Springer Series in
Surface Sciences

6

W0234494

Editor: Gerhard Ertl

Springer Series in **Surface Sciences**

Editors: Gerhard Ertl and Robert Gomer

M.A. Van Hove
W.H. Weinberg   C.-M. Chan

Low-Energy Electron Diffraction

Experiment, Theory and Surface Structure Determination

With 213 Figures

Springer-Verlag Berlin Heidelberg New York
London Paris Tokyo

Dr. Michel A. Van Hove

Materials and Molecular Research Division, Lawrence Berkeley Laboratory, and
Department of Chemistry, University of California, Berkeley, CA 94720, USA

Professor William H. Weinberg

California Institute of Technology, Pasadena, CA 91125, USA

Dr. Chi-Ming Chan

Raychem Corp., 300 Constitution Dr.,
Menlo Park, CA 94025, USA

Series Editors

Professor Dr. Gerhard Ertl

Fritz-Haber-Institut der Max-Planck-Gesellschaft, Faradayweg 4–6
D-1000 Berlin 33

Professor Robert Gomer

The James Franck Institute, The University of Chicago, 5640 Ellis Avenue,
Chicago, IL 60637, USA

ISBN-13:978-3-642-82723-5   e-ISBN-13:978-3-642-82721-1
DOI: 10.1007/978-3-642-82721-1

Library of Congress Cataloging-in-Publication Data. Hove, M. A. Van (Michel André), 1947-. Low-energy
electron diffraction. (Springer series in surface sciences ; 6). Bibliography p. Includes index. 1. Surfaces
(Physics) 2. Low-energy electron diffraction. 3. Surface chemistry. I. Weinberg, W. H. (William Henry),
1944-. II. Chan, C.-M. (Chi-Ming), 1955-. III. Title. IV. Series. QC173.4.S94H68 1986 530.4'1 86-20351

© Springer-Verlag Berlin Heidelberg 1986
Softcover reprint of the hardcover 1st edition 1986

2153/3150-543210

Preface

Surface crystallography plays the same fundamental role in surface science which bulk crystallography has played so successfully in solid-state physics and chemistry. The atomic-scale structure is one of the most important aspects in the understanding of the behavior of surfaces in such widely diverse fields as heterogeneous catalysis, microelectronics, adhesion, lubrication, corrosion, coatings, and solid-solid and solid-liquid interfaces.

Low-Energy Electron Diffraction or LEED has become the prime technique used to determine atomic locations at surfaces. On one hand, LEED has yielded the most numerous and complete structural results to date (almost 200 structures), while on the other, LEED has been regarded as the "technique to beat" by a variety of other surface crystallographic methods, such as photoemission, SEXAFS, ion scattering and atomic diffraction. Although these other approaches have had impressive successes, LEED has remained the most productive technique and has shown the most versatility of application: from adsorbed rare gases, to reconstructed surfaces of semiconductors and metals, to molecules adsorbed on metals. However, these statements should not be viewed as excessively dogmatic since all surface-sensitive techniques retain untapped potentials that will undoubtedly be explored and exploited. Moreover, surface science remains a multi-technique endeavor. In particular, LEED never has been and never will be self-sufficient.

LEED has evolved considerably and, in fact, has reached a watershed. The technique is now well established and can therefore be extended confidently into new directions. Consequently, this is an appropriate time to put the entire field into perspective in the form of a book. Recent trends in LEED that are reflected in this book also point toward the future shape of the field. These include the following:

1. A base of confidence in the method has been established since past controversies have been resolved.

2. Diversification is taking place. For example, high-accuracy refinements are being obtained in structures that were previously considered to be simple (e.g. multilayer relaxations); complex structures are being analyzed (e.g. surface reconstructions and molecular adsorption); the variety of materials under study is continually increasing (e.g. metal alloys, metals

on semiconductors, carbides, oxides, layered compounds, coadsorption systems, and multilayer growth); and disordered systems are beginning to be investigated beyond two-dimensional ordering characteristics (i.e. in terms of local bonding geometry).

3. A variety of theoretical approaches have been developed to deal efficiently with many different situations: strong, intermediate, or weak multiple scattering; long or short mean free paths; and large or small interlayer spacings. Developments are continuing to take place.

4. New experimental techniques are rapidly being adapted to automate and render the LEED measurement more efficient, as well as to enhance its reliability.

5. Theory and experiment are increasingly being compared by reproducible means, namely through R-factors which remove part of the human subjectivity and contribute importantly to the automation of the search for the correct structure.

6. The competition between LEED and other surface crystallographic techniques has spurred mutually beneficial influences, such as efforts to obtain higher accuracy and higher reliability, questioning of certain assumptions (e.g. the assumption of highly symmetrical structures, or of the need for well-ordered surfaces in LEED), and the injection of theoretical concepts from LEED into other fields (especially treatments of electron-atom scattering and multiple scattering).

With the above observations in mind, we felt that it was desirable in composing this book to discuss a wide variety of topics in the field of LEED, rather than delve deeply into selected aspects. At the same time, we have attempted to preserve a certain unity of approach in order to emphasize the links between the various topics. An important concern has also been the didactic value of the text, since it is our hope that more researchers will find LEED useful in their work. To that end, we introduce gradually the various concepts used in LEED, starting with the kinematic limit, before presenting one-by-one the complications that arise from multiple scattering. The text is self-contained in that no previous knowledge of LEED is assumed. As a consequence, this book should be accessible to many readers who have any level of familiarity with the subject or with related fields, such as solid-state physics or diffraction physics. At the same time, we anticipate that this text can serve as a reference book for the more established practitioner of LEED. This is facilitated by a variety of tables and examples, and, in the case of structural results, by a comprehensive list of structures with a bibliography.

Our text starts with a historical sketch of the development of LEED (Chap. 1), followed by a description of experimental issues and techniques (Chap. 2). A general discussion of surface structures and their notations is given in Chap. 3, together with an analysis of the information content of the LEED pattern. The kinematic and dynamical theories of LEED are treated in Chaps. 4 and 5, respectively, while their application to structural determinations appears in Chap. 6. Several diverse examples of such surface structural determinations are presented next (Chap. 7). Chaps. 8, 9, and 10 deal with LEED studies of phase transitions: order-disorder effects, chemical reactions and island formation, respectively. Probable future developments and the connection between LEED and other surface crystallographic techniques are addressed in Chap. 11. Finally, a comprehensive table and a bibliography of surface structural results are given in Chap. 12.

We are very grateful for the generous help of those who have contributed in one way or another to the completion of this book. The hospitality of the Institute of Physical Chemistry at the University of Munich is acknowledged by one of the authors (WHW) during a visit that was largely devoted to work on this book. Segments of this book are based on unpublished reports of S. L. Cunningham, who is hereby thanked. Important parts of the manuscript were critically read by A. C. Sobrero, S. Y. Tong, S.-W. Wang and E. D. Williams, but remaining errors and misrepresentations are the authors' exclusive responsibility. The typing of the entire manuscript represented a demanding word-processing task which was accomplished efficiently and expertly by Ms. Kathy Lewis at the California Institute of Technology. Further typing assistance was graciously provided by Raychem Corporation. The Graphic Arts Department of the California Institute of Technology was most helpful in preparing the many figures. Finally, we wish to acknowledge our colleagues who allowed us to reproduce figures from their publications.

Berkeley, Menlo Park,   *M. A. Van Hove, C.-M. Chan,*
and Pasadena, April 1986   *and W. H. Weinberg*

Contents

List of Reference Tables and Figures

This list is intended to help the reader locate tables and figures in this book that contain valuable data

1. The Relevance and Historical Development of LEED

1.1 The Relevance of Surface Crystallography

Of utmost importance in understanding both the physical and the chemical properties of any state of matter is a detailed knowledge of the relative positions of the atoms. This simple fact cannot be overemphasized. The geometric structure is arguably the most fundamental property of the surface of a solid. Even more than this, it is one of the most fundamentally important properties of a solid since the bulk of the solid "communicates" inevitably with its environment via its surface. Furthermore, there is an increasing trend in a myriad of technologically important areas to maximize the ratio of the surface area of a material to its volume.

A prime example is that of electronic devices. Their size is continually being decreased to reduce power losses from resistive heating, to shorten the transit time of electrical signals, and to decrease space requirements. The result is a large ratio of surface to volume, so that surface-related issues appear in the conceptual design, the manufacture, and the use of such small devices. Another very important example concerns heterogeneous catalysis, which often employs expensive metals as the catalysts. Since heterogeneous catalysis is purely a surface phenomenon, a large ratio of surface to volume is obtained by "dispersing" the metal into small particles which are "supported" on an inexpensive material such as alumina or silica. Here again, the understanding of the surface properties is crucial to the improvement of practical catalysts.

Although it certainly has not been fully appreciated historically, it is now clear that the "surface region" of a solid (the last few atomic layers) at a solid-gas, solid-liquid or solid-solid interface may have a structure that is quite different from the bulk. Consequently, the electronic and vibrational character of a surface may differ substantially from the bulk properties, leading to unique physical and chemical properties of surfaces.

For example, consider first the electronic properties of a clean surface. The very existence of surface states, in which electron density is localized near the surface, can depend upon whether the surface layer has relaxed or not, i.e. whether it has contracted or expanded relative to the bulk interlayer spacing. In addition, the work function, which is a property of the surface, is quite sensitive to the dipole layer at the surface, which in turn is connected intimately to the surface geometry. If the surface is reconstructed with a periodic disturbance, which may be due to buckling or to missing atoms, for example, then the surface Brillouin zone is changed completely. This can cause electronic states that were originally at the Brillouin zone edge to lie in the zone center, and this can affect the surface conductivity, for instance. Furthermore, properties that depend on averages over the entire Brillouin zone (such as the surface electronic specific heat) depend strongly on the detailed structure of the surface.

Second, the vibrational properties of the clean surface should be considered. The surface phonons at short wavelengths depend very strongly on the atomic positions and upon whether the interatomic force constants change upon relaxation. The mere existence of the surface reduces the symmetry of the lattice, and forces that are symmetry-forbidden in the bulk may be present at the surface. Detailed knowledge of atomic positions in the surface region is needed in order to ascertain to what extent the symmetry is broken. These new surface-symmetry-induced forces can cause the surface to reconstruct. The reconstruction changes the surface Brillouin zone, and this affects many vibrational properties, such as the Debye-Waller factor and the surface vibrational specific heat.

Third, the structure of atoms or molecules adsorbed on surfaces is of extreme importance. This issue is relevant, for example, to physical adsorption, chemisorption, heterogeneous catalysis, and epitaxial growth (indeed solid-solid interfaces generally). The nature of the chemical bond (including both the ground state and all excitation spectra) of an adsorbate to an extended surface cannot be calculated without knowledge of the exact position of the adsorbate. This requires not only general information, such as the type of adsorption site that is occupied, but also specific information regarding bond lengths and bond angles. In addition, the position of the substrate atoms must be known both before and after chemisorption occurs, since chemisorption-induced relaxation or reconstruction of the substrate can affect profoundly the nature of the chemisorption bond. Furthermore, determination of the structure of adsorbates is critical to assess the accuracy and the validity of ab initio calculations for finite-sized clusters. The atomic positions determined from experiment are also required to test the total-energy band-structure calculations that are now predicting surface geometries from first principles. These are only a few of the many reasons why it is fundamentally important that the precise position of the surface atoms be known.

At least for model systems, Low-Energy Electron Diffraction (LEED) has been demonstrated to be effective at providing fundamental information of use in technological applications. Although the use of model (e.g. single crystalline) surfaces has frequently prohibited a one-to-one correspondence between the scientific result and the technological application, the model systems have, at the very least, provided invaluable insight to the technologist.

Moreover, experience has shown that much of the "chemistry" or the "physics" which is of interest to the technologist is embodied in the scientific results from the "model" systems. Two industries that have profited immeasurably from surface crystallography are the microelectronics and the chemical industries. In the former case, a better understanding of surface states at semiconductor surfaces, and interface states at metal-semiconductor interfaces (which dictate the height of the Schottky barrier, for instance) have been of fundamental importance. The major impact of surface crystallography on the chemical industry has been related to the ubiquitous field of heterogeneous catalysis. The contributions of surface crystallography here have been multifareous, for example the determination of the geometry of various stable chemical intermediates (in adsorption, decomposition, or synthesis reactions) at surfaces. The determination of the stoichiometry and the structure of the intermediate is of immense value because it provides a precedent for the occurrence of such species. This knowledge is of fundamental importance since such an intermediate might well occur, but not be stable (isolatable), in other catalytic reactions, the mechanisms of which can be put on a firmer theoretical foundation by the mere precedence of the existence of such a species on the surface. Furthermore, the systematics of the bonding of various ligands of chemical importance on various surfaces of various symmetries provides very important chemical insight. Many ligands could be cited as examples here, not the least of which in practical importance would be hydrogen atoms, carbon monoxide. and innumerable saturated and unsaturated hydrocarbons.

1.2 The Historical Development of LEED

1.2.1 The Period Before Wave Mechanics

Prior to the discovery of LEED — indeed prior to the recognition of the wave nature of the electron — extremely elegant and relevant experimental work was carried out concerning the reflectivity of low-energy (≤ 100 eV) electrons from polycrystalline samples of nickel [1.1,2], copper [1.3], and silver and iron [1.4]. The first significant conclusion to be drawn from these measurements was that the elastic reflectivity, proportional to the probability that an electron is reflected from the surface with (essentially) no loss in kinetic energy, is a strong function of the incident energy of the electron (i.e. the wavelength of the electron) and the angle of emergence. Second, maxima were observed in the reflectivity for incident electron energies below approximately 25 eV. Although unrecognized at the time, these variations were probably due to electron diffraction.

An impressive component of these early measurements is that, in retrospect, they were performed under ultrahigh vacuum (UHV) conditions (p $\leq$ 2 x 10^{-10} Torr). Such an environment is necessary when working with clean surfaces of low area. If the probability of adsorption of the ambient gas on the surface is near unity, then the surface would become saturated with contamination from the background ambient gas in approximately one second at a pressure of 10^{-6} Torr. In these early measurements, UHV was attained by liquid-nitrogen-trapped mercury diffusion pumps, augmented by evaporated metallic film getters and high-temperature (≤ 700 K) bakeout of the apparatus. Although the ultimate pressure that could be measured then was approximately 10^{-8} Torr, due to the "X-ray limit" of the available ionization gauges, subsequent measurements with Bayard-Alpert ionization gauges verified that the base pressures were on the order of 10^{-10} Torr.

1.2.2 The Discovery of Electron Diffraction

The theoretical possibility of the occurrence of electron diffraction was a consequence of the wave mechanics proposed by de Broglie in 1924 [1.5]. This theory extended to all particles the coexistence of waves and particles discovered by Einstein in 1908 for light and photons. Einstein's wave-particle duality had led to von Laue's discovery of X-ray diffraction from crystals in 1912 [1.6] and to the use of X-ray diffraction for crystallography, beginning with the work of Bragg in 1913 [1.7].

Electron beams had been discovered in 1897 by J. J. Thompson, who measured the ratio of charge to mass for the electron. Consequently, the particle nature of the electron had been well established. De Broglie postulated that the wavelength of a particle which has a linear momentum p is given by h/p, where h is Planck's constant. For electrons with a kinetic energy of 100 eV, this corresponds to a wavelength of approximately 1 Å Consequently, such low-energy electrons should diffract from a grating with a periodicity on the order of atomic dimensions, e.g. from a crystal.

The experimental observation of LEED was made by Davisson and Germer as a consequence of a laboratory accident at Bell Laboratories. These events are related in detail by Gehrenbeck [1.8]. The glass vacuum chamber, which contained a polycrystalline nickel sample, cracked while the sample was being heated, resulting in severe oxidation of the nickel. After repairing the apparatus, the sample was cleaned by high-temperature chemical reduction with hydrogen. This severe oxidation-reduction cycle resulted in an extensive recrystallization of the polycrystalline nickel into (111) oriented microfacets, i.e. the polycrystal simulated a single crystal of (111) orientation. The experimental observations caused Davisson and Germer to realize that the angular dependence of the elastic electron scattering is due to crystal effects rather than intra-atomic effects as they had believed previously. The possible connection between wave mechanics and earlier, related results by Davisson and Kunsman was proposed by Max Born at an Oxford meeting in August 1926. This suggestion came as a surprise to Davisson, who was present. It led Davisson and Germer to focus their efforts on the search for well-defined electron-diffraction effects that could be related to de Broglie's wavelength-momentum relationship. Not until early 1927 was success achieved with the clear observation of an off-specular intensity peak for a Ni(111) single-crystal surface and of threefold azimuthal symmetry corresponding to the structural symmetry of the crystal. Their results were first published in *Nature* [1.9] and in *Physical Review* [1.10]. A further series of experiments led to the often-quoted key publication in *Physical Review* in late 1927 [1.11]. This paper establishes the wave-particle duality of electrons on a firm experimental basis. However, some unresolved discrepancies between experiment and theory remained. In particular, "anomalous beams" were observed, the introduction of an ad hoc "contraction factor" of the crystal lattice was found to be necessary, and the observed peak widths seemed too large.

Davisson and Germer's article in *Nature* preceded by only one month another article in *Nature* by Thompson and Reid [1.12], which also revealed the wave nature of electrons. Thompson, at the same Oxford meeting, had realized that his positive-ray scattering apparatus could easily be adapted to search for electron diffraction. Thompson and his student Reid benefited from higher kinetic energies (tens of keV rather than Davisson and Germer's tens of eV), requiring a less stringent vacuum, and chose celluloid as the diffracting object. The resulting diffraction patterns, taken in the transmission mode, are the core of their *Nature* article. Later experiments used the better-known lattices of metal samples in the form of films and produced further justification of the de Broglie postulate. In 1937, Davisson and Thompson shared the Nobel Prize for Physics for their pioneering work in electron diffraction.

1.2.3 The Aftermath of the Discovery of Electron Diffraction

Within a few years of the discovery of electron diffraction, the wave-particle duality was also demonstrated for atoms and molecules. Indeed, helium atoms and hydrogen molecules were found to diffract from (100) surfaces of LiF, NaF, and NaCl [1.13], while hydrogen atoms were diffracted from LiF surfaces [1.14]. Neutron diffraction was also exhibited only a few years later [1.15].

Meanwhile, Davisson and Germer [1.16] confirmed that LEED could not be described quantitatively by the kinematic (single-scattering) theory used in the analysis of X-ray diffraction data [1.17]. After Davisson and Germer's observations, Bethe studied the question of the mismatch in peak positions between theory and experiment. He proposed correctly that a negative potential could resolve this issue [1.18], and today this is called the inner potential. Actually, Bethe's first publication on this subject mentions a *positive* potential, which was obtained erroneously with a different indexing of the orders of reflection [1.19]. Bethe then published the first "dynamical theory" of LEED [1.20], which was inspired by Ewald's theory of X-ray diffraction [1.21]. It was followed by the first one-electron energy-band theory applied to LEED by Morse in 1930 [1.22], which was based on the Bloch-wave theory introduced in 1928. However, success was still not forthcoming in matching the experimental data.

It took over 40 years from its discovery until LEED could be used to determine atomic positions. By contrast, Thompson's High-Energy Electron Diffraction (HEED) developed within a few years to the stage of determining bond lengths in small gas-phase molecules, initially CCl_4 [1.23], by using high energies that allowed kinematic scattering to be assumed. HEED was also soon used extensively to study a variety of properties of sufficiently small crystallites of solid materials, including high-accuracy bond-length determinations, and, of course, it also spawned Electron Microscopy [1.24]. The slow progress of LEED contrasts with the rapid development of X-ray diffraction as well. After the discovery of X-ray diffraction in 1912 by von Laue [1.6], Bragg [1.7] published the first X-ray crystallographic result in 1913!

1.2.4 The Period 1930-1965

Rather little experimental LEED work was carried out during the period between its original discovery and the early 1960s. This was presumably due both to the technological complexity of the measurement and the lack of an adequate theory. Nevertheless, at least two important technical advances were introduced: the post-acceleration display LEED system which allows the entire spot diffraction pattern to be viewed on a fluorescent screen [1.25-27], as well as a high-speed, movable Faraday cup collection scheme for the diffracted beams [1.28].

The most notable exception to the lack of significant experimental activity in LEED during this time was the work of Farnsworth [1.29,30]. Three examples of the fundamentally significant work of Farnsworth and his co-workers will suffice to document his contributions. First, he demonstrated by the epitaxial growth of silver onto a (100) surface of gold that those electrons which are reflected elastically (diffracted) penetrate only a very few atomic layers into the "bulk". For example, at an incident energy of 60 eV, on the order of 90% of the diffracted electrons sample no more than two atomic layers [1.31]. Second, he showed that clean surfaces may be prepared under UHV conditions via argon ion bombardment (i.e. the sputtering of impurities from the surface), followed by annealing to remove the lattice damage induced by the sputtering [1.32]. This technique, chemical cleaning, or a combination of both, remain to this day the most popular and most reliable ways of obtaining clean surfaces of low area solids, especially for use in a UHV environment. Third, and perhaps most important from a scientific point of view, Farnsworth discovered that clean (100) and (111) surfaces of silicon and germanium are reconstructed, i.e. the primitive unit cell of the surface is larger than that which would be expected based on a simple termination of the bulk structure [1.33,34].

1.2.5 The Renaissance of LEED: Experimental Advances in the Mid-1960s

During the 1960s there was an astounding increase in the amount of experimental LEED work. This was sparked in particular by renewed efforts by Germer [1.26] and contributions by Lander [1.27,35], and was fueled by the rise of semiconductor device technology and the aerospace industry. The spread to other laboratories was made possible by commercialization: the commercial availability of ion-pumped UHV systems, the commercial availability of LEED optics compatible with the UHV hardware, and finally the commercial availability of Auger spectrometers which could monitor the chemical composition of a surface and insure its cleanliness. The addition of Auger Electron Spectroscopy illustrates a trend that was to become increasingly evident, namely surface science was to become a multi-technique effort. In particular, LEED would be much less useful if it were not supplemented by other techniques which give additional information concerning the state of the surface under investigation.

Since a quantitative theory of LEED had not then been developed, a detailed determination of surface structures, including adsorption sites, bond lengths and bond angles, was not possible in the 1960s. However, the LEED pattern is a most helpful observable, almost universally used to this day to monitor the surface condition and establish its reproducibility. In addition, there was and there remains an interest in monitoring two-dimensional surface phase transitions by LEED [1.36]. An understanding of these phenomena and the related phenomenon of the ordering of adsorbates into islands did not require a full multiple-scattering LEED theory. It did require, however, a quantification of the LEED "instrumental response function", and this was provided by Park et al. [1.37].

1.2.6 The Theoretical Solution: The Late 1960s and Early 1970s

The experimental advances of the 1960s rekindled the interest in developing a successful theory of LEED. McRae in 1964 [1.38] produced evidence that multiple scattering is a necessary ingredient of such a theory. He then proceeded in 1966 to formulate an approach which generalized Darwin's two-beam dynamical X-ray theory of 1914 [1.17] to an N-beam theory [1.39]. In 1968 he incorporated [1.40] the Bloch-wave concept introduced in 1928, together with wave matching across the solid-vacuum interface.

In the meantime, many of these theoretical techniques had become commonplace in the band-structure theory of three-dimensional crystals, which applies primarily to electrons below the Fermi level. Another such technique was the pseudopotential. It became an ingredient of a LEED theory developed by Boudreaux and Heine [1.41], which also incorporated Bloch waves. Soon afterward, in 1968, Marcus and Jepsen introduced [1.42] a LEED theory based on Fourier-transforming the wave field parallel to the surface and which included the propagation-matrix variant of Bloch-wave theory.

In the same year, Beeby presented [1.43] a Green's-function t-matrix formalism that makes fewer assumptions about periodicity parallel to the surface and allows a more refined treatment of the electron-atom scattering. This approach was utilized by Holland [1.44] to exhibit how the effect of thermal vibrations could be introduced by means of an atomic Debye-Waller factor and temperature-dependent phase shifts. A related theory was proposed in 1969 by Duke in the s-wave limit [1.45], which introduced another important ingredient, namely the effect of inelastic scattering. The consequence of inelastic scattering is that electrons with kinetic energies above the Fermi level can lose energy, and fewer electrons diffract elastically. This is simulated by a mean free path or an imaginary component of the scattering potential, as had been introduced previously in the bulk by Slater [1.46], Moliere [1.47] and Yoshioka [1.48], and at surfaces by Hirabayashi [1.49]. The need for such a theoretical ingredient in LEED had been predicted earlier by Webb [1.50]. Duke's theory was generalized in 1970 [1.51] to include proper atomic scattering following Beeby's method and to incorporate a self-consistent solution for the electron scattering by an atomic monolayer, developed in 1967 by Kambe [1.52] within the t-matrix formalism. Duke also inserted temperature corrections similar to Holland's. In addition, a solution of LEED for disordered atomic layers was proposed by Duke [1.53], followed [1.54] by a theory of inelastic LEED (ILEED).

In the meantime, Heine's Bloch-wave approach had been modified in 1969 by Pendry [1.55] to include atomic scattering in the t-matrix formalism, as well as inelastic effects. Pendry emphasized the importance of ion-core scattering. He then replaced [1.56] the Bloch-wave method by a much more efficient perturbation expansion (Renormalized Forward Scattering) in 1971 and by the intermediate Layer Doubling method [1.57,58] in 1974.

In 1971, Tong produced a perturbation version of Beeby's theory, called the third-order τ-matrix method [1.59,60], which includes the mean-free-path effect. This was followed by the third-order t-matrix method in 1972 [1.61]. Also in 1971, Jepsen and Marcus improved [1.62] their previous theory by inclusion of the layer-KKR technique, the t-matrix formalism of Kambe and inelastic effects. Several other theoretical methods were also proposed [1.63-66] during this same period.

The results of all these approaches were compared with experimental curves of diffracted intensity versus energy or intensity versus incident angle for simple metal surfaces. These "I-V curves" or "I-θ curves" involved principally low-Miller-index surfaces of aluminum, copper, and nickel and were measured by Jona [1.42], Andersson [1.67], Somorjai [1.68] and a number of others. The various physical ingredients mentioned above (an atomic scattering potential represented by a t-matrix, an inelastic mean free path and a temperature correction) were found to be essential to provide a good description of the experimental data [1.57,60]. On the other hand, the various formalisms which included those ingredients yielded essentially the same results, which were in reasonable agreement with experiment [1.57,60].

It should be noted that a completely different approach has also been developed for the LEED problem to try to avoid the mathematical complexities due to multiple scattering and to retain only a kinematic theory. In 1971, Lagally et al. [1.69] proposed a data-averaging scheme which would reduce the importance of multiple-scattering effects in the data. This direction of investigation was only partly successful.

1.2.7 The Era of Structural Determination: The 1970s and 1980s

At this time, a streamlining of the number of LEED formalisms took place, as the calculations began to be applied to actual structural determinations of surfaces. The first reliable structural determinations came in 1971 as by-products of the above-mentioned comparisons between theory and experiment. These initial surface structural determinations concerned simple clean metal surfaces. Earlier comparisons between simpler theories and experiment might be called preliminary structural determinations.

Thereafter, the theoretical developments were aimed at reducing the computational effort needed in the structural determination. Some of these have already been mentioned above. Van Hove [1.58] systematically applied symmetries and other efficiency features to overlayer systems with superlattices, starting in 1974. In 1975, Zimmer and Holland [1.70] produced a perturbation version of Beeby's formalism, called reverse scattering peturbation. It was then combined by Tong and Van Hove with some of Tong's and Pendry's methods into a combined-space theory [1.71,72]. This theory allows an optimum use of both the plane-wave representation of the layer-based methods and the spherical-wave representation of the Green's-function methods in cases of some structural complexity. At about the same time, Pendry produced the "chain method" [1.73], which is designed for the higher energies of Medium-Energy Electron Diffraction (MEED) and Reflection High-Energy Electron Diffraction (RHEED).

A large number of surface structures were solved during the 1970s and into the 1980s with all these methods, initiating the "era of structural determination by LEED". It was thereby shown that LEED can be applied not only to clean metal surfaces, but also to atomic overlayers on these metals, to semiconductors, to ionic compounds, and even to adsorbed molecules [1.72,74]. R-factors were also introduced in this period to help in performing structural determinations.

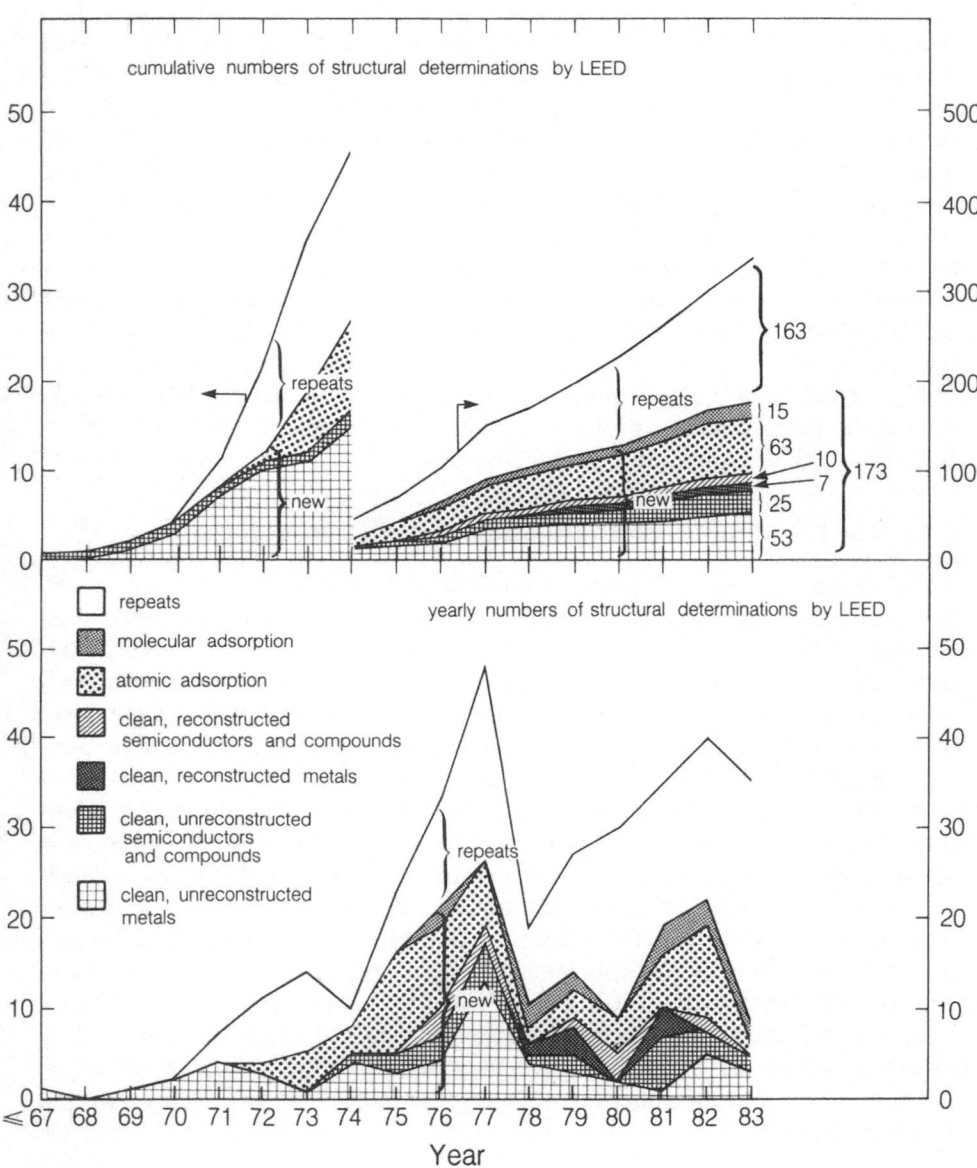

Fig. 1.1. Yearly (*bottom panel*) and cumulative (*top panel*) numbers of structural determinations by LEED

Figure 1.1 illustrates the yearly number of structural determinations by LEED up to the end of 1983, broken down into first ("new") determinations of a surface structure and duplications or refinements ("repeats"). The new determinations are in turn broken down by type of surface. Figure 1.1 also shows cumulative numbers of structural determinations. A few observations can be made on the basis of Fig. 1.1. First, it is quite obvious which are the easier structures to solve, namely clean, unreconstructed metal surfaces (53 of them up to 1983) and atomic adsorbates on these (63 structures). Second, the trend toward greater diversity in the types of surfaces is clear (a further breakdown of categories after 1980 would enhance that impression). Third, despite wide fluctuations, the average yearly number of new structures has remained steady around 15 since 1975, in the face of growing competition from other surface-crystallography techniques. That figure is doubled to 30 per year when one includes the "repeat" studies, which comprise many refinements made possible by the increase in the accuracy of LEED determinations. The spike-and-dip of 1977-78 in Fig. 1.1 appears to be due more to a statistical fluctuation than to a scientific event. However, it signals a leveling off in the rate of structural determinations, as Fig. 1.1 shows. This is probably due in part to the involvement of some LEED specialists in other competitive techniques.

Increasingly complex structures have been tackled recently. For instance, reconstructions of some clean semiconductor surfaces require the optimization of a dozen structural parameters. Similar situations arise with refinements of earlier results, leading for instance to the discovery of multilayer relaxation at clean metal surfaces. Furthermore, large molecules are being investigated, using appropriate theoretical procedures.

On the experimental side, much has been learned about maintaining reproducible surface conditions. Moreover, improved intensity measurement techniques have been introduced, in particular photography, and use of the video camera and position-sensitive detectors.

As a consequence, LEED has arguably reached the stage of being a mature scientific discipline. However, there is still considerable room for what might best be considered as evolutionary as opposed to revolutionary improvements in LEED and, in particular, in surface structural determinations by LEED. Representative examples of areas where progress will surely be made include structure determinations of larger unit cells, of unit cells with a greater number of atoms in the basis, of molecular overlayers with larger and more complicated admolecules, of overlayers which contain a significant amount of disorder, and an elucidation of the properties of magnetic surfaces.

The large effort put into developing LEED over the years, on both the experimental and theoretical fronts, has paid dividends also in other areas. Since electron scattering and diffraction are inherent to many surface-sensitive spectroscopies, techniques developed for LEED have assisted these other spectroscopies. These include Angle-Resolved Photoelectron Spectroscopy (ARPES), Surface Extended X-ray Absorption Fine Structure (SEXAFS), Extended Appearance Potential Fine Structure (EAPFS), High-Resolution Electron Energy Loss Spectroscopy (HREELS) and many others.

At the same time, these newer spectroscopies, as well as others that do not involve electron scattering (e.g. Rutherford Back Scattering), have become competitors to LEED. The net effect has been beneficial both to LEED and to surface science as a whole.

In the following chapters, all of the concepts introduced either explicitly or implicitly in this Introduction are developed fully in a self-contained fashion. In addition, LEED is compared with other experimental techniques that can determine some aspects of surface structure. In this sense, this text provides a useful access to surface science more generally, and not just LEED.

2. The LEED Experiment

As was described in the historical overview of Chap. 1, a number of important technological and procedural developments have enabled reliable LEED data to be measured. This chapter describes the main experimental methods presently in use. Many of these are of interest to any surface experimentalist who wishes to observe the LEED pattern, whether for the purpose of measuring diffracted beam intensities or simply to monitor the state of the surface for other investigations. Some of these methods are more specifically relevant to those researchers interested in other surface-sensitive spectroscopies that involve electron impact and/or emission (Sect. 11.4). This is especially true when accuracy is important in such variables as impact and emission angles.

The first step in LEED experiments is common to virtually all surface experiments: the surface preparation. From a single crystal, one needs to cut a slice of approximately 1 mm, or less, in thickness and of approximately 1 cm in diameter. One or both faces of the resulting sample should be oriented well within 1° of the desired crystallographic orientation. This is easily achieved with materials that cleave along the desired plane, but most materials must be cut with great orientational precision. By a succession of finer polishing steps, the surface can be brought to optical flatness. Atomic-scale flatness is achieved by annealing, often at temperatures near the melting point of the sample for many hours. During this process, a sharp LEED pattern becomes visible. Chemical purity of the surface may be obtained by this annealing through desorption into the vacuum, but often nonvolatile bulk impurities are found to segregate to the surface. These may be removed by chemical cleaning methods (e.g. oxidation and reduction cycles) or by ion sputtering (followed by an additional annealing cycle). Once a satisfactorily cleaned and ordered surface has been obtained, one may wish to adsorb monolayer or submonolayer concentrations of atoms or molecules onto it. In many cases, this is accomplished by exposure of the surface to molecular gases. In that case atomic adsorption is obtained by molecular decomposition. The variety of approaches is too large to be detailed here. In any case, the state of the surface is often characterized by Auger Electron Spectroscopy, Thermal Desorption Mass Spectrometry, observation of the LEED pattern, X-ray Photoelectron Spectroscopy, or other suitable techniques (Sect. 11.4).

2.1 General Features of LEED Experiments

All definitive LEED measurements are performed in an ultrahigh vacuum (UHV) system which is maintained at pressures below 10^{-10} Torr. It is quite important to maintain a sufficiently low pressure in the experimental chamber to keep the surface under study free of impurities, since residual gases which adsorb on the surface of a crystal can cause a significant error in the measurement of experimental data. For example, a surface can be contaminated with a monolayer of a gas with a probability of adsorption of unity in approximately one second at an ambient pressure of 10^{-6} Torr. Thus, at a pressure of 10^{-10} Torr, it will take at least 10^4 seconds before the surface is contaminated with one monolayer of the background gases. Typical experimental measurement times are approximately 15 to 20 minutes, using a Faraday-cup collector, for obtaining one LEED intensity-voltage beam profile with an energy range of 200 eV in intervals of 2 eV. Therefore, it is necessary to maintain a low pressure in order to provide sufficient time for obtaining good experimental data. To monitor the concentration of impurities on the surface, an Auger-Electron Spectroscopy system is usually installed along with the LEED system.

Low-energy electrons can be deflected by the magnetic field of the earth. To facilitate the data collection, this magnetic field is either compensated by a set of Helmholtz coils or isolated by a μ-metal shield. A typical, completely equipped LEED system consists of

1. An UHV chamber.
2. A high-precision crystal manipulator.
3. An electron gun with a movable Faraday cup for collection of the diffracted beams.
4. A display system for visual observation of the LEED pattern.
5. Sets of Helmholtz coils to compensate for the magnetic field of the earth.
6. A sputtering gun for cleaning of the surface.
7. An Auger-Electron Spectroscopy system to monitor surface impurities.

2.2 Sample Mounting

The high-precision manipulator used commonly in an UHV environment is a goniometer capable of precisely positioning a sample for different types of surface analyses. A typical manipulator, shown in Fig. 2.1, can position a sample in three orthogonal axes (the z-direction is along the axis of the manipulator). Rotation of 360° about the z-axis is provided as shown in Fig. 2.1. A flip mechanism is often used for rotation about an axis perpendicular to the z-axis to vary the angle of incidence of the electron beam.

Fig. 2.1. A commercial (Varian) high-precision manipulator with flip mechanism

Most commercial manipulators have no mechanism for rotation about the axis normal to the crystal surface. Nevertheless, this is adequate for most LEED measurements, since such measurements are often performed at normal incidence in order to maintain the highest symmetry among the diffraction beams. This choice can also facilitate the dynamical calculations [2.1]. Otherwise, and in some special cases [2.2], LEED measurements must be made at an off-normal incidence angle and at an arbitrary azimuthal angle, which is determined by the way the crystal is mounted. The incidence direction is then determined less reliably. In general, if no azimuthal angle rotation is possible, there is no symmetry among the diffraction beams. Consequently, the calculations may become needlessly time-consuming and expensive. To avoid such a situation, a mirror plane of symmetry can be maintained by rotating the crystal about an axis perpendicular to a mirror plane of the surface structure, as in Fig. 2.2. Designs for manipulators that allow azimuthal rotation are available [2.3,4].

The manipulator flange usually has a variety of "feedthroughs" mounted on it as well. This allows for heating and cooling of the sample and measuring its temperature by means of a thermocouple. The sample can be heated either resistively or by electron bombardment [2.5]. The sample can also be maintained below 100 K by passing liquid nitrogen through cooling coils which are connected to a coolant reservoir that is in contact with the sample.

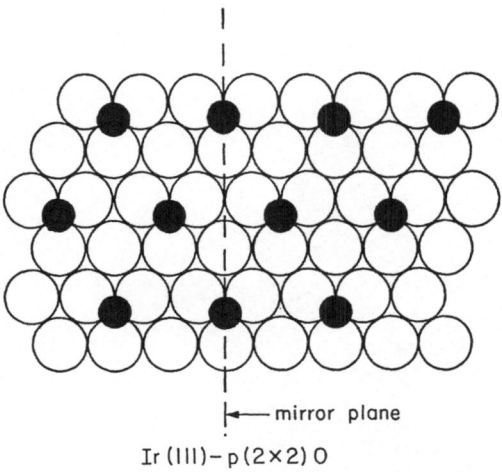

Ir (III) – p(2×2) O

Fig. 2.2. A Schematic diagram showing the maintenance of a mirror plane of symmetry for an Ir(111)-p(2×2)O overlayer structure by tilting the crystal around an axis perpendicular to a mirror plane of the surface structure

A special manipulator, called the Wilson manipulator, allows crystal translation and rotation as well as temperature control in the range between 25 and 2500 K. It is an important device due to its versatility and simplicity [2.6]. It consists of two parts: the differentially pumped rotary feedthrough shown in Fig. 2.3 and the crystal holder shown in Fig. 2.4. The rotary

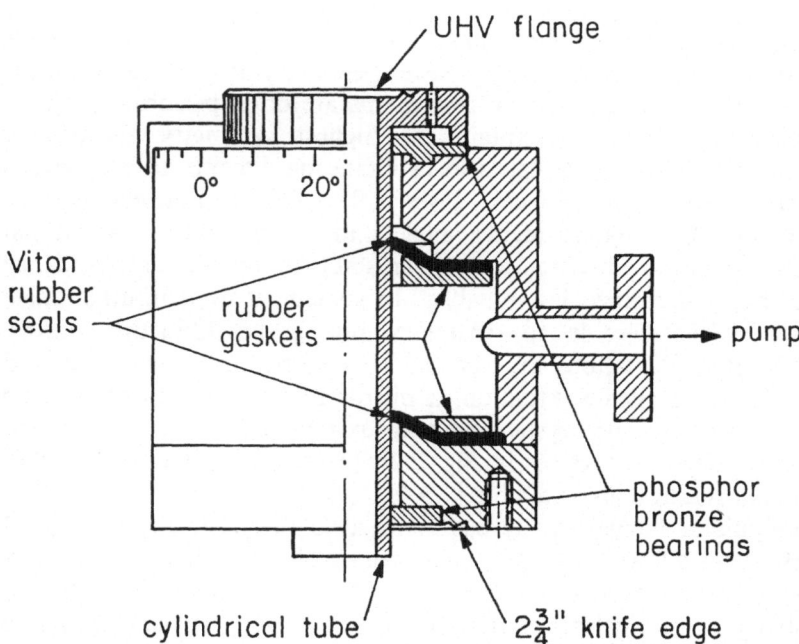

Fig. 2.3. Schematic of a differentially pumped rotary feedthrough. After [2.6]

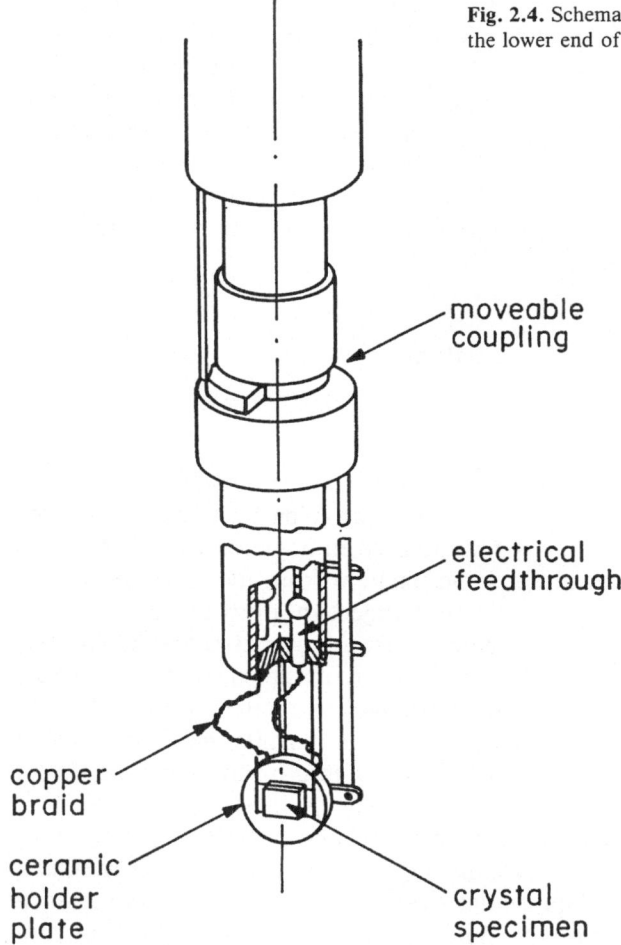

Fig. 2.4. Schematic of a crystal holder mounted at the lower end of a cold-finger. After [2.6]

moveable
coupling

electrical
feedthrough

copper
braid

ceramic
holder
plate

crystal
specimen

feedthrough can be attached to the top of a commercial manipulator. A long cylindrical stainless steel tube, which functions as a cold-finger, is inserted through the central part of the feedthrough. The upper end of the tube is welded to a small flange. Cooling the sample to approximately 80 K or 20 K can be accomplished by passing liquid nitrogen or liquid helium, respectively, through the cylindrical tube. The sample can be heated to a temperature of 2500 K by passing a current through two current leads inserted through the cylindrical tube. The problem of sealing a rotatable cold-finger has been solved by employing a differentially pumped double Wilson seal [2.7]. The cylindrical tube slides through a thin sheet of Viton rubber in which a hole has been cut that is considerably smaller in diameter than the tube. This Viton rubber seal separates the atmosphere from the differential pumping stage, which is at a pressure of approximately 10^{-2} Torr. Another similar Viton rubber seal is placed so that it separates the UHV chamber from the differential pumping stage. This double-seal design allows a base pressure of 5 x 10^{-11} Torr to be reached in the UHV chamber [2.6]. The crystal mounting that allows a certain degree of rotation of the crystal about its surface normal is shown in Fig. 2.4. The rotation is possible by controlling an axial feedthrough situated near the top of the manipulator. The motion is transferred to a movable coupling which moves the connector between the coupling and sampling mount either up or down depending on the desired direction of rotation.

2.3 Electron Gun and Display System

2.3.1 Electron Gun

The main features of a commercial electron gun are shown in Fig. 2.5. An electron-beam current from 10^{-8} to 10^{-4} A is provided by a heated tungsten filament. The filament assembly is mounted at an angle of 13° off the main gun axis to provide optical baffling of the filament from the exit aperture. A deflector assembly then brings the electron trajectory onto the main axis of the gun. The energy range of the electrons, from approximately 30 to 1000 eV for LEED, is determined by the potential maintained across the anode and the cathode. The electron beam is brought to focus approximately 10 cm from the end of the gun. The gun lens system permits the electron beam to remain in focus with a diameter of less than 1 mm on the sample over the entire energy range.

Most commercial filament control units are not capable of maintaining the steady, low beam current (approximately 10^{-9} A) necessary for measurement of I-V beam profiles of molecular overlayer structures that are dissociated easily by a high beam flux. Modification of a commercial filament con-

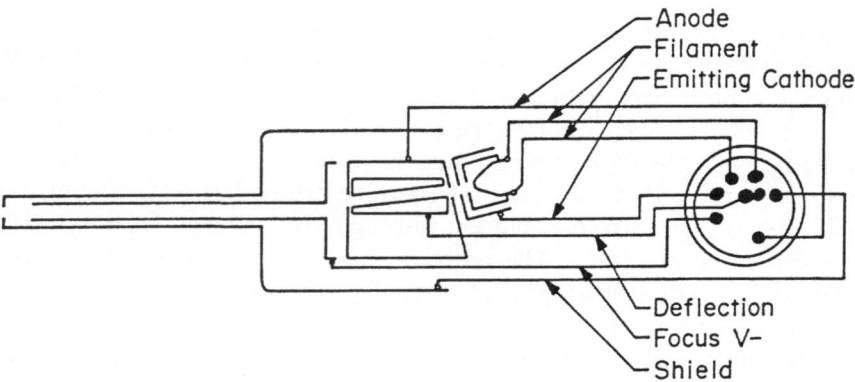

Fig. 2.5. A commercial (Varian) electron gun wiring diagram

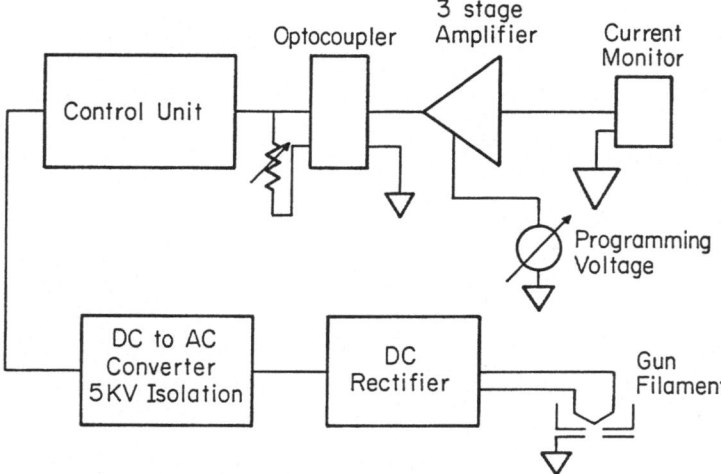

Fig. 2.6. Schematic diagram showing the control unit modification to maintain the beam current near 10^{-9} A. After [2.8]

trol unit to supply a steady beam current of 30 x 10^{-9} A or less has been achieved by Feulner and Menzel [2.8]. A schematic diagram of this modification is shown in Fig. 2.6. In normal operating conditions, the filament control unit is in the constant beam-current mode. The voltage developed between the cathode and the earth ground by the beam current is amplified by the error amplifier. The amplified output is applied to a voltage regulator as an error signal. The error signal is used to control the filament voltage and thus the beam current. For a very low beam current, the error amplifier does not have sufficient sensitivity or gain to control the voltage regulator and thus the beam current. Modification of the circuit was made under a constant beam-voltage mode of operation. During standard operation, a voltage, which is determined by the setting of a variable resistor, is applied to control the voltage regulator in order to control the filament voltage and thus the beam current. However, for a very low beam current, the voltage determined by the variable resistor is not sufficient to control the voltage regulator. Another regulating voltage is needed to control and modify this voltage. A voltage that is proportional to the beam current is obtained from a monitor output of the LEED power supply.

The major problem with the voltage is that it contains a high-frequency ac noise which has to be filtered and the dc component has to be amplified. The latter is accomplished with a three-stage amplifier (shown in Fig. 2.7). Stage 1 is a high-input-impedance noninverting amplifier, which uses an AD517L IC operational amplifier. It has a low-pass frequency characteristic, and its output provides a voltage proportional to beam current at the conversion ratio of 1 V/μA. Stage 2 is a unity-gain inverting buffer. Stage 3 has two inputs, one from stage 2 and one from a programming voltage source.

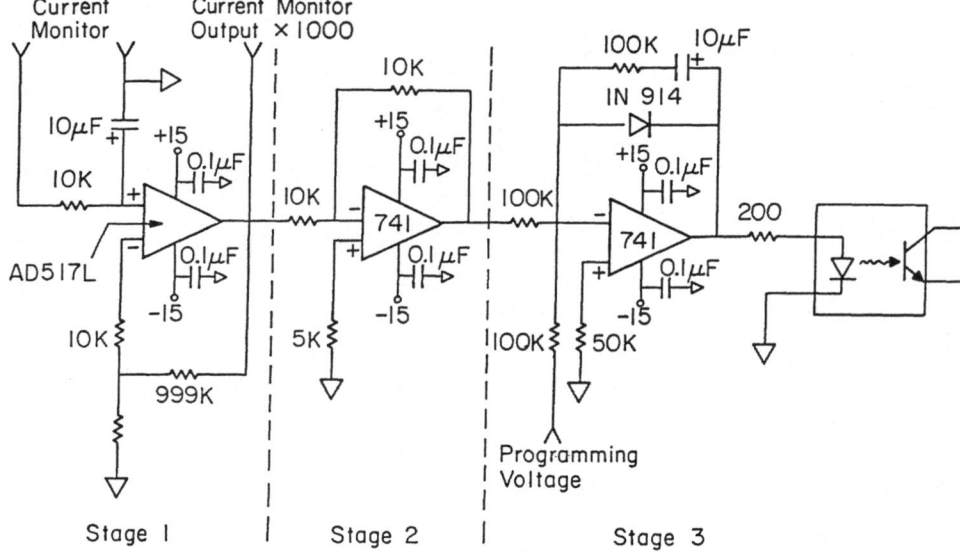

Fig. 2.7. Schematic of the three-stage amplifier used in Fig. 2.6

The programming voltage sets the quiescent beam current, which is ordinarily determined by the variable resistor. Stage 3 has a high dc gain and rolls off with a frequency due to the ac feedback loop provided by the combination of a 100 kΩ resistor and a 10 µF capacitor. Stage 3 drives the light emitting diode (LED) input of the optocoupler (Fig. 2.6). The LED in the optocoupler controls a phototransistor which is in parallel with the variable resistor. The voltage developed across the variable resistor in the usual operation is modified and controlled by the phototransistor. The resulting voltage controls the voltage regulator which subsequently controls the beam current.

2.3.2 Display System

The apparatus most commonly used to display the LEED pattern is the retarding-field energy analyzer. It consists typically of four hemispherical concentric grids and a fluorescent screen, each containing a central hole through which the LEED gun is inserted as shown in Fig. 2.8. The first grid is connected to earth ground to provide an essentially field-free region between the sample and the first grid. This minimizes an undesirable electrostatic deflection of diffracted electrons. A suitable negative potential is applied to the second and third grids (suppressor grids) to allow a narrow range of the elastically scattered electrons to be transmitted to the fluorescent screen. The fourth grid is usually grounded to reduce field penetration of the

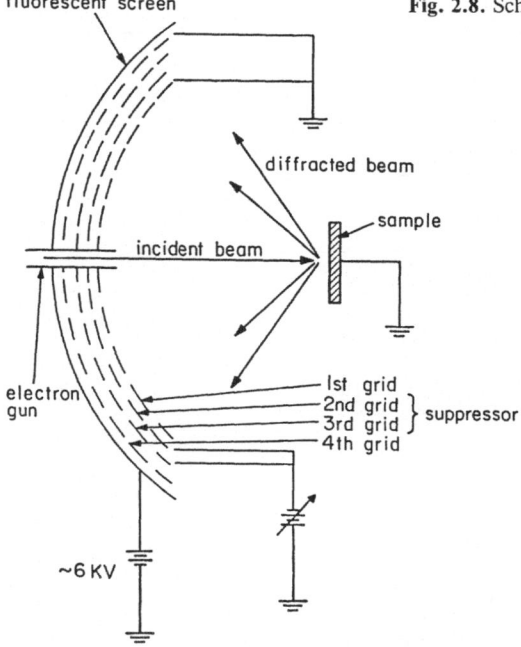

fluorescent screen

Fig. 2.8. Schematic of a four-grid LEED display system

diffracted beam

sample

incident beam

electron gun

1st grid
2nd grid
3rd grid } suppressor
4th grid

~6 KV

suppressor grids by the screen voltage when a potential of a few kilovolts is applied to the screen in order to render the diffraction beams visible.

2.4 Methods of Data Acquisition

There are several different types of LEED intensity measurements which may be made, depending on the purpose of the experiment: for example, intensity versus energy or accelerating voltage (I-V) of the incident electron beam for determination of surface structures, intensity versus polar angle of emergence (I-θ) for determination of the instrumental response function and surface perfection, and intensity versus temperature of the crystal surface (I-T) for determination of the Debye temperature of the crystal. Since the most important measurement is that of the I-V beam profiles, the discussion below is devoted to different methods of collecting this type of data.

2.4.1 Faraday-Cup Collector and Spot Photometer

The Faraday-cup collector [2.9,10] and the spot photometer [2.11] are the most commonly used detector systems in a LEED measurement. A commercial Faraday-cup assembly is shown in Fig. 2.9. The intensity of the backscattered electrons is measured by a movable Faraday cup, which col-

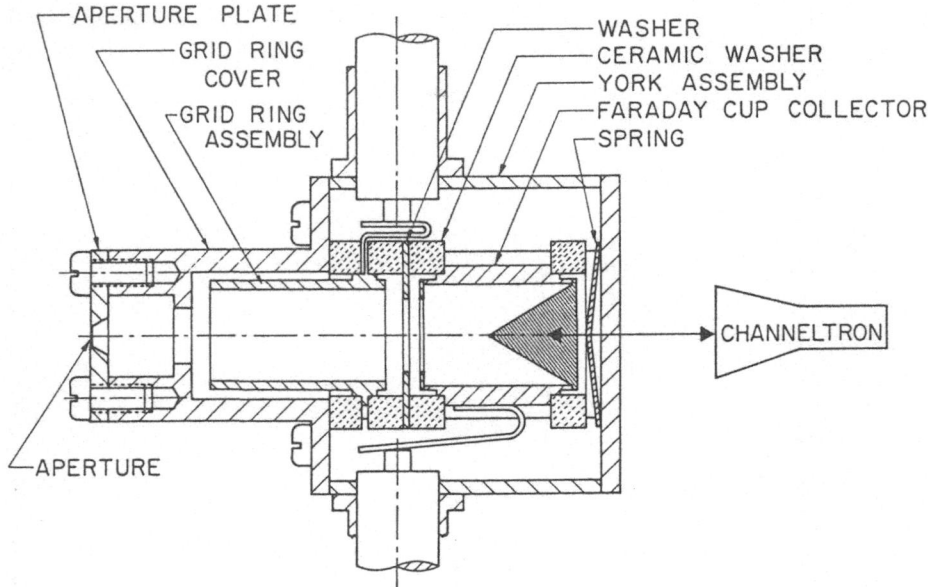

Fig. 2.9. A commercial (Varian) Faraday-cup assembly. The Faraday-cup collector can be modified to detect low beam currents by the addition of a channeltron electron multiplier

lects only the elastic electrons by having a retarding potential between two tungsten grids near the aperture. The energy spread in the elastic electrons depends on the bias potential applied at the grids. If the electron-gun control unit has been modified to supply a very low beam current, as discussed in Sect. 2.3.1, the collector of the Faraday cup can be modified by the addition of a channeltron electron multiplier [2.12]. This modification enables measurement of I-V beam profiles of molecular overlayers easily dissociated by a high incident beam current. The aperture diameter of the Faraday cup is typically 0.02" to 0.06"; the Faraday cup has two orthogonal degrees of angular motion as shown in Fig. 2.10. During the LEED I-V measurement, the Faraday cup can be positioned at any diffracted beam and can be controlled to follow the movement of that beam as the energy of the incident electron beam changes. A spot photometer may also be used to collect LEED intensities by measuring the brightness of the spots due to the diffraction beams. Both methods have been employed successfully in measuring LEED I-V profiles, but they suffer from the disadvantage of long and tedious data-collection times. This is typically 15-20 minutes when using a Faraday cup or 5-10 minutes when using a photometer to measure a single I-V beam profile of energy range from 20 to 200 eV at 2 eV intervals. For overlayer surface structures with a large unit cell, the number of diffracted beams is several times the number of diffracted beams obtained from the clean sub-

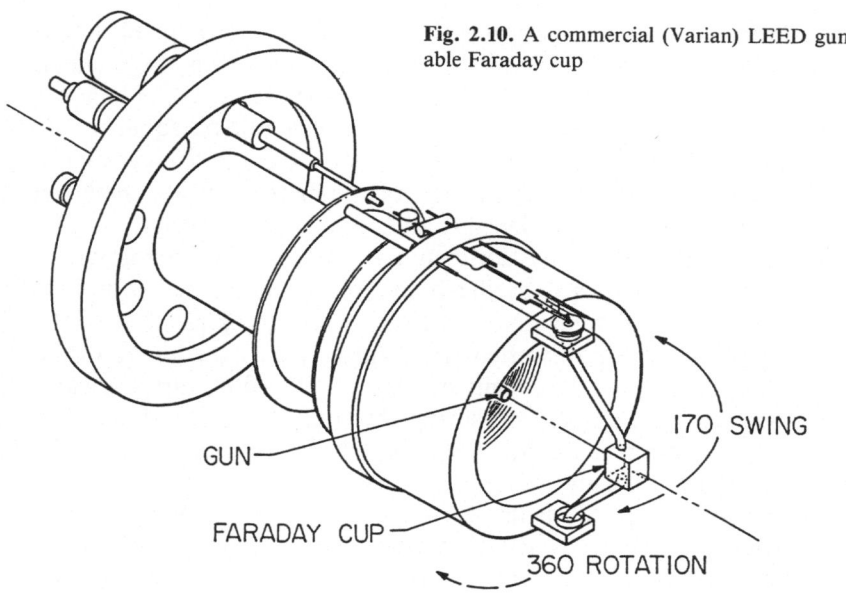

Fig. 2.10. A commercial (Varian) LEED gun with a movable Faraday cup

strate for the same energy range. For example, the formation of a $(\sqrt{3} \times \sqrt{3})R30°$ overlayer structure on an fcc(111) surface increases the number of diffracted beams compared to that for the clean fcc(111) surface by a factor of three [2.13]. Situations like this make the collection of a complete set of data quite time-consuming.

Recently, several experimental developments have occurred which permit both a rapid and an accurate collection of LEED intensities. Stair et al. have developed a method to collect LEED intensities using a photographic technique [2.14]. Heilmann et al. have developed another procedure which measures LEED intensities by scanning the LEED screen with a vidicon-camera connected directly to a processing computer [2.15]. Further improvements in the speed of data collection have been achieved by Frost et al. [2.16], as well as by Tommet et al. [2.17] by combining both the photographic technique and the computer-controlled vidicon camera method.

2.4.2 Photographic Technique

In the method developed by Stair et al. [2.14], the diffraction intensities are measured by analyzing photographs of the diffracted beams displayed directly on the fluorescent screen; a fast film of suitable sensitivity (e.g. Kodak Tri-X emulsion No. 40-265) is used. An exposure time of one second is chosen, with an aperture of either f/4.0 or f/2.8, depending upon whether the diffracted feature of interest is the (00) beam or a nonspecular beam, respectively. Before the intensities of the diffraction beams can be computed from the negative images on the film, a calibration which converts the optical density to the diffracted beam intensity must be established. For fixed conditions of film emulsion, developing, and spectral distribution in the incident light, the optical density is a monotonic function of exposure. If the incident light intensities remain constant in time, the exposure will be equal to the product of the light intensity and time. Keeping the exposure-time constant, the optical intensity is a monotonic function of the light intensity. Hence a calibration between the optical density and the light intensity can be achieved. Film from the same emulsion batch as that used in photographing the diffraction pattern is exposed to the image of a calibrated, continuous neutral density wedge. The wedge is calibrated so that the logarithm of the relative intensity of the light transmitted varies linearly over the length of the wedge. Thus, the logarithm of the light incident on the film varies linearly over the length of the image. By measuring the optical density of the film along this image, a relation is obtained between the optical density and the relative diffracted beam intensity.

The negatives are scanned using a computer-controlled digital-output microdensitometer. A table of measured optical densities is generated by the computer. The intensity of each spot is then computed by converting the optical density of each point to an intensity, subtracting a locally determined background, and summing the intensities for all points having intensities greater than the background. To determine the spot size from the negatives, the number of points used in the summation is counted, since the spot size should be proportional to this number of points. Finally, a computer listing of the location and the intensity of various diffracted beams for each frame is obtained. An I-V profile for a particular beam is obtained by scanning the output of a sequence of frames taken at different energies. This method requires a short data-acquisition time (one second per frame or about ten minutes for an I-V profile of energy range 20-200 eV in 2 eV intervals) and produces a hard copy of all the diffracted beam intensities simultaneously. The disadvantage of this method is the long and tedious analysis of the photographic negatives.

2.4.3 Vidicon Camera Method

Another procedure, developed by Heilmann et al. to measure I-V beam profiles, is to use a system which consists of a vidicon camera, a commercial TV unit, a processing computer, and a time base [2.15]. The same system was modified by Lang et al. with the addition of a video tape recorder [2.18] to provide shorter data-collection times. The principal features of the modified system are shown in Fig. 2.11. The LEED pattern is displayed on a TV screen through the vidicon camera. The intensities of all the beams are stored in the video tape recorder. A particular diffracted beam can be delineated by cursors. Its intensity can be computed from a mean profile obtained by averaging slit profiles of the beam prior to the subtraction of a

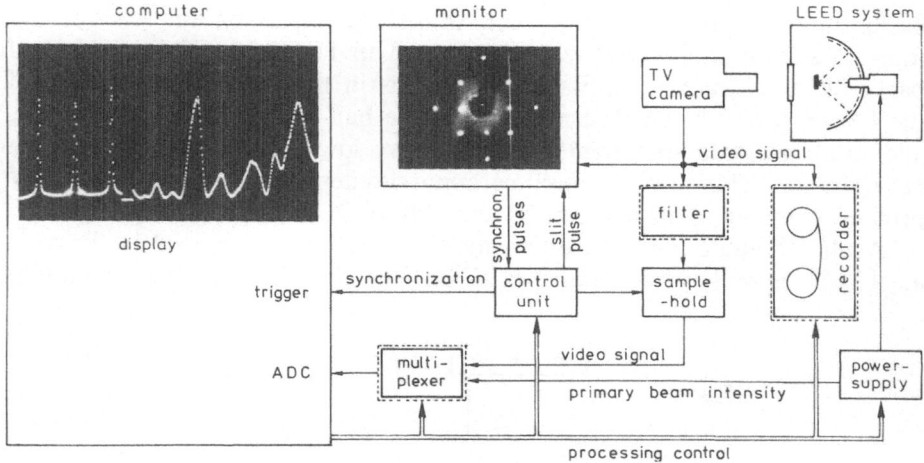

Fig. 2.11. Schematic diagram of a modified computer-controlled TV system [2.18]. The elements modified in [2.18] with respect to [2.15] are enclosed by broken lines

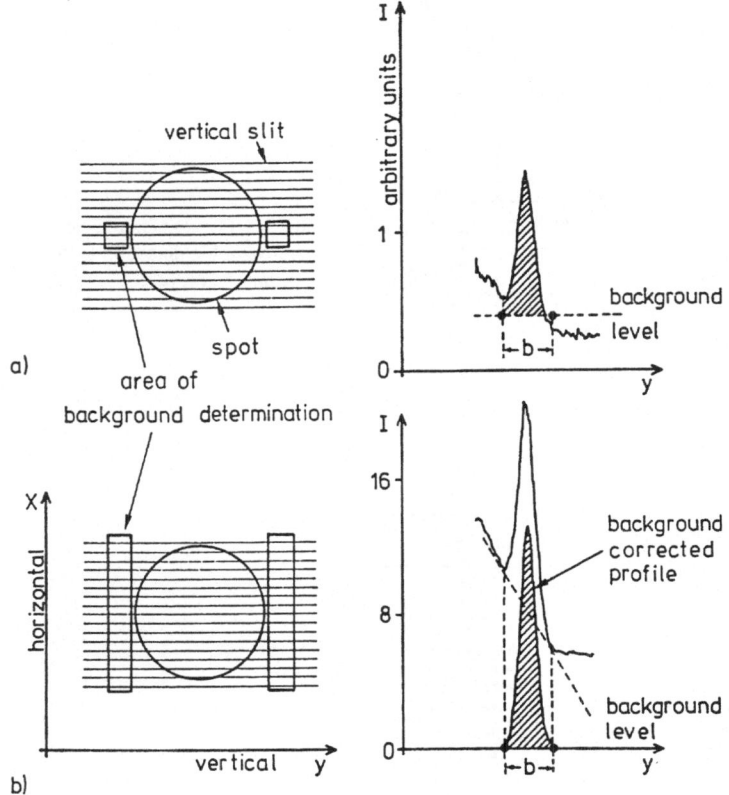

Fig. 2.12a, b. Background treatment in the vidicon camera method of Heilmann et al. [2.18]: **(a)** Background determination; **(b)** background subtraction

background intensity, which is determined by averaging the background intensities around the cursors, as illustrated in Fig. 2.12. This method of background subtraction allows improvements in the signal to noise ratio of the I-V spectra, a better determination of the half-width of the beam profile, and more accurate spot tracing even at very low beam intensity and high background. This method requires approximately ten minutes for 20 I-V profiles of energy range between 20 and 300 eV in intervals of 1 eV [2.18].

A photographic vidicon-camera method has been developed by Frost et al. [2.16], as well as by Tommet et al. [2.17]. The LEED diffraction pattern

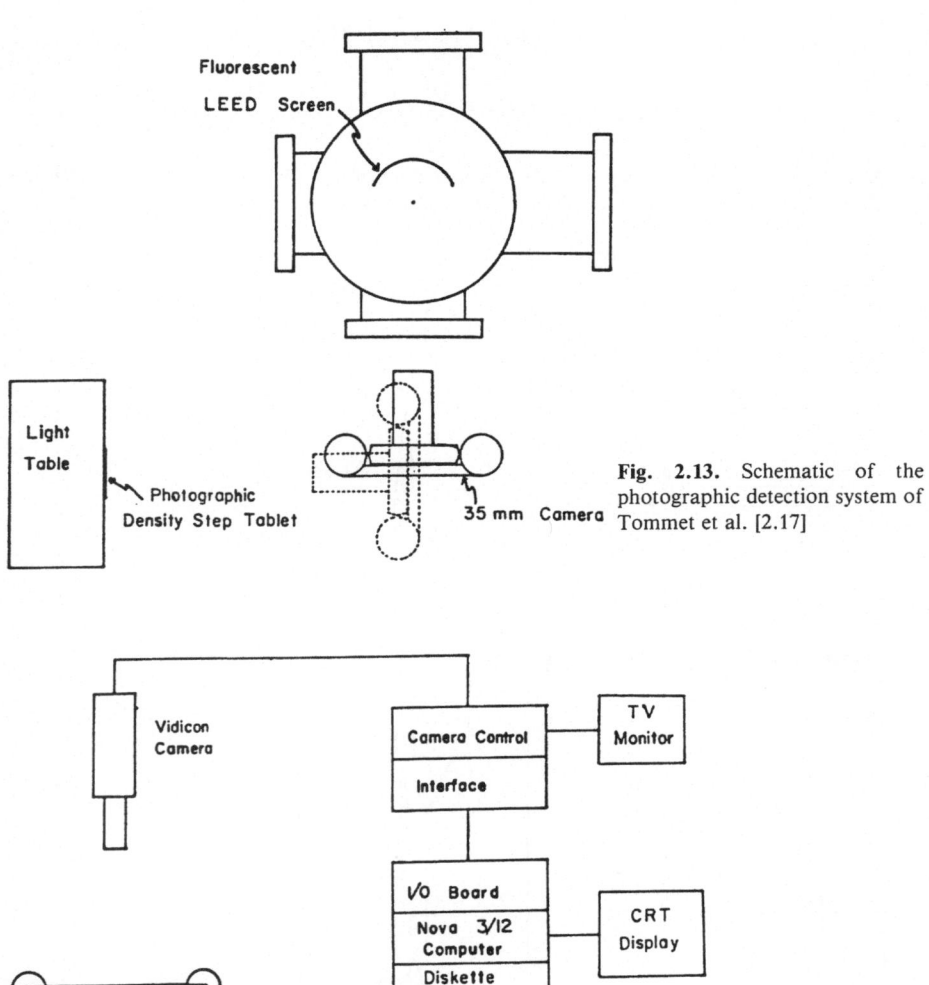

Fig. 2.13. Schematic of the photographic detection system of Tommet et al. [2.17]

Fig. 2.14. Schematic of the vidicon detection system of Tommet et al. [2.17]

is photographed by a 35-mm camera, a procedure similar to that used by Stair et al. The intensities of the diffracted beams are computed by scanning the negatives with a computer-controlled vidicon camera. The vidicon camera is interfaced to the processing computer, which includes a dual-diskette drive and a cathode-ray-tube display. Schematic diagrams of the photographic system and the vidicon system are shown in Figs. 2.13 and 2.14. This method is comparable to the modified version of the TV camera-computer system in many respects, however, the latter method does not require photographic film and a calibration between the light intensity and the optical density.

2.4.4 Position-Sensitive Detector

In a recent design, the spherical retarding-field grid optics are coupled to a two-stage channel electron multiplier array and a resistive-anode image converter [2.19]. The back-to-back channel plates amplify the electron current by a factor of approximately 10^6. The electron beam is then accelerated further onto the resistive anode that records the position of the electron pulse. Although the pulse shape is distorted, its centroid can still be determined accurately. The resistive anode acts as a current divider where the fractional charge that moves to one of two opposing sides is inversely proportional to the distance the charge must travel to reach that side. Thus, by measuring the charge collected at the corners of the resistive anode, the position as well as the charge of an arriving electron pulse is determined. Since the detection electronics can measure each event within 5-10 μsec, a 50-kHz count rate can be achieved without a severe pulse pileup. In this way, a full set of I-V profiles (approximately 20 beams) containing 100 intensity points can be measured in less than 10 minutes.

The major advantage of this approach is the extremely low incident beam current (approximately 10^{-12} A) that can be achieved, which results in negligible beam damage of the surface or any adlayer. The pin-cushion distortion caused by the flat electron detector can easily be corrected in the beam-intensity-evaluation programs.

2.5 Instrumental Response Function

2.5.1 Basic Concepts

The observed LEED diffraction pattern would be infinitely sharp if the experiment were performed on a perfect, rigid, and infinite crystal with a perfect instrument. In practice, the diffraction spots are of finite size due to instrumental distortion [2.20-22] and surface imperfections [2.23,24]. The effect of the instrumental distortion on the measured LEED diffraction pattern from a perfect, rigid, and infinite crystal can be estimated through the use of an instrumental response function [2.20-25].

The experimentally measured function will always be broadened by the instrumental response function. In this case, the measured intensity function $J(\bar{s})$, which the instrument would measure using a diffracted beam from a perfect lattice, can be expressed as the convolution product of two terms: the intensity function $I(\bar{s})$, which is a measurement obtained from a perfect instrument, and the instrumental response function $T(\bar{s})$, which represents the response of the instrument to a diffracted beam from a perfect crystal. This convolution product depends on the scattering vector $\bar{s}$ (i.e. on the momentum transfer $\bar{s}$):

$$J(\bar{s}) = T(\bar{s})*I(\bar{s}) \equiv \int_{-\infty}^{\infty} T(\bar{s}')I(\bar{s} - \bar{s}')d\bar{s}' \ . \tag{2.1}$$

It is well known from the kinematic theory of scattering that the Fourier transform of the intensity function gives the autocorrelation function of the crystal structure. The Fourier transformation of (2.1) yields

$$F\{J(\bar{s})\} = F\{T(\bar{s})*I(\bar{s})\} = t(\bar{r})\Phi(\bar{r}) \ . \tag{2.2}$$

The autocorrelation function $\Phi(\bar{r})$ is a measure of the number of pairs of scatterers that are connected by a real space vector $\bar{r}$. Consequently, it provides structural information concerning the crystal. In principle, $I(\bar{s})$ can be determined accurately if $T(\bar{s})$ and $J(\bar{s})$ can be determined accurately. However, in practice, both $T(\bar{s})$ and $J(\bar{s})$ have uncertainties associated with them, and the function $I(\bar{s})$ cannot be determined with a width less than the sum of the uncertainties in the widths of $T(\bar{s})$ and $J(\bar{s})$. The Fourier transform of the instrumental response function $t(\bar{r})$ is defined as the transfer function [2.21]. The full-width at half-maximum (FWHM) of the transfer function, $w[t(\bar{r})]$, is defined as the transfer width. The autocorrelation function is modulated by the transfer function. The correlation between two scatterers cannot be detected over the range for which $t(\bar{r})$ is zero. Hence, the transfer width is regarded as the effective coherent range of the instrument, however, it does not represent a limit on the region over which phase correlation exists or can be detected. The transfer width determined by measuring the angular profile of a diffracted beam varies typically between 20 and 120 Å depending on diffraction conditions [2.22].

To evaluate the instrumental response function, several experimental factors have to be considered. If each experimental contribution is assumed to be a Gaussian function, then the overall instrumental response of the system is the convolution product of these separate response functions, i.e.

$$T(\bar{s}) = T_1(\bar{s})*T_2(\bar{s})*T_3(\bar{s})*T_4(\bar{s})* \ \cdots \ . \tag{2.3}$$

Then the FWHM of $T(\bar{s})$, which is the response width, is given by the square root of the sum of the squares of the FWHM's of these separate response functions,

$$W[T(\overline{s})] = \{\sum_i W^2[T_i(\overline{s})]\}^{1/2} . \qquad (2.4)$$

Since $t(\overline{r})$ is the Fourier transform of $T(\overline{s})$, the equivalent width of $t(\overline{r})$ is the reciprocal equivalent width of $T(\overline{s})$, i.e.

$$w[t(\overline{r})] = \frac{1}{W[T(\overline{s})]} .$$

2.5.2 Contributions to the Response Width

The instrumental response function as a function of $\overline{s}_\parallel$, which is the component of the momentum transfer of the electron in a direction parallel to the crystal surface, can be determined by measuring the angular profiles of diffracted beams. The transformation of $J(\overline{s}_\parallel)$ to $T(\overline{s}_\parallel)$ can be achieved through the relationship governing the conservation of linear momentum of the electron parallel to the crystal surface, i.e.

$$|\overline{s}_\parallel| = [(2mE)^{1/2}/\hbar](\sin\theta - \sin\theta_o) , \qquad (2.5)$$

where E is the incident electron energy in the vacuum, m is the mass of the electron, $2\pi\hbar$ is Planck's constant, θ_0 is the angle of incidence with respect to the surface normal, and θ is the angle of emergence in the same azimuth. (Note that θ and θ_0 may be different due to parallel momentum exchange in the amount of a reciprocal-lattice vector, cf. Chapter 3.) A more general expression is needed if the azimuthal angles for θ and θ_0 have different

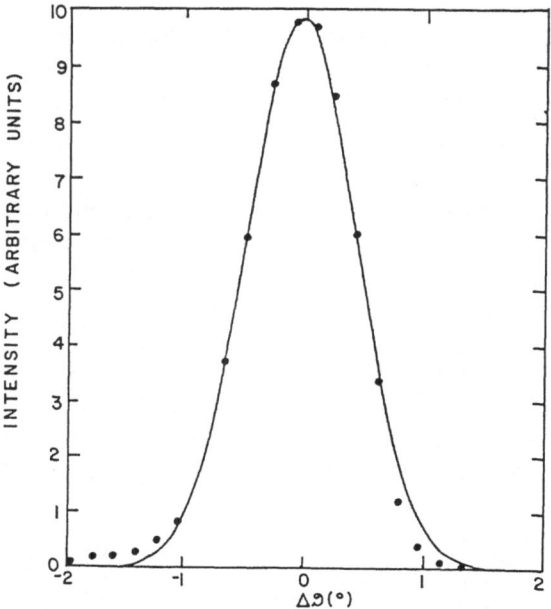

Fig. 2.15. Comparison between an experimentally measured beam profile $J(\theta)$ of the (20) beam for the W(110) surface at E = 97.5 eV, $\theta_0 = 17°$, and $\varphi = 35.25°$ (*solid line*) and a Gaussian fit (*filled circles*) [2.22]. $\Delta\theta$ measures the deviation from the ideal spot direction θ_0

values. An example of a measured angular profile, $J(\vec{s}_{\parallel})$, for the (20) beam of the W(110) surface, is shown in Fig. 2.15. The angular profile is approximately a Gaussian function. Application of (2.1,2) gives

$$J(\vec{s}_{\parallel}) = T(\vec{s}_{\parallel})*I(\vec{s}_{\parallel}) \; , \; \text{and} \; F\{J(\vec{s}_{\parallel})\} = t(\vec{r}_{\parallel})\Phi(\vec{r}_{\parallel}) \; . \tag{2.6}$$

The instrumental response width of a LEED system consists of four major contributions: the energy spread of the incident electron beam, ΔE; the aperture width of the detector, d; the source extension, γ; and the electron beam diameter, D. Application of (2.3) gives

$$T(\vec{s}_{\parallel}) = T_{\Delta E}(\vec{s}_{\parallel})*T_d(\vec{s}_{\parallel})*T_\gamma(\vec{s}_{\parallel})*T_D(\vec{s}_{\parallel}) \; . \tag{2.7}$$

The response width may be regarded as the uncertainty associated with the measurement of position in reciprocal space. Each of the above terms contributes individually to this overall uncertainty, $|\Delta\vec{s}_{\parallel}|$. If each contribution is assumed to be a Gaussian function, the response width is given by

$$W[T(\vec{s}_{\parallel})] = |\Delta\vec{s}_{\parallel}| = \{\sum_i W^2[T_i(\vec{s}_{\parallel})]\}^{1/2} = (\sum_i |\Delta\vec{s}_{\parallel}|_i^2)^{1/2} \; . \tag{2.8}$$

In addition, $|\Delta\vec{s}_{\parallel}|$ is related to the FWHM $\Delta\theta_{1/2}$ of the angular profile $J(\vec{s}_{\parallel})$ by

$$|\Delta\vec{s}_{\parallel}| = [(2mE)^{1/2}/\hbar]\Delta\theta_{1/2}\cos\theta \; . \tag{2.9}$$

We shall discuss next how each of these factors contributes individually to $|\Delta\vec{s}_{\parallel}|$.

(a) *Energy Spread* ΔE. To estimate $|\Delta\vec{s}_{\parallel}|$ due to the energy spread ΔE, the partial derivative of $\vec{s}_{\parallel}$ with respect to E must be evaluated. One finds that

$$|\Delta\vec{s}_{\parallel}|_{\Delta E} = \frac{1}{\hbar}\left[\frac{m}{2E}\right]^{1/2}\Delta E|\sin\theta - \sin\theta_o| \; . \tag{2.10}$$

The energy spread is obtained by measuring the intensity of the diffracted beam as a function of the bias potential applied to the Faraday cup. In the system studied by Wang and Lagally [2.22], the energy spread is given by

$$\Delta E \text{ (eV)} = 0.35 + 0.66 \times 10^{-3}E \text{ (eV)} \; .$$

(b) *Aperture Width* d. The detector has a finite aperture which causes an uncertainty in the measurement of the angle of diffraction, as may be seen in Fig. 2.16a. To estimate $|\Delta\vec{s}_{\parallel}|$ due to the uncertainty in $\Delta\theta$ caused by the finite aperture width, the partial derivative of (2.5) with respect to θ must be evaluated. One finds that

$$|\Delta\vec{s}_{\parallel}|_d = \frac{1}{\hbar}(2mE)^{1/2}(\frac{d}{R})\cos\theta \; , \tag{2.11}$$

where $d/R = \Delta\theta$, d is the geometrical diameter of the aperture, and R is the distance from the sample to the collector.

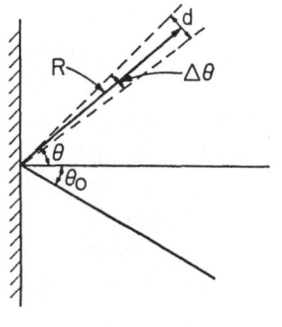

(a) APERTURE WIDTH

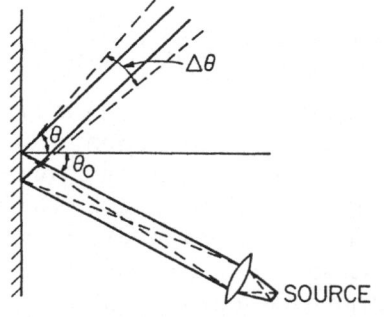

(b) SOURCE EXTENSION

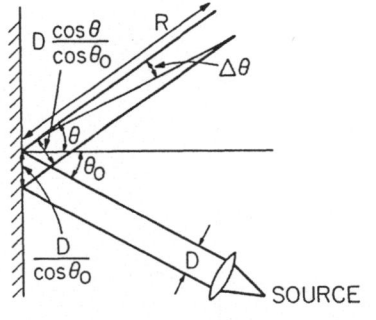

(c) ELECTRON BEAM DIAMETER

Fig. 2.16a – c. Uncertainty in $|\Delta \vec{s}_{\parallel}|$ caused by uncertainty in $\Delta\theta_0$ or $\Delta\theta$ as the result of (a) aperture width d, (b) source extension γ, and (c) beam diameter D. After [2.22]

(c) Source Extension γ. The angle of incidence of the electron beam can be determined accurately if the electron source is a point source. In practice, the electron source has a finite size which causes an uncertainty in the angle of incidence of the beam, as may be seen in Fig. 2.16b. The source extension $\gamma = \Delta\theta_0$ is defined as the angle subtended by the source as viewed from the crystal through the lens of the gun. Its contribution to $\Delta\vec{s}_{\parallel}$ is given by the partial derivative of (2.5) with respect to θ_0, with the result that

$$|\Delta\vec{s}_{\parallel}|_\gamma = [(2mE)^{1/2}/\hbar\,]\,\gamma\cos\theta_0 \ . \tag{2.12}$$

From the partial derivatives of $\vec{s}_{\parallel}$ with respect to θ and θ_0, one finds that

$$\Delta\theta_0\cos\theta_0 = \Delta\theta\cos\theta \ . \tag{2.13}$$

Hence, the source extension can be estimated by measuring the angular profile of a diffracted beam at very large angles [2.26].

(d) Beam Diameter D. The electron beam has a finite size that causes an uncertainty in $\Delta\theta$ which, in turn, causes an uncertainty in $\vec{s}_{\parallel}$, Fig. 2.16c. It is given by

$$|\Delta \vec{s}_{\parallel}|_D = [(2mE)^{1/2}/\hbar \,] \left(\frac{D}{R}\right) \frac{\cos^2\theta}{\cos\theta_o} \,, \tag{2.14}$$

with $\Delta\theta = (D\cos\theta)/(R\cos\theta_o)$, where D is the diameter of the electron beam. The diameter of the electron beam may be measured by scanning the incident beam across a knife edge on a goniometer and measuring the resulting secondary-electron current [2.22]. The beam diameter increases rapidly as the energy decreases below 50 eV, as shown in Fig. 2.17. Usually, D is sufficiently small so that the beam diameter is not an important limiting factor for the instrumental response function.

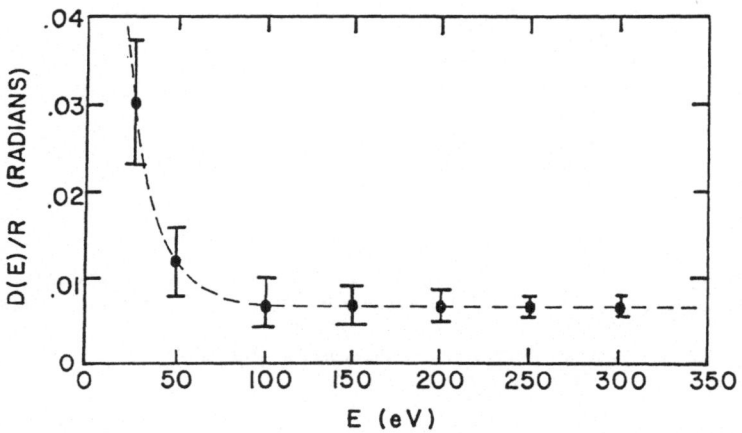

Fig. 2.17. Ratio of beam diameter D to sample-to-detector distance R = 70 mm as a function of electron beam energy [2.22]

The FWHM $\Delta\theta_{1/2}$ of the angular profile $J(\vec{s}_{\parallel})$ and the transfer width of the (20) beam for the W(110) surface at normal incidence have been calculated by employing the above equations [2.22]. The results are shown in Fig. 2.18. The source extension and the aperture width are the dominant factors in determining $\Delta\theta_{1/2}$.

The measured FWHM of the angular profiles of the (20) beam as a function of the incident electron beam energy is shown in Fig. 2.19. The oscillations in the $\Delta\theta_{1/2}$ profiles are due to the presence of steps on the surface [2.24,25]. In addition to the instrumental distortion, the presence of steps on the surface causes a further broadening of the diffraction spots due to constructive and destructive interference from the terraces at different heights. The minima in the $\Delta\theta_{1/2}$ profile can be regarded as the result of coherent scattering from an ordered surface. The broadening beyond the minimum $\Delta\theta_{1/2}$ gives an indication of step disorder. The transfer width is calculated by using the minimum values in the $\Delta\theta_{1/2}$ profiles. The calculated values for the transfer width using the geometrical aperture do not fit the

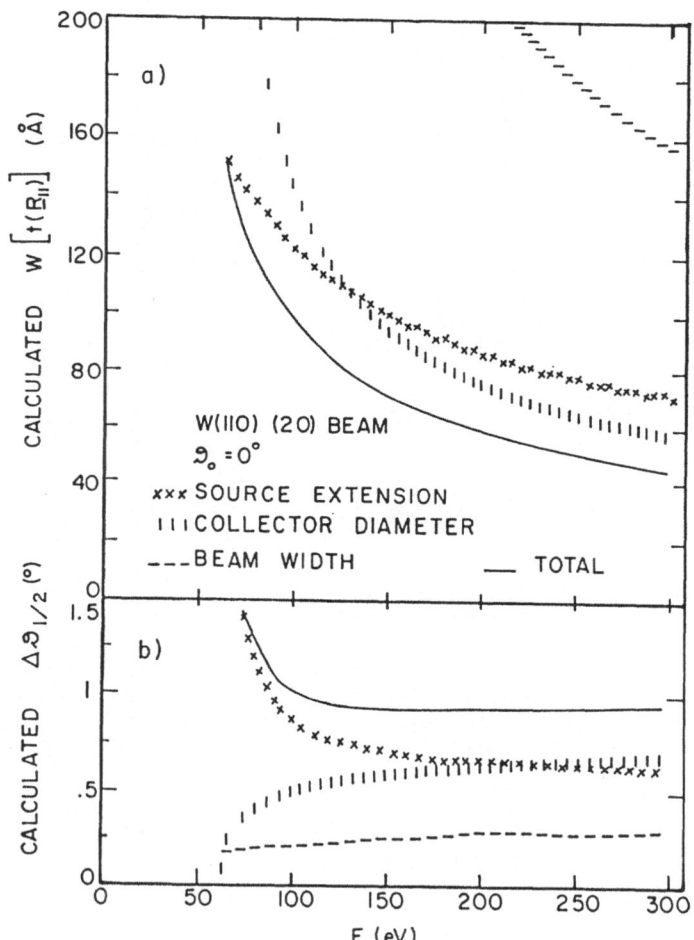

Fig. 2.18. (a) Calculated transfer width of the (20) beam for the W(110) surface at normal incidence as a function of energy. The contributions to the transfer width from different factors are plotted individually. The parameters used in the calculation are: source extension $\gamma = 0.01$, detector aperture $d/R = 0.014$, and beam diameter D/R as shown in Fig. 2.17 [2.22]. **(b)** Same as **(a)** for the angular width

experimental transfer width satisfactorily. However, the geometrical aperture may be different from the actual aperture due to the bias on the Faraday cup [2.22]. The effective aperture can be obtained by a best fit to the experimental transfer width. It may be seen in Fig. 2.19 that the best fit occurs when a value of d/R of 0.10 is used. Hence, the effective aperture is smaller than the geometrical aperture. The effective aperture can be decreased by using a proper bias, so that its contribution to the instrumental response can be minimized. The source extension is probably the most important limitation to an increased transfer width. It can be minimized by having a large angle

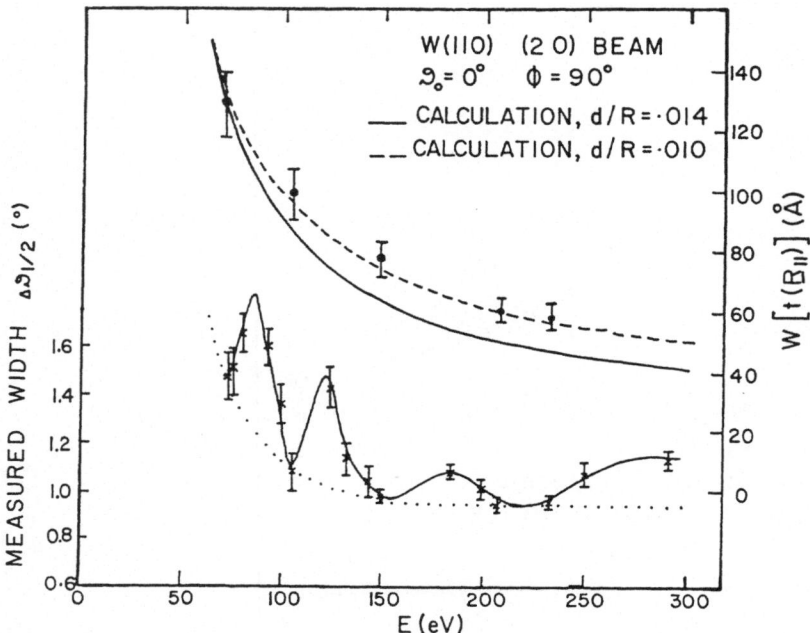

Fig. 2.19. Measured FWHMs of the (20) beam for W(110) surface at normal incidence as a function of energy. The corresponding transfer widths (the filled circles) are calculated by inverting the minima in the measured FWHMs. The calculated transfer widths using the parameters of Fig. 2.18 are shown as solid curves. The dashed curve is the same as the solid curve except for d/R = 0.01. After [2.22]

of incidence, but measurement at an angle near normal incidence has smaller uncertainty which is more desirable experimentally.

2.6 Determination of Angle of Incidence

2.6.1 Different Methods

Currently, there are four methods in use for finding the angle of incidence in LEED. The most common approach is to calibrate the manipulator outside the vacuum system [2.27]. This is done, for example, by correlating a dial setting on the manipulator with the angle of reflection of a laser beam. A second method is to mount a protractor and vernier near the movable sample which can be read during the experiment (usually with the aid of a telescope) [2.28]. A third method is to mount a protractor on the Faraday-cup assembly. By centering the Faraday cup on the specular beam, the angle of incidence can be related to the angle of the Faraday cup [2.29]. The fourth method, which will be discussed in detail, was developed by Cunningham and Weinberg [2.30] and is more convenient to use than the methods mentioned above.

The method developed by Cunningham and Weinberg involves initially taking a photograph of the LEED pattern as it appears on a standard fluorescent display screen. Then, a clear plastic sheet marked with angles (e.g. a transparent photocopy of circular graph paper) is used to measure the angle between the line from the (00) spot to the center of the screen and the line from the center of the screen to the (hk) spot. Finally, a simple program to be described below determines the two angles of incidence.

The advantages of this method are the following. First, no special equipment other than a camera is needed. Second, no work with the manipulator is needed outside the vacuum system. Third, the analysis of the photograph is simple, does not take very much time and does not need to be extremely accurate. Fourth, the angle determination is made from a large number of measurements rather than from just one measurement (such as manipulator setting or protractor reading). Consequently, the statistics of the measurement make it possible to obtain the desired angles quite accurately. Fifth, the crystal does not have to be located at the center of curvature of the LEED screen as is necessary when using the Faraday cup to determine the angle. Finally, the method can be applied to all experimental situations in which a fluorescent screen is used to display the diffraction pattern.

2.6.2 Theory

We proceed as follows. First, the relationship between the wave vector of the incident and scattered electron in both the laboratory reference frame and the crystal reference frame will be developed. This relationship depends on the incidence angle θ (the angle of the incident beam with respect to the surface normal), and the azimuthal angle φ (the angle between the crystal x_c-axis and the projection of the incident beam on the crystal surface). Second, the relationship between the wave vectors and the angle measured on the photograph will be determined. A description of the program that solves the equations numerically is given in Appendix B.

The laboratory reference frame, shown in Fig. 2.20, is defined such that the z_l-axis extends along the axis of the electron gun, and this, as well as the specular LEED beam, defines the $y_l z_l$ plane. In a typical experimental system, the y_l-axis is nearly vertical. The flip mechanism of the manipulator causes the specular spot to move down the screen in the $-y_l$ direction. The crystal reference frame, also shown in Fig. 2.20, is defined such that the z_c-axis extends into the crystal and the x_c-axis points along one of the unit cell vectors. At normal incidence the crystal x_c-axis is rotated by an angle ψ with respect to the laboratory x_l-axis, and the z-axes in the two reference systems are collinear. The reason for defining the crystal z_c-axis pointing into the crystal is that this allows the labeling of the LEED spots which are in the laboratory frame and the labeling of the reciprocal lattice vectors which are in the crystal frame both to use right-hand coordinate systems.

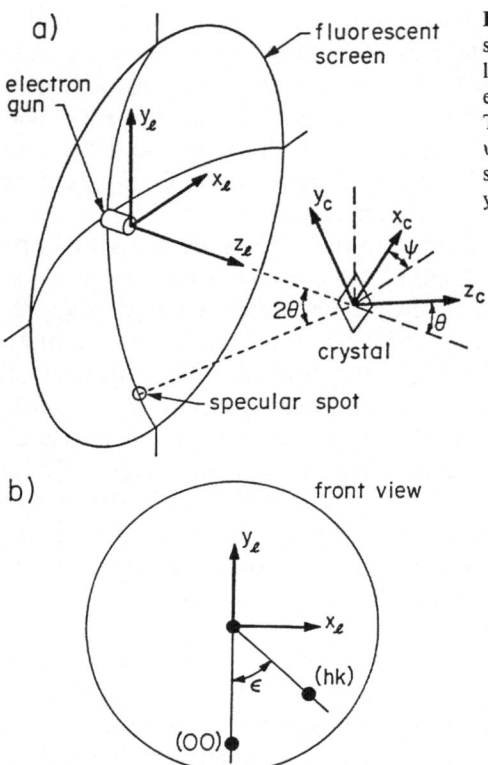

a)

electron gun

fluorescent screen

specular spot

b) front view

Fig. 2.20. (a) Schematic of the LEED experiment showing the laboratory-coordinate system (as related to the hemispherical fluorescent screen and electron gun) and the crystal-coordinate system. The two systems are related by the angles θ and $\psi = -\varphi - \pi/2$. **(b)** Front view of the fluorescent screen showing the definitions of the laboratory y-axis and the angle ε for the (hk) LEED spot

The transformation matrix which relates a vector $\vec{k}_c$ in the crystal frame to a vector $\vec{k}_l$ in the laboratory frame is given by

$$\vec{k}_c = \begin{pmatrix} -\sin\varphi & -\cos\theta\cos\varphi & \sin\theta\cos\varphi \\ \cos\varphi & -\cos\theta\sin\varphi & \sin\theta\sin\varphi \\ 0 & \sin\theta & \cos\theta \end{pmatrix} \vec{k}_l , \tag{2.15}$$

where $\varphi = -\psi - \pi/2$. The incident wave vector in the laboratory frame $\vec{k}_l^i$ has only a z_l-component and is given by

$$k_{lz}^i = k = (2mE)^{1/2}/\hbar , \tag{2.16}$$

where E is the electron energy relative to vacuum zero, m is the free-electron mass, and $2\pi\hbar$ is Planck's constant. In the crystal frame, the incident wave vector is given by

$$k_{cx}^i = k \sin\theta \cos\varphi ,$$
$$k_{cy}^i = k \sin\theta \sin\varphi , \tag{2.17}$$
$$k_{cz}^i = k \cos\theta .$$

Upon scattering from the crystal, the (hk) diffracted beam has the wave-vector components

$$k_{cx}^s = k_{cx}^i + g_x(hk) \ ,$$

$$k_{cy}^s = k_{cy}^i + g_y(hk) \ , \tag{2.18}$$

$$k_{cz}^s = -[k^2 - (k_{cx}^s)^2 - (k_{cy}^s)^2]^{1/2} \ .$$

Here, $\bar{g}(hk)$ is the (hk) reciprocal lattice vector, and the z_c-component is determined by energy conservation. Finally, using the transpose of (2.15), the scattered wave components in the laboratory frame are given by

$$k_{lx}^s = -k_{cx}^s \sin\varphi + k_{cy}^s \cos\varphi \ , \tag{2.19}$$

$$k_{ly}^s = -k_{cx}^s \cos\theta\cos\varphi - k_{cy}^s\cos\theta\sin\varphi + k_{cz}^s\sin\theta \ .$$

From the LEED photograph, the angle $\varepsilon(hk)$ between the $-y_l$ axis and the (hk) spot (see Fig. 2.20b) is related to the wave vector components by

$$\tan\varepsilon(hk) = k_{lx}^s/(-k_{ly}^s) \ . \tag{2.20}$$

For a single spot, (2.20) is a single equation with two unknowns. Therefore, in principle, any two spots on the photograph [other than the (00) spot which defines the $-y_l$ axis] can be used to determine the angles θ and φ.

For any two chosen spots, labeled n = 1 and 2, Eq. 2.20 represents two highly nonlinear equations in two unknowns. These may be solved numerically by using Newton's method. The equations can be written

$$f_n(\theta,\varphi) = k_{lx}^s + k_{ly}^s \tan\varepsilon(hk) = 0 \qquad n = 1, 2 \ . \tag{2.21}$$

For the ith iteration, the (2x2) matrix equation

$$\begin{bmatrix} J_{1\theta} & J_{1\varphi} \\ J_{2\theta} & J_{2\varphi} \end{bmatrix} \begin{bmatrix} \Delta\theta \\ \Delta\varphi \end{bmatrix} = \begin{bmatrix} -f_1(\theta_i,\varphi_i) \\ -f_2(\theta_i,\varphi_i) \end{bmatrix} \tag{2.22}$$

is solved numerically for $\Delta\theta$ and $\Delta\varphi$, where J is the Jacobian matrix given by (for example)

$$J_{1\theta} \equiv \frac{\partial f_1(\theta,\varphi)}{\partial\theta}\Bigg|_{\theta_i,\varphi_i} \ , \text{ etc.} \tag{2.23}$$

New values of θ and φ are determined by

$$\theta_{i+1} = \theta_i + \Delta\theta \ ,$$

$$\varphi_{i+1} = \varphi_i + \Delta\varphi \ , \tag{2.24}$$

and the procedure is repeated until $\Delta\theta$ and $\Delta\varphi$ are less than, for example, 10^{-3} rad. This numerical procedure is quite rapid due partly to the fact that the Jacobian is analytic. A computer program which solves this problem is given in Appendix B.

One advantage of this technique is that the angle $\varepsilon(hk)$ does not depend on the location of the crystal. The angle $\varepsilon(hk)$ is invariant as the crystal is moved along the axis of the hemispherical screen (assuming that the electron

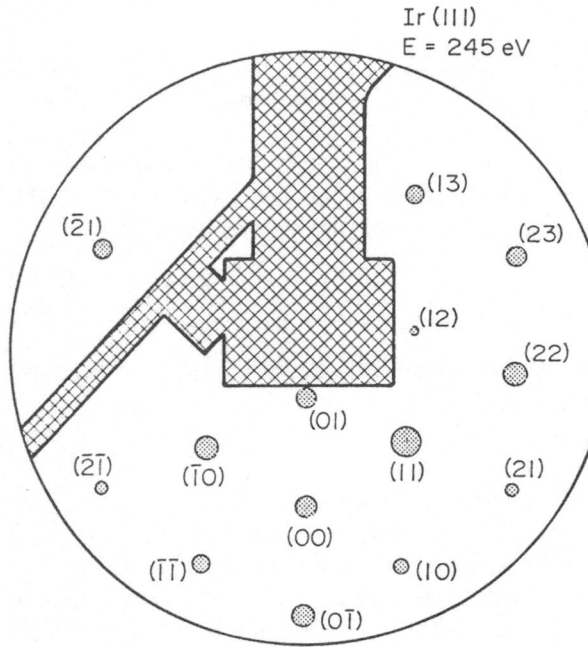

Ir (III)
E = 245 eV

Fig. 2.21. Schematic of a LEED pattern obtained from Ir(111) at E = 245 eV along with the labeling of the diffraction spots

beam is colinear with this axis). It is important, however, that the camera is also located on this axis, so care must be taken in this alignment.

2.6.3 An Example

A schematic of a LEED pattern taken from an Ir(111) surface is shown in Fig. 2.21 [2.30]. The energy of the incident electron is E = 245 eV. There are 13 discernable beams other than the specular beam. Many of the spots are large, but the angles of the lines through the centers of the spots can be estimated easily to ±2°, perhaps even ±1°. The angle ε for each of the beams is given in Table 2.1.

Any two spots (other than the specular spot) are sufficient to determine θ and φ. For n spots, there are n(n − 1)/2 spot combinations. Thus, for the 13 spots in Table 2.1, there are 78 independent determinations of θ and φ. Examples of two of these determinations, one for the (01) and (10) combination and the other for the (23) and (21) combination are shown in Table 2.2. The changes in θ and φ which result from an uncertainty in the measured angles ε of ±2° are also shown in Table 2.2. These two examples illustrate that the error in θ and φ is smaller than the error in ε. This means that these particular spot combinations are quite sensitive to θ and φ. Small changes in (θ,φ) lead to large changes in the value of ε for these spots. On the other hand, some of the spot combinations give large errors in θ and φ

Table 2.1. Angle ε for the spots shown in Fig. 2.21

h	k	ε [°]
0	0	0 defined
1	0	24
1	1	46
$\overline{0}$	1	0
$\overline{1}$	0	-46
$\overline{1}$	$\overline{1}$	-26
0	$\overline{1}$	-1
2	1	55
2	2	83
2	3	115
1	2	100
$\overline{1}$	3	145
$\overline{2}$	$\overline{1}$	-116
$\overline{2}$	1	-56

Table 2.2. Examples of angles of incidence θ and φ (in degrees) determined from the (01) and ($\overline{1}$0) spots and from the (23) and ($\overline{2}$1) spots along with the errors resulting from an assumed uncertainty of ±2° in ε

ε(01)	Δε	ε($\overline{1}$0)	Δε	θ	Δθ	φ	Δφ
0		-46		13.18		-90.00	
	+2		-2		-0.56		0.64
	-2		+2		0.60		-0.87
	-2		-2		-0.64		-0.62
	+2		+2		0.68		0.89
ε(23)	Δε	ε($\overline{2}$1)	Δε	θ	Δθ	φ	Δφ
115		-116		12.99		-89.58	
	+2		-2		-0.77		0.00
	-2		+2		0.75		0.00
	-2		-2		0.02		-1.73
	+2		+2		0.01		1.72

(approximately 10°) for the same ±2° uncertainty in ε. The reason why some spots are more sensitive than others is due to the way in which the angle ε is measured. For example, a small increase in the tilt angle θ moves all the spots the same distance down the screen. However, the angle ε for the (01) spot does not change at all, the angle ε for the (21) spot changes an intermediate amount, and the angle ε for the (12) spot changes considerably.

In obtaining the resultant θ and φ for this photograph, the values are averaged over the various spot combinations. In this average, however, those spot combinations are omitted for which the uncertainty in the determination of the spot position leads to an excessive uncertainty in the angular determination (namely, for an uncertainty $\delta\varepsilon$ in the spot position ε, the resulting uncertainties $\delta\theta$ and $\delta\varphi$ are required to satisfy $|\delta\theta| \leqslant 2|\delta\varepsilon|$ and $|\delta\varphi| \leqslant 3|\delta\varepsilon|$). For the example shown in Fig. 2.21, this eliminates 24 of the spot combinations. The straight average of the remaining 54 values gives $\theta_{ave} = 13.17 \pm 0.54°$ and $\varphi_{ave} = -89.22 \pm 1.23°$. The errors given here are the standard deviations from the respective means. It should be noted that the solutions to (2.20) are weakly dependent on the initially (guessed) solutions, as in any iterative scheme.

Since an error can be associated with each determination of θ and φ, it is possible to perform a weighted average. It is here that statistics work to the advantage of this method. If σ_i is the standard deviation of the ith measurement, then the weighted mean is given by

$$x_m = \sigma_m^2 \sum_i x_i/\sigma_i^2 \ , \qquad \text{where} \tag{2.25}$$

$$\frac{1}{\sigma_m^2} = \sum_i \frac{1}{\sigma_i^2} \ , \tag{2.26}$$

and where x_i is the ith measurement, and σ_m is the standard deviation of the mean.

It is assumed that the maximum $\Delta\theta$ and $\Delta\varphi$ due to the error in ε is the standard deviation for that spot combination. Thus, for example, the standard deviations for the first value in Table 2.2 would be taken to be 0.68° and 0.89° for θ and φ, respectively. The weighted averages for the example in Fig. 2.21 using 54 values are $\theta_m = 13.14 \pm 0.12°$ and $\varphi_m = -89.80 \pm 0.25°$. If an error of $\pm 1°$ in ε is assumed, then the weighted averages of 63 points (15 being omitted) are $\theta_m = 13.18 \pm 0.06°$ and $\varphi_m = -89.81 \pm 0.11°$. This shows that this method, coupled with care in making the measurements and the use of a photograph with many spots, is capable of producing very accurate values for the angles of incidence.

This method of determining the angles of incidence in a LEED experiment is both accurate and simple. The method is accurate because the angles are determined a large number of times from a single photograph. Absolute accuracies of less than 0.1° should be attainable routinely. The method is simple in that it requires only a camera. In addition, there is no need for a calibration point (such as normal incidence) to be established. However, it should be noted that the above procedure is applicable only when the electron gun is collinear with the axis of the camera. A related method was developed by Price [2.31] to determine the incident and azimuthal angles by treating the position of the electron gun and the electron energy as unknowns and using at least three LEED spots, including the specular beam. This method, however, requires the crystal to be positioned precisely at the center of curvature of the LEED screen. In addition, both methods require that the incident electron beam be aligned with the center of curvature of the LEED screen. Under special circumstances when the crystal is off-center or the electron gun is misaligned, a more general method developed by Sobrero and Weinberg [2.32] should be used. This method can be used to check the alignment of the electron gun and the position of the crystal in addition to determining the polar and azimuthal angles of incidence.

2.7 Determination of the Debye Temperature

Theoretical calculations predict a larger root-mean-square (rms) displacement of surface atoms compared to bulk atoms [2.33-35]. Moreover, the parallel component is predicted to have a smaller rms displacement than the perpendicular component. This is consistent with the idea that the creation of a free surface leads to a greater reduction of the total forces affecting the perpendicular motion than the parallel motion.

To determine the vibrational properties of the surface atoms, a surface-sensitive probe is necessary. LEED is the most commonly employed probe since low-energy electrons, which interact strongly with matter, are sensitive to only a limited number of atoms near the surface. Within the kinematic approximation, if all atoms had identical vibrational amplitudes, the intensities of the diffracted beams would decrease exponentially with an increase in the temperature of the crystal surface, as shown in Fig. 2.22. By measuring the intensities of the diffracted beams as a function of the temperature of the

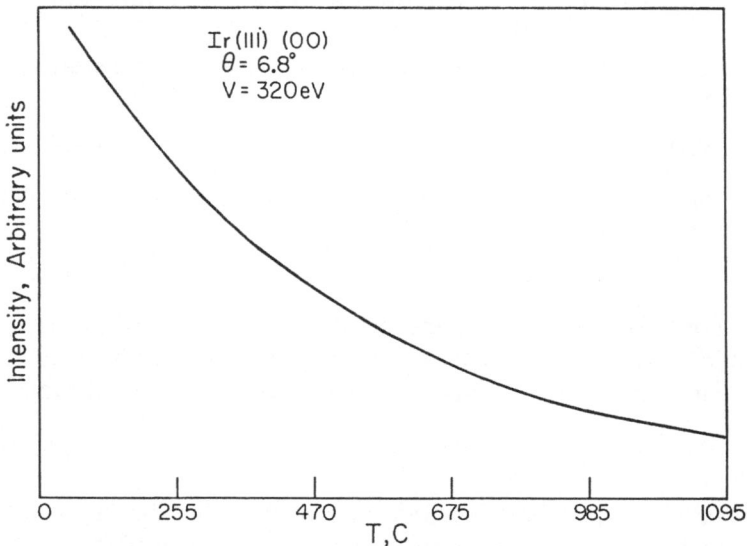

Fig. 2.22. The temperature dependence of the intensity of the (00) beam diffracted from the Ir(111) surface at an angle of incidence of 6.8° and at a kinetic energy of 320 eV

crystal surface, the vibrational properties of the surface atoms can be investigated. Even though LEED is a surface-sensitive probe, the electron beam still penetrates a few layers of atoms into the bulk (dependent upon its incident energy). The measured "effective" Debye temperatures, which are designated as θ_D^{eff}, are an average over these few layers. At electron energies on the order of 50 eV, θ_D^{eff} can be taken as the surface Debye temperature, and θ_D^{eff} may be taken to be the bulk Debye temperature at much higher electron energies, e.g. above a few hundred eV. A plot of θ_D^{eff} as a function of the incident electron energy demonstrates a gradual change in the vibrational properties of atoms from the surface to the bulk. However, multiple-scattering effects can severely distort the values of θ_D^{eff}, as will be discussed in Sects. 4.7 and 5.11.

2.7.1 The Debye Temperature Normal to the Crystal Surface

To determine the surface Debye temperature experimentally is relatively straightforward. The crystal should be positioned so that the (00) diffraction spot is visible on the fluorescent screen. The beam intensity at a particular electron energy corresponding to a peak maximum in the I-V profile is measured with a Faraday-cup collector (or any other method discussed in Sect. 2.3) as the crystal cools from an initially elevated temperature. Ten to twelve such beam intensities should be obtained for a particular angle of incidence as a function of temperature at different electron energies corresponding to different peak maxima in the I-V beam profile. Sets of such data at different angles of incidence are then obtained. The angle of incidence of the LEED beam can be determined easily by the procedure of Cunningham and Weinberg [2.30].

Within the kinematic approximation, the intensity of the backscattered electrons I may be written as

$$I = I_o \exp(-2M) , \tag{2.27}$$

where exp(-2M) is the Debye-Waller factor. For the specular beam, the Debye-Waller exponent is given by

$$2M = |s_\perp|^2 < (\Delta r_\perp)^2 > , \tag{2.28}$$

where $s_\perp$ is the normal component of the momentum transfer of the electron, and $<(\Delta r_\perp)^2>$ is the root-mean-square displacement of the surface atom perpendicular to the crystal surface. In the case of the specular beam,

$$|s_\perp|^2 = \frac{8m}{\hbar^2} (E\cos^2\theta + V_o) , \tag{2.29}$$

where m is the mass of the electron, $2\pi\hbar$ is Planck's constant, E is the kinetic energy of the electron in vacuum, θ is the angle of incidence with respect to the surface normal, and V_o (>0) is the inner potential. In the high temperature limit,

$$<(\Delta r_\perp)^2> = \frac{3\hbar^2 T}{m_a k_B (\theta_{D\perp}^{eff})^2} ,$$ (2.30)

where m_a is the mass of the surface atom, k_B is the Boltzmann constant, and $\theta_{D\perp}^{eff}$ is the effective Debye temperature for the perpendicular component of the atomic vibration. Consequently,

$$2M = \frac{24 m T (E \cos^2\theta + V_o)}{m_a k_B (\theta_{D\perp}^{eff})^2} .$$ (2.31)

The effective Debye temperature can be evaluated from the slope of a plot of $\ln(I/I_o)$ as a function of surface temperature. However, the experimentally measured intensity of a beam does not necessarily correspond to the true intensity, which results from elastic scattering, due to the presence of the background intensity from phonon-assisted inelastic scattering. To determine the effective Debye temperature correctly, the background intensity must be subtracted from the experimentally measured intensity [2.36]. Hence, (2.27) should be rewritten as

$$I - I_b = I_o \exp(-2M) ,$$ (2.32)

where I_b is the background intensity.

Experimentally, it is difficult to determine an accurate background intensity, especially at high incident electron beam energies ($E > 500$ eV) where the diffraction spots are closely spaced. Therefore, in a least-squares fitting procedure [2.36] for the plot of $\ln(I - I_b)$ as a function of T, the value of I_b should be varied within physically reasonable limits (between the asymptotic value of the intensity at high temperatures and zero) until the best straight line for the representation of $\ln(I - I_b)$ as a function of T is obtained. In the least-squares fitting procedure, flexibility is provided to assign a weighting factor to each data point. The data taken at low temperatures can be measured more accurately and are less sensitive to the variation of the background intensity. Hence, the value of $\theta_{D\perp}^{eff}$ is determined from the slope of the best straight line in the plot of $\ln(I - I_b)$ as a function of T with a weighting factor for each data point proportional to its experimentally measured intensity. (Note, however, that other weighting schemes can be used as well.) In the calculation of $|s_\perp|^2$, the value for the inner potential can be obtained from the results of dynamical LEED calculations, or, even better, from a kinematic analysis, Sects. 4.6 and 6.1.

As an example [2.37], in Fig. 2.23, $\theta_{D\perp}^{eff}$ as a function of $s_\perp$ for the (00) beam of the clean (110)-(1x2) surface of Ir is shown at three different angles of incidence with respect to the surface normal: $\theta = 8.5°$, $\theta = 16.5°$, and $\theta = 25.3°$. The solid line is a second-degree least-squares polynomial fit to all data points. The value of $\theta_{D\perp}^{eff}$ at the lowest value of $s_\perp$ (6.5 Å^{-1}) obtained experimentally is taken as the "surface" value of θ_D which is 150 ± 12 K. The high-energy asymptote of the polynomial fit to the data gives a value for

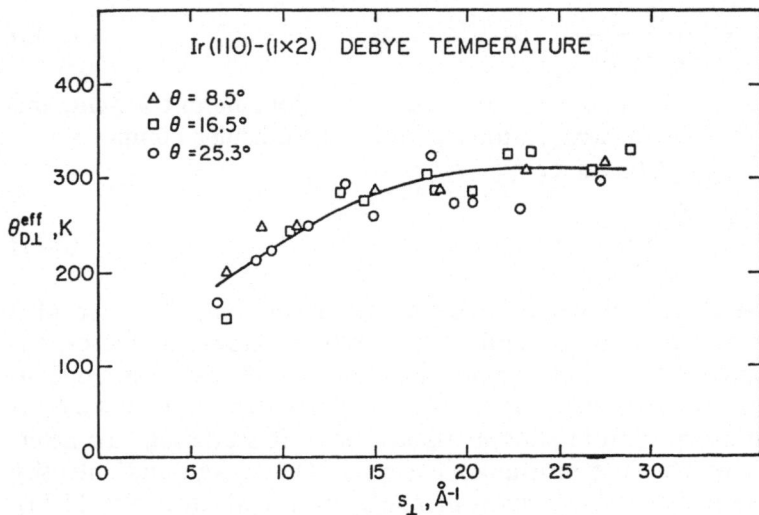

Fig. 2.23. Plot of $\theta_D^{eff}{}_\perp$ as a function of $s_\perp$ for the (00) beam of the (110)-(1 × 2) reconstructed surface of Ir at three different angles of incidence: $\theta = 8.5°$, $\theta = 16.5°$, and $\theta = 25.3°$. The solid line is a second-degree least-squares polynomial fit to the data

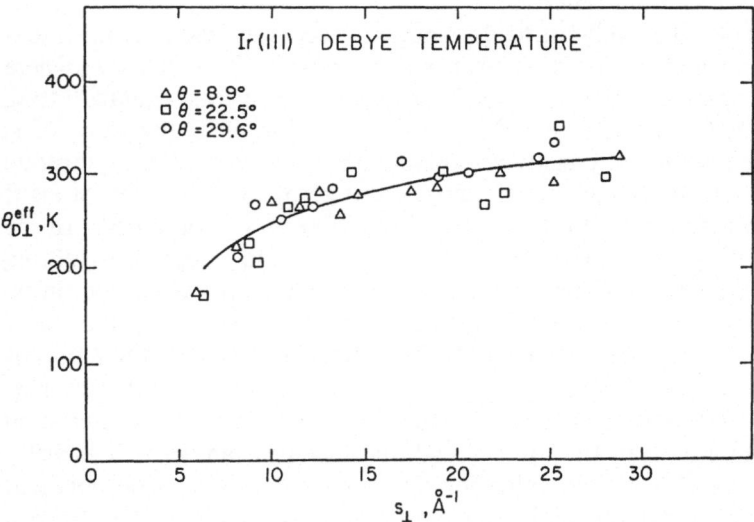

Fig. 2.24. Same as Fig. 2.23 except the (111) surface of Ir at angles of incidence: $\theta = 8.9°$, $\theta = 22.5°$, and $\theta = 29.6°$

the bulk θ_D of 310 ± 22 K. The errors were estimated from the standard deviation of the measured data. The corresponding results for the Ir(111) surface at angles of incidence $\theta = 8.9°$, $\theta = 22.5°$, and $\theta = 29.6°$ are shown in Fig. 2.24. The "surface" $\theta_{D\perp}$ and the bulk θ_D were found to be 170 ± 12 K and 315 ± 22 K, respectively.

Surface	Surface θ_D^{eff} [K]	Bulk θ_D^{eff} [K]	Reference
Al(100)	189	370	[2.38]
V(100)	250	380	[2.39]
Cr(100)	175	440	[2.40]
Ni(110)	220	390	[2.41]
Nb(100)	106	281	[2.42]
Rh(111)	197	380	[2.36]
Pd(100)	140	274	[2.43]
Pd(111)	140	274	[2.43]
Mo(100)	239	380	[2.40]
Ag(111)	155	225	[2.43]
Cu(111)	244	322	[2.44]
Cu(100)	210	322	[2.44]
W(110)	200	280	[2.40]
W(100)	150	280	[2.45]
Ir(100)	175	285	[2.46]
Ir(110)	150	310	[2.37]
Ir(111)	170	315	[2.37]
Pt(100)	118	234	[2.47]
Pt(111)	111	234	[2.47]
Pt(110)	107	234	[2.47]
Th(111)	109	143	[2.48]
Cu$_3$Au(100)	164	216	[2.49]
Zn(10$\bar{1}$0)	200	360	[2.50]

Table 2.3. Effective surface and bulk Debye temperatures for several elements and alloys, as determined by LEED

The results presented in Figs. 2.23,24 are indistinguishable within experimental uncertainties. This indicates that the difference in the packing of the surface atoms in this case does not affect the perpendicular component of the effective "surface" Debye temperature at a level detectable experimentally, and the effective bulk Debye temperatures obtained from these two surfaces are also identical.

The surface θ_D^{eff} and bulk θ_D^{eff} for several elements and alloys are listed in Table 2.3. They demonstrate that the surface θ_D^{eff} is approximately half that of the bulk. There appears to be little difference in the surface θ_D^{eff} for different low-index crystal planes.

2.7.2 The Debye Temperature Parallel to the Crystal Surface

To measure the Debye-Waller factor parallel to the crystal surface, nonspecular beams are needed. Consider a simple case as an example. For nonspecular beams at normal incidence, the Debye-Waller factor for a surface with vibrations that are isotropic in the surface plane is given by

$$2M = |s_\perp|^2 <(\Delta r_\perp)^2> + |s_\parallel|^2 <(\Delta r_\parallel)^2 > , \tag{2.33}$$

where

$$|s_\perp| = |\vec{k}_o|(1 + \cos\gamma_{hk}) , \tag{2.34}$$

and

$$|\vec{s}_\parallel| = |\vec{g}_{hk}|, \tag{2.35}$$

Here $|\vec{k}_o| = [(2m/\hbar^2) (E + V_o)]^{1/2}$ is the incident wavevector inside the surface, $\vec{g}_{hk}$ is the reciprocal-space-lattice vector for the (hk) beam, and γ_{hk} is the angle between the surface normal and the diffracted (hk) beam. An expression for this angle is

$$\cos\gamma_{hk} = \left(\frac{|\vec{k}_o|^2 - |\vec{g}_{hk}|^2}{|\vec{k}_o|^2} \right)^{1/2} . \tag{2.36}$$

Consequently,

$$2M = \frac{6mT}{m_a k_B} (E + V_o) \left[\frac{(1 + \cos\gamma_{hk})^2}{(\theta_{D_\perp}^{eff})^2} + \frac{\sin^2\gamma_{hk}}{(\theta_{D_\parallel}^{eff})^2} \right] . \tag{2.37}$$

Thus, using the value of $\theta_{D_\perp}^{eff}$ obtained previously via the (00) diffracted beam, the value of $\theta_{D_\parallel}^{eff}$ can be calculated using (2.37). However, as pointed out by Somorjai and Farrell [2.46], the two experimentally determined slopes for the specular and nonspecular beams are of comparable magnitude. Thus, an uncertainty in either experimentally determined value would be amplified causing a large uncertainty in the determination of $\theta_{D_\parallel}^{eff}$. Possible ways to circumvent this problem include either using a grazing angle of incidence of the low-energy electron beam or a diffracted beam of high order, in order to minimize the term in (2.37) which contains $\theta_{D_\perp}^{eff}$.

3. Ordered Surfaces: Structure and Diffraction Pattern

Low-energy electrons can diffract from many assemblies of atoms, including molecules [3.1], liquid surfaces [3.2], and otherwise disordered surfaces [3.3]. However, under the name LEED, this technique has been applied mainly to the study of single-crystal surfaces with a high degree of ordering of the surface atoms, especially as a tool for determining surface structures [3.4]. Therefore we shall be very much concerned with the detailed atomic structure of single-crystal surfaces. This chapter discusses the general properties and the nomenclature of surface structures and their connection with the LEED diffraction pattern that is observed experimentally. In particular, we explore the structural information that the diffraction pattern can provide without consideration of beam intensities.

3.1 Two-Dimensional Periodicity and the LEED Pattern

The ideas that are familiar to solid-state physicists and X-ray crystallographers concerning three-dimensional symmetry operations can easily be adapted to the two-dimensional case of an ordered, solid surface. Here the discussion will concern a surface that is free of both defects and any disorder. Such imperfections are considered explicitly in Chaps. 8-10.

3.1.1 Miller and Miller-Bravais Indices

First, we should recall the concept of Miller indices (h,k,l), or (hkl) for short when no confusion can arise by dropping the commas, and Miller-Bravais indices (h,k,i,l) or (hkil), where h, k, i, and l are all integers. These indices serve to define a particular crystallographic orientation in a three-dimensional crystal. In particular, Miller and Miller-Bravais indices are used to specify the crystallographic orientation of a surface. Indices in the notation [h,k,l] or [hkl] indicate a direction, i.e. a vector in a lattice. The collection of such directions that are equivalent by symmetry is labeled <h,k,l> or <hkl>. The Miller indices of a given crystallographic plane are denoted (h,k,l) or (hkl): the index values are determined by the intercepts of the given plane with the three crystallographic axes [3.5,6]. The collection of such planes that are equivalent by symmetry is labeled {h,k,l} or {hkl}.

Thus, {100} stands for the collection (100) (-1,0,0), (010), (0,-1,0), (001) and (0,0,-1), if these six planes are equivalent.

Miller indices are simplest to work with in cubic lattices, including simple cubic (sc), face-centered cubic (fcc), body-centered cubic (bcc), zinc blende-like, diamond-like lattices, because then the vector [hkl] is perpendicular to the plane (hkl). For example, the (100) plane is perpendicular to the [100] lattice vector, which is parallel to a cube edge in a cubic lattice. Therefore, one can easily see that the (100) surface has a square two-dimensional lattice when the bulk has a cubic three-dimensional lattice.

For the case of hexagonal lattices, such as the hexagonal close-packed (hcp) lattice, Miller-Bravais indices (hkil) have been introduced for convenience [3.5]. Briefly, an additional (redundant) crystallographic axis is defined (corresponding to the index i) within the hexagonal basal plane. The first two crystallographic axes (corresponding to the indices h and k) also lie within the hexagonal basal plane, separated by an angle of 120° from each other and from the additional axis. The fourth axis (corresponding to the index l) is perpendicular to the basal plane. One may have threefold or sixfold rotational symmetry around this axis. The simple relation i = -h-k holds, which can be proved from the concept of intercepts mentioned above. Therefore, the notation (hk.l) is frequently also used, since the index i is redundant. For example, the (0001) or (00.1) surface is simply the basal plane, which could be labeled (001) in the conventional Miller-index notation, by dropping the index i. Other commonly occurring (low-index) surfaces of hexagonal lattices are the (10$\bar{1}$0) = (10.0) = (100) and the (11$\bar{2}$0) = (11.0) = (110) surfaces. Both of these surfaces (planes) are perpendicular to the basal plane, since l = 0. It should be clear that the (100) and (001) planes are far from equivalent. Also note the bar notation for negative indices: $\bar{1}$ and $\bar{2}$ stand for -1 and -2, respectively.

3.1.2 Lattice and Basis

When considering the symmetry of a perfectly ordered surface, there are two concepts of fundamental importance, that of a *lattice* and that of a *basis* (as in three-dimensional crystals). The lattice consists of a two-dimensional array of points which possess translational symmetry. The basis, on the other hand, designates the arrangement of the atoms with respect to the lattice points of the surface, for example by specifying all atomic positions within one unit cell. Consequently, the basis together with the lattice define the atomic geometry of the surface: either alone is insufficient.

Perfect surfaces are characterized not only by a translational symmetry but also by a point group symmetry, the latter consisting of rotation operations about the surface normal, and reflection and glide operations with respect to planes perpendicular to the surface. Taken together, these space group operations limit the possible geometries of the surface to five two-

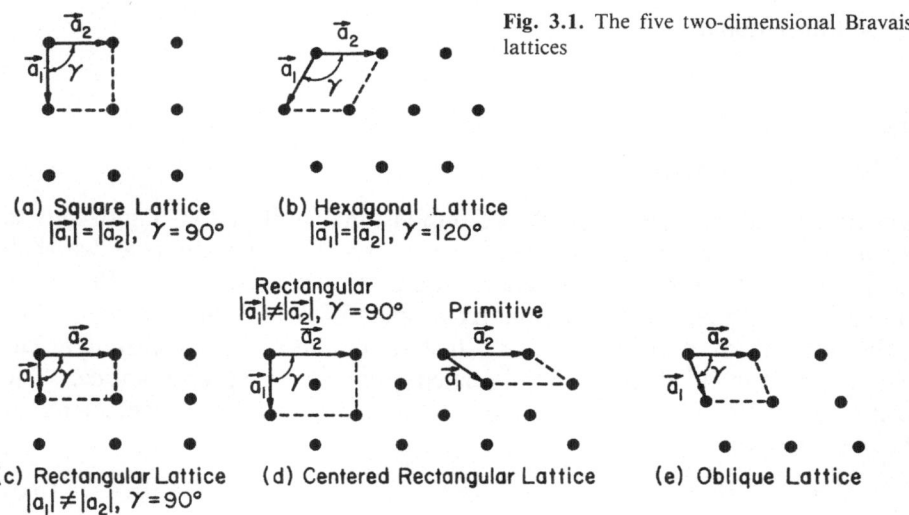

Fig. 3.1. The five two-dimensional Bravais lattices

(a) Square Lattice
$|\vec{a_1}| = |\vec{a_2}|$, $\gamma = 90°$

(b) Hexagonal Lattice
$|\vec{a_1}| = |\vec{a_2}|$, $\gamma = 120°$

Rectangular
$|\vec{a_1}| \neq |\vec{a_2}|$, $\gamma = 90°$ Primitive

(c) Rectangular Lattice
$|a_1| \neq |a_2|$, $\gamma = 90°$

(d) Centered Rectangular Lattice

(e) Oblique Lattice

dimensional Bravais lattices [3.5], as shown in Fig. 3.1. The point group symmetries which are allowed are: rotations of $2\pi/n$ (for n = 1, 2, 3, 4 or 6), a rotation of this type combined with mirror reflections, and/or a translation (glide). We shall discuss further aspects of symmetry at surfaces in Sect. 3.4.

The most common procedure for obtaining an ordered surface is to prepare a planar cut through a three-dimensional crystalline sample. The ideal truncation of a crystal can be described simply by specifying the Miller indices of the plane chosen for the cut [3.6]. Although real surfaces can deviate in various ways from the ideal truncation of the bulk structure (by relaxation, reconstruction, and atomic or molecular adsorption), most surfaces can still be characterized at least in part by the Miller indices defining the cut orientation. Attention has focused primarily on low-Miller-index surfaces of face-centered cubic and body-centered cubic metal crystals because of their structural simplicity. These surfaces tend to have the highest degree of symmetry and the smallest unit cells. Examples are fcc(111), fcc(100), and fcc(110) surfaces, which have threefold, fourfold and twofold rotational symmetry, respectively; and bcc(110), bcc(100), and bcc(111) surfaces, which have twofold, fourfold, and threefold rotational symmetry, respectively. (All these surfaces also have mirror planes in addition to the rotation axes.)

3.1.3 Direct and Reciprocal Lattices

A two-dimensional solid surface contains a primitive unit cell (with an associated minimum basis), which is defined by the translational vectors $\vec{a_1}$ and $\vec{a_2}$. Defining unit vectors in Cartesian coordinates $\hat{x}$ and $\hat{y}$, the unit cell of the surface is described by the matrix **A**, the elements A_{ij} of which are related to the translational vectors

$$\vec{a}_1 = A_{11}\hat{x} + A_{12}\hat{y} , \tag{3.1}$$

and

$$\vec{a}_2 = A_{21}\hat{x} + A_{22}\hat{y} . \tag{3.2}$$

Associated with the (real space) two-dimensional unit cell is a unit cell in *reciprocal space* which is described by a matrix $\mathbf{A}^*$. The latter, as we shall see in the next chapter, is extremely useful and pertinent in LEED. The matrix $\mathbf{A}^*$ is defined so that the matrix product of $\mathbf{A}$ and the transpose of $\mathbf{A}^*$ is the unit matrix multiplied by 2π. Just as for the real-space primitive lattice, the matrix $\mathbf{A}^*$ defines an associated primitive reciprocal lattice. The reciprocal lattice consists of linear combinations with integral coefficients of the two vectors

$$\vec{a}_1^* = A_{11}^*\hat{x} + A_{12}^*\hat{y} , \tag{3.3}$$

and

$$\vec{a}_2^* = A_{21}^*\hat{x} + A_{22}^*\hat{y} . \tag{3.4}$$

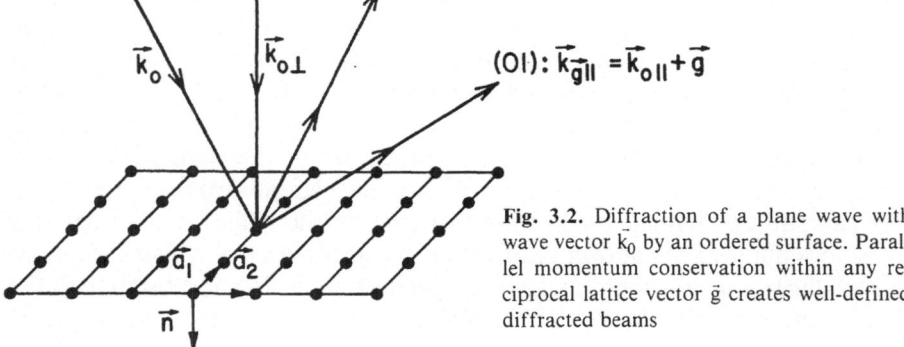

Fig. 3.2. Diffraction of a plane wave with wave vector $\vec{k}_0$ by an ordered surface. Parallel momentum conservation within any reciprocal lattice vector $\vec{g}$ creates well-defined diffracted beams

The diffraction of low-energy electrons at a surface is governed by the translational symmetry of the surface. The angle of emergence of the diffracted beams is determined by conservation of linear momentum ($\vec{p} = \hbar\vec{k}$ where $\vec{k}$ is the wavevector of the electron) parallel to the surface. As depicted schematically in Fig. 3.2, this conservation of the parallel component of the momentum ($\vec{p}_\parallel = \hbar\vec{k}_\parallel$) can occur in two ways. After the diffractive scattering (denoted by a prime), the parallel component of the momentum can be equal to that of the incident electron beam, i.e.

$$\vec{k}'_\parallel = \vec{k}_\parallel , \tag{3.5}$$

or it can be equal to $\vec{k}_\parallel$ to within a reciprocal lattice vector $\vec{g}$, i.e.

$$\vec{k}'_\parallel = \vec{k}_\parallel + \vec{g} \ , \tag{3.6}$$

where $\vec{g}$ is any of the reciprocal lattice vectors that are associated with the primitive Bravais lattice of the surface. The reciprocal lattice vectors $\vec{g}$ are related to the vectors $\vec{a}_1^*$ and $\vec{a}_2^*$, defined earlier, by

$$\vec{g} \equiv h\vec{a}_1^* + k\vec{a}_2^* \ . \tag{3.7}$$

Thus each diffracted beam corresponds to a different reciprocal lattice vector $\vec{g}$, and, in fact, each beam is labeled by the values h and k as the beam (h,k).

The relationships between the primitive translation vectors of the surface, $\vec{a}_1$ and $\vec{a}_2$ (cf. Fig. 3.2), and the primitive translation vectors of the reciprocal lattice, $\vec{a}_1^*$ and a_2^*, may be written as

$$\vec{a}_1^* = 2\pi \left(\frac{\vec{a}_2 \times \vec{n}}{\vec{a}_1 \cdot (\vec{a}_2 \times \vec{n})} \right) , \quad \text{and} \tag{3.8}$$

$$\vec{a}_2^* = 2\pi \left(\frac{\vec{n} \times \vec{a}_1}{\vec{a}_2 \cdot (\vec{n} \times \vec{a}_1)} \right) , \tag{3.9}$$

when $\vec{n}$ is taken to be a vector normal to the surface. From these relations, it is clear that $\vec{a}_1 \cdot \vec{a}_1^* = \vec{a}_2 \cdot \vec{a}_2^* = 2\pi$ and $\vec{a}_1 \cdot \vec{a}_2^* = \vec{a}_2 \cdot \vec{a}_1^* = 0$ are satisfied, as we required previously.

We are now in a position to discuss the LEED pattern of a perfect surface in terms of the direct and reciprocal translational lattice vectors. We note in passing that the reciprocal lattice vector $\vec{g}$ lies in a direction that is orthogonal to the plane of the direct lattice that is defined by (hk) Miller indices, and the magnitude of $\vec{g}$ is equal to the product of 2π and the reciprocal of the spacing between atomic planes of (hk) orientation. The relation between the direct lattice and the reciprocal lattice for the five two-dimensional Bravais nets is shown explicitly in Fig. 3.3. When h and k are both equal to zero, then the parallel component of the linear momentum of the electron is conserved absolutely, see (3.5), and we are considering the specularly reflected or (00) diffraction beam. When h and k are not both equal to zero, so-called "higher-order" diffraction beams are under discussion, e.g. the (01) beam is shown explicitly in Fig. 3.2. The relationship between the translation vectors of the direct lattice and those of the reciprocal lattice should be noted in Fig. 3.3.

3.2 Superlattices at Surfaces

After the discussion in the previous sections, it is obvious that the possible observation of "fractional-order" LEED beams is indicative of an ordered superlattice (or superstructure) with a periodicity larger than that of the

DIRECT LATTICE   RECIPROCAL LATTICE

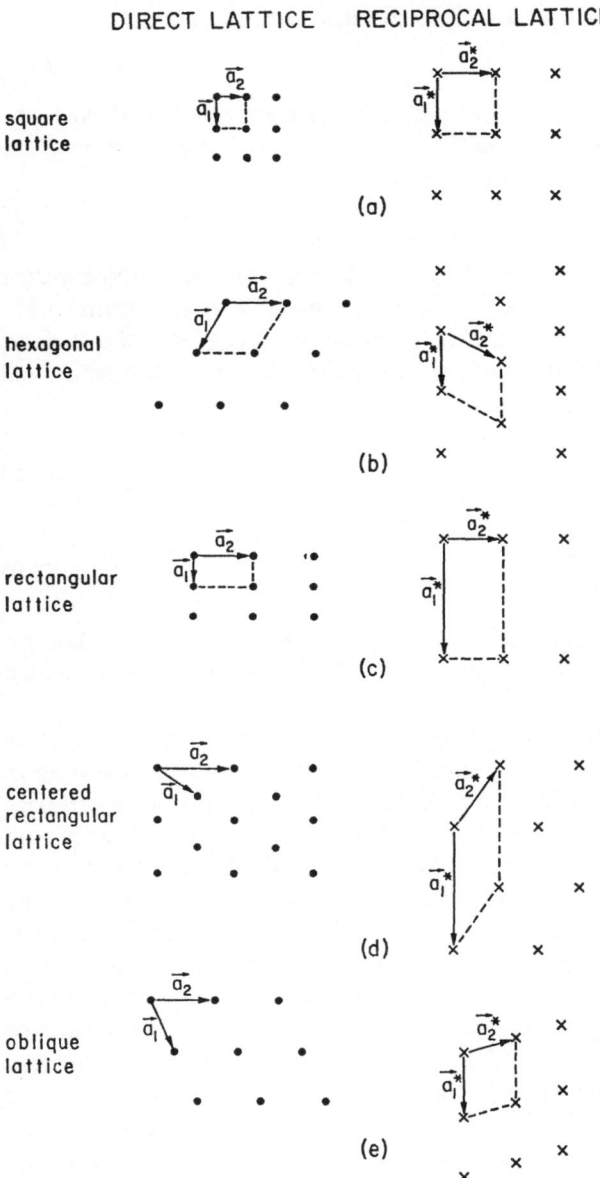

Fig. 3.3. Direct (*left*) and reciprocal (*right*) lattices for the five two-dimensional Bravais lattices

primitive unit cell of the substrate. This additional periodicity could be due to an intrinsic reconstruction of the clean surface (Sect. 7.2), the presence of an ordered atomic overlayer (Sect. 7.3), or the presence of an ordered molecular overlayer (Sect. 7.4). It is fortunate indeed that the occurrence of ordered overlayers is quite common, otherwise detailed LEED calculations to determine the geometrical structure of the overlayer, i.e. the basis of the superlattice, might be much more complex than they are. The structural determination of atomic or molecular adsorbates is one of the fundamental contributions of LEED to furthering a microscopic understanding of all phenomena that occur at surfaces of solids.

Following Wood [3.7], we describe the primitive unit cell of the ordered surface layer by a matrix $\mathbf{B}$, the elements of which, B_{ij}, are related to the translational unit-cell vectors $\vec{b}_1$ and $\vec{b}_2$ by [cf. (3.1,2)]

$$\vec{b}_1 = B_{11}\hat{x} + B_{12}\hat{y} \ , \tag{3.10}$$

and

$$\vec{b}_2 = B_{21}\hat{x} + B_{22}\hat{y} \ . \tag{3.11}$$

The important relationship between the unit cells of the substrate and the surface layer is embodied in the matrix $\mathbf{M}$ where

$$\mathbf{M} = \mathbf{B}\mathbf{A}^{-1} \ . \tag{3.12}$$

The matrix $\mathbf{M}$ allows a convenient classification of the unit cell of the surface layer vis-à-vis that of the substrate. In particular, there are three pertinent cases:

1. When all matrix elements M_{ij} are integers, the unit cells of the surface layer and substrate are simply related, and the structure of the surface layer is termed simple.

2. When all matrix elements M_{ij} are rational numbers, the two unit cells are rationally related, the surface is said to have a coincidence structure, and the superlattice is termed commensurate.

3. When at least one matrix element M_{ij} is an irrational number, the two unit cells are irrationally related, and the superstructure is termed incommensurate.

In the first and second cases, the combined surface layer and substrate is characterized also by a Bravais lattice, and the surface layer is in registry with the substrate. In each of the three cases, the total system (substrate and surface layer) can be described by

$$S(hkl) - \mathbf{M} - \eta A \ , \tag{3.13}$$

where $S(hkl)$ is the (hkl) crystallograhic orientation of the substrate with chemical composition S, while the $\mathbf{M}$ matrix is defined by (3.12). The chemical stoichiometry of the atomic or molecular overlayer is given by A, and η is the number of adspecies in the overlayer unit cell. (ηA is dropped in the case of reconstruction involving no adsorbates.) For example, CO adsorbed molecularly on the Ni(100) surface at a fractional surface coverage of one-

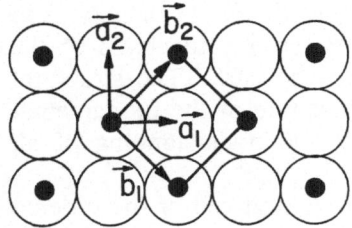

Fig. 3.4. A simple superlattice on the fcc(100) surface: $\begin{pmatrix} 1 & -1 \\ 1 & 1 \end{pmatrix}$, c(2×2) or $(\sqrt{2} \times \sqrt{2})R45°$

half forms the overlayer shown in Fig. 3.4. In this case, (3.13) becomes

$$Ni(100) - \begin{pmatrix} 1 & -1 \\ 1 & 1 \end{pmatrix} - CO \ ,$$

and it is clear that the **M** matrix displays the translational vectors of the direct lattice of the overlayer: (1,-1) and (1,1), see Fig. 3.4.

A most useful relationship is the fact that the determinant of **M** (det **M**) is simply the ratio of the unit-cell areas of the superlattice and the substrate lattice. In the above example, $\det\begin{pmatrix} 1 & -1 \\ 1 & 1 \end{pmatrix} = 2$ indicates that the CO over-layer has a unit cell twice as large in area as that of the substrate.

As an alternative to the above *matrix notation*, the more transparent *Wood notation* [3.7] may be used in many cases as a labeling scheme for the superlattice. In this case, the total system consisting of the substrate and the surface layer is described by

$$S(hkl) - i\left| \frac{b_1}{a_1} \times \frac{b_2}{a_2} \right| R\alpha - \eta A \ , \tag{3.14}$$

where $S(hkl)$, A, and η are as defined above. In (3.14), i is either "p" (for primitive) or "c" (for centered) according to the way in which the unit cell of the overlayer is defined. When the letter "p" is dropped, the primitive nota-tion is understood implicitly. The quantity

$$\left| \frac{b_1}{a_1} \times \frac{b_2}{a_2} \right|$$

indicates the ratios of the magnitudes of the unit-cell vectors of the surface layer to those of the substrate, and $R\alpha$ includes the possibility of a rotation of the unit cell of the overlayer by α degrees with respect to the unit cell of the substrate, i.e. the angle between $\vec{b}_1$ and $\vec{a}_1$. If α is zero, then $R\alpha$ is omit-ted from (3.14). In the example of CO chemisorbed on the (100) surface of Ni shown in Fig. 3.4, the Wood notation would be either

$$Ni(100) - p(\sqrt{2} \times \sqrt{2})R45° - CO \ ,$$

or, equivalently,

$$Ni(100) - c(2 \times 2) - CO \ .$$

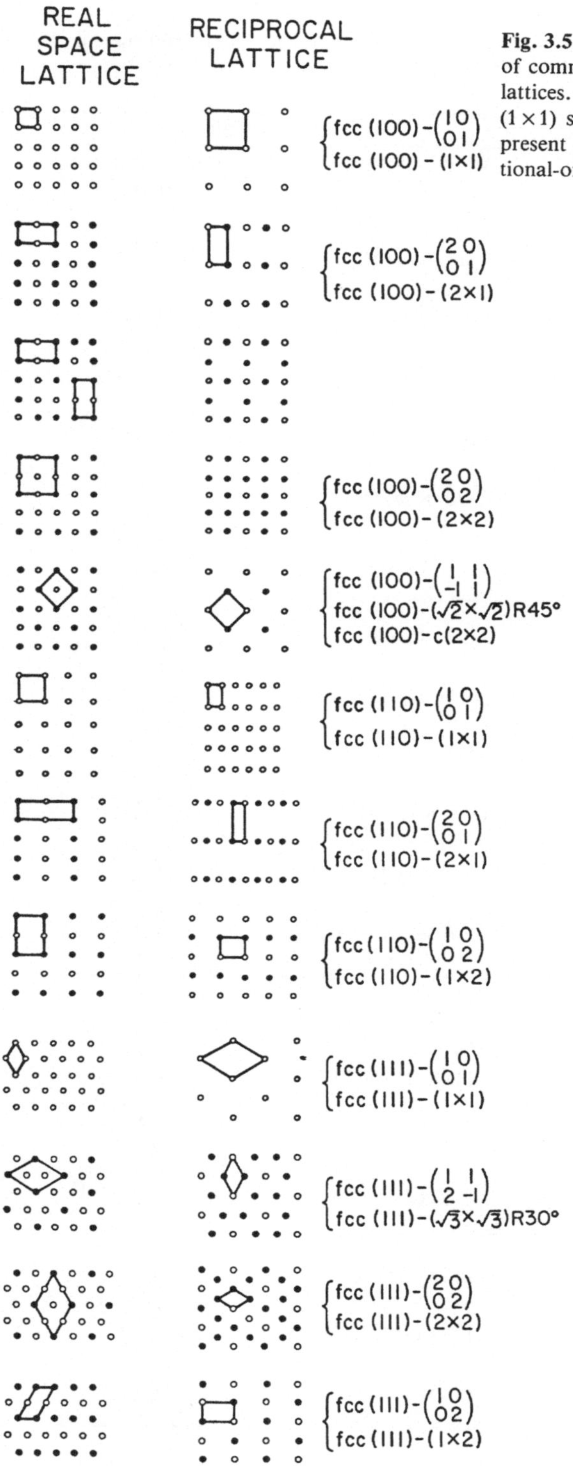

REAL
SPACE
LATTICE

RECIPROCAL
LATTICE

Fig. 3.5. Direct and reciprocal lattices of a series of commonly occurring two-dimensional superlattices. Open circles correspond to the ideal (1×1) surface structure, while filled circles represent adatoms in the direct lattice and fractional-order beams in the reciprocal lattice

$$\begin{cases} \text{fcc (100)} - \begin{pmatrix} 1 & 0 \\ 0 & 1 \end{pmatrix} \\ \text{fcc (100)} - (1 \times 1) \end{cases}$$

$$\begin{cases} \text{fcc (100)} - \begin{pmatrix} 2 & 0 \\ 0 & 1 \end{pmatrix} \\ \text{fcc (100)} - (2 \times 1) \end{cases}$$

$$\begin{cases} \text{fcc (100)} - \begin{pmatrix} 2 & 0 \\ 0 & 2 \end{pmatrix} \\ \text{fcc (100)} - (2 \times 2) \end{cases}$$

$$\begin{cases} \text{fcc (100)} - \begin{pmatrix} 1 & 1 \\ -1 & 1 \end{pmatrix} \\ \text{fcc (100)} - (\sqrt{2} \times \sqrt{2}) R45° \\ \text{fcc (100)} - c(2 \times 2) \end{cases}$$

$$\begin{cases} \text{fcc (110)} - \begin{pmatrix} 1 & 0 \\ 0 & 1 \end{pmatrix} \\ \text{fcc (110)} - (1 \times 1) \end{cases}$$

$$\begin{cases} \text{fcc (110)} - \begin{pmatrix} 2 & 0 \\ 0 & 1 \end{pmatrix} \\ \text{fcc (110)} - (2 \times 1) \end{cases}$$

$$\begin{cases} \text{fcc (110)} - \begin{pmatrix} 1 & 0 \\ 0 & 2 \end{pmatrix} \\ \text{fcc (110)} - (1 \times 2) \end{cases}$$

$$\begin{cases} \text{fcc (111)} - \begin{pmatrix} 1 & 0 \\ 0 & 1 \end{pmatrix} \\ \text{fcc (111)} - (1 \times 1) \end{cases}$$

$$\begin{cases} \text{fcc (111)} - \begin{pmatrix} 1 & 1 \\ 2 & -1 \end{pmatrix} \\ \text{fcc (111)} - (\sqrt{3} \times \sqrt{3}) R30° \end{cases}$$

$$\begin{cases} \text{fcc (111)} - \begin{pmatrix} 2 & 0 \\ 0 & 2 \end{pmatrix} \\ \text{fcc (111)} - (2 \times 2) \end{cases}$$

$$\begin{cases} \text{fcc (111)} - \begin{pmatrix} 1 & 0 \\ 0 & 2 \end{pmatrix} \\ \text{fcc (111)} - (1 \times 2) \end{cases}$$

Table 3.1. Some common unit cells at surfaces and their notations (different but equivalent notation matrices can be obtained by choosing different sets of basis vectors)

Substrate	Surface Unit Cell	
	Wood Notation	Matrix Notation
	p(1x1)=(1x1)	$\left(\begin{smallmatrix} 1 & 0 \\ 0 & 1 \end{smallmatrix}\right)$
fcc(100),bcc(100),	p(2x1)=(2x1)	$\left(\begin{smallmatrix} 2 & 0 \\ 0 & 1 \end{smallmatrix}\right)$
diamond(100),	p(1x2)=(1x2)	$\left(\begin{smallmatrix} 1 & 0 \\ 0 & 2 \end{smallmatrix}\right)$
zincblende(100)	c(2x2)=($\sqrt{2} \times \sqrt{2}$)R45°	$\left(\begin{smallmatrix} 1 & \bar{1} \\ 1 & 1 \end{smallmatrix}\right)$
	p(2x2)=(2x2)	$\left(\begin{smallmatrix} 2 & 0 \\ 0 & 2 \end{smallmatrix}\right)$
	($2\sqrt{2} \times \sqrt{2}$)R45°	$\left(\begin{smallmatrix} 2 & 2 \\ \bar{1} & 1 \end{smallmatrix}\right)$
	c(4x2)	$\left(\begin{smallmatrix} 2 & \bar{1} \\ 0 & 2 \end{smallmatrix}\right)$
	p(1x1)=(1x1)	$\left(\begin{smallmatrix} 1 & 0 \\ 0 & 1 \end{smallmatrix}\right)$
fcc(111),hcp(0001),	p(2x1)=c(2x2)=(2x1)	$\left(\begin{smallmatrix} 2 & 0 \\ 0 & 1 \end{smallmatrix}\right)$
diamond(111),zincblende(111),	p(2x2)=(2x2)	$\left(\begin{smallmatrix} 2 & 0 \\ 0 & 2 \end{smallmatrix}\right)$
graphite(0001)	($\sqrt{3} \times \sqrt{3}$)R30°	$\left(\begin{smallmatrix} 1 & \bar{1} \\ 1 & 2 \end{smallmatrix}\right)$
	c(4x2)	$\left(\begin{smallmatrix} 2 & \bar{1} \\ 0 & 2 \end{smallmatrix}\right)$
	($\sqrt{7} \times \sqrt{7}$)Rarctan($\sqrt{3}/5$)	$\left(\begin{smallmatrix} 2 & 1 \\ \bar{1} & 3 \end{smallmatrix}\right)$
	p(1x1)=(1x1)	$\left(\begin{smallmatrix} 1 & 0 \\ 0 & 1 \end{smallmatrix}\right)$
fcc(110),diamond(110),	p(2x1)=(2x1)	$\left(\begin{smallmatrix} 2 & 0 \\ 0 & 1 \end{smallmatrix}\right)$
zincblende(110)	p(1x2)=(1x2)	$\left(\begin{smallmatrix} 1 & 0 \\ 0 & 2 \end{smallmatrix}\right)$
	c(2x2)	$\left(\begin{smallmatrix} 1 & \bar{1} \\ 1 & 1 \end{smallmatrix}\right)$
	p(1x1)=(1x1)	$\left(\begin{smallmatrix} 1 & 0 \\ 0 & 1 \end{smallmatrix}\right)$
bcc(110)	p(2x1)=(2x1)	$\left(\begin{smallmatrix} 2 & 0 \\ 0 & 1 \end{smallmatrix}\right)$
	p(2x2)=(2x2)	$\left(\begin{smallmatrix} 2 & 0 \\ 0 & 2 \end{smallmatrix}\right)$

A number of other direct superlattices and the corresponding reciprocal-space lattices (i.e. the LEED patterns) are shown in Fig. 3.5 for the (100), (110), and (111) surfaces of a face-centered cubic crystal with the appropriate matrix and Wood notations shown explicitly in the figure. A more extensive list appears in Table 3.1. Note that the Wood notation cannot always be applied, since it assumes that both basis vectors $\vec{a}_1$ and $\vec{a}_2$ are rotated by the same angle α to yield $\vec{b}_1$ and $\vec{b}_2$.

Unfortunately, in the Wood notation, there is no standard way of labeling the two unit-cell vectors. For example, the structures labeled as fcc(110)-(2x1) and fcc(110)-(1x2) in Fig. 3.5 could equally well be labeled as fcc(110)-(1x2) and fcc(110)-(2x1), respectively. This should lead to no confusion as long as the system that is adopted is defined explicitly.

3.3 Stepped and Kinked Surfaces

Surface scientists have chosen to work with single crystals to avoid the structural complexities of more natural surfaces. To provide the greatest experimental and theoretical simplicity, the focus has been on low-Miller-index surfaces, since these are relatively flat and close-packed, at least for most metals. However, defects at surfaces are thought to play an important role in many processes, one example of which is heterogeneous catalysis [3.8]. Steps and kinks are defects that can be created reproducibly and studied at surfaces. Therefore they have become models for the complex variety of structures found at more natural surfaces. Not only are steps and kinks reproducible, but their nature and concentration can also be controlled by varying the cutting angle (i.e. the orientation) of a single crystal. Thus, we need to be able to describe accurately the structure of stepped and kinked surfaces, collectively called high-Miller-index surfaces.

The unit-cell notation used in the previous section is not sufficient to describe the detailed structure of a high-Miller-index surface. As a result, an illustrated atlas of stepped and kinked surfaces was produced to help relate the Miller indices to the detailed atomic surface structure [3.9]. It was assumed that the termination of the bulk crystalline structure is ideal, i.e. it involves no reconstruction, faceting, etc. This assumption appears to have been justified experimentally for many clean surfaces of this type. We shall now introduce two notations designed specifically to describe step and kink structures in a simple fashion, one being the "step notation", the other the "microfacet notation". The latter has the ability to directly correlate the Miller indices with the detailed structure.

3.3.1 The Step Notation

A stepped or high-Miller-index surface is obtained by cutting a crystal along a plane the Miller indices of which are not all small. The simplest case is

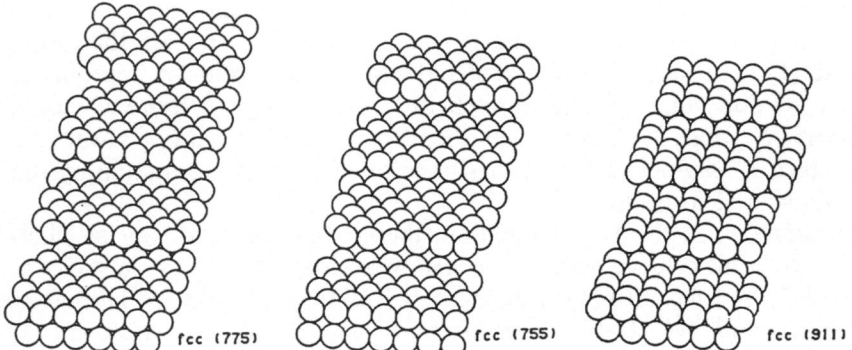

Fig. 3.6. Stepped surfaces with Miller indices (775), (755), and (911) for a face-centered cubic material

that of low-Miller-index terraces of constant width separated by steps of single-atom height, and these steps are segments of low-Miller-index faces. For example, the (775) surface of an fcc crystal consists of (111) terraces, six atoms wide, separated by steps of (11$\bar{1}$) orientation and single-atom height, Fig. 3.6. Since this surface consists largely of (111) segments, it is called a surface that is "vicinal" to the (111) surface, i.e. related to it by a small difference in cutting angle. Another (111) vicinal surface is the (755) surface, which has the same type of terrace but a step of (100) orientation, as shown in Fig. 3.6. One can also prepare vicinal (100) surfaces, such as the (911), which has (100) terraces five atoms wide and steps of (111) orientation, Fig. 3.6.

The step notation devised by Lang et al. [3.10] simply compacts this type of information into the general form

$$n(h_t k_t l_t) \times (h_s k_s l_s) , \qquad (3.15)$$

where $(h_t k_t l_t)$ and $(h_s k_s l_s)$ are the Miller indices of the terrace plane and the step plane, respectively, while n is the number of atoms that are counted in the width of the terrace, including the step-edge atom and the in-step atom. Thus, the fcc(775) surface is denoted by 7(111) × (11$\bar{1}$), or also by 7(111) × (111) for simplicity. The fcc(755) surface has the notation 6(111) × (100), and the fcc(911) surface is denoted by 5(100) × (111).

A stepped surface which has steps that are themselves high-Miller-index faces is termed a kinked surface. In this case, the steps themselves have steps, called kinks. These "kink atoms" generally have a lower coordination than other atoms. An example is shown in Fig. 3.7 of such a surface with Miller indices (10,8,7). It is a vicinal (111) surface with (310) step faces and a step notation, according to (3.15), of 7(111) × (310). The number n of atoms across the step is not well defined in this case, since it depends on the row of atoms one chooses for counting. If n were allowed to be nonintegral, the notation of (3.15) could be retained rigorously, but it would become

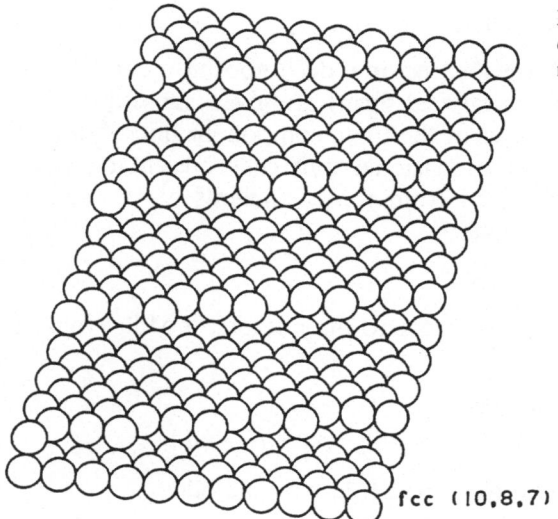

Fig. 3.7. Kinked surface with Miller indices (10,8,7) for a face-centered cubic material

fcc (10,8,7)

somewhat inconvenient because n would not be easy to evaluate. The step notation is, of course, equally applicable to surfaces of bcc, hcp, and other crystals, in addition to surfaces of fcc crystals. However, the overwhelming majority of experimental research on high-Miller-index surfaces has utilized fcc crystals.

3.3.2 The Microfacet Notation for Cubic Materials

The justification for a further notation in addition to the step notation is twofold. First, the steps of the fcc (10,8,7) surface of Fig. 3.7, which have a (310) orientation, could be decomposed in turn into "terraces" and "steps", following the idea of the step notation to its logical conclusion. Second, the number n in the notation of (3.15) does not have an obvious relationship to the Miller indices. The microfacet notation developed by Van Hove and Somorjai [3.11] remedies these shortcomings. It is based on the simple idea that any vector (hk$l$) can be decomposed in terms of three linearly independent vectors such as (111), (11$\bar{1}$), and (100), that are chosen to correspond to low-Miller-index faces. This is easily accomplished for cubic materials (simple cubic, fcc, bcc, diamond, NaCl, CsCl, etc.) where the [hkl] vector is always perpendicular to the (hkl) plane. For noncubic materials (including hcp materials), this is not the case and complications arise. Consequently, the microfacet notation and its use have thus far only been developed for cubic materials.

Let $(h_1k_1l_1)$, $(h_2k_2l_2)$, and $(h_3k_3l_3)$ be three linearly independent Miller-index vectors that represent three low-Miller-index faces, which are termed "microfacets". Then an arbitrary surface (of a cubic material), given by the Miller indices (hk$l$), can be denoted as

$$(hkl) = a^1_\lambda(h_1k_1l_1) + a^2_\mu(h_2k_2l_2) + a^3_\nu(h_3k_3l_3) \ , \tag{3.16}$$

where the factors a^1, a^2, and a^3 are the vectorial-decomposition coefficients. These coefficients (even when in numerical form) are given subscripts λ, μ, and ν which indicate the relative sizes of the three component microfacets in terms of the number of basic two-dimensional unit cells present on each. The integers λ, μ, and ν are obtained from the coefficients a^1, a^2, and a^3 according to the following proportionality:

$$\lambda{:}\mu{:}\nu = a^1p_1{:}a^2p_2{:}a^3p_3 \ , \tag{3.17}$$

where p_1, p_2, and p_3 are defined by reference to the microfacet Miller indices $(h_ik_il_i)$ $(i = 1,2,3)$:

$$p_i = \begin{cases} 1, \ \text{for} \ \begin{cases} \text{sc lattices} \ , \\ \text{bcc lattices when } h_i + k_i + l_i = \text{odd,} \end{cases} \\[2mm] 2, \ \text{for} \ \begin{cases} \text{fcc lattices when } h_i,k_i,l_i \text{ not all odd,} \\ \text{bcc lattices when } h_i + k_i + l_i = \text{even,} \end{cases} \\[2mm] 4, \ \text{for} \ \ \ \text{fcc lattices when } h_i,k_i,l_i \text{ all odd} \ . \end{cases} \tag{3.18}$$

Whenever fractional numbers appear on the right-hand side of (3.17), these numbers should be scaled to become integers; or, whenever they have a common divisor, they should be divided by that divisor to yield λ, μ, and ν.

These definitions are best clarified with specific examples, for which we first consider the fcc(hkl) = fcc(10,8,7) surface. Let us choose as microfacets the $(h_1k_1l_1) = (111)$, $(h_2k_2l_2) = (11\bar{1})$, and $(h_3k_3l_3) = (100)$ faces. It is easy to decompose (10,8,7) on this basis, yielding $(10,8,7) = 15/2 \ (111) + 1/2 \ (11\bar{1}) + 2 \ (100)$, so that $a^1 = 15/2$, $a^2 = 1/2$ and $a^3 = 2$. (Note the decreasing magnitude from left to right of the three Miller indices for each surface or microfacet; this choice makes the decomposition quite convenient.) To obtain λ, μ, and ν we need p_1, p_2, and p_3, which, according to (3.18), are $p_1 = 4$, $p_2 = 4$, and $p_3 = 2$, because we are using fcc(111), fcc(11$\bar{1}$), and fcc(100) microfacets, respectively. Thus, $\lambda{:}\mu{:}\nu = 15/2 \times 4{:}1/2 \times 4{:}2 \times 2 = 30{:}2{:}4 = 15{:}1{:}2$. As a consequence, we have the notation

$$\text{fcc}(10,8,7) = \text{fcc}[(\frac{15}{2})_{15}(111) + (\frac{1}{2})_1(11\bar{1}) + 2_2(100)] \ . \tag{3.19}$$

The subscripts 15, 1, and 2 indicate that one unit cell of the (10,8,7) surface contains fifteen (111) unit cells, one (11$\bar{1}$) unit cell and two (100) unit cells, as can be verified in Fig. 3.7. By (111), (11$\bar{1}$) and (100) unit cells, we mean the basic rhombus and square containing one surface atom on these surfaces.

We could have chosen different microfacets if we had preferred, for example (110) rather than (11$\bar{1}$). Indeed, any arbitrary faces could be used as microfacets, depending on individual preference or on the application. In terms of the microfacets (111), (110) and (100), the fcc(10,8,7) surface has the microfacet notation

$$fcc(10,8,7) = fcc[7_{14}(111) + 1_1(110) + 2_2(100)] . \tag{3.20}$$

In these terms the (10,8,7) unit cell contains fourteen unit cells of the (111) microfacet, one unit cell of (110), and two unit cells of (100).

We can similarly describe the fcc(775), fcc(755), and fcc(911) surfaces as

$$fcc(775) = fcc[6_6(111) + 1_1(11\bar{1})] = fcc[5_5(111) + 2_1(110)] ,$$

$$fcc(755) = fcc[5_5(111) + 2_1(100)] , \tag{3.21}$$

$$fcc(911) = fcc[8_4(100) + 1_1(111)] .$$

The fact that only two microfacets at a time are needed in these examples corresponds to the absence of kinks on these surfaces on which only terraces and low-Miller-index step faces occur. A list of further examples of both the step notation and the microfacet notation is given in Table 3.2. Note the different counting methods in the two notations, e.g. fcc(755) = fcc[5_5 (111) + 2_1 (100)] = fcc[6(111) x (100)] with factors 5_5 and 6, respectively. The microfacet notation counts unit cells between atom centers, while the step notation counts the atom centers themselves.

3.3.3 Unit Cells of Stepped and Kinked Surfaces

To recognize a particular high-Miller-index surface from its LEED pattern or, conversely, to predict the LEED pattern for a surface with given Miller indices, we need to determine the surface unit cell from the Miller indices.

Table 3.2. Comparison of notations for several classes of stepped or kinked surfaces. The Miller indices of a surface are proportionally reduced to the smallest set of integers, e.g. (220) is written as (110), and this affects the microfacet notation in some cases, producing apparent irregularities

n	Miller Indices (hk$l$)	Microfacet Notation	Step Notation
general	(n+1,n+1,n-1)	$n_n(111)+1_1(11\bar{1})$	(n+1)(111)x(111)
1	(110)	$(\frac{1}{2})_1(111)+(\frac{1}{2})_1(11\bar{1})$	2(111)x(111)
2	(331)	$2_2(111)+1_1(11\bar{1})$	3(111)x(111)
3	(221)	$(\frac{3}{2})_3(111)+(\frac{1}{2})_1(11\bar{1})$	4(111)x(111)
general	(n+2,n,n)	$n_n(111)+2_1(100)$	(n+1)(111)x(100)
1	(311)	$1_1(111)+2_1(100)$	2(111)x(100) or 2(100)x(111)
2	(211)	$1_2(111)+1_1(100)$	3(111)x(100)
3	(533)	$3_3(111)+2_1(100)$	4(111)x(100)

Table 3.2 (continued)

n	Miller Indices (hk$l$)	Microfacet Notation	Step Notation
general	(n+3,n+1,n)	$n_{2n}(111)+1_1(310)$	n(111)x(310)
1	(421)	$1_2(111)+1_1(310)$	1(111)x(310)
2	(532)	$2_4(111)+1_1(310)$	2(111)x(310)
3	(643)	$3_6(111)+1_1(310)$	3(111)x(310)
general	$(2n+1,1,\bar{1}1)$	$2n_n(100)+1_1(111)$	(n+1)(100)x(111)
1	$(31\bar{1})$	$2_1(100)+1_1(11\bar{1})$	2(100)x(111) or
			2(111)x(100)
2	$(51\bar{1})$	$4_2(100)+1_1(11\bar{1})$	3(100)x(111)
general	(n+1,1,0)	$n_n(100)+1_1(110)$	(n+1)(100)x(110)
1	(210)	$1_1(100)+1_1(110)$	2(100)x(110) or
			2(110)x(100)
2	(310)	$2_2(100)+1_1(110)$	3(100)x(110)
3	(410)	$3_3(100)+1_1(110)$	4(100)x(110)
general	(2n+1,2n+1,1)	$2n_n(110)+1_1(111)$	(n+1)(110)x(111)
1	(331)	$2_1(110)+1_1(111)$	2(110)x(111)
2	(551)	$4_2(110)+1_1(111)$	3(110)x(111)
3	(771)	$6_3(110)+1_1(111)$	4(110)x(111)
general	(n+1,n,0)	$n_n(110)+1_1(100)$	(n+1)(110)x(100)
1	(210)	$1_1(110)+1_1(100)$	2(110)x(100) or
			2(100)x(110)
2	(320)	$2_2(110)+1_1(100)$	3(110)x(100)
3	(430)	$3_3(110)+1_1(100)$	4(110)x(100)

The surface unit cell can be defined by two vectors in the plane of the surface, $\vec{e}$ and $\vec{w}$, where $\vec{e}$ is parallel to a step edge, while $\vec{w}$ joins equivalent atoms on two adjacent steps and gives the width of the terraces (but $\vec{w}$, in general, will not be perpendicular to the step edges). We shall now determine $\vec{e}$ and $\vec{w}$ for given Miller indices [3.11].

If $\vec{u} = (hkl)$ is the given surface orientation and $\vec{u}_t = (h_t k_t l_t)$ is its terrace orientation (easily deduced from the microfacet notation as being the largest microfacet), then, using the vector product, $\vec{e}$ becomes

$$\vec{e} = \pm \frac{q_t}{p} \vec{u}_t \times \vec{u} . \tag{3.22}$$

Here p is given by (3.18), using the indices (hkl), and

$$q_t = \begin{cases} 1 \text{ for } \begin{cases} sc, \\ fcc, \text{ when } h_t, k_t, l_t \text{ all odd}, \\ bcc, \text{ when } h_t + k_t + l_t = \text{even}, \end{cases} \\ 2 \text{ for } \begin{cases} fcc, \text{ when } h_t, k_t, l_t \text{ not all odd}, \\ bcc, \text{ when } h_t + k_t + l_t = \text{odd}. \end{cases} \end{cases} \tag{3.23}$$

The vector $\vec{w}$ is generated in three steps. First, the vector $\vec{w}' = \vec{u} \times \vec{e}$ is obtained. Second, this vector is scaled to $\omega\vec{w}'$ in order to satisfy the relation

$$|\omega\vec{w}' \times \vec{e}| = |\vec{u}|/p ,$$

which specifies ω; ω becomes the inverse of an integer $1/\omega = n$. Third, the vector $\vec{w}$ is obtained as $\vec{w} = \omega(\vec{w}' + m\vec{e})$, where the integer m is varied from unity to $n = 1/\omega$ until $\vec{w}$ is a lattice vector of the crystal structure.

The reciprocal lattice of the surface can now be generated as in Sect. 3.1.3. Its basis vectors $\vec{e}^*$ and $\vec{w}^*$ are

$$\vec{e}^* = 2\pi \frac{\vec{w} \times \vec{u}}{\vec{e} \cdot (\vec{w} \times \vec{u})} , \quad \vec{w}^* = 2\pi \frac{\vec{u} \times \vec{e}}{\vec{e} \cdot (\vec{w} \times \vec{u})} . \tag{3.24}$$

The resulting vector $\vec{e}^*$ is perpendicular to $\vec{w}$, which allows one to observe the step-to-step relationship in the LEED pattern. The vector $\vec{w}^*$ is perpendicular to the step edges and describes the "spot splitting" normally observed in LEED patterns for stepped and kinked surfaces. The resulting diffraction pattern is sketched in Fig. 3.8 for the case of the fcc(10,8,7) surface. The disappearance of spots far removed from the reciprocal-lattice points of the terraces is due to the predominance of the terraces, see Sect. 3.5.7.

The length of $\vec{w}^*$ is important, since it is related directly to the terrace width w_s through $w_s = 2\pi/|\vec{w}^*|$. This width is measured parallel to the macroscopic surface (not in the terrace plane) and perpendicular to the step edges.

The inclination angle and the step height are also of interest in high-Miller-index surfaces. The inclination angle α_i between a stepped or kinked surface $\vec{u} = (hkl)$ and any of its microfacets $(h_i k_i l_i)$, in particular the terraces, is given by

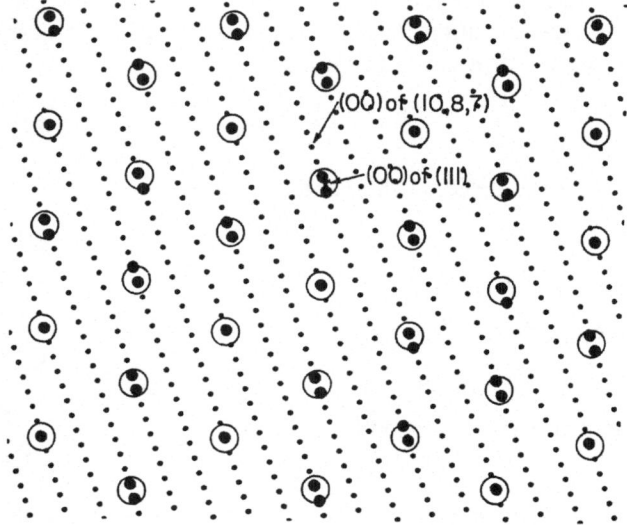

Fig. 3.8. LEED pattern for a fcc(10,8,7) kinked surface consisting of beams (shown as large dots) determined by the conjunction of the circles [which represent broadened beams due to the finite width of the (111) terraces] and the dots (which represent the periodicity of the steps). As a function of increasing energy, the dots converge toward the immobile (00) specular beam of the (10,8,7) surface. At the same time, the circles, that define which beams are intense, converge toward the immobile (00) specular beam position of the (111) terraces. As a result of these two motions, different beams in the pattern are alternately split or narrow as a function of energy

$$\cos\alpha_i = \frac{\vec{u}\cdot\vec{u}_i}{|\vec{u}||\vec{u}_i|} \ . \tag{3.25}$$

The step height d is just the interplanar distance of bulk layers parallel to the terraces:

$$d = \frac{1}{q_t} \frac{1}{|\vec{u}_t|} \ , \tag{3.26}$$

making use of (3.23).

3.4 Symmetries and Domains at Surfaces

3.4.1 Symmetries in Two Dimensions

Space Groups

Structures with periodicities in three-space directions, such as three-dimensional crystals, can be classified into 230 independent space groups, when all possible combinations of symmetry elements are included: center of symmetry, two-, three-, four- and sixfold rotational axes, three-, four-, and sixfold inversion axes, screw axes, and mirror and glide planes. These space groups are listed in the International Tables for X-Ray Crystallography [3.5]. Planar structures with periodicity in two dimensions and a nonzero extent in the third dimension, such as films (with no substrate) can be classified into 80 diperiodic space groups. Included here is the possibility of symmetry operations in the third dimension, such as mirroring across the plane of the film or 180° rotation about an axis parallel to the plane of the film. These 80 groups are discussed and listed by Wood in [3.7].

At surfaces, one can define the structure of interest as that accessible to LEED electrons, typically a slab of 5-10 Å thickness. In this situation, the symmetry operations in the third dimension, perpendicular to the surface, can be assumed to be absent. Then only 17 space groups are possible, namely those that can exist in a strictly two-dimensional space with periodicities in both dimensions. These space groups are also listed in [3.5], and they are illustrated in Fig. 3.9.

Most low-Miller-index surfaces of fcc, bcc, and hcp lattices have a high structural symmetry, at least when they are clean and unreconstructed. Thus, fcc(100) and bcc(100) have the p4m space group, while fcc(111), bcc(111), and hcp(0001) have the p3m1 space group, fcc(110) has the pmm space group, and bcc(110) has the cmm space group. Note that layer spacing changes normal to the surface do not affect the space group, whereas layer registry changes do.

In the presence of layers of adsorbates or in the case of clean surfaces reconstructions, the space group can change compared to the unreconstructed clean surface, especially when a superlattice is present. For example, a (2x1) adsorbate superlattice on an fcc(100) surface might have a pmm space group, assuming adsorption sites of suitable symmetry. Adsorption sites of lower symmetry, whatever their superlattice arrangement, will of course lower the overall surface symmetry. The degree of symmetry is necessarily lowered or at least unchanged, and never increased, by reconstruction or adsorption, unless a surface layer of such thickness is formed that the substrate is no longer accessible to the LEED electrons. In that case, any new space group can appear at the surface. For example, a thick oxide surface layer might appear after oxidation of a metal surface. The oxide can have its own lattice constant, orientation, and symmetry independent of the undetectable substrate.

Symmetries and the LEED Pattern

We shall now discuss the symmetries that appear in a LEED pattern, including spot intensities, not just spot positions, and how these symmetries relate to those of the structure of the diffracting surface. The translational symmetries of the surface, of course, produce well-defined spot positions (i.e.

The 17 two-dimensional symmetry groups

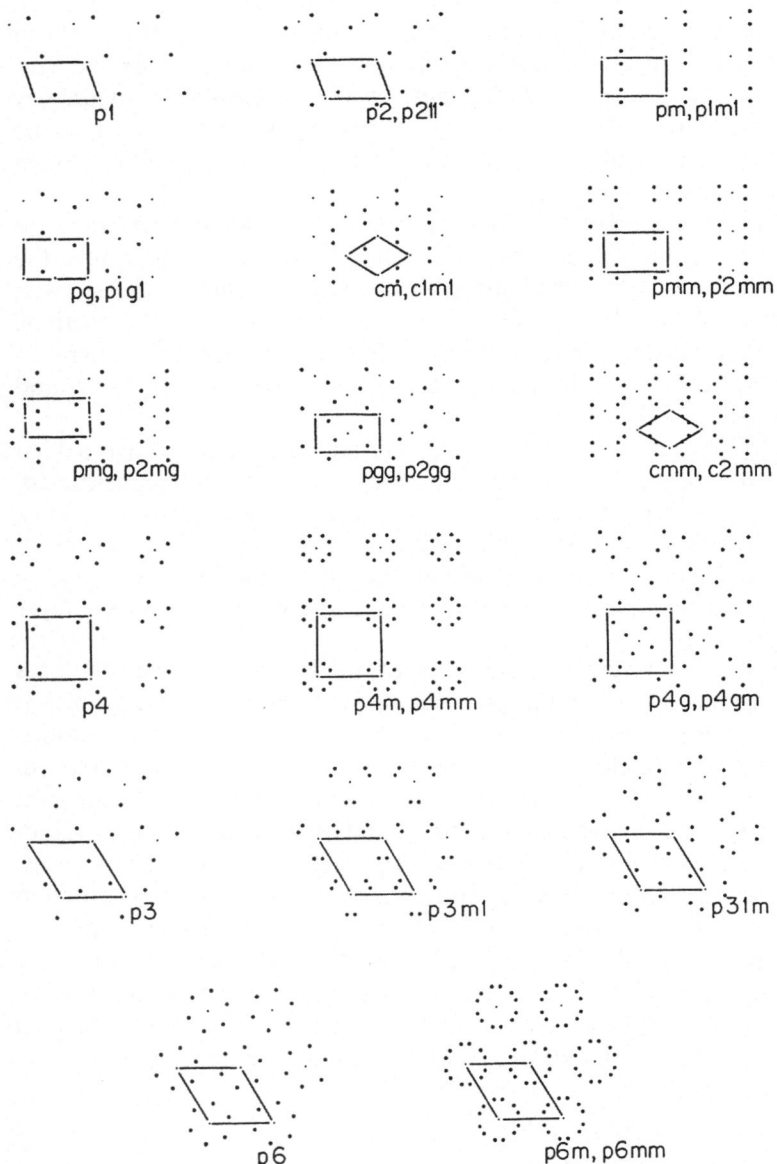

Fig. 3.9. The 17 two-dimensional space groups, represented as parts of lattices satisfying the symmetries of those space groups. The small dots are at the corners of the unit cells or at midpoints, for reference. In each case, one large dot is positioned at an arbitrary nonsymmetrical location within the unit cell, and the other large dots are obtained from this one by applying all the relevant symmetry operations. The "short" and "long" notations of the groups are given

beam directions), as discussed in Sect. 3.1. The other symmetry properties of the surface structure (namely the point symmetries such as rotation axes, mirror planes, etc.) affect the intensities of the spots. However, one should make a clear distinction between these structural symmetries of the surface and the symmetries observed among spot intensities in a LEED diffraction pattern. This is quite obvious when one varies the incidence direction of the LEED beam. At normal incidence one may observe a highly symmetric pattern of beam intensities, which can become totally asymmetric at off-normal incidence. As a general rule, the observed symmetry elements in the spot intensities are those symmetry elements that are common to the surface structure and to the incident electron beam. Thus, if the incident beam has a direction parallel to a mirror plane of the surface structure (so that both the surface and the incident beam have this mirror plane as a symmetry element), then the spot intensities will exhibit this mirror plane as well. At normal incidence, the spot intensities will have exactly the same symmetries as the surface structure (not considering domain effects, which are discussed in Sect. 3.4.2).

It is important to note that the contents of the unit cell, both the number and positions of atoms within the unit cell, do not affect the positions of diffraction spots, i.e. the beam directions. Only the spot intensities are affected by these considerations. However, one special case is of particular interest, namely that of structures with glide-planes (a glide plane can be obtained by inserting additional atoms at suitable positions of the unit cell). Glide planes are manifest by the absence of some spots in the LEED pattern at all electron energies. The general rule is that if the incident beam direction is parallel to a glide plane, then extinctions occur for alternate spots in that row of spots parallel to the glide-plane that contains the specular (00) spot, which is not extinguished. At normal incidence, the presence of more than one glide plane of different orientations extinguishes spots in more than one such row of spots. These extinctions make the glide plane symmetry unique in being the only structural symmetry that can be identified unambiguously in a LEED pattern. (However, domain effects may cancel these extinctions, as discussed below.)

Another important point is that the LEED pattern is determined by the translational and other symmetries of the surface, regardless of the physical interaction laws between electrons and surfaces. Thus, multiple scattering does not affect spot positions or the symmetries among spots, although multiple scattering can make some spots visible that otherwise would be too weak to be detected.

3.4.2 Domains

An important consideration in the pattern symmetries is the occurrence of domains (also called mosaics). Domains are a form of long-range imperfec-

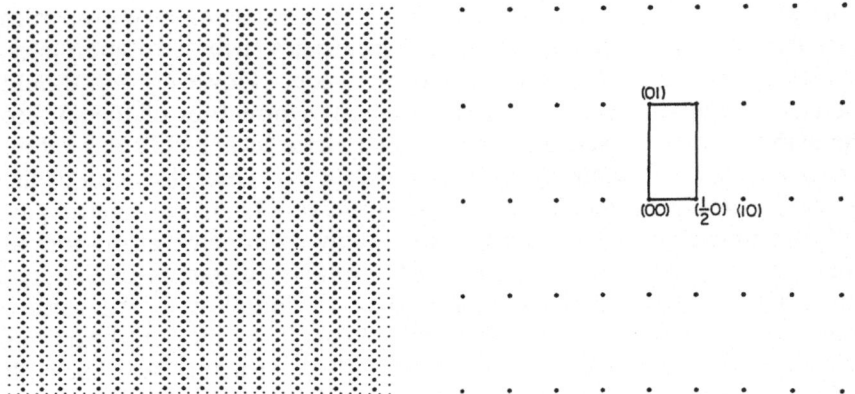

Fig. 3.10. On the left, a (2×1) superlattice (*large dots*) on a square lattice (*small dots*) with translational domains. On the right, the corresponding diffraction pattern in the limit of very large domains

tion in which distinct patches of the surface coexist. These different patches have mutually identical structures, but they do not mesh together without breaks in the periodicity. Thus, domain boundaries occur across which the periodicity is broken. Domains are prevalent in the presence of superlattice overlayers and reconstructions. There are several kinds of domains: translational, rotational, mirrored, and combinations thereof. An example of translational domains is given in Fig. 3.10 for a (2x1) superlattice overlayer. Each domain is related to each other domain by a substrate lattice vector that is not a superlattice vector. In the (2x1) case, there are antiphase domains, since half-order diffraction spots will receive contributions from both sides of a domain boundary that are exactly out-of-phase. (The term *antiphase domains* is often used to denote all translational domains or even all kinds of domains, although the strict antiphase relationship exists only in a few cases.) An example of rotational domains is shown in Fig. 3.11. Now the mutual rotation prevents meshing without breaks. Mirrored domains are obtained in a similar way by mirroring one domain onto another. Glide-plane symmetry, which combines translation and mirroring, and other combined symmetries can also produce domains. In general, a surface structure with a symmetry lower than that of the substrate can produce domain structures by applying all the symmetry operations of the substrate.

We shall assume here that any domains present at a surface are large compared to the lateral coherence length of the incident electron beam as determined by the instrumental response function (which is typically of the order of a few hundred Ångstroms). Then each domain can be considered to contribute independently to the diffraction pattern, which becomes simply the sum of the individual diffraction patterns of each domain, as illustrated in Figs. 3.10,11. Smaller domains of irregular sizes produce disorder effects

Fig. 3.11. As in Fig. 3.10 with rotational domains, showing additional spots in the diffraction pattern on the right

in the diffraction pattern, such as weakened and broadened fractional-order spots and diffuse intensity distributions. These effects will be discussed in Chaps. 8-10.

An important aspect of domains is that different domains with the same internal structure should be found with equal probability at a surface, since they have the same structural energy (unless the substrate has asymmetric features such as well-oriented steps, or directional defects). A LEED beam with a typical diameter of approximately 1 mm should therefore sample essentially equal areas of each domain type, thereby giving approximately equal contributions to the diffraction pattern from each domain type. It follows that whatever the symmetries or lack thereof that one domain may have, the combined contributions from equal domains at normal incidence produce a pattern and spot intensities that have the same symmetry as the substrate. For example, at normal incidence the pattern and spot intensities in Fig. 3.11 have the same symmetries as the substrate structure, if equal domain sizes are assumed, namely fourfold rotation and mirror planes. This is true even if the basis structure within one domain has either low or no symmetry. If in Fig. 3.11 the adatoms had an arbitrary position in the unit cell, rather than being centered as drawn, the various domains obtained by rotations, mirrorings, and translations would, taken together, still give a pattern with the full symmetry of the substrate.

As a very important consequence, *the observation of a highly symmetrical LEED pattern at normal incidence of the electron beam does not in any way imply a highly symmetrical surface structure.* Indeed, it is often experimentally possible to eliminate certain domains selectively (for example, by the presence of steps in the substrate surface), and then observe the reduced symmetry of individual types of domains.

3.5 Interpretation of LEED Patterns

Numerous sharp LEED patterns occur during experiments with ordered surfaces. Some are simple, as for overlayers with small unit cells, while others are complex, as for large unit cells or for multiple domain structures. In either case, given the pattern, one wishes to know what the corresponding surface unit cell is, since that is important information concerning the surface structure. As will become clear, often much more than the size and shape of the unit cell can be learned from a pattern, namely some information concerning the contents of the unit cell.

In this section, we shall discuss the interpretation of patterns in a general fashion, ranging over many possible kinds of patterns. Generally, sharp patterns will be considered, since disorder-induced patterns can be even more complicated and varied. The latter are treated in Chaps. 8-10. We shall also exclude patterns that are due to independent superlattice structures coexisting simultaneously at a surface, such as (2x1) and c(2x2) structures existing simultaneously on different parts of an fcc(100) surface. Such coexisting structures are quite common in the course of surface preparation, and we shall assume that the observer can identify this situation without further help.

First, let us discuss a few general issues concerning LEED patterns. Many patterns are due to superlattices that have formed at the surface, enlarging the unit cell from that expected from ideal truncation of the substrate [3.12]. Apart from asking which surface unit cell corresponds to the observed pattern, one may ask whether the pattern is due to reconstruction of the clean substrate or rather to an overlayer of different atoms (or even to an underlayer of different atoms). Unfortunately, the LEED pattern itself often does not help in answering this kind of qualitative question. The main reason is that most atoms have comparable scattering strengths, so that one cannot, for example, conclude that fractional-order spots appearing in the presence of an adsorbate of low atomic number imply a substrate reconstruction. Even with the adsorption of hydrogen, the weakest scatterer among all atoms, weak fractional-order spots could be explained as either small substrate atom displacements, or as a hydrogen superlattice on top of or within an unmodified substrate. In addressing these questions, it is more helpful, in addition to analysis of the surface composition by Auger-Electron Spectroscopy and other techniques, to observe sequences of LEED patterns as a function of coverage or temperature. These can indicate the most probable kind of surface structure. Nevertheless, there is no guaranteed, simple way to answer this kind of structural question, although we shall show that the pattern itself can in certain cases give considerable structural information.

As another general point, it is always useful to study LEED patterns at many electron energies, at different angles of incidence, and with as large a range of reciprocal space as possible. Varying the electron energy allows one to avoid overlooking some spots that happen to be almost extinguished at a certain energy. Of great importance are certain spots that are missing at all energies, since they probably indicate the presence of glide-plane symmetry, as discussed in Sect. 3.5.3. The use of a large range of reciprocal space, in particular by the observation of patterns at sufficiently high energies to expose many spots, provides a better overview of the patterns, and avoids false conclusions based on insufficient information in patterns that contain only a few spots. We shall give examples of this.

We now turn to the pattern interpretation and classify patterns according to increasing complexity, starting with the simplest category. We shall then conclude this section with a brief description of the use of laser diffraction patterns as models of LEED patterns for the study of surface structures.

3.5.1 Patterns with a Bravais Array of Spots

The Bravais array patterns are characterized by the fact that all spots are part of a single reciprocal lattice, and that there are no missing spots in that lattice. (A Bravais lattice can be defined as a lattice the unit cell of which contains only one lattice point.) An example may be found in Fig. 3.10, whereas the pattern of Fig. 3.11 does not have a Bravais array of spots. In our category of Bravais array patterns, no multiple domain orientations are allowed if new spots are generated by different domains. Such new spots usually fall outside the spot array of the single domain. Thus, on the fcc(100) surface, a (2x1) structure can coexist with a (1x2) structure in different domains, giving a combined pattern that is not a Bravais array of spots (Fig. 3.11); a (2x1) domain by itself does produce a Bravais array pattern (Fig. 3.10).

The Bravais array patterns include many simple patterns, such as those for the (1x1), c(2x2), (2x2) and $(\sqrt{3} \times \sqrt{3})R30°$ structures on substrates of appropriate symmetry. They also include less simple patterns, namely those that correspond to

$$\begin{pmatrix} m_{11} & m_{12} \\ m_{21} & m_{22} \end{pmatrix}$$

structures (using the matrix notation), for arbitrary integers m_{ij}. There is an infinite number of such patterns, since each element m_{ij} can take any integral value, as long as the determinant $|m_{11}m_{22} - m_{12}m_{21}|$ is nonzero. (This determinant is the area of the superlattice unit cell, given in units of the substrate unit-cell area.)

In order to determine the matrix

$$\begin{pmatrix} m_{11} & m_{12} \\ m_{21} & m_{22} \end{pmatrix}$$

describing the surface structure (which, of course, also includes the simpler cases like (1x1), c(2x2), etc.), the general procedure is to identify which among all the observed spots are the "substrate spots", i.e. those that an ideal termination of the substrate would produce. Let the substrate spots have the basis vectors $\vec{a}_1^*$, $\vec{a}_2^*$ in reciprocal space, and let the superlattice spots have the basis vectors $\vec{b}_1^*$, $\vec{b}_2^*$ in reciprocal space. Using the pattern, it is then sufficient to determine the integers m_{ij}^* in the equation

$$\begin{pmatrix} \vec{a}_1^* \\ \vec{a}_2^* \end{pmatrix} = \begin{pmatrix} m_{11}^* & m_{12}^* \\ m_{21}^* & m_{22}^* \end{pmatrix} \begin{pmatrix} \vec{b}_1^* \\ \vec{b}_2^* \end{pmatrix}, \tag{3.27}$$

expressing the substrate reciprocal lattice in terms of the reciprocal superlattice. From this relation it follows that

$$\begin{pmatrix} \vec{b}_1^* \\ \vec{b}_2^* \end{pmatrix} = \frac{1}{m_{11}^* m_{22}^* - m_{12}^* m_{21}^*} \begin{pmatrix} m_{22}^* & -m_{12}^* \\ -m_{21}^* & m_{11}^* \end{pmatrix} \begin{pmatrix} \vec{a}_1^* \\ \vec{a}_2^* \end{pmatrix}. \tag{3.28}$$

We can now reconstruct the direct-space relationship between the surface basis vectors $\vec{a}_1, \vec{a}_2$ for the substrate and $\vec{b}_1, \vec{b}_2$ for the superlattice (these are reciprocal to $\vec{a}_1^*, \vec{a}_2^*$ and $\vec{b}_1^*, \vec{b}_2^*$, respectively):

$$\begin{pmatrix} \vec{b}_1 \\ \vec{b}_2 \end{pmatrix} = \begin{pmatrix} m_{11}^* & m_{21}^* \\ m_{12}^* & m_{22}^* \end{pmatrix} \begin{pmatrix} \vec{a}_1 \\ \vec{a}_2 \end{pmatrix}. \tag{3.29}$$

In this conveniently simple relationship, note the exchange of the elements m_{12}^* and m_{21}^* compared with (3.27).

As an illustration of this unit-cell determination, we consider first the case of a c(2x2) structure on a surface of square symmetry, such as the fcc(100) surface shown in Fig. 3.12. The observed pattern can be analyzed as follows in terms of its basis vectors:

$$\begin{pmatrix} \vec{a}_1^* \\ \vec{a}_2^* \end{pmatrix} = \begin{pmatrix} 1 & 1 \\ \bar{1} & 1 \end{pmatrix} \begin{pmatrix} \vec{b}_1^* \\ \vec{b}_2^* \end{pmatrix},$$

which implies

$$\begin{pmatrix} \vec{b}_1 \\ \vec{b}_2 \end{pmatrix} = \begin{pmatrix} 1 & \bar{1} \\ 1 & 1 \end{pmatrix} \begin{pmatrix} \vec{a}_1 \\ \vec{a}_2 \end{pmatrix},$$

thereby yielding the matrix notation of the c(2x2) direct lattice. Slightly less trivial is our second example, which concerns the $(\sqrt{3} \times \sqrt{3})R30°$ lattice on a surface of hexagonal symmetry, as illustrated in Fig. 3.13. In this case, we find from the pattern that

$$\begin{pmatrix} \vec{a}_1^* \\ \vec{a}_2^* \end{pmatrix} = \begin{pmatrix} 2 & \bar{1} \\ 1 & 1 \end{pmatrix} \begin{pmatrix} \vec{b}_1^* \\ \vec{b}_2^* \end{pmatrix},$$

so that

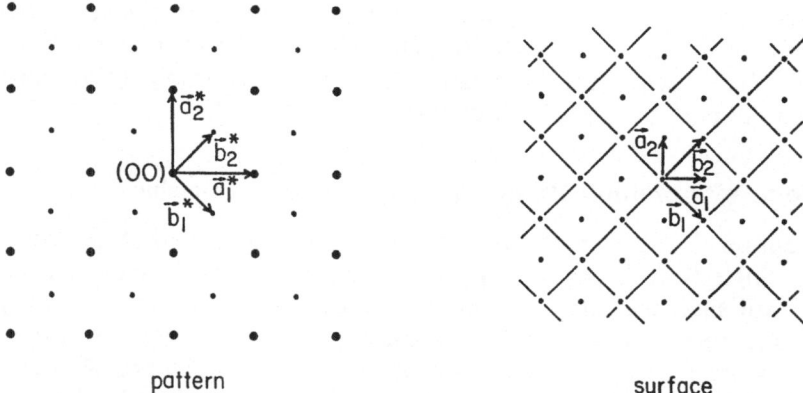

Fig. 3.12. LEED pattern (*left*) for a c(2×2) surface structure on a square substrate (*right*). In the pattern, large dots represent "substrate spots". In the surface structure diagram, the lines indicate the superlattice

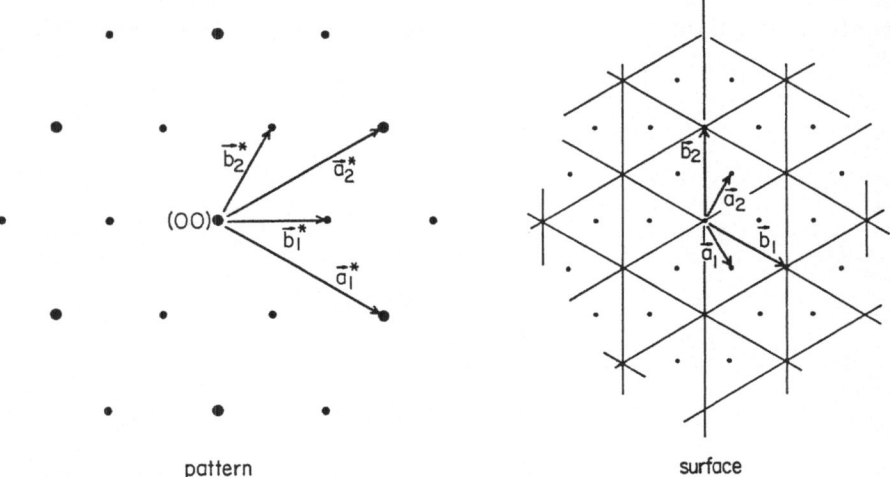

Fig. 3.13. LEED pattern (*left*) for a $(\sqrt{3} \times \sqrt{3})$R30° surface structure on a hexagonal substrate (*right*). Notation as in Fig. 3.12

$$\begin{pmatrix} \vec{b}_1 \\ \vec{b}_2 \end{pmatrix} = \begin{pmatrix} 2 & 1 \\ 1 & 1 \end{pmatrix} \begin{pmatrix} \vec{a}_1 \\ \vec{a}_2 \end{pmatrix} .$$

The matrix in this relation is just that corresponding to the $(\sqrt{3} \times \sqrt{3})$R30° superlattice. This example is instructive for practicing the transformation between direct and reciprocal lattices when the basis vectors are not mutually orthogonal to each other.

A special case of some ambiguity and importance is the observation of (2x2) patterns with hexagonal surfaces. These can be interpreted as Bravais array patterns, which yield a (2x2) surface structure. However, they can also be interpreted as due to (2x1) surface structures with different domain orientations. We shall discuss this point and solutions to this ambiguity in connection with the next class of LEED patterns.

3.5.2 Patterns with Multiple Bravais Arrays of Spots — Domains

As was mentioned above, patterns occur frequently that cannot be interpreted as a single Bravais array of spots, i.e. no pair of basis vectors can be found that will span the array of observed spots. The coexistence of (2x1) and (1x2) structures on the fcc(100) surface provides a simple example. Such patterns are recognized easily through the apparent absence of many spots, as may be seen in Fig. 3.11. In the case of complex unit cells, the patterns can be quite unintelligible at first, although frequently of artistic beauty. Examples are shown in Fig. 3.14.

The principal practical difficulty in interpreting patterns with multiple Bravais arrays of spots, especially when they are of some complexity, is to identify to which domain each spot belongs. The difficulty is most extreme

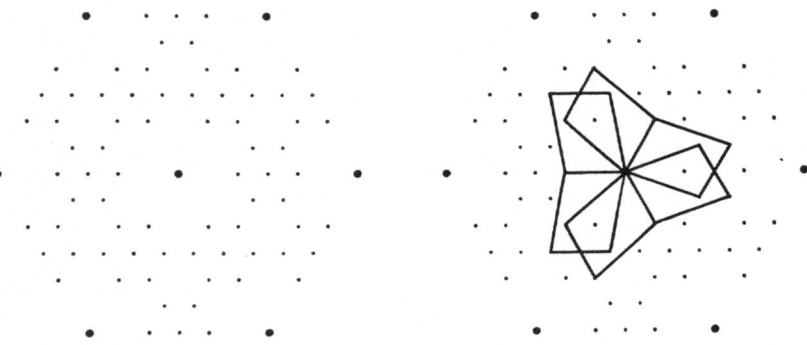

Fig. 3.14. Complex LEED patterns for a hexagonal substrate. Top right shows the unit cells for different domains present in the pattern at top left. Bottom shows the pattern with a slightly different unit cell that avoids coincidence of spots. Large dots represent substrate spots

in cases of intermediate complexity where spots from different domains coincide, so that a spot can belong to different domains at the same time. Moreover, spots from different domains can be aligned in such a way that the observer is led to believe that they belong to the same domain, as in the top pattern of Fig. 3.14. For simpler patterns, for example (2x1) + (1x2), the number of possible solutions of the pattern is so small that these factors are no great inconvenience in determining the surface structure. For patterns of greater complexity, as the bottom pattern of Fig. 3.14, different domains often produce no spot coincidences, and it is easier to identify the domain to which a spot belongs. In the latter case, one can usually solve the pattern by first searching for lines of equidistant spots passing through the specular spot (or any other substrate spot). Such lines with the smallest possible spot-to-spot distances are the most useful ones. Similar lines of spots parallel to these should be found at equidistant spacings on either side of the initial ones. All these lines should clearly define equivalent unit cells of the pattern for different domain orientations. (One should not only include the possibility of rotational domains, but also of mirrored and glided domains, as was discussed in Sect. 3.4.2). Once the single-domain pattern is identified, its interpretation proceeds as described in the previously discussed class of Bravais array patterns. This procedure can be exercised on the patterns of Fig. 3.14.

In such a pattern analysis, it is important to avoid normal incidence of the electron beam, because of the possibility of the presence of glide-plane symmetry in the surface structure. Glide-plane symmetry can extinguish spots on lines of spots passing through the specular spot, and this would severely complicate the task of determining the unit cell, or even mislead one into an erroneous result.

As mentioned previously, in some special cases such as with a (2x2) pattern for a hexagonal surface, the possibility of domains can produce an ambiguity in the pattern interpretation. Both a (2x2) surface structure and domains of (2x1) surface structure can generate a (2x2) pattern. It is not an easy matter to resolve such an ambiguity, but there are several experimental methods that can be helpful. It may be possible to favor one domain orientation over another and thus produce marked asymmetries in the spot intensities due to different domains, at normal incidence. A domain orientation may be favored by the presence of oriented steps on the surface, and it may also be favored by other anisotropic surface damage, such as that due to off-normal ion bombardment. Another way to resolve the ambiguity relies on streaking of the spots due to disorder. The direction of spot streaking may distinguish between the different possibilities. Other less direct arguments for resolving the ambiguity could be used as well, such as adatom coverage, atomic or molecular size and shape, etc.

3.5.3 Patterns Exhibiting Extinctions Due to Glide-Plane Symmetry

As was mentioned in Sect. 3.4, systematic spot extinctions can occur along lines of spots passing through the specular spot. In particular, alternate spots, excluding the specular spot, are absent at all energies. This occurs for special incidence directions, namely normal incidence and for some off-normal incidence directions that are parallel to glide planes of the surface structure. Glide-plane symmetry is the only possible cause of this kind of extinction, as will be demonstrated in Sect. 3.5.8. In some patterns there exist consistently weak spots that are systematically invisible, such as with split spots, which belong to complete strings of generally invisible spots. This will be discussed in Sect. 3.5.7. There can rarely be confusion, however, between this kind of missing spots and the glide-plane extinctions, because they occur at very different positions in the pattern and have a different dependence on the angle of incidence.

In the case of a single domain orientation, the direction of a line or lines of missing spots indicates directly the orientation of the glide plane or planes. When different domain orientations coexist, one must first ascertain to which domain orientation a line of missing spots belongs.

More difficult to analyze, and especially to recognize, is the situation where, even though a single domain orientation produces missing spots, other domain orientations give nonzero intensities in those same positions. The net effect is that no spots appear to be missing. This can occur for example with the pmg space group, which applies to the W(100)-c(2x2) reconstruction [3.13]. At normal incidence, one orientation of a surface with that symmetry produces one and only one line of alternate missing spots, while a domain of orientation rotated by 90° produces a line of missing spots perpendicular to that of the first orientation of the domain, and vice versa. Thus, no spots are normally missing. To recognize such a situation requires that the total sizes of differently oriented domains be substantially different, so that asymmetries occur in the spot intensities at normal incidence, with the extreme case of the complete absence of one domain orientation being ideal, since it leads to truly missing spots.

As may be seen in Fig. 3.9, no space group based on hexagonal lattices has an overall glide-plane symmetry. (Although individual glide planes may exist, they are parallel to mirror planes that prevent spot extinctions.) Therefore, if an overlayer structure on a hexagonal surface does produce extinctions, one can conclude immediately that a nonhexagonal unit cell has been formed, namely a rectangular unit cell. In certain cases, this can even provide a criterion for distinguishing between (2x1) and (2x2) unit cells on hexagonal surfaces, or other similar ambiguities [3.14].

Finally, it should be emphasized again that it is most desirable to observe patterns both at normal incidence and at off-normal incidence, and to examine a large range of reciprocal space to render many spots visible. This per-

mits a more reliable extraction of easily obtained information from the
LEED pattern.

3.5.4 Rationally Related Lattices and Coincidence Lattices

The idea of rationally related lattices is simple, but leads to practical
subtleties in the LEED pattern that we must treat in some detail. A one-
dimensional example is given in Fig. 3.15: a substrate with period a is combined with an overlayer of period 3a/2, hence the term "rationally related lattices". The key observation in this kind of structure is that the overlayer
atoms are not all identical to each other. First, their environment is different
in terms of distances and coordination number from neighboring atoms.

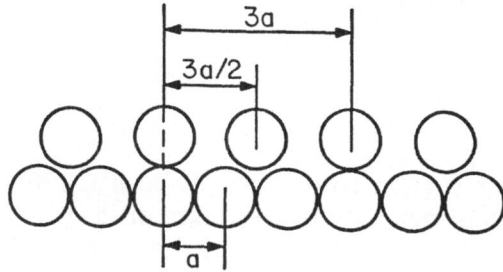

Fig. 3.15. One-dimensional surface with
rationally related lattices of atoms,
viewed parallel to the surface

This difference may be manifest in diffraction with multiple scattering.
Second, one may expect positional differences in these atoms due to the
different environments. For example, some overlayer atoms may sink closer
to the substrate than the others as a result of the different "registry". After
all, the very existence of a relation between the lattices, even if it is only
rational, implies some physical interaction between them, which may act not
only parallel to the surface but also perpendicular to it. The resulting buckling (nonplanarity) of the overlayer can be strong, if the adsorbate-substrate
interactions dominate the situation as in chemisorption, or weak, if the
adsorbate-adsorbate interactions dominate the situation as in compact physical adsorption. Of course, the substrate layers can buckle as well for the
same reasons. Again, diffraction will detect this difference, and, as a result,
the inequivalence of a' priori equivalent atoms leads to the conclusion that
the unit cell of the combined substrate-overlayer system does not, in our
example, have the length 3a/2 but rather the length 3a, and that it includes
not one overlayer atom but two inequivalent ones. This larger unit cell generates the so-called *coincidence lattice*, which has the true periodicity of the
combined substrate-overlayer system.

In general, on two-dimensional surfaces, the unit cell of the coincidence
lattice can be defined as the smallest unit cell that contains an integral
number of substrate unit cells and of overlayer unit cells. An example of a

fcc (100) : buckled hexagonal top layer

two-bridge   top/center

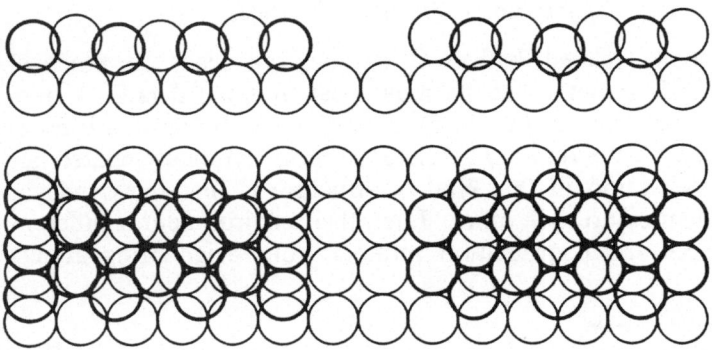

Fig. 3.16. (*Bottom*) Top view of Ir(100)-(5 × 1) surface with quasi-hexagonal top layer in two registries. (*Top*) Side view of the same structures, including buckling

coincidence lattice is provided by the (5x1) reconstruction of clean Ir(100), assuming the model in which the top metal monolayer takes on a quasi-hexagonal lattice, which matches the square substrate lattice with a (5x1) coincidence lattice, as shown in Fig. 3.16. This figure also illustrates the buckling of the top layer that one may expect in such a situation.

Now, we turn to the LEED patterns that can arise in the case of coincidence lattices. According to our discussion, the only true unit cell from the point of view of diffraction is the unit cell of the coincidence lattice rather than that of the substrate or of the overlayer. In these terms, the diffraction pattern is simply the one which is constructed from the reciprocal lattice of the coincidence lattice, as exemplified in Fig. 3.17 for the Ir(100)-(5x1) surface. Although this reasoning is correct, another point of view is often used since it permits more physical insight.

Consider a planar (i.e. nonbuckled) overlayer with the (5x1) lattice of Fig. 3.16. In the kinematic limit, one can easily be convinced that it is possible to construct the diffraction pattern as the sum of the patterns of the substrate and of the overlayer. In this way, one obtains spots due to the square lattice and spots due to the quasi-hexagonal lattice, denoted in Fig. 3.17 as square and triangular symbols, respectively. This explains some but not all of the (5x1) spots. There are two ways to generate the additional (5x1) spots. The first, valid both within the kinematical approximation and in the multiple-scattering case, is the introduction of buckling, as shown in Fig. 3.16. This produces immediately a (5x1) unit cell with six inequivalent overlayer atoms arranged in nonregular positions within the unit cell, and hence all (5x1) spots will appear. (In other words, the form factor of the unit cell gives nonzero intensities in all fifth-order spots.)

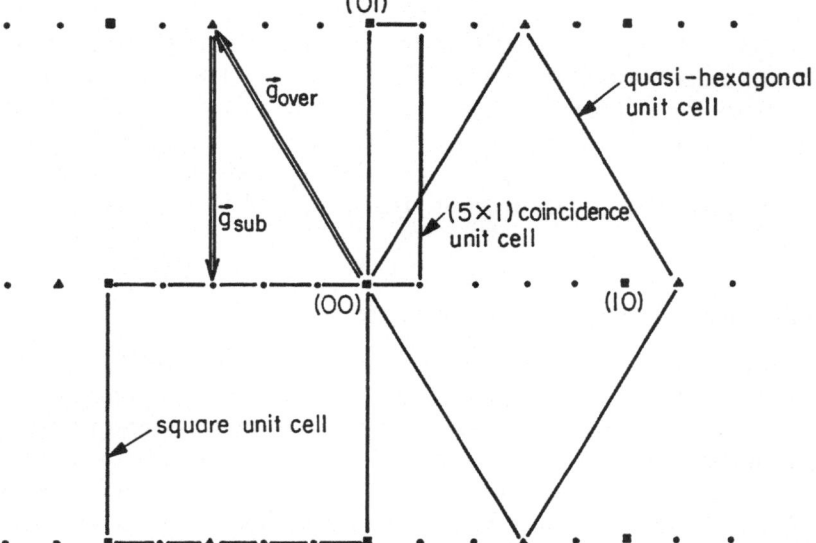

Fig. 3.17. LEED pattern for the Ir(100)-(5 × 1) surface (one domain orientation only), indicating various unit cells in reciprocal space

The second explanation of the extra "nonsquare" and "nonhexagonal" spots relies on multiple scattering. Figure 3.17 shows how in two scatterings one can reach any arbitrary spot of the pattern. Starting from the (00) beam, a scattering by the overlayer through one of its reciprocal lattice vectors $\vec{g}_{over}$, followed by a scattering by the substrate through one of its reciprocal lattice vectors $\vec{g}_{sub}$, leads to a fifth-order spot labeled $\vec{g}_{sub} + \vec{g}_{over}$. By a suitable choice of $\vec{g}_{sub}$ and $\vec{g}_{over}$, any fifth-order spot can be reached with $\vec{g}_{sub} + \vec{g}_{over}$. By triple and higher-order multiple scattering, one merely translates from one of those spots to another of those spots, without modifying the pattern.

Thus, one can explain the observed LEED pattern either in terms of overlayer buckling, whether in the kinematical or dynamical case, or by multiple scattering, whether with planar or buckled layers. The real situation contains contributions from both mechanisms, buckling and multiple scattering. Is it then possible to distinguish whether buckling is present or not by merely inspecting the diffraction pattern? It turns out that to some extent this is possible, but it is necessary to consider the relative (energy-averaged) intensities of the spots.

As a general rule in LEED, a diffraction spot due entirely to double scattering (i.e. not to buckling) is approximately an order of magnitude weaker than a single-scattering (kinematic) spot. There is an additional interesting factor influencing the average spot intensity: the larger the reciprocal lattice vectors $\vec{g}_{sub}$ and $\vec{g}_{over}$ (which often implies that evanescent plane waves are involved), the weaker the spot at $\vec{g}_{sub} + \vec{g}_{over}$. Thus, if the quasi-hexagonal overlayer on Ir(100)-(5x1) were planar, the "square" and "hexagonal" spots in Fig. 3.17 would on the average be strong compared to the other fifth-order spots due to multiple scattering. The experimentally observed pattern for Ir(100)-(5x1) does not indicate this, but rather shows approximately equal intensities in all spots. We can conclude that the over-layer is not a planar quasi-hexagonal layer, but perhaps a buckled quasi-hexagonal layer, as is borne out by a full dynamical LEED intensity analysis [3.15].

3.5.5 An Instructive Example of Pattern Interpretation

We shall analyze now a LEED pattern that involves many of the factors dis-cussed thus far in this section and that therefore will serve to recapitulate some of the important points. It is an unusual case in that the interpretation of the pattern determines the surface structure to a large extent, without a need to resort to dynamical LEED calculations. The pattern occurs when a half-monolayer of carbon monoxide is adsorbed without dissociation on a clean Pd(100) surface [3.16]; it is depicted in Fig. 3.18.

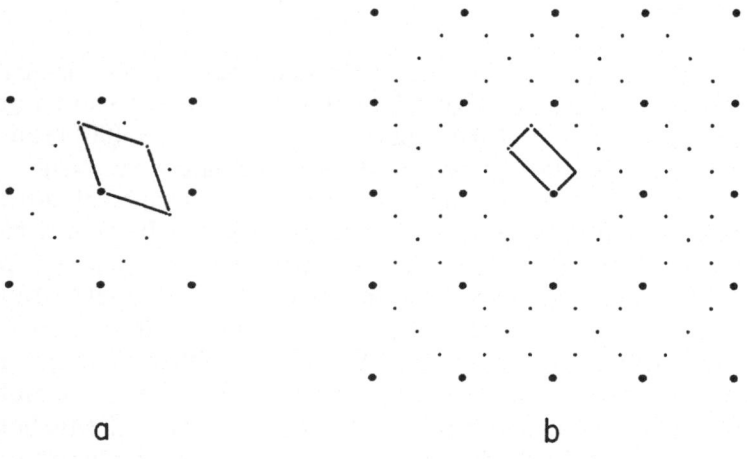

a   b

Fig. 3.18a, b. LEED pattern for Pd(100)-(2$\sqrt{2}$×$\sqrt{2}$)R45°-2 CO: small portion (**a**) and larger portion (**b**). Large dots represent substrate spots

Panel **a** of Fig. 3.18 shows a limited region of reciprocal space, as would be visible at low electron energies, and normal incidence is assumed. The spots do not form a Bravais array, because there are too many missing spots, most of them not explainable by glide-plane symmetry. If one tentatively chooses the drawn rhombus as the unit cell in reciprocal space, all observed extra spots are explained with two domain orientations at 90° with respect to each other. This unit cell, yielding a c($2\sqrt{2} \times \sqrt{2}$)R45° superlattice, is tempting because one molecule per unit cell yields the desired coverage of 1/2. However, there is an immediate inconsistency. If some integral-order spots are not part of the overlayer lattice, as is the case here with the (10) spot, then there should exist multiple-scattering spots $\vec{g}_{sub} + \vec{g}_{over}$; but some of these are not present in our example. Another flaw appears when a larger range of reciprocal space is observed, as at higher electron energies, cf. Fig. 3.18b. A number of fractional-order spots are not explained by our choice of unit cell, such as the (5/4,3/4) spot. However, since we could obtain these by multiple scattering, there is still some hope that our rhombic unit cell contains some element of truth. Off-normal incidence would produce yet another problem: new spots appear experimentally at the locations (1/4,1/4), (3/4,3/4), etc., showing that those were extinguished at normal incidence only. This observation indicates the presence of glide-plane symmetry and actually also suggests the proper unit cell, the rectangle drawn in Fig. 3.18b. This unit cell, corresponding to a p($2\sqrt{2} \times \sqrt{2}$)R45° lattice, solves all the above problems, but creates a new one: it implies a coverage of 1/4. A solution to this problem is to allow two molecules per unit cell. This solution is actually also imposed by the glide-plane symmetry, which requires pairs of identical objects in the unit cell. These identical objects must also have identical environments, since they are related by the glide-plane symmetry.

It is now easy to draw a model for the surface structure in our example. We need a ($2\sqrt{2} \times \sqrt{2}$)R45° unit cell containing two molecules in identical environments. The only possible solution is given in Fig. 3.19, where the "bridge sites" are implied by the symmetry requirements.

Consider now the CO lattice of Fig. 3.19 taken in isolation from the substrate. One recognizes a regular layer with a quasi-hexagonal lattice. It has a smaller unit cell than the rectangle, namely the rhombus shown in Fig. 3.19; this unit cell corresponds to a c($2\sqrt{2} \times \sqrt{2}$)R45° lattice with respect to the substrate. We now recognize the superlattice that we had at first

Pd(100) – ($2\sqrt{2} \times \sqrt{2}$) R45°–2CO

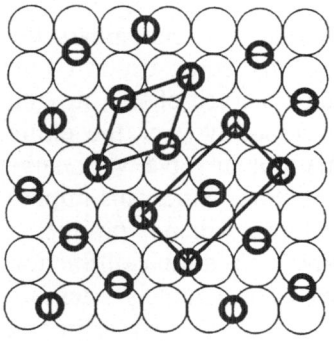

Fig. 3.19. Surface structure of Pd(100)-($2\sqrt{2} \times \sqrt{2}$)R45°-2 CO. The admolecules are represented by thick circles

hypothesized and then abandoned: it turns out to have been partly correct. What has happened is that a coincidence lattice is present. The (1x1) substrate lattice and the $c(2\sqrt{2} \times \sqrt{2})R45°$ overlayer lattice coincide as a $p(2\sqrt{2} \times \sqrt{2})R45°$ lattice. Both smaller lattices can be identified within the pattern, but additional "combined spots" are generated, by multiple scattering in this case since there is no need for buckling in the layers. (Buckling would also violate the glide-plane symmetry.)

This structure has been confirmed by vibrational frequency measurements [3.17,18] and by a dynamical LEED analysis [3.18], the latter having in addition determined bond lengths and bond angles. However, as an exercise, the reader may wish to analyze the orientation of the CO molecules (which are believed to be perpendicular to the surface) on the basis of the pattern alone, to see how much additional information can be extracted.

3.5.6 Incommensurate Lattices

Thus far, we have dealt with combinations of commensurate lattices, i.e. cases where the combination of substrate and overlayer lattices still produce a finite unit cell that is simply related to the substrate lattice. When the overlayer and substrate lattices have a matrix relationship involving not only integral or rational numbers, but also irrational numbers, one speaks of incommensurate lattices. An example would be the case where an overlayer has lattice vectors $\vec{b}_1 = \pi \vec{a}_1$ and $\vec{b}_2 = \pi \vec{a}_2$, where $\pi = 3.14159...$ is irrational, and $\vec{a}_1$ and $\vec{a}_2$ are substrate lattice vectors. [Note that a $(\sqrt{2} \times \sqrt{2})R45° = c(2x2)$ lattice on the fcc(100) surface is not irrational even though $\sqrt{2}$ is irrational, because its matrix notation contains no irrational numbers.]

Essentially, we are dealing with the combination of two independent lattices, one defined by the lattice vectors $\vec{a}_1$ and $\vec{a}_2$, the other by lattice vectors $\vec{b}_1$ and $\vec{b}_2$. The diffraction pattern can be constructed as we did in discussing rationally-related lattices. We start with single scattering by nonbuckled layers, which superimposes two independent patterns: that for the substrate based on the reciprocal lattice vectors $\vec{a}_1^*$ and $\vec{a}_2^*$, and that for the overlayer based on the reciprocal lattice vectors $\vec{b}_1^*$ and $\vec{b}_2^*$.

By double scattering, we obtain all possible combinations $k\vec{a}_1^* + l\vec{a}_2^* + m\vec{b}_1^* + n\vec{b}_2^*$ with integral coefficients. In principle, we can encompass the entirety of reciprocal space with all these combinations, since any arbitrary point of reciprocal space can be reached by a suitable choice of k, l, m, and n. Consequently, intensity would be distributed smoothly over the entire LEED screen. However, most of those combinations involve very large values of k, l, m, and n, i.e. large reciprocal lattice vectors. Since this implies that evanescent beams are involved, such combinations will not produce a detectable intensity. It follows that in practice only some double-diffraction spots will become visible. Moreover, they will have approximately a tenfold reduced intensity compared to single-scattering spots. Third-order scattering can in turn add a small amount of intensity to spots that double scattering could not adequately illuminate, but they would rarely be observed.

This situation is illustrated by the reconstruction of the clean Au(100) surface, which shows a rather complex pattern. On the basis of recent high-resolution LEED patterns measured by Zehner at a variety of energies (at normal incidence) [3.15,19], we sketch this pattern in Fig. 3.20. It is reminiscent of the Ir(100)-(5x1) pattern of Fig. 3.17, but with small displacements in many of its spots and additional satellite spots. This pattern can be understood, as with Ir(100)-(5x1), by assuming a quasi-hexagonal top metal layer, but with an incommensurate lattice. As in Fig. 3.17, we show in this pattern "square" and "hexagonal" spots. The hexagonal spots now no longer fall exactly in fifth-order positions, but nearly so. This indicates slightly modified lattice constants in the hexagonal layer, specifically in this case approximately 4% contraction with respect to bulk Au-Au interatomic distances. The combinations $\vec{g}_{sub} + \vec{g}_{over}$ can be used to predict the positions of other spots, and it is gratifying to find that in this way one can reproduce the observed pattern. This constitutes very strong support for the hexagonal reconstruction model, because it is very difficult to imagine another structure that would reproduce all the observed spots and only the observed spots. It was in fact the Au(100) pattern that gave rise to the first suggestion, by Fedak and Gjostein [3.20], that the reconstruction involves a quasi-hexagonal top layer.

To obtain good relative spot intensities in modeling the Au(100) pattern of Fig. 3.20, one must include some buckling in the "fivefold direction" along

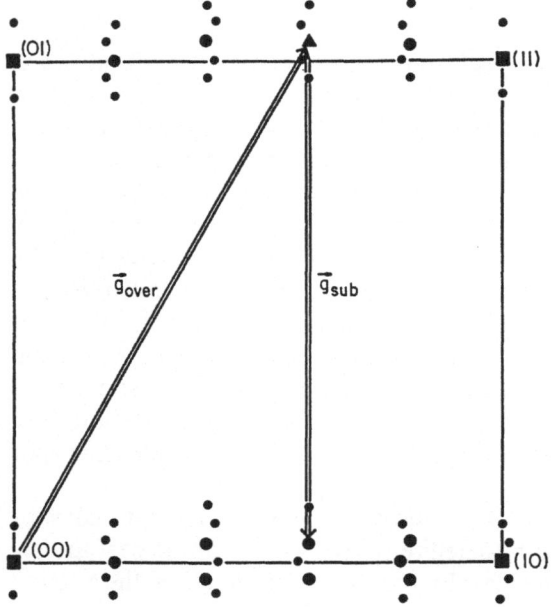

Fig. 3.20. Portion of LEED pattern for the reconstructed Au(100) surface. Triangles mark spots due to single scattering from the quasi-hexagonal top layer. The size of the symbols is approximately proportional to the observed intensities

the surface, similar to that concluded for the Ir(100) surface. Otherwise many spots would be an order of magnitude weaker than observed. In the surface direction perpendicular to that fivefold direction, one can understand the spot splittings in terms of a structural waviness with a long wavelength of approximately 25 substrate lattice constants. This will be justified in our discussion of the next class of patterns.

If the quasi-hexagonal top layer of Au(100) were perfectly planar, one would see a much simpler pattern in which the single-scattering spots would dominate. Such simpler patterns occur frequently with hexagonally close-packed monolayers of rare gases on many substrates [3.21] and with thin oxide films obtained by oxidation of a metal substrate [3.22]. In these cases, the overlayer and the substrate essentially diffract independently, and it can be concluded that there is little or no structural waviness or buckling.

3.5.7 Split Spots

One speaks of split spots in a LEED pattern when isolated clusters of spots appear that are obviously derived in some sense from single spots. Thus the reconstructed Au(100) pattern shown in Fig. 3.20 has satellite spots around each substrate and each overlayer spot, and each of the other fifth-order spots has been replaced by a few spots in the same neighborhood. In general, one has to distinguish between what happens to substrate spots and what happens to superlattice spots. We shall discuss the former first.

Splitting of Integral-Order Spots

A very characteristic form of spot splitting, which occurs for all substrate spots simultaneously, is manifest as follows when the electron energy is varied continuously. A continuous stream of spots passes through the area of each substrate spot, which itself is not visible, as if one were seeing a moving string of lights through a small orifice that allows only a few lights to be seen simultaneously. This behavior indicates that the surface is stepped, i.e. not cut along a low-Miller-index plane. We have already described the diffraction pattern of stepped surfaces in Sect. 3.3, and Fig. 3.8 illustrates the behavior described above, especially when combined with the Ewald construction to be given in Sect. 4.2.11. In short, as a function of increasing energy, the (10,8,7) spots of Fig. 3.8 converge toward the specular spot of the macroscopic (10,8,7) surface, while the terraces continue to generate intensity in the neighborhood of (111) spots which converge toward the specular spot of the (111) surface.

We now turn to another class of patterns, those with undisplaced substrate spots that are accompanied by satellite spots. These can be best understood in terms of Fourier transforms. In the kinematic limit of the theory, each spot with label $\bar{g}$ has an intensity proportional to the square of the two-dimensional Fourier transform of the surface structure with wave vector $\bar{g}$. Thus the LEED pattern is, approximately, a Fourier map of the surface: a spot at reciprocal lattice vector $\bar{g}$ indicates the presence of a structural feature with the corresponding direction and wavelength $2\pi/|\bar{g}|$. Consequently, if the specular spot is accompanied by a spot with a small value of $|\bar{g}|$, one may conclude that a structural feature with a long wavelength $2\pi/|\bar{g}|$ is present. The simplest example is a long-wavelength charge-density wave that induces a waviness in the atomic positions, as if a long-wavelength phonon had been frozen as in a snapshot. This, incidentally, is just the way that a real phonon affects electron diffraction, since phonon vibrational times are large compared to the LEED electron transit time. See also the discussion for modulated structures in Sect. 4.2.9.

Such a waviness in a surface layer can also be induced by a lattice mismatch to the substrate, which forces a long-wavelength buckling. This is the most probable explanation for the Au(100) reconstruction pattern shown in Fig. 3.20.

Splitting of Fractional-Order Spots

We turn now to patterns that show splittings of fractional-order spots. These occur particularly often for overlayers when the adatom coverage is varied, or during annealing of overlayer structures. Generally, split fractional-order spots are considered a nuisance that indicates undesirable disorder. However, valuable information can be obtained easily from the study of such patterns.

The general cause of the splitting of fractional-order spots is the coexistence of antiphase domains. This is not necessarily true, however, if the integral-order spots are also split, since this merely implies that the surface is stepped. The case of ordered arrangements of antiphase domains is easiest to

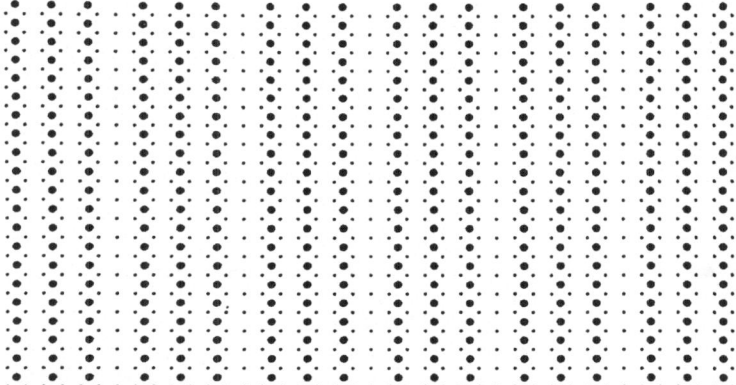

Fig. 3.21. A (7×1) array of (2×1) domains

understand. Figure 3.21 gives an example of ordered overlayer domains. A (2x1) structure exists within each domain, and, since this is a dominant surface structure, it should also in some strong way appear in the diffraction pattern, namely as intensity at least in the neighborhood of half-order locations. Because of the regular antiphase arrangement, a true surface lattice exists, shown in Fig. 3.21, with a (7x1) unit cell. Therefore the diffraction pattern should also exhibit this feature in some way, namely as seventh-order spots. One can already guess the outcome: only those seventh-order spots near half-order positions will have strong intensities, and it will appear as if the half-order spots have split into pairs.

To put the above guess on a sounder footing, we proceed as follows. If we isolate one domain with its actual width and construct its diffraction pattern, we do not obtain a simple (2x1) pattern with sharp spots but rather one with wide spots, as illustrated in the spot profiles at the top of Fig. 3.22. The spot width is due to the limited extent of the domains: this follows simply from the standard theory of diffraction by a grating composed of a finite number of identical elements. As a general rule the spot width (at half-height) for a domain size of N units is 1/N of the spot-to-spot distance,

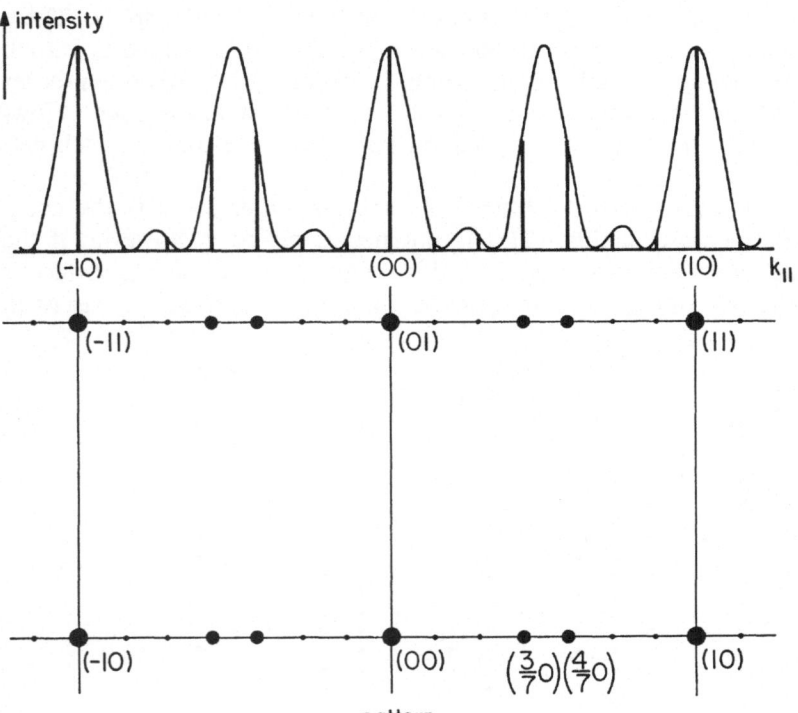

Fig. 3.22. (*Top*) Intensity profile for diffraction from one domain of the (2 × 1) superlattice of Fig. 3.21 (*continuous curve*) and delta-functions for (7 × 1) periodicity. (*Bottom*) Resulting diffraction pattern (dot size indicates intensity)

which becomes 1/3 in this case. This (2x1) pattern with broad spots is just the "form factor" of the single domain, using the well-known concept from X-ray Diffraction. This form factor must be multiplied (at least in the kinematic limit) by the atomic form factor, which we ignore in the illustration for simplicity, since it only modulates the end result. We now have a form factor for the basic repeating unit of the full surface. This in turn must be multiplied by the structure factor of the full surface, namely the pattern due to the (7x1) periodicity. For an infinitely extended surface, we obtain delta functions at seventh-order positions, as shown in Fig. 3.22. The product of the form factor and the structure factor yields the expected result of pairs of dominant fractional-order spots around the half-order positions.

Notice that the amount of splitting, 1/7th of the integral-order-spot separation in Fig. 3.22, corresponds to the size of the domain, seven substrate unit cells in this case. This is true in general: a 1/N splitting indicates an N-fold repeat distance.

It can be left as a simple exercise for the reader to show that the spot splitting and the spot width of the single-domain pattern are always proportional to each other. Consequently, as the number N in (Nx1) becomes large, one will still see only the pairs of Nth-order spots, not triplets or quadruplets. The other Nth-order spots are not totally absent, but in practice their intensity is near to or below the level of detectability of standard LEED equipment. Another simple exercise consists of showing that if one adds a fourth row of adatoms in each domain of Fig. 3.21, only the spot intensities can change.

An interesting aspect of antiphase-domain structures is that as the surface unit cell changes, while the internal-domain structure remains constant, the adatom coverage must change. As one goes from (7x1) to (8x1) to (9x1), etc. in the example above, the coverage must take on a sequence of discrete values. Therefore it is often possible to monitor the coverage through the fractional-order-spot splitting.

In this connection, an intriguing question is how the pattern transforms from one discrete state to the next as the coverage changes continuously between the two corresponding discrete values, since some disorder must exist at intermediate coverages. Experimentally, one usually observes a continuous transition between the patterns in the form of a continuous change of the spot splitting. This is in fact quite understandable: the diffraction by a statistical mixture of the two discrete states has been shown theoretically to behave in just the observed manner [3.23], as long as the domain sizes of the two discrete states are sufficiently small.

We have not emphasized the significance of yet another factor: the orientation of spot splitting indicates directly the orientation of the strips of domains. In addition, it follows that if the strips of domains have a rather random orientation, the result is diffuse spot splitting. Furthermore, because of the possibility of different orientations of strips of domains, one must take into account that a pattern with split spots may often be simplified into the individual contributions of differently oriented domains. Consequently, if a c(2x2) structure on the fcc(100) surface develops splittings in the half-order spots, these can occur in different directions at the same time by a simple superposition of rotated patterns.

However, there is also the possibility that domains form in small patches rather than in long narrow strips, in which case an ordered array of domains can exist in both surface directions, and spot splittings can occur in both directions as well. Further aspects of these topics will be treated in Chaps. 8-10.

3.5.8 An Example: Compact Structures vs. Antiphase Domain Structures of Adsorbed Carbon Monoxide Overlayers

On many metal surfaces, overlayers of CO for fractional coverages above approximately 1/2 produce diffraction patterns that have often been interpreted as due to an incommensurate overlayer that is quasi-hexagonally close-packed [3.24]. This overlayer is then thought to be compacted gradually and uniformly as the coverage is increased. However, there are some unsatisfactory features in this interpretation. First, some of the patterns are inconsistent with the quasi-hexagonal lattice if one looks at a sufficiently large range of reciprocal space: some necessary spots are just not there. Second, the patterns show no satellite spots indicative of buckling in the incommensurate layer, which would imply weak bonding to the substrate. However, the bonding is known to be strong from Thermal-Desorption Mass Spectrometry and from vibrational electron energy loss measurements. Third, with incommensurate lattices one would expect simultaneously many different bonding arrangements on the substrate, yielding many different vibrational loss frequencies. However, at most, one or two frequencies are observed experimentally [3.17,25]. In other words, the actual CO overlayers appear to involve at most one or two different binding sites, but they nevertheless produce a continuous sequence of LEED patterns as a function of coverage.

This last feature is reminiscent of our previous antiphase-domain example. There, only one adsorption site was involved, where the bonding could be of any strength, and yet a continuous series of patterns could be generated as a function of coverage. We can apply the same idea of antiphase-domain structures to CO overlayers. The basic internal domain structure is then, for example, the well-known c(2x2) structure of CO on Ni(100) at 1/2-monolayer coverage, or the $(\sqrt{3} \times \sqrt{3})R30°$ structure of CO on Ru(0001) at 1/3-monolayer coverage. With simple antiphase-domain structures at higher coverages, one can reproduce all the published LEED patterns for such CO overlayers and at the same time satisfy the requirements of coverage and number of adsorption sites [3.26]. In many cases the resulting structure is not much different from the proposed quasi-hexagonal model (after small relaxations of the individual CO-CO distances), but in a number of cases the new solution is entirely different and inescapable.

This example illustrates several issues. A large region of reciprocal space should be used. The nonplanarity of an overlayer can often be detected in the pattern itself. The concept of spot splitting can include large splittings, not just small ones. (One reason that the compact CO structures were not treated initially in terms of antiphase domains is that the splittings are not so small as one might at first expect.)

3.5.9 Patterns with Multiple Specular Spots

One occasionally encounters patterns in which, as a function of energy, more than one spot is immobile. Immobility of a spot implies that it is due to specular reflection from a flat surface. The only way to get more than one direction of specular reflection is to have more than one flat surface. Therefore, such patterns are characteristic of faceting, i.e. the formation of multiple surfaces large compared to the instrumental response width, and thus large compared to approximately 100 Å The individual orientations of the different facets can be obtained easily from the pattern by observing the specular-beam direction and the pattern symmetries.

3.5.10 Laser Simulation of LEED Patterns

The process of LEED can be simulated with the diffraction of a laser beam by a two-dimensional grating the lattice constants of which are comparable to the wavelength of the light. All one needs is an image with dots approximately 50 μm apart, such as a reduced photographic transparency of a macroscopic drawing or of a perforated sheet of paper, where each perforation represents an atom [3.15,27]. Particularly convenient are computer-generated plots, output directly on microfiche or 35-mm film, which serve directly as diffraction gratings [3.15].

The third dimension perpendicular to the real surface cannot be represented easily in laser diffraction. Since it generally affects spot intensities rather than the existence of spots, one can still make many useful simulations. In fact, some three-dimensional effects can be simulated. For example, the buckling of a surface layer can be introduced with dot displacements parallel to the surface rather than perpendicular to it. Even the presence of multiple-scattering spots can be simulated. When the dots representing two atoms overlap, the resulting image is usually not just the sum of two dots in terms of the gray level, and this will actually create proper additional spots when these would be expected from multiple scattering.

Laser simulation of LEED can be quite useful in checking an interpretation of complex patterns, since almost all effects that we have discussed in this section can be included. In addition, it can often discriminate between different structural models. For example, when one is faced with positioning several atoms in a unit cell, there are usually many possibilities that can be distinguished to some extent by laser diffraction [3.15,26]. Finally, laser simulation of LEED is an instructive activity for those who need to analyze many LEED diffraction patterns, since it exercises one's mental ability to relate direct and reciprocal lattices, as well as one's ability to perform Fourier transformations mentally.

4. Kinematic LEED Theory and Its Limitations

In the previous chapter, we discussed how much information concerning a surface can be extracted from a LEED pattern without attempting to analyze the intensities of the diffracted beams in any detail. To obtain additional crystallographic data concerning the surface, namely atomic coordinates defining bond lengths, bond angles, adsorption sites, etc., it is necessary to study the beam intensities. In order to make the link between beam intensities and atomic positions, we must understand both the basic electron scattering mechanism and the nature of electron diffraction at LEED energies.

The theoretically most convenient situation arises when only "single scattering" occurs, in which case an electron that has been scattered once by a surface atom will not be scattered again by another surface atom. This describes the kinematic limit of diffraction and leads to a relatively simple theory which has been quite successful in the interpretation of the diffraction of X-rays [4.1-3], neutrons [4.4], and even high-energy electrons (HEED) with sufficiently small crystallites [4.5,6]. As we shall discuss in this chapter, the kinematic theory has been less successful in LEED since "multiple scattering" usually plays an important role that makes a kinematic interpretation less straightforward, although by no means always impossible. In fact, multiple scattering, which leads to the "dynamical theory", is much stronger in LEED than in HEED or in Transmission Electron Microscopy.

Multiple scattering and the small penetration depth of LEED electrons into surfaces are both caused by the large cross section for electron-atom collisions prevalent at low energies. Therefore, multiple scattering is difficult to avoid if one desires surface sensitivity.

First it is necessary to define more precisely what is meant by single scattering or kinematic theory, since one can adopt different levels of sophistication even in this relatively simple limit. Since kinematic methods can occasionally be applied in LEED and since many concepts of kinematic diffraction are central to the more complete dynamical description of LEED, we shall discuss these in some detail in this chapter. However, we shall also point out here where a purely kinematic description of LEED fails.

4.1 Definition of Kinematic Theory

We shall define kinematic diffraction to be the situation in which each wave packet representing one electron incident on a surface is diffracted elastically only once by that surface.

4.1.1 Atomic Scattering Factor

The above definition of kinematic diffraction puts no restrictions on the manner in which the scattering by a single atom occurs. We avoid such restrictions because the atomic scattering of electrons at LEED energies (unlike in HEED) is in reality always a multiple-scattering process within the individual atom. One may ignore such complications and use a constant atomic scattering factor f as in X-ray Diffraction. This choice corresponds to isotropic (i.e. s-wave) scattering, since it produces no angular dependence, and it can also be kept energy independent for convenience. This simplest of cases may be adequate in LEED for some purposes, such as the prediction of the LEED pattern, the angular dependence of disorder-induced diffuse scattering, or kinematic peak positions in intensity curves. However, in kinematic LEED one often includes the actual angle and energy dependence of the atomic scattering factor $f(\vec{s})$, where $\vec{s} = \vec{k}_{out} - \vec{k}_o$ is the momentum transfer between an incident and a diffracted plane wave with wave vectors $\vec{k}_o$ and $\vec{k}_{out}$, respectively, since it is such a strong dependence.

The variations of f with $\vec{s}$ are strongly affected by multiple scattering within the atom. This may be considered to be strong diffraction of the electrons by the complicated "electrostatic lens" that one can view an atom to be. (In HEED, f is also a strong function of $\vec{s}$, although multiple scattering within the atom is very weak.) We shall return to this topic in Sect. 4.5 in some detail, where it will be shown how the scattering factor can be described with "phase shifts" for partial waves. Another source of structure in the scattering factor is thermal vibrations, the effect of which on kinematic diffraction from ordered surfaces can be described conveniently by the familiar Debye-Waller factor of X-ray Diffraction. This subject will be discussed in Sect. 4.7 for the case of kinematic LEED.

4.1.2 Elastic Scattering

Our definition of kinematic diffraction specifies *elastic* scattering. This qualification must be explained more precisely. Experimentally, the so-called "elastic" peak in the energy distribution of diffracted electrons is found to have a width ΔE of at least 0.25 eV, due in large part to instrumental effects. This width encompasses the small inelastic losses to phonons at the surface. For experimental convenience, this entire peak is included in the LEED intensity measurement, with the assumption that diffraction at the energy E

− ΔE is essentially the same as that at the energy E. (A similar criterion is used in setting the acceptance angle for measurement of the beam intensities.) Although for most of today's applications these assumptions are justified, one rapidly exceeds their range of validity when carrying out more precise work. Thus, LEED data are in practice integrated over the range of energies of high-resolution electron energy-loss spectroscopy, whereas the theory is kept strictly elastic.

4.1.3 Amplitude of Diffraction

Now we may write down expressions for the diffracted electron waves of kinematic LEED. Let us consider a monoenergetic electron beam impinging on a surface. The beam is represented by a plane wave which is described by

$$A_i = A_o \exp(i\vec{k}_o \cdot \vec{r}) \ , \tag{4.1}$$

where A_i is the amplitude of the incident wave, A_o is a constant, $\vec{k}_o$ is the incident wave vector, and $\vec{r}$ is a space vector. If multiple scattering is neglected, the amplitude of a diffracted beam is represented by

$$A_s = A_o \left[\sum_n \alpha f_n(\vec{s}) \exp(i\vec{s} \cdot \vec{r}_n) \right] \exp(i\vec{k}_{out} \cdot \vec{r}) \ , \tag{4.2}$$

where $f_n(\vec{s})$ is the atomic scattering factor for the nth atom located at position $\vec{r}_n$, $\vec{s} = \vec{k}_{out} - \vec{k}_o$ is the momentum transfer, and $\vec{k}_{out}$ is the wave vector of the scattered wave (α is a constant to be determined later). The wave vectors $\vec{k}_o$ and $\vec{k}_{out}$ are related to the constant energy by

$$E = \frac{\hbar^2}{2m} |\vec{k}_o|^2 = \frac{\hbar^2}{2m} |\vec{k}_{out}|^2 \ . \tag{4.3}$$

Electrons passing from vacuum into a surface are actually accelerated to a higher kinetic energy by surface dipole layers, so that the kinetic energy inside the surface is $E + V_o$, where V_o (>0) is called the "inner potential" in kinematic LEED. The effect is to cause refraction with a change in both $\vec{k}_o$ and $\vec{k}_{out}$ at the surface. This issue will be discussed in more detail in Sect. 4.2.3. Notice that so far we have not assumed any particular crystal structure nor even any form of order.

4.1.4 Surface Sensitivity

Before dealing with the lattice structure, we must first introduce the surface sensitivity of LEED. As it stands, (4.2) does not favor surface atoms over bulk atoms. The inelastic effects that contribute strongly to the surface sensitivity can be represented by a mean free path through a factor that decays exponentially in the direction of wave propagation. We shall justify and describe this approach in Sect. 4.3; the result will be that the components of the wavevectors $\vec{k}_o$ and $\vec{k}_{out}$ perpendicular to the surface, and therefore also of the momentum transfer $\vec{s} = \vec{k}_{out} - \vec{k}_o$, are given imaginary parts, which dampen the waves with the proper length of the mean free path (within the surface, not in the vacuum). In this way, atoms deeper in the surface give progressively smaller contributions to the diffraction.

4.1.5 From Amplitudes to Intensities of Diffraction

We now need to link the amplitude of a diffracted wave with its intensity, the latter being proportional to the beam current measured experimentally.

When an incident plane wave $A_o \exp(i\vec{k}_o \cdot \vec{r})$ diffracts into a plane wave $B \exp(i\vec{k}_{out} \cdot \vec{r})$, the quantities $k|A_o|^2$ and $k|B|^2$, where $k = |\vec{k}_o| = |\vec{k}_{out}|$, are the respective fluxes in the two beams. These fluxes measure the number of electrons per unit time passing unit area oriented perpendicular to the direction of electron motion, assuming beams with an infinite cross section. In the experiment, however, the beams have finite cross sections (approximately 1 mm^2 for the incident beam and a comparable area for the diffracted beams), and the entire cross section of a diffracted beam normally fits within the detection aperture of the spot photometer, the Faraday cup, or other measuring device. The fluxes integrated over the actual beam cross sections are then, in the theory, $Ck|A_o|^2$ and $C(\cos\theta_{out}/\cos\theta)k|B|^2$, where C is the cross section of the incident beam and $C\cos\theta_{out}/\cos\theta$ that of the diffracted beam, given the polar angles of incidence θ and of emergence θ_{out} with respect to the surface normal. This is illustrated in Fig. 4.1. The theoretical reflectivity R of the surface thus becomes

$$R^{th} = \frac{\cos\theta_{out}}{\cos\theta} \frac{|B|^2}{|A_o|^2} .\tag{4.4}$$

(A_o is usually set equal to unity in the theory, for simplicity). The equivalent experimental quantity, to which the theoretical reflectivity of (4.4) should be

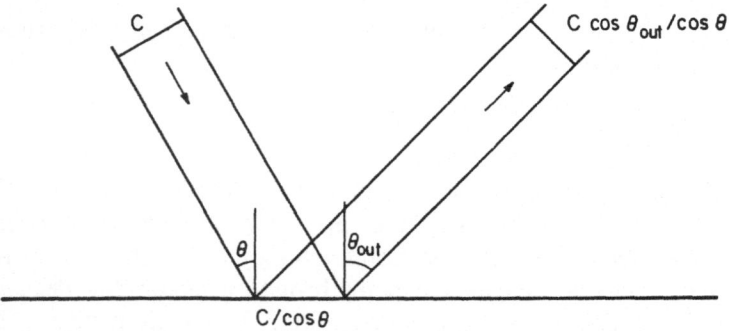

Fig. 4.1. The cross section of a diffracted beam due to an incident beam of finite cross section C

compared, is the ratio of the measured intensity (current) of a diffracted beam I_{out} and that of the incident beam I_o:

$$R^{exp} = \frac{I_{out}}{I_o} \; .$$
(4.5)

Equations (4.4,5) are valid not only in the kinematic limit but also when multiple scattering is included.

4.2 The Kinematic Structure Factor for Ordered Surfaces

4.2.1 Two-Dimensional Bragg Conditions

The sum

$$\sum_n f_n \exp(i\vec{s}\cdot\vec{r}_n)$$

that appears in (4.2) is generally known as the structure factor S in diffraction theory. If the diffracting specimen has a crystalline structure, the structure factor gives rise to specific conditions for the appearance of diffracted beams. For three-dimensional cases, as in the diffraction of X-rays [4.1-3], neutrons [4.4] and high-energy electrons [4.5], Bragg conditions appear as follows.

Let the atomic positions $\vec{r}_n$ be represented by $\vec{r}_n = \vec{R}_p + m_1\vec{a}_1 + m_2\vec{a}_2 + m_3\vec{a}_3$. Here $\vec{a}_i$ (i = 1,2,3) are the basis vectors of the lattice; m_1,m_2,m_3 are integers; and $\vec{R}_p$ (p = 1,2,...,N) are the locations of the atoms within one unit cell. Then, in the three-dimensional case, the structure factor is given by

$$S^{(3)} = \left[\sum_p f_p \exp(i\vec{s}\cdot\vec{R}_p)\right]\left\{\sum_{m_1 m_2 m_3} \exp\left[i\vec{s}\cdot(m_1\vec{a}_1 + m_2\vec{a}_2 + m_3\vec{a}_3)\right]\right\} .$$
(4.6)

The sum over lattice vectors is proportional to the Dirac delta function $\delta(\vec{s} - \vec{g}^{(3)})$, where $\vec{g}^{(3)}$ is any of the three-dimensional reciprocal lattice vectors of the given lattice $(\vec{a}_1,\vec{a}_2,\vec{a}_3)$:

$$\vec{g}^{(3)} = h\vec{g}_1 + k\vec{g}_2 + l\vec{g}_3 \quad (\text{h,k,}l \text{ integers}) ,$$
(4.7)

with

$$\vec{g}_1 = 2\pi \frac{\vec{a}_2 \times \vec{a}_3}{\vec{a}_1\cdot(\vec{a}_2 \times \vec{a}_3)} , \quad \vec{g}_2 = 2\pi \frac{\vec{a}_3 \times \vec{a}_1}{\vec{a}_1\cdot(\vec{a}_2 \times \vec{a}_3)} ,$$

$$\vec{g}_3 = 2\pi \frac{\vec{a}_1 \times \vec{a}_2}{\vec{a}_1\cdot(\vec{a}_2 \times \vec{a}_3)} .$$
(4.8)

The delta functions define the Bragg conditions for the existence of diffracted beams,

$$\vec{s} = \vec{g}^{(3)} = h\vec{g}_1 + k\vec{g}_2 + l\vec{g}_3 \text{ for any integers } h,k,l \ . \tag{4.9}$$

In the case of LEED, the periodicity of the diffracting lattice is only retained parallel to the surface: $\vec{r}_n = \vec{R}_p + m_1\vec{a}_1 + m_2\vec{a}_2$ (m_1,m_2 integers), where $\vec{R}_p$ includes all atoms in the two-dimensional unit cell of the surface. The unit cell now extends to $+\infty$ and $-\infty$ in the direction perpendicular to the surface, which in practice means down into the surface to the depth reached by the electrons. The structure factor now becomes, in the two-dimensional case:

$$S \equiv S^{(2)} = \left[\sum_p f_p \exp(i\vec{s}_\parallel \cdot \vec{R}_p)\right]\left\{\sum_{m_1 m_2} \exp\left[i\vec{s}_\parallel \cdot (m_1\vec{a}_1 + m_2\vec{a}_2)\right]\right\} \ . \tag{4.10}$$

Here, the two-dimensional lattice sum is proportional to $\delta(\vec{s}_\parallel - \vec{g}^{(2)})$, where $\vec{s}_\parallel$ is the component of $\vec{s}$ parallel to the surface, and $\vec{g}^{(2)}$ is any of the *two*-dimensional reciprocal lattice vectors of the surface lattice ($\vec{a}_1,\vec{a}_2$)

$$\vec{g} \equiv \vec{g}^{(2)} = h\vec{g}_1 + k\vec{g}_2 \quad (\text{h,k integers}) \ , \tag{4.11}$$

with

$$\vec{g}_1 = 2\pi\frac{\vec{a}_2 \times \vec{n}}{\vec{a}_1 \cdot (\vec{a}_2 \times \vec{n})} = 2\pi\frac{\vec{a}_2 \times \vec{n}}{\vec{n} \cdot (\vec{a}_1 \times \vec{a}_2)} = 2\pi\frac{(a_{2y}, -a_{2x})}{a_{1x}a_{2y} - a_{2x}a_{1y}} \ ,$$

$$\vec{g}_2 = 2\pi\frac{\vec{n} \times \vec{a}_1}{\vec{a}_1 \cdot (\vec{a}_2 \times \vec{n})} = 2\pi\frac{\vec{n} \times \vec{a}_1}{\vec{n} \cdot (\vec{a}_1 \times \vec{a}_2)} = 2\pi\frac{(-a_{1y}, a_{1x})}{a_{1x}a_{2y} - a_{2x}a_{1y}} \ . \tag{4.12}$$

In these equations, $\vec{n}$ is a unit vector normal to and pointing *out* of the surface, replacing $\vec{a}_3$ in (4.8). We have used a *left*-handed Cartesian-coordinate system with the x- and y-axes in the plane of the surface and the z-axis pointing *into* the surface. Examples of reciprocal lattices appear in Sects. 3.1-3.

Thus, the Bragg conditions have become, for the case of LEED,

$$\vec{s}_\parallel = \vec{k}_{\text{out}_\parallel} - \vec{k}_{o_\parallel} = \vec{g}^{(2)} = \vec{g} \ . \tag{4.13}$$

In the case of a three-dimensional crystal, the Bragg conditions can be explained in terms of the condition that the difference in path-length between specular reflections from successive planes of atoms is a multiple of the wavelength. In the same fashion, in the case of a two-dimensional crystal, one can interpret the two-dimensional Bragg conditions in terms of a corresponding path-length difference between rows of surface atoms that is equal to a multiple of the wavelength.

The physical origin of the various diffracted beams is best illustrated with the Huygens construction of the diffracted wave field, as shown in Fig. 4.2. One simply draws spherical wave fronts originating from each scatterer and

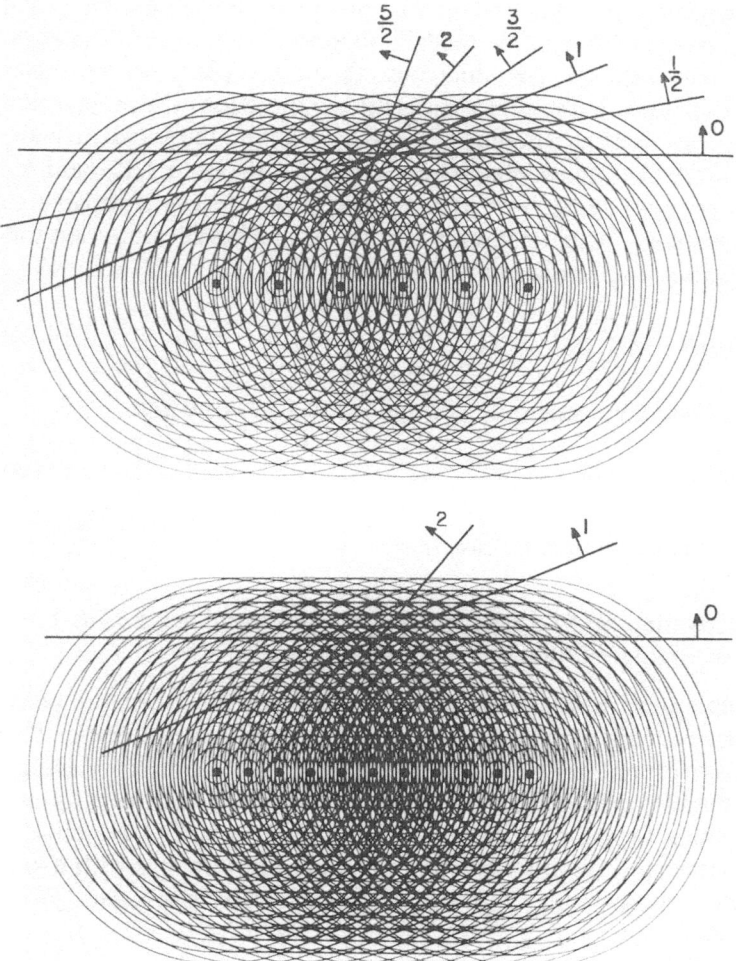

Fig. 4.2. Huygens construction of planar wave fronts in diffraction by a string of scatterers. Two values of the lattice constant are used, exhibiting the multiplication of the number of plane-wave diffraction directions when the lattice constant is enlarged. The wave fronts are best seen by viewing the figure at a glancing angle. The labels indicate the order of the various waves

observes that at a sufficient distance from the scatterers the wave fronts emanating from different scatterers tend to merge into planar wave fronts that have only a few well-defined orientations. These orientations correspond to the diffracted beam directions.

In LEED, no Bragg condition restricts the momentum transfer perpendicular to the surface. However, the perpendicular component of $\vec{s}$ is restricted in another way, as it is also in X-ray Diffraction, by the condition of energy conservation, given by Eq. (4.3) (the inner-potential correction V_0 is to be included only inside the surface and not in the vacuum). Equation (4.13) states that $\vec{k}_{out_\parallel} = \vec{k}_{o_\parallel} + \vec{g}$, and we shall therefore adopt the notation $\vec{k}_{out} = \vec{k}_{\bar{g}}$, where the negative sign indicates that these plane waves travel back toward the vacuum, namely in the negative z-direction. In preparation for the future, we also use the notation $\vec{k}_o^+ = \vec{k}_{\bar{o}}^+ = \vec{k}_o$ for the incident wave vector. Furthermore, we allow plane waves with wave vectors $\vec{k}_{\bar{o}}^-$ and $\vec{k}_{\bar{g}}^+$ that travel in the opposite direction with respect to the surface compared to $\vec{k}_o^+$ and $\vec{k}_{\bar{g}}^-$, respectively, while maintaining the same parallel momentum component. Equation (4.3) now implies that

$$\vec{k}_o^\pm = \left\{ k_{ox}, k_{oy}, \pm \left[\frac{2m}{\hbar^2} (E + V_o) - |\vec{k}_{o\parallel}|^2 \right]^{1/2} \right\} , \tag{4.14}$$

(again, V_o is dropped in the vacuum). Equations (4.13,3) also give

$$\vec{k}_{\bar{g}}^\pm = \left\{ k_{ox} + g_x, k_{oy} + g_y, \pm \left[\frac{2m}{\hbar^2}(E + V_o) - |\vec{k}_{o\parallel} + \vec{g}|^2 \right]^{1/2} \right\} , \tag{4.15}$$

which includes (4.14) as the special case of $\vec{g} = \vec{0}$.

4.2.2 General Derivation of Two-Dimensional Bragg Conditions in LEED from the Schrödinger Equation

We now derive the same results given above in a more general fashion, based on the Schrödinger equation for the scattering of electrons by a surface. This has the advantage of not making any assumption of kinematicity (single scattering) and thereby proves that the previous results are both general and of fundamental value for the case of multiple scattering in LEED as well.

When a monoenergetic, collimated beam of electrons interacts with a perfectly well-ordered surface, the total wave function $\psi(\vec{r})$ must satisfy the Schrödinger equation

$$\left[-\frac{\hbar^2}{2m} \nabla^2 + V(\vec{r}) - E \right] \psi(\vec{r}) = 0 , \tag{4.16}$$

where m is the mass of the electron, $2\pi\hbar$ is Planck's constant, $\vec{r}$ is a three-dimensional real-space vector, $V(\vec{r})$ is the one-electron crystal potential, and E is the total electron energy. The wave function may be written as $\psi(\vec{r}) = \psi_i(\vec{r}) + \psi_{in}(\vec{r}) + \psi_s(\vec{r})$, where $\psi_i(\vec{r})$ and $\psi_s(\vec{r})$ are the wave functions of the incident and scattered electron in vacuum, respectively, and $\psi_{in}(\vec{r})$ is the wave function of the electrons within the surface. The wave function of the incident electron can be represented simply by the wave function of the free electron in vacuum, namely a plane wave with wavevector $\vec{k}_o$,

$$\psi_i(\vec{r}) = A_o \exp(i\vec{k}_o \cdot \vec{r}) , \tag{4.17}$$

where $|k_o| = (2mE)^{1/2}/\hbar$ and A_o is a constant.

A perfect crystal surface has two-dimensional periodicity so that the crystal potential is invariant under a two-dimensional translational operation $\vec{T}$,

$$V(\vec{r}) = V(\vec{r} + \vec{T}) \ . \tag{4.18}$$

With $\vec{a}_1$ and $\vec{a}_2$ as the basis vectors of the surface lattice, $\vec{T}$ can be written as

$$\vec{T} = m_1\vec{a}_1 + m_2\vec{a}_2 \ , \tag{4.19}$$

where m_1 and m_2 are integers.

According to Bloch's theorem, since the crystal potential is periodic, the wave function of the electrons within the surface can be expressed as the product of a plane wave, $\exp(i\vec{k}_{o\parallel}\cdot\vec{r})$, and a Bloch function, $U_{\vec{k}_{o\parallel}}(\vec{r})$, with the periodicity of the crystal surface:

$$\psi_{in}(\vec{r}) = \exp(i\vec{k}_{o\parallel}\cdot\vec{r})U_{\vec{k}_{o\parallel}}(\vec{r}) \ , \tag{4.20}$$

where $\vec{k}_{o\parallel}$ is the component of the incident wave vector parallel to the surface. Since the Bloch function is periodic, it can be expressed in a Fourier series as

$$U_{\vec{k}_{o\parallel}}(\vec{r}) = \sum_{\vec{g}} A_{\vec{k}_{o\parallel},\vec{g}}(z)\exp(i\vec{g}\cdot\vec{r}) \ , \tag{4.21}$$

where now the quantity $\vec{g}$ is a two-dimensional vector determined as follows. To satisfy the periodicity of $U_{\vec{k}_{o\parallel}}(\vec{r})$, it is necessary that

$$\exp(i\vec{g}\cdot\vec{a}_1) = 1 \quad \text{and} \quad \exp(i\vec{g}\cdot\vec{a}_2) = 1 \ . \tag{4.22}$$

Hence,

$$\vec{g}\cdot\vec{a}_1 = 2\pi\cdot \text{integer} \quad \text{and} \quad \vec{g}\cdot\vec{a}_2 = 2\pi\cdot \text{integer} \ . \tag{4.23}$$

Equations (4.23) are satisfied only by vectors $\vec{g}$ of the kind given in (4.11), namely

$$\vec{g} = h\vec{g}_1 + k\vec{g}_2 \ ,$$

with h and k integers, where $\vec{g}_1$ and $\vec{g}_2$ are defined by (4.12). Thus we have obtained again the reciprocal lattice. From (4.21) it also follows that the same plane waves with wave vectors $\vec{k}_{\vec{g}}$ are generated in the general dynamical case allowed here as in the kinematic limit assumed previously.

4.2.3 Plane Waves, Beams, and the LEED Pattern

The plane waves corresponding to the wave vectors $\vec{k}_{\vec{g}}$ form well-defined beams that may emanate from the crystal and travel toward the detector. In particular, if a luminescent screen is employed to intercept them, each beam can produce a light spot. Unlike the case of X-ray Diffraction or HEED, however, each beam that reaches the LEED screen has nonzero intensity in general, because one of the three Bragg conditions for nonextinction is absent. Each spot is labeled by the values of h and k in $\vec{g} = h\vec{g}_1 + k\vec{g}_2$, for example, $(1,0)$, $(1,\bar{1})$. The $(0,0)$ spot is the specularly reflected spot and often plays a special role.

The symmetry of the diffraction pattern directly reflects the symmetry of the surface structure, which makes it easy to verify the type of surface with which one is dealing. For example, a perfect fcc(111) surface, which has a hexagonal surface lattice, a threefold rotation axis and three mirror planes, produces a pattern consisting of a hexagonal array of spots with those same symmetries when normal incidence is used. A more detailed discussion of symmetries in patterns is given in Sect. 3.4.

The angular directions of the diffracted beams can be calculated as follows. Let (θ,φ) define the incidence direction in vacuum ($\theta = 0$ is along the $-z$-axis, i.e. along the outward surface normal, while $\varphi = 0$ is along the x-axis, cf. Fig. 4.3), and let $(\theta_{\vec{g}}^{\pm},\varphi_{\vec{g}})$ define the emergence direction of beam $\vec{g}$ in vacuum ($\varphi_{\vec{g}}^{+} = \varphi_{\vec{g}}^{-} = \varphi_{\vec{g}}$). Then it follows from the parallel-momentum relations

$$k_{gx}^{\pm} = k\sin\theta_{\vec{g}}^{\pm}\cos\varphi_{\vec{g}} = k_{ox} + g_x = k\sin\theta\cos\varphi + g_x \ , \quad \text{and}$$

$$k_{gy}^{\pm} = k\sin\theta_{\vec{g}}^{\pm}\sin\varphi_{\vec{g}} = k_{oy} + g_y = k\sin\theta\sin\varphi + g_y \ , \quad \text{that}$$

$$\theta_{\vec{g}}^{-} = \arccos\left[\frac{k^2 - (k_{ox} + g_x)^2 - (k_{oy} + g_y)^2}{k^2}\right]^{1/2}$$

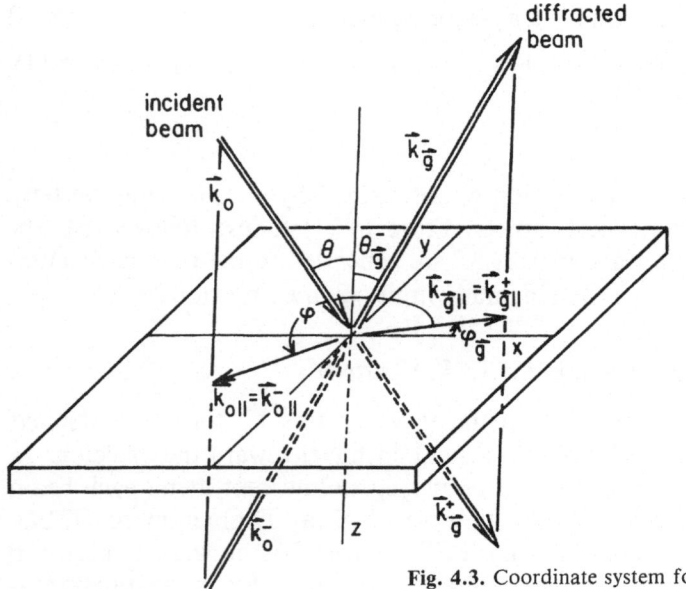

Fig. 4.3. Coordinate system for diffraction by a surface

$$= \arcsin\left[\frac{(k_{ox} + g_x)^2 + (k_{oy} + g_y)^2}{k^2}\right]^{1/2},$$

$$\theta_{\bar{g}}^+ = \pi - \theta_{\bar{g}}^-,$$

(4.24)

and

$$\varphi_{\bar{g}} = \arccos\left[\frac{(k_{ox} + g_x)^2}{(k_{ox} + g_x)^2 + (k_{oy} + g_y)^2}\right]^{1/2}$$

$$= \arcsin\left[\frac{(k_{oy} + g_y)^2}{(k_{ox} + g_x)^2 + (k_{oy} + g_y)^2}\right]^{1/2},$$

where the positive square root is taken. In (4.24), $k = (2mE)^{1/2}/\hbar$ is the wave number outside the surface. Within the surface, the kinetic energy is $E + V_o$ due to the inner potential, which causes refraction at the plane of onset of the inner potential V_o. Since this refraction changes the directions of propagation of all beams (except those with normal incidence or normal exit), the directions inside the surface are different from those outside the surface. These different directions can also be obtained from (4.24) by using $k = [2m(E + V_o)]^{1/2}/\hbar$ rather than $k = (2mE)^{1/2}/\hbar$, and by noting that $k_{ox} + g_x$ and $k_{oy} + g_y$ are not affected by refraction (this last assumption is discussed in Sect. 4.4.3). The situation is illustrated in Fig. 4.4.

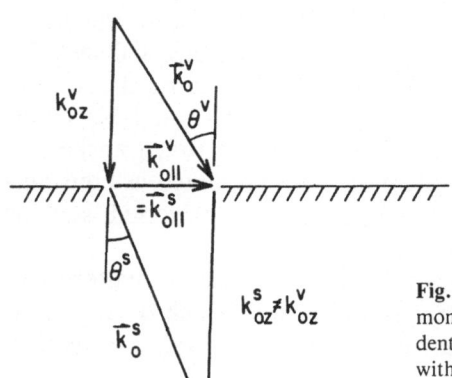

Fig. 4.4. Refraction at a surface with conservation of momentum parallel to the surface. A plane wave $\vec{k}_0^v$ incident from vacuum is refracted into the plane wave $\vec{k}_0^s$ within the surface, causing a change of propagation direction from θ^v to θ^s

Since $|g_x|$ and $|g_y|$ can be arbitrarily large, namely for large indices $|h|$ and $|k|$, the expressions for $\theta_{\bar{g}}^-$ may not have real solutions. This occurs when $|\vec{k}_{\bar{g}\parallel}|^2 = (k_{ox} + g_x)^2 + (k_{oy} + g_y)^2 > k^2$. The situation can be best understood from (4.15) in which $V_o = 0$ is chosen in the vacuum. For large $|g_x|$ and $|g_y|$, the square root in the z-component of $\vec{k}_{\bar{g}}^-$ does not have a real solution, but rather an imaginary solution $k_{\bar{g}z}^- = -i\kappa$, where κ is real and positive. The corresponding plane wave then becomes

$$\exp(i\vec{k_g}\cdot\vec{r}) = \exp(\vec{k}_{g\parallel}\cdot\vec{r}_\parallel)\exp(\kappa z) \ ,$$

namely, a damped wave which decays away from the surface ($z < 0$). Such a wave cannot reach a detector located at macroscopic distances from the surface. As a result, only a finite number of diffracted beams can emerge from the surface and be detected, namely those with sufficiently small $|g_x|$ and $|g_y|$.

4.2.4 I-V, I-θ, I-φ, and Other Collections of Data

In X-ray Diffraction and in HEED, one normally uses a fixed wavelength and measures the intensity of discrete beams at a variety of suitable crystal orientations that satisfy the Bragg conditions, producing a discrete collection of reflection intensities. As we have seen, in the case of LEED, all those beams that can emerge from the surface based on momentum and energy arguments have nonzero intensities in general, whatever the electron energy and the crystal orientation may be. There is no need to select particular crystal orientations to obtain nonzero intensities. Such a collection of intensities can be called I-g data, consisting of one intensity for each beam. It may include from approximately 5 to over 100 intensity points for one crystal orientation in the more common cases.

More data are normally desired, however, and these are obtained by varying any of the available parameters: energy (i.e. wavelength), or polar or azimuthal angles of incidence. The corresponding curves of beam intensities are called, respectively, I-V curves (V stands for accelerating voltage), I-θ curves, and I-φ curves (also called rocking curves). By far the most common form of data collection used in structural determinations are I-V curves. Occasionally they are given as I-s curves, using the momentum transfer s = $|\vec{s}|$ rather than the energy. Occasionally, data are also presented as constant-intensity contour plots in which two of these parameters are varied, such as energy and polar angle.

For different purposes, the angular profile of the diffraction pattern can be measured for constant incidence energy and direction. Such data, including beam profiles and diffuse-background variations, are of value in analyzing disorder phenomena at surfaces.

One may also vary nongeometrical parameters. The most obvious among these is the temperature, the effects of which can be used to study thermal vibrations through the Debye-Waller factor or to investigate phase transitions at surfaces. The latter subject may also be studied by observing diffraction intensities as a function of adsorbate coverage, for instance. This topic is discussed in detail in Chap. 8.

4.2.5 Kinematic Diffraction by Bravais Lattices of Atoms

Assume at first the simple case of diffraction by a single two-dimensional layer of atoms arranged in a regular ordered array, where each unit cell contains only one atom. The last condition makes this array a Bravais lattice. The one-dimensional analogue of this lattice and its effect on diffraction is illustrated in Fig. 4.2. It can be shown that the plane waves diffracted by the two-dimensional Bravais lattice have, in the kinematic limit, the complex amplitude

$$M_{out,o} = \frac{i}{Ak_{oz}} \, f(\vec{s})\delta(\vec{k}_{\|out} - \vec{k}_{\|o} - \vec{g}) \, , \tag{4.25}$$

where $i = (-1)^{1/2}$, A is the area of the unit cell, $\vec{s} = \vec{k}_{out} - \vec{k}_o$ is the momentum transfer from $\vec{k}_o$ to $\vec{k}_{out}$, and $f(\vec{s})$ is the atomic scattering factor used in (4.2). For simplicity $f(\vec{s})$ can be taken as $\vec{s}$-independent, but more frequently its $\vec{s}$ dependence is taken into account [$f(\vec{s})$ will be discussed in more detail in Sect. 4.5, where, in particular, it will be expressed in terms of phase shifts]. The complex prefactor i/Ak_{oz}, although simple in appearance, is the result of a nontrivial integration over the contributions by all scatterers in the layer. It can be derived with the concept of Fresnel zones following principles of optical diffraction [4.7], or by direct integration [4.7-10]. The factor 1/A is the density of atoms in the layer, to which the diffraction amplitudes should obviously be proportional. Thus, as a function of energy and of incidence or detection angles, $M_{out,o}$ varies in the same way that the atomic scattering factor does, apart from the smoothly varying factor $(k_{oz})^{-1}$.

If our atomic layer now has more than one atom per unit cell, at locations $\vec{R}_p$, see (4.10), the diffraction amplitude of (4.25) is simply generalized to

$$M_{out,o} = \frac{i}{Ak_{oz}} \left[\sum_p f_p(\vec{s})\exp(i\vec{s}\cdot\vec{R}_p) \right] \delta(\vec{k}_{\|out} - \vec{k}_{\|o} - \vec{g}) \, , \tag{4.26}$$

where the unit-cell form factor has been included in square brackets.

Next we introduce periodicity perpendicular to the surface in this atomic layer, i.e. we think of a regular lattice of the Bravais type that extends deep below the surface. Then $\vec{R}_p = (p - 1)\vec{a}_3$, $p = 1,2,...$, where $\vec{a}_3$ is not necessarily perpendicular to the surface, and the summation over p can be carried out analytically, setting $f_p = f$ for all p.

$$\sum_p f_p(\vec{s})\exp(i\vec{s}\cdot\vec{R}_p) = \frac{f}{1 - \exp(i\vec{s}\cdot\vec{a}_3)} \, . \tag{4.27}$$

If $\vec{s}$ were real, corresponding to the absence of inelastic losses and therefore infinite penetration into the lattice, we would recover sharp diffraction peaks of the delta-function type at the "third" Bragg conditions

$$\vec{s}\cdot\vec{a}_3 = n2\pi \quad \text{(n integer)} \, , \tag{4.28}$$

as in X-ray Diffraction and HEED. Inelastic effects lead us to give $\vec{s}$ a (relatively small) imaginary component which broadens the peaks, but leaves them essentially centered on the Bragg conditions (4.28). This is a situation

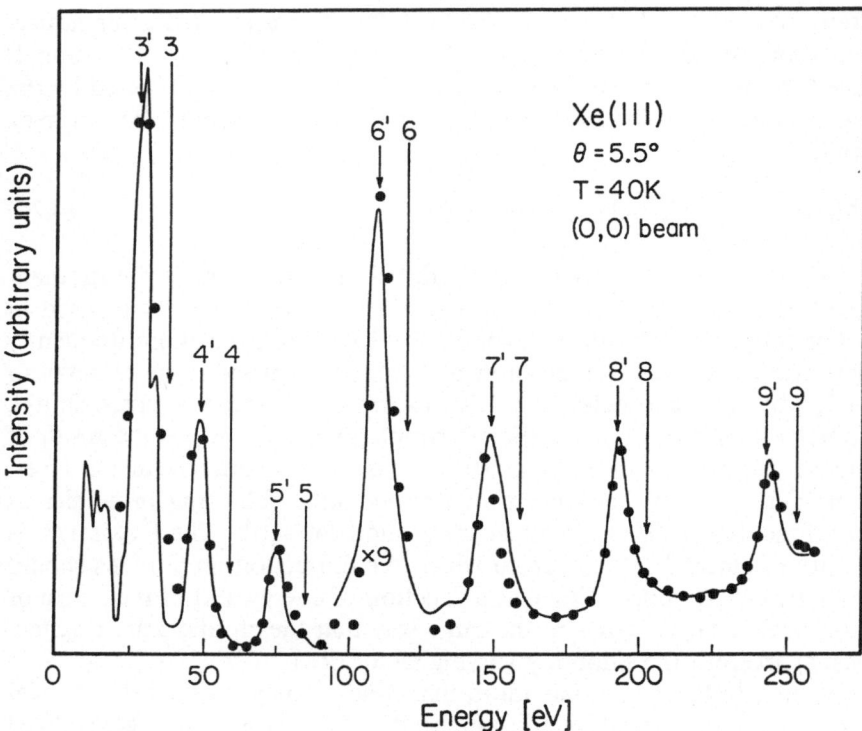

Fig. 4.5. Specular I-V curve for Xe(111) showing Bragg peaks satisfying (4.28) (after [4.11]). Arrows indicate Bragg-peak positions expected for an ideally-terminated bulk xenon lattice, both before and after a downward inner-potential shift of 10 eV

that some real surfaces approximate rather well (except for an inner-potential shift which can easily be taken into account). For example, the (111) surface of the fcc xenon crystal produces I-V curves [4.11] that show "Bragg peaks" at energies compatible with the Bragg condition, [4.28], as shown in Fig. 4.5. Many metal surfaces have Bravais lattices as assumed here (if they can be thought to have an ideal truncation of the bulk lattice). All unreconstructed fcc and bcc surfaces fall into this category, whether they have low or high Miller indices.

One can modify the Bragg condition, (4.28), using the energy-conservation rule, in order to show explicitly at which energies, called Bragg energies, the kinematic condition for maximum intensity is satisfied for any given diffraction beam, $\vec{g}_{hk} = h\vec{g}_1 + k\vec{g}_2$. To that end, it is convenient to express the lattice vector $\vec{a}_3$ as

$$\vec{a}_3 = a\vec{a}_1 + b\vec{a}_2 + d(0,0,1) \ , \tag{4.29}$$

where d is the interlayer spacing and (a,b) represents the "registry shift" from one atomic plane to the next. Then, in atomic units (bohrs for distances, 1 bohr = 0.529 Å, hartrees for energies, 1h = 27.18 eV), the Bragg energies E^B follow from

$$2(E^B + V_o) - |\vec{k}_{o\parallel}|^2 = 2(E^B + V_o)\cos^2\theta$$

$$= \left[\left(2\vec{k}_{o\parallel}\cdot\vec{g}_{hk} + |\vec{g}_{hk}|^2\right)\frac{d}{4\pi(n - ah - bk)} + \frac{\pi}{d}(n - ah - bk)\right]^2 . \quad (4.30)$$

This equation is valid for any Bravais lattice, for any incidence direction, and for any beam. Specializing to normal incidence ($\theta = 0$, $\vec{k}_{o\parallel} = \vec{0}$) yields

$$2(E^B + V_o) = \left[|\vec{g}_{hk}|^2\frac{d}{4\pi(n - ah - bk)} + \frac{\pi}{d}(n - ah - bk)\right]^2 . \quad (4.31)$$

Taking in addition a rectangular lattice, i.e. $\vec{g}_1 = (2\pi/a_1, 0) \perp \vec{g}_2 = (0, 2\pi/a_2)$, gives

$$2(E^B + V_o)$$

$$= \left\{\left[\left(\frac{h}{a_1}\right)^2 + \left(\frac{k}{a_2}\right)^2\right]\frac{\pi d}{n - ah - bk} + \frac{\pi}{d}(n - ah - bk)\right\}^2 . \quad (4.32)$$

For a general angle of incidence and a general Bravais lattice, the specularly reflected beam ($h = k = 0$) has maxima at the following Bragg energies:

$$2(E^B + V_o) - |\vec{k}_{o\parallel}|^2 = 2(E^B + V_o)\cos^2\theta = n^2\frac{\pi^2}{d^2} . \quad (4.33)$$

It is often of some value, even in dynamical cases, to attempt to identify Bragg peaks on the basis of these equations, even though there are usually many more "multiple-scattering" peaks present. As we shall discuss at the end of this chapter, however, Bragg peaks can be shifted in energy by several electron volts due to dynamical effects, in addition to the inner-potential shift. If nothing else, this exercise gives a good idea of the relative importance of multiple scattering in a given material. If, as for xenon and aluminum surfaces, one finds strong peaks close to the Bragg energies, the material is relatively kinematic (Fig. 4.5). However, if no obvious connection exists between Bragg energies and strong peaks, multiple scattering is very strong, as in platinum surfaces [4.12], see Fig. 4.6.

4.2.6 The Case of Non-Bravais Lattices

The kinematic diffraction by a non-Bravais lattice is easily obtained from (4.25). If the three-dimensional unit cell, below the surface and spanned by the vectors $\vec{a}_1, \vec{a}_2, \vec{a}_3$, contains atoms with scattering factors f_q at positions $\vec{R}_q$ ($q = 1,...,N$), then the reflected amplitude becomes

$$M_{out,o} = \frac{i}{Ak_{oz}}\left[\sum_{q=1}^{N} f_q\exp(i\vec{s}\cdot\vec{R}_q)\right]\frac{1}{1 - \exp(i\vec{s}\cdot\vec{a}_3)} \delta(\vec{k}_{\parallel out} - \vec{k}_{\parallel o} - \vec{g}) . \quad (4.34)$$

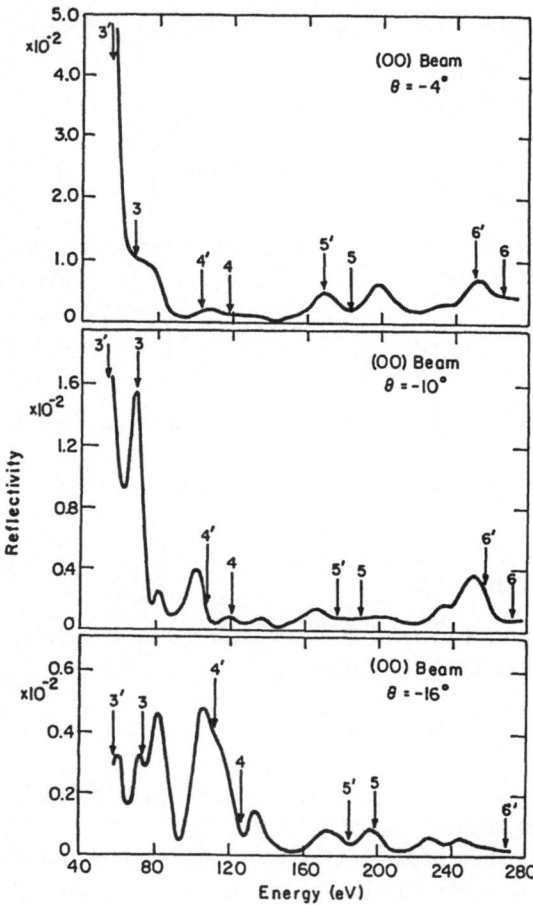

Fig. 4.6. As Fig. 4.5. for Pt(111), using an inner-potential shift of 15 eV. After [4.12]

We have simply added a form factor for the contents of the three-dimensional bulk unit cell.

This expression still has Bragg energies as given by (4.30-33), but the new form factor in special cases may produce extinctions at some of those Bragg energies. Consequently, some expected peaks in I-V curves may be absent. The special extinctions can be predicted easily when the atomic positions in the bulk unit cell are known.

Non-Bravais lattices include surfaces of hcp metals and of substances with a diamond lattice, a wurtzite lattice, a NaCl lattice, etc., and (4.34) assumes that they have an ideal bulklike truncation at the surface. Silicon has a diamond lattice and is relatively kinematic. Hence most of the diffraction peaks in I-V curves expected for the ideally truncated surface can be predicted based on (4.30-33), together with additional conditions due to the contents of the unit cell.

4.2.7 Surface Structures Deviating from the Bulk Structure

Most real surfaces show deviations of the atomic positions from the bulk structure or include atoms of a different chemical type. To calculate the kinematic diffraction by such surfaces, it is convenient to consider separately the scattering by the substrate and the scattering by the deviating surface atoms, which we shall label by the index s = 1,...,M. If we assume the same two-dimensional surface periodicity as exists in the bulk, namely a (1×1) surface, then the diffraction amplitude might appear as follows, generalizing (4.34),

$$M_{out,o} = \frac{i}{Ak_{oz}} \delta(\vec{k}_{\|out} - \vec{k}_{\|o} - \vec{g})$$

$$\times \left\{ \left[\sum_{s=1}^{M} f_s \exp(i\vec{s}\cdot\vec{R}_s) \right] + \left[\sum_{q=1}^{N} f_q \exp(i\vec{s}\cdot\vec{R}_q) \right] \frac{\exp(i\vec{s}\cdot\vec{R}_{sub})}{1 - \exp(i\vec{s}\cdot\vec{a}_3)} \right\}, \qquad (4.35)$$

where $\vec{R}_{sub}$ points to the first atom below the surface that fits in the substrate lattice. Simplifying to M=1 and N=1 [one deviating atom per (1×1) unit cell at the surface and a Bravais lattice in the bulk], with $\vec{R}_s = \vec{R}_q = \vec{0}$, we obtain

$$M_{out,o} = \frac{i}{Ak_{oz}} \delta(\vec{k}_{\|out} - \vec{k}_{\|o} - \vec{g}) \left[f_{surf} + f_{sub}\frac{\exp(i\vec{s}\cdot\vec{R}_{sub})}{1 - \exp(i\vec{s}\cdot\vec{a}_3)} \right]. \qquad (4.36)$$

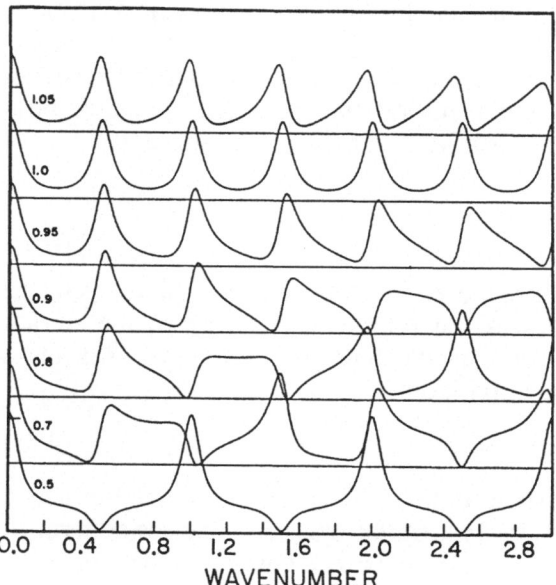

0.0   0.4   0.8   1.2   1.6   2.0   2.4   2.8
WAVENUMBER

Fig. 4.7. Kinematic specular I-V curve calculated at normal incidence for the (100) surface of a simple cubic lattice, in which the spacing of the top layer of atoms above the next layer takes on the fractions 0.5, 0.7, 0.8, 0.9, 0.95, 1.0 and 1.05 of the bulk spacing (from bottom to top in the figure). Typical numerical values are used for the layer spacing and the damping factor

If $f_{surf} \cong f_{sub}$ and $\vec{R}_{sub} \cong \vec{a}_3$, i.e. if the surface is almost bulklike, the term f_{surf} only slightly changes the diffraction amplitude due to the substrate (apart from an overall phase factor that drops out in forming the intensity $|M_{out,o}|^2$). The simple I-V profiles with peaks at Bragg energies corresponding to the ideal truncation at the surface are then only slightly distorted by the surface deviation, namely the peaks become skewed, as illustrated in Fig. 4.7 in a kinematic model calculation. On the other hand, for large deviations of the surface structure from the bulk structure, the I-V curves can clearly lose all resemblance with the initial I-V curves of the ideal bulklike surface, illustrating the surface sensitivity of LEED.

4.2.8 Surfaces with Superlattices

A more radical departure from the bulk structure is presented by surfaces with superlattices, in which the surface atoms have a two-dimensional unit cell different from, but often related to, that of the ideally truncated (1×1) surface. We have given examples of such surface structures in Sect. 3.2.

With superlattices there appear "extra" or "fractional-order" beams that are absent for the (1×1) surface. Since, in the kinematic limit, the (1×1) substrate cannot contribute to the extra beams, these derive all their intensity from the deviating surface layer, and they contain no information whatsoever concerning the substrate, not even concerning the relative position of the deviating layer with respect to the substrate. (With multiple scattering, this statement is no longer true, and in that case all beams contain information concerning the entire surface structure.)

On the other hand, for those superlattice beams that coincide with substrate beams (these are the "substrate" or "integral-order" beams), one simply adds the amplitudes diffracted by the different parts of the surface, including interference, as in (4.35,4.36). Therefore these special integral-order beams, even in the kinematic limit, contain information concerning relative positions of any atoms in the deviating surface layer and in the substrate.

4.2.9 Modulated Structures

For many surface structures of some complexity, it is useful to view the kinematic diffraction amplitudes as the Fourier coefficients of the surface structure, as implied by (4.2). We can illustrate this as follows. Assume a one-dimensional string of atoms at positions

$$R_p = pa + r \cos(qpa), \qquad p = 0, \pm 1, \pm 2, ..., \pm \infty \ . \tag{4.37}$$

If $r=0$, this is a string of equidistant atoms, with spacing a, which diffracts a wave into beams labeled by $g = m \, 2\pi/a$ (m an integer). The cosine term in (4.37) modulates the regularly spaced positions with a wave vector q (as for a phonon with wavelength $2\pi/q$, for example). The diffraction amplitude will then be proportional to the structure factor

$$S = f\sum_p \exp(isR_p) = f\sum_p \exp(ispa)\exp[(isr)\cos(qpa)] \ . \tag{4.38}$$

We expand the second exponential

$$\exp[(isr)\cos(qpa)] = 1 - isr\cos(qpa) - \frac{s^2r^2}{2}\cos^2(qpa) + \cdots$$

$$= 1 - \frac{isr}{2}\left[\exp(iqpa) + \exp(-iqpa)\right]$$

$$- \frac{s^2r^2}{8}\left[2 + \exp(i2qpa) + \exp(-i2qpa)\right] + \cdots ,$$

and thus (4.38) becomes

$$S = f\left\{\left(1 - \frac{s^2r^2}{8}\right)\left[\sum_p \exp(ispa)\right]\right.$$

$$- f(\frac{isr}{2} + \cdots)\left\{\sum_p \exp[i(s+q)pa] + \sum_p \exp[i(s-q)pa]\right\}$$

$$- f\left(\frac{s^2r^2}{8} + \cdots\right)\left\{\sum_p \exp[i(s+2q)pa] + \sum_p \exp[i(s-2q)pa]\right\} + \cdots \tag{4.39}$$

Each summation over exponentials in (4.39) produces a delta function. The first one is $\delta(s - g)$, the next two are $\delta(s + q - g)$ and $\delta(s - q - g)$, respectively, and the following two are $\delta(s + 2q - g)$ and $\delta(s - 2q - g)$, respectively. Thus one sees that "satellite" diffractions at $s = g \pm q$, $s = g \pm 2q$, ... appear adjacent to the main beams at $s=g$ (q need not necessarily be small for this description to hold). The terms of (4.39) proportional to r, which are the first ones to become intense as the modulation r increases, produce the satellite beams at $s = g \pm q$. The next beams to appear as r increases are those at $s = g \pm 2q$. More and more satellite beams, farther and farther away from the main beams at $s=g$ appear as the modulation r becomes larger. Thus the number of visible satellites is a measure of the amplitude of the structure modulation. Note that the factor $1 - s^2r^2/8 + \ldots$ in (4.39) can be approximated by $\exp(-s^2r^2/8)$, representing an intensity loss from the main beams to the satellite beams. This factor gives rise to the Debye-Waller factor in the case of surface vibrations.

Thus the kinematic LEED pattern becomes a map of two-dimensional structural wave vectors $\bar{q}$, together with their weights which affect the spot intensities. We have already made use of this fact in Sect. 3.5.6 in the interpretation of LEED patterns that arise in the presence of a long-wavelength modulation of the surface structure (as in a charge-density wave, for instance, or in substrate-induced buckling of an overlayer which has a unit cell that is slightly mismatched to that of the substrate).

4.2.10 The Simple Effect of Multiple Scattering on LEED Patterns

Multiple scattering generally has little qualitative effect on the LEED pattern. Spot positions are never modified by multiple scattering, as our general derivation at the beginning of this section has shown. Therefore, the above kinematic considerations are of real practical value in analyzing actual LEED patterns.

Nevertheless, in some cases multiple scattering can produce qualitative effects on the LEED pattern, namely by adding spots to it. Fortunately, kinematic considerations can be extended easily to the case of multiple scattering for the purpose of understanding LEED patterns. One merely allows a wave that has already diffracted once, say into the beam labeled $\vec{g}$, to diffract more often, by adding at each additional diffraction another reciprocal lattice vector. The label becomes $\vec{g} + \vec{g}'$ after the second diffraction, followed by $g + \vec{g}' + \vec{g}''$ after the third diffraction, etc. In practice the third diffraction is already sufficiently weak to be ignored.

The position in the LEED pattern of the spot labeled $\vec{g} + \vec{g}'$ can be predicted easily by vector addition of the reciprocal lattice vectors defining the spot $\vec{g}$ and the spot $\vec{g}'$. For example, if $\vec{g} = (1,1)$ and $\vec{g}' = (0,1/5)$, then $\vec{g} + \vec{g}' = (1,6/5)$ represents a spot that receives a double-scattering contribution by first diffracting from the incident beam $\vec{0} = (0,0)$ to the beam $(1,1)$ and by subsequently diffracting from the beam $(1,1)$ to the beam $(1,1) + (0,1/5) = (1,6/5)$.

This method is particulary useful when the beams $\vec{g}$ belong to the substrate and the beams $\vec{g}'$ belong to a superlattice surface layer (or vice versa), in such a way that the combination $\vec{g} + \vec{g}'$ is new, i.e. is not already present in the set of all vectors $\vec{g}$ or in the set of all vectors $\vec{g}'$. In this case, the beam $\vec{g} + \vec{g}'$ is a multiple-scattering beam that did not exist in the kinematic limit. We have already used these ideas in a number of examples of pattern interpretation in Sect. 3.5, and those can serve as more detailed illustrations of these concepts.

4.2.11 The Ewald Sphere

In X-ray Diffraction, the Ewald sphere has served as an extremely convenient concept to visualize the occurrence of Bragg conditions [4.1-5]. The following conditions (with three-dimensional periodicity) must be satisfied: the momentum transfer $\vec{s} = \vec{k}_{out} - \vec{k}_o$ must be a reciprocal lattice vector $\vec{g}^{(3)}$, and $|\vec{k}_o| = |\vec{k}_{out}| = k$ must hold. This can be represented graphically in reciprocal space by the condition that a reflection will occur only when a reciprocal lattice point falls on the Ewald sphere, the center of which is at the origin of the vector $\vec{k}_o$ and the radius of which is $|\vec{k}_o| = k$.

The Ewald Sphere in LEED

This visualization can be adapted to the case of LEED by noting that the three-dimensional reciprocal lattice is replaced by a two-dimensional one with no condition being placed on the third coordinate of $\vec{k}_{out}$. The effect is to replace reciprocal lattice points by reciprocal lattice "rods" perpendicular to the surface and to demand that $\vec{s} = \vec{k}_{out} - \vec{k}_o$ be a vector linking any two points on the rods. Since the energy conservation requirement $|\vec{k}_o| = |\vec{k}_{out}|$

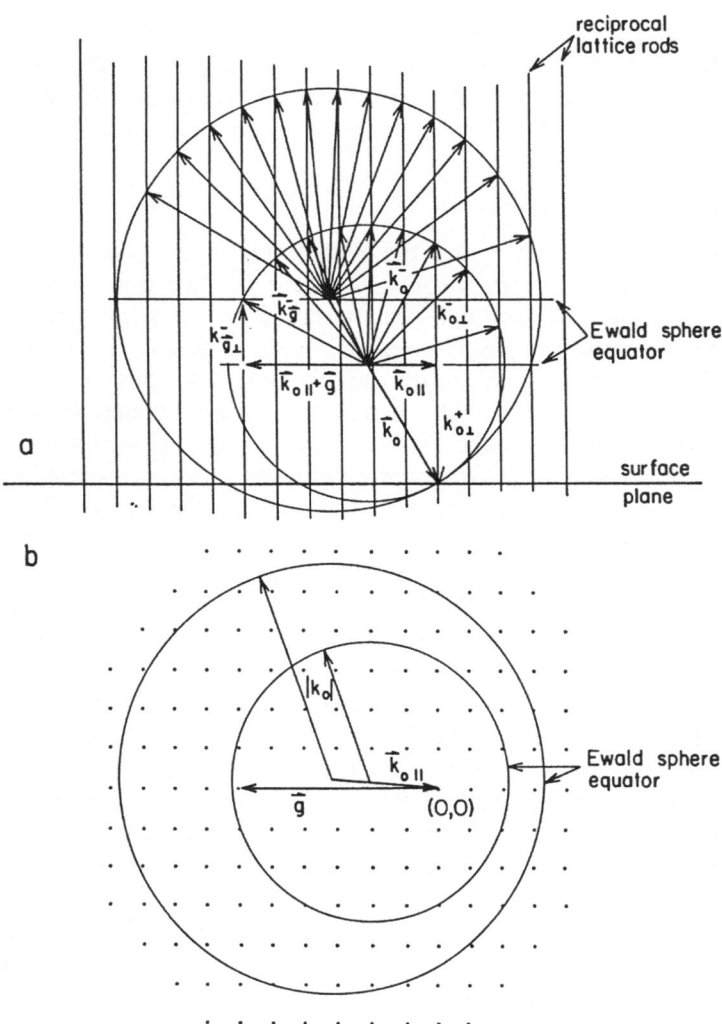

Fig. 4.8a, b. The Ewald-sphere construction in the case of LEED. The Ewald-sphere is shown at two energies for the same incidence direction: (a) view parallel to the surface, and (b) view perpendicular to the surface showing the equator of the Ewald sphere

= k must still be fulfilled, an Ewald sphere still exists, and the result in reciprocal space is as drawn in Fig. 4.8a. The end of the vector $\vec{k}_0$ is shown located on a rod drawn perpendicular to the surface. This may be called the "specular rod" (by convention, the end of $\vec{k}_0$ is also located on the "surface plane", imagined drawn into the reciprocal lattice). Other parallel rods exist and are separated by any of the reciprocal lattice vectors $\vec{g}^{(2)} = \vec{g}$. Thus each rod corresponds to a reciprocal lattice vector $\vec{g}$. The Ewald sphere is again drawn centered on the origin of $\vec{k}_0$. Now any $\vec{k}_{out}$ must point from the origin of $\vec{k}_0$ to one of the intersections of the sphere with a rod. This determines a finite and discrete set of outgoing beams. Compared to the case of X-ray Diffraction, however, many more opportunities for nonextinct beams exist, since the reciprocal lattice rods are not broken up into reciprocal lattice points.

Energy Dependence of the LEED Pattern

As the electron energy is increased and the Ewald sphere increases in size and shifts its center, one sees from the Ewald-sphere construction that the diffracted beams smoothly change their orientation, converging toward the specular beam (the orientation of which does not change). At the same time, new beams appear at discrete energies in the grazing emergence directions.

Such beam appearances occur as the equator of the Ewald sphere (which is parallel to the surface) increases and intersects new reciprocal lattice rods, i.e. as new values of (h,k) become compatible with the energy-conservation requirement. This observation leads to Fig. 4.8b, where the Ewald-sphere equator is drawn projected onto the surface, in which case the rods are viewed end on. Those two-dimensional reciprocal lattice vectors $\vec{g} = h\vec{g}_1 + k\vec{g}_2$ that fall within the Ewald-sphere equator in this diagram correspond to beams that can emerge from the surface and can be detected. In the case of a hemispherical LEED screen centered on the crystal, the Ewald-sphere equator corresponds to the edge of the screen. In addition, the projection is such that the pattern on the screen is an undistorted representation of the reciprocal lattice, if viewed from infinity along the axis of the screen. For example, a square reciprocal lattice is manifest as a square lattice of spots. Therefore one can easily predict how the spot pattern on the screen varies with energy. Note that the (00) spot (which is the end of the vector $\vec{k}_{0||}$) does not move with energy. The spot pattern then merely contracts toward the (00) spot as the energy is increased. The variation with incidence direction can also be visualized easily, as the projection $\vec{k}_{0||}$ changes its direction; but now the (00) spot is in general not immobile.

Note the important point that the Ewald-sphere construction remains valid in the presence of multiple scattering, since it makes use of only the two-dimensional periodicity of the surface and energy conservation.

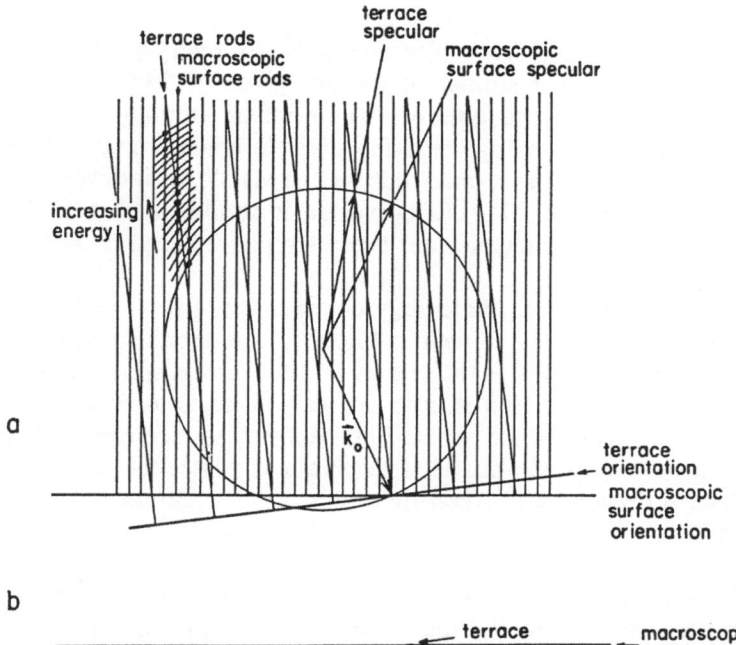

Fig. 4.9a, b. The Ewald-sphere construction for a stepped surface. (**a**) The Ewald sphere and two sets of reciprocal lattice rods in a view parallel to the steps. Segments of the sphere at various energies are shown, and strong beams are indicated by dots on those segments. (**b**) Profile of the stepped surface in question

The LEED Pattern of Stepped Surfaces

Perhaps the most elegant application of the Ewald sphere in LEED lies in the understanding of the LEED pattern of stepped surfaces. Figure 4.9b shows the profile of a stepped surface. In Fig. 4.9a the Ewald sphere is drawn, together with the reciprocal lattice rods of the stepped surface, which are packed in inverse proportion to the terrace length. The intersection of the Ewald sphere with each of the many rods yields an allowed beam. However, we can include additional information. We can consider the diffraction by a single isolated terrace and multiply that as a form factor by the structure factor corresponding to the repetition of large unit cells, each containing one terrace. The unit-cell form factor is related to that of an infinitely extended terrace, inclined with respect to the so-called "macroscopic surface". Thus, its diffraction properties can be approximated by inclined reciprocal lattice rods as drawn in Fig. 4.9a. However, these rods are not infinitely "thin" because they do not arise from a sharp delta-function condition, but rather they have a nonzero width due to the finite width of the terrace. Therefore, any macroscopic surface beam that falls near the terrace rods has a strong intensity, while a beam that falls between terrace rods has a very small intensity. (This intensity argument is similar to the one given in Sect. 3.5.7 for the splitting of extra spots in the presence of antiphase domains.) As a result, we can understand that among the many possible macroscopic surface beams, only those that lie near the terrace rods will be visible, as indicated by dots in Fig. 4.9a. Consequently, the LEED pattern is similar to that of the nonstepped surface, with each integral-order beam simply replaced by two or three closely spaced spots.

In addition, we can understand the energy dependence of the LEED pattern for stepped surfaces. As may be seen in Fig. 4.9a, the terrace rods favor different macroscopic surface rods as the energy and the Ewald-sphere radius increase. In particular, a given macroscopic surface rod starts to be selected when its intersection with the Ewald sphere approaches a terrace rod. Then its intensity goes through a maximum as it reaches and passes the center of the rod, after which it decays again when the intersection is too far removed from that terrace rod. (Incidentally, the passage through the center of the terrace rod corresponds to a Bragg condition for the momentum transfer perpendicular to the surface and is therefore related to the spacing between layers parallel to the surface.) In the meantime, the next macroscopic surface rod has approached and gained intensity prior to decaying in turn, and this sequence will be repeated indefinitely as the energy increases. Since both the width of the terrace rods and the spacing between macroscopic surface rods are inversely proportional to the distance between the steps, at any one energy either two or three neighboring macroscopic surface rods are visible on a LEED screen, rather than a longer string of closely spaced spots.

One can also see from this Ewald-sphere construction that the individual split-off spots must converge with increasing energy toward the specular direction of the macroscopic surface, while the locations of the groups of bright spots (specified by the terrace rods) converge toward the specular direction of the terraces. Although we have already described this situation in other words in Sects. 3.3 and 3.5.7, the Ewald-sphere construction gives a much more intuitive feeling for it than was possible there.

4.2.12 Further Applications of the Kinematic Theory of LEED

The mathematical simplicity of the kinematic LEED theory is sufficient reason to attempt to apply it in spite of the frequent occurrence of multiple scattering in real situations. The hope is that multiple scattering is not so strong as to overshadow the single-scattering effects completely. In fact, there are many proven cases where this hope is justified. In the majority of situations, however, the kinematic approximation generates unacceptably large errors, in particular in the important area of structural determination of binding sites, bond lengths, and bond angles. Therefore, this approximation has not gained general acceptance, despite a reasonable amount of effort to make it suitable for the extraction of surface structural information.

There is one major exception to the decline in popularity of the kinematic theory, which is its application to the interpretation of the LEED pattern, especially in cases of defect or disordered surface structures (step formation, island formation, domain formation, two-dimensional phase transitions, and the like). We shall discuss these in some detail in Chaps. 8-10.

Due to the attractiveness of a kinematic LEED theory for structural determination of atomic positions, several approaches have been developed that have proved useful in some circumstances. We shall devote Sects. 6.1-3 to these methods, but we shall summarize their principles here.

Of course, the simplest kinematic approach to structural determinations consists in applying the kinematic formulae for the diffraction amplitude (and hence intensity), in order to simulate directly actual experimental data such as I-V curves. In a structural determination, one can effectively and inexpensively test many different trial surface structures this way. For substances with little multiple scattering (xenon and silicon, for example), this is a good approximation. However, even when one cannot reach a very accurate structural determination in this fashion, due to multiple scattering, at least one should be able to exclude many unpromising structures on the basis of complete disagreement between kinematic theory and experiment. This is most helpful in reducing the cost of a more refined structural search with dynamical calculations.

A more sophisticated kinematic approach consists of attempting to average out multiple-scattering effects from the experimental data, leaving mainly kinematic features, and then to apply kinematic formulas to interpret the kinematically enhanced data. The best known of such methods is the Constant-Momentum-Transfer-Averaging (CMTA) scheme, in which one averages over intensities measured at different incidence directions and energies, but at constant momentum transfer $\vec{s} = \vec{k_g} - \vec{k_o}$, yielding I-s curves. In principle, this kind of averaging favors the kinematic Bragg peaks (the positions of which depend only on $\vec{s}$), while discriminating against multiple-scattering peaks, the positions of which vary in a more complicated fashion with incidence direction and energy, and which thus tend to be averaged into a smooth background. However, it has been shown that the averaging can introduce unwanted distortions due to dynamical effects that cannot be averaged out properly [4.8].

Another approach is based on Fourier transforming the LEED intensity data in momentum-transfer space. In the kinematic limit, this should yield the interatomic vectors of the surface structure after suitable deconvolution (apart from the complication, well-known in diffraction theory, due to the unknown phase). To some extent, the Fourier transformation to some extent averages out some multiple-scattering effects as well, by attempting to relegate them to the noise frequency range.

The method of Fourier-transform convolution (as opposed to deconvolution) has been developed to gain more reliability in structural determination, by acknowledging the nonuniqueness of deconvolution and allowing a more refined parametrization of the kinematic process.

Both of these Fourier-transform methods have been able to extract acceptable structural information, even in cases where strong multiple scattering is present. However, they have only been applied to rather simple structures (specifically clean and unreconstructed metal surfaces), and it remains an open question whether they can be applied successfully to overlayers and reconstructions.

4.3 The Scattering Processes in LEED

We now turn our attention to the physical ingredients that can be taken into account in a kinematic theory.

4.3.1 Inelastic Scattering Processes

Electrons diffracting from a surface are scattered by all electric charges present in their paths: nuclei and electrons. The localized core electrons as well as the less localized valence electrons and conduction electrons (when present) should be included in this interaction [4.8].

We shall be only indirectly concerned with interactions between the diffracting electrons and the collective excitations of the surface. This is because a significant energy loss (or gain) is often involved, such as with a plasmon excitation (or de-excitation), leading to "inelastic" LEED (ILEED) [4.13], which we do not consider here. All these inelastically scattered electrons will be treated in the theory through a decrease of elastic beam intensities. This is achieved by introducing a mean free path, or alternatively, an imaginary part of the electron-surface interaction potential [4.14]. Furthermore, interactions with collective excitations can give rise to momentum transfers that are filtered out in the LEED intensity measurement, when these interactions deflect electrons into the diffuse background between spots. Interactions with phonons, with their smaller energy exchange and their small momentum transfer, are a borderline case. Most experiments include in the beam-intensity measurement a number of (quasi-elastic) electrons that have undergone interactions with phonons. This number depends strongly on the energy and angle acceptances of the experiments (i.e. on the biasing voltage and on the angular aperture of the spot photometer, the Faraday cup, etc.). However, it appears after numerous comparisons between experiment and theory, that it is appropriate for the purpose of surface structural analysis to consider all phonon interactions in the theory as being strictly inelastic and therefore excluded from the elastic intensity. (A Debye-Waller factor should be included, though.)

Other inelastic processes, such as electronic excitations (including secondary-electron emission) are only of concern in elastic LEED in that they influence the mean free path of the diffracting electrons.

The justification for representing all the inelastic processes by a single number (the mean free path or the imaginary part of the potential) is that in elastic LEED we need only know what fraction of the LEED electrons lose energy, not why or how. As will become clear shortly, this information is obtained easily from the experimental peak widths in I-V curves. Note that the mean free path is usually taken to be position independent in calculations, although occasionally a layer-dependent value may be chosen. The inelastic effects, of course, decrease gradually into the vacuum, but in the theory they are often made to terminate abruptly at some surface plane. When the effects of a potential step are included, a more realistic decay can be used easily, but it has been found to be of little consequence, except for "resonances" (cf. Sect. 5.12).

In general, the electronic mean free path is energy dependent, since the inelastic processes are energy dependent [4.8]. However, the mean free path is relatively independent of the material, so that one may speak of a universal curve (Fig. 4.10). This curve shows that LEED electrons can probe a surface region that extends to no more than 10-20 Å in depth for the usual energies (10-300 eV). The mean-free-path lengths shown in Fig. 4.10 correspond to an imaginary part of the potential of approximately 1 to 5 eV.

4.3.2 Modeling the Effect of the Mean Free Path

We shall now illustrate the effect of the mean free path in LEED with a very simple one-dimensional model. A plane wave exp(ikx), where k is the wave

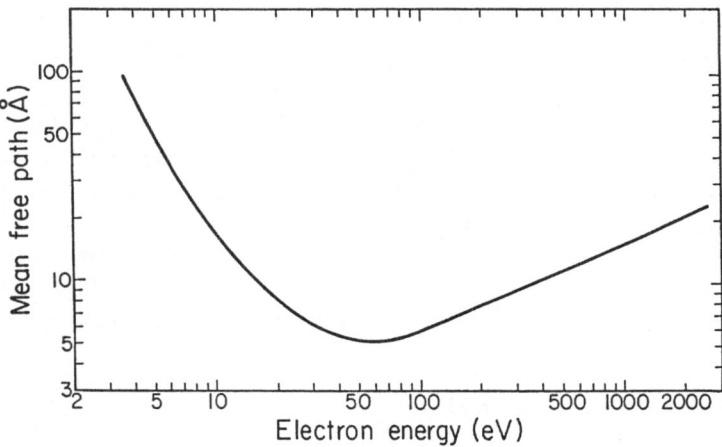

Fig. 4.10. "Universal Curve" of electron mean free path as a function of electron kinetic energy, as fit to experimental data for many metallic surfaces

number and x is the only dimension, is incident on a chain of atoms extending from $x = 0$ to $x = \infty$. In the vacuum, i.e. for $x < 0$, the wave number k and the kinetic energy are related by $k^2/2 = E$ (the factor $1/2$ stands for $\hbar^2/2m$, using atomic units). After penetrating the surface, but ignoring the scattering by the atomic cores, this wave has the form $\exp(ik'x)$ with $k'^2/2 = E + V_o = E + V_{or} + iV_{oi}$. Here, V_{or} is the inner potential (or muffin-tin constant), while $V_{oi} > 0$ is the imaginary part of the potential, used to represent all inelastic effects. Consequently, k' must be complex, and we may write

$$\tfrac{1}{2}k'^2 = \tfrac{1}{2}(k'_r + ik'_i)^2 = \tfrac{1}{2}(k'^2_r - k'^2_i) + ik'_r k'_i = E + V_{or} + iV_{oi} , \quad (4.40)$$

so that

$$\tfrac{1}{2}(k'^2_r - k'^2_i) = E + V_{or} , \quad \text{and} \quad k'_r k'_i = V_{oi} .$$

A better intuitive feeling for the implications of this result may be obtained by neglecting k'^2_i with respect to k'^2_r (which is justified when V_{oi} is small compared to $E + V_{or}$). This leads to

$$\tfrac{1}{2}k'^2_r \cong E + V_{or} , \quad \text{and} \quad k'_i = V_{oi}/k'_r \cong V_{oi}/[2(E + V_{or})]^{1/2} . \quad (4.41)$$

Thus, k'_i is easily related to V_{oi}, while k'_r (and thus the wavelength $2\pi/k'_r$) is hardly affected by the presence of a small V_{oi}. The latter point implies that the conditions for constructive and destructive interference in diffraction are barely affected by V_{oi}.

Now the incident wave within the surface has the form

$$\exp(ik'x) = \exp(ik'_r x)\exp(-k'_i x) = \exp(ik'_r x)\exp(-x/\lambda_e) , \quad (4.42)$$

where the damping effect is represented by the mean-free-path length λ_e. It follows that

$$\lambda_e = 1/k'_i \cong [2(E + V_{or})]^{1/2}/V_{oi} , \quad (4.43)$$

which relates λ_e to V_{oi} and thereby shows the equivalence of the concepts of the mean free path and the imaginary part of the potential.

We shall now relate V_{oi} (and λ_e) to the peak width in I-V curves, again within the one-dimensional model used above. Damping masks the deeper parts of the crystal from the electrons. Therefore, a mean free path λ_e can be simulated qualitatively by taking a surface with a thickness of only λ_e (and by neglecting the damping otherwise). To understand the effect of this truncation of the crystal into a thin slab, let us first scatter the wave $\exp(ik'x)$ from just two atoms located at $x=0$ and $x=a$. This produces a reflected wave amplitude which is equal to

$$1 + \exp(2ik'a) = 2\cos(k'a)\exp(ik'a) , \quad (4.44)$$

and thus a reflected wave intensity of $4\cos^2 k'a$. This is shown schematically in Fig. 4.11a, and the peaks that are visible there are Bragg peaks. With three equidistant atoms one would obtain the result of Fig. 4.11b, and with

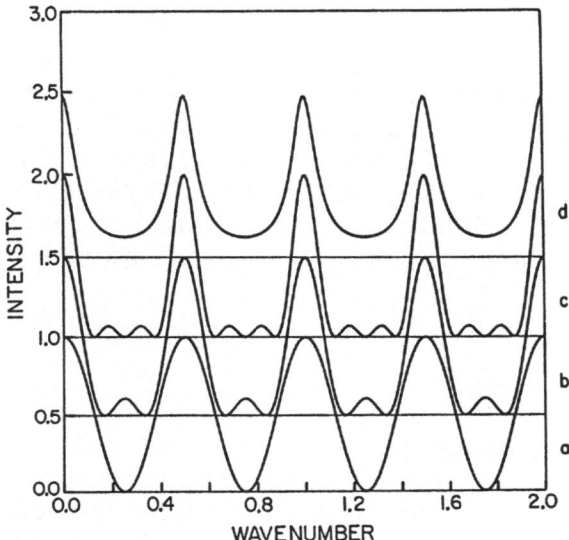

Fig. 4.11. Curves of normalized kinematic intensity as a function of wave number in a one-dimensional space occupied by a string of atoms of varying length: (*a*) only two atoms, no damping; (*b*) three atoms, no damping; (*c*) four atoms, no damping; (*d*) semi-infinite string of atoms with damping giving equivalent penetration of approximately three atoms

four atoms that of Fig. 4.11c. The width of the dominant peaks in Fig. 4.11a-c is $2\Delta k'_r \sim 2/(Na)$, where N is the number of equidistant atoms involved in the scattering. If we equate Na and λ_e, we find that

$$2\Delta k'_r \sim 2/\lambda_e \cong 2V_{oi}/[2(E + V_{or})]^{1/2} \; . \tag{4.45}$$

The width in energy of these dominant peaks is then

$$2\Delta E \cong 2k'_r \Delta k'_r \cong 2V_{oi} \; , \tag{4.46}$$

a well-known and very useful result, since it allows an easy estimation of V_{oi} simply by measuring the experimental peak widths. In fact, (4.46) is more exactly satisfied than our simplified derivation would imply. Incidentally, the secondary peaks that are shown in Figs. 4.11a-c are not present in reality when the truncated crystal slab with no damping is replaced by a semi-infinite crystal in which damping provides the limited electron penetration. The electrons then see an ill-defined thickness of the surface, the reachable depth of which is weighted in an exponentially decaying fashion. This is manifest in the reflected intensities as ill-defined secondary peaks at 1/2, 1/3, 1/4, etc., positions between the dominant diffraction peaks. Consequently, this smears out the secondary peaks and leaves an intensity similar to that shown in Fig. 4.11d.

4.3.3 Spin Effects

Diffracting electrons can interact with a surface through their spin in addition to their charge. This leads to relativistic effects in the case of spin-orbit interactions and to magnetic effects in the case of spin-spin interactions. The relativistic effects are observable in LEED intensities with surface atoms of large atomic number (e.g. $Z \geqslant 50$). In Spin-Polarized LEED (SPLEED), where the electron spin is monitored in addition to the diffracted intensity, both relativistic and magnetic effects are prominent. We shall discuss spin effects more fully in Sect. 5.13.

4.4 The Elastic Scattering Potential

The interaction potentials responsible for the elastic scattering of electrons at surfaces correspond to various forces acting between the diffracting electrons and the surface [4.8]. There is the electrostatic force between the diffracting electrons and all the point charges present at the surface. This force attracts the diffracting electrons toward the nuclei, producing a periodic potential of the form $-Ze/|\vec{r} - \vec{r}_n|$ near each nucleus $\vec{r}_n$, as illustrated in Fig. 4.12. The bound electrons in the atomic cores manage to screen the nuclei completely only outside the atomic radius. An additional contribution to the interaction potential is derived from the Pauli exclusion principle, which tends to keep electrons away from each other. This "exchange" contribution is important numerically but does not alter the qualitative picture given above. Note that in positron diffraction [4.15], no exchange term appears, since the Pauli exclusion principle does not apply. This is the major difference between LEED and Low-Energy Positron Diffraction (LEPD).

Assuming an antisymmetrical product wave function for the system of all N electrons of the surface, with wave functions given by $\psi_i(\vec{r}_i)$ (i = 1,...,N), and a LEED electron, with wave function $\varphi(\vec{r})$, and ignoring spin, the relevant Schrödinger equation becomes the following Hartree-Fock equation:

$$\left[-\frac{\hbar^2}{2m} \vec{\nabla}_j^2 - \sum_j \frac{Z_j e^2}{|\vec{r} - \vec{r}_{nj}|} + V_{sc}(\vec{r}) + \sum_i \int \frac{e^2 |\psi_i(\vec{r}_i)|^2}{|\vec{r} - \vec{r}_i|} \, d^3 r_i \right] \varphi(\vec{r})$$

$$- \sum_i \left[\int \frac{e^2 \psi_i^*(\vec{r}_i)\varphi(\vec{r}_i)}{|\vec{r} - \vec{r}_i|} \, d^3 r_i \right] \psi_i(\vec{r}) = E\varphi(\vec{r}) \ . \tag{4.47}$$

Here Z_j is the nuclear charge at position $\vec{r}_{nj}$ and $V_{sc}(\vec{r})$ is the potential due to the charge that screens the LEED electron (it is usually neglected), while the terms between the second set of square brackets are the exchange terms.

Since $\varphi(\vec{r})$ appears in the exchange terms (which are therefore nonlocal), (4.47) must be solved self-consistently, for example, iteratively. This involves rather large computations, common in Hartree-Fock molecular calculations.

This approach has been adopted in LEED by Pendry [4.8]. However, the computations can be reduced substantially by making the exchange terms local through the Slater approximation [4.16]:

$$V_{ex}(\vec{r}) = -3\alpha \left[\frac{3\rho(\vec{r})}{8\pi} \right]^{1/3} , \tag{4.48}$$

where $\rho(\vec{r})$ is the local density of bound electrons:

$$\rho(\vec{r}) = e\sum_i |\psi_i(\vec{r})|^2 . \tag{4.49}$$

The factor α in (4.48) can adopt various values between 2/3 and unity, depending both on the situation and on the author [4.16]. There is no general agreement on the "best" value. This so-called $X\alpha$ exchange potential is the one most widely used in LEED. Many other approximations and combinations of approximations have been proposed to describe the Hamiltonian of a LEED electron [4.17]. However, the issue of the best choice of a calculational scheme for the elastic scattering potential is far from having been solved completely and satisfactorily.

4.4.1 Atomic Potentials

Equation (4.47), as it stands or with any of the approximations referred to above, is not yet suitable for computation, partly because the wave functions ψ_i and the charge distribution ρ are not known. Since accurate wave functions for the free atoms are available and should be transferable to the surface environment, at least for the nonvalence electrons, one wishes to use these free-atom wave functions as a starting point for the LEED problem. It has been found that LEED is sufficiently sensitive to the valence electrons that the free-atom wave functions do not suffice to describe these in the surface environment. Various approaches have been used to obtain an improved description. Pendry [4.8] has proposed an increase in the population of the higher occupied free-atom orbitals to account for the overlap of electrons from neighboring atoms. Loucks [4.18], within the $X\alpha$ approximation in band-structure calculations, has constructed ρ by summing overlapping, spherically symmetrical charge densities from many nearby atoms, for the exchange term, and similarly summing overlapping electrostatic potentials from surrounding atoms. A more sophisticated approach is that of Moruzzi et al. [4.19], who, within the local-density functional theory of exchange, have performed self-consistent band-structure calculations for all crystal electrons. This produces a one-electron potential that can be used for the LEED electrons. In general, band-structure potentials have performed well in LEED.

One additional approximation has been applied in virtually all LEED work to date to make the computation more tractable: spherical symmetry has been assumed around each atom within a sphere that does not overlap similar spheres surrounding the neighboring atoms. With this assumption, (4.47) can be reduced to a one-dimensional differential equation when solved in terms of "partial waves". These are solutions consisting of the product of a radial function $R_l(r)$ (to be obtained by numerical integration of the radial differential equation) and an angular function that is a spherical harmonic $Y_{lm}(\theta,\varphi)$ of angular momenta l and m. The radial differential equation is

$$-\frac{\hbar^2}{2m}\left(\frac{1}{r^2}\right)\frac{d}{dr}\left(r^2\frac{dR_l(r)}{dr}\right) + \frac{\hbar^2 l(l+1)}{2mr^2}R_l(r)$$

$$+\left(-\frac{Ze^2}{r} + V_{sc}(r) + V_{ex}(r)\right)R_l(r) = ER_l(r) \ . \tag{4.50}$$

In an ionic material an additional potential term corresponding to a Madelung correction should be included in this equation.

The assumption of atomic spherical symmetry can be risky at surfaces. Although it has worked well in many LEED investigations, this remains a controversial assumption, especially for non-close-packed layers of adsorbed atoms and molecules. However, LEED electrons, with kinetic energy large compared to the variations in the potential energy they can encounter in bonding regions of a crystal surface (i.e. away from the atomic cores), have been found to be relatively insensitive to the electronic distribution in those regions. This insensitivity requires LEED energies of at least 10 to 30 eV, depending on the openness of the lattice. At these energies the wavelength is small compared to the distance scale of potential variations.

We have now obtained the following model of the LEED problem. Each atom of the surface is represented by a spherical potential within a sphere (called the muffin-tin sphere) that does not overlap neighboring spheres. The idea is to solve the Schrödinger equation (4.50) within each sphere separately, and then to match these solutions to the local solutions in the remaining parts of space. Consequently, we must turn our attention to these other parts of space: the interstitial regions between spheres, and the transition regions between layers of different composition or between the surface and vacuum.

4.4.2 The Muffin-Tin Constant

In relatively close-packed materials (rather more compact than, for example, silicon), the interstitial regions have potential variations that are small compared to LEED energies, namely at most a few electron volts in the worst cases. It is then quite justifiable to assume a constant value of the potential in those regions. This assumption, together with the spherically symmetrical atomic potentials, constitutes the well-known muffin-tin model and is illus-

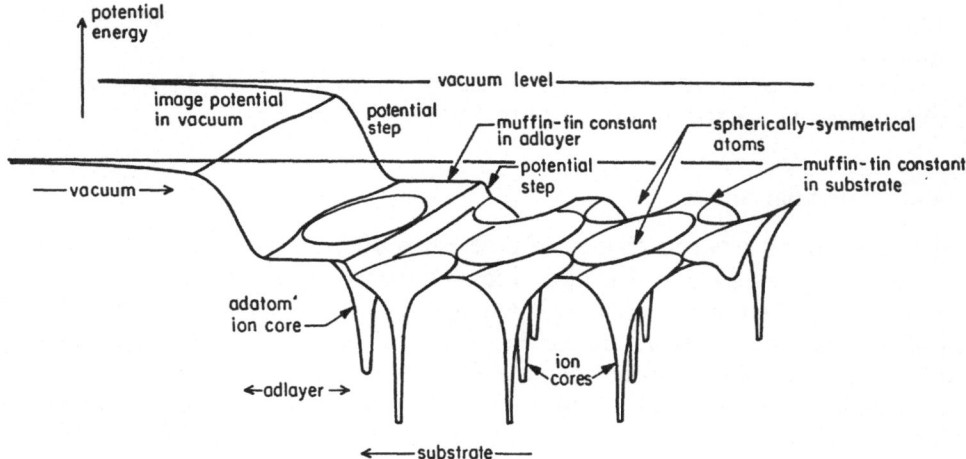

Fig. 4.12. Sketch of muffin-tin potential at a surface, linked through a layer of adatoms to the vacuum. Various terms in common usage are indicated

trated in Fig. 4.12. The great advantage of the constant interstitial potential (or "muffin-tin constant") is derived from the fact that waves propagating in these regions have a simple form, for example, that of plane waves, and this property is very much exploited in the theory. The value of the muffin-tin constant can be calculated together with the spherical potentials, with respect to an internal reference energy, such as electronic-energy levels.

However, in LEED the muffin-tin constant must be referred to an external-reference energy, such as the zero-energy level in vacuum, since this relation determines the change in kinetic energy that a LEED electron undergoes on entering or leaving the solid through the surface. As indicated earlier, this energy difference is influenced by the charge distribution at the surface. Present theories are not routinely capable of calculating such effects sufficiently accurately to be used as input to LEED calculations (an accuracy of approximately 1 eV is required). Rather, the muffin-tin constant is generally treated as an adjustable parameter that is fit to the experiment during a structural optimization. An exception occurs when the muffin-tin constant can be inferred from that of another surface through measurement of a work-function change. Thus, in the adsorption of an overlayer, the new muffin-tin constant in the substrate is just the sum of the "theoretical" value fit before adsorption and the experimental work-function change. Typical values of the muffin-tin constant range from 5 to 15 eV, depending on the material and on the method of construction of the spherical potentials. Little work has been carried out with the aim of optimizing the muffin-tin constant independently in different layers (it has been done occasionally for the case of one atomic monolayer adsorbed on a metal substrate), since it appears that at the usual LEED energies, layer-dependent muffin-tin values are of limited importance. This issue remains largely untested, however, for molecular overlayers.

The concept of the muffin-tin constant is closely related to that of the inner potential, as will be discussed more fully in Sect. 4.6.

4.4.3 Potential Steps

The transitions between regions of different muffin-tin constants give rise to "potential steps", see Fig. 4.12, that influence propagating electrons in several ways. First, there is a change in kinetic energy at a potential step. If such a potential step has no structure parallel to the surface, this results in *refraction* of the propagating electronic plane waves. Consequently, there is momentum conservation parallel to the surface, but not perpendicular to the surface. Due to its importance, all LEED calculations include refraction of the LEED electronic waves. Second, there is *reflection* at any potential step and therefore also reduced transmission. When the kinetic energy of the electron is sufficiently large compared to the potential-step height, the reflection is negligible. This occurs for energies larger than typically 30 eV, assuming that the potential step is not too abrupt, namely that it extends over at least one Ångstrom in the direction perpendicular to the surface (which seems to be a realistic assumption, judging from theoretical estimates and electron-resonance studies, cf. Sect. 5.12). A number of LEED calculations have included reflection and transmission corrections for potential steps, but for structural studies these corrections are usually omitted. Third, if the potential step has structure parallel to the surface (this structure would have the periodicity of the surface lattice), one would obtain *diffraction* of incident plane waves into various directions just as with any other periodic scattering layer. There is no evidence yet that this effect should be included in any LEED calculation, although it may well turn out to be important in the case of non-close-packed surfaces or overlayers at lower kinetic energies.

4.5 Atomic Scattering

With the muffin-tin model of the elastic scattering potential of a crystal surface, it is natural to think in terms of plane waves and spherical waves in the interstitial regions. These are the waves which should be matched to the solutions of the radial Schrödinger equation (4.50) within the muffin-tin spheres. Using plane waves is advantageous in connection with the fact that a plane wave will be incident on the surface, and also in connection with the fact that the two-dimensional periodicity of the surface will diffract this incident wave again into plane waves. The spherical waves are the natural choice to describe the scattering by the individual atoms, since these are assumed spherical. We shall therefore need to understand the scattering properties of both spherical and plane waves by spherical potentials, and that will be the subject matter of this section.

4.5.1 Spherical-Wave Scattering

First we examine the scattering of a spherical wave by a spherical potential of finite radius (the muffin-tin radius), outside of which the potential is constant. In the absence of this scattering potential, i.e. with a constant potential throughout, a solution of (4.50) is the Bessel function j_l for any positive integer l. (The complete solution, of course, also contains spherical harmonics Y_{lm} as factors multiplying the Bessel functions.) Such a solution may be written as

$$j_l(kr) = \tfrac{1}{2}\left[h_l^{(1)}(kr) + h_l^{(2)}(kr)\right] .$$ (4.51)

Here, $h_l^{(1)}$ and $h_l^{(2)}$ are the spherical Hankel functions of the first and second kind, respectively, and $k = [2(E + V_{or})]^{1/2}$ (E being the kinetic energy of the incoming electron, and atomic units are used). The asymptotic form of this solution is

$$j_l(kr) \xrightarrow[r\to\infty]{} i^{-(l+1)}\frac{\exp(ikr)}{kr} + i^{l+1}\frac{\exp(-ikr)}{kr} ,$$ (4.52)

which exhibits the fact that j_l contains an incoming wave and an outgoing wave of equal magnitude, thus satisfying current conservation.

The switching on of a spherical scattering potential (it is real since we ignore inelastic effects here) will not violate current conservation, and, consequently, the outgoing wave in the region outside the muffin-tin radius retains the same magnitude and thus can only be phase-shifted. The solution j_l of (4.51) becomes

$$\tfrac{1}{2}\left[\exp(i2\delta_l)h_l^{(1)}(kr) + h_l^{(2)}(kr)\right] ,$$ (4.53)

where the real number δ_l is called the phase shift. Note that the outgoing wave has the same value of l as the incoming wave, since the spherical nature of the scattering potential cannot change the angular momentum of the incident spherical wave.

The reason for the inclusion of the phase shift is easy to understand qualitatively. For convenience, we may think of the analogy of a plane wave passing over a square well of an arbitrary depth. Since within the well the wave has a shorter wavelength than outside, the phase of the emerging wave has advanced faster than would be the case in the absence of the well. The excess in phase over the unperturbed case is the phase shift. A resonance effect may and does occur as well, as is familiar with the square well. At certain energies the wave reflects back and forth between the inner walls of the well and is reinforced by constructive interference or is weakened by destructive interference, thereby producing maxima and minima in the amplitude of the emerging wave, with corresponding variations in the phase shift. Thus, phase shifts can be strongly energy-dependent, and, in addition, they include multiple scattering effects internal to the scatterer, i.e. internal to a single atom in our case. For comparison, at HEED energies, multiple-scattering within the atoms is negligible (one can use the Born approximation and Fourier transform the potential between incoming and outgoing plane waves), but the scattering amplitude is not a constant. In particular, it depends on $|\vec{s}| = |\vec{k}_{out} - \vec{k}_o|$ [4.5].

Returning to (4.53), the scattering potential has generated the scattered wave

$$\tfrac{1}{2}\left[\exp(i2\delta_l) - 1\right]h_l^{(1)}(kr) , \qquad (4.54)$$

which is just the difference between (4.53) and (4.51). The prefactor of $h_l^{(1)}$ in (4.54) is the amplitude of the scattered wave for unit unscattered wave j_l. The amount of scattering is more commonly expressed as the t-matrix element, namely

$$
\begin{aligned}
t_l &= -\frac{\hbar^2}{2m}\left(\frac{1}{2ik}\right)\left[\exp(2i\delta_l) - 1\right] \\
&= -\frac{\hbar^2}{2m}\left(\frac{1}{k}\right)\sin\delta_l\exp(i\delta_l) .
\end{aligned}
\qquad (4.55)
$$

4.5.2 Plane Wave Scattering

To determine how a plane wave $\exp(i\vec{k}\cdot\vec{r})$ with $k = |\vec{k}| = [2m(E + V_{or})]^{1/2}/\hbar$ is affected by the same potential, we use the spherical-wave expansion [4.20]

$$\exp(i\vec{k}\cdot\vec{r}) + \sum_{l=0}^{\infty}\sum_{m=-l}^{l} 4\pi i^l j_l(kr)Y_{lm}^*(\vec{k})Y_{lm}(\vec{r}) . \qquad (4.56)$$

Replacing the unscattered waves j_l of (4.56) by the perturbed waves of (4.53), in order to represent the effect of turning on the scattering potential, we obtain the resulting wave

$$
\begin{aligned}
&\exp(i\vec{k}\cdot\vec{r}) + \sum_l i^{l+1}(2l + 1)\sin\delta_l\exp(i\delta_l)P_l(\cos\theta)h_l^{(1)}(kr) \\
&\quad = \exp(i\vec{k}\cdot\vec{r}) - \frac{2mk}{\hbar^2}\sum_l i^{l+1}(2l + 1)t_l P_l(\cos\theta)h_l^{(1)}(kr) ,
\end{aligned}
\qquad (4.57)
$$

where we have obtained Legendre polynomials $P_l(\cos\theta)$ involving the scattering angle θ between $\vec{k}$ and $\vec{r}$, by applying the relation

$$P_l(\cos\theta) = \frac{4\pi}{2l + 1}\sum_{m=-l}^{l} Y_{lm}^*(\vec{k})Y_{lm}(\vec{r}) . \qquad (4.58)$$

The asymptotic form of the wave of (4.57) for $r \to \infty$ is

$$\exp(i\vec{k}\cdot\vec{r}) + f(\theta)\frac{\exp(ikr)}{r} , \qquad (4.59)$$

i.e. it is composed of the original incident plane wave and a scattered spherical wave with the angular distribution

$$f(\theta) = -4\pi\sum_{l}(2l + 1)t_l P_l(\cos\theta) \; . \tag{4.60}$$

An interesting feature of (4.57) is that, given the phase shifts δ_l, we can calculate the scattering by the muffin-tin sphere of any wave incident on it, without having to solve the Schrödinger equation *inside* the sphere repeatedly for different incident waves. Therefore, in practice, one separates the problem of obtaining phase shifts from that of calculating the scattering in the surface lattice. The phase shifts can be obtained once for a given atomic element and tabulated for multiple use in many different LEED calculations. Thus, in the LEED calculations themselves, one does not determine the form of the wave function within the muffin-tin spheres, but only its form between these spheres. In other problem areas, such as Angular-Resolved Photoelectron Spectroscopy, Auger-Electron Spectroscopy and High-Resolution Electron Energy-Loss Spectroscopy, the form of the electronic wave function *within* the muffin-tin spheres is required, since matrix elements between initial and final electronic states have to be evaluated. For such purposes this form can be recovered easily by summing all the incident partial waves, as determined by the LEED-like lattice scattering calculation, using the radial functions obtained from (4.50).

4.5.3 Phase Shifts

In this discussion, we have not yet shown how the phase shifts of (4.53-60) are obtained from the solution of (4.50). The procedure is simple: we require that the solution of the radial equation inside the muffin-tin sphere match smoothly to the solution outside. This can be achieved by equating the logarithmic derivatives just inside and just outside the muffin-tin radius r_m. If $R_l(r)$ is the solution inside, as obtained by numerical integration of (4.50) from the origin r=0 (with the boundary condition that the solution is not singular at the origin), and if (4.53) describes the radial wavefunction outside the muffin-tin radius, we obtain the condition [4.21]

$$\frac{R'_l(r_m)}{R_l(r_m)} = \frac{\exp(i2\delta_l)h_l^{(1)'}(kr_m) + h_l^{(2)'}(kr_m)}{\exp(i2\delta_l)h_l^{(1)}(kr_m) + h_l^{(2)}(kr_m)} \; . \tag{4.61}$$

This yields

$$\exp(2i\delta_l) = \frac{L_l h_l^{(2)}(kr_m) - h_l^{(2)'}(kr_m)}{h_l^{(1)'}(kr_m) - L_l h_l^{(1)}(kr_m)} \; , \quad \text{with} \quad L_l = \frac{R'_l(r_m)}{R_l(r_m)} \; . \tag{4.62}$$

Inelastic Effects

The results quoted thus far apply to the case where there are no inelastic effects, since the atomic potential in (4.50) was chosen to be real. Inelastic effects can be included through an imaginary part of this potential, without changing any of the above formulas, except that k becomes complex. However, the phase shifts will acquire an imaginary part as well. In particular, in (4.50) the outgoing wave will not only be phase-shifted by $\exp(i2\text{Re}\delta_l)$, but will also be damped by $\exp(-2\text{Im}\delta_l)$. Although this is the proper approach, in practice the inelastic effects are often not included in the phase shifts, but rather in the interatomic-wave propagation. The main reason is that as a consequence of this choice, the value of the imaginary part of the potential V_{oi} can remain unspecified until later and can thus be revised without having to recompute the phase shifts. The justification is that the choice of the treatment of V_{oi} is not critical; we shall return to this point later.

The Number of Phase Shifts

Before proceeding, it may be useful to give an idea of the numerical values of the phase shifts. The first question, however, is how many phase shifts are needed in practice. Equation (4.60) involves a sum over l from zero to infinity, which, hopefully, can be truncated at a tractable limit l_{max}. One way to estimate the value of l_{max} is to use the direct analogy of the scattering of a plane wave by a finite-sized object with diffraction through an aperture in a screen (Babinet's principle). An aperture of diameter d produces diffraction rings with rings of maximum intensity separated by an angle on the order of λ/d, where λ is the wavelength. Thus the entire polar range from $\theta=0$ to π will have approximately $\pi d/\lambda$ maxima. If we are to describe this distribution of intensity, we need Legendre polynomials $P_l(\cos\theta)$ with l ranging up to at least the number of maxima in this distribution, since a polynomial of order l has approximately l full oscillations. Thus, $l_{max} \sim \pi d/\lambda$. For an atom with a diameter of three Ångstroms and for electrons with a wavelength of one Ångstrom (corresponding to a kinetic energy of approximately 100 eV), we find $l_{max} \sim 9$. This value can be reduced somewhat in practice, because the outer part of the atom scatters relatively weakly, reducing the effective atomic diameter. Most LEED calculations can be performed satisfactorily at 100 eV with $l_{max} \sim 4$.

It also follows from our considerations concerning the maximum value of l that one needs a greater l_{max} for atoms with a larger radius. A relatively small l_{max} is sufficient for hydrogen atoms, whereas a large value is needed for alkali atoms. The transition-metal atoms all have rather similar radii and consequently require similar values of l_{max}.

It is obvious that l_{max} increases with energy, since the decreasing wavelength causes the diffraction pattern due to the single atom to be more finely structured. However, this is partly compensated by the fact that the effective atomic diameter becomes gradually smaller with increasing kinetic energy, since valence electrons become less and less influential.

The Magnitude of Phase Shifts

Concerning the magnitude of the phase shifts, we recall that resonances can occur in the scattering since the potential is sufficiently deep. Resonances imply that phases can vary by π or more. Note that the scattering amplitude of (4.55) is π-periodic in each phase shift (i.e. $\delta_l + \pi$ gives the same amplitude as δ_l). As a consequence, only the value $\delta_l \mathrm{mod} \pi$ is relevant (i.e. the reduced value in, say, the interval $[-\pi/2, \pi/2]$). Thus, the large possible variations in the phase shifts can produce all possible scattering strengths up to the theoretical limit: we are in the energy range of the maximum interaction between the electron and the atom. As a consequence, no approximation, such as the Born approximation, can be applied in calculating the phase shifts. Another consequence is that every atom in the periodic table requires its own independent set of phase shifts (not counting the additional differences that varying bonding arrangements may imply).

A quantity of some importance is the overall strength of the atomic scattering, namely the cross section for elastic scattering, since this will in large part determine whether the surface scattering is kinematic or dynamical, i.e. whether multiple scattering between atoms is negligible or not. There is a general, but not systematic, increase in the cross section as the atomic number Z increases. Thus, aluminum with $Z = 13$ is considered a relatively kinematic metal, whereas tungsten with $Z = 74$ exhibits considerably more multiple scattering. There are exceptions and also other considerations, however. Iridium with $Z = 77$ is noticeably more kinematic than platinum with $Z = 78$; there is no obvious simple reason for this particular difference, just that the partial waves tend to cancel each other for iridium but not for platinum. Strong scatterers with a low density, such as xenon in its bulk lattice, produce little interatomic multiple scattering because the interatomic distances are large compared to the mean free path, which induces relatively strong overall inelastic damping in the multiple-scattering path lengths. Conversely, there can be considerable multiple scattering between two small adjacent atoms, such as between carbon atoms in graphite or organic molecules, although carbon by itself is not a very strong scatterer.

The Angular Dependence of the Scattering Amplitude

Finally, we draw attention to the angular dependence of the scattered spherical wave. Figure 4.13 shows this in a polar plot of the scattering amplitude squared $|f(\theta)|^2$, (4.60), in the case of platinum. The scattering amplitude typically has a few lobes (these are essentially diffraction rings, or cones, due to the finite size of the atom, as discussed above), the prominent lobe occurring usually in the forward direction. As will be discussed later, this forward

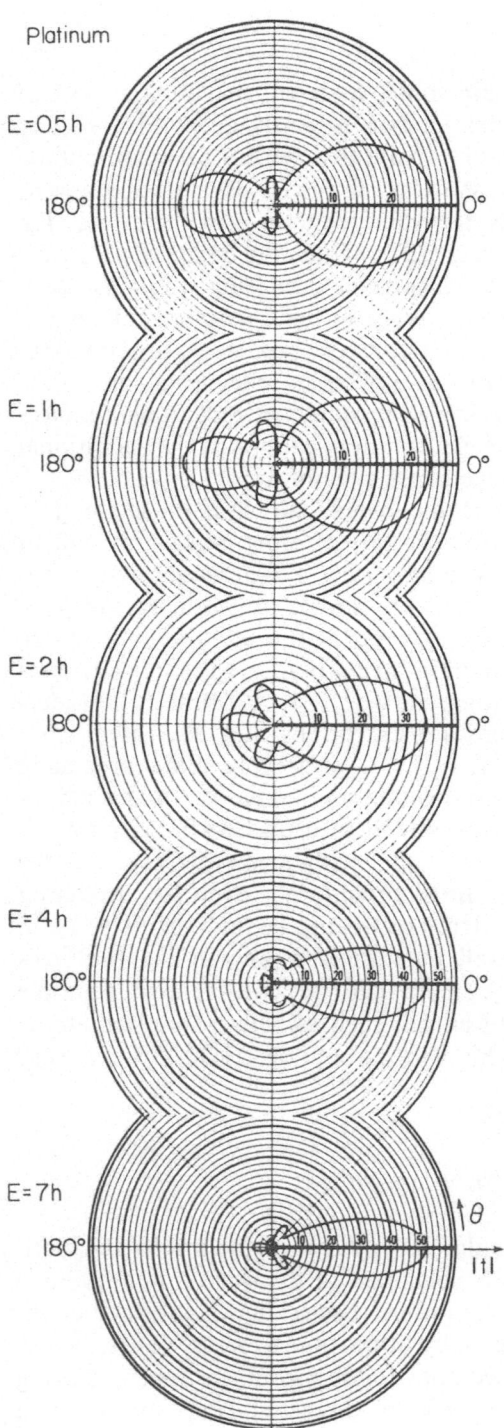

Platinum

E = 0.5 h

180°   0°

E = 1 h

180°   0°

E = 2 h

180°   0°

E = 4 h

180°   0°

E = 7 h

180°   0°

Fig. 4.13. Polar plot of the atomic scattering amplitude squared, $|f(\theta)|^2$, for platinum at different electron kinetic energies (1 hartree $\cong 27.18$ eV). Note the varying radial scale

scattering is quite important. There is also usually a smaller lobe pointing in the backward direction where most LEED observations are made.

4.5.4 Atoms as Point Scatterers

We have seen how the matching of the LEED wave functions inside and outside the muffin-tin spheres has led to a scattering description of the LEED problem. After calculating the phase shifts, we know how spherical and plane waves are scattered by the surface atoms, and we need no longer be concerned about the wave function inside the spheres. Furthermore, there is no explicit mention of the atomic size any more; the size is only implicitly present in the phase shifts. This fact allows us to adopt the following convenient picture of the scattering geometry. Let us assume that the atoms are now represented by *point* scatterers endowed with the scattering phase shifts derived previously for finite atoms. Everywhere between the point scatterers we may assume a constant potential (except at potential steps between atomic layers of different composition). Then the elementary waves in the surface can be everywhere regarded as free-space plane waves and spherical waves (with Bessel or Hankel radial functions), that change abruptly at the nuclear sites. This picture of point scatterers is not an approximation. Rather it is a description that produces results identical to those of the finite muffin-tin sphere representation. We have not changed the wave functions between the muffin-tin spheres, but only replaced the wave functions within the spheres by more convenient ones that produce the same effect. Figure 4.14 illustrates the procedure explictly.

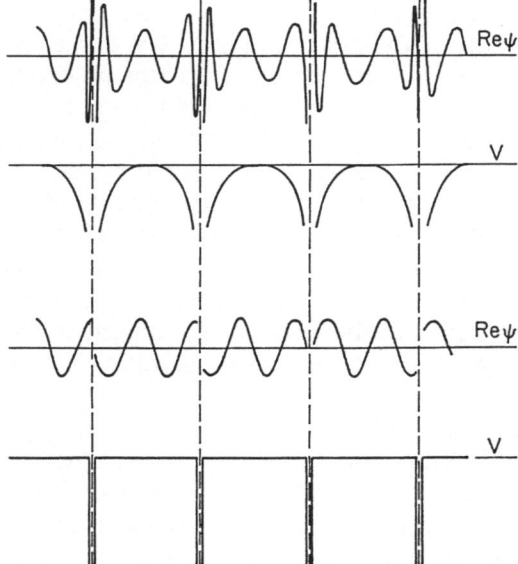

Fig. 4.14. (*Top pair of plots*) Realistic electronic wave function and atomic potential for a LEED electron in a periodic lattice (one-dimensional representation). (*Bottom pair of plots*) Equivalent wave function and potential in the point-scatterer model, showing the same wave function behavior as with the realistic wavefunction away from the atomic cores

Most LEED theories use the point-scatterer model for the atoms, since it is the natural result of scattering theory. This implies, for example, that the elementary spherical or plane waves will be propagated all the way from one nucleus to another nucleus using their simple mathematical form, such as $\exp(i\vec{k}\cdot\vec{r})$ for plane waves, rather than having to give these waves their actual complicated functional form inside the muffin-tin spheres.

Within the point-scatterer model, it is easy to justify that inelastic effects are usually not included in the atomic phase shifts, but rather in the spherical- and plane-wave propagation between atoms. For that purpose one imagines that the point scatterers take only the elastic effects into account, so that the inelastic effects must be included elsewhere. Since these take the form of a mean free path (or, equivalently, an imaginary part of the potential), it is natural to introduce the inelastic effect as a decay of propagating waves, achieved simply by allowing any vector $\vec{k}$ to have an imaginary part. In this fashion we allow damping of a wave to take place all the way from one nucleus to another, as desired.

4.6 The Inner Potential and the Muffin-Tin Constant

We have previously described the inner potential to be equal to the kinetic energy that an electron gains upon entering a surface. We have also noted that the muffin-tin constant is that value of the constant potential between muffin-tin spheres. These two quantities are not identical. In principle, the inner potential can be measured experimentally for any given surface: for example, through the shift in position of the Bragg peaks in I-V curves at high energies, assuming a kinematic model (in practice, the inner potential will be found to be energy dependent). The inner potential is a sort of spatial average over the actual potential felt by the LEED electrons, including the region inside the atoms. The muffin-tin constant, on the other hand, is not related to the interior of the atoms at all, and it is not measurable. It is rather a theoretical construct, the value of which actually depends on the method of construction of the muffin-tin potential. Another difference is that the inner potential is always referred to vacuum, whereas the theoretically calculated muffin-tin constant can only be referred to vacuum if the surface properties (surface dipole layer, etc.) are taken into account, as was pointed out earlier.

However, the two phrases *inner potential* and *muffin-tin constant* have been used interchangeably in the LEED literature. Therefore, we shall do so here as well. However, it must be remembered that, for example, the inner potential used in a kinematic fit to experiment bears no relationship to the muffin-tin constant derived from a dynamical fit to the same experiment, and in fact can differ by 10 eV or more.

As mentioned previously, one of the physical effects of the inner potential on the LEED electron that enters or leaves the solid is a change in kinetic energy. This is manifest as a more or less rigid shift of the energy scale. Therefore, since the value of the inner potential (or muffin-tin constant) is often not known a priori and is not easy to calculate, it is normally fit to experiment by simply shifting the theoretical energy scale until the diffraction peaks are aligned.

4.7 Temperature Effects

The thermal vibrations of surface atoms have a number of effects on electron diffraction. Several of these can be discussed easily within a kinematic

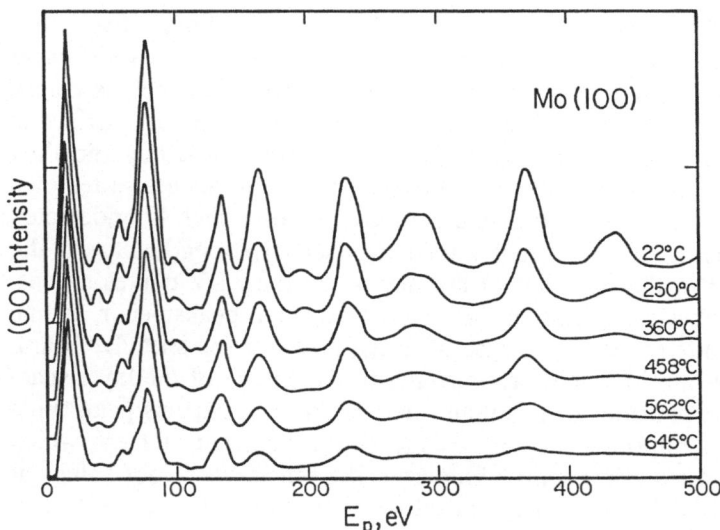

Fig. 4.15. Experimental I-V curves for Mo(100) at different temperatures (baselines are shifted for clarity). After [4.22]

framework. As the surface temperature is increased in a LEED experiment, the most obvious changes are that the diffraction spots (beams) broaden gradually, the spot intensities decrease, and the background intensity that is visible between spots increases.

These effects are due directly to positional disorder of the vibrating surface atoms. The spot broadening can be described as due to momentum exchange of the electrons with phonons. (Phonon-energy losses, although they are strictly speaking inelastic, are considered to be elastic in this context.) The decrease of spot intensities illustrated in Fig. 4.15 is readily understood as due to destructive interference effects between waves scattered by slightly disordered atoms, as described, for example, by the Debye-Waller factor that is familiar from X-ray Diffraction [4.1-3]. The background increase can be explained as thermal-diffuse scattering and corresponds to the intensity that is lost through the Debye-Waller factor. It can equally well be described as an increase of incoherent diffraction at the expense of the coherent diffraction into sharp beams.

These thermal effects are particularly evident in the case of order-disorder phase transitions of overlayers that produce superlattice spots when ordered. As a consequence of their importance, the study by LEED of this and other kinds of disorder will receive close attention in Chaps. 8-10. In this section we shall concentrate on the intensity decrease in the coherent diffraction that gives rise to sharp beams, i.e. on the Debye-Waller factor. This will provide a firmer basis for our earlier discussion of the Debye-Waller factor in Sect. 2.7.

Prior to considering the Debye-Waller factor, we can discuss two other observations that have been made in careful experiments. It has been found that the small thermal lattice expansion due to anharmonic effects is barely observable with LEED [4.22]. The lattice expansion parallel to a surface is expressed as a contraction of the two-dimensional reciprocal lattice and thus as a contraction of the spot pattern. This contraction is on the order of a fraction of one percent for practical temperature ranges, which is undetected in most experiments. The thermal lattice expansion perpendicular to the surface is manifest in the positions of diffraction peaks in I-V curves. A lattice expansion implies a decrease of the energies at which peaks occur, as one can most easily understand in the case of Bragg peaks in the specular beam, since their energies are inversely proportional to the square of the layer spacing. With lattice expansions of less than one percent, one expects peak shifts at energies of a few hundred electron volts to be on the order of 1 eV, which have been observed experimentally. However, the sensitivity is not sufficient to examine reliably differences between lattice expansions at surfaces and in the bulk material.

4.7.1 The Debye-Waller Factor

We shall summarize here the principle of the Debye-Waller factor in the kinematic limit: multiple-scattering effects will be treated in Sect. 5.11. The physical basis of the Debye-Waller factor is quite simple [4.8]. Suppose the

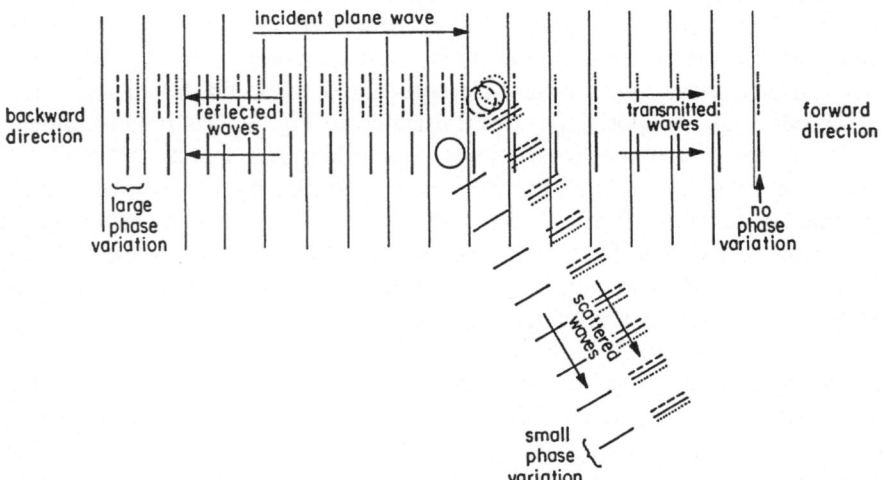

Fig. 4.16. Wave fronts in kinematic diffraction from two atoms. One atom is rigid for reference, while the other takes thermally-random positions around its equilibrium site. Waves scattered in Bragg directions are phase shifted by the random atomic positions. The magnitude of the phase shift depends on the scattering angle. The lines in the diagram show wave fronts, including phase-shifted wave fronts

two atoms in Fig. 4.16 form part of a periodic lattice and are at rest, and suppose a plane wave strikes them in such a way that Bragg reflections occur in the three scattering directions that are indicated, i.e. in the forward direction, near the backward direction (near 180° deflection) and in a third "sideways" direction. In this case, in each of the three directions the scattered waves from both atoms (and from the others in the solid) are in phase.

Now let one of these two atoms move slightly away from its equilibrium position, as in the presence of thermal vibrations. (We may assume a static picture of the lattice because atomic vibrational times are orders of magnitude longer than electron scattering times.) The forward-scattered wave is not influenced by this movement, whatever its direction, because the distance traversed by the wave is not changed. Consequently, the waves emanating from different atoms remain perfectly in phase. In other scattering directions, however, the wave emanating from the displaced atom is somewhat phase shifted. The phase shift is, on the average, largest for the backscattering direction because it implies larger scattering-path differences due to atomic displacements. The phase shifts reduce the coherence of the diffraction in the nonforward directions, thereby reducing the intensities measured in the Bragg directions (the intensity between Bragg directions increases as a result). Clearly, the effect is stronger for larger vibrational amplitudes and for smaller wavelengths. This decrease in intensities is quantified by the Debye-Waller factor $\exp(-2M)$ multiplying the intensities as $I = I_o\exp(-2M)$. The Debye-Waller factor is well known to have the form

$$\exp(-2M) = \exp[-|\vec{s}|^2 <(\Delta\vec{r})^2>] \ , \tag{4.63}$$

for small vibrational amplitudes, with the assumption of independent atomic vibrations, if $\vec{s}$ is the three-dimensional momentum transfer of the beam under consideration. Here the mean-square deviation $<(\Delta\vec{r})^2>$ may be anisotropic. With the Debye model of thermal vibrations and in the case of isotropic vibrations, (4.63) becomes

$$\exp(-2M) = \exp\left(-\frac{3\hbar^2|\vec{s}|^2T}{m_a k_B \theta_D^2}\right) \ , \tag{4.64}$$

where T is the temperature, m_a the atomic mass of the surface atoms, k_B the Boltzmann constant, and θ_D the Debye temperature, which measures the rigidity of the lattice with respect to vibrations. Notice that the Debye-Waller factor is unity in the forward direction ($\vec{s}$=0) and smallest in the backward direction ($|\vec{s}|$ maximal). This reinforces the relative predominance of forward scattering in atomic scattering, and generally strongly reduces LEED intensities, since in LEED $|\vec{s}|$ is usually largely due to near-backward scattering.

Since the Debye model is a simplified picture of lattice vibrations, the Debye temperature θ_D is not a well-defined measurable quantity. Thus, its value depends on the type of experiment performed to measure it (e.g.

specific-heat measurement, temperature dependence of diffraction intensity, see Sect. 2.7); and, within the Debye-Waller formalism, it also depends on the temperature itself. In the case of surfaces, one can imagine the further complication of a layer-dependent Debye temperature, since vibrations in surface layers certainly behave differently from those in the bulk material: enhanced and anisotropic vibrational amplitudes occur [4.8]. In addition, multiple scattering in LEED may seriously affect any attempt to measure θ_D at a surface. These complications exist in reality and have rendered the extraction of detailed information concerning thermal vibrations at surfaces unreliable, as several attempts have demonstrated [4.8,24,25].

It is true that the logarithm of the measured intensities of LEED diffraction spots, when plotted against the temperature T, produces remarkably straight lines from whose slopes θ_D can be estimated. This would seem to support the validity of (4.64), despite the approximations built into it. However, in so doing one immediately encounters a serious inconsistency. The value obtained for θ_D depends strongly on the electron energy used, in contradiction to the nature of θ_D as a supposed material constant. This can be seen clearly in Fig. 4.15 where some peaks, e.g. at 160 and 290 eV, decrease with increasing temperature much more rapidly than nearby peaks, leading to much smaller values of θ_D for those peaks. Multiple scattering is the major reason for this apparently anomalous behavior, while a layer dependence of θ_D also plays a role. We shall return to these issues in some detail in Sect. 5.11, equipped with multiple-scattering formalisms.

4.8 From Kinematic to Dynamical LEED

It is easily shown that a kinematic model of LEED is inadequate. For example, if the kinematic model were correct, an azimuthal I-φ plot of the specular (0,0) beam intensity would exhibit no dependence of I on the incidence azimuth φ. However, such an I-φ plot can show very large intensity variations. An I-V curve for a surface structure that is the ideal termination of a simple crystal lattice, such as a face-centered cubic lattice, should in the kinematic limit produce a well-ordered sequence of Bragg peaks at predictable energy intervals (Figs. 4.5,6). However, experimental measurements show the presence of many more peaks in addition to the Bragg peaks, which themselves are found at somewhat erratic energies. Finally, in the kinematic model it should be possible to extract from the temperature dependence of LEED intensities a Debye temperature for surface atoms, possibly anisotropic and layer dependent. Unfortunately, the strong energy dependence of measured "effective Debye temperatures" shows that other complications play an important role.

It should not come as a surprise that multiple scattering is the cause of the observed deviations from kinematic behavior. At one extreme of the energy scale, electron scattering at energies relevant to valence and conduction electrons was studied extensively long before the solution of the LEED problem, for example in band-structure calculations. At those very low energies, the existence of multiple scattering is so evident that the self-consistent solution in terms of Bloch waves (rather than simple plane waves) leading to band structures has long been well established [4.16]. In this context, self-consistency is equivalent to the exact inclusion of multiple scattering. At the other extreme, multiple scattering has long been known to be of serious concern at high electron energies [4.5,6], as in Electron Microscopy and gas-phase or crystal electron diffraction.

In this section we wish to introduce the dynamical effects in a gradual and simple fashion, starting from the kinematic principles that we have already presented, in order to allow the concepts and implications of multiple scattering to become clear. This can be accomplished while avoiding the obscurity that comes from mathematical formalisms. (The next chapter will discuss multiple scattering in a more rigorous fashion.) It will emerge that a simple qualitative understanding of the apparently complicated LEED intensities is eminently possible.

We shall begin by considering a very simple clean surface, in one dimension. It will soon become clear that the kinematic model, as used in X-ray Diffraction, violates current conservation and naturally forces the effects of multiple scattering upon us. Bragg peaks, for example, will be seen to be intimately related to the familiar band gaps of periodic solids. The erratic positions of Bragg peaks will also become understandable in terms of phase shifts, i.e. intra-atomic scattering. The appearance of extra peaks will be shown to be a three-dimensional effect due to the opening up of new scattering channels that are not allowed in the single-scattering limit.

4.8.1 Clean Crystals and Bragg Reflections in One Dimension

Let us first consider a one-dimensional space, one-half of which ($x>0$) is occupied by a crystal composed of a semi-infinite row of identical, equally spaced "atoms" (this represents a clean, ideal surface). An undamped quantum-mechanical wave $\exp(ikx)$, where k is the wave vector, is incident on the crystal from $x = -\infty$. This wave is scattered (toward $x = -\infty$ and $x = +\infty$) by each atom, as described by complex reflection and transmission coefficients r and t.

The weak-scattering, kinematic limit is characterized by a very small value of $|r|$ and essentially total transmission $|t| = 1$ [in that limit the phase $\arg(t)$ of the transmission coefficient must vanish for wave function continuity]. Our one-dimensional crystal reflects a part of the wave $\exp(ikx)$ into a wave given by

$$\sum_{j=0}^{\infty} r\exp(ik2aj)\exp(-ikx) = \frac{r\exp(-ikx)}{1 - \exp(ik2a)} \; , \qquad (4.65)$$

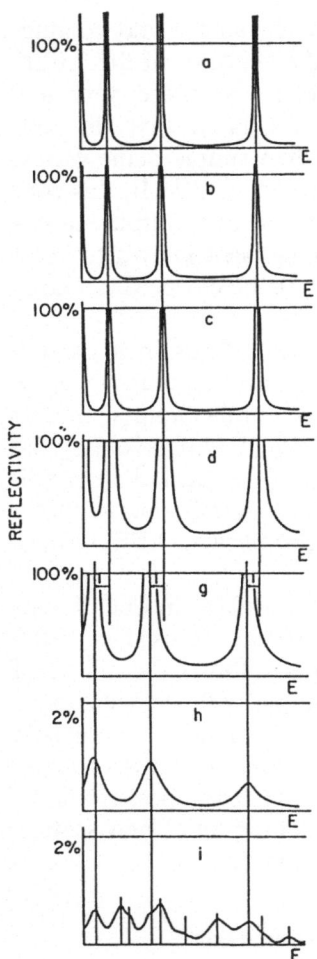

REFLECTIVITY

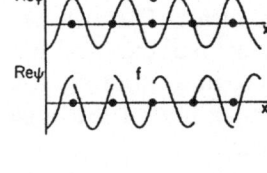

Fig. 4.17a – i. LEED I-V curves as a function of increasing exactness from kinematic to dynamical, for a clean surface in a one-dimensional space. See text for details

where a is the lattice parameter, and x=0 defines the first atom of the surface. Generally, the amplitude of the reflected wave is small, but strong reflection takes place when the Bragg condition $k2a = n2\pi$ (n integer) is satisfied, because constructive interference occurs only then. An illustration of the energy dependence of this reflection amplitude (squared) is given in Fig. 4.17a [energy E and wave vector k being related by $E = (\hbar k)^2/2m$]: infinitely sharp Bragg reflection peaks occur. These peaks are infinitely sharp because the number of contributing atoms is infinite.

Incomplete Transmission

Moreover, the reflection amplitude is infinite at each Bragg condition, clearly violating current conservation. Current conservation is actually always violated due to the assumption of total transmission ($|t| = 1$) through a scatterer. For an improved physical description, we must accept that $|t| < 1$. Equation (4.65) may be rewritten easily for this case, if we recognize that $r \cdot \exp(ik2aj)$ represents the round-trip amplitude factor for an excursion from the crystal surface to atom $j + 1$ and back. This quantity should be replaced by $rt^{2j}\exp(ik2aj)$ because the wave is transmitted twice through each intervening atom. In this case, we have, rather than (4.65)

$$\sum_{j=0}^{\infty} rt^{2j}\exp(ik2aj)\exp(-ikx) = \frac{r\exp(-ikx)}{1 - t^2\exp(ik2a)} \ . \tag{4.66}$$

The new reflection amplitude, given by (4.66), still has reflection maxima at the same energies, but they are not of infinite height, since $|t| < 1$. (However, they may still exceed 100%!.) Nor are these peaks infinitely sharp any more. Assuming that t is real [arg(t) = 0], the resulting energy dependence of the reflection amplitude has the behavior shown in Fig. 4.17b.

Multiple Reflections

Current conservation has, however, not yet been introduced quite correctly in (4.66). At each of the transmissions mentioned above, the lost current goes into a new reflected wave, which in turn can reflect, etc. In short, multiple reflections take place, at each of which current conservation should be respected. In fact, current conservation will be satisfied only if all multiple reflections (to infinite order) are included. This cannot be conveniently expressed in a relation like (4.66). The result is given by the Bloch wave theory, which treats the multiple scattering self-consistently. One consequence that follows from the Bloch wave theory is that multiple scattering leaves the reflection peaks essentially unshifted with respect to the kinematic case and does not appreciably change their widths (cf. Fig. 4.17c). However, the peaks are now flat-topped, with a height of unity; the multiple reflections conspire exactly to produce total reflection in the neighborhood of each Bragg condition. This merely corresponds to the familiar band gaps of a crystalline material. At electron energies within a band gap no propagation can take place, and total reflection must occur (whether in the bulk or at the surface). Within the bands an electron can penetrate a crystal infinitely far (if we ignore inelastic scattering), and hence there is only partial reflection. Thus we see clearly that Bragg reflections are simply connected with band gaps of the bulk material, and we are reminded that band gaps are simply due to complete constructive interference in reflection, forcing all current to change direction. The peak widths must increase as the scattering strength is increased, corresponding to wider band gaps, as shown in Fig. 4.17d.

Phase-Shifted Transmission

Although we have assumed a real transmission coefficient t, actually it is normally complex. This is necessary in order to satisfy wave function continuity. Inserting this fact in (4.66) does not affect the reflection peak shapes, but the condition for a reflection maximum (constructive interference) changes to

$$k2a + 2\arg(t) = n2\pi \quad \text{(n integer)} .\qquad\qquad (4.67)$$

What has happened is that the "optical path" of the wave (i.e. the path length measured with the wavelength as a unit) has been modified by a phase shift arg(t) at each transmission through an atom. This may be understood, in the case of attractive atomic potentials, as a temporary acceleration of the electron through the atom, causing the electron to emerge behind the atom somewhat ahead of its original phase (the wavelength is momentarily shortened through an increased kinetic energy, thereby making the phase of the scattered electron advance faster). The effect is illustrated in Fig. 4.17e,f. For an attractive atomic potential, the electron wave effectively is "telescoped" to some extent due to the shortened wavelength, resulting in an effective-average-wavelength reduction in the crystal. (The effect is dramatized here by making the wave discontinuous at the nuclei rather than smoothly varying in wavelength; see also Fig. 4.14.) To keep the Bragg condition satisfied, this shortened wavelength must be increased through a decrease of the energy, and hence Bragg peaks are normally found at energies substantially below the kinematic prediction (see Fig. 4.17g). Note that the phase of the reflection coefficient r has no effect on this result, apart from a relatively small indirect effect through multiple scattering.

The physical origin of this "transmission phase shift" can be split into two components. The first is the familiar inner potential, the effect of which can be more directly described as a rigid shift of the energy axis (insofar as the small, gradual energy dependence of the inner potential is ignored) by approximately 5 to 15 eV, depending on the material. The second is due to the scattering of an electron within a single atomic core, i.e. the diffraction of the electron through the complicated "electrostatic lens" which an atom is. Multiple scattering within an atom can take place, and in fact strong resonance effects can and do occur (Sect. 4.5.3), giving rise to substantial transmission phase shifts. These resonances are directly analogous to those one finds when a wave passes over a square potential well. The resonance effects in atoms are energy dependent, and therefore the transmission phase shifts are also energy dependent. As a result, peaks in I-V curves occur not only well below their kinematically expected energies (due to the inner potential) but also scatter about their inner-potential-corrected positions (due to intra-atomic resonances). This scatter is largely toward even lower energies. In dynamical LEED calculations, the effects of intra-atomic multiple scattering are contained in the partial-wave phase shifts. Calculated I-V curves exhibiting these effects are shown in Fig. 4.18.

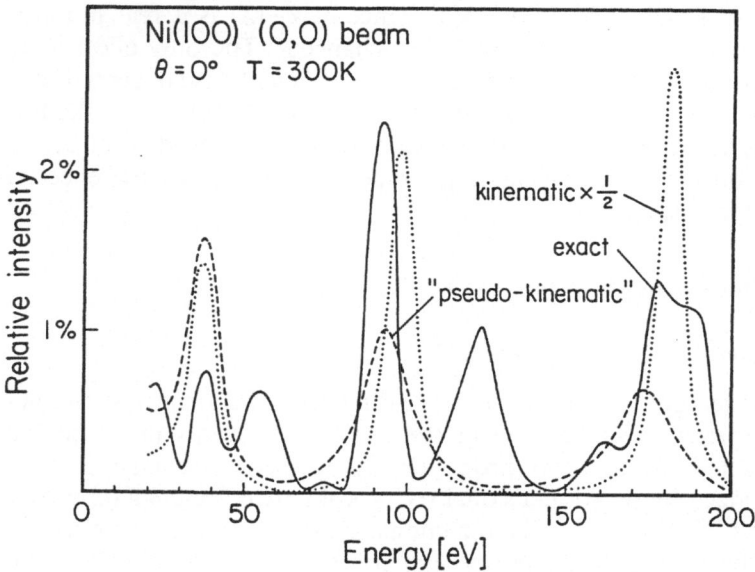

Fig. 4.18. Calculated I-V curves for the Ni(100) specular beam at normal incidence (T = 300 K) at three levels of accuracy: in the kinematic limit, with forward-scattering correction (otherwise kinematic: "pseudokinematic"), and exact. The same inner potential is used in the three cases

Transmission Phase and Perturbation Methods

The transmission-phase effect of the intra-atomic resonances is one of the major causes of the difficulty that has been experienced in LEED theory with perturbation methods based on the number of scatterings undergone by an electron. A good perturbation method should have a starting approximation that needs only small corrections in subsequent terms of higher order. Figures 4.17e and f show the problem clearly in the case of LEED. Without correction for the transmission phase shift, the wave exp(ikx) of Fig. 4.17e is, after a few lattice spacings, a very poor approximation to the actual wave, which is more like the one shown in Fig. 4.17f. The lowest order of a simple perturbation theory in LEED includes only reflection and no transmission phase factor. Consequently, it produces peaks at incorrect energies (even after an inner-potential correction). The next orders of perturbation must both destroy those peaks and create new peaks at new energies, since that is what is involved in shifting a peak; hardly a perturbation.

Temperature Effects

Some temperature effects may also be understood easily within our simple picture of the electron-diffraction process. The Debye-Waller factor reduces the reflection coefficient r to r·exp(−M). Since exp(−M) is a real number, the phase of the reflection coefficient is not changed. The only effect is an intensity reduction. As far as the transmission coefficient t is concerned, it is not changed at all by the Debye-Waller factor since exp(−M) = 1 in the forward scattering direction, see (4.64). Thus arg(t) is not changed, and, consequently, peaks do not shift with varying temperature (excepting the effect of anharmonic lattice expansions).

4.8.2 Three-Dimensional Effects

The LEED Pattern is Always "Kinematic"

The major new feature introduced by three-dimensional (in contrast to one-dimensional) crystals is an increase in the number of beams in which the elastically diffracted electrons can propagate. At surfaces, one has the set of beams characterized by two-dimensional reciprocal lattice vectors $\vec{g}$ (observed as spots on the LEED screen). The directions of these beams are determined uniquely by the two-dimensional periodicity and the wavelength. The scattering mechanism, whether kinematic or dynamical, does not affect the directions of diffracted beams. (This answers a frequently-asked question.) This is because the beam directions are determined solely by the relative phase of reflected waves emanating from scattering atoms equivalent under the periodic translations of the surface. This relative phase depends on the unit-cell dimensions, but not on the scattering mechanism. The scattering mechanism, no matter how complicated, influences only the absolute phase (as well as the intensity) of the reflected waves, a quantity that disappears when relative phases are considered. Therefore, LEED spots always have positions that can be determined kinematically.

Generalized Bragg Conditions

Each layer in the clean three-dimensional crystal surface diffracts a beam $\vec{g}$ into a beam $\vec{g}'$ with a reflection coefficient $r_{\vec{g}'\vec{g}}$ or a transmission coefficient $t_{\vec{g}'\vec{g}}$. Each beam has its particular wave vector $\vec{k}_{\vec{g}}$, and $k_{\vec{g}z}$ will denote the component of $\vec{k}_{\vec{g}}$ perpendicular to the layers. Now (4.66) generalizes immediately to the following form, if we consider scattering only from $\vec{g}$ to $\vec{g}'$:

$$\sum_{j=0}^{\infty} r_{\vec{g}'\vec{g}}(t_{\vec{g}'\vec{g}'})^{j}(t_{\vec{g}\vec{g}})^{j}\exp(ik_{\vec{g}'z}aj)\exp(ik_{\vec{g}z}aj)\exp(-i\vec{k}_{\vec{g}'}\cdot\vec{r})$$

$$= \exp(-i\vec{k}_{\vec{g}'}\cdot\vec{r})r_{\vec{g}'\vec{g}}\left[1 - t_{\vec{g}'\vec{g}'}t_{\vec{g}\vec{g}}\exp\Big[i(k_{\vec{g}'z} + k_{\vec{g}z})a\Big]\right]^{-1}, \tag{4.68}$$

where "a" is the layer separation.

We can now, as in one dimension, generalize the familiar Bragg reflection conditions used in X-ray Diffraction by including transmission phases induced by the strong potentials in the atomic layers, i.e.

$$k_{\bar{g}z}a + \arg(t_{\bar{g}\bar{g}}) + k_{\bar{g}'z}a + \arg(t_{\bar{g}'\bar{g}'}) = n2\pi \quad \text{(n integer)} \tag{4.69}$$

gives conditions for maximum reflection from beam $\bar{g}$ into beam $\bar{g}'$. This equation is rigorous for kinematic (i.e. single) scattering that includes zero-angle forward scattering $t_{\bar{g}\bar{g}}$. In the presence of multiple scattering it holds inasmuch as, in diffraction from a single atomic layer, the zero-angle forward-scattered beam is much stronger than all other diffracted beams. Due to the terms of (4.69) involving transmission, the reflection maxma are shifted substantially from the weak-scattering positions by amounts that are, as a consequence of resonances, very much energy dependent, just as in our one-dimensional example.

Intermediate Beams and "Multiple-Scattering" Peaks

In the limit of weak multiple scattering, the various reflected beams have intensity maxima only when (4.69) is satisfied, $\bar{g}$ being the incident beam, i.e. $\bar{g} = \bar{0}$. These are essentially the Bragg peaks. However, strong multiple scattering can, however, generate additional maxima that have the same general appearance and often similar intensities as the single-scattering ones. This is easily visualized in terms of intermediate beams. The incident beam $\bar{0}$ can scatter into an intermediate beam $\bar{g}_1$, which in turn can scatter into an emerging beam $\bar{g}'$. More successive intermediate beams are of course possible, but they tend to be increasingly weaker due to the additional scatterings that are involved. Either of the two scatterings, $\bar{0}$ to $\bar{g}_1$ and $\bar{g}_1$ to $\bar{g}'$, will at particular energies satisfy the condition of (4.69) with the proper $\bar{g}$'s substituted, thereby generating peaks in the emerging beam $\bar{g}'$ at energies different from those predicted by the single-scattering conditions (directly from $\bar{0}$ to $\bar{g}'$). Therefore, we may say that a peak occurs whenever, in a chain of scatterings, one of the scatterings satisfies the "Bragg condition" of (4.69). Hence, the abundant set of "multiple-scattering" peaks that is so characteristic of LEED I-V curves is explained. This is also responsible for some of the difficulties encountered in treating these I-V curves with kinematic (single-scattering) methods of data averaging and data reduction.

4.8.3 Overlayer Effects

To this point, we have analyzed only "clean" surfaces. The additional effects resulting from a more complicated termination of the bulk structure at the surface can be illustrated with the case of an overlayer adsorbed on a clean surface.

We consider an overlayer of monatomic thickness adsorbed at a distance d from the clean substrate. The total electron reflection coefficient R can be regarded as being composed of interfering reflections from the overlayer and the substrate. In one dimension especially, this is a simple situation. The substrate reflection coefficient R_s and the overlayer reflection and transmission coefficients r and t combine to give

$$R = r + \exp(2ikd)tR_s[1 - \exp(2ikd)rR_s]^{-1}t \ . \tag{4.70}$$

An analogous matrix equation holds for three dimensions. The factor $[1 - \exp(2ikd) rR_s]^{-1}$ describes multiple scattering between substrate and overlayer, as proved by a geometric expansion. This factor usually plays no qualitative role in practice due both to inelastic effects and the not so large values of $|r|$ and $|R_s|$ (typically 0.1 and 0.5 to 0.1, respectively). Neglecting this multiple-scattering factor, (4.70) has interference maxima between overlayer and substrate when

$$2kd + 2\arg(t) + \arg(R_s) - \arg(r) = n2\pi \quad \text{(n integer)} \ . \tag{4.71}$$

Again, we have energy-dependent shifts of the maxima away from simple geometric Bragg-like conditions. These maxima are superimposed on the strong energy dependence of the substrate reflection R_s (a complex number) in such a way that only by including the relevant scattering phase shifts can one reproduce experimental data faithfully. This is true with one- as well as with three-dimensional crystals. Interference also produces intensity minima in addition to maxima, and total destruction can occur easily. Thus, zero intensities can be observed, and Figs. 4.17h and i illustrate this effect. Figure 4.17h represents the substrate reflectivity $|R_s|^2$, with peaks smoothed compared to Fig. 4.17g, due to the inclusion of inelastic effects. Addition of the overlayer and inclusion of the three-dimensional effects produces Fig. 4.17i, showing additional maxima due to new diffraction channels.

The interference between overlayer and substrate is of course strongly geometry dependent and explains the sensitivity of I-V curves to the surface-layer position. The maxima due to this interference generally have a larger width than those due to the substrate alone, since the width is inversely proportional to the number of interfering components, only two in this case. However, the general appearance of the I-V curves remains essentially the same.

With more complicated surface structures the above considerations still apply, but it is clear that it becomes rather difficult to predict the result of all the different interference conditions, especially in three dimensions, without doing the actual calculations. Thus, I-V curves become difficult to understand directly. Although the basic mechanisms of diffraction are simple, the multitude of simultaneous events obscures that simplicity.

5. Dynamical LEED Theory

The last chapter has shown the limitations of the kinematic approximation in LEED. In this chapter we shall describe the dynamical or multiple-scattering formalism that has explained successfully the experimentally observed intensities.

Dynamical theory has had its most profound impact in the area of surface-structure determination, and, consequently, LEED has been used to determine a large number of surface structures for a wide variety of materials as will be described in Chap. 7. The dynamical theory of LEED has also been the basis for subsequent parallel developments in other electron spectroscopies of surfaces, most notably in photoelectron emission.

5.1 Multiple Scattering

Electron diffraction, whether at the high energies of Electron Microscopy or at the low energies of LEED, is termed a dynamical process to describe the fact that the forces acting between the crystalline material and the diffracting electrons influence the diffraction characteristics markedly more than in the kinematic limit of X-ray [5.1] or Neutron [5.2] Diffraction. Dynamical effects are also familiar in High-Energy Electron Diffraction [5.3], in addition to Electron Microscopy [5.4]. The term *dynamical* has its origin in the study of X-ray Diffraction and describes the situations where the response of the crystal to the incident wave is included self-consistently, for example through a dielectric constant, and where multiple scattering occurs.

Two main features are commonly implied when Low-Energy Electron Diffraction is termed dynamical. First, the electron scattering by an individual atom, taken in isolation, is relatively complicated (Chap. 4). Multiple scattering of the diffracting electron occurs within the individual atom (this can induce resonance effects), and the scattering must be described by several parameters (e.g. the phase shifts), which are often strongly energy dependent. Thus, a relatively large effort must go into calculating phase shifts, since these cannot be obtained from experiment. By contrast, this atomic scattering is relatively simple at HEED and electron-microscopy energies.

Second, a substantial amount of multiple scattering of the diffracting electrons occurs between the atoms of the crystal surface. This multiple scattering removes much of the simplicity of X-ray, Neutron, and even High-Energy Electron Diffraction and is responsible, together with the multiparameter atomic scattering problem, for the much greater computational requirements of an accurate LEED theory. This multiple scattering is to a lesser extent also present at HEED and electron-microscopy energies [5.3,4], but it can be avoided by using very small crystals or very thin specimens.

As a result of the complications due to dynamical effects, it is not normally possible in LEED to "invert" experimental data directly (in the sense of the Patterson function [5.1], for example) in order to obtain the atomic positions in the crystal surface, even for the simplest structures. Rather, one must postulate a series of plausible surface structures, make LEED intensity calculations for each of these, and ascertain which structure produces the best fit with experiment. This procedure will be discussed in more detail in Chap. 6.

In this chapter we shall describe theoretical methods employed in dynamical LEED. The aim is to calculate the intensities of diffraction of a plane wave which represents the incident electron beam that impinges on a periodic and defect-free crystal surface. Although real crystal surfaces are known to be often imperfect, diffraction has the welcome ability to filter out a large part of these imperfections. Our approach will be first to reduce the problem to its basic elements and then to construct the solution of the general problem by discussing in turn the electron-diffraction characteristics of single atoms, of small collections of atoms, of individual layers of atoms and of stacks of such layers. The discussion will then turn to the treatment of superlattices, domains, symmetries, thermal effects, spin effects and surface resonances. A review of further approaches to and extensions of LEED theory will also be provided.

5.2 Diffraction in Crystalline Lattices

In Chap. 4 we discussed the electron scattering properties of several ingredients of any crystal surface: the elastic scattering by individual atoms and by individual potential steps taken in isolation, as well as the inelastic scattering that removes electrons from the elastic LEED current. The next stage is to put these ingredients together in the form of a crystalline surface and to describe the scattering of an incident plane wave, representing the incident LEED beam, by the entire surface.

With that in mind, we must first discuss the ways in which the total electron wave function can be represented at a surface. There are three basic approaches to this problem: expansion in spherical waves, expansion in plane waves, and expansion in Bloch waves. In practice these three choices are often combined to take advantage of the virtues of each individual expansion whenever possible. In particular, plane waves and spherical waves are generally used side-by-side in most calculational schemes. It will emerge that it is advantageous first to obtain the atomic scattering properties in the spherical-wave representation; then the multiple scattering within individual atomic layers (parallel to the surface) can also be treated with spherical waves. At that point, one can in many cases switch to the plane-wave representation to obtain the scattering properties of a stack of layers representing the surface. In some cases, namely when interlayer spacings are too small, the plane-wave representation is not convergent, and spherical waves remain favorable. Bloch waves (which are expanded in plane waves) are useful when a substantial part of the subsurface region, normally the substrate, is composed of a simple periodic sequence of layers, and especially when the mean free path is very large (such as at energies below approximately 10 eV).

5.2.1 Expansion in Spherical Waves

As we have seen in treating the scattering of electrons by individual atoms of the surface lattice (Sect. 4.5), spherical waves occur as a consequence of the near-spherical symmetry of most of these atoms. Since atomic scattering is the most important ingredient in electron-surface diffraction, we shall find the spherical-wave representation to be at least part of each calculational method. As a consequence, in Sect. 5.3 we shall begin our detailed treatment of multiple scattering by focusing on the spherical-wave representation. We have already introduced the basic concepts of this representation when dealing with atomic scattering in Sect. 4.5, and we refer the reader there for details.

Spherical-wave expansions have the virtue of converging well in LEED in terms of the number $l_{max} + 1$ of partial waves used. The only particular caution that should be recommended concerns the use of temperature-dependent phase shifts (Sect. 5.11): because of the enhancement of forward scattering over backward scattering due to the Debye-Waller factor, more partial waves, i.e. a larger l_{max}, may be required when including temperature effects through the phase shifts.

5.2.2 Expansion in Plane Waves

In the commonly used point-scatterer model of crystal surfaces (see Sect. 4.5.4), plane waves are natural solutions of the Schrödinger equation between the scatterers, where the potential is assumed constant. Furthermore, the two-dimensional surface periodicity provides a convenient natural selection of a limited number of plane waves to be used as a basis set for an expansion. This remains true in the multiple-scattering case, as we demonstrated in Sect. 4.2.2. (It is useful to recall also that multiple scattering does not change the LEED pattern: it changes only the spot intensities.) As described in Chap. 4, the relevant plane waves are those of a given energy E that have wave vectors $\vec{k}_{\bar{g}}^{\pm}$ such that

$$\vec{k}_{\bar{g}}^{\pm}{}_{\parallel} = \vec{k}_{0_\parallel} + \vec{g} \; , \tag{5.1}$$

where $\vec{k}_{0_\parallel}$ is the projection onto the surface of the incident wave vector $\vec{k}_0$, and $\vec{g}$ is any of the two-dimensional reciprocal lattice vectors of the surface periodicity, including any superlattice, if present (+ and − refer to propagation toward the bulk and toward the vacuum, respectively). The energy E in vacuum specifies the component $k_{\bar{g}z}^{\pm}$ perpendicular to the surface, i.e.

$$\frac{\hbar^2}{2m} |\vec{k}_{\bar{g}}^{\pm}|^2 = \frac{\hbar^2}{2m} (|\vec{k}_{0_\parallel} + \vec{g}|^2 + |k_{\bar{g}z}^{\pm}|^2) = E + V_o + iV_{oi} \; , \tag{5.2}$$

where V_o (>0) is the local constant potential (muffin-tin constant), V_{oi} (>0) is the local imaginary part of the potential, and $k_{\bar{g}z}^{\pm}$ is complex if V_{oi} is nonzero, providing wave damping due to inelastic processes. When $|\vec{k}_{0_\parallel} + \vec{g}|$ is large, $k_{\bar{g}z}^{\pm}$ acquires a large imaginary part (independent of the presence of V_{oi}) corresponding to exponentially decaying (evanescent) waves. When $|\vec{k}_{0_\parallel} + \vec{g}|$ is not too large, we obtain "propagating waves", many of which can emerge from the crystal and reach a detector, producing, for example, spots on a LEED screen. Some of these waves can propagate within the crystal but cannot emerge into vacuum because of the potential step. These waves are of special interest because they suffer total internal reflection at the potential step, and they are important in connection with "resonances" and surface states, cf. Sect. 5.12.

There are an infinite number of reciprocal lattice vectors $\vec{g}$, but only a finite number that give plane waves $\exp(\pm i\vec{k}_{\bar{g}}^{\pm} \cdot \vec{r})$ which carry flux over a distance that matters. If a wave cannot reach from one atomic layer to the next, it plays no role in the diffraction process. It can then be ignored, and this provides the criterion for reducing the infinite number of vectors $\vec{g}$ to a limited number.

Two plane waves have special significance. First, of course, is the incident beam itself, $\vec{k}_0^+$, characterized by $\vec{g} = 0$. This is also the plane wave that can be transmitted without deviation through any atomic layer. The other special plane wave is the specularly reflected plane wave $\vec{k}_0^-$, also characterized by $\vec{g} = 0$. This is the only reflected plane wave that does not change its direction as a function of electron energy or azimuthal crystal orientation, and thus it gives a fixed spot on the LEED screen.

5.2.3 Expansion in Bloch Waves

In any periodic medium the Bloch waves are the natural eigenfunctions of the Schrödinger equation. Surfaces have a two-dimensional periodicity parallel to the surface, and, consequently, Bloch waves with vectors $\vec{k}$ parallel to the surface can be envisaged. This choice has not, however, been adopted in LEED, because it does not provide a convenient matching with the incident plane wave.

Bloch waves, on the other hand, have been extensively used in LEED to describe the wave function in the direction perpendicular to the surface. In the bulk below the surface, one normally encounters a periodic stacking of atomic layers (each layer being parallel to the surface), and in that part of the crystal Bloch waves can be used as a basis set of eigenfunctions to describe the total wave function. To determine the amplitudes of these Bloch waves, one has to match the total wave function across the surface to the wave function in the vacuum, which consists of one incident plane wave of known amplitude and a set of emerging plane waves of unknown amplitudes. The non-bulklike part of the surface can be arbitrarily complicated: it may consist, for example, of a potential step, of a surface layer with a deviating layer spacing to the next layer, of a reconstructed layer, of an overlayer of adsorbates, or of any combination of these. The complexity of this surface affects the complexity of the matching equations between Bloch waves and vacuum plane waves.

The Bloch waves must be obtained from an eigenvalue problem (involving a matrix diagonalization), as prescribed by the Schrödinger equation combined with the Bloch condition. To solve this eigenvalue problem it is necessary to know the scattering properties of the individual layers of the periodic subsurface region. These are discussed most easily in connection with spherical and plane waves. Furthermore, we need a convenient representation of each Bloch wave, and plane waves provide the most suitable basis set in which these Bloch waves can be expanded. It is then natural to delay our treatment of Bloch waves until we have developed the use of plane waves more fully (see Sect. 5.5).

5.2.4 Forward vs. Backward Scattering

It is important in LEED to appreciate the relative importance of forward and backward scattering. As shown in Fig. 4.13, atomic scattering tends to "bunch" diffracted electrons in the forward direction, especially at higher energies. This tendency is strongly enhanced by the Debye-Waller factor, the effect of which also increases with energy. Only at low energies, on the order of 0-30 eV, is backward scattering similar in strength to forward scattering.

In addition, the absolute strength of forward scattering is important. Most electrons transmitted elastically through an atomic layer are either undeflected (i.e. undergo zero-angle forward "scattering") or undergo a small-angle forward scattering. Thus, most LEED electrons that penetrate a surface by a few layers and return to the vacuum have experienced several forward scatterings and one or at most a few backward scatterings. The important feature here is that a successful computational scheme must attempt to include as many forward scatterings as possible, while the number of backward scatterings that are included can remain small (except at the lowest energies).

5.3 Multiple Scattering in the Spherical-Wave Representation – Self-Consistent Formalism

5.3.1 Scattering by Two Atoms

We shall solve the full multiple-scattering problem at a crystal surface in gradual steps, following the ideas of the Korringa-Kohn-Rostoker (KKR) approach [5.5] and their application to the LEED problem [5.6-20]. The scattering by a single atom has been described fully in Chap. 4 in terms of a t-matrix (4.59,60). The nature of the spherical waves, e.g. $h_l^{(1)}(kr)Y_{lm}(\theta,\varphi)$, that we shall use as a basis set for the wave function in the surface has also been described.

The Green's Function

To include the effect of a subsequent scattering by another atom, it is necessary to express how a spherical wave of given angular momentum $L' = (l'm')$ centered on the first atom at $\bar{r}_1$ propagates to the second one, and how it decomposes into spherical waves centered on that second atom at $\bar{r}_2$. The result is given by a Green's function, which, assuming a constant potential between the atoms, has the following form in our case:

$$\bar{G}_{LL'}^{21} = -4\pi i \frac{2m}{\hbar^2} k \sum_{L_1} i^{l''} a(L,L',L_1) h_l^{(1)}(k|\bar{r}_2 - \bar{r}_1|) Y_{L_1}(\bar{r}_2 - \bar{r}_1) . \tag{5.3}$$

Here, $k = |\bar{k}_{\bar{g}}^{\pm}|$, and $L_1 = (l_1 m_1)$ extends over all values of l_1 and m_1 compatible with $L = (lm)$ and $L' = (l'm')$, namely $|l - l'| \leqslant l_1 \leqslant l + l'$ and $m + m' = m_1$. This compatibility is manifest as nonzero values of the Clebsch-Gordan (or Gaunt) coefficients

$$a(L,L',L_1) = \int Y_L^*(\Omega) Y_{L'}(\Omega) Y_{L_1}^*(\Omega) d\Omega , \tag{5.4}$$

where Ω ranges over all values of solid angle.

Equation (5.3) gives the amplitude of the spherical wave $L = (lm)$ centered on and arriving at the second atom at $\bar{r}_2$ due to a spherical wave $L' = (l'm')$ centered on and leaving the first atom at $\bar{r}_1$. The Hankel function $h_l^{(1)}$ and the spherical harmonic Y_{L_1} describe the propagation, while the Clebsch-Gordan coefficients are due to the expansion of a spherical wave centered on one atom into spherical waves centered on the second atom. Note that no scattering but only free-space propagation is included in this Green's function. Note also the implicit assumption of a point-scatterer picture, in which a constant potential extends completely between the atomic centers (see Sect. 4.5.4).

Scattering Paths

The t-matrix expresses the amplitude of the scattering of a spherical wave by an atom [the value of the angular momentum (lm) does not change in the scattering by a spherical potential]. Therefore, the product $\overline{G}^{21}_{LL'}t_{l'}$ expresses the fact that a spherical wave $L' = (l'm')$ *incident* on atom 1 scatters into the wave L' departing from atom 1 which in turn induces an incident wave $L = (lm)$ arriving at the second atom. The product $\overline{G}^{21}_{LL'}t_{l'}$ in fact is just the amplitude of this wave incident on the second atom ($t_{l'}$ should be considered to be a diagonal element of a diagonal matrix t). Clearly, it is now possible to combine such scattering events sequentially. If t^1_L and t^2_L are the respective t-matrices of the first and second atoms, one may express the amplitude of a succession of scatterings as, for example,

$$t^2\overline{G}^{21}t^1\overline{G}^{12}t^2\overline{G}^{21}t^1$$

(dropping the angular-momentum indices in matrix notation). This string of symbols is to be read from right to left in the fashion of matrix multiplication. As depicted in Fig. 5.1, it indicates that a spherical wave is incident on atom 1, from which it scatters (t^1) and propagates to atom 2 ($\overline{G}^{21}$), to be scattered again (t^2) and propagate back to atom 1 ($\overline{G}^{12}$). There it scatters again (t^1) and propagates to atom 2 ($\overline{G}^{21}$), where it scatters a final time (t^2). In this way, any multiple-scattering path of any finite length can be represented explicitly. (The matrix notation used here implies summation over intermediate spherical waves of all possible angular momenta, whereas explicit indices would indicate one particular choice of angular momenta.)

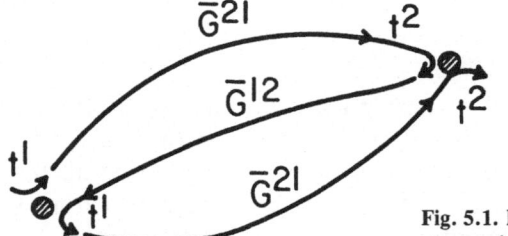

Fig. 5.1. Multiple scattering of a spherical wave by two atoms, showing Green's function notation

If we wanted to develop a perturbation theory on the basis that only scattering paths with a few scatterings will contribute to the diffraction, it could be done conveniently with this formalism by summing up contributions from the short scattering paths. However, since experience has shown that such a perturbation approach often does not converge properly because of numerous forward scatterings, it is necessary to be able to include paths involving many scatterings. A self-consistent (i.e. closed) formalism that includes all paths up to an infinite length can be developed as follows.

Self-Consistency

First we write down side-by-side the sums of the amplitudes of all scattering paths that terminate at atom 1 and at atom 2, respectively, denoting these sums by T^1 and T^2.

$$T^1 = t^1 + t^1 \overline{G}^{12} t^2 + t^1 \overline{G}^{12} t^2 \overline{G}^{21} t^1 + t^1 \overline{G}^{12} t^2 \overline{G}^{21} t^1 \overline{G}^{12} t^2 + \cdots , \qquad (5.5)$$

$$T^2 = t^2 + t^2 \overline{G}^{21} t^1 + t^2 \overline{G}^{21} t^1 \overline{G}^{12} t^2 + t^2 \overline{G}^{21} t^1 \overline{G}^{12} t^2 \overline{G}^{21} t^1 + \cdots . \qquad (5.6)$$

Before proceeding, we have to consider a phase problem in these two expressions. In T^1, for example, both paths originating at atom 1 and paths originating at atom 2 are included. As the expression for T^1 stands, the amplitude (modulus and phase) of whatever wave reaches these two atoms from the outside source of LEED electrons is assumed to be the same at the two atoms. In general this is not correct, however. If the incident wave is $\exp(i\vec{k}_{in} \cdot \vec{r})$, there is an amplitude ratio of $\exp[i\vec{k}_{in} \cdot (\vec{r}_1 - \vec{r}_2)]$ between the amplitudes arriving from the outside at atoms 2 and 1. We shall insert this correction within the definition of $\overline{G}^{21}$, which now becomes

$$\overline{G}^{21}_{LL'} = -4\pi i \frac{2m}{\hbar^2} k \sum_{L_1} i^{l''} a(L,L',L_1) h^{(1)}_{l_1}(k|\vec{r}_2 - \vec{r}_1|)$$

$$Y_{L_1}(\vec{r}_2 - \vec{r}_1) \exp[-i\vec{k}_{in} \cdot (\vec{r}_2 - \vec{r}_1)] . \qquad (5.7)$$

The same expression with $\vec{r}_1$ and $\vec{r}_2$ exchanged applies for $\overline{G}^{12}_{LL'}$. With this definition, (5.5,6) are correct. As pointed out by Beeby, [5.8], these expressions have the interesting property that they are closely connected, namely

$$T^1 = t^1 + t^1 \overline{G}^{12} T^2 , \qquad (5.8)$$

$$T^2 = t^2 + t^2 \overline{G}^{21} T^1 . \qquad (5.9)$$

Equation (5.8) implies that the set of scattering paths terminating at atom 1 consists, on one hand, of the once-scattered event t^1 and, on the other, of a set of multiple scattering paths. The latter can be considered to be all the paths which terminate at atom 2 (given by T^2) and then propagate to atom 1 ($\overline{G}^{12}$), finally to scatter there (t^1). Equation (5.9) has the same meaning for scattering paths terminating at atom 2.

Equations (5.8,9) constitute a self-consistent set of equations that one can solve easily in closed form for T^1 and T^2. Using matrix notation, one obtains

$$\begin{pmatrix} T^1 \\ T^2 \end{pmatrix} = \begin{pmatrix} I & -t^1 \overline{G}^{12} \\ -t^2 \overline{G}^{21} & I \end{pmatrix}^{-1} \begin{pmatrix} t^1 \\ t^2 \end{pmatrix} , \qquad (5.10)$$

where I is a unit matrix. The dimension of the matrices I, G, t, and T is determined by the maximum number of phase shifts required to describe the atomic scattering: if $l \leqslant l_{max}$, and all values of m such that $-l \leqslant m \leqslant l$ are needed, then this dimension is $(l_{max} + 1)^2$. Note that the geometrical series obtained by expanding the large-matrix inversion of (5.10) reproduces the expressions given by (5.5,6), yielding an expansion in terms of the number of scatterings. In fact, (5.10) is rigorously valid even when (5.5,6) do not converge well, because (5.8,9) are rigorous (self-consistent).

5.3.2 Scattering by N Atoms

It is straightforward to generalize our results to a set of N atoms, located at positions $\vec{r}_1, \vec{r}_2, \cdots, \vec{r}_N$. Equations (5.8) and (5.9) generalize immediately to

$$
\begin{aligned}
T^1 &= t^1 & + & t^1\overline{G}^{12}T^2 & +t1\overline{G}^{13}T^3 & + \cdots + t^1\overline{G}^{1N}T^N \\
T^2 &= t^2\overline{G}^{21}T^1 + & & t^2 & + t^2\overline{G}^{23}T^3 & + \cdots + t^2\overline{G}^{2N}T^N \\
T^3 &= t^3\overline{G}^{31}T^1 + & t^3\overline{G}^{32}T^2 & + t^3 & & + \cdots + t^3\overline{G}^{3N}T^N \\
&\;\;\vdots & \vdots & \vdots & & \vdots
\end{aligned}
$$

$$
T^N = t^N\overline{G}^{N1}T^1 + t^N\overline{G}^{N2}T^2 + t^N\overline{G}^{N3}T^3 \;\; + \cdots + \;\; t^N
$$

(5.11)

These equations are solved by

$$
\begin{pmatrix} T^1 \\ T^2 \\ T^3 \\ \vdots \\ T^N \end{pmatrix} = \begin{pmatrix} I & -t^1\overline{G}^{12} & -t^1\overline{G}^{13} & \cdots & -t^1\overline{G}^{1N} \\ -t^2\overline{G}^{21} & I & -t^2\overline{G}^{23} & \cdots & -t^2\overline{G}^{2N} \\ -t^3\overline{G}^{31} & -t^3\overline{G}^{32} & I & \cdots & -t^3\overline{G}^{3N} \\ \vdots & \vdots & \vdots & & \vdots \\ -t^N\overline{G}^{N1} & -t^N\overline{G}^{N2} & -t^N\overline{G}^{N3} & \cdots & I \end{pmatrix}^{-1} \begin{pmatrix} t^1 \\ t^2 \\ t^3 \\ \vdots \\ t^N \end{pmatrix}.
$$

(5.12)

It remains to recombine the scattering paths that were separated into the different sets $T^1, T^2, \cdots, T^N$. The quantities $T^1, T^2, \cdots, T^N$ can be simply added together, since they are the amplitudes of scattered waves. However, phase differences have to be taken into account. If we keep the incident plane wave $\exp(i\vec{k}_{in}\cdot\vec{r})$ used up to this point and wish to consider the amplitude of the outgoing plane wave $\exp(i\vec{k}_{out}\cdot\vec{r})$, it follows that the total scattering amplitude will be (still in the spherical-wave representation)

$$
T_{LL'} = \sum_{i=1}^{N} \exp[i(\vec{k}_{in} - \vec{k}_{out})\cdot\vec{r}_i]T^i_{LL'} .
$$

(5.13)

We have chosen an outgoing *plane* wave in view of the subsequent application of this formalism to periodic surface structures with their well-defined outgoing plane waves. For a disordered set of atoms, plane waves might not be the preferred representation of the diffracted wave, and (5.13) would then be modified accordingly. [Note also that a plane incident wave was assumed in (5.7)].

5.3.3 One Periodic Plane of Atoms

The results expressed in (5.12,13) have the serious disadvantage, in the case
of a surface, that too many atoms (on the order of $N = 100$) have to be
included in the formalism to provide a reasonable description of the
diffracted electrons: the large matrices become far too large. However, we
can reduce this computational problem enormously by exploiting the periodicity of the *ordered* surface.

First, let us consider a single plane of identical atoms that extends to
infinity in all directions of the plane and has a so-called Bravais lattice, i.e.
there is only one atom in each unit cell. Later in our discussion, this plane
will represent a plane of equivalent atoms in a crystal surface. These identical atoms now all have the same t-matrix t. By periodic symmetry, the
diffraction of a spherical wave incident on any one of these atoms is identical
to the diffraction of an identical spherical wave incident on any other of
these atoms, even after inclusion of all multiple scattering. Consequently, if
T^1, T^2, ... represent the scattering from atoms 1,2,... as used in (5.11), then
these quantities must be equal, i.e.

$$T^1 = T^2 = \cdots \equiv \tau \ . \tag{5.14}$$

We shall henceforth reserve the symbol τ for the scattering by a Bravais-
lattice plane of atoms. From any of the equations (5.11), it follows that

$$\tau = t + t \left(\sum_n{}' \overline{G}^{in} \right) \tau \ , \tag{5.15}$$

where n extends over all atoms other than the atom i under consideration.
(This exclusion is indicated by the prime on the summation symbol.) We
define the new Green's function

$$G^{ii}_{LL'} = \sum_n{}' \overline{G}^{in}_{LL'} = -4\pi i \ \frac{2m}{\hbar^2} \ k$$
$$\times \sum_{L_1} \sum_{\vec{P}}{}' \ i^{l_1} a(L,L',L_1) h^{(1)}_{l_1}(k|\vec{P}|) Y_{L_1}(\vec{P}) \exp(-\vec{k}_{in} \cdot \vec{P}) \ , \tag{5.16}$$

which includes a sum over all atoms at positions $\vec{P}$, except $\vec{P} = \vec{0}$. (This
exclusion is again denoted by the prime on the summation symbol.) The
superscripts ii of $G^{ii}_{LL'}$ have become meaningless since the right-hand side of
(5.16) no longer depends on i, but we shall retain them for later generalization. Now, (5.15) implies that

$$\tau = (1 - tG^{ii})^{-1} t = t(1 - G^{ii}t)^{-1} \ . \tag{5.17}$$

Equation (5.17) essentially solves the multiple-scattering problem in a
Bravais-lattice plane. Although an infinite number of atoms is involved, the
matrices have a relatively small dimension, namely $(l_{max} + 1)^2$, and we have

a self-consistent solution valid whatever the amount of multiple scattering may be. We are still employing the spherical-wave representation, and, consequently, τ has matrix elements $\tau_{LL'}$. The quantity $\tau_{LL'}$ indicates that an incident spherical wave L' of amplitude 1 centered on any of the layer atoms eventually produces an amplitude $\tau_{LL'}$ in a departing spherical wave L centered on the same atom.

The lattice sum in (5.16) is expected to converge only if sufficient damping is present in the form of the imaginary part of k in the Hankel function $h^{(1)}$. Thus, when inelastic effects are weak, such as at kinetic energies near or below the Fermi level, a different summation procedure is required. Kambe has formulated an alternative method based on an Ewald summation scheme [5.6]. Basically it consists of a Fourier summation over the far part of the atomic lattice and a direct summation over the near part of the atomic lattice. Even for normal LEED cases where the kinetic energy is far above the Fermi level, the Kambe approach is computationally slightly faster than the direct-space summation of (5.16) (at the expense of more complicated programming).

5.3.4 Several Periodic Planes of Atoms

The next step is to combine a pair of Bravais-lattice planes that have the same unit cell, but possibly different atoms, characterized by the t-matrices t^1 and t^2. The single planes are characterized by the τ-matrices τ^1 and τ^2, determined by (5.17). We shall derive the multiple-scattering paths of the two planes in exactly the way we followed in the case of two individual atoms. Let T^1 and T^2 represent all multiple-scattering paths that terminate at an atom of planes 1 and 2, respectively. Then T^1 can be considered to consist, on one hand, of all paths internal to plane 1 (giving τ^1, i.e. paths that do not scatter from any atom of plane 2), and, on the other, of all paths that terminate at any atom i of plane 2 (T^2) and then propagate to any atom of plane 1 ($\overline{G}^{12(i)}$), where they may multiply scatter and terminate at some atom of plane 1 (τ^1), without ever returning to plane 2. Similar reasoning applies to T^2, and we find that

$$T^1 = \tau^1 + \tau^1 \sum_{(i)} \overline{G}^{12(i)}T^2 \ , \tag{5.18}$$

$$T^2 = \tau^2 + \tau^2 \sum_{(i)} \overline{G}^{21(i)}T^1 \ . \tag{5.19}$$

With the definitions

$$G^{12} = \sum_{(i)} \overline{G}^{12(i)} \quad \text{and} \quad G^{21} = \sum_{(i)} \overline{G}^{21(i)} \ , \tag{5.20}$$

(5.18,19) yield a solution very reminiscent of (5.10) with, in fact, identical matrix dimensions:

$$\begin{pmatrix} T^1 \\ T^2 \end{pmatrix} = \begin{pmatrix} I & -\tau^1 G^{12} \\ -\tau^2 G^{21} & I \end{pmatrix}^{-1} \begin{pmatrix} \tau^1 \\ \tau^2 \end{pmatrix} . \tag{5.21}$$

Here, as with individual atoms, a generalization to N Bravais-lattice planes with identical unit cells is straightforward:

$$\begin{pmatrix} T^1 \\ T^2 \\ T^3 \\ \cdot \\ \cdot \\ \cdot \\ T^N \end{pmatrix} = \begin{pmatrix} I & -\tau^1 G^{12} & -\tau^1 G^{13} & \cdots & -\tau^1 G^{1N} \\ -\tau^2 G^{21} & I & -\tau^2 G^{23} & \cdots & -\tau^2 G^{2N} \\ -\tau^3 G^{31} & -\tau^3 G^{32} & I & \cdots & -\tau^3 G^{3N} \\ \cdot & & & & \cdot \\ \cdot & & & & \cdot \\ \cdot & & & & \cdot \\ -\tau^N G^{N1} & -\tau^N G^{N2} & -\tau^N G^{N3} & \cdots & I \end{pmatrix}^{-1} \begin{pmatrix} \tau^1 \\ \tau^2 \\ \tau^3 \\ \cdot \\ \cdot \\ \cdot \\ \tau^N \end{pmatrix} , \tag{5.22}$$

with the Green's functions,

$$G^{ji}_{LL'} = -4\pi i \, \frac{2m}{\hbar^2} \, k \sum_L \sum_{\vec{P}}{}' \, i^{l'} a(L,L',L_1) h^{(1)}_{l_1}(k|\vec{r}_j - \vec{r}_i + \vec{P}|)$$

$$\times \, Y_{L_1}(\vec{r}_j - \vec{r}_i + \vec{P}) \exp[-i\vec{k}_{in} \cdot (\vec{r}_j - \vec{r}_i + \vec{P})] . \tag{5.23}$$

In this expression, $\vec{r}_i$ and $\vec{r}_j$ are atomic sites in planes i and j, respectively, while $\vec{P}$ extends over the lattice points of any of the planes, except the point $\vec{r}_j - \vec{r}_i + \vec{P} = \vec{0}$. Note that (5.16) follows from (5.23) when j = i. The total scattering $T_{LL'}$ by the N layers is obtained by applying (5.13), which can then be used again unchanged with the new values of $T^i_{LL'}$ determined from (5.22).

Equation (5.22) solves all multiple scattering within a surface, if N planes of atoms are sufficient to describe that surface. The number N is determined by the condition that all Bravais-lattice planes of a surface be included that are closer to the surface than a few times the electron-mean-free-path length.

5.3.5 Change to Plane-Wave Amplitudes

Since a periodic surface scatters an incoming electron beam into a set of well-defined departing electron beams, we wish ultimately to obtain the intensity of these outgoing beams. The treatment of the preceding sections has already been biased in this direction, as is most visible in (5.13) which, together with (5.22), represents the closest we have advanced toward this goal.

Now we need only a relation between an arbitrary t-matrix $t_{LL'}$, giving the diffraction amplitude between two spherical waves L' and L, and a plane-wave diffraction matrix $M_{out,in}$, giving the diffraction amplitude between two plane waves $\vec{k}_{in}$ and $\vec{k}_{out}$. For a lattice with two-dimensional periodicity of the type we have been assuming for surfaces, such a relation is given by

$$M_{out,in} = -\frac{8\pi^2 i}{Ak_{outz}}\frac{2m}{\hbar^2}\sum_{LL'}Y_L(\vec{k}_{out})t_{LL'}Y_L^*(\vec{k}_{in}) ,\qquad (5.24)$$

where A is the area of the two-dimensional unit cell ($1/A$ is the density of atoms in each atomic plane). This relation is formally equivalent to (4.26), the delta function being omitted. Derivations of (5.24) can be found in the literature [5.7]. Thus, for the incident beam $\vec{k}_o$ and the outgoing beams $\vec{k}_{\bar{g}}^-$, we obtain, by inserting (5.13) into (5.24), the reflected amplitudes

$$M_{\bar{g},0}^{-+} = -\frac{16\pi^2 im}{Ak_{\bar{g}z}^-\hbar^2}\sum_{LL'}Y_L(\vec{k}_{\bar{g}}^-)T_{LL'}Y_L^*(\vec{k}_o) ,\qquad (5.25)$$

where $T_{LL'}$ is calculated from the results of (5.22), via (5.13).

Equation (5.25) gives the amplitudes of the plane waves $\exp(i\vec{k}_{\bar{g}}^-\cdot\vec{r})$ diffracted from a complete surface, provided the number N of atomic planes that are included is sufficient. In the absence of inelastic effects, this number must be infinite since the LEED electrons can then penetrate the surface to an infinite depth. Actual electron-mean-free-path lengths, however, limit the number of required planes severely and can make (5.25) tractable. Nevertheless, even in the most favorable case, such as a clean low-index fcc, hcp, or bcc metal surface, the number of required planes is at least approximately five. With a value of four for l_{max}, which is sufficient in most cases up to an energy of 100 eV, this number of planes produces a matrix dimension of $5(4 + 1)^2 = 125$ for the large matrix that is to be inverted in (5.22). Since this matrix inversion must be repeated for each new geometry and for each new value of the energy (or angle of incidence), in practice this approach is prohibitively expensive in any structural determination of more complicated surfaces (e.g. atomic or molecular overlayers with superlattices, surface reconstruction, etc.).

However, it will prove to be possible to apply this formalism to restricted, thin slices of the surface, in conjunction with more efficient methods for combining these slices with the remaining parts of the surface. Therefore, we shall summarize the result of this spherical-wave approach in the case of a thin slice of the surface, which we shall call a layer, each layer being composed of any number of atomic Bravais-lattice planes.

5.3.6 Layer Diffraction Matrices for Plane Waves

A layer will be understood to consist of one or more atomic planes (we shall also use the word "subplane" synonymously with "plane"). Henceforth, we shall allow any plane wave $\vec{k}_{\bar{g}}^\pm$, not just the primary incident beam $\vec{k}_o$, to be incident onto such a layer, and we wish to determine the amplitude $M_{\bar{g}'\bar{g}}^{\pm\pm}$ of any of the diffracted plane waves $\vec{k}_{\bar{g}'}^\pm$ due to such an incident wave.

Case of a Bravais-Lattice Layer

First, for a layer with only one atomic plane, which we may therefore call a Bravais-lattice layer, we can apply (5.24) with the τ-matrix (5.15) for that layer:

$$M_{\vec{g}'\vec{g}}^{\pm\pm} = -\frac{16\pi^2 im}{Ak_{\vec{g}z}^+\hbar^2} \sum_{LL'} Y_L(\vec{k}_{\vec{g}}^{\pm})\tau_{LL'}Y_{L'}^*(\vec{k}_{\vec{g}}^{\pm}) + \delta_{\vec{g}'\vec{g}}\delta_{\pm\pm} \ . \qquad (5.26)$$

The Kronecker deltas (equal to unity when $\vec{g}' = \vec{g}$ and when the signs are the same) that have been added to this expression represent the unscattered plane wave which is transmitted without change of direction through the layer. This corresponds to the term $\exp(i\vec{k}\cdot\vec{r})$ of (4.59), which was omitted from our discussion in favor of the scattered part of the wave. Note the right-to-left logical order of the indices of $M_{\vec{g}'\vec{g}}^{\pm\pm}$.

Equation (5.26) can be obtained in various ways other than that described here. Our derivation (including the parts not reproduced here) is based partially on and therefore parallels that of Beeby [5.8] and that of Tong [5.7]. Pendry [5.9] has arrived at a similar result by a somewhat different route. His final expression is sufficiently different in appearance (although mathematically equivalent, as has been proved [5.10]) that we reproduce it here as well. Equation (5.26) is formally equivalent to

$$M_{\vec{g}'\vec{g}}^{\pm\pm} = \frac{8\pi^2 im}{Akk_{\vec{g}'z}^+\hbar^2} \sum_{lml'm'} \left[i^l(-1)^m Y_{l-m}(\vec{k}_{\vec{g}}^{\pm}) \right] (1 - X)_{lm,l'm'}^{-1}$$

$$\times \left[i^{-l'} Y_{l'm'}(\vec{k}_{\vec{g}'}^{\pm}) \right] \sin\delta_{l'}\exp(i\delta_{l'}) + \delta_{\vec{g}'\vec{g}}\delta_{\pm\pm} \qquad (5.27)$$

with

$$X_{lm,l'm'} = \sum_{l'+m'=\text{even}} C^l(l'm',l''m'')_{l'm'}\sin\delta_{l'}\exp(i\delta_{l'}) \ , \qquad (5.28)$$

$$C^l(l'm',l''m'') = 4\pi(-1)^{(l-l'-l'')/2}(-1)^{m'+m''}Y_{l'-m'}(\theta = \frac{\pi}{2},\varphi = 0)$$

$$\times \int Y_{lm}Y_{l'm'}Y_{l''-m''}d\Omega \ , \quad\text{and} \qquad (5.29)$$

$$F_{l'm'} = \sum_{\vec{P}}{}' \exp(i\vec{k}_{\parallel}\cdot\vec{P})h_{l'}^{(1)}(k|\vec{P}|)(-1)^{m'}\exp[-im'\varphi(\vec{P})] \ . \qquad (5.30)$$

Here, $C^l(l'm',l''m'')$ is a set of Clebsch-Gordan coefficients that is different from the set $a(lm,l'm',l''m'')$ of (5.4). However, these two sets are equivalent under suitable permutations, apart from different prefactors. In (5.30), $\varphi(\vec{P})$ is the azimuth in the surface plane of the lattice vector $\vec{P} \neq \vec{0}$. (One minor discrepancy occasionally appears in some versions of these formulae: the complex factor k in the denominator of (5.27) is replaced occasionally by its absolute value or by its real part.)

Other workers, including Kambe [5.6], Beeby [5.8], Duke and Tucker [5.11], McRae [5.12], and Jepsen et al. [5.13], have derived similar formulas by other approaches. Their results are very likely equivalent to the ones shown here, although to our knowledge this equivalence has not been proven (other than by the close correspondence of calculated I-V curves).

Case of a Layer with N Planes

The general case of diffraction by a layer with N planes, i.e. with N atoms at positions $\vec{r}_i$ in the surface unit cell, may be obtained from (5.24) in the same way that (5.25) was derived.

$$
M_{\vec{g}'\vec{g}}^{\pm\pm} = -\frac{16\pi^2 im}{Ak_{\vec{g}'z}^+\hbar^2} \sum_{LL'} Y_L(\vec{k}_{\vec{g}'}^{\pm})
$$

$$
\times \sum_{i=1}^{N} \left\{ \exp[i(\pm\vec{k}_{\vec{g}}^{\pm} \mp \vec{k}_{\vec{g}'}^{\pm})\cdot\vec{r}_i]T_{LL'}^i \right\} Y_L^*(\vec{k}_{\vec{g}}^{\pm}) + \delta_{\vec{g}'\vec{g}}\delta_{\pm\pm} . \tag{5.31}
$$

One small modification is needed here in the Green's function of (5.23): the incident wave is now $\vec{k}_{\vec{g}}^{\pm}$ rather than $\vec{k}_{in}$. Simultaneously with the final form of $G_{LL'}^{ji}$, we now give an alternate form derived by Tong [5.7], based on a reciprocal-lattice rather than a direct-lattice summation:

$$
G_{LL'}^{ji} = \exp\left[-i\vec{k}_{\vec{g}}^{\pm}\cdot(\vec{r}_j - \vec{r}_i)\right]\hat{G}_{LL'}^{ji}, \quad \text{with} \tag{5.32}
$$

$$
\hat{G}_{LL'}^{ji} = -8\pi\frac{ikm}{\hbar^2} \sum_{L_1} \sum_{\vec{P}} {}'i^{l_1}a(L,L',L_1)h_{l_1}^{(1)}(k|\vec{r}_j - \vec{r}_i + \vec{P}|)
$$

$$
\times Y_{L_1}(\vec{r}_j - \vec{r}_i + \vec{P})\exp(-i\vec{k}\cdot\vec{P}) , \quad \text{or} \tag{5.33}
$$

$$
\hat{G}_{LL'}^{ji} = -\frac{16\pi^2 im}{A\hbar^2} \sum_{\vec{g}_1} \frac{\exp\left[i\vec{k}_{\vec{g}_1}^{\pm}\cdot(\vec{r}_j - \vec{r}_i)\right]}{k_{\vec{g}_1z}^+} Y_L^*(\vec{k}_{\vec{g}_1}^{\pm})Y_L(\vec{k}_{\vec{g}_1}^{\pm}) . \tag{5.34}
$$

Equation (5.34) has computational advantages over (5.33) under certain circumstances, namely when the atomic planes of the layer are not spaced too closely.

Note that (5.26) and (5.27) assume a reference point (origin of coordinates) at the center of an atom of the layer under consideration: $M_{\vec{g}'\vec{g}}^{\pm\pm}$ is the ratio of outgoing to incident plane-wave amplitudes measured in that atomic center (or at least extrapolated to that atomic center as if we had point scatterers). In (5.31) the corresponding reference point is the origin of the vectors $\vec{r}_i$ and thus can be an arbitrary point. However, it is important to be consistent in the use of a particular reference point in each layer when carrying out a calculation, as we shall discuss later, especially in connection with the utilization of symmetries.

Reflection and Transmission Matrices

Finally, we introduce a notation that will be more convenient later when we discuss the stacking of layers in a surface. We distinguish between reflection and transmission at a layer, following normal practice. Transmission occurs when electrons emerge ultimately on the side of the layer opposite to that from which they impinge on it, reflection when they emerge ultimately on the same side, with an appropriate generalization in the case of evanescent waves. This definition implies, for example, that grazing specular diffraction is treated as a reflection rather than as a transmission, even though the scattering angle may be close to 0°. Thus we define reflection and transmission matrices

$$r^{+-} = M^{+-}, \quad r^{\mp} = M^{\mp}, \quad t^{++} = M^{++} \quad \text{and} \quad t^{--} = M^{--}, \tag{5.35}$$

where $M^{\pm\pm}$ is derived from (5.26,27, or 31).

5.3.7 One-Center Expansion

A calculational method related closely to the Green's-function method described in this section uses a one-center partial-wave expansion. This recent approach, proposed by Pendry [5.14], focuses on *clusters* of surface atoms rather than on *planes* (subplanes) of surface atoms. Thus a cluster of atoms (such as an adsorbed molecule or a larger fragment of the surface) is imagined isolated from the rest of the surface, and its complete scattering properties are calculated before the surface periodicity is taken into account. First, an equation very similar to (5.12) is obtained for the cluster, with the following difference: rather than the incoming plane wave being expanded in spherical waves centered on each atom, as was done for (5.12), the incoming wave is expanded about a single, arbitrary reference point, best chosen near the cluster center. One thereby obtains a t-matrix $T_{LL'}$ that does not depend on the propagation direction of the initial plane wave, unlike the case of (5.12) [see, for example, (5.7,5.13)]. As a consequence, it is easy to rotate the cluster in space without having to recalculate the multiple scattering within it. This is accomplished by a simple rotation matrix R given by Slater [5.15], which simulates the spatial rotation by changing the t-matrix T into $R^{-1}TR$. The periodic repetition of this cluster can then be taken into account according to (5.17) where t is replaced by $R^{-1}TR$, followed by the application of (5.24), yielding (5.26) with suitable modifications to t and τ.

The cluster t-matrix $T_{LL'}$ obtained with the one-center expansion can be used in other interesting ways. It can be combined with other clusters (besides the periodic counterparts) into large clusters, allowing large clusters to be constructed stepwise from smaller ones. The matrix $T_{LL'}$ also can be used conveniently, in conjunction with a Debye-Waller factor, to represent rigid vibrations of the entire cluster with respect to the surrounding medium.

This is useful, for example, in the case of a strongly coherent molecule that is loosely bound to a substrate.

Another feature of the one-center expansion is that the number of spherical waves to be included is determined, for a given cluster, by its diameter rather than by the atomic properties as was the case previously in this section. Consequently, l_{max} is usually rather large in the one-center expansion.

5.4 Perturbation Expansion of Multiple Scattering in the Spherical-Wave Representation: Reverse-Scattering Perturbation (RSP) Method

The matrix inversions of (5.17,22) ensure that multiple scattering is included to infinite order at the price of a large but finite number of computational operations. One can avoid the matrix inversion by an expansion as in

$$\tau = (1 - tG^{ii})^{-1}t = t + tG^{ii}t + tG^{ii}tG^{ii}t + \cdots ,$$

where each term corresponds to a different order of multiple scattering: first single (kinematic) scattering, then double scattering, etc. (5.17). In practice, however, this particular expansion has been found to converge poorly [5.7] in simple close-packed metal layers (in the sense that matrix inversion is computationally at least as rapid as the expansion), and therefore it has not been employed frequently.

5.4.1 The Principle of RSP

The expansion of the large-matrix inversion of (5.22) can be an advantageous choice, however, if the atoms involved do not scatter too strongly. (Some heavy metals like tungsten and platinum produce too strong multiple scattering even for this expansion.) Toward this end, Zimmer and Holland proposed a Reverse-Scattering Perturbation (RSP) method [5.16] in which expansion terms are grouped in an efficient manner that is formally equivalent to the Gauss-Seidel-Aitken matrix-inversion procedure [5.17]. We shall describe here a variant of the Zimmer-Holland method, used by Tong and Van Hove [5.18], that reflects the preponderance of forward scattering over backward scattering. The approach is to allow a plane wave incident on the atomic layer under consideration to interact first with the subplane nearest the vacuum. Spherical waves scattered from that first subplane (after the inclusion of any multiple scattering within that subplane) propagate to and reach the next subplane into the surface, which is also simultaneously reached by the original unscattered incident plane wave. This is depicted in Fig. 5.2, where a layer of four subplanes is assumed. The result is a new collection of scattered spherical waves that propagate from the

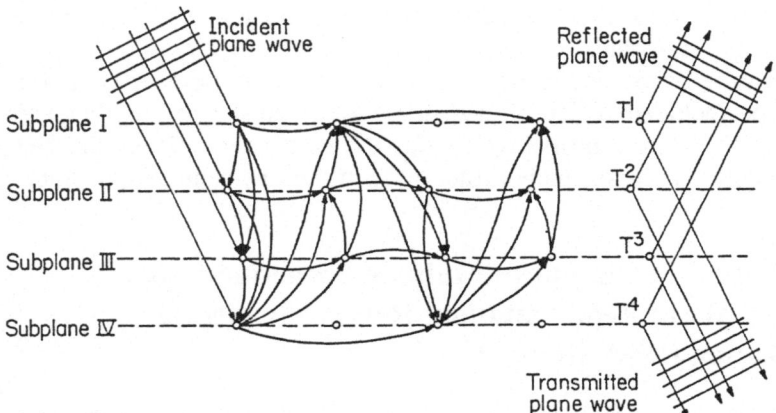

Fig. 5.2. Diagram of the Reverse-Scattering Perturbation scheme for the case of a composite layer with four subplanes. See text for details

second subplane to the third subplane, which receives, in addition, not only the original plane wave but also spherical waves emanating from the first subplane. In turn, the fourth subplane receives the plane wave and spherical waves emanating from the first, second, and third subplanes. Note that the waves reaching this fourth subplane have already been scattered by zero, one, two or three other subplanes, in such a way that all possible multiple *forward* scattering events have been included.

This procedure continues until the deepest subplane, numbered N, of the surface layer under consideration has been reached. At that point, the procedure is reversed, and one propagates various waves to the (N-1)th subplane, namely all those waves that have previously reached subplane N. To this, one also adds the waves that had already reached subplane N-1 on their way into the surface. Next, subplane N-2 receives all waves that have previously reached the subplanes N, N-1, and N-2 on their way in, and this is repeated from subplane to subplane until the topmost one has been reached again. At this point, all multiple-scattering paths involving just one back-scattering from a subplane have been produced, and their combined contribution to an exiting plane wave is recorded.

Now a second in-and-out pass through the layer is carried out, using the waves computed in the first pass (but not the primary incident plane wave). This generates all multiple-scattering paths with two backscatterings on the inward journey and those with three backscatterings on the outward journey. The latter give the next contribution to reflection by the layer, while the two-backscattering paths contribute to the transmission (together with the zero-backscattering paths obtained in the first pass). This procedure can be iterated to convergence and repeated for other incident plane waves, including any plane waves incident from below the layer under consideration, if applicable.

5.4.2 The Formalism of RSP

The procedure described above can be formalized as follows [5.16,18]. First, a simplification is realized by defining the quantities

$$T^i_L(\vec{g}^{\pm}) = \sum_{L'} T^i_{LL'} Y^*_L(\vec{k}^{\pm}_{\vec{g}}), \qquad i = 1,2,\cdots,N, \quad \text{and} \qquad (5.36)$$

$$\tau^i_L(\vec{g}^{\pm}) = \sum_{L'} T^i_{LL'} Y^*_L(\vec{k}^{\pm}_{\vec{g}}), \qquad i = 1,2,\cdots,N. \qquad (5.37)$$

The right-hand side of (5.36) can be recognized in (5.31), while the right-hand side of (5.37) appears in (5.26). The term $T^i_L(\vec{g}^{\pm})$ contains all multiple-scattering paths terminating at subplane i (from an incident beam $\vec{k}^{\pm}_{\vec{g}}$), while $\tau^i_L(\vec{g}^{\pm})$ contains all paths within subplane i. Then $T^i_L(\vec{g}^{\pm})$ is expanded according to the number of backscattering events, i.e.

$$T^i_L(\vec{g}^{\pm}) = \tau^i_L(\vec{g}^{\pm}) + \sum_{n=0}^{\infty} \left[T^{i+(n)}_L(\vec{g}^{\pm}) + T^{i-(n)}_L(\vec{g}^{\pm}) \right]. \qquad (5.38)$$

Here, $T^{i+(n)}$ contains those paths terminating at subplane i that arrive from the substrate side of this subplane and that have experienced n reversals of direction (backscatterings), not counting the possible change of direction at the first scattering by a subplane; $T^{i-(n)}$ is identical to $T^{i+(n)}$ except that it includes only paths arriving from the vacuum side of subplane i.

The following iteration relations describe the subplane-to-subplane propagation scheme outlined above:

$$T^{i+(n)}_L(\vec{g}^{\pm}) = \sum_{L_1 L_2} \tau^i_{LL_1} \sum_{j>i} G^{ij}_{L_1 L_2} \left[T^{j+(n)}_{L_2}(\vec{g}^{\pm}) + T^{j-(n-1)}_{L_2}(\vec{g}^{\pm}) \right], \qquad (5.39)$$

and

$$T^{i-(n)}_L(\vec{g}^{\pm}) = \sum_{L_1 L_2} \tau^i_{LL_1} \sum_{j<i} G^{ij}_{L_1 L_2} \left[T^{j-(n)}_{L_2}(\vec{g}^{\pm}) + T^{j+(n-1)}_{L_2}(\vec{g}^{\pm}) \right]. \qquad (5.40)$$

The iteration scheme is initiated with the conditions

$$T^{1-(n)}_L(\vec{g}^{\pm}) = 0, \qquad n = -1,0,1,2,\ldots, \qquad (5.41)$$

$$T^{N+(n)}_L(\vec{g}^{\pm}) = 0, \qquad n = -1,0,1,2,\ldots, \quad \text{and} \qquad (5.42)$$

$$T^{i-(-1)}_L(\vec{g}^{\pm}) = T^{i+(-1)}_L(\vec{g}^{\pm}) = \tau^i_L(\vec{g}^{\pm}), \qquad i = 1,2,\ldots,N. \qquad (5.43)$$

The iterations of (5.39,40) compute in succession $T^{1-(0)}$, $T^{2-(0)}$, ..., $T^{N-(0)}$, $T^{N+(0)}$, ..., $T^{2+(0)}$, $T^{1+(0)}$ and repeat this sequence with n = 1,2,... until convergence of (5.38).

5.4.3 The Use of RSP

The RSP method has proved valuable for cases of many atoms per unit cell. Zimmer and Holland [5.16] found the conversion to the plane-wave representation in (5.31) time-consuming and therefore chose to compute only the matrix elements $M_{\vec{g}\vec{0}}^{\mp}$, i.e. they chose to use only the incident LEED beam $\vec{k}_0^+$ and only the diffracted beams $\vec{k}_{\vec{g}'}^-$ for those $\vec{g}'$ available experimentally. The price of this choice is that a thick surface layer is needed to represent the entire depth of the surface, as determined by the electron-mean-free-path length. Tong and Van Hove [5.18] chose to accept the cost of conversion to the plane-wave representation with all incident waves $\vec{k}_{\vec{g}}^{\pm}$ and all emergent waves $\vec{k}_{\vec{g}'}^{\pm}$ for thinner surface layers with fewer atoms per unit cell in order to benefit from the plane-wave representation whenever possible in the remainder of the surface depth. (This benefit is substantially enhanced when symmetries are present and exploited.) It was found, however, that for a number of typical calculations, the use of RSP rather than the large-matrix inversion of (5.31) did not substantially reduce the computation time. Rather, the major benefit accrued from the reduced core-size requirements due to the absence of the large matrix. However, RSP should prevail for the more complex structures, since the computation time is proportional to $N^2(l_{max} + 1)^4$, while matrix inversion is proportional in time to $N^3(l_{max} + 1)^6$, where N is the number of atoms per unit cell in the layer.

5.5 Diffraction by a Stack of Layers: Transfer-Matrix and Bloch-Wave Method

The Bloch-wave method and its close relative, the transfer-matrix method, together with Beeby's large-matrix-inversion method (5.12,5.13), were the earliest solutions of the LEED problem. The Bloch-wave and transfer-matrix methods have the virtues both of providing a connection with the familiar band-structure theory of bulk materials and of giving a simple interpretation of the electron-diffraction process at elementary surfaces. However, these advantages have not been particularly useful in treating more complex surfaces involving adsorbates or reconstructions.

5.5.1 The Bloch Condition

The basic idea is to apply the Bloch condition to the electronic wave functions in the direction perpendicular to the surface for that part of the subsurface that has a well-defined periodicity perpendicular to the surface. An eigenvalue problem then provides the Bloch functions, the relative amplitudes of which are determined by a matching across the surface to the plane waves in the vacuum. The method has been described by McRae [5.12], Boudreaux and Heine [5.19], Kambe [5.6], Pendry [5.9], and Jepsen et al. [5.13].

Let a Bloch wave have a plane-wave decomposition

$$\varphi_i(\vec{r}) = \sum_{\vec{g}} \left[b_{i\vec{g}}^+ \exp(i\vec{k}_{\vec{g}}^+ \cdot \vec{r}) + b_{i\vec{g}}^- \exp(-i\vec{k}_{\vec{g}}^- \cdot \vec{r}) \right] \tag{5.44}$$

midway between layers i and i+1. Note the inclusion of both penetrating and emerging plane waves in this expansion. The expansion coefficients depend on the index i because each layer intermixes the plane waves that are scattered by it. We shall need this mixing information and use it to relate the components of the Bloch waves φ_i and φ_{i+1} on either side of layer i+1. This mixing is described by the layer-diffraction matrices $M_{\vec{g}'\vec{g}}^{\pm\pm}$ of (5.26,31), which we shall write here in a form similar to that defined in (5.35),

$$R_{\vec{g}'\vec{g}}^{-+} = p_{\vec{g}'}^- r_{\vec{g}'\vec{g}}^{\mp} p_{\vec{g}}^+ = p_{\vec{g}'}^- M_{\vec{g}'\vec{g}}^{\mp} p_{\vec{g}}^+ \; ,$$

$$R_{\vec{g}'\vec{g}}^{+-} = p_{\vec{g}'}^+ r_{\vec{g}'\vec{g}}^{+-} p_{\vec{g}}^- = p_{\vec{g}'}^+ M_{\vec{g}'\vec{g}}^{+-} p_{\vec{g}}^- \; ,$$

$$T_{\vec{g}'\vec{g}}^{++} = p_{\vec{g}'}^+ t_{\vec{g}'\vec{g}}^{++} p_{\vec{g}}^+ = p_{\vec{g}'}^+ M_{\vec{g}'\vec{g}}^{++} p_{\vec{g}}^+ \; ,$$

$$T_{\vec{g}'\vec{g}}^{--} = p_{\vec{g}'}^- t_{\vec{g}'\vec{g}}^{--} p_{\vec{g}}^- = p_{\vec{g}'}^- M_{\vec{g}'\vec{g}}^{--} p_{\vec{g}}^- \; , \tag{5.45}$$

where

$$p_{\vec{g}}^\pm = \exp(\pm i\vec{k}_{\vec{g}}^\pm \cdot \vec{a}/2) \; , \tag{5.46}$$

if $\vec{a}$ is the layer-repetition vector. With this notation, we can relate the components of φ_i and φ_{i+1}, as illustrated in Fig. 5.3, i.e.

$$b_{i+1}^+ = T^{++} b_i^+ + R^{+-} b_{i+1}^- \; , \tag{5.47}$$

and

$$b_i^- = T^{--} b_{i+1}^- + R^{\mp} b_i^+ \; , \tag{5.48}$$

where matrix/vector notation is used.

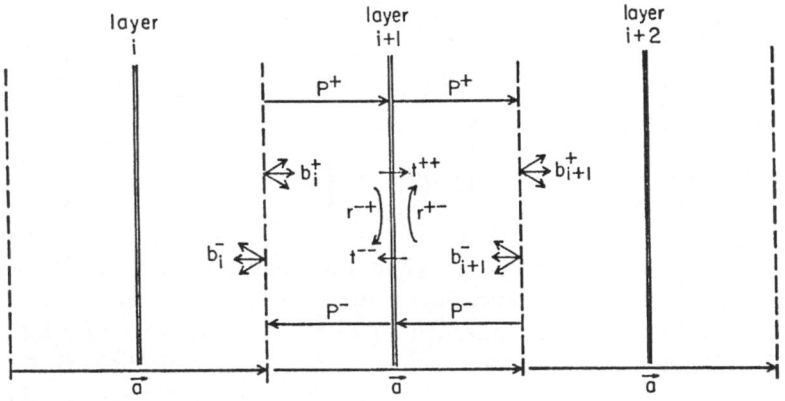

Fig. 5.3. Notation for components of Bloch waves in a periodic lattice

The Bloch condition implies that the relations

$$b_{i+1}^+ = b_i^+ \exp(i\vec{k}_B \cdot \vec{a}) \;, \tag{5.49a}$$

and

$$b_{i+1}^- = b_i^- \exp(i\vec{k}_B \cdot \vec{a}) \tag{5.49b}$$

are valid for each plane-wave amplitude of the Bloch wave φ_i. Here $\exp(i\vec{k}_B \cdot \vec{a})$ is an eigenvalue to be determined, thereby specifying the value of the Bloch wavevector $\vec{k}_B$.

5.5.2 The Bloch Functions

Equations (5.47-49) can be combined to yield the following eigenvalue equation, using a notation with double-length vectors:

$$\begin{pmatrix} T^{++} & R^{+-} \\ (T^{--})^{-1}R^{-+}T^{++} & -(T^{--})^{-1}R^{-+}R^{+-} + (T^{--})^{-1} \end{pmatrix} \begin{pmatrix} b_i^+ \\ b_{i+1}^- \end{pmatrix}$$

$$= \exp(i\vec{k}_B \cdot \vec{a}) \begin{pmatrix} b_i^+ \\ b_{i+1}^- \end{pmatrix} . \tag{5.50}$$

The eigenvectors of (5.50) are the Bloch wave functions that we sought. These wave functions have a simple behavior from one layer to the next, as given by (5.49). Consequently, it is sufficient to solve (5.50) for only one value of the layer index i.

5.5.3 The Transfer Matrix

An equivalent approach to the above problem uses the transfer matrix S that relates the plane waves from one layer to the next via

$$\begin{pmatrix} b_{i+1}^+ \\ b_{i+1}^- \end{pmatrix} = S \begin{pmatrix} b_i^+ \\ b_i^- \end{pmatrix} . \tag{5.51}$$

From (5.47,48), S is readily found to be

$$S = \begin{pmatrix} T^{++} - R^{+-}(T^{--})^{-1}R^{-+} & R^{+-}(T^{--})^{-1} \\ -(T^{--})^{-1}R^{-+} & (T^{--})^{-1} \end{pmatrix} . \tag{5.52}$$

According to (5.49,51), the matrix S, when diagonalized, also yields the Bloch wave functions and the eigenvalues. Whether in the transfer-matrix approach or in the Bloch-wave approach of (5.50), the number of Bloch wave functions obtained is 2g, where g is the number of $\bar{g}$-values used in the expansion of (5.44). Half of these Bloch wave functions propagate into the crystal, while the other half propagate out of the crystal toward the surface.

Only the former half are relevant, since no net electron current flows from the crystal interior to its surface in the LEED wave function. (This would require an electron source of the same energy E at the back of the crystal.) The reflected LEED electrons belong to the Bloch waves that propagate into the surface, i.e. have $\vec{k}_B$ pointing into the surface, and are represented by the second term in the brackets in (5.44).

5.5.4 Wave Matching at the Surface

Once the desired Bloch wavefunctions φ^m (m = 1,2,...,g) are known, the next step is to find the total wave function inside the surface, i.e. that linear combination $\Sigma_m a_m \varphi^m$ which applies. The simplest case is that of a periodic bulk material which is truncated suddenly at the surface plane, where the Bloch waves can be matched directly to the plane waves in the vacuum. These plane waves consist of the incident plane wave $\exp(i\vec{k}_0^+ \cdot \vec{r})$, arbitrarily given an amplitude of unity, and the reflected plane waves $\Sigma_g c_g^- \exp(-i\vec{k}_g^- \cdot \vec{r})$, of unknown amplitudes c_g^-. (We should correct the wave vectors $\vec{k}_g^\pm$ here for the surface-potential step which affects the electron kinetic energy, but, for simplicity, we shall ignore this detail at present). Near the surface plane, the combination $\Sigma_m a_m \varphi^m$ contains the incident set of plane waves $\Sigma_m a_m b_{mg}^+ \exp(i\vec{k}_g^+ \cdot \vec{r})$, which should be equal to the actual incident plane wave $\exp(i\vec{k}_0^+ \cdot \vec{r})$. Setting $\vec{r}=\vec{0}$ at the surface for convenience, this yields $\Sigma_m a_m b_{mg}^+ = \delta_{g\vec{0}}$, which is a system of g linear equations that can be solved for the g coefficients a_m. The emergent beam amplitudes are then simply $c_g^- = \Sigma_m a_m b_{mg}^-$. Finally, beam intensities (or reflectivities) follow from c_g^- as in (4.4).

The general case, however, includes surface layers that deviate from the bulk structure. Suppose there are s such layers, each characterized by known reflection and transmission coefficients. Then there are s+1 matching planes, namely s-1 between these layers, one on their vacuum side and one on their bulk side. Also, there are s-1 sets of 2g unknown plane-wave amplitudes near those layers [the plane waves on either side of each layer are directly related by expressions such as (5.47,48)], as well as the g unknown Bloch-wave amplitudes and the g unknown reflected plane-wave amplitudes in the vacuum, i.e. a total of 2gs unknowns. Each of the s+1 matching planes provides 2g conditions (counting both the plane-wave amplitudes and their derivatives) for a total of 2g(s+1). Of these conditions, 2g are superfluous since boundary conditions at both infinities (z = ±∞) have already been taken into account to reduce the number of unknowns to 2gs, i.e. the incident plane waves have already been specified to contain just the beam $\exp(i\vec{k}_0^+ \cdot \vec{r})$, and half the Bloch waves have been eliminated. Thus a system of 2gs linear equations must be solved to produce the desired wave amplitudes c_g^-, which solves the LEED problem in the general case.

5.5.5 Small Layer Spacings

We mention briefly a recent development by Jepsen of the transfer-matrix method that is designed to treat those cases where the spacing between adjacent atomic layers is relatively small [5.20]. In this case the plane-wave representation converges poorly, which affects the Bloch-wave representation equally, since it is expressed in terms of plane waves. The basic idea is to express the waves *incident* on a layer in terms of plane waves, while expanding the *scattered* waves in terms of spherical waves. The formalism is somewhat involved, but it allows a halving of the lower limit for tractable layer spacings. This is especially valuable for treating stepped surfaces since these have small layer spacings, presuming that layers are defined to be parallel to the macroscopic surface rather than parallel to the terraces of a stepped surface.

5.5.6 Relation to Band Structure

The Bloch-wave picture of LEED is convenient in visualizing the diffraction in terms of known concepts of band-structure theory. In the absence of inelastic processes, the Bloch waves have the property, familiar in band-structure theory, of propagating with constant magnitudes (but varying phases) through a crystalline lattice, i.e. the Bloch wave vector $\vec{k}_B$ is a real vector. This is true as long as the energy does not happen to fall within a band gap. If this occurs, no such propagating electron state exists, and the magnitude of the Bloch wave varies exponentially with position in the direction of the Bloch wavevector. In this case, $\vec{k}_B$ is complex, with an imaginary part that causes the Bloch waves either to decay or to increase exponentially. Waves that vary exponentially cannot be accommodated in an infinite medium, since they would have infinite amplitude somewhere, and therefore they are forbidden in the bulk of perfect crystalline solids. However, near the surface of a crystalline medium (or at other interfaces), such waves can exist, because the surface prevents them from growing to infinite values.

It is pedagogic to consider how band gaps arise in terms of the behavior of a plane wave impinging upon a crystalline sample from the outside (see also Sect. 4.8). If the incident plane wave has an energy and a direction such that a Bragg condition is satisified, we know from kinematic theory that constructive interference creates a relatively strong diffraction into certain other plane waves. This constructive interference, due to successive layers of atoms, builds up more and more intensity in the diffracted waves as the incident wave penetrates deeper into the surface. As a result, the incident wave loses intensity and decays exponentially. This occurs in X-ray diffraction as well as in electron diffraction, but much more strongly in the latter case. In fact, at LEED energies below approximately 30 eV, where the atomic scattering strength is quite large, this decay can take place over a distance of only a few atomic layers, sometimes well within the inelastic mean free path of the electron. This is nothing but a band gap: the incident wave cannot penetrate through the lattice, but is deflected into other directions, for example back out of the crystal. Thus, crystals with ideally-truncated surfaces (i.e. without deviating surface structures that can redistribute electrons among beams) give a strong reflectivity into some beams whenever a band-gap energy is chosen, each band gap corresponding to a Bragg condition. For completeness, one must include multiple scattering, which allows additional Bragg conditions between diffracted waves, not just between the primary incident wave and a diffracted wave. This allows more instances of strong diffraction. These concepts explain the relative complexity of LEED I-V curves, although the basic physical principles involved remain as simple as described earlier (Sect. 4.8).

The preceding discussion has assumed the absence of inelastic effects. When one includes these effects, the only changes in the preceding arguments are (1) even within the allowed bands, the Bloch waves decay exponentially as a result of current loss; and (2) the inelastic effects reduce the number of reflected electrons and smooth its dependence on energy and incident direction, removing, for example, the sharp band-gap edges that would otherwise create abrupt changes in the reflectivity as a function of energy or incidence direction (see Fig. 4.17 g versus h). More extensive discussions and illustrations of these issues can be found in the literature [5.9].

5.6 Diffraction by a Stack of Layers: Layer-Stacking and Layer-Doubling Method

5.6.1 The Case of Two Layers

To calculate exactly the reflection by two layers, including all multiple scattering, we may sum explicitly all orders of multiple scattering occurring between the individual layers [(the multiple scattering within layers is assumed solved, see (5.26 or 31 and 35)]. If A and B label two adjacent layers, and $P_1^{\pm}$, $P^{\pm}$, and $P_2^{\pm}$ are plane-wave propagators (for every $\bar{g}$) between planes 1 and A, A and B, and B and 2, respectively, see (5.46) and Fig. 5.4, then we may write in matrix/vector notation, for the reflectivity of the layer pair A+B,

$$R^{\mp} = P_1^- r_A^{-+} P_1^+$$
$$+ P_1^- t_A^{--} P^- r_B^{-+} P^+ t_A^{++} P_1^+$$
$$+ P_1^- t_A^{--} P^- r_B^{-+} P^+ r_A^{+-} P^- r_B^{-+} P^+ t_A^{++} P_1^+ + \cdots$$
$$= P_1^- [r_A^{-+} + t_A^{--} P^- r_B^{-+} P^+ (I - r_A^{+-} P^- r_B^{-+} P^+)^{-1} t_A^{++}] P_1^+ . \qquad (5.53)$$

The infinite sum over all paths has produced a geometric series leading to an

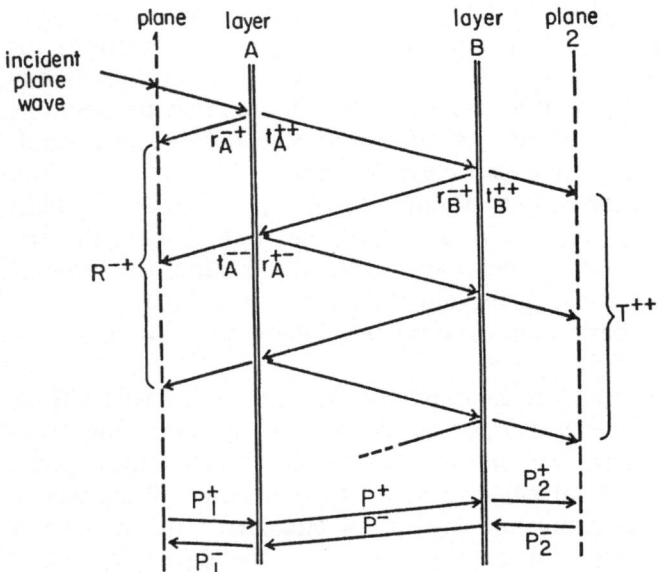

Fig. 5.4. Diagram of the Layer-Stacking scheme for one pair of layers

exact expression (here I is the unit matrix). The same reasoning applies to transmission through the double layer, in which case

$$T^{++} = P_2^+ t_B^{++} P^+ t_A^{++} P_1^+$$
$$+ P_2^+ t_B^{++} P^+ r_A^{+-} P^- r_B^{-+} P^+ t_A^{++} P_1^+ + \cdots$$
$$= P_2^+ [t_B^{++} P^+ (I - r_A^{+-} P^- r_B^{-+} P^+)^{-1} t_A^{++}] P_1^+ . \tag{5.54}$$

One obtains R^{+-} and T^{--} (for incidence from the bulk side of the double layer) from $R^{\mp}$ and T^{++} by interchanging + and -, A and B, and 1 and 2 everywhere. It is convenient for the following discussion to make the planes 1 and A coincide, as well as to make the planes B and 2 coincide. Then the reflection and transmission matrices for the layer pair A+B are

$$R^{-+} = r_A^{-+} + t_A^{--} P^- r_B^{-+} P^+ (I - r_A^{+-} P^- r_B^{-+} P^+)^{-1} t_A^{++} , \tag{5.55a}$$

$$T^{++} = t_B^{++} P^+ (I - r_A^{+-} P^- r_B^{-+} P^+)^{-1} t_A^{++} , \tag{5.55b}$$

$$R^{+-} = r_B^{+-} + t_B^{++} P^+ r_A^{+-} P^- (I - r_B^{-+} P^+ r_A^{+-} P^-)^{-1} t_B^{--} , \tag{5.55c}$$

$$T^{--} = t_A^{--} P^- (I - r_B^{-+} P^+ r_A^{+-} P^-)^{-1} t_B^{--} , \tag{5.55d}$$

with

$$P_{\vec{g}}^{\pm} = \exp(\pm i \vec{k}_{\vec{g}}^{\pm} \cdot \vec{r}_{BA}) , \tag{5.56}$$

if $\vec{r}_{BA}$ links reference points on the layers A and B.

5.6.2 The Case of Many Layers

There are no equivalent formulas for more than two layers, but we can construct a crystal by iterating the two-layer idea, namely by adding stepwise individual layers to a growing stack of layers and by the iterative use of (5.55). The reflection matrix R^{-+} of the growing stack will converge due to the limited electronic mean free path. Typically, ten metal layers produce convergence.

As a result of bulk periodicity perpendicular to the surface, which applies to many clean surfaces and substrates, the iteration can be rendered considerably more rapid when identical layers are being stacked. In this case, at each step the iteration can double the thickness of the slab of layers by repeatedly combining two identical slabs as depicted in Fig. 5.5. Thus, the nth iteration step combines two identical slabs, consisting of 2^{n-1} layers each, into one slab consisting of 2^n layers. This is the "Layer-Doubling" method developed by Pendry [5.9]. For a stack of eight layers, three iterations are sufficient; a fourth iteration would produce a stack of 16 layers in only 33% more time, which illustrates the effectiveness of this method for treating relatively deep electron penetration.

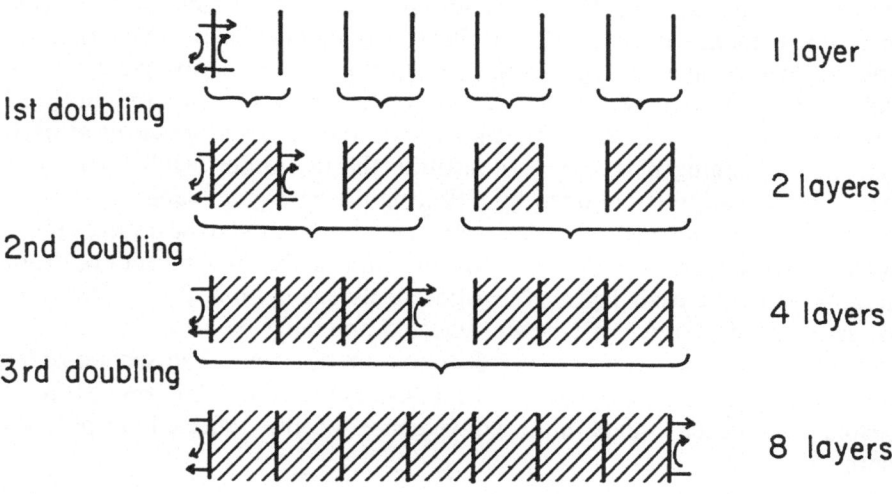

Fig. 5.5. Diagram of the Layer-Doubling scheme with the surface at the left

The Layer-Doubling method yields a full reflection matrix for the stack of equal layers. Other layers different from bulk layers can then be added to the surface at will, using the first of (5.55a) again. For efficiency, the bulk reflection can be stored and used repeatedly for many different configurations of surface layers (such as different overlayer spacings and registries), a convenient feature in a surface structure search.

5.7 Diffraction by a Stack of Layers: Renormalized-Forward-Scattering (RFS) Perturbation Method

5.7.1 The Principle of RFS

The Renormalized-Forward-Scattering (RFS) perturbation method intro-duced by Pendry [5.9] follows the principle that transmission through any layer should not be described by unperturbed plane waves, but rather by plane waves modified by forward scattering through that layer together with all other plane waves transmitted with various scattering angles. It is the *reflection* by any layer that is considered to be weak, and therefore the per-turbation is based on an expansion of the total reflectivity of the surface in terms of the number of reflections. The lowest order contains all paths that have been reflected only once, but transmitted any number of times; the next order contains only triple-reflection paths (an odd number of reflections is needed to bring electrons back out of the surface), and so on. This is the same approach that was employed in the reverse-scattering perturbation scheme described in Sect. 5.4. (In fact, RSP was modeled after RFS.)

A convenient way to produce the scattering paths for the lowest order of perturbation is to follow the plane waves generated by the incident beam as they are forward-scattered from layer to layer into the surface until inelastic effects make them die out. This is illustrated in Fig. 5.6a. Then, starting from the deepest layer reached in the penetration, the emerging plane waves, obtained by reflecting the penetrating ones, are forward-scattered outward from layer to layer, collecting reflections from the penetrating waves at each layer. The emerging plane waves constitute the first-order result. The next order of perturbation is obtained by reflecting the emerging plane waves back into the crystal, starting from the top layer, and forward-scattering them inward as in the first order, but now collecting additional reflections from emerging waves at each layer. Inelastic effects again limit the penetration, and, as in the first order, the newly reflected emerging plane waves are pro-pagated to the surface, collecting reflections on the way. The second-order result emerges, which is added to the first-order result. This procedure is repeated for higher orders until convergence of the reflected amplitudes occurs.

5.7.2 The Formalism of RFS

Each transmission through one layer, with the accompanying collection of reflections, is performed in the following way. The plane-wave amplitudes $a_{(i)\bar{g}}$ in the ith interlayer spacing are computed iteratively using two expres-sions, one for penetration (Fig. 5.6b),

$$a_{(i)\bar{g}}^{new} = \sum_{\bar{g}'}(t_{\bar{g}\bar{g}'}^{++}P_{\bar{g}'}^{+(i-1)}a_{(i-1)\bar{g}'} + r_{\bar{g}\bar{g}'}^{+-}P_{\bar{g}'}^{-(i)}a_{(i)\bar{g}'}) \,, \tag{5.57}$$

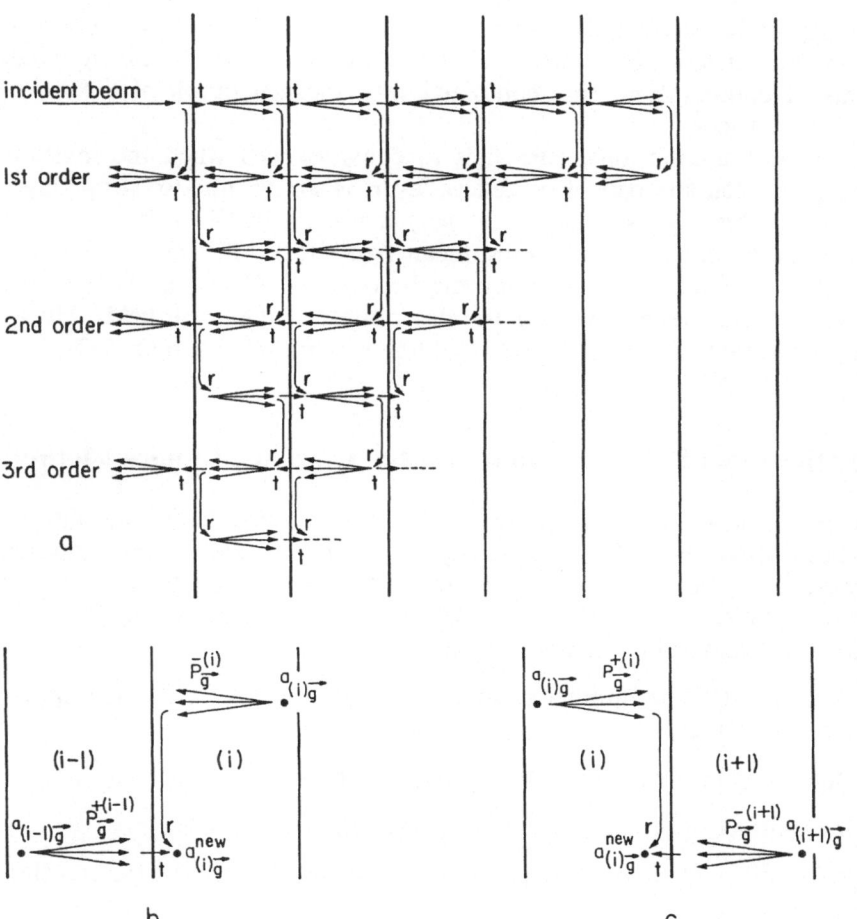

Fig. 5.6. Diagram of the Renormalized-Forward-Scattering scheme with the surface at the left

and one for emergence (Fig. 5.6c),

$$a_{(i)\bar{g}}^{new} = \sum_{\bar{g}'}(t_{\bar{g}\bar{g}'}^{--}P_{\bar{g}'}^{-(i+1)}a_{(i+1)\bar{g}'} + r_{\bar{g}\bar{g}'}^{-+}P_{\bar{g}'}^{+(i)}a_{(i)\bar{g}'}) , \tag{5.58}$$

where $P_{\bar{g}}^{\pm(i)}$ are plane-wave propagators between appropriate reference points on successive layers, and $a_{(i)\bar{g}}^{new}$ should be considered as constantly overwriting $a_{(i)\bar{g}}$. Note that we have arbitrarily chosen to use the amplitudes just past each scattering layer, see Figs. 5.6b,c, and not, for example, halfway between layers. Counting the surface plane (where the inner potential and damping begin) as i=1, the initial values for the iteration of (5.57,58) are

$$a_{(i)\bar{g}} = 0 \quad \text{for all } i,\bar{g}, \text{ except } a_{(1)\bar{0}} = 1 . \tag{5.59}$$

The RFS scheme typically uses 12 to 15 layers and three to four orders of iteration for convergence. The RFS computation requires relatively small amounts of computation time compared with the generation of the layer-diffraction matrices.

By its very assumptions, RFS fails to converge well when the multiple scattering between any pair of successive layers is strong (due to strong layer reflections). This occurs especially at very low energies (E ~ 10 eV) when the electron damping is small (long mean free path). The RFS method is also unstable when an interlayer spacing becomes small ($\lesssim 1.0$ Å), requiring many evanescent plane waves for the description of the wave field. Under all these conditions, Layer Stacking/Doubling is more reliable than RFS.

5.8 Efficiency of Computation and the Combined-Space Method

In the previous sections, several methods for calculating LEED amplitudes have been discussed. It is necessary to compare their computational efficiency, since the calculations must be performed repeatedly in structural searches in surface crystallography. We shall compare their scaling behavior with the most important variables, namely:

$L = (l_{max} + 1)^2$, the number of spherical waves needed for the description of atomic scattering;

N, the number of subplanes or layers included;

g, the number of beams (after symmetrization, if applicable);

s, the number of surface matching planes in the Bloch-wave (or transfer-matrix) approach.

The computational time for calculating all elements of the diffraction matrices for an atomic layer with N atoms per unit cell by matrix inversion is approximately proportional to $L^3 N^3 g^2$, whereas, if RSP is used, this time scales as $L^2 N^2 g^2$. The Bloch-wave (or transfer-matrix) method is approximately proportional to $(2g)^3 + (2gs)^3$. Layer Stacking scales as $g^3 N$, Layer Doubling scales as $g^3 \ln(N)$, while RFS scales as $g^2 N$. If matrix inversion or RSP is applied to a thick layer that avoids the need for Layer Stacking in the plane-wave representation, one need only calculate g matrix elements (corresponding, for example, to the measured beams only). In this case, the time scales as $L^3 N^3 g$ or $L^2 N^2 g$, respectively.

Thus, the most effective combination of methods consists of using layers that have one atom per unit cell and assembling these with RFS. However, this neglects the complication that the plane-wave representation is poor for closely spaced layers. This effect can be estimated with the following formula that gives the number of plane waves needed as a function of the unit cell area A, the kinetic energy E, and the smallest interlayer spacing d_{min}:

$$g = \frac{A}{4\pi} \left[\frac{2m}{\hbar^2} E + \frac{c}{d_{min}^2} \right],$$

(5.60)

where c defines the cutoff for evanescent waves and is determined by the desired accuracy. Thus, the computation times of the affected calculation schemes scale, for small layer spacings d_{min}, as d_{min}^{-4} or d_{min}^{-6}, not to speak of convergence problems that are then encountered. Consequently, one is forced to use the time-consuming spherical-wave representation for treating small layer spacings.

Nevertheless, one may compromise quite effectively by combining the plane-wave and spherical-wave representations whenever possible. This occurs when *some* layer spacings are small while others are large. For example, an overlayer that approaches very closely the top substrate layer, while the substrate layers are not mutually close together, is a good candidate for the "Combined-Space Method". A surface composed of well-separated layers, each layer containing more than one atom per unit cell, which may be coplanar or nearly so, is another example. The Combined-Space Method, as described by Tong and Van Hove [5.18], has several advantages. (1) The number of plane waves g can be reduced substantially, (2) the use of plane waves is especially attractive when one must make repeated calculations for many surface structures in a structural search, and (3) one can easily exploit the advantages of superlattices (block-diagonalization of substrate layer-diffraction matrices, to be described in Sect. 5.9.1) and of symmetries present in the surface structure.

5.9 Superlattices and Domains

5.9.1 Diffraction and Superlattices

Many ordered surfaces, whether they are clean and reconstructed or have an adsorbate overlayer, have superlattices, that is, two-dimensional unit cells occur that are larger than the unit cell expected from an ideal truncation of the substrate [5.21]. Often, both the unit cell of the substrate and that of the surface occur simultaneously in LEED in different layers, because the LEED electrons penetrate below the deviating surface layers into the ideal substrate. This is the usual case for adsorption with coverages below a monolayer or for reconstructions that are limited to one or two atomic layers. Thus, it will be necessary to consider electron diffraction by a combination of layers with different unit cells. Types of superlattices occurring at surfaces and their notations have been described in a general fashion in Sect. 3.2. Kinematic diffraction by surfaces with superlattices has been discussed in Sect. 4.2.8.

Here, in the dynamical case, we shall limit ourselves to commensurate super-lattices, the unit-cell area of which is an integral multiple of the substrate unit cell. This includes many common surface structures such as (2x1), c(2x2), p(2x2), and $(\sqrt{3} \times \sqrt{3})R30°$.

Dynamical Effects

A convenient way to visualize the dynamical effect on diffraction by a combination of different lattices in different layers is to follow a plane wave as it propagates from one layer to the next. Assume that $\vec{g}^{(S)}$ and $\vec{g}^{(B)}$ are the reciprocal lattice vectors of the surface superlattice and the substrate lattice, respectively. Then a plane wave $\exp(i\vec{k}_0 \cdot \vec{r})$ incident on the surface will give rise first to reflected and transmitted plane waves $\{\vec{k}_{0\parallel} + \vec{g}^{(S)}\}$, where the braces denote the full set of vectors obtained by taking all values of $\vec{g}^{(S)}$. The transmitted collection of plane waves $\{\vec{k}_{0\parallel} + \vec{g}^{(S)}\}$ encounters next the second layer of atoms. If this layer has the same superlattice, it will merely redistribute the plane-wave amplitudes among the same plane waves $\{\vec{k}_{0\parallel} + \vec{g}^{(S)}\}$, without generating new plane waves. The first substrate layer that is reached, however, can diffract all the incident plane waves $\{\vec{k}_{0\parallel} + \vec{g}^{(S)}\}$ into any of the plane waves $\{\vec{k}_{0\parallel} + \vec{g}^{(S)} + \vec{g}^{(B)}\}$, and any subsequent substrate layer will again merely redistribute the amplitudes among the plane waves $\{\vec{k}_{0\parallel} + \vec{g}^{(S)} + \vec{g}^{(B)}\}$. Our assumption of commensurability and an integral ratio of unit-cell areas implies that the vectors $\{\vec{g}^{(B)}\}$ are a subset of the vectors $\{\vec{g}^{(S)}\}$. In the LEED pattern, this simply means that the "integral-order" spots also belong to the reciprocal lattice of the superlattice. For example, in a p(2x2) pattern the p(2x2) spots include the integral-order spots, whereas in an incommensurate situation the substrate-induced spots do not coincide with any of the superlattice spots, except for the specular beam. Thus, the set of waves $\{\vec{k}_{0\parallel} + \vec{g}^{(S)} + \vec{g}^{(B)}\}$ is not different from the set $\{\vec{k}_{0\parallel} + \vec{g}^{(S)}\}$. As the plane waves $\{\vec{k}_{0\parallel} + \vec{g}^{(S)}\}$ continue their journey from layer to layer and emerge at the surface, no new plane waves are generated: only a redistribution of amplitudes occurs at each layer. Consequently, it is sufficient to include for an entire LEED calculation just the set of plane waves $\{\vec{k}_{0\parallel} + \vec{g}^{(S)}\}$.

Block-Diagonalization of Substrate Diffraction Matrices

One simplification has not been mentioned above for substrate layers. This is that a substrate layer can scatter a given wave $\vec{k}_{0\parallel} + \vec{g}_i^{(S)}$ only to a certain restricted part of the set of waves $\{\vec{k}_{0\parallel} + \vec{g}^{(S)}\}$, namely to those waves $\vec{k}_{0\parallel} + \vec{g}_j^{(S)}$ for which $\vec{g}_j^{(S)} - \vec{g}_i^{(S)} = \vec{g}^{(B)}$ for any bulk vector $\vec{g}^{(B)}$. In other words, in the substrate a surface vector $\vec{g}_i^{(S)}$ can be scattered only through a bulk vector $\vec{g}^{(B)}$. An example for a p(2x2) overlayer is shown in Fig. 5.7. The effect of this is that the plane waves $\{\vec{k}_{0\parallel} + \vec{g}^{(S)}\}$ fall into subsets such that the waves within any given subset are mutually related by a *bulk* reciprocal lat-

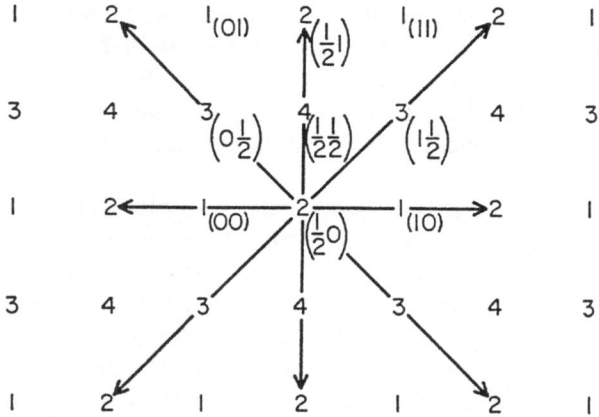

Fig. 5.7. Reciprocal lattice of a (2 × 2) structure on a square substrate. Arrows are reciprocal lattice vectors $\vec{g}^{(B)}$ of the substrate. Beams are labeled by numbers (1 through 4) that indicate to which subset of beams they belong (see text)

tice vector $\vec{g}^{(B)}$ and that waves in different subsets cannot scatter into each other in the substrate. This implies that the diffraction-matrix element $M_{\vec{g}'\vec{g}}^{\pm\pm}$ is only nonzero for substrate layers when $\vec{g} - \vec{g}'$ is a substrate vector $\vec{g}^{(B)}$. Then, by grouping vectors $\vec{g}$ into subsets, the matrices $M_{\vec{g}'\vec{g}}^{\pm\pm}$ block-diagonalize as shown in Fig. 5.8. The number of subsets and therefore the number of blocks in the block-diagonalization is simply equal to the ratio P of unit-cell areas, e.g. four for a p(2x2) superlattice. The computational benefit of such block-diagonalization is large, both in storage requirements

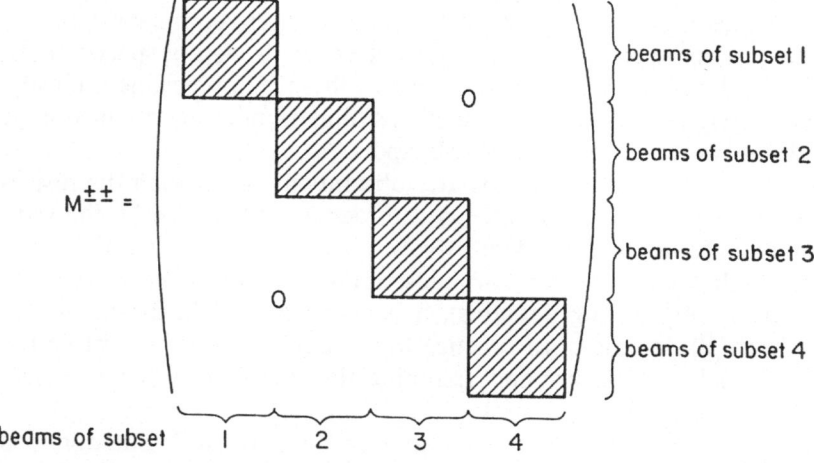

Fig. 5.8. Block-diagonalized substrate-layer-diffraction matrix for the case of four subsets of beams, as with a (2 × 2) structure

and in speed. This is true whatever the computational method used, so long as it is based on plane waves, such as the Bloch-wave method, Layer Stacking/Doubling or RFS. Another important benefit that results is the fact that the substrate layers can be treated as layers with one atom per unit cell, even though the surface unit cell would imply P atoms per unit cell in the substrate layers.

Such a block-diagonalization does not occur if the spherical-wave representation is used throughout the surface, as in matrix inversion and in RSP, where no benefit arises from the separation of beams into subsets. In addition, with the spherical-wave representation, each of the substrate layers must be considered to have P atoms per unit cell, a serious drawback. The use of the Combined-Space Method avoids this drawback, when the plane-wave representation is convergent in the substrate: it allows the beam subsets and block-diagonalization to be used in the substrate and confines the matrix inversion or RSP to the surface layer.

5.9.2 Domains

With superlattices at surfaces, one has to keep in mind that domain structures are possible, as discussed in Sect. 3.4. In the case of domains due to periodicity degeneracy, illustrated in Fig. 3.10, each domain of the surface has the proper periodicity but is mismatched across a domain boundary to neighboring out-of-phase domains. If the domains are smaller than the instrumental transfer width (which is often on the order of a few hundred Ångstroms), the diffraction is complicated because different domains contribute coherently to the diffraction. For example, with domains of random size and shape, certain LEED spots can broaden. With a periodic array of domains of constant size and shape with regular domain-to-domain relationships, one obtains a new, larger unit cell (containing one complete domain). This produces a corresponding diffraction pattern with closely spaced spots, see Sect. 3.5.7. Only a fraction of these spots will have appreciable intensity, namely those spots located near the spot positions implied by a single large domain: this gives the appearance of split spots.

On the other hand, if the domains are substantially larger than the instrumental transfer width, it is legitimate to consider the diffraction by the individual domains to be independent. Now, rather than adding diffracted amplitudes (with the relative phases describing the interference between amplitudes from neighboring domains), it is sufficient to add the intensities emanating from all domains. In practice this is equivalent to neglecting the presence of domains altogether and assuming that the surface has a perfect two-dimensional periodicity.

In the case of rotational degeneracy among domains, different domains can produce different sets of spots as discussed in Sect. 3.4 (see Fig. 3.11). Then some spot intensities are due to one domain orientation, while others are due to another domain orientation. Therefore the spot intensities are proportional to the fractional surface area that has the corresponding domain orientation. The remaining spots, which are usually the integral-order spots, contain contributions from all domain orientations, with weights in proportion to the relative domain sizes (assuming incoherence due to sufficiently large domains). In general, therefore, each such spot has an I-V curve that is the weighted sum of independent I-V curves.

The assumption of large domains giving incoherent contributions has always been made in LEED calculations. It has presumably worked well because the corresponding experiments have always attempted to achieve the sharpest possible diffraction spots, i.e. the largest possible domain size (at the same time as the best possible order within the domains).

5.10 Symmetries

5.10.1 Types of Symmetry

Symmetries often occur in the electron-diffraction pattern and, when they occur, may be utilized to reduce the computational effort in a calculation. Types of symmetry that often occur are (see also Sect. 3.4):

1. Structural symmetries particular to the atomic positions in the surface, such as two-, three-, or fourfold axes of rotational symmetry (perpendicular to the surface), mirror planes and glide planes (also perpendicular to the surface), and mirror planes parallel to the surface for individual atomic layers. Such symmetries can be used to reduce the computational effort regardless of the presence or absence of any wave field symmetries indicated below.

2. Wave field symmetries that arise out of a privileged incidence direction of the LEED electron beam. Most of the structural symmetries of a surface can appear in the LEED wave field (whatever the contribution of multiple scattering) and can then be utilized to render the computation more efficient. On the other hand, the wave field may at times appear to be more symmetrical than the structure itself. This is the case when equivalent rotated or mirrored domains of surface structure sum their contributions to the diffraction pattern and yield a higher symmetry in the pattern (e.g. a fourfold rotational axis of symmetry) than the individual domains may have (e.g. a twofold axis of structural symmetry).

An inconvenience of utilizing symmetry is that the variety of possible surface symmetries requires separate programming for each type of symmetry. This can be especially cumbersome in the spherical-wave representation, and therefore few computations utilize symmetry within that representation, despite potentially large gains. The programming is much more manageable in the plane-wave representation, and proportionally more computations are performed in this way. The potential benefit is also very large here: it suffices to recall that typical computation times scale as the square or the cube of the number of plane waves involved and that, for example, an n-fold rotational axis effectively divides that number of plane waves by approximately n.

5.10.2 The Formalism of Symmetrization

For all but the glide-plane symmetries, the mathematics of symmetrization are familiar and quite straightforward. However, since the case of glide-plane symmetries is not so commonly used and since it presents some features of particular interest at surfaces, we shall discuss the general formalism of symmetrization in LEED here, limiting ourselves to its application to plane-wave diffraction by layers of atoms.

First we exclude glide-plane symmetry from our considerations. Then, if an electron beam is incident on a crystal surface along an axis or a plane of symmetry of the surface structure, the diffracted plane waves will have correspondingly symmetrical amplitudes. Standard group theory can be applied in this case to reduce greatly the required computational effort, both in storage and in time.

Two mutually symmetrical plane waves a_1 and a_2 traveling between atomic layers can be combined to produce waves $b_s = (1/\sqrt{2})(a_1 + a_2)$ and $b_a = (1/\sqrt{2})(a_1 - a_2)$ that are symmetrical and antisymmetrical, respectively, about the axis or plane of symmetry relating a_1 and a_2. If the two waves a_1 and a_2 have equal amplitude due to the primary incident beam being parallel to the symmetry axis or plane, then the antisymmetrical combination will have zero amplitude. In the same symmetry situation a layer-diffraction matrix

$$M = \begin{pmatrix} m_{11} & m_{12} \\ m_{21} & m_{22} \end{pmatrix}$$

operating on the vector

$$\begin{pmatrix} a_1 \\ a_2 \end{pmatrix}$$

has the symmetries $m_{11} = m_{22}$ and $m_{12} = m_{21}$ (at least if the origin of coordinates used in defining M lies on the symmetry axis or plane). Under the basis set transformation

$$\begin{pmatrix} a_1 \\ a_2 \end{pmatrix} \rightarrow \begin{pmatrix} b_s \\ b_a \end{pmatrix} = T \begin{pmatrix} a_1 \\ a_2 \end{pmatrix},$$

where

$$T = (1/\sqrt{2})\begin{pmatrix} 1 & 1 \\ 1 & -1 \end{pmatrix},$$

M changes into

$$M' = TMT^{-1} = \begin{pmatrix} m_{11} + m_{21} & 0 \\ 0 & m_{11} - m_{21} \end{pmatrix}. \tag{5.61}$$

Due to the block-diagonal form of (5.61), the antisymmetrical wave b_a is independent of the symmetrical wave b_s and can be dropped from the calculation because it is not excited by the symmetrical wave field incident on the crystal surface. Thus the matrix dimensions are reduced accordingly, since only the submatrix $(m_{11} + m_{21})$ is retained.

The basic symmetry utilization described above extends to mirror planes, two-, three-, four-, and sixfold symmetry axes and combinations thereof. In general, matrix reduction is performed in the following way. Let us describe a plane wave $a_{\bar{g}}$ together with the group of I-1 plane waves related to it by symmetry by the quantities $a_{\bar{g}i}$ (i = 1,2,..., I). We obtain a symmetrical wave function by constructing

$$b_{\bar{g}} = I^{-1/2} \sum_{i=1}^{I} a_{\bar{g}i};$$

nonsymmetrical wave functions need not be considered, since they are not excited by the incident beam. We designate the diffraction-matrix elements by $M_{\bar{g}'j,\bar{g}i}$ (i = 1,2,..., I; j = 1,2,..., J; and I and J can be different). Here, J represents any group of plane waves related to each other by the same symmetry. The matrix element for diffraction from the symmetrical wavefunction $b_{\bar{g}}$ into the symmetrical wavefunction

$$b_{\bar{g}'} = J^{-1/2} \sum_{j=1}^{J} a_{\bar{g}'j}$$

becomes, generalizing the steps leading to (5.61),

$$M'_{\bar{g}'\bar{g}} = \sqrt{J/I} \sum_{i=1}^{I} M_{\bar{g}'1,\bar{g}i} = \sqrt{I/J} \sum_{j=1}^{J} M_{\bar{g}'j,\bar{g}1}. \tag{5.62}$$

5.10.3 Glide-Plane Symmetry

We now consider glide-plane symmetries, taking as a representative example just one glide plane. First we must consider the effect of this glide plane on diffraction. Suppose a plane wave $\bar{g}$ is incident on a layer which has a glide plane parallel to the yz plane (z being perpendicular to the surface). The propagation direction of the plane wave shall be parallel to this glide plane, but not necessarily at normal incidence. We shall use a reference point for plane-wave amplitudes that is located in the glide plane (the choice of reference point is important for the details of symmetrization).

One can then show easily that the layer-diffraction matrix $M_{\vec{g}'\vec{g}}$ has the following properties [see also, for example, (7.13) of Sect. 7.2.3]. If $\vec{g} = (g_x,g_y) = (0,g_y)$ for the incident beam and $\vec{g}' = (g'_x,g'_y)$ for the diffracted beams, then

$$M_{(-g'_x,g'_y),(0,g_y)} = \exp[i(g'_y - g_y)a_y/2]M_{(g'_x,g'_y),(0,g_y)} , \tag{5.63}$$

where a_y is the lattice constant in the direction parallel to the glide plane. The exponential factor is alternately +1 and -1 for successive values of g'_y. Therefore the incident beam is diffracted into pairs of symmetrical beams (g'_x,g'_y) and $(-g'_x,g'_y)$ which are alternately in phase and in antiphase with each other as g'_y varies. Note that when $g'_x = 0$, the antiphase relationship makes the diffracted intensity vanish for every other value of g'_y: these are the extinctions characteristic of glide-plane symmetry. As a result, these vanishing beams can simply be dropped from the formalism from the outset.

The next step is to generalize our result to a pair of symmetrical incident plane waves (g_x,g_y) and $(-g_x,g_y)$ that will diffract to a pair of waves (g'_x,g'_y) and $(-g'_x,g'_y)$, among others. Now the relevant layer-diffraction-matrix elements, in the simplified notation m_{ij}, satisfy

$$m_{22} = \exp[i(g'_y - g_y)a_y/2]m_{11} ,$$
$$m_{21} = \exp[i(g'_y - g_y)a_y/2]m_{12} , \tag{5.64}$$

with the same exponential factor as encountered in the previous case. Consequently, $m_{22} = \pm m_{11}$ and $m_{21} = \pm m_{12}$. Using the same transformation T that led to (5.61) for non-glide-plane symmetry, we now find

$$M' = \begin{pmatrix} m_{11} + m_{21} & 0 \\ 0 & m_{11} - m_{21} \end{pmatrix} \quad \text{for} \quad \exp[i(g'_y - g_y)a_y/2] = 1 , \tag{5.65}$$

and

$$M' = \begin{pmatrix} 0 & m_{11} + m_{21} \\ m_{11} - m_{21} & 0 \end{pmatrix} \quad \text{for} \quad \exp[i(g'_y - g_y)a_y/2] = -1 . \tag{5.66}$$

Equation (5.65) shows that the symmetrical and antisymmetrical combinations of the beams $(\pm g_x,g_y)$ can diffract only into the symmetrical and antisymmetrical combinations, respectively, of the beams $(\pm g'_x,g'_y)$ when the exponential factor is +1. Likewise (5.66) shows that, when the exponential factor is -1, the symmetrical combination diffracts only into the antisymmetrical combination and vice versa. Consequently, in the presence of glide-plane symmetry we need to follow simultaneously both symmetrical and antisymmetrical combinations of symmetrical beams, depending on the value of $g'_y - g_y$, since they interact alternately; by contrast, for non-glide-plane symmetries, only the symmetrical combinations occur. Nevertheless, the same reduction of matrix dimensions results.

The question immediately arises how this actually functions at surfaces where some atomic layers, e.g. an overlayer, have glide-plane symmetry, while others have mirror-plane symmetry (a commonly expected case). We must assume that the full surface, with all layers taken together, satisfies the given glide-plane symmetry, since otherwise there is no symmetry of interest here. This is possible because mirror-plane layers can also satisfy glide-plane symmetry, if properly positioned. The general answer is that layers with glide-plane symmetry behave as described above (i.e. symmetrical and antisymmetrical combinations can interact), while layers with mirror-plane symmetry diffract symmetrical combinations of beams only into symmetrical combinations and antisymmetrical combinations only into antisymmetrical combinations. Consequently, even in layers with mirror symmetry, one encounters not only symmetrical combinations, but also antisymmetrical combinations due to a layer with glide-plane symmetry. For this description to be valid, it is important to maintain the reference points for plane waves on the glide plane. Moreover, one should realize that layers with glide-plane symmetry usually have a larger unit cell than the adjoining layers with mirror planes and thus create fractional-order beams that fall into independent subsets insofar as the mirror-plane layers are concerned. (In this case, block-diagonalization occurs in the mirror-plane layers.)

We have sketched the case of one glide plane: it leads to the same reduction in diffraction-matrix dimensions as in the case of a mirror plane, while the additional computational cost in dealing with the antisymmetrical combination of beams is minimal. More generally, one can have, for example, one glide plane and one mirror plane perpendicular to it, or two orthogonal glide planes, or two glide planes and a fourfold rotation axis, with correspondingly larger reductions in matrix dimensions. All these symmetries can clearly also be utilized in angular-momentum space.

5.11 Thermal Effects

In Sect. 4.7 we discussed thermal effects within a kinematic framework, in particular with the use of the Debye-Waller factor exp(-2M) familiar in X-ray Diffraction [5.1]. We pointed out that complications arise from multiple scattering, and these complications shall be discussed here.

The observed intensity decrease with increasing temperature is understood easily as due to thermal structural disorder, which causes disruption of the phase coincidence in the constructive interference responsible for peaks (Fig. 4.16). This is the basis for the explanation of the corresponding X-ray Diffraction behavior as given by Debye and Waller; each diffracted beam is on the average effectively weakened by the Debye-Waller factor. Application of this idea to LEED is not straightforward because multiple scattering makes the wave field incident on any atom depend on the instantaneous position of the other vibrating atoms, unlike the X-ray case where the incident beam is not noticeably affected by the crystal. As pointed out by Holland [5.22], this implies that we cannot simply time-average the wave field as it reaches the observer (the Debye-Waller idea), but that we must time-average the wave field scattered by each atom before it reaches another atom. The full formalism for calculating LEED intensities can then be applied to the waves that have been time-averaged "in situ".

With this in mind, let us consider the following model. After scattering by a vibrating atom, the intensity of a LEED wave is reduced from the non-vibrating case by the Debye-Waller factor exp(-2M), where M contains the momentum transfer appropriate for that scattering. Then a second scattering is allowed to occur in which one can again apply a Debye-Waller factor exp(-2M'), where M' now contains the momentum transfer of the second scattering. Further scatterings can be dealt with similarly. In this model the thermal effects of subsequent scatterings are assumed to be independent, as if different atoms in a lattice were vibrating completely independently of each other. This is clearly not correct, especially for near-neighbor atoms. However, it is difficult to include these correlations between atomic vibrations in a diffraction formalism, and therefore it has to our knowledge not been done in any existing LEED formalism.

In the above model, an important factor has been ignored, namely the phase of the scattering wave as it propagates from atom to atom. This phase can be restored easily if we follow the wave amplitude rather than its intensity. The corresponding Debye-Waller factor is then simply exp(-M) rather than exp(-2M). Thus each atomic scattering amplitude f(θ), see Sect. 4.5, is replaced by f(θ)exp(-M) in the presence of atomic vibrations, while the remaining wave propagation between atoms is unaffected. In other words, the multiple-scattering formalisms discussed earlier in this chapter are applied with just the substitution of f(θ)exp(-M) for f(θ). This model of thermal effects in the multiple-scattering case has been shown theoretically by Holland [5.22] and by Duke and Laramore [5.23] to be correct with the assumption of independent atomic vibrations. It also has the virtue of fitting very naturally into the multiple-scattering formalism, at least for isotropic atomic vibrations. This is because the scattering amplitude f exp(-M) can be expanded in partial waves almost as easily as f itself can be. We obtain "temperature-dependent" phase shifts $\delta_l(T)$ rather than the traditional phase shifts δ_l and can then reduce the whole problem of temperature effects to the simple prescription of replacing δ_l by $\delta_l(T)$. This is the choice made in all present LEED theories. How these temperature-dependent phase shifts are obtained is the subject of the next section.

5.11.1 Temperature-Dependent Phase Shifts

We recall (4.60) for the atomic scattering amplitude due to a spherical atom:

$$f(\theta) = -4\pi \sum_l (2l + 1)t_l P_l(\cos\theta) , \qquad (5.67)$$

with t_l given by (4.55)

$$t_l = - \frac{\hbar^2}{2m} \left(\frac{1}{2ik}\right) [\exp(2i\delta_l) - 1] = - \frac{\hbar^2}{2m} \left(\frac{1}{k}\right) \sin\delta_l \exp(i\delta_l) \ . \quad (5.68)$$

In the presence of thermal vibrations, we can express the *effective* atomic scattering amplitude in a similar way, if we assume isotropic vibrational amplitudes,

$$f(\theta)\exp(-M) = -4\pi\sum_l (2l + 1)t_l(T)P_l(\cos\theta) \ , \quad (5.69)$$

where $t_l(T)$ is to be determined. The expansion in l of (5.69) is valid because isotropic vibrations have the effect that the angular-momentum component m is conserved. For the case where

$$\exp(-M) = \exp(-\frac{1}{2} |\vec{s}|^2 <(\Delta\vec{r})^2>) \ , \quad (5.70)$$

as in (4.63), the quantities $t_l(T)$ of (5.69) can be shown to have the following values [5.9]

$$t_l(T) = \sum_{l'l''} i^{l'}\exp[-2\alpha(E + V_{or})]j_{l'}[-2\alpha(E + V_{or})]t_{l''}$$

$$\times \left[\frac{4\pi(2l' + 1)(2l'' + 1)}{2l + 1}\right]^{1/2} \int Y_{l''0}(\Omega)Y_{l'0}(\Omega)Y_{l0}(\Omega)d\Omega \ , \quad (5.71)$$

where $\alpha = (m/\hbar^2) < (\Delta\vec{r})^2>$. The phase shifts $\delta_l(T)$ can then be obtained, according to (5.68), as

$$\delta_l(T) = \frac{1}{2i} \ln[1 - \frac{4kim}{\hbar^2} t_l(T)] \ . \quad (5.72)$$

One word of caution is in order here: the number of significant temperature-dependent phase shifts increases with increasing temperature since the Debye-Waller factor makes the atomic scattering amplitude more and more sharply forward peaked, requiring more partial waves for its description.

5.11.2 Illustrations of Multiple-Scattering Effects in Temperature-Dependent LEED

In Fig. 4.15, we noted that different peaks in an experimental I-V curve can yield markedly different Debye temperatures θ_D if one applies the standard Debye-Waller factor. This is illustrated again in Fig. 5.9, where the major peak intensities are shown for the same data at the lowest temperature, together with the "effective Debye temperature" obtained from their temperature dependence. Some interesting observations can be made.

First, the effective Debye temperatures generally fall below the bulk value, especially at the lower energies. Second, there is a tendency for the effective

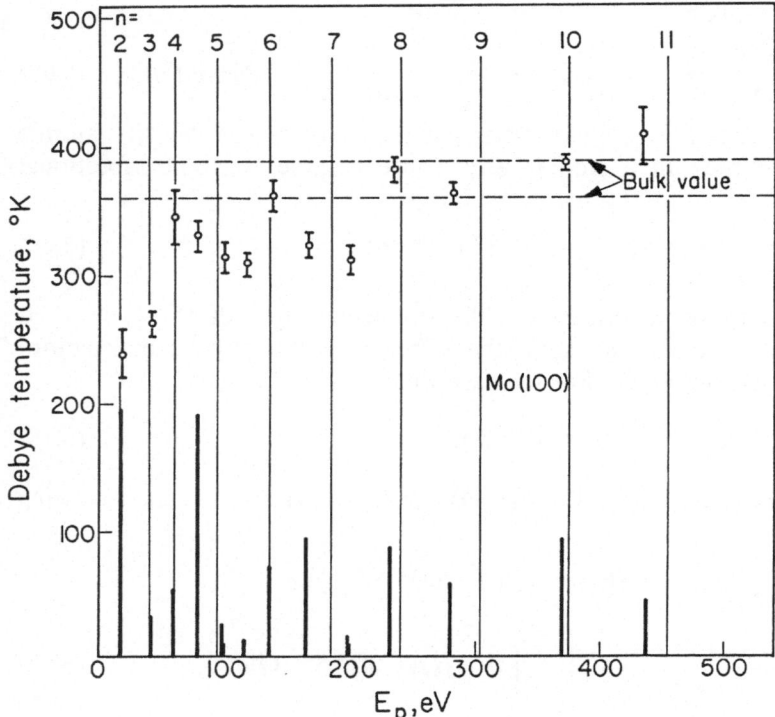

Fig. 5.9. Experimental LEED intensities (*vertical bars*) and effective Debye temperatures (*small circles with error bars*) of peaks in the specular Mo(100) I-V curve shown in Fig. 4.15. The range of Debye temperatures obtained from nonsurface measurements is shown. Bragg peaks have even values of n. After [5.24]

Debye temperatures of Bragg peaks (those labeled n=even in Fig. 5.9) to be larger than those of intermediate peaks, which are multiple-scattering peaks. Not shown are effective Debye temperatures measured at energies between the peaks. These would fluctuate widely both above and below the values that are shown. In fact, there are isolated cases where the intensity increases with increasing temperature, yielding a negative θ_D^2, and thus an imaginary θ_D.

The low values of effective Debye temperatures at low energies are due both to multiple scattering and to increased vibrational amplitudes at a surface. These effects are so intertwined that the "surface Debye temperature" can be obtained only by trial and error through full LEED calculations at different temperatures. Unfortunately, attempts to do this have usually failed, presumably because of the various approximations involved and because of uncertainties in other physical parameters.

Figure 5.10 shows calculated I-V curves for a Ni(100)-c(2x2)-Na surface in which only the sodium overlayer is allowed to vibrate. Clearly illustrated

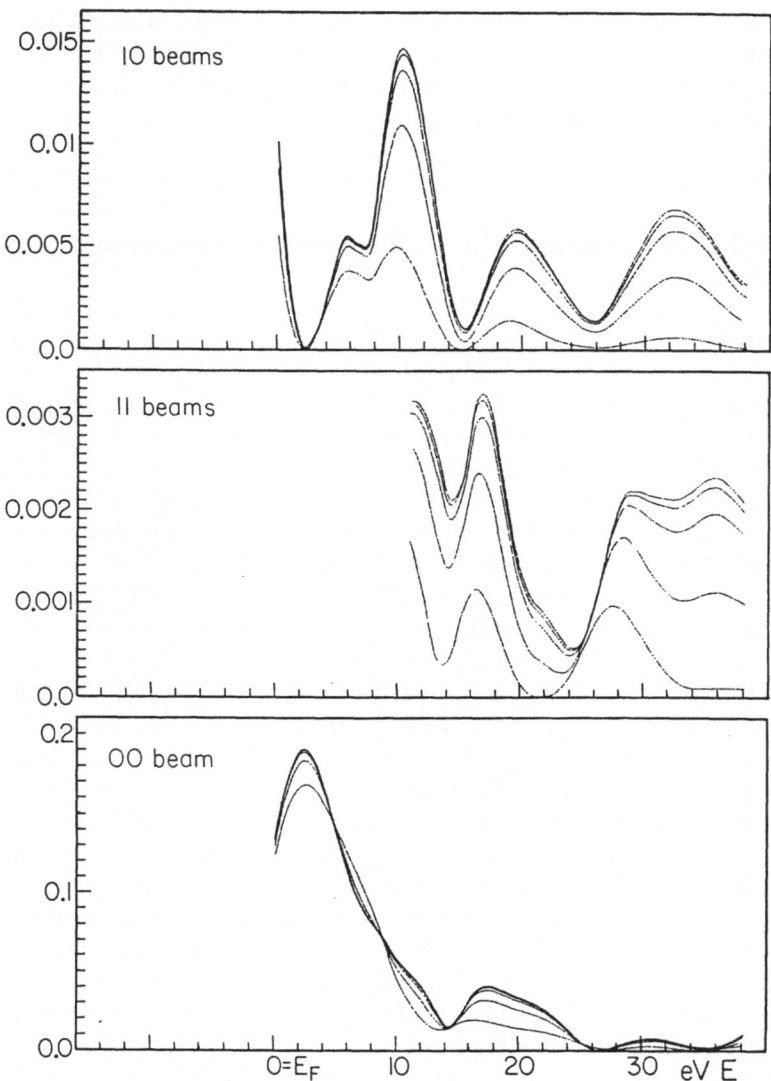

Fig. 5.10. Calculated I-V curves for Ni(100)-c(2×2)-Na with varying Na vibrational amplitudes (the substrate is rigid). Rms amplitudes of 0.0, 0.1, 0.2, 0.4 and 0.8 atomic units (1 a.u. = 0.529 Å) are used. Intensities are shown in fractions of 1 (for total reflection)

are various features mentioned previously. Peak positions are normally not affected by temperature changes (apart from the small possible effect of a lattice expansion). An exception is observed near 30 eV where a peak shifts by approximately 1 eV due to changes in the intralayer multiple scattering.

Peak widths remain unchanged with heating, except as a result of changes in overlap between neighboring peaks of variable heights. Neighboring peaks can have a very different temperature dependence, i.e. very different effective Debye temperatures. An example of *increasing* intensity with *increasing* temperature is included.

5.12 Potential Steps, Surface States, Surface Resonances and LEED Fine Structure

5.12.1 Potential Steps

In this section we address effects due to the change in potential that occurs at a surface between the inner potential and the vacuum zero level. An introductory discussion has been given in Sect. 4.4.3.

Partial reflection and partial transmission of plane waves occur at a potential step of any shape with nonzero height. To calculate this effect requires a detailed knowledge of the step profile. Only for clean, close-packed metal surfaces is this potential profile known to some extent [5.25-27]. It is characterized by a smooth transition from the vacuum constant to the muffin-tin constant over a distance of approximately 2 Å outside the outermost nuclear plane. For a metal surface, the asymptotic form of the potential profile away from the surface is that of the image potential, proportional to the inverse of the distance from the surface. The potential step probably has very little structure parallel to the surface for close-packed surfaces. The effect of such a potential step on a plane wave can be obtained by one-dimensional numerical integration through the step region [5.9]. (This is a one-dimensional problem if we ignore structure parallel to the surface.) By assuming relatively simple mathematical forms for the potential profile, one can also obtain analytical formulas for the reflection and transmission coefficients of such a potential step. Examples are given by Jennings and Price [5.28] and by Dietz et al. [5.29]. From various investigations of these effects, one may conclude the following.

For electron wavelengths smaller than the step width of approximately 2 Å i.e. for energies above approximately 30 eV, the potential step has only minor effects that can be ignored in structural determinations by LEED. Among these effects are weak "resonances" that are of interest in their own right, and we shall return to these. At energies on the order of 10 eV or lower, the potential step strongly influences LEED intensities, affecting the positions of diffraction peaks or even creating new peaks. A square step, i.e. a discontinuous potential jump, produces an excessive effect at all energies: it is usually better to ignore the step than to include a square step in the theory. Most LEED calculations to date ignore the potential step (except for the change in kinetic energy and the resulting refraction), since, for this and other reasons, the energy range below 20 eV is rarely used in structural determinations.

To date, no work has investigated the shape of the potential step for surfaces that are not close packed, such as for stepped surfaces, semiconductor surfaces, or adsorbate-covered surfaces, but calculated charge-density profiles give an idea of what shape to expect. This shape would undoubtedly be complicated, including significant structure parallel to the surface, and its effect on transmitted or related plane waves would be more difficult to evaluate. The periodic structure along the surface would produce diffracted beams like any periodic layer of atoms. However, we may apply the rule that for wavelengths short compared to the scale of variations in the surface potential, i.e. probably again for energies above approximately 30 eV, there should be no significant effect on the diffraction intensities of the surface.

5.12.2 Surface States, Surface Resonances and LEED Fine Structure

The conjunction of potential steps and multiple scattering gives rise to the interesting phenomena of surface states, surface resonances, and LEED fine structure. These effects have been discussed by Forstmann [5.30], Baribeau and Carette [5.31], McRae and Kane [5.32], Echenique and Pendry [5.33], and Rundgren and Malmström [5.34], among others. They consist of the permanent or temporary trapping of electron waves between the outermost atomic cores of the surface and the potential step.

Consider a single plane wave propagating outward from the outermost atomic cores to the potential step in one dimension as shown in Fig. 5.11. If this plane wave has an energy below that of the vacuum zero level, the potential step will completely reflect it back into the surface. On reaching the atomic cores, it will be partly reflected outward again. This process can be repeated, building up a standing wave at the surface. This standing wave will have appreciable amplitude if the phase change for the round trip of the wave between the ion cores and the potential step is an integral multiple of 2π. In that case, we obtain a wave that consists of a standing wave with a large amplitude at the surface, linked to a small-amplitude wave extending into the substrate (and consisting of components propagating both inward and outward). This is a surface "resonance". The word "resonance" exaggerates the effect somewhat, since usually only a very few round trips of the wave occur between the ion cores and the potential step.

The amplitude of such a surface resonance can be reinforced when the reflection by the substrate is strong, as can happen at an energy that falls within a band gap. In this case the wave is perfectly trapped between two impenetrable walls, and we have a "surface state". Surface states, as well as surface resonances, have been observed experimentally in photoemission involving the photoexcitation of electrons out of such trapped states [5.30]. This can take place when the trapped states occur below the Fermi energy,

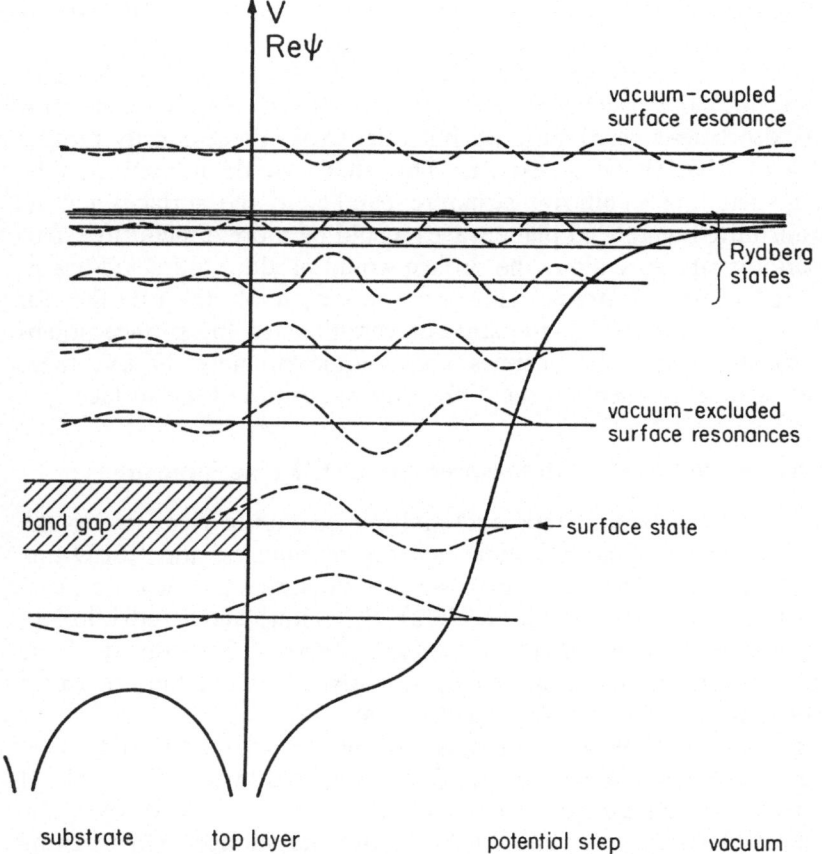

Fig. 5.11. Surface states and "resonances" shown as the real part of the wave function (*dashed lines*) at each corresponding energy level. The surface-potential profile is shown as a continuous line. Note the increased wave amplitudes in the surface region

so that they are occupied by electrons. Occupied surface states and surface resonances are of great importance for the electrical properties of surfaces, especially in the case of semiconductor surfaces.

Rydberg States

The condition that the phase change of a resonating wave be an integral multiple of 2π in a roundtrip between the ion cores and the potential step gives rise to a series of possible states very close to the vacuum level. There the long-range image potential dominates and allows the electron waves to extend farther and farther, but still a finite distance, into the vacuum as the energy approaches the vacuum level. For each distinct multiple of 2π, a surface resonance exists, and it can be shown that the sequence of such resonances forms a Rydberg series of energies. As pointed out by Echenique and Pendry [5.33], and by Rundgren and Malmström [5.34], this is analogous to the bound electron states of a hydrogen atom. Rydberg series have been observed experimentally in LEED at high energy resolution [5.31]. The sharpness necessary to resolve the peaks in a Rydberg series is provided by the reduced inelastic effects away from the surface, where the electrons in a Rydberg state spend most of their time.

LEED Fine Structure

To a limited extent, surface "resonances" (we are still restricting ourselves to one dimension) can also occur at energies above the vacuum level. However, the reflectivity of the potential step decreases rapidly with increasing energy and makes resonances weak. In this case, the term "LEED fine structure" is more appropriate than the term "resonance".

Observation of Surface States, Resonances, and LEED Fine Structure

It might be deduced from the preceding discussion that surface states, resonances, and LEED fine structure are difficult to detect in LEED, because low energies near or below the vacuum level are required. This neglects the influence of the two dimensions parallel to the surface that allow additional waves to exist at any given energy. In reality, a surface resonance of any kind will often be coupled by diffraction through a reciprocal lattice vector $\vec{g}$ to a plane wave (of the same energy) that can propagate from the surface into the vacuum. This can be seen most readily as follows. An incident plane wave $\vec{k}_0$ can diffract into waves $\vec{k}_{\vec{g}}$ with $k_{\vec{g}z} = [(2m/\hbar^2)(E + V_0) - |\vec{k}_{\vec{g}_{\parallel}}|^2]^{1/2}$. Since a potential step that has no structure parallel to the surface conserves $\vec{k}_{\vec{g}_{\parallel}}$ when the wave $\vec{k}_{\vec{g}}$ interacts with it, only $k_{\vec{g}z}$ matters in the scattering by the step. The situation for this wave reduces to a one-dimensional problem with the effective electron energy $E_{eff} + V_0 = (\hbar^2/2m)k_{\vec{g}z}^2 = E + V_0 - (\hbar^2/2m)|\vec{k}_{\vec{g}_{\parallel}}|^2 \leqslant E + V_0$. Thus, an incoming wave can sample the potential step through diffracted waves with lower effective energies $E_{eff} < E$. Since E_{eff} may be arbitrarily small or even negative, the incident wave can couple to surface resonances that exist below the vacuum level, which can be relatively strong. Such coupling appears as small ripples on I-V curves, especially just below energies at which any beam emerges from the surface, since at these energies rapid changes occur in the diffraction condition within a surface resonance [5.32]. In these circumstances, the surface resonance can propagate nearly parallel to the surface and monitor the presence of irregularities (steps, defects, etc.) over thousands of Ångstroms.

In contrast to a surface resonance, a surface state, by definition, cannot be detected in LEED. This is because, in order to be detected, the surface state would have to be coupled to a wave that emanates from the surface and would therefore provide a finite lifetime for electrons residing in that surface state, making it a surface resonance.

5.13 Relativistic and Spin-Dependent Effects in LEED

Thus far we have only briefly mentioned (Sect. 4.3.3) the fact that electrons possess spin, which in LEED gives access to information concerning a surface in the form of a variable spin polarization. This subject has been reviewed by Pierce and Coletta [5.35]. Spin can be controlled experimentally to produce polarized incident electron beams, and spin can be measured. Most importantly, the electron spin can interact with other spins resident in the surface, as in a magnetic surface, through exchange as a result of the Pauli exclusion principle. In addition the spin is sensitive to relativistic effects through spin-orbit coupling. Other relativistic effects which are not due to spin-orbit coupling exist also, such as the relativistic mass effect. All of these factors influence the intensities of diffracted LEED beams so that in principle they can be detected without measuring spin polarizations. However, in general, intensities are affected only marginally, while the spin polarization is very sensitive to both the relativistic and exchange effects. The spin polarization is also more sensitive to structural and other physical parameters of the surface than conventional "intensity-only" LEED; see, for example, the work of Feder and Kirschner [5.36]. With the recent development of efficient polarized electron sources and polarization detectors, this field has gained considerably in importance. In particular, the hitherto neglected study of surface magnetism has now become possible.

Detecting Spin Polarization

The measurement of spin polarization is usually carried out in one of two ways [5.35]. In surface-related experiments, the method used first historically consists of an unpolarized incident electron beam and a measurement of the polarization P of a diffracted beam, with

$$P = \frac{N\uparrow - N\downarrow}{N\uparrow + N\downarrow} \ . \tag{5.73}$$

Here $N\uparrow$ and $N\downarrow$ are the numbers of electrons in the beam with spin parallel and antiparallel, respectively, to a given reference direction. This approach requires the use of a cumbersome and inefficient Mott detector. The experiment is simplified greatly if a source of electrons with known polarization is available. Such a polarized electron beam can be generated by diffraction of an unpolarized beam by a separate crystal that produces polarized diffraction [5.36], or by photoemission from a surface such as GaAs using polarized photons [5.37]. In that case, one can measure the quantity

$$S = \frac{1}{|P_o|} \frac{I_{\uparrow\uparrow} - I_{\uparrow\downarrow}}{I_{\uparrow\uparrow} + I_{\uparrow\downarrow}} ,$$ (5.74)

where $I_{\uparrow\uparrow}$ and $I_{\uparrow\downarrow}$ are intensities of the beam of interest when the incident beam is polarized parallel and antiparallel, respectively, to a given reference direction, while $|P_o|$ is the polarization of the incident beam. The quantities S and P are equal only in some cases of particular symmetry, but either quantity can be used for comparison between theory and experiment.

The Dirac Equation

The general form of the theory for relativistic and spin-polarized LEED is based on the Dirac equation, where each conventional electron wave function that is a solution of the Schrödinger equation is replaced by four component spinors. However, for the case of interest at surfaces, it is believed that the Pauli equation is sufficient. The latter takes into account the two large spinors, neglecting the two small ones. Thus, each conventional wave function of the Schrödinger equation acquires two components, one associated with up spin, the other with down spin. Let us assume that all the relativistic and exchange effects take place inside the muffin-tin sphere. Then, following Jennings [5.38], the relevant Pauli equation takes the following form, with r the radial distance from the nucleus (atomic units are used):

$$\left[-\nabla^2 + V(r) - \frac{\alpha^2}{4} [E - V(r)]^2 + \frac{\alpha^2}{8} \nabla^2 V(r) \right.$$

$$\left. + \left(\frac{\alpha^2}{4} \right) \left(\frac{1}{r} \right) \frac{\partial V}{\partial r} \vec{L} \cdot \vec{\sigma} \right] \psi(r) = E\psi(r) .$$ (5.75)

Here, $\alpha = e^2/(4\pi\varepsilon_0 \hbar c) \cong 1/137$ is the fine-structure constant, which has been used as an expansion parameter, truncating all terms of order larger than α^2, $\vec{\sigma}$ are the Pauli spin matrices, and $\vec{L} = -i(\vec{r} \times \vec{\nabla})$.

Relativistic Effects

All terms in α^2 in (5.75) are relativistic, the third one (involving $\vec{L} \cdot \vec{\sigma}$) resulting from spin-orbit coupling. These three terms can become important when the potential V(r) and its derivatives are large, which happens near the atomic nucleus for heavy elements, especially in the fifth and subsequent rows in the periodic table. These contributions are known to be important for the heavy elements in band-structure calculations, and hence they are also manifest in conventional "intensity-only" LEED. For $Z \geq 60$ one should use relativistic atomic potentials, even when spin polarizations are not measured.

The only term in (5.75) that can flip a spin (up to down or vice versa) is the spin-orbit term. It is responsible for the change in spin polarization that an electron beam undergoes as it diffracts from a surface. Thus, the two spin orientations are coupled, leading to matrix equations with dimensions double those of conventional LEED. Hence, computation times for LEED calculations including spin-orbit effects are approximately eightfold those of conventional LEED calculations. In gas-phase electron-atom collisions, the spin polarization changes most when the scattered intensity is small, since then the relative intensity changes for spin up and spin down are greatest [5.38]. In the surface case, however, multiple scattering can cause considerable polarizations (occasionally in excess of 50%) even when the intensities are not small. This is illustrated in Fig. 5.12 where experiment is compared with various theoretical calculations. It should be noted in Fig. 5.12 that the spin polarization is much more sensitive to the theoretical parameters (whether structural or nonstructural) than is the intensity.

Magnetic Effects

We have not yet noted that the potential V(r) in (5.75) can be spin dependent. This arises in magnetic materials where individual atoms have a net spin due to unpaired electrons. Different spin-spin interactions can then take place between the incident electrons and the atomic electrons, the only one of importance in the present context being the exchange interaction resulting from the Pauli principle. To make the calculation tractable, an "exchange potential" is designed and included in V(r) of (5.75). The exchange interaction is a nonlocal phenomenon resulting from the requirement of antisymmetry the total wave function of the system. It is simplified into a local interaction by assuming an exchange potential. Neglecting the spin-orbit term (and other terms in α^2) of (5.75), this Pauli equation decouples into two independent equations for the two spin orientations (relative to the direction of magnetization), since no spin-flip mechanism is operative. Thus one can use conventional LEED programs by executing them once with the exchange potential for parallel spins and once with the exchange potential for antiparallel spins, as performed by Wang [5.40].

The application of spin-polarized LEED to the study of the properties of magnetic surfaces is just beginning, since the experimental challenge of combining the magnetic fields due to the sample and well-defined electron beam trajectories is only now being attacked seriously.

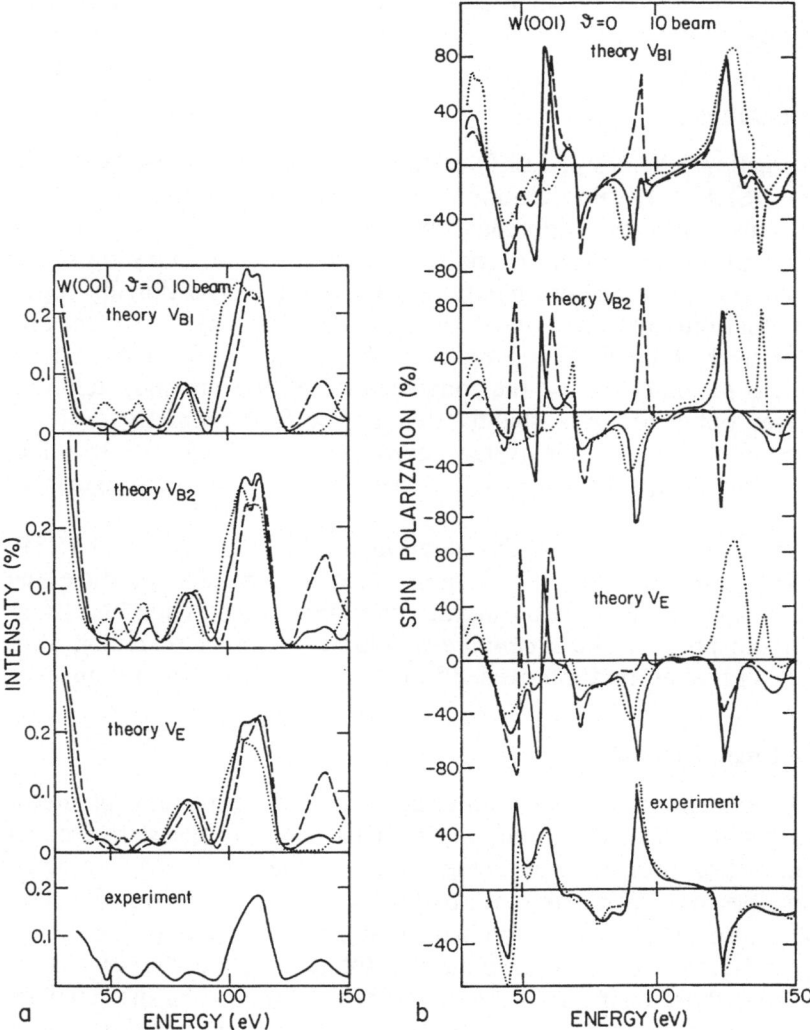

Fig. 5.12. Intensity (**a**) and spin polarization (**b**) of the (10) beam from W(100) at normal incidence. The theoretical results for a top interlayer spacing contracted by 0% ($\cdots\cdots$), 5% (——), and 10% (– – –) are given for three different atomic potentials labeled V_{B1}, V_{B2}, and V_E. The experimental spin polarization is also shown for the (10) beam ($\cdots\cdots$) in addition to the (10) beam (——). After [5.36]

5.14 Some Other Theoretical Techniques

Thus far in this chapter we have described several theoretical methods that have been used in LEED calculations. We chose these because of their popularity or pedagogical value. A large number of modifications of these methods exist, while substantially different approaches have also been employed. In this section, we shall describe some of these modified or alternative methods.

5.14.1 Bootstrapping

In the perturbation schemes of LEED calculations such as Reverse-Scattering Perturbation (Sect. 5.4) and Renormalized-Forward Scattering (Sect. 5.7), one attempts, by iteration, to approach the desired solution as rapidly as possible. Critical in this respect is the choice of a guess for the initiation of the first iteration. The two schemes mentioned above are conventionally begun by assuming that only the unscattered incident wave exists at the surface and must be iteratively refined to convergence. It has been shown by Adams [5.17] that it is preferable to use the converged result of a previous iteration, even if it was obtained for a slightly different surface geometry or for a slightly different energy. On the average, convergence is speeded up by a factor of two in that case. This idea is especially useful when convergence is slow.

Only minor modifications to the iteration formulas of Sects. 5.4 (RSP) and 5.7 (RFS) are required. For example, these formulae produce the *differential* change to the solution at each iteration, whereas for bootstrapping one should produce the *accumulated* solution in order to retain, after convergence, all the information needed to begin the next iteration sequence.

5.14.2 The Chain Method

In Sect. 5.3 we encountered a two-dimensional lattice summation for multiple scattering within a single atomic layer. This is one of the steps leading from the diffraction properties of a single atom to that of a complete layer. It is also possible to split this step into two substeps. First, one would obtain the diffraction properties of a linear chain of atoms from those of the individual atom, and then one would generate the diffraction properties of a complete layer from those of its consitituent chains of atoms. The intermediate step of linear chains gives rise to an additional set of basis functions, apart from the spherical waves, the plane waves, and the Bloch waves. These additional functions are cylindrical waves. This "chain method" has been developed by Masud et al. [5.41].

The chain method is especially designed for relatively high energies that extend into the medium- and high-energy electron diffraction regimes (from 500 to 1000 eV and above 1000 eV, respectively). At these higher energies, grazing incidence is often used to ensure surface sensitivity through limited penetration orthogonal to the surface, compensating for the increased electron-mean-free-path length. Furthermore, since the atomic scattering factor becomes very strongly forward-peaked at such energies, the multiple scattering becomes rather channeled along chains of atoms parallel to the

surface close to the direction of incidence. The conventional LEED calculation schemes for single-layer diffraction do not benefit from this effect,
whereas the chain method does. Finally, the chain method is computationally competitive with the conventional approaches (the direct lattice summation or the Kambe lattice summation) at LEED energies.

5.14.3 Multiple Scattering in Disordered Systems

The calculational methods that have been described up to this point are
designed for perfectly ordered surfaces. Not only does one benefit from the
existence of a well-defined set of reciprocal lattice vectors $\bar{g}$, but also one
essentially only has to calculate the multiple scattering within one unit cell,
the rest of the surface being included in a relatively simple fashion. In disordered systems both of these great advantages disappear. One has to consider
all diffracted directions involved in broadened or streaked beams (if there are
any) and the diffuse background between beams. Furthermore, in principle,
the "unit cell" has become infinitely large. This problem was first addressed
in LEED by Duke and Laramore [5.23] and later by Moritz et al. [5.42].
We shall describe the more recent approach of the latter authors.

In the kinematic limit, the diffracted amplitude is just the sum over wave
amplitudes emanating from the individual atoms at their particular positions.
The randomness in their positions is treated by suitable averaging (which is a
nontrivial task in general). In the multiple-scattering case one can reduce the
problem to the kinematic one by summing similarly over the waves emanating from each surface atom. However, each of these waves has been multiply scattered prior to reaching the surface atom from which it will leave the
surface. To take this into account, one can consider the *vicinity* of each surface atom, let the incident plane wave multiply scatter through this cluster of
atoms, and extract just the amplitude that finally leaves the surface atom
under consideration. This can be done with the methods of Sect. 5.3. For
this approach to be adequate, one should select a finite but representative set
of surface clusters and average the scattering over these. Moreover, one
should choose a sufficiently large vicinity to simulate properly the various
multiple-scattering paths. The computational effort involved in this aproach
is very large unless one makes the approximation that only the nearest and
possibly next-nearest neighbors are treated exactly, whereas for more distant
neighbors a uniform, averaged multiple scattering is assumed.

An interesting observation has been made on the basis of such multiple-
scattering calculations for disordered surfaces. The angular profile of the
diffracted intensity is not qualitatively modified by multiple scattering relative to the kinematically expected profile. As a fortunate consequence, spot
broadening and streaking and the diffuse background distribution can be
understood in kinematic terms [5.41]. Furthermore, I-V curves for well-
defined beams do not suffer appreciably from point defects (vacancies,

impurities, etc.) below a defect concentration of approximately 10%. This is a reassuring conclusion since real surfaces necessarily have defects [5.43].

5.14.4 Pseudopotentials

As we have seen in Sect. 4.5, the scattering of electrons by a spherical atomic potential is described fully by a set of phase shifts δ_l. In fact only $\delta_l \bmod(\pi)$ matters, a feature which may be explained as follows. If the strength of the atomic scattering potential is changed in some fashion so that δ_l for some value of l changes by $+\pi$ or $-\pi$, we see from (5.68) that the new scattered wave is indistinguishable from the old one, since $\exp[i2(\delta_l \pm \pi)] = \exp(i2\delta_l)$. What has happened is that the number of oscillations of the radial wave function near the atomic core has changed by exactly one as a result of the change in the potential. Thus the scattered wave has changed its phase by exactly $\pm 2\pi$. This illustrates the fact that different potentials can produce the same scattering amplitude, at least for a given value of l, as illustrated in Fig. 5.13. In particular, for a given atomic potential there must exist a "pseudopotential", the phase shift of which is just $\delta'_l = \delta_l \bmod(\pi)$. Such a pseudopotential will be the "weakest" in the sense that it induces the fewest oscillations of the radial wave function near the atomic core. Clearly, such a pseudopotential must be l dependent, with different values for different partial waves.

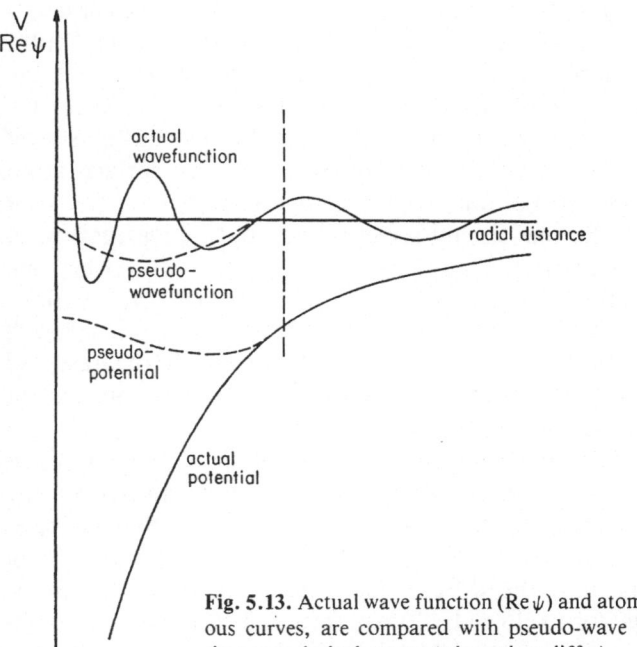

Fig. 5.13. Actual wave function ($\operatorname{Re}\psi$) and atomic potential, shown as continuous curves, are compared with pseudo-wave function and pseudopotential, shown as dashed curves (where they differ)

The existence of pseudopotentials is of great importance in simplifying the treatment of electron scattering in solids, particularly when one uses plane-wave expansions rather than spherical-wave expansions. The reason lies in the rapid spatial variation of the wave function in the atomic cores near the nuclei, which when represented in terms of a plane-wave expansion requires a large number of Fourier components $\bar{g}$. Therefore, matrices with very large dimensions are needed (typically 1000x1000) with unpleasant computational consequences. A pseudopotential, on the other hand, replaces the rapidly varying potential and wave functions of the atomic core with smooth functions that can be represented by considerably fewer Fourier components, leading to tractable numerical calculations.

Pseudopotentials are used extensively in band-structure calculations near the Fermi level, both in the bulk [5.44] and at surfaces [5.45], and they have been applied also to the higher energies present in the LEED case [5.9]. However, the extensive use of the spherical-wave representation and the trivially simple appearance of the pseudopotential effect through $\exp[i2(\delta_l \pm \pi)]$ $= \exp(i2\delta_l)$ have largely eliminated the use of pseudopotentials in LEED. Nevertheless, the pseudopotential method deserves attention both as a major technique involved in the early development of LEED and as a viable alternative that may be revived in the future.

5.14.5 A Semiclassical Theory of LEED

Success has been obtained with a semiclassical formalism in Atom Diffraction from surfaces at thermal kinetic energies. This is because scattering atoms probe parts of the surface where the interaction potential is not a rapid function of position on the scale of the atomic wavelength. This idea may be applied to LEED as well and would be expected to perform well whenever the wavelength of the LEED electron is small compared with the distances over which the scattering potential varies appreciably. Thus one can expect success at high energies (because of short wavelengths), where "high" in this case means above approximately 50 eV.

The semiclassical theory of LEED has thus far been formulated and applied only in a mixed mode by Jauho et al. [5.46]. Conventional quantum-mechanical LEED theory is used to stack layers, for example with Renormalized Forward Scattering, given layer-diffraction matrices obtained with the semiclassical approach. Within an atomic layer, the few multiply scattered, classical trajectories that lead from a given incident direction to a given diffracted direction are found and their amplitudes evaluated, including the phase factor. The summation of these complex amplitudes provides the interference needed to simulate the more exact quantum-mechanical description.

In its present state, the semiclassical LEED theory yields I-V curves in very good agreement with experiment and with quantum-mechanical calculations above approximately 100 eV, but it produces poor results at lower energies. The computational effort is reportedly much smaller than for conventional LEED formalisms. Further development of this approach is likely, which is also necessary to assess the potentialities of the method.

5.15 Outstanding Theoretical Problems in LEED

Although much progress has been achieved in the development of various LEED theories, there remain a number of basic, unresolved problems that affect either the accuracy of the calculations or their feasibility.

Such a problem is the generation of suitable atomic potentials and, from them, scattering amplitudes. For example, as has been discussed by Echenique and Titterington [5.47], even in the limited context of metals, no unique prescription for obtaining an atomic potential has been found that would apply to all metals studied to date, although metals provide potentially the easiest case because of a relatively dense atomic packing and a relatively uniform spatial distribution of valence electrons. The problem becomes even worse with adsorbed atoms and adsorbed molecules. These have highly anisotropic environments, and the spherical muffin-tin potential is of doubtful quality. Such concerns are especially important at the lowest LEED energies: below approximately 20 eV, the electrons are particularly sensitive to details of the surface electronic charge distribution away from the atomic cores, namely in the bonding regions and in the surface-step region. With less smooth surfaces (as with open surfaces and overlayers of fractional monolayer coverage) this is manifest also in a potential step that is corrugated in the directions parallel to the surface.

Another weakness in present LEED theories is the treatment of temperature effects. The Debye-Waller factor in a kinematic picture already has its limitations, well known in X-ray scattering [5.1]. The multiple-scattering effects complicate this issue considerably, bringing in factors such as correlations between the thermal vibrations of neighboring atoms. The fact that surface vibrations are different from their bulk counterparts introduces additional uncertain parameters.

Computational limits arise when the surface unit cell becomes large, multiplying the number of beams that has to be considered. Moreover, in that case one usually has to consider many atoms in the unit cell, even if one takes only one layer at a time. This produces a large number of inequivalent partial waves in the spherical-wave representation. Matrix dimensions and computation times then rapidly exceed the capabilities of existing computers. As a consequence, for example, at present we are not in a position to perform LEED calculations for stepped surfaces with high Miller indices, a type of surface of particular interest because it provides quite different types of adsorption sites.

5.16 Application of LEED Theory to Other Electron Spectroscopies

The propagation of low-energy electrons through surfaces is not confined to LEED. All other electron spectroscopies used to study surfaces also involve this propagation. It follows that the theoretical methods that have been developed in LEED can also be applied to these other techniques, which has in fact been done successfully in the past few years.

The techniques that resemble LEED most closely from a theoretical point of view are those that use angle-resolved measurements of the electrons emanating from the surface. Included in this category are Angle-Resolved Photoelectron Spectroscopy (ARUPS in the case of incident ultraviolet light, providing sensitivity mainly to valence levels in the surface electronic structure, and ARXPS in the case of incident X-rays, probing mainly the core levels of surface atoms) [5.48], Angle-Resolved Auger-Electron Spectroscopy (ARAES) [5.49,50], Inelastic LEED (ILEED) [5.23], and High-Resolution Electron Energy-Loss Spectroscopy (HREELS) [5.51,52]. The close analogy with LEED is seen easily if one follows the outgoing electrons, which may be represented by a plane wave at the detector, in reverse time. The plane wave at the detector must originate from plane waves with the usual k-vectors $\bar{k}_{\bar{g}}$ by diffraction through the two-dimensional periodic surface layers. One can backtrack in this fashion from plane wave to plane wave using conventional LEED reasoning until one reaches the point where the electron has undergone a different kind of process, not present in LEED: photoexcitation from an occupied level in the case of ARUPS and ARXPS, Auger excitation from an occupied level in the case of ARAES, inelastic scattering by a phonon, plasmon, surface electron, or other quasiparticle or collective excitation in the case of ILEED, and in particular inelastic scattering by a surface vibrational mode (often a vibration in adsorbed molecules) in HREELS. If one regresses in time one more step, one finds that several of these techniques again involve electrons propagating through the lattice like LEED electrons, although usually with energy or parallel momentum different from that of emerging electrons. This is the case, for example, for electron-stimulated ARAES, for ILEED, and for HREELS. In the case of ARUPS the initial state of the ejected electron may also have to be calculated (since it is needed to compute the atomic photoexcitation-matrix elements), and this can be considered a LEED problem as well, one which is equivalent to the surface electronic structure problem.

Thus, the angle-resolved surface electron spectroscopies essentially consist of two parts. First, a matrix element for the excitation or type of scattering under consideration must be evaluated. Second, the electrons must be propagated through the surface lattice just like any LEED electron. Therefore, these techniques can be used to study both the excitation mechanisms and the surface structure, together with the electronic states of the surface in the case of photoemission.

In another class of surface electron spectroscopies, the electron current emitted from the surface is integrated over the accessible half-sphere of emission directions. This integration, which averages over the directions involved, has the effect of reducing somewhat the relative contribution of multiple scattering, compared to the angle-resolved techniques. Included in this class of techniques are Surface Extended X-ray Absorption Fine Structure (SEXAFS) [5.53] and Extended Appearance Potential Fine Structure (EAPFS) [5.54]. In both cases, electrons that are emitted from a surface atom can scatter back from the neighboring atoms to the emitting atom and modulate the overall excitation cross section by wave interference. This modulation contains the information concerning the interatomic distances, although atomic scattering phase shifts must be considered here also.

5.17 Computer Programs

A number of computer programs have been developed to perform LEED computations. They differ in the theoretical methods used, in the effects that are included, in the complexity of surface structure that they can treat, in their ease of use, in their documentation, in their portability from one computer to another, and in their availability. There are too many variables to present a complete picture of the various computer programs, and, consequently, we shall limit ourselves to a brief description of those programs that are in frequent use and/or are readily available. In addition, some ancillary programs will be mentioned, dealing primarily with the calculation of phase shifts (Chapt. 4) and R-factor (Chap. 6). Some related programs developed for other electron spectroscopies will also be mentioned. In the discussion, the alphabetical order of the principal author of the individual programs will be followed.

Davis [5.55] has written a LEED program based on the Beeby matrix-inversion method for atomic layers with one or more atoms per unit cell, and the RFS and Layer-Doubling methods for layer stacking. Also, a program for ARAES has been implemented [5.50].

Duke and co-workers [5.36] also base their present LEED programs on the Beeby matrix-inversion method for atomic layers, combined with RFS between layers.

Feder has written spin-polarized LEED programs [5.36]. The Bloch-wave method and the Layer-Doubling method are applied. He also has a non-spin-polarized program, in which more than one atom per unit cell in a layer is allowed. A version of this program has been implemented for ARUPS [5.57]. The spin-polarized LEED program is equivalent, although not identical, to one written independently by Jennings [5.38], who also has a non-spin-polarized program that includes a potential step at the surface [5.28].

Holland and co-workers [5.16], in addition to a one-atom per unit cell program, have a LEED program based on RSP. This treats the entire surface as a single thick layer in the spherical-wave representation. Holland has also developed an ARUPS program [5.58].

Jona and Zanazzi [5.59] have available an R-factor program that is in wide use for the comparison of theory and experiment in LEED crystallography, see Sect. 6.5.

Jepsen and Marcus have two principal programs in use at present. The more commonly used one, called HEX [5.13], is based on the Beeby matrix-inversion method for atomic layers with one or more atoms per unit cell, combined with the transfer-matrix approach. The other program, called THIN [5.20], is designed for small interlayer spacings, such as those occurring on stepped surfaces. These programs have the particularity that their main programs are written in PL1 (the subprograms being written in FORTRAN), allowing dynamical storage allocation, but reducing portability. The documentation is nearly nonexistent, but the programs are easy to use. Included is a surface-potential step. Jepsen has also developed an ARUPS program [5.60].

Moritz [5.42] has written a multiple-scattering LEED program for disordered surfaces, using Beeby matrix-inversion as a basic ingredient. He has also written a program for periodic surfaces, embodying Beeby matrix-inversion, RFS, and Layer Doubling, allowing more than one atom per unit cell.

Pendry and co-workers (principally Kinniburgh, Titterington, Masud, and Aers) have built up a package of programs, the CAVLEED package originating at the Cavendish Laboratory, which is distributed widely. It is particularly broad in scope, including many methods and useful ancillary programs. The basic LEED program of the CAVLEED package is written for one atom per unit cell in each layer, employing either the Bloch-wave method, Layer Doubling, or RFS. The Kambe lattice summation is available, and a potential step may be included. Another program implements the one-center expansion method [5.14] for layers with more than one atom per unit cell. A further program is built around the chain method [5.41], which can be used in LEED, MEED, and RHEED. Also available are a very useful phase shift program as well as an R-factor program. The CAVLEED package is very well documented and easy to use for those surfaces which fit available programs. Further developments include an ARUPS [5.61] and an HREELS program [5.52].

Rundgren and Salwen [5.62] have written a program for one atom per unit cell in each layer, based on RFS. This program can utilize any available symmetry, both in the plane-wave and in the spherical-wave parts and therefore is very efficient for those surfaces to which it can be applied. It is readily available and well documented.

Tong and Van Hove [5.62] provide a package of LEED programs based on a flexible building-block structure. The user can select the desired combination of theoretical methods (provided as subroutines) and write a main program accordingly. Tens of sample programs are provided that include most common applications. Atomic layers with any number of atoms per unit cell are treated by the Beeby matrix-inversion method or RSP (or even a combination of the two). Layers are stacked with Layer Doubling or RFS. Symmetry can be used in the plane-wave representation. These features are designed to allow the most efficient use of computer resources. These very portable programs are documented in detail, and their use is easy when a sample program already suits the problem at hand. An R-factor program is also available that combines ten different R-factor definitions. In addition, Tong and co-workers have also developed ARUPS [5.48] and HREELS [5.64] programs based on the LEED programs.

6. Methods of Surface Crystallography by LEED

The major impetus for developing the use of LEED has been the prospect of surface crystallography, i.e. the determination of atomic positions at crystalline surfaces. In addition to the use of LEED for surface characterization (especially the use of diffraction patterns to monitor the state of the surface), surface crystallography has indeed been the major application of LEED. This chapter will describe the specific methods that have been developed for the determination of surface structures through the interpretation of intensity data. (The interpretation of the LEED pattern has been discussed extensively in Chap. 3.) These methods fall into two classes: the kinematic and the dynamical methods.

The kinematic methods of surface crystallography (such as kinematic simulation of I-V curves, Constant-Momentum-Transfer Averaging and Fourier-Transform techniques) are relatively simple and therefore provide a computationally efficient approach, similar to the case of X-ray Diffraction. However, few actual surfaces diffract low-energy electrons kinematically. Consequently, the approximations involved in the kinematic approach generally lead to a loss of accuracy and reliability. For detailed and reliable structural determinations, one must usually resort to the dynamical-scattering formalism and attempt to reproduce the experimental data theoretically by a trial-and-error search for the correct structure. The computational cost can be correspondingly high.

To give an idea of this computational cost, which in fact can be rather lower than many people believe, we give a few typical examples. The full dynamical LEED calculation, for a complete energy range of 20 to 200 eV, for the simplest type of surface, such as the unreconstructed fcc(111) or (100), can easily cost as little as US $1 for each structure, at normal computer-usage rates, especially if symmetry is used at normal incidence. In this case, one often finds that an R-factor analysis and the mere plotting of the data cost more than the dynamical calculation! The corresponding calculation for a simple overlayer, such as Ni(100)-p(2x2)-S, may cost approximately US $10 per structure. As soon as more than one atom per unit cell has to be considered in any atomic layer, the cost increases, because one has to rely more on the use of spherical waves (Sect. 5.8). Thus, the more complicated structures investigated in recent years yield costs on the order of US

$100 to $500 per structure. As a result, a complete structural analysis in the three quoted situations could cost in practice approximately US $ 50, $ 500, and $ 2,000 to $ 10,000, respectively. The actual costs to the researcher are often considerably lower due to preferential rates, low priorities in computer-job queuing, etc. On the other hand, by reducing the labor of setting up efficient LEED calculations, one may also easily end up with much higher computational costs.

6.1 The Kinematic Approach to Surface Crystallography

6.1.1 Kinematic Simulation of Intensity Data

The simplest approach to the extraction of structural information from experimental LEED intensities consists of applying the kinematic formulas of Sect. 4.2 with assumed atomic positions. By varying the atomic positions until the agreement between theory and experiment is maximized, one hopes to find the correct surface structure.

It became clear at the very outset of the use of LEED that a kinematic approach can fail to produce theoretical results that match experimental data [6.1]. Nevertheless, this approach has been attempted repeatedly ever since for a variety of surfaces. For example, intriguing reconstructions of semiconductor surfaces as well as other surface structures were investigated in this manner [6.2-6] with only limited success. This situation contributed strongly to the subsequent development of the dynamical LEED theory. However, cases have emerged where relatively "kinematic" intensities were found. A good example is the Xe(111) surface [6.7] which, for not too low electron energies, produces nearly kinematic I-V curves with well-resolved Bragg peaks (Fig. 4.5). By comparing the peak *shapes* of Fig. 4.5 with those in I-V curves calculated with a varying top-layer spacing (Fig. 4.7), one can clearly tell that the Xe(111) surface is essentially bulklike: one may estimate that the top-layer spacing deviates from the bulk value by at most a few percent. The "kinematicity" of the Xe(111) data is due to the large interatomic distances in that crystal lattice, which, except at the lowest energies, prevent multiple scattering from becoming important, despite the rather large scattering strength of each xenon atom.

At higher electron energies, I-V curves for any material often appear to become relatively kinematic, in the sense that Bragg peaks become dominant features. However, it must be understood that strong multiple scattering in a narrow forward cone still takes place in most materials in those circumstances. Hence, in reality, the "Bragg peaks" are not simply kinematic, but contain multiple-forward-scattering contributions as well. What is happening is that the intensity of multiple-scattering peaks distant from Bragg energies is strongly reduced by temperature effects, see Sect. 5.11. This

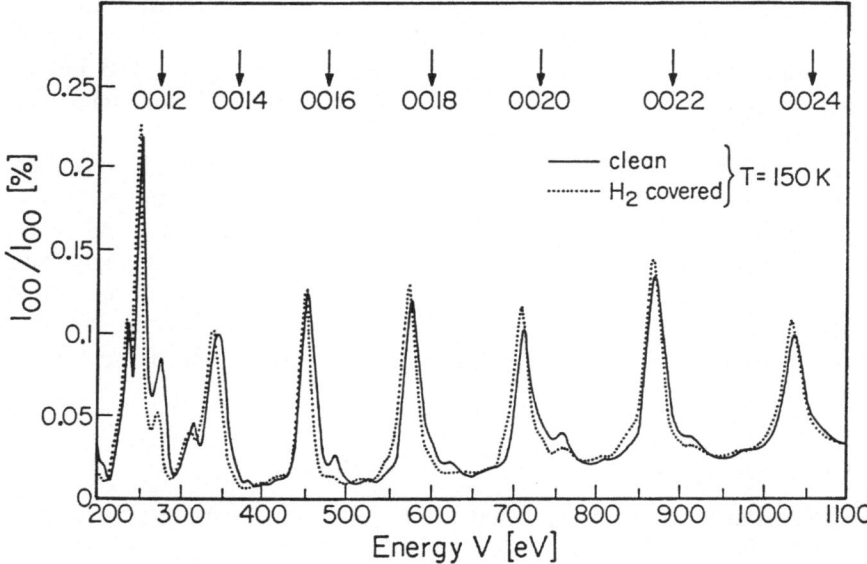

Fig. 6.1. I-V curves for the specular beam diffracted from a clean Pt(111) surface (*full line*) and from a hydrogen-covered Pt(111) surface (*dotted line*). Arrows indicate the kinematic Bragg peak positions without an inner-potential correction [6.10]

apparent kinematicity has even been observed with such strongly scattering surfaces as Pd(111) [6.8] and Pt(111) [6.9], Fig. 6.1. Particularly illustrative is the fact that adsorption of hydrogen on these surfaces slightly shifts the positions of the "Bragg peaks", as shown in Fig. 6.1. This indicates a hydrogen-induced change in the spacing between the near-surface metal layers. By simple fitting of kinematically calculated I-V curves, one can estimate that the three or four topmost metal-layer spacings expand by about 1% at these surfaces due to the adsorbed hydrogen [6.11]. The hydrogen atoms themselves are weak scatterers, and they are probably disordered at these surfaces. Consequently, their position is difficult to determine by LEED, but in special circumstances, hydrogen-atom positions can be found, as is illustrated in Sect. 8.2.4. Note that the diffraction peaks in Fig. 6.1 show very little skewness. A skewness would be indicative of an expansion of only the topmost layer spacing. Note also that the peak shifts at the higher energies are not larger than those at lower energies suggesting that the bulk layer spacing at a depth of several atomic layers (reached by the higher-energy electrons) has not expanded.

Another example in which kinematic I-V curves agree reasonably well with experimental data is provided by silicon surfaces. In the case of the reconstructed Si(100)-(2x1) surface, the structure of which was long under debate, a decisive step towards its resolution came with a kinematic I-V curve calculation for a theoretically predicted dimer model [6.12], as described in more detail in Sect. 7.2.2. Although this structure was later refined with the help of multiple-scattering calculations, it was shown that a kinematic approach could yield the basic structural features for this lightly scattering material. In addition, this example shows that relatively complicated structures may be treated in this way.

Therefore, one should expect the applicability of the kinematic approach to the surface crystallography of any surface of lightly scattering materials, including possibly aluminum, beryllium, organic crystals, films, and overlayers. Furthermore, with materials that scatter electrons more strongly, one may expect the possibility of an approximate structural determination with a kinematic approach. This can also provide an inexpensive way to rule out many a priori plausible structures before a more detailed, multiple-scattering analysis is undertaken.

6.1.2 Layer Spacings from Sequences of Bragg Peaks

For simple surfaces where Bragg peaks are the dominant features, the examples of Xe(111), Pt(111), and Pd(111) in the previous section suggest that one could use the energies at which the Bragg peaks occur to determine the spacing between layers parallel to the surface, i.e., the layer spacing and the Bragg energies are directly related, as was discussed in Sect. 4.2.5. For example, the "third" Bragg condition for the specularly reflected beam can be written in atomic units as (Eq. 4.33)

$$2(E^B + V_o)\cos^2\theta = \frac{\pi^2 n^2}{d^2} , \qquad (6.1)$$

where E^B (in hartrees) is the energy of the incident electron beam, V_o is the inner potential, θ is the angle of incidence of the electron beam, and d (in bohr) is the interlayer spacing. The plot of E^B as a function of n^2 gives a straight line, and V_o as well as d can be extracted from the intercept and the slope of this straight line, respectively. Christmann et al. [6.8,9], Andersson and Kasemo [6.13] as well as Stern and Sinharoy [6.14] have determined the surface interlayer spacing for Pt(111), Pd(111), Ni(100), Cu(100), and W(110) surfaces from such plots using specular LEED beam data. Mark et al. [6.15] have also determined the surface interlayer spacing for several semiconductor surfaces including ZnO(11$\bar{2}$0), ZnO(10$\bar{1}$0), CdS(11$\bar{2}$0), CdS(10$\bar{1}$0), and GaAs(110). Within the limits of accuracy, the results obtained by this method were found to agree with the bulk interlayer spacings determined by the X-ray method. However, the surface interlayer spacing determined by this method should be interpreted as the average interlayer spacing for the top several layers defined by the electron mean free path, since only one layer-independent value of d is considered in (6.1). In addition, a contraction or expansion which occurs only at the top layer of a surface, shows up more as peak skewness than as peak shifts in I-V curves

(Fig. 4.7). Therefore, this method is not very sensitive to small changes in the value of the topmost surface layer spacing, such as a small expansion or contraction from the bulk value, which are often detected by dynamical methods.

The accuracy and reliability of this approach to structural determination depend on the correct measurement of the locations of the Bragg peaks in the I-V curves. The positions of the Bragg peaks can be shifted significantly from the kinematically expected positions when multiple scattering is strong, even after the correction for the inner potential (Sect. 4.8). Furthermore, the extra non-Bragg peaks generated by multiple scattering can cause additional difficulties in the identification of the Bragg peaks. Hence, this simple kinematic method of determining surface interlayer spacings is not very reliable in the case of LEED data showing strong multiple scattering.

6.2 Averaging Methods

6.2.1 Constant-Momentum-Transfer Averaging (CMTA)

Since multiple scattering adversely affects the kinematic approaches to surface crystallography that we have discussed thus far in this chapter, one must seek ways to reduce these adverse effects. To that end, a method called Constant-Momentum-Transfer Averaging (CMTA) was proposed by Lagally et al. [6.16] to produce averaged experimental data which exhibit enhanced kinematic characteristics. It was claimed without rigorous theoretical justification that if a set of I-V data, obtained for a wide range of polar and azimuthal angles of incidence, are averaged in a suitable fashion, then the contributions from multiple scattering would average out to a constant background, giving rise to experimental data with a kinematic appearance.

According to the kinematic theory, the reflected plane wave amplitude $M_{out,o}$ is given by, from (4.34),

$$M_{out,o} = \frac{i}{Ak_{oz}} \left[\sum_{q=1}^{N} f_q \exp(i\vec{s}\cdot\vec{R}_q) \right] \frac{1}{1 - \exp(i\vec{s}\cdot\vec{a}_3)} \, \delta(\vec{k}_{\|out} - \vec{k}_{\|o} - \vec{g}) \, , \quad (6.2)$$

for a bulk-like surface with N atoms per three-dimensional unit cell at locations $\vec{R}_q$ (q = 1,2,...,N), repeated by the vector $\vec{a}_3$ into the surface. This amplitude has maxima when the "third Bragg condition" is satisfied, i.e. when

$$\vec{s}\cdot\vec{a}_3 = n2\pi \quad \text{(n integer)} \, . \tag{6.3}$$

This condition for the occurrence of maxima in the kinematic expression depends upon the beam energy and directions only through the momentum transfer, $\vec{s} = \vec{k}_{out} - \vec{k}_o$. By contrast, the multiple-scattering contributions to the amplitude depend explicitly on both $\vec{k}_{out}$ and $\vec{k}_o$ separately. Lagally et al. assume that the multiple-scattering effects modify the kinematic intensity in a statistically random fashion and thus produce no systematic effects. With this assumption, it is postulated that if a set of I-V data taken for a wide range of incidence directions are averaged at constant values of the momentum transfer $\vec{s}$, the averaged data should show enhanced kinematic features that are unaffected by multiple scattering. However, it was shown by Pendry [6.17] that systematic effects can still exist in the averaged experimental data (apart from the effects of incomplete averaging over an insufficient range of incidence directions).

First, multiple scattering decreases the beam intensity systematically. When multiple scattering is strong, electrons travel a longer, more tortuous path than in the kinematic case. Consequently, peak intensities are reduced due to a higher probability of inelastic processes. Another effect that reduces peak intensities is "beam sharing". A Bragg-reflected beam will lose part of its intensity to other beams which are averaged out into the background intensity. Moreover, flux can be removed from the incident beam into other beams not necessarily in Bragg conditions, thus reducing the flux available in the incident beam for kinematic processes. These effects increase the difficulties in the matching of the peak intensities between experiment and theory. Second, the inner potential is somewhat energy-dependent, which can introduce systematic effects. Third, multiple scattering can shift the kinematic peaks up or down along the energy axis. Consequently, even if one assumes that the mean positions of the peaks remain constant, the averaged data would contain kinematic peaks that are broader than those of truly kinematic peaks.

In spite of these effects, it has been shown by Ngoc [6.18] that the averaged data do exhibit enhanced kinematic features as shown in Fig. 6.2 for a Ni(111) surface. A set of data was taken at 12 polar angles of incidence between $\theta = 62°$ and $84°$ in steps of $2°$, and six azimuthal angles of incidence between $\varphi = 0°$ and $50°$ in steps of $10°$. All the I-V data were first plotted as a function of the reduced scattering vector s/s_o, where $s = |\vec{s}|$, $s_o = 2\pi/d_{111}$, and d_{111} is the bulk value of the interplanar spacing for the Ni(111) surface. Then all curves were averaged at constant values of the reduced scattering vector. Comparison between the averaged experimental data and the calculated kinematic data shows reasonably good agreement. The results, which indicate that the Ni(111) surface is neither expanded nor contracted appreciably, are consistent with the findings of later dynamical calculations [6.19,20]. However, an inelastic damping as large as 9 eV was used in the calculations to fit the peaks that were broadened in the averaging process. Before averaging, the peaks have a natural width corresponding to a damping of 4-5 eV. This method has also been applied to the Ag(111) [6.18] and Ni(100) [6.21] surfaces with satisfactory results. However, in the case of Cu(100), nonkinematic features were observed in an averaged (00) beam

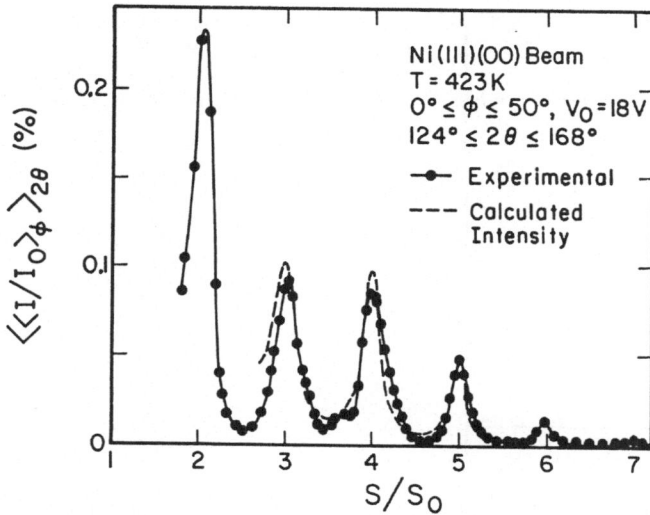

Fig. 6.2. Experimental average of intensity data for the Ni(111) (00) beam and a comparison with kinematically calculated intensity. A constant inner potential of 18 eV has been taken into account in the experimental data. The calculation was normalized to the n = 5 Bragg peak height. Polar and azimuthal angles of incidence are θ and φ, respectively. After [6.18]

[6.22]. This arouses doubt about the use of this technique to identify unknown structures.

The CMTA technique has also been applied to analyze LEED data from overlayer structures, including Cu(100)-c(2x2)-O [6.22], Ni(100)-(2x2)-2C [6.23], and Ni(110)-(1x2)-H [6.23]. Such attempts resulted in very limited success, due mainly to the presence of residual strong multiple-scattering features in the averaged data. The failure of the CMTA to average away the multiple-scattering features is due in part to the differences in the scattering factor between the adsorbate and the substrate atoms [6.24]. Such differences can cause the kinematic peaks to be shifted to different momentum transfers when intensity data are taken at different incidence angles. Kinematic approaches in general tend to have problems with the simultaneous presence of different kinds of atoms. Although one may explicitly take different atomic scattering factors into account in a kinematic theory, the subsequent manipulations (such as averaging or Fourier transformation and deconvolution) can often no longer be carried out rigorously.

6.2.2 CMTA with Azimuthal Averaging at Constant Energy

Based on the ideas formulated by Lagally et al. [6.16] in their CMTA technique, Aberdam et al. [6.25] have proposed another method to produce averaged data with enhanced kinematic features. The procedure adopted by Lagally et al. is to obtain experimentally a large set of I-V profiles at various

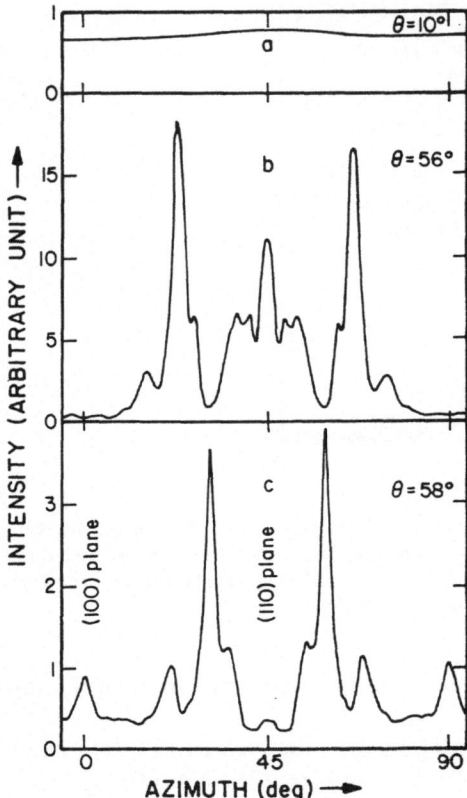

Fig. 6.3. Rotation diagrams for the specular beam reflected from the Al(001) surface at electron energy E = 380 eV, for angles of incidence θ = 10°, 56° and 58°. After [6.25]

polar and azimuthal angles of incidence. Then, the I-V data are averaged at constant values of the momentum transfer $\bar{s}$. Two critical factors, which are considered as drawbacks in this approach, are discussed by Aberdam et al. [6.25]. First, the I-V spectra are so sensitive to the polar and azimuthal angles of incidence that it is difficult (time-consuming) to obtain a set of I-V curves with sufficient continuity to give a realistic angular average. This effect is particularly pronounced at high electron energies as shown in Fig. 6.3. Second, the conversion of the kinematic energy of the incident electron to the momentum transfer requires a known value of the inner potential. For the simple case of a specular beam at a polar angle of incidence θ, s is given by

$$s = \frac{2}{\hbar} [2m(E\cos^2\theta + V_o)]^{1/2} \quad . \tag{6.4}$$

The value of the inner potential is usually not known a priori. In addition, intensities measured at different energies and polar angles of incidence corresponding to a given value of $\bar{s}$ generally involve different values of the atomic scattering factor. This is because the atomic scattering factor does not depend only on $\bar{s}$, as assumed in the kinematic approximation, but on E and θ separately, see (4.60). Thus, in the averaging process, different I-$\bar{s}$ spectra have different weighting factors, determined by the respective scattering factors. This problem can be avoided by maintaining E and θ constant during the averaging process, which leaves the azimuthal angle as the only averaging variable. As an added benefit, the incident electron current (which varies with the energy) remains constant. Thus, no further normalization of the data is necessary so long as only one energy value is used.

In light of the above arguments, Aberdam et al. [6.25] proposed that averaging should be done as follows. Rotation diagrams, which are the beam intensities measured as a function of the azimuthal angle, are obtained at a constant energy and different polar angles of incidence. The intensity of each rotation diagram is averaged continuously with respect to the azimuthal angle. To obtain an averaged polar-rotation diagram, as shown in Fig. 6.4, the azimuthally averaged intensity is plotted as a function of the polar angle of incidence.

At this point, one can apply the same kinematic formulas used in the CMTA technique described in Sect. 6.2.1. However, the authors of the azimuthal-averaging CMTA approach have chosen to improve on that treatment. In particular, a pseudo-kinematic calculation based on Darwin's two-

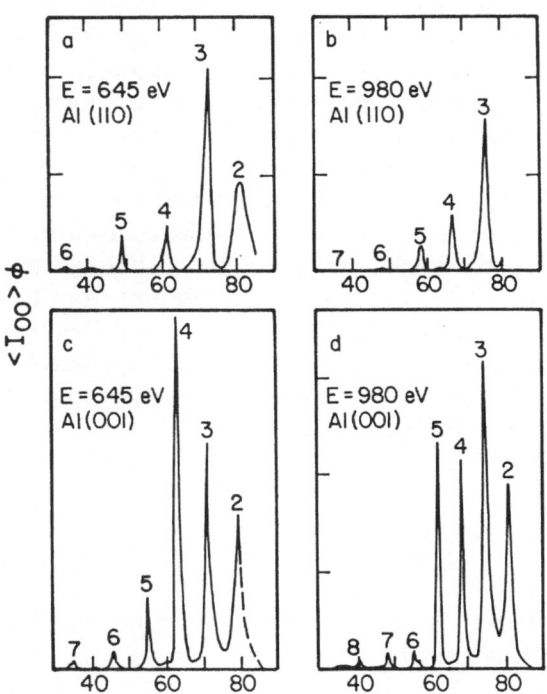

Fig. 6.4a – d. Averaged rotation diagrams for Al(001) and Al(110) surfaces and electron energies E = 980 and 645 eV. Here, $\langle I_{00} \rangle_{\varphi}$ denotes the intensity of the specular beam, averaged with respect to the azimuthal angle φ, and θ denotes the polar angle of incidence. Bragg orders are indicated at the tops of the peaks. The measurements were made at room temperature. After [6.25]

beam dynamical theory [6.26,27] is used to analyze the averaged rotation diagrams. This involves taking into account atomic phase shifts and kinematic reflection and transmission coefficients for each individual layer. To a limited extent, some multiple scattering is thereby included between successive layers, namely between the beams $\vec{k}_{\vec{g}}^{+}$ and $\vec{k}_{\vec{g}}^{-}$ for any given $\vec{g}$. Each Bragg peak in the rotation diagram is fit with a set of optimum values of surface relaxation, optical potential, and inner potential.

This method has been applied to analyze averaged rotation diagrams for Al(001) and Al(110) surfaces, as shown in Fig. 6.4. The results indicate that the (001) surface is not significantly relaxed, while the (110) surface is contracted by a few percent. However, the contraction of the (110) surface cannot be determined very accurately due to the large dispersion of the results based on data sets taken at different energies and on different assumed parameter values.

6.3 Fourier-Transform Methods

In the kinematic limit, the plane-wave reflection amplitude takes the form of a Fourier series involving the atomic positions (Sect. 4.2). Therefore, one could recover the atomic positions by Fourier-transforming measured reflection amplitudes (at all momentum transfers), which consist of a reflection magnitude and a phase. However, the phase is not directly measurable, so that part of the information needed for the determination of a crystal structure is lacking. The more convenient, although clearly approximate, choice is to Fourier-transform the measured intensities, which are the reflection magnitudes squared. This approach has been used quite successfully in X-ray Diffraction [6.28,29], and we shall now describe its use in LEED. An advantage that can be expected from Fourier-transforming LEED data is a "filtering" out of multiple-scattering effects. Although the hope is that these effects are separated out in the Fourier transform from the structural information, we shall see that this hope is realized only under certain circumstances. Our discussion will move from simple and relatively unsuccessful approaches to more sophisticated and productive methods of structural determination based on Fourier transformation.

6.3.1 The Patterson Function

General Principles

The Fourier transform of the intensity is called the Patterson function $P(\vec{r})$, defined for X-ray Diffraction as [6.28,29]

$$P(\vec{r}) = \frac{1}{V}\sum_{hkl}I_{hkl}\exp[2\pi i(hx + ky + lz)] \ , \tag{6.5}$$

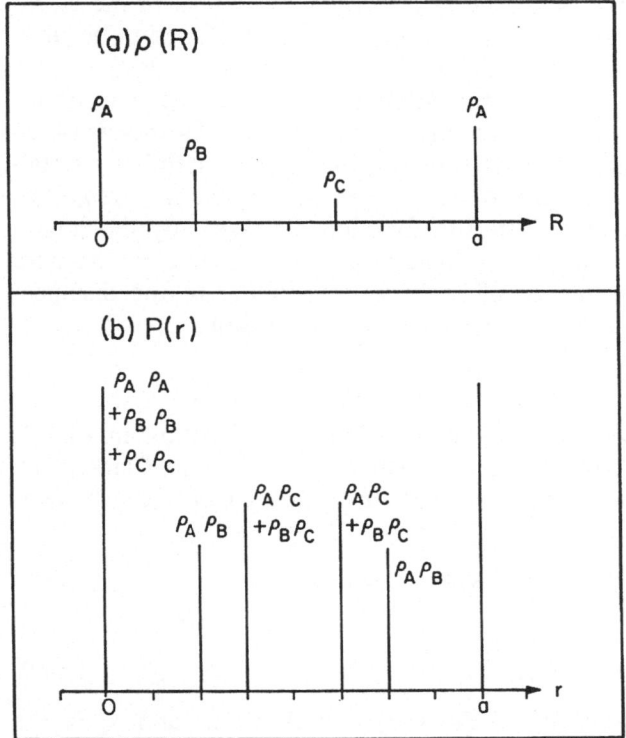

Fig. 6.5. (a) A schematic diagram of a one-dimensional unit cell containing three atoms A, B, and C, with total electron densities ρ_A, ρ_B, and ρ_C, respectively. The electron densities are treated as delta functions. (b) The Patterson function corresponding to (a)

where $\vec{r} = (x,y,z)$ is a vector from an arbitrary origin in the crystal, V is the volume of a unit cell, and I_{hkl} is the intensity of the (hkl) diffracted X-ray beam. This function can also be shown to be the autocorrelation function of the electron density $\rho(\vec{r})$:

$$P(\vec{r}) = \int \rho(\vec{R})\rho(\vec{r} + \vec{R})d\vec{R} \ . \tag{6.6}$$

The relationship between the electron density and the Patterson function for a one-dimensional case is illustrated in Fig. 6.5. The asymmetric one-dimensional electron density contains, within each unit cell, three peaks which correspond to three atoms A, B, and C with total electron densities, ρ_A, ρ_B, and ρ_C, respectively, which we shall treat as delta functions, for convenience. The corresponding Patterson function

$$P(r) = \sum_{i,j=A,B,C} \int \rho_i(R)\rho_j(r + R)dR \ , \tag{6.7}$$

contains nine peaks, three of which lie at the origin and represent the atoms overlapping with themselves. Other peaks in the Patterson function occur at distances from the origin that correspond to the interatomic distances (or interatomic vectors, in the two- or three-dimensional cases). For example, the peak at a/4 from the origin represents the interatomic distance between atoms A and B. Although the Patterson function is very useful in crystal-structure determination, there are disadvantages preventing its straightfor-ward interpretation. First, the Patterson function displays only the intera-tomic vectors without showing the individual atomic coordinates. Second, the possibility of the coincidence of peaks corresponding to the distances between different pairs of atoms can be a source of confusion.

Applications in LEED

Several attempts have been made to use the Patterson function in LEED, notably in the early work of Tucker and Duke [6.30-32] and Clark et al. [6.33-35]. In analogy to (6.5), the Patterson function for LEED can be con-structed as

$$P(x,y,z) = \sum_{hk} P_{hk}(z) \exp[2\pi i(hx + ky)] \ , \quad \text{where} \tag{6.8}$$

$$P_{hk}(z) = \int_{0}^{\infty} I_{hk}(s_z) \exp(is_z z) ds_z \ . \tag{6.9}$$

In (6.9), $I_{hk}(s_z)$ is the intensity of the (hk) LEED beam, and s_z is the momentum-transfer component perpendicular to the surface. Here $P_{hk}(z)$ represents a one-dimensional line projection containing peaks corresponding to interatomic vectors shifted to a common origin and projected onto the z-axis. In the work of Tucker and Duke, the aim was the determination of the structures of carbon monoxide chemisorbed on Pt(100) [6.30] and on Rh(100) [6.31], as well as oxygen adsorbed on Rh(100) [6.32]. $P_{hk}(0)$ was obtained by simple integration of the intensity of the I-V profiles of the (hk) beams, and P(x,y,0) was constructed. This gives the x-y structure of the first layer but not the registry, i.e. not its position parallel to the surface relative to the underlying layers. However, with just this information, possible over-layer structures can be deduced.

In the work of Clark et al. [6.33-35], additional attempts have been made to determine the interlayer spacing. As an illustration, the Fourier transform P(z) of the (00) intensity profile of the Ni(100)-c(2x2)-Na overlayer structure is shown in Fig. 6.6. It was claimed that peaks in this curve provide the Ni-Ni (1.8 Å) and Ni-Na (2.7 Å) interlayer spacings. However, the Ni-Na spac-ing was later found to be rather different with full dynamical calculations [6.36,37], namely approximately 2.23 Å The authors of the Patterson-function analysis did recognize two weaknesses in their work. First, the Fourier transform was taken without correcting for the inner potential, and the structural peaks can be shifted significantly depending on the value used

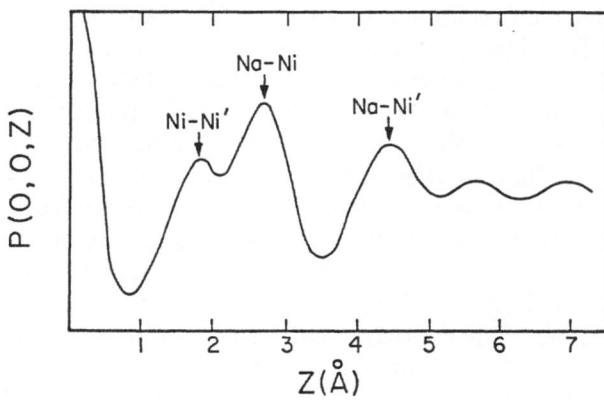

Fig. 6.6. The Fourier transform $P(z)$ of the (00) beam for the Ni(100)-c(2 × 2)-Na overlayer surface. After [6.35]

for the inner potential correction. A difficulty with the Patterson function is indeed the unknown value of the inner potential V_o, which enters the following equation, where one converts kinematic energy to momentum transfer s_z:

$$s_z = \left[\frac{2m}{\hbar^2}(E + V_o) - |\vec{k}_{o\parallel} + \vec{g}|^2\right]^{1/2} + \left[\frac{2m}{\hbar^2}(E + V_o) - |\vec{k}_{o\parallel}|^2\right]^{1/2} .(6.10)$$

The momentum transfer is needed inside the surface, while E is the measured kinetic energy outside the surface. Hence the inner potential is involved. Second, the finite data range used in the analysis causes the appearance of spurious peaks, known as Gibbs oscillations, in the Patterson function. These oscillations are partly due to the abrupt truncation of the data at both ends of their range. How to distinguish between the structural

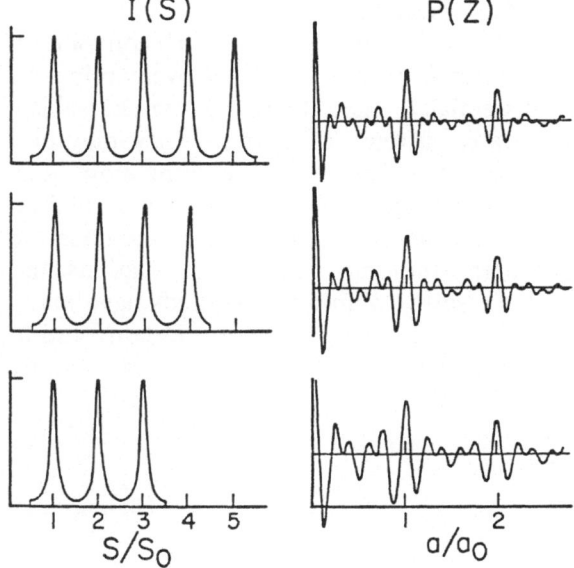

Fig. 6.7. A series of diagrams showing, for a set of idealized data, that the Gibbs oscillations ("noise" in the right-hand graphs) increase as the data range used in the Fourier transform decreases (left-hand graphs)

peaks and the spurious peaks is an additional problem in the straightforward application of the Patterson function. Figure 6.7 shows, for a set of idealized data, that the Gibbs oscillations increase as the data range used in the Fourier transform decreases. In addition, broadening of the structural peaks is also observed. To minimize the Gibbs oscillations, the usual procedure is to "apodize" the data, i.e. the data are weighted with a function that goes smoothly to zero at both ends of the range of data. This procedure has been applied by Buchholz et al. [6.38] to individual curves and constant momentum transfer averaged curves for the Ni(111) surface. While the Gibbs oscillations are reduced by the apodization of the data, at the same time the peaks are also broadened. Hence, this approach reduces the position resolution of the transform method. The results of Buchholz et al. from the analysis of the averaged data indicate that the Ni(111) surface is neither expanded nor contracted. However, inconclusive results are obtained from the analysis of the individual I-V profiles due to the presence of spurious peaks among the structural peaks. The spurious peaks are the result of the multiple-scattering features in the individual I-V profiles.

The effects of multiple scattering can indeed be an important perturbation in the transform methods. The Fourier transformation is intended in part to filter out or average out the multiple-scattering effects, in a way analogous to the averaging methods discussed previously. In fact, averaging and Fourier transformation can both be applied in succession to the same data, as was done by Buchholz et al. [6.38] and Cohen et al. [6.39], see Sect. 6.3.4. However, the uncertainties introduced by the inner potential correction, the data truncation, and the residual effects of multiple scattering have severely restricted the applicability of the simple transform methods. Nevertheless, with relatively kinematic data, one can expect some success with these computationally efficient approaches.

The Convolution-Transform [6.40-42] and Transform-Deconvolution [6.43-46] methods have been developed with the purpose of overcoming the difficulties mentioned above. Essentially, they follow a more realistic kinematic formalism based on iterative fitting procedures. They may also include the angle and energy dependence of the atomic scattering amplitude, as well as its temperature dependence through a Debye-Waller factor, and an attenuation coefficient to represent the inelastic losses. These methods are still relatively simple compared to the multiple-scattering calculations and remain computationally efficient. These two transform methods have many common features. Consequently, although the Transform-Deconvolution method was applied earlier, we shall first describe the somewhat simpler Convolution-Transform method.

6.3.2 The Convolution-Transform Method

An Expression for the Kinematic Intensity

In order to understand the Convolution-Transform method of analyzing LEED data, we first present a general expression for the kinematic scattering intensity of the specular beam diffracted from the surface of an elemental single crystal in which the surface layer has an arbitrary relaxation perpendicular to the surface. We will then Fourier-transform this analytic expression to derive the expected Patterson function for this ideal case.

The scattering amplitude A can be written in general as

$$A(E) = f(E)\sum_{n} \alpha(n,E)\exp[-M(n)] \sum_{j} \exp[i\vec{s}\cdot\vec{r}_j(n)] \sum_{l_1 l_2}\exp[i\vec{s}\cdot\vec{r}(l_1 l_2)] \ . \quad (6.11)$$

Here, E is the energy of the incident electron, $\exp[-M(n)]$ is the square root of the Debye-Waller factor for the nth layer, $\alpha(n,E)$ is the attenuation coefficient which is the amplitude at energy E that makes the round trip from the surface to the nth layer and back to the surface, $\vec{r}_j(n)$ is the relative position of the jth basis atom in the nth layer, and $\vec{r}(l_1 l_2)$ is the absolute position of the two-dimensional surface unit cell labeled by integers l_1 and l_2. The quantity $\vec{s}$ is the momentum transfer to the surface and is given by

$$\vec{s} = \vec{k}_{out} - \vec{k}_o \ , \quad (6.12)$$

where $\vec{k}_{out}(\vec{k}_o)$ is the wave vector of the electron after (before) the scattering event. The prefactor $f(E)$ contains all contributions other than those explicitly included in the sums, such as the atomic scattering amplitude, which is assumed to be identical for all scatterers, and the transmission coefficient of the surface step. These contributions do not affect the analysis and can be absorbed into an energy-dependent prefactor.

We now specialize the derivation to the scattered intensity of the specular beam for a model structure with the bulk interlayer spacing denoted by d and the fractional expansion of the topmost layer spacing relative to d denoted by t (t<0 for a contraction). First, we assume that the attenuation coefficient α is a constant (not dependent on energy) which may be written as

$$\alpha = \exp\left(-\frac{d}{\lambda_e \cos\theta}\right) \ , \quad (6.13)$$

where λ_e is the inelastic mean free path of the electron, and θ is the polar angle of the incident beam relative to the surface normal. Second, we assume that the Debye-Waller factor for the surface atoms is the same as for the bulk atoms. With these assumptions, the scattered intensity can be written in terms of a sum over the amplitudes scattered from each layer:

$$I(s_z) = |A(E)|^2 = |f(E)|^2 \exp(-2M)\left\{ 1 - \frac{\alpha^{2t}\alpha^2}{1 - \alpha^2} + \frac{2\alpha^{2t}\alpha^2}{1 - \alpha^2} \sum_{n=0}^{\infty} \alpha^n \cos(ns_z d) \right.$$

$$\left. + 2\alpha^t\alpha \sum_{n=0}^{\infty} \alpha^n \cos[(n + 1 + t)s_z d] \right\} . \tag{6.14}$$

The prefactor $|f(E)|^2$ is a gradual function of energy and forms a smooth envelope to the terms in the braces, which constitute an interference function depending on s_z only through the cosine terms.

An Analytic Patterson Function

Assuming that the range of momentum transfer over which the intensity is measured is $0 < s < +\infty$, then the expression for the intensity, (6.14), can be Fourier-transformed analytically to give the Patterson function:

$$P(z) = \int_0^{\infty} I(s_z) \exp(is_z z) ds_z . \tag{6.15}$$

If we consider only the interference terms in braces in Eq. 6.14, then the Patterson function becomes

$$P(z) = 2\pi \left\{ \frac{1 - \alpha^2 - \alpha^2\alpha^{2t}}{1 - \alpha^2} \delta(z) + \frac{\alpha^{2t}\alpha^2}{1 - \alpha^2} \sum_{n=0}^{\infty} \alpha^n [\delta(z + nd) + \delta(z - nd)] \right.$$

$$\left. + \alpha^t\alpha \sum_{n=0}^{\infty} \alpha^n [\delta(z + (n + 1 + t)d) + \delta(z - (n + 1 + t)d)] \right\} . \tag{6.16}$$

This function has a relatively simple form, i.e. it is a series of delta functions, symmetrically placed about z=0. The first term, which is the largest, represents the coefficient of the delta function at the origin. The second term represents a series of delta functions placed at multiples of the bulk spacing. The last term is similar to the second one, but now the delta functions are at multiples of the bulk spacing plus or minus the surface-layer spacing. We shall prove later that, as z becomes larger, the coefficients of all the delta functions become smaller with an exponential decay.

To illustrate this behavior, we show in panels (a) and (b) of Fig. 6.8 the interference function of (6.14) and the Patterson function of (6.16), respectively, for the case of the specular beam at normal incidence, choosing $t = 15\%$ and $\lambda_e = 1.12d$ (which corresponds to a constant $\alpha = 0.41$). We present only a part of the interference function, which is assumed to extend to $+\infty$, and we show only the coefficients of the delta functions for positive z. Note that the Patterson function shows a doublet structure near every multiple of the bulk spacing even though only the surface layer is assumed to relax.

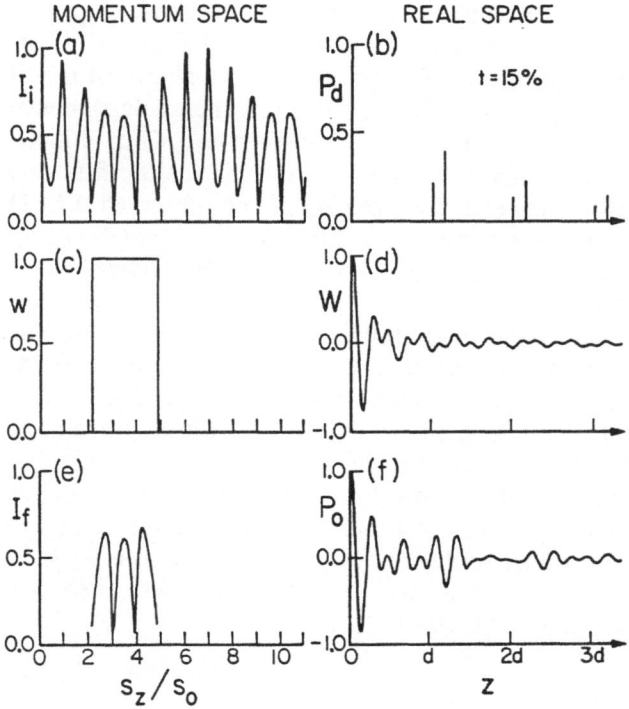

MOMENTUM SPACE   REAL SPACE

Fig. 6.8a – f. Schematic illustration of idealized intensities in momentum space along with the corresponding Fourier transforms in real space. (a) Interference function for 15% relaxation of the surface. (b) Fourier transform of (a) assuming the data extend to infinity. This shows the type of doublet structure expected for a relaxed surface. There is a single delta function of unit magnitude at the origin. (c – d) Schematic of the window function limiting the data range along with the real part of its Fourier transform. (e – f) Appearance of a finite data set along with the real part of the Patterson function showing spurious features not associated with structure. Curve (e) is the product of (a) and (c). Curve (f) is therefore the convolution product of (b) and (d)

Limitations of the Patterson Function

There are five important differences between the interference function shown in Fig. 6.8a and experimental LEED data. Specifically, the calculated function *(i)* assumes a constant electron mean free path, *(ii)* equalizes the Debye temperatures of bulk and surface layers, *(iii)* omits all multiple-scattering structure, *(iv)* ignores the energy-dependent envelope $|f(E)|^2 \exp(-2M)$, and *(v)* extends in momentum space to infinity.

In the Convolution-Transform method of analysis, the last two differences are dealt with exactly. This is demonstrated by multiplying the infinite interference function $I_i(s_z)$ by a window function $w(s_z)$, as shown in Fig. 6.8c, to give data $I_f(s_z)$ of a finite range, as shown in Fig. 6.8e. The window function need not be a constant, as shown, but can be any function of s_z over the range where it is nonzero. In particular, the window function $w(s_z)$ can include all the energy-dependent effects in the first factor of Eq. 6.14. This term cannot, however, explicitly account for the energy dependence of α and a layer dependence of the Debye-Waller factor, since the energy dependence of these two quantities cannot be factored. In fact, in the Convolution-Transform method of analysis, there is no way to account for the first two differences mentioned above between the calculated and observed LEED intensities. That the method works nonetheless is because, for a small energy range, it is reasonable to assume that α is constant, and the Debye-Waller factor is layer independent.

If we use the finite range of the interference function as shown in Fig. 6.8e and calculate the Patterson function, (6.15), we obtain a complex quantity, the real part of which is shown in Fig. 6.8f and is labeled $P_o(z)$. This quantity is symmetric in z. In Fig. 6.8d, we show the Fourier transform, labeled $W(z)$, of the window function $w(s_z)$ in Fig. 6.8c. The structure is due entirely to the finite nature of the window function: these are the Gibbs oscillations mentioned previously. The convolution theorem states that since the curve of Fig. 6.8e is the product of the curves of Figs. 6.8a and 6.8c, then the curve of Fig. 6.8f is the convolution product of those of Figs. 6.8b and 6.8d (before taking the real part), indicated as

$$P_o(z) = P_d(z)*W(z) \ . \tag{6.17}$$

The problem of determining the structure of a relaxed surface can now be succinctly stated as: Given the Fourier transform $P_o(z)$ of the data, and of a chosen window $W(z)$, what is the series of delta functions $P_d(z)$ which gives the structure? This is the problem of deconvoluting two known functions to get a third one which is unknown.

Extraction of the Surface Structure

To solve this problem, Landman and Adams [6.43-46] have presented a direct deconvolution procedure based on the Southwell method (Sect. 6.3.3). The nonuniqueness of the deconvolution scheme is a major complication in this method. As a consequence, a different method has been developed by Cunningham et al. [6.40-42] for determining the set of delta functions $P_d(z)$ which gives the structure. The method is as follows. Assume that $P_o(z)$ and $W(z)$ are known and that $P_d(z)$ has the form

$$P_d(z) = \sum_i g_i[\delta(z + z_i) + \delta(z - z_i)] \ , \tag{6.18}$$

where the factors g_i are the relative weights of the delta functions. From (6.16), these weights are

$$\left. \begin{array}{l} g_1(z = 0) = g_0[1 - \alpha^2(1 + \alpha^{2t})] \ , \\[4pt] g_{2n}(z = nd) = g_0\alpha^{n+2}\alpha^{2t} \ , \\[4pt] g_{2n+1}[z = (n + t)d] = g_0(1 - \alpha^2)\alpha^n\alpha^t \ , \end{array} \right\} \qquad n = 1,2,... \qquad (6.19)$$

where

$$g_0 = \frac{1}{1 - \alpha^2} \ . \qquad (6.20)$$

The convolution product of $W(z)$ and the assumed $P_d(z)$ can be performed to give a calculated Patterson function $P_c(z)$ given by

$$P_c(z) = \int_{-\infty}^{\infty} W(z')P_d(z - z')dz'$$

$$= \sum_i g_i[W(z + z_i) + W(z - z_i)] \ . \qquad (6.21)$$

This calculated Patterson function can now be compared with the corresponding Patterson function $P_0(z)$ deduced from experiment. Thus, the outermost layer spacing of the crystal surface can be determined by varying the position of the delta functions in the calculated Patterson function until one obtains the best fit between the "observed" and the calculated Patterson functions. This best fit is determined by finding the minimum of a parameter called the residual, the square of which is given by

$$R^2 = \int_0^{\infty} [P_0(z) - P_c(z)]^2 dz \ . \qquad (6.22)$$

The Convolution-Transform method can now be summarized by indicating its seven basic steps.

1. A window function is chosen which delimits the energy range and defines the envelope of the experimental data.
2. A value of the inner potential V_0 (which is one of the two variable parameters) is chosen, since V_0 is needed to convert from energy space to momentum-transfer space.
3. The "observed" Patterson function $P_0(z)$ is obtained by Fourier-transforming the experimental intensity from momentum space into real space.
4. The position of a set of trial delta functions, $\{z_i\}$, representing interlayer distances, is chosen. This set is completely specified by the known bulk interlayer spacing and one parameter t, the percentage of relaxation of the surface layer spacing. The relative weights of these delta functions, $\{g_i\}$, are considered to be unknown.
5. The set of relative weights $\{g_i\}$ which results from a least-squares fit between $P_0(z)$ and $P_c(z)$ is determined by minimizing the residual with respect to the unknown weights.

6. The residual in (6.22) is evaluated using the determined weights and normalized by dividing by the value of R obtained from (6.22) with $P_c(z) = 0$.

7. The procedure is repeated with other values of V_0 and other sets of interlayer spacings, and the correct surface structure appears as a minimum of the residual as a function of V_0 and t. The minimum in the residual surface is characterized by its depth, which we describe by the parameter

$$\Delta = 100(R_{ave} - R_{min})/R_{ave} , \tag{6.23}$$

where R_{ave} is the average value of the residual in the residual surface, and R_{min} is its value at the minimum. The deepest possible minimum ($R_{min} = 0$), and therefore the best possible result, gives $\Delta = 100\%$.

The Window Function

Among these seven steps, all are well-defined except step (1) which involves the choice of the window function. We shall now discuss the method used for choosing the window function and the results of various, different choices. In the work of Landman and Adams [6.43-46], the window function is calculated by carefully evaluating the first factor in (6.14). This requires much detailed information concerning the surface, such as the scattering phase shifts and the Debye-Waller factor. In the Convolution-Transform method, this step is avoided. The window function used is an approximate, hand-drawn envelope of the data, touching or approaching the I-V curve at the positions where Bragg peaks may occur. This window has a degree of arbitrariness about it which has been shown to be of no importance to the results of the analysis [6.40]. The reason why this drawn window is the desired window function rather than the calculated prefactor in (6.14) can be understood when we consider what it is to accomplish. The Patterson function of the data contains spurious structure due to truncation effects, which appear as high-frequency oscillations in z-space. On the other hand, low-frequency features in z-space are due to the gross features of the intensity in z-space: in particular, the envelope of the data. By making the window function in z-space mimic the envelope of the data as well as the cutoff, we have a better chance of accounting for the extra peaks in the Patterson function.

Examples

The Convolution-Transform method has been applied to various clean, unreconstructed surfaces [6.41,42,47]. The analysis performed on the Ir(111) surface [6.42] is one of the examples. In this analysis, the specular beam at two different angles of incidence, $\theta = 7°$ and $\theta = 25°$, was used. The window functions that were chosen are shown as the drawn envelopes in Fig. 6.9.

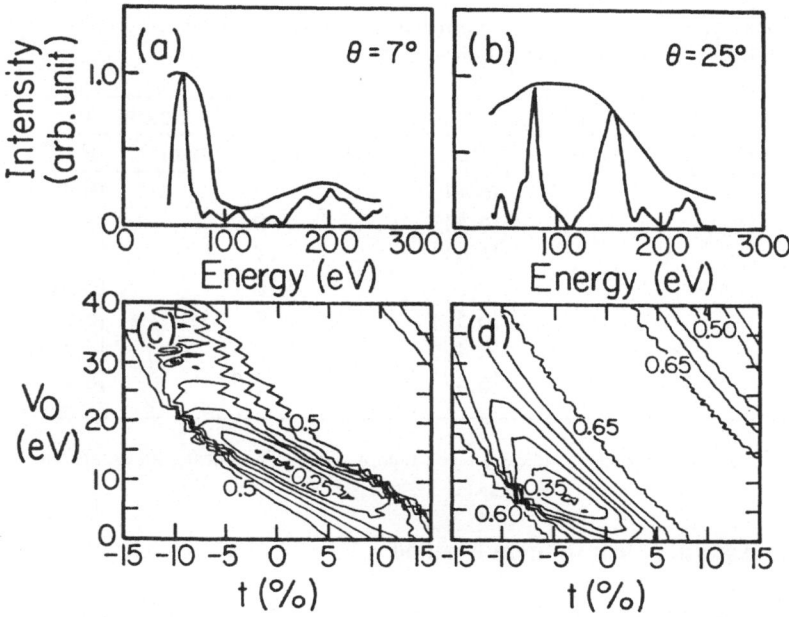

Fig. 6.9. (a – b) I-V curves and drawn windows for $\theta = 7°$ and $\theta = 25°$, respectively, of the specular beam from the Ir(111) surface. **(c – d)** Topographic plots for the residual surface R, as a function of inner potential (V_0) and % relaxation (t) for $\theta = 7°$ and $\theta = 25°$, respectively. The ragged contour lines are due to the limited number of grid points used in plotting the graphs. In reality, they should be smooth

Contour plots for the residual surface R as a function of inner potential V_0 and relaxation t are shown in Figs. 6.9c,d. The minimum in Fig. 6.9c occurs at $V_0 = 13$ eV and t = 0% with a relative depth of $\Delta = 60\%$; while the minimum in Fig. 6.9d occurs at $V_0 = 7$ eV and t = −4% with a relative depth of $\Delta = 43\%$. From these results, it may be concluded that the surface layer of the Ir(111) crystal is contracted from 0% to 4% of its bulk value of d = 2.217 Å These results are consistent with those obtained from dynamical calculations which indicate that the Ir(111) surface contracts by less than 5% of its bulk value [6.42].

The Convolution-Transform method was also applied to other clean metal surfaces: Ni(110), Al(110), Ag(110) [6.41], and Rh(111) [6.47]. A summary of results obtained by the Convolution-Transform method and a comparison of these results with those obtained by dynamical calculations are given in Table 6.1. Comparison between results obtained by these two methods shows good agreement.

Thus, the Convolution-Transform method seems to be able to determine the degree of relaxation, at least of clean metal surfaces. The method has not been applied to other systems.

Table 6.1. Summary of results obtained by the convolution-transform method and comparison of these results with those obtained by dynamical calculations

Surface	Convolution-transform method		Dynamical calculations	
	Surface relaxation[a]	Ref.	Surface relaxation[a]	Ref.
Ni(110)	-5	[6.41]	-5	[6.48]
Al(110)	-4	[6.41]	-10	[6.49]
Ag(110)	-6	[6.41]	-7	[6.50]
			-10	[6.51]
Pt(111)	-3	[6.52]	0	[6.52]
Ir(111)	+3	[6.42]	0	[6.42]
Rh(111)	-4	[6.47]	-2.5	[6.47]

[a] In % of bulk interlayer spacing. $(-)$ indicates a contraction while $(+)$ indicates an expansion

6.3.3 The Transform-Deconvolution Method

The Convolution-Transform method described in the previous section is very similar to the Transform-Deconvolution method proposed earlier by Landman and Adams [6.43-46]. The major difference between these two methods is the procedure used to obtain $P_d(z)$, the set of delta functions that characterizes the surface structure. In the Convolution-Transform method, $P_d(z)$ is a trial function, whereas in the Transform-Deconvolution method, $P_d(z)$ is obtained by applying the Southwell deconvolution scheme [6.53] to (6.17). Within the formalism of the Transform-Deconvolution method, $P_o(z)$ is still the Fourier transform of the experimental data, but $W(z)$ is no longer an arbitrary window function as in the case of the Convolution-Transform method. Rather, it is obtained by explicitly calculating the atomic scattering factor, including the initial guess for the Debye-Waller factor. Since $P_o(z)$ and $W(z)$ are now known functions, $P_d(z)$ can be obtained by a deconvolution scheme. However, it is not possible to generate a unique solution of $P_d(z)$ by a deconvolution scheme, due to the finite range of data used in the transform. A unique determination of $P_d(z)$ by a deconvolution procedure requires the intensity data over a complete range of s from 0 to ∞. The problem of the nonuniqueness is treated by using the Southwell deconvolution scheme. Although it is not guaranteed that a unique solution of $P_d(z)$ will be produced, it has been shown that this scheme usually produces a solution of the required form.

The Convolution-Transform method can be summarized by indicating its six major steps.

1. Assuming a value of V_o, $P_o(z)$ is obtained by Fourier-transformation the experimental data and $W(z)$ is obtained by Fourier-transforming the cal-

culated scattering factor over the same range of s as the experimental data with the initial guess for the Debye-Waller factor, determined by the Debye temperature θ_D.

2. The Southwell deconvolution scheme is applied to (6.17) and yields a set of delta functions $\{z_i\}$ that specifies the interlayer spacings, the attenuation factor α that is calculated by using the relative amplitudes of the delta functions, and a scaling constant c.

3. The series of delta functions $P_d(z)$ is constructed with $\{z_i\}$ and α. Then $P_c(z)$ is obtained by convoluting $P_d(z)$ and $W(z)$,

$$P_c(z) = [P_d(z) + P_n(z)]*W(z) \ , \tag{6.24}$$

where $P_n(z)$ is an arbitrary function which is included to account for the input errors and effects of multiple scattering. The construction is on a uniform grid of z with typical increments of 0.1 Å over a range between 0 and 10 Å

4. The residue of the fit between $P_o(z)$ and $P_c(z)$ is calculated by

$$R = \sum_i |P_o(z_i) - P_c(z_i)| / \sum_i |P_o(z_i)| \ . \tag{6.25}$$

5. Steps (3) and (4) are iterated with variation over $\{z_i\}$, α, c, and $P_n(z)$ until, e.g. $R \leqslant 0.1\%$.

6. Finally, steps (1)-(5) are repeated with other values of V_o and θ_D until a maximum of the signal S to noise N ratio

$$\frac{S}{N} = \frac{\sum_i P_d(z_i)}{\sum_i P_n(z_i)} \tag{6.26}$$

is obtained.

An example of the output from the Transform-Deconvolution analysis on the Cu(100) surface is shown in Table 6.2. Computer deconvolutions corresponding to the best and the worst cases in terms of S/N for this surface are shown in Fig. 6.10. Landman and Adams have applied this method to the analysis of other systems including the Ni(100), Al(111), and Al(100) surfaces. The results obtained for these surfaces are consistent with those obtained by dynamical calculations, and they suggest that this method can adequately treat the analysis of clean metal surfaces.

6.3.4 Fourier Transform of Intensity Beats from Overlayer and Substrate

The Fourier Transform approach has been applied by Cohen et al. [6.39] to the analysis of intensity beats from a monolayer of xenon on the Ag(111) surface. The work of Cohen et al. successfully demonstrates that CMTA fol-

Table 6.2. Deconvolution output parameters for Cu(100) for the specular beam at a sequence of directions (θ, φ) of the incident beam, together with mean values weighted according to the signal-to-noise ratio [6.46]

θ [°]	φ [°]	d [Å]	V_o [eV]	θ_D [K]	μ [Å^{-1}]	S/N
10	0	1.81	12.0	181	0.24	59
10	5	1.81	11.8	180	0.25	62
10	10	1.80	13.0	198	0.32	65
10	45	1.78	10.0	190	0.57	50
12	0	1.81	12.0	170	0.32	57
12	10	1.81	12.0	176	0.40	54
12	45	1.78	12.1	195	0.49	47
Weighted mean		1.80 ±0.01	11.9 ±0.8	184 ±10	0.36 ±0.11	

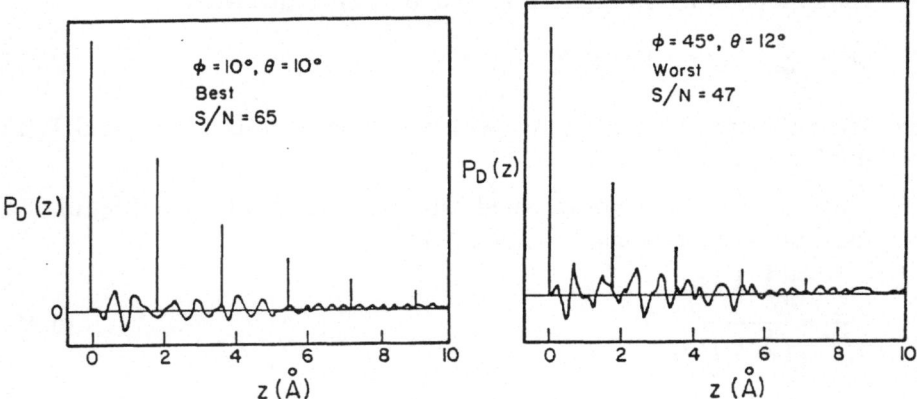

Fig. 6.10. Deconvolutions of P(z) functions of two specular intensity spectra for Cu(100) at different incidence directions, corresponding to the best and worst cases in terms of signal-to-noise ratio of the seven specular intensity spectra which were analyzed. The deconvolutions contain delta functions at multiples of the uniform interlayer spacing in the Cu(100) surface, together with a random-noise component. The exponential decay of the delta function series is due to the attenuation of the electron flux in the crystal, as characterized by the attenuation factor. After [6.46]

lowed by Fourier transformation can be applied to experimental data to determine an overlayer-substrate interlayer spacing.

Xenon forms an incommensurate hexagonal overlayer lattice on the Ag(111) surface [6.39,54]. Due to the incommensurability, this system was not amenable to dynamical LEED calculations until the later work of Stoner et al. [6.55]. The CMTA method was applied to ten specular intensity profiles obtained at polar angles of incidence ranging from 0° to 18° in steps

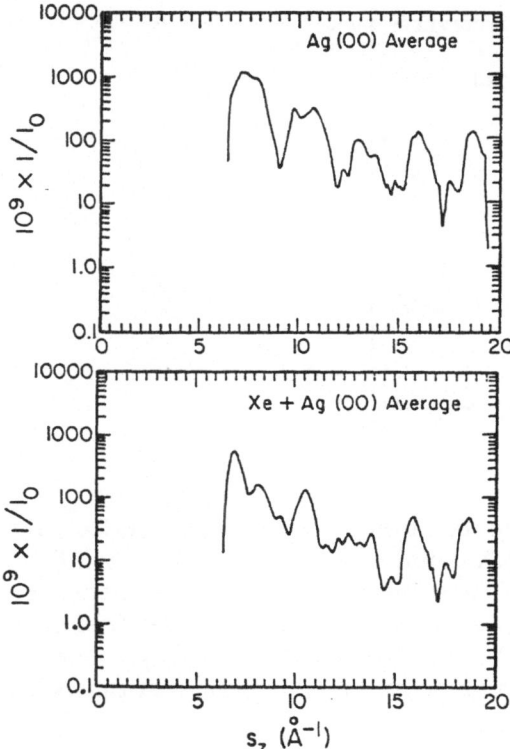

Fig. 6.11. Averages at constant momentum of the (00) beam for both clean Ag(111) and Xe-covered Ag(111) surfaces. Curves taken at angles of incidence from 0° to 18° from the normal were used in the average. After [6.39]

of 2° for both the clean and the Xe-covered Ag(111) surfaces. It was found that the averaged intensity profiles of the clean and the Xe-covered surfaces are very similar, as shown in Fig. 6.11. Consequently, to obtain the Xe-Ag interlayer spacing, the data analysis will have to be sensitive to small differences between these averaged intensity profiles. It was suggested by Cohen et al. that the superposition of the amplitude scattered by the overlayer and the amplitude scattered by the entire substrate should produce beats in the diffracted intensity with a periodicity that is related to the overlayer spacing. It was also shown that beats would be amplified if the ratio of the intensities from the Xe-covered and the clean surfaces is taken. This ratio for the experimental, averaged intensities for xenon on Ag(111) is shown as a function of s_z in Fig. 6.12.

Within the kinematic approximation, dividing the intensity of the overlayer-covered surface by the intensity of the clean surface gives

$$\left(\frac{I}{I_{Ag}}\right) \propto a_{Xe}^2 + \frac{I_{Xe}}{I_{Ag}} + 2a_{Xe}\frac{|f_{Xe}|}{|f_{Ag}|}\left\{\cos(s_z d + \delta_{Xe} + \delta_{Ag})\right.$$

$$\left. - a_{Xe}\cos[s_z(d-c) + \delta_{Xe} + \delta_{Ag}]\right\} \ , \tag{6.27}$$

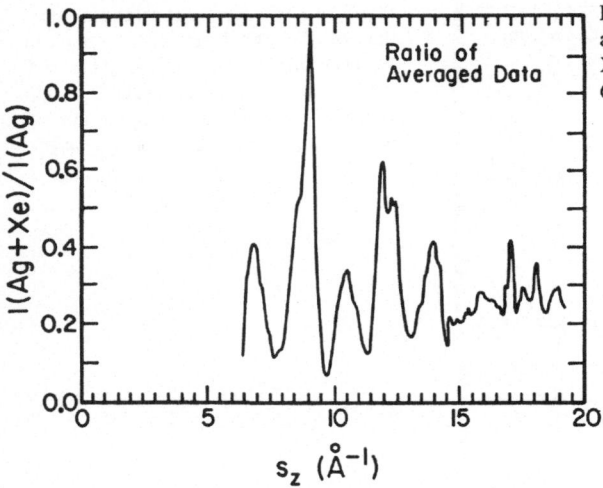

Fig. 6.12. The ratio of the experimentally averaged intensities from the clean and Xe-covered Ag(111) surfaces of Fig. 6.11. After [6.39]

where I, I_{Ag}, and I_{Xe} are the intensities from the Xe-covered surface, the clean Ag substrate, and an isolated Xe monolayer, respectively; |f| and δ are the magnitude and phase of the atomic scattering factor; α is the attenuation factor; and d and c are the Xe-Ag spacing and the bulk interlayer spacing of Ag(111), respectively. In (6.27), the dominant oscillating term is $\cos(s_z d + \delta_{Xe} + \delta_{Ag})$. Within the cosine term, $s_z d$ is the term that causes the

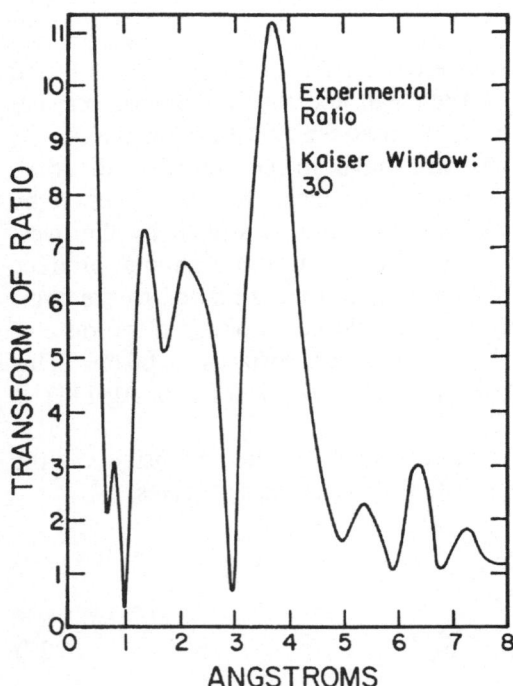

Fig. 6.13. The Fourier transform of the experimental ratio of Fig. 6.12. A Kaiser window of strength 3.0 was used to apodize the data. After [6.39]

most rapid oscillation. As a result, the Fourier transform of the ratio of the averaged profiles, shown in Fig. 6.13, shows a major peak in the transform at 3.71 Å which is associated with the Xe-Ag spacing and attributed to the term $\cos(s_z d + \delta_{Xe} + \delta_{Ag})$. However, the Xe-Ag spacing is still unknown due to the unknown contributions of δ_{Xe} and δ_{Ag}. This problem is solved by evaluating the theoretical ratio I/I_{Ag} using the δ's and |f|'s of xenon and silver calculated for free atoms. Then the transform of the experimental I/I_{Ag} is compared with the transforms of the calculated I/I_{Ag} assuming various Xe-Ag spacings. The best fit is obtained for a Xe-Ag interlayer spacing of 3.6 Å without the correction for the inner potential. With a correction for an inner potential of 7 eV, equal to the average of the inner potentials measured for clean Ag [6.56], the Xe-Ag interlayer spacing is found to be 3.5 Å These results are in favorable agreement with the results obtained in later dynamical calculations [6.55], which indicate an optimum Xe-Ag spacing of 3.55 Å

Shaw et al. [6.57] also applied the above method to extract structural information for incommensurate argon and krypton overlayers adsorbed on the graphite basal plane. However, they used only individual intensity profiles rather than averaged data. In the case of a monolayer of argon adsorbed on graphite, the Ar-graphite interlayer spacing is determined to be 3.2 Å The results for krypton are not well defined, but a suggested value of 3.3 Å is given for the Kr-graphite interlayer spacing. Since averaged data were not used in this work, it is possible that the intensity profile for the Kr-covered surface has significant contributions from multiple scattering, giving less well-defined results. It should be pointed out that this method in no way assumes an incommensurate overlayer lattice. In fact, with commensurate overlayers, other beams in addition to the specular beam become available for the analysis. The only requirement is that those beams receive, in the kinematic limit, interfering contributions from both the overlayer and the substrate.

6.4 The Dynamical Approach to Surface Crystallography

The accuracy and reliability of a structure determination is adversely affected by the approximations involved in the kinematic approaches described thus far in this chapter, except of course in those cases where almost no multiple scattering occurs. For increased accuracy and reliability, one turns to methods that incorporate dynamical effects. First, we should review briefly in what way dynamical effects and especially multiple scattering are manifest in LEED intensities, in order to make the best use of measured intensity data.

6.4.1 Dynamical Effects on Intensity Data

We first recapitulate some of the findings of Chaps. 3 and 4 and, in particular, of Sect. 4.8. Dynamical effects do not normally affect the presence or absence of spots or their position in LEED patterns. An exception is the case of multiple-scattering spots, which occur in cases of some complexity that we do not discuss here.

I-V Curves

It is the intensities of diffracted beams rather than their directions that are affected by dynamical effects. Thus, in an I-V curve, where the energy is varied, one finds shifts in the energies at which Bragg peaks occur. This also applies to any other kinematically expected peak when the surface structure is not a simple termination of the bulk structure. Furthermore, additional peaks appear that are due entirely to multiple scattering. The latter contain as much structural information as the original Bragg peaks and can therefore be considered as useful data for structural determinations.

I-θ Plots

An I-θ plot for any given beam, where the polar angle of incidence θ is varied, would in the kinematic limit have essentially the shape of the atomic scattering amplitude $|f(\theta')|^2$, where θ' is the scattering angle that varies with θ. This shape is modulated by Bragg and other kinematic peaks, as in the case of I-V curves (the Ewald construction can help in visualizing this). Multiple scattering also strongly distorts this shape in a way that depends sensitively on the surface structure, making I-θ plots equally useful sources of structural information. Examples of (φ-averaged) I-θ plots are shown in Fig. 6.4.

I-φ Plots

From kinematic theory, an azimuthal I-φ plot for a given beam should also reflect $|f(\theta')|^2$ modulated by a kinematic structure factor, as an Ewald construction will show. (A variation of φ implies, for nonspecular beams, a variation of the scattering angle θ'.) Multiple scattering again induces strong deviations in I-φ curves from the kinematic form, and thus these azimuthal curves also contain much structural information. Examples of I-φ plots are shown in Fig. 6.3.

I-ḡ Data

One can refrain from varying any incidence parameter, whether the energy, the polar angle or the azimuthal angle and limit oneself to the intensities of all the beams $\bar{g}$ diffracted under one incidence condition. We may label these I-$\bar{g}$ data. Such data, of course, contain the same type of information available in I-V, I-θ, and I-φ curves. However, the amount of information present in these different kinds of data could be different, a matter to which we turn our attention now.

6.4.2 Information Content of Measured Data

The amount of experimentally available information influences the reliability of a structural determination, so that one should look for data containing as much structural information as possible. In the case of I-$\bar{g}$ data, it is straight-forward to count the amount of available information, which is equal to the number of beams for which the intensity has been measured. The only assumption here is that the intensities of different beams are independent. Aside from symmetry effects, there is no reason to doubt the validity of this assumption. (We mean here experimental independence rather than theoretical independence. In principle, only a small number of theoretical parameters suffice to describe the entire surface structure and LEED process completely, whereas one can make independent measurements of many more data points than that small number.) Symmetry-degenerate beam intensities presumably are to be counted as a single data point for which a relatively higher accuracy is obtained due to better statistics. These considerations will also apply to I-V, I-θ and I-φ curves, where symmetry-degenerate data can also be measured. The issue of weighting such data arises especially in computing R-factors.

I-$\bar{g}$ data, however, have a weakness [6.58]. They are only of use if one can assume that the relative intensities of different beams are determined mainly by the ordered surface structure and are affected little by nonstructural factors such as thermal effects and other forms of disorder. Unfortunately, this assumption does not necessarily hold, especially with overlayers where extra superlattice spots are easily dimmed by adsorbate disorder or incomplete coverage. Even on simple, clean metal surfaces, this assumption is known not to hold in general, presumably because of surface defects. Of course, I-V, I-θ, and I-φ curves suffer from these effects as well, but with them one is more interested in the presence and position of peaks, i.e. in the local intensity variations around a certain energy or angle, rather than in relative heights of distant peaks. Disorder does not affect the presence and positions of peaks nearly so much as it affects the relative intensities of different beams. Hence, I-$\bar{g}$ data, in addition to containing relatively little information compared to a set of I-V, I-θ, or I-φ curves, are also relatively sensitive to disorder, making them less attractive for structural determination.

The amount of information available in I-V curves is less easily counted than with I-$\bar{g}$ data. It is perhaps best (though not entirely satisfactory) to count each energy interval equal to the local peak width as containing one unit of information. This is justified since I-V curves usually contain as many peaks as can possibly be fit into the available energy range, especially when multiple scattering occurs. Thus, a typical I-V curve extending from 20 to 220 eV and having peaks of average width 10 eV would contain 20 units of information, essentially corresponding to 20 peaks. One can similarly count the information in I-θ and I-φ curves in units of peak widths, taking into account the fact that the peak widths in these cases can depend on both θ and φ, as one can imagine by considering an Ewald construction and using the known peak widths as a function of the perpendicular component of $\bar{s}$. In I-θ curves, peak widths are on the average of the order of 5° to 10°; while in I-φ curves the peak widths are of the order of 15° for $\theta \sim$ 45°, and smaller (larger) at larger (smaller) values of θ for typical LEED energies.

In I-θ and I-φ curves, the information content is limited by the natural range of the independent variables: $0 \leqslant \theta \leqslant 90°$ and $0 \leqslant \varphi \leqslant 360°$. In I-V curves the energy is in practice limited to the range from about 20 eV to a few hundred electron volts, the upper limit depending primarily and inversely on the unit-cell size since it affects the cost of LEED calculations. It appears that these three types of data are equivalent in information content, when they are measured in terms of peak widths. Furthermore, there does not seem to be much difference in the total amount of information available in these types of data. In any case, the distinction becomes blurred in practice because, for example, I-V curves are usually taken at a series of different incidence angles, while I-θ curves may be measured at a series of different energies or azimuths, in addition to the presence of a collection of diffracted beams emerging in different directions.

Not only I-$\bar{g}$ data, but also I-V, I-θ, and I-φ data are unreliable for structural determination under certain circumstances. The main situations to be avoided are those of low energies E $\leq$ 20 eV and grazing angles of incidence or emergence. In those cases, the data are too sensitive to poorly understood details of the scattering mechanism, such as the effects of the valence and conduction electrons and of the surface-potential step. The electron beams are then also more difficult to control, especially at low energies. Furthermore, high-energy data that are weak because of thermal effects are unreliable because they are difficult to separate from the diffuse background intensity.

6.4.3 Extraction of Structural Information from Dynamical LEED Intensities

The Trial-and-Error Approach

The theoretical multiple-scattering formalism, as described in Chap. 5, is too complex to allow a direct inversion of the type presented in previous sections of this chapter. One is forced to adopt the trial-and-error approach, aided by a suitable search strategy. Thus, one should make intensity calculations for all plausible surface geometries and hope that the one which gives the best agreement with experiment is indeed the correct geometry. There is no way to ensure that the correct geometry has not been missed in the list of plausible geometries. Just as with X-ray Diffraction, one is never sure that the "best" structure found cannot be improved upon by a completely different structure.

Need for Independent Information

It follows that any independent information that can restrict or direct the structural search is very valuable. Such information can include the elemental composition, reasonable bond lengths and bond angles, the coverage (giving the number of atoms or molecules in the unit cell), whether an overlayer or underlayer is formed, the nature of a molecular surface species (such as whether a molecule dissociates or to what extent an adsorbed molecule reacts with the surface and changes its stoichiometry) or many other kinds of evidence concerning the surface, obtainable with a variety of other surface-sensitive techniques.

Structural-Search Strategies

Thus far, search strategies have been developed only for the simplest kinds of surfaces. Consequently, for a clean unreconstructed fcc(100) surface, a structural search would allow only one or a few layer spacings to vary, no other change from the bulk structure being considered likely. The problem of finding the maximum agreement between theory and experiment when more structural parameters are free to vary has not been addressed adequately in LEED yet. Most well-known minimization or maximization procedures (such as the methods of least-squares or steepest descent) are rather wasteful in LEED, because they make no multiple use of partial results (e.g. layer-diffraction matrices) that have been obtained at relatively large expense. It should be recalled that it is computationally inexpensive to vary an interlayer spacing when the plane-wave representation is used, whereas a variation of the buckling of a layer, for example, is computationally expensive. In such a case one should, for each amount of buckling, let the calculations range over a substantial number of values of an interlayer spacing to benefit from the availability of the expensive diffraction matrices for the buckled layer. A search strategy should exploit such differences in the cost of varying different structural parameters.

Switch between Calculation Methods

A further complication arises in the not unusual case where the variation of a positional parameter beyond a certain value requires a switch to a different calculation method. For example, as the distance of an adsorbate from a substrate is gradually decreased in a sequence of calculations, there comes a point where the plane-wave representation fails and the spherical-wave representation must take over. To incorporate the option of such a switch of methods in one single program is often quite wasteful of computer resources, because an unused set of subroutines can occupy large amounts of core space.

Yet another factor of importance is the use of symmetry for reducing the computing needs. By assuming a given symmetry, the search is restricted to those structures that satisfy that symmetry. Again, it would be inconvenient to allow a switch to a different and/or lower symmetry within the same calculation.

As a result of these various factors, at present no general search strategy exists in LEED. Each surface-structural search is handled in its own manner, varying the "cheap parameters" (mainly layer spacings and to some extent layer registries) as much as possible compared to the relative atomic positions within individual composite layers. A priori knowledge, of course, also enters into the choice of search strategy. For example, one may wish to use knowledge concerning bond lengths and bond angles and only vary the structure accordingly.

A Trend: Insensitivity to Registry

Nevertheless, two trends often emerge in structural searches that are of rather general value. First, if better agreement between theory and experiment is obtained at a certain layer spacing compared to neighboring layer spacings, there is a good chance that the same layer spacing at other registries (i.e. displacements parallel to the surface) will also produce relatively good results. This follows from the greater sensitivity of LEED to layer spacings than to registries, at least for momentum transfers $\vec{s}$ not too far from the surface normal.

Another Trend: Multiple Coincidences

Second, in the same case of a relatively good layer spacing, it will often be found that layer spacings differing from this one by a (positive or negative) multiple of 0.5 to 0.7 Å also produce rather good agreement with experiment. This "multiple-coincidence" effect [6.36] occurs because, in the kinematic limit, a change in layer spacing by half a wavelength (approximately 0.6 Å in most practical cases), only changes by 2π the relative phases of the waves reflected from the different layers, thereby restoring the original interference conditions. The multiple-coincidence effect is best illustrated with the system Ni(100)-c(2x2)-O. The first published LEED analysis indicated a Ni-O layer spacing of 1.5 Å [6.59], which was modified in later work to 0.9 Å [6.60] (the first search had not ranged to such small values of the spacing).

Subsequently, a further extension of the search [6.61], prompted by other work [6.62], showed that the spacing could in fact also be approximately 0.3 Å A more extensive data base is needed to resolve the issue.

6.5 Reliability Factors (R-Factors)

Surface crystallography by LEED depends heavily on the optimization of the agreement between measured and calculated intensity curves. For various reasons it is desirable to judge this agreement both in an objective manner and in a quantitative manner rather than by a visual evaluation that otherwise must be used. The latter is to some extent irreproducible and subjective. Various quantifications, variously called reliability factors, R-factors, residues, or residuals, have been proposed in the past for these purposes [6.40,63-73]. R-factors, as we shall call them here, are single numbers summarizing the level of agreement between sets of curves. Consequently, they can be used as figures of merit for trial structures.

R-factors have been proposed to satisfy several needs in LEED in analogy with their application in X-ray crystallography [6.28]. Ideally, an R-factor should be sensitive to atomic positions at the surface and insensitive to nonstructural surface properties (electron-atom scattering potential, electron damping, thermal state, disorder, etc.). Such a figure of merit can then be used to discriminate between different surface geometries and hopefully indicate which is the correct geometry.

A related need arises from the trial-and-error approach that a structural determination by LEED is forced to adopt. A structure is usually determined by calculating I-V curves for a variety of plausible geometries until the agreement with experimental I-V curves is satisfactory. Thus, typically hundreds of curves must be compared. This approach clearly becomes unreliable when left to visual comparison. An R-factor can also provide a basis for a structural-search strategy (possibly automated) in which it is desirable to converge to the correct structure by using the results of a preliminary search. Furthermore, an R-factor would allow a comparison of the quality of different structural determinations and enable other workers to judge the quality of a particular result.

An often-stated function of R-factors is to satisfy the need for an objective criterion to compare theoretical and experimental curves. In the sense of treating all curves equally, R-factors provide more objectivity than a visual comparison does. However, there is no consensus among LEED crystallographers concerning the relative importance in I-V curves of such features as peak positions, relative peak heights, peak skewness, and peak widths. Consequently, it is not possible to design an R-factor that will satisfy everyone, let alone be objective. In addition, the relative importance that one may give various features can depend on the particular surface under consideration and on its condition. For example, in a substance with relatively kinematic scattering, calculated intensities should in general be more reliable than in a substance with strong multiple scattering. However, if the thermal motion happens to be much more complicated in the kinematic substance, the reverse might be true. Any R-factor that one may propose for LEED is based on a subjective choice of the relative importance of the various features of I-V curves.

The subjective aspects of choosing an R-factor suitable for LEED have led to the coexistence of a variety of R-factors that differ in their sensitivity to different features of intensity curves. Thus far, R-factors have been applied only to I-V curves, but they could equally well be used to compare experimental and theoretical I-θ and I-φ curves (for I-$\bar{g}$ data one can simply use the X-ray R-factors [6.28]). We shall now review the R-factors that have been proposed and then discuss their use.

6.5.1 Various R-Factors

The R-Factors RPOS1 and RPOS2

Since peak positions in I-V curves constitute the feature most sensitive to atomic locations, the earliest R-factors in LEED concentrated on these positions. For example, if corresponding peaks in the theory and the experiment differ in their positions by ΔE_i, then one can define the R-factors [6.63,64]

$$RPOS1 = \frac{1}{N}\sum_i |\Delta E_i| \quad \text{and} \tag{6.28}$$

$$RPOS2 = \frac{1}{N}\sum_i |\Delta E_i|^2 \ , \tag{6.29}$$

where the index i labels N peaks (cf. Fig. 6.14). A difficulty with this kind of R-factor is that it is well-defined only when each peak in the theory exists in a one-to-one relationship with a peak in the experiment, which often fails to happen. Furthermore, peak positions are sometimes difficult to locate, for example for peaks with shoulders. A consequence is the difficulty of programming these R-factors for automatic computation.

The R-Factor ROS

These difficulties and uncertainties are removed in an R-factor that measures the fraction of the energy range over which the theoretical and experimental curves have energy derivatives dI/dE of opposite signs [6.65] (illustrated in Fig. 6.14):

ROS = fraction of energy range with slopes of opposite signs . (6.30)

It should be pointed out that, although the above R-factors consider only the

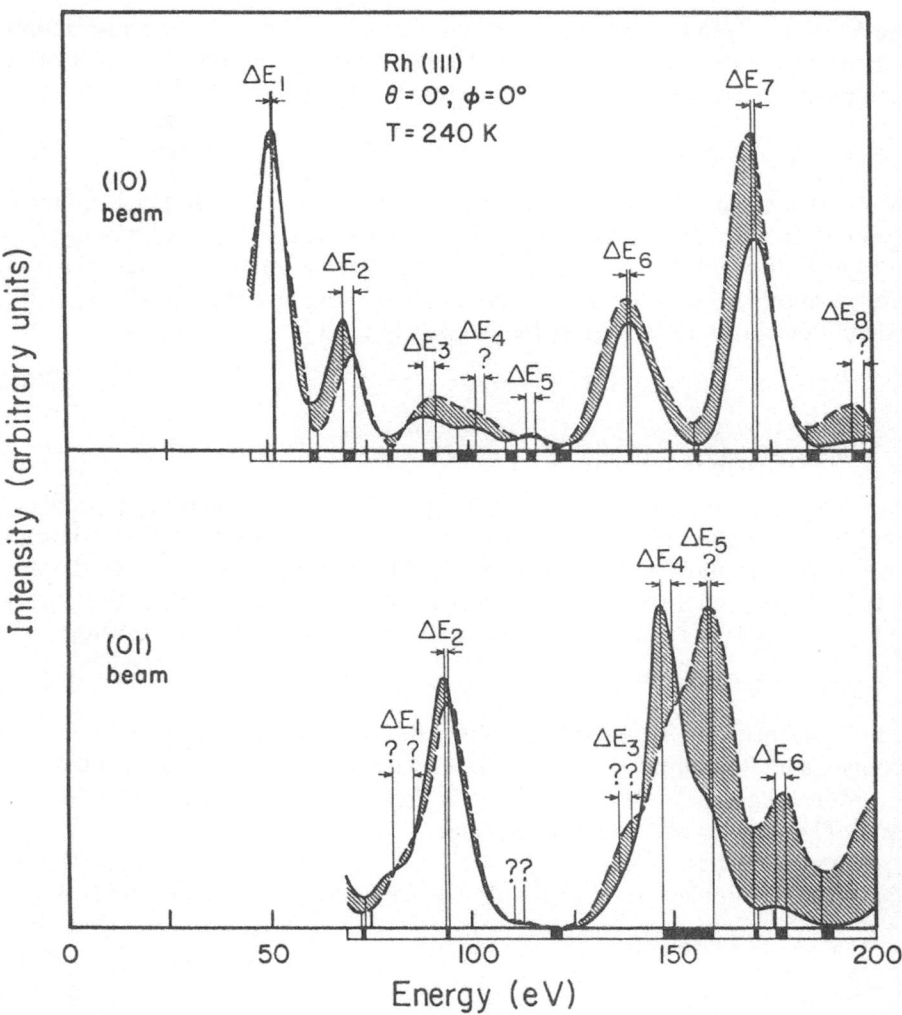

Fig. 6.14. An illustration of the definition of some R-factors, with the example of two beams diffract-ed from the clean, unrelaxed Rh(111) surface, comparing experiment (*full lines*) with theory (*dashed lines*). The peak mismatches ΔE_i enter into the R-factors RPOS1 and RPOS2. Question marks indi-cate cases where ΔE_i is ill-defined. The bars below the energy axes show in black those regions of the energy range where the experimental and theoretical intensity slopes (derivatives with respect to ener-gy) have opposite signs. The ratios of these regions to the full energy range define the R-factor ROS. The shaded areas define the R-factor R1 (except that in this figure the curves have not been properly normalized by the factor c)

positions of intensity extrema, they are somewhat sensitive to nonstructural parameters, since phase shifts and relative heights of component peaks of a composite peak affect the positions of these extrema.

The R-Factors R1 and R2

Much structural information is contained in the relative peak heights, largely ignored in the above R-factors. In analogy with X-ray crystallography (where individual Bragg reflection intensities rather than continuous curves are measured), the following R-factors have been proposed [6.40,65-68,70], which depend on relative peak heights (cf. Fig. 6.14):

$$R1 = A_1 \int |I_e - cI_t| dE , \quad \text{and} \tag{6.31}$$

$$R2 = A_2 \int (I_e - cI_t)^2 dE . \tag{6.32}$$

Here the subscripts e and t stand for experiment and theory, respectively, and the integration (or summation) ranges over the energy intervals common to experiment and theory. The constant c is inserted to account for the usually unknown relative intensity scales in the experiment and in the theory, and it serves to normalize the curves to each other, for example through

$$c = \int I_e dE / \int I_t dE . \tag{6.33}$$

Thus, insensitivity to the relative intensity scales is explicitly built in. Of course, some assumptions are involved, such as the energy independence of c. Moreover, there is the question whether c should be beam-dependent or not. These issues will be discussed later.

The prefactors A_1 and A_2 in (6.31,32) are designed to render the R-factors dimensionless and to provide normalization. The usual choice is

$$A_1 = 1/\int I_e dE \quad \text{and} \quad A_2 = 1/\int I_e^2 dE . \tag{6.34}$$

A χ^2-factor has been proposed [6.71] that is very similar to R2 of (6.32). It differs in the choice of A_2, using the variance of the fit between experiment and theory, i.e.

$$A_2 = 1/< \int (I_e - \bar{c}I_t)^2 dE > ,$$

where the average is taken over all available beams, and $\bar{c}$ is also the weighted average of c over all available beams.

The R-Factors RP1 and RP2

A drawback of using the intensities in (6.31,32) is that any unfiltered background intensity present in the experiment can contribute to the R-factors. The background, although it has little structure, is difficult to eliminate reliably. However, using the derivatives of the intensities with respect to energy will largely remove the background contribution [6.65]

$$RP1 = A_{p1} \int |I_e' - cI_t'| dE \; , \tag{6.35}$$

and

$$RP2 = A_{p2} \int (I_e' - cI_t')^2 dE \; , \tag{6.36}$$

with

$$A_{p1} = 1/\int |I_e'| dE \quad \text{and} \quad A_{p2} = 1/\int (I_e')^2 dE \; . \tag{6.37}$$

The R-Factors RPP1 and RPP2

One may argue that RP1 and RP2 put too much emphasis on the flanks of the peaks, where the derivatives are largest, and too little emphasis on the tops of the peaks or on the minima. To correct these deficiencies, one may use the second derivatives, since they are largest at the peaks and minima, i.e.

$$RPP1 = A_{pp1} \int |I_e'' - cI_t''| dE \; , \tag{6.38}$$

and

$$RPP2 = A_{pp2} \int (I_e'' - cI_t'')^2 dE \; , \tag{6.39}$$

with

$$A_{pp1} = 1/\int |I_e''| dE \quad \text{and} \quad A_{pp2} = 1/\int (I_e'')^2 dE \; . \tag{6.40}$$

The Zanazzi-Jona R-Factor RRZJ

Another R-factor, which has been used extensively, also uses the second derivatives. This R-factor, proposed by Zanazzi and Jona [6.68] (we mention only their "reduced" R-factor), combines first and second derivatives to obtain the advantages of both

$$RRZJ = A_{RZJ} \int [|I_e'' - cI_t''||I_e' - cI_t'|/(|I_e'| + \max|I_e'|)] dE \; , \tag{6.41}$$

with

$$A_{RJZ} = 1/(0.027 \int I_e dE) \; . \tag{6.42}$$

The reduction factor 0.027 is a typical average value that RRZJ would have in the absence of this factor for unrelated LEED curves. In addition, this factor makes RRZJ dimensionless. The dimension can be eliminated more naturally by using a modified normalization

$$A_{MZJ} = 1/\int |I_e''| dE \; . \tag{6.43}$$

Because of this simplification, which avoids the need for a reduction factor, one can use a modification of the Zanazzi-Jona R-factor,

$$RMZJ = A_{MZJ} \int [|I_e'' - cI_t''||I_e' - cI_t'|/(|I_e'| + \max|cI_t'|)]dE . \quad (6.44)$$

We have also inserted here $\max|cI_t'|$ in the place of $\max|I_e'|$ of [6.41]. This change avoids a normalization problem that occurs when an unusually large $|I_t'|$ could dominate the integrand.

The Pendry R-Factor RPE

In all the R-factors in (6.37-44), the peaks are weighted in proportion to their heights. Whether this is reasonable is debatable. It is true that a small peak is less reliably measured, but it is also true that its mere existence and its position imply equally valuable geometrical information as that of a large peak. This is especially true of the higher-energy parts of I-V curves, where thermal and atomic scattering effects often make all intensities much smaller than at the lower energies. For these reasons, Pendry has suggested an R-factor [6.72] that attempts to treat all peaks (and minima) with equal weight. It is based on the logarithmic derivative L of the I- V curves,

$$L = I'/I , \quad (6.45)$$

arranged so as to avoid singularities when I = 0 by defining the function

$$Y = L/(1 + V_{oi}^2 L^2) , \quad (6.46)$$

where $2V_{oi}$ can be taken to be the average peak width of single peaks (as opposed to overlapping peaks). Pendry's R-factor is then

$$RPE = \int (Y_e - Y_t)^2 dE / \int (Y_e^2 + Y_t^2) dE . \quad (6.47)$$

Although, in principle, large and small peaks have equal weights in this R-factor, there are exceptions. For example, overlapping peaks are not treated equally if they have different heights.

Metrics as R-Factors

A class of R-factors that has more global sensitivity to the entire intensity curve, has been proposed by Philip and Rundgren [6.73]. These R-factors are based on the integral functions of the experimental and theoretical intensity curves, namely

$$J_{e,t}(E) = A_{e,t} \int_{E_{min}}^{E} I_{e,t}(E')dE' \quad \text{and} \quad A_{e,t} = 1/\int_{E_{min}}^{E_{max}} I_{e,t}(E')dE' , \quad (6.48)$$

where (E_{min}, E_{max}) is the available energy range common to experiment and

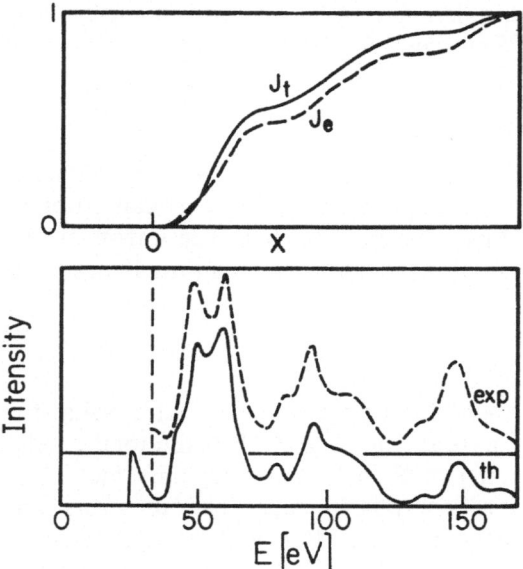

Fig. 6.15. The functions J_e and J_t (*top diagram*) corresponding to the I-V curves shown in the bottom diagram for the (01) beam of Al(111) at normal incidence. After [6.73]

theory, and the subscript e or t is chosen. For normalization to the interval (0,1), the energy E is replaced by a variable x such that

$$E = (E_{max} - E_{min})x + E_{min} \quad .\tag{6.49}$$

Note that $0 \leqslant J_{e,t} \leqslant 1$ holds, so that a plot of J_e and J_t might look like that of Fig. 6.15. Then one may measure the discrepancy between the experimental and theoretical curves $J_e(x)$ and $J_t(x)$ as

$$RPR1 = \int_0^1 |J_e - J_t| dx \quad ,\tag{6.50}$$

which represents the area between the curves J_e and J_t. Another R-factor starts by intersecting the curves J_e and J_t with a straight line of slope -1 and projecting the distance between the two intersections onto the intensity axis: the largest such projection for all positions of the intersection line is the value of this R-factor, which is also called the Levy distance. Yet another R-factor takes for each point on the curves J_e or J_t the shortest distance to the other curve: the largest of these shortest distances is the value of this R-factor, also called the Hausdorff distance.

The basic advantage of these R-factors is that they are metrics [6.73], which guarantees among other things that, when they are smaller, the corresponding curves are necessarily closer together and that convergence of such an R-factor to zero implies that the experimental and theoretical curves are necessarily converging to each other. (This cannot be said of other R-factors.) Another advantage is that the effect of noise is considerably reduced, because integration rather than differentiation of the intensity curves is used. On the other hand, differences in background intensities and in overall intensity envelopes can affect these R-factors adversely.

6.5.2 Reliability of Reliability Factors

To obtain a nonempirical estimate of the significance of a given R-factor value, Pendry has proposed the following approach [6.72]. One can estimate the effect on an R-factor of random errors made in the experiment and in the theory. A variance varR due to these random fluctuations is obtained that satisfies the approximate relation

$$RR = varR/\overline{R} \sim (8V_{oi}/\Delta E)^{1/2} \ , \tag{6.51}$$

valid for any LEED I-V curve, where $\overline{R}$ is an average R-factor value for uncorrelated data, $2V_{oi}$ is the average peak width, and ΔE is the total energy range being considered. The meaning of R, a "double-reliability factor", is explained by the statement that for uncorrelated data, random errors will cause the R-factor value to be within the range $\overline{R}(1 \pm RR) = \overline{R} \pm varR$ with 68% probability. Equivalently, an R-factor value falling outside the range $\overline{R}(1 \pm 1.96 \ RR) = \overline{R} \pm 1.96 \ varR$ has a probability of only 10% of being caused by random fluctuations, for uncorrelated data. A significance factor S can now be defined as

$$S = 1 - (2\pi)^{-1/2} \int_{-T}^{\infty} \exp(-t^2/2)dt \ , \tag{6.52}$$

where

$$T = (\overline{R} - R)/RR \ . \tag{6.53}$$

The significance factor S is the probability of a random fluctuation of an R-factor to a value less than R, and by implication 1-S gives the probability that a structural choice producing the value R is correct [6.72].

6.5.3 Dealing with Different Experiments and Different Beams

Experimental LEED data may come in the form of separate I-V curves, measured for different diffraction beams, or for different experimental measurements performed under similar conditions, or for different angles of incidence. One may average all equivalent I-V curves, such as those measured in different experiments or those degenerate by symmetry. After this averaging, one is left with inequivalent curves the R-factors of which can then be averaged in turn to produce an overall R-factor (although we shall be interested in the individual beam-dependent R-factors as well). Usually, weights proportional to the energy range ΔE_i of each beam are used to produce the average

$$\bar{R} = (\sum_i \Delta E_i R_i)/(\sum_i \Delta E_i) \ . \tag{6.54}$$

Less obvious is the proposal to weight the overall R-factor to reflect the number of independent curves used to calculate it [6.68,69]. This serves the justified purpose of penalizing results based on a small amount of data. However, it is not true that the amount of data is simply proportional to the number of independent I-V curves. The length of the energy range or, even better, the number of peaks is a better measure of the amount of information available and used in I-V curves. Even so, it is not obvious that very dynamical I-V curves with their relatively large number of peaks and thus large amount of information are preferable to very kinematic I-V curves with their smaller amount of information, since multiple scattering is inherently more difficult to reproduce theoretically than single scattering. The choice of an appropriate weighting factor thus becomes uncertain.

In addition, the use of such a global weighting factor is only meaningful if all other considerations of reliability are built into the final R-factor as well. For example, the number of surface structures tested affects the reliability of a result and could equally well be included in the R-factor. However, it is very difficult to quantify such considerations, and therefore it does not appear fruitful to take all the available information into account in R-factors. A further consequence of this is that it is not strictly meaningful to compare R-factor values obtained in independent structural determinations, a fact which is true also in X-ray crystallography [6.28,74].

When one considers different beams measured in one experiment, the question arises whether the intensity scales of the different beams should be kept proportional to each other or may be independently adjusted. Various factors favor independent adjustment, producing a beam-dependent factor c in [6.31-44]. These are, among others, solid-angle effects, projection-angle effects, varying screen response, surface disorder (especially with adsorbate-induced fractional-order beams), and angle-dependent effects of inelastic damping and of thermal vibrations.

6.5.4 Noise and Smoothing

Experimental noise is always present in measured intensities and is of great concern when taking first and especially second derivatives. A noise-to-signal ratio of up to 1:20 is not uncommon in practice. In this case, it is necessary to smooth the experimental curves. A filtering of high frequencies is not necessarily unambiguous since the experimental energy step (often 1 or 2 eV) is not always very small compared with peak widths (5 to 10 eV). Not only is it not clear where the filtering cutoff should occur, but, in addition, this smoothing affects the underlying shape of the curves (i.e. the signal), since the lowest noise frequencies can easily overlap and be indistinguishable from the highest signal frequencies. One way to circumvent this difficulty would be to smooth the theoretical curves in the same manner so that the underlying curve shapes of the experiment and the theory are distorted in the same way. (In this discussion, we neglect the complications of fine structure in I-V curves that arises near beam emergence conditions, see Sect. 5.12.)

In addition to filtering, one can also smooth with simpler procedures. An approach that has been used successfully [6.69] smoothes point-by-point with a three-point formula. Each intensity I(E) is replaced by

$$I_{new}(E) = \frac{1}{4} \left[I(E - dE) + 2I(E) + I(E + dE) \right] , \tag{6.55}$$

where dE is the energy grid interval. Smoothing each curve once or twice in this way produces well-behaved first and second derivatives. It should be pointed out, however, that this method of smoothing also changes the underlying shape of the curves to some extent. In particular, it decreases all curvatures.

6.5.5 The Use of R-Factors

Multiple Minima in R-Factors

The primary use of R-factors is to aid in identifying the most promising surface model in a structural search, especially in the case where a large amount of data is involved that renders visual comparison between theory and experiment cumbersome and unreliable. Thus, one looks for an R-factor minimum as a function of atomic positions. Unfortunately, several "local" R-factor minima usually exist for different surface geometries, as in the case of multiple coincidences found when interlayer spacings are varied. The first consequence of this situation is that it is not sufficient to make a random guess at the structure and then locally minimize the R-factor by adjusting the structural parameters. This would merely bring the search to a nearby, possibly incorrect minimum. One must also investigate the possible existence of other minima. The second consequence is that one must learn to discriminate against minima that correspond to incorrect structures, especially when the R-factor values at the minima are comparable.

One solution to this problem is to increase the data base and to add corresponding calculations. A more cost-effective approach comes from contrasting the behavior of the R-factors calculated for individual beams, as opposed to the R-factors averaged over all beams. It appears that in the neighborhood of a minimum corresponding to an incorrect structure, these "beam R-factors" disagree between themselves relatively strongly in their predictions of the best structure, i.e. the "best structure" is very dependent on the beam used to determine it [6.74-76]. This is illustrated in Figs. 6.16b and 6.17. Large disagreements between beams occur, that may be only weakly reflected in the beam-averaged R-factor. A similar, though less strik-

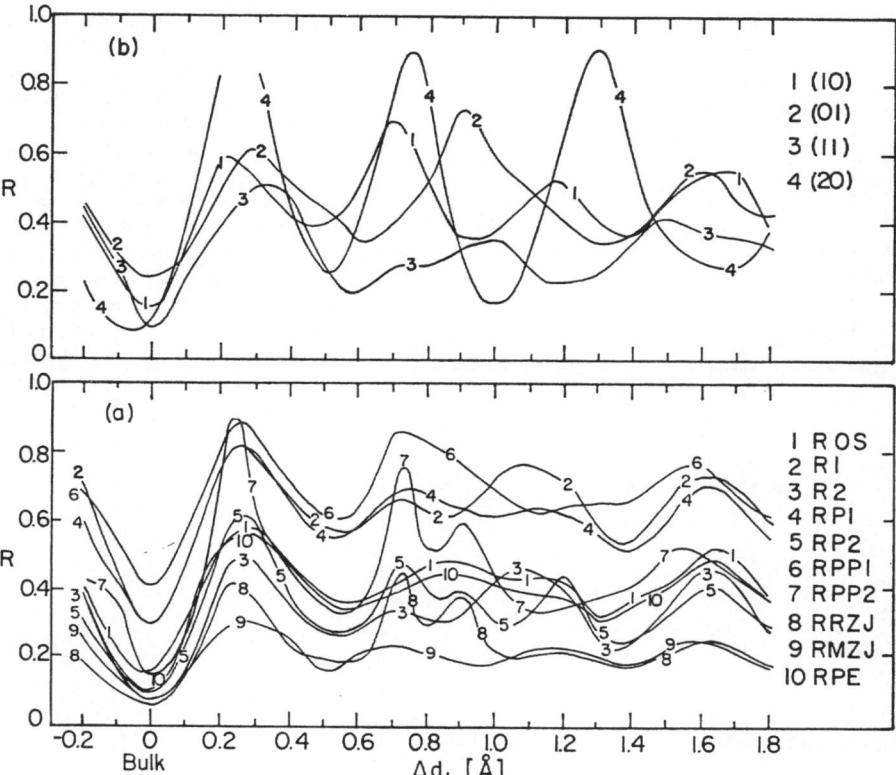

Fig. 6.16a,b. Various R-factors applied to several beams diffracted from Rh(111) ($\theta = 0°$, T = 240 K). The top-layer spacing is varied from a 0.2 Å contraction to a 1.8 Å expansion. **(a)** For each of ten R-factors defined in the text, the beam-averaged value is plotted as a function of the top-layer relaxation. **(b)** For each beam, the value averaged over the ten R-factor definitions is plotted. Note the good agreement between R-factors near the correct structure (unrelaxed spacing), while all other R-factor minima vary from one R-factor to another R-factor, or from one beam to another beam

ing effect is found when one contrasts different R-factors. They also disagree in their structural prediction in the neighborhood of an incorrect R-factor minimum (Fig. 6.16a). The use of different R-factors also helps to establish whether one minimum is reliably deeper than another minimum. If most R-factors make the same minimum deepest, then there should be no doubt that it corresponds to a better structural choice [6.74].

In addition, a correct R-factor minimum should not shift with varying diffraction geometry and varying wavelength (due to varying energy). By contrast, incorrect R-factor minima can shift with such variations, since they are due to accidentally favorable interference conditions. This is clear for the case of "multiple coincidences" (Sect. 6.4.3), which occur at layer spacings

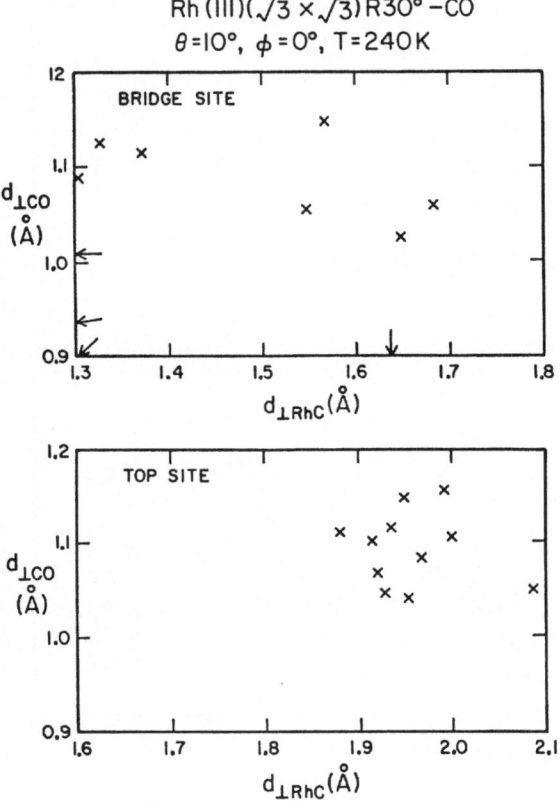

Rh (III)($\sqrt{3} \times \sqrt{3}$) R30° –CO
$\theta = 10°$, $\phi = 0°$, T = 240 K

Fig. 6.17. In a structural analysis of Rh(111)-($\sqrt{3} \times \sqrt{3}$)R30°-CO [6.75], the optimum Rh-C and C-O layer spacings were determined separately for each of 11 beams, both for the on-top adsorption site (*bottom panel*) and for the bridge site (*top panel*). Each cross indicates the optimum layer spacings for one beam (arrows are used when the optimum structure lies outside the range of the search). The optimum spacings are markedly clustered for the on-top site, while they scatter widely for the bridge site, indicating a preference for the on-top site

that differ by a multiple of half a wavelength from a correct layer spacing: the incorrect minima shift with varying wavelength, while the correct minimum remains stationary. However, it remains to be demonstrated, in the general case, that one can recognize an incorrect R-factor minimum by its shifting with varying incidence angles and with varying wavelength.

Graphical Representations of R-Factors

It is often convenient to represent R-factors graphically in the course of a structural determination. When only one parameter p is varied, the representation of an R-factor R(p) yields a smooth curve, as illustrated in Fig. 6.16, which can be used to determine the optimal value p_0 of the parameter p that minimizes R(p) by simple interpolation between the parameter grid points. Moreover, the sensitivity of R to p, or lack thereof, is easily visualized in the form of a sharp or a shallow minimum, respectively.

However, often more than a single parameter must be varied independently, such as a layer spacing and the inner potential, or various atomic coordinates. If only two parameters are varied, a two-dimensional contour plot is convenient (Figs. 6.9c,d). Again, the minimum can be located by interpolation between the grid points used for the parameters in question, and the sensitivity to the parameter variation can be gauged. In addition, correlation between the two varied parameters can be recognized when the contours near a minimum have an elliptical shape with major axes not parallel to the parameter axes (uncorrelated parameters would produce major axes parallel to the parameter axes). For example, a layer spacing and the inner potential are always found to be correlated (Figs. 6.9c,d). This is not surprising, since the interference conditions for reflections by different layers depend not only on the interlayer spacing, but also simultaneously on the local wavelength, which is affected by the inner potential.

A useful graphical procedure to locate the position of a minimum in a two-parameter contour plot of the R-factor $R(p,q)$ is the following [6.71]. One first determines the locations where the partial derivative $(\partial R/\partial p)$ ($q =$ constant) vanishes for the available values of $q =$ constant (Fig. 6.18). These locations define a curve that must go through the minimum of $R(p,q)$. Similarly, one determines the locations where the partial derivative $(\partial R/\partial q)$ ($p =$ constant) vanishes. These locations define another curve that must also pass through the minimum of $R(p,q)$. The minimum is now simply found graphically as the intersection of these two curves. If $R(p,q)$ is a quadratic function of p and q, the two curves in question are straight lines. In practice, this is a good approximation in the proximity of the minimum. Additional information concerning the location of the minimum can be found in the same manner by taking partial derivatives not parallel to the p-axis $(\partial R/\partial p)$ or the q-axis $(\partial R/\partial q)$, but along lines of grid points with other orientations, see Fig. 6.18. Again, the curves joining points of vanishing partial derivatives must go through the minimum. This can be useful when the contours have a strong elongation that makes an angle with the p or q axes. The curves in question can then be chosen to lie approximately parallel and perpendicular to the elongation axis. This helps to define their intersection precisely, whereas the curves obtained from $\partial R/\partial p = 0$ and $\partial R/\partial q = 0$ are almost tangent, yielding a poorly defined intersection.

R-Factors as Absolute Measures of Reliability

In addition to being used as guides in a structural search, R-factors are also used to summarize the reliability of a structural result in an absolute sense and to provide a basis for comparison with other structural determinations of the same surface and with results obtained for other surfaces. These uses, though very tempting, are controversial, because there is a large step between formulas such as those proposed for R-factors and an absolute measure of reliability. Nevertheless, for lack of a better, yet succinct alternative, it has become customary to at least accompany each new structural result with its

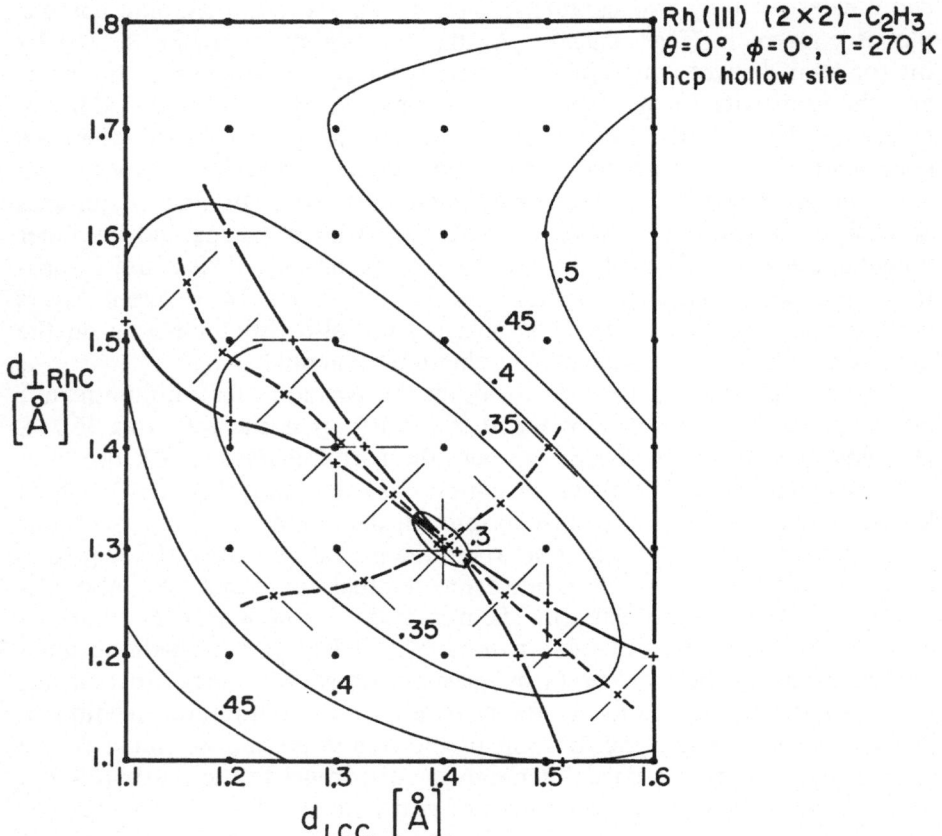

Fig. 6.18. A representative R-factor contour plot for the variation of two structural parameters: the Rh-C and C-C layer spacings in Rh(111)-(2 × 2)-C_2H_3 (ethylidyne) [6.76], which were varied over the grid points shown as dots. Four lines of vanishing partial first derivatives are drawn, two for the partial derivatives with respect to $d_{\perp RhC}$ and $d_{\perp CC}$ (*continuous lines linking* +'*s*) and two for the partial derivatives with respect to $d_{\perp RhC} + d_{\perp CC}$ and $d_{\perp RhC} - d_{\perp CC}$ (*dashed lines linking* × '*s*). Because of the elliptical elongation of the contour lines near the minimum, the intersection of the two continuous lines is poorly defined, since they are nearly tangent to each other. By contrast, the two dashed lines intersect each other at nearly 90°, giving a well-defined location for the minimum. Note that the four lines under discussion also determine the local direction of the contours: the straight lines $d_{\perp RhC} =$ constant, $d_{\perp CC} =$ constant, $d_{\perp RhC} + d_{\perp CC} =$ constant, and $d_{\perp RhC} - d_{\perp CC} =$ constant are tangent to contours at these four lines. This is helpful in hand drawing contours, as was done here

Zanazzi-Jona R-factor. Results with a Zanazzi-Jona R-factor of approximately 0.2 or below are called "good", those with a value of approximately 0.35 are called "mediocre" and those with a value of approximately 0.5 "bad". The corresponding reliabilities are termed "very probable", "probable" and "doubtful", respectively [6.68].

6.6 Accuracy and Precision of Structural Determination

Accuracy

No a priori method of establishing the accuracy and the precision of a structural determination by LEED is known, because the errors involved both in the experiment and in the theory are difficult to estimate. Systematic errors inherent to our limited understanding of the LEED mechanism can be evaluated only by comparison of structural predictions with those of very different techniques, since some of these can produce results with comparable accuracy (e.g. High-Energy Ion Scattering and Surface Extended X-ray Absorption Fine Structure). Within the quoted uncertainties, these techniques appear to agree with LEED results, although a few discrepancies exist [6.77] [that may be due, however, to differing surface preparations [6.78]. Thus, one may estimate that the inherent precision of LEED is, at the worst, on the order of 0.1 Å for interlayer spacings of simple surfaces. This value clearly improves with the amount of effort expended on optimizing structural and nonstructural parameters of the theory and with the quality and size of the experimental data base.

Precision

The precision, as opposed to accuracy, of structural determinations by LEED can be gauged by comparing studies based on independent LEED measurements and/or independent LEED calculations, or by comparing structural predictions based on separate subsets of the available data (e.g. separate beams, see Fig. 6.17, or data taken at different incidence directions). Here again one finds that a value of 0.1 Å is a conservative estimate for the uncertainty in interlayer spacings, with values down to 0.01 Å being quoted occasionally. Once more, a smaller uncertainty results from better optimization of parameters and the availability of a more reliable or larger data base.

Atomic positions parallel to the surface are usually not determined with the same precision or accuracy as interlayer spacings: a reduction by a factor of two to three is often encountered in the quoted accuracies of such results compared to those of layer spacings. This is primarily due to the fact that the momentum transfer $\vec{s}$ is close to the surface normal in most data. This gives the highest sensitivity to position components perpendicular to the surface. Not enough experience has been accumulated yet in the case of adsorbed molecules to estimate the uncertainties in their structural determinations, although one should expect layer spacings again to be relatively more accurate than other quantities.

Standard Deviation

The precision of structural determinations by LEED can be quantified by using standard statistical methods as demonstrated by Watson et al. [6.79] and Jona et al. [6.80]. Let us take the case where two unknown parameters are to be determined: the topmost interlayer spacing d and the muffin-tin zero level V_o. After finding the minimum in the overall R-factor, the standard errors in these two parameters can be determined as follows. The standard error in the value of d at the optimum value of V_o can be defined as [6.79]

$$\varepsilon_d = \{[\sum_i \Delta E_i(d^i_{min} - \bar{d}_{min})^2]/[(n-1)\sum_i \Delta E_i]\}^{1/2} , \quad \text{where} \tag{6.56}$$

$$\bar{d}_{min} = (\sum_i \Delta E_i d^i_{min})/(\sum_i \Delta E_i) . \tag{6.57}$$

Here d^i_{min} is the value of d which minimizes the R-factor for the ith beam, ΔE_i is the energy range of the ith beam, and n is the total number of beams included in the analysis. Similarly, the standard error for V_o at the optimal value of d is defined as

$$\varepsilon_{V_o} = \{[\sum_i \Delta E_i(V^i_{o,min} - \bar{V}_{o,min})^2]/[(n-1)\sum_i \Delta E_i]\}^{1/2} , \quad \text{where} \tag{6.58}$$

$$\bar{V}_{o,min} = (\sum_i \Delta E_i V^i_{o,min})/(\sum_i \Delta E_i) . \tag{6.59}$$

The quantities ε_d or ε_{V_o} can be used as the sample standard deviations (also called root-mean-square deviations) of a normal distribution although such usage is strictly justified only if $n \geq 30$. The result for the interlayer spacing can then be quoted as $\bar{d}_{min} \pm \varepsilon_d$ (for 68% probability), $\bar{d}_{min} \pm 2\varepsilon_d$ (for 95% probability), etc. For example, the topmost interlayer spacing of the Cu(111) surface was shown to be contracted by 4.1 ± 0.6% of the bulk interlayer spacing with a muffin-tin zero at $V_o = 9.4 \pm 0.6$ eV (for 68% probability) [6.79].

However, most LEED analyses are performed with a smaller number of beams, $n < 30$. In that case, it is more proper to use the Student's t-distribution [6.81], which was chosen by Jona et al. [6.80] in determining the standard errors. One groups the observed values of the parameter Y into sets labeled $j = 1,2,...,k$ (this way, the internal consistency between the sets can be judged as well). For example, the sets could be small groups of beams, where each beam determines separately the optimized values of the parameters Y=d and $Y=V_o$. The standard error ε_s at a chosen confidence interval is then defined as

$$\varepsilon_s = t_{\alpha,\nu}(\frac{s}{k})\left[\sum_{j=1}^{k}\sum_{i=1}^{l_j}(w_{ij})^2\right]^{1/2} , \quad \text{where} \tag{6.60}$$

$$s^2 = \left[\sum_{j=1}^{k}\sum_{i=1}^{l_j}(Y_{ij} - \bar{Y}_j)^2\right]/\nu , \tag{6.61}$$

$$w_{ij} = \Delta E_{ij} / \sum_{i=1}^{l_j} \Delta E_{ij} \ , \tag{6.62}$$

$$\nu = \sum_{j=1}^{k} l_j - k \ , \tag{6.63}$$

$t_{\alpha,\nu}$ is a value obtained from a table of the Student's t-distribution for the confidence level α (e.g. $\alpha = 95\%$) and the degree of freedom ν, l_j is the number of beams in set j, k is the total number of sets, and $\overline{Y}_j$ is the average value of the parameter Y_{ij} in set j with the beams labeled by i. For example, the top interlayer spacing and the muffin-tin zero level of the Al(111) surface were determined to be 2.39 ± 0.03 Å and 8.7 ± 0.6 eV, respectively, for 95% confidence using the Student's t-distribution [6.80].

7. Results of Structural Analyses by LEED

In the last three chapters, we have treated the various methods that have been employed to extract surface structural information from LEED data. We now review briefly the structural results that have been obtained and discuss several specific examples of structural determinations in detail, in order to illustrate how the procedure is carried out in practice.

Thus far, LEED has been applied to the determination of over 150 structures, with over 110 additional studies performing redeterminations or refinements. A comprehensive listing of these studies and of their results is given in Chap. 12, which also includes an extensive list of references. The present discussion is concerned with the four major categories of structures: clean unreconstructed surfaces, clean reconstructed surfaces, atomic adlayers, and molecular adlayers. For each category, we shall introduce first the general concepts and structural trends and turn then to specific examples.

7.1 Clean Unreconstructed Surfaces

The use of LEED as a technique for surface structural determination began with the study of clean surfaces. Low-Miller-index unreconstructed surfaces of metallic single crystals were commonly chosen because they have the lowest surface energy; hence, they are the most stable ones. The (100), (110), and (111) surfaces of face-centered cubic (fcc) and body-centered cubic (bcc) crystals as well as the (0001) surface of the hexagonal close-packed (hcp) crystals have frequently been the subject of structural determinations. These surfaces are illustrated in Fig. 7.1. The study of these surfaces serves several purposes. (1) They serve as the prime candidates for testing the LEED theory, including, in particular, nonstructural parameters that may be needed in determining related structures of more complexity, such as reconstructions and adsorbate structures, (2) The structural results of these studies are important for the subsequent investigation of ordered overlayers of atoms and molecules adsorbed on the clean surfaces, and (3) These studies provide surface structural information for the calculation of the electronic structure and the total energy of the surface.

A surface represents a unique environment for atoms, which is intermediate between the bulk crystalline phase and the atomic or molecular gas

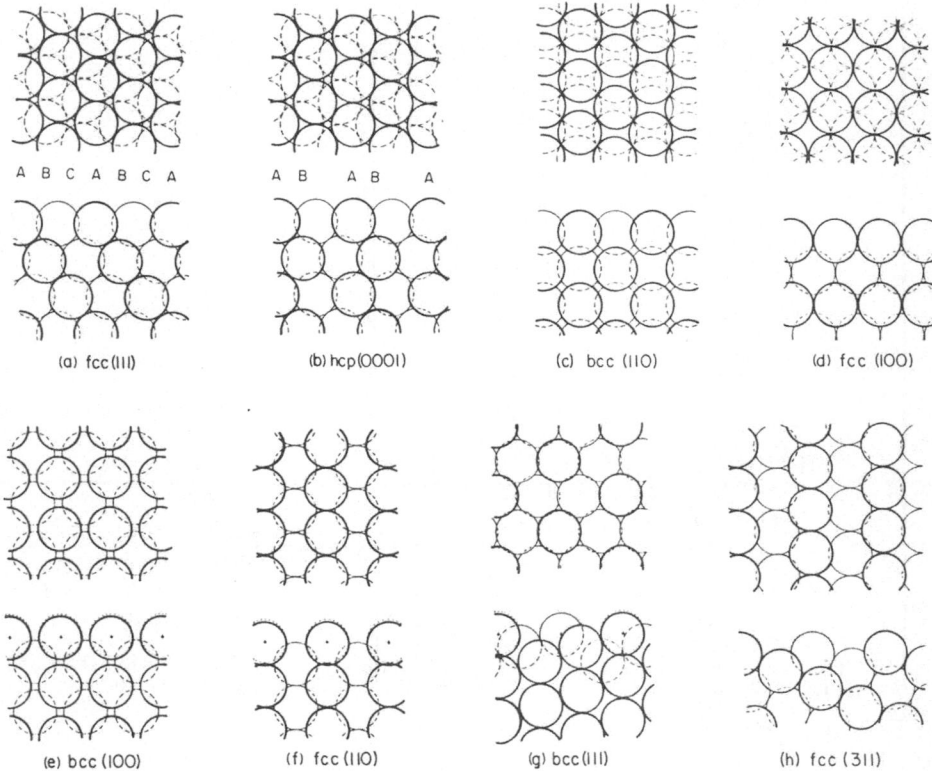

Fig. 7.1a – h. Surface structures of low-Miller-index unreconstructed metal surfaces. In each panel a top view and a bottom view are shown. Dashed circles indicate ideal bulk positions of surface atoms

phase. An atom at a surface is located in a highly asymmetric position, with bonds to a number of nearest neighbors on the side of the substrate and no bonds at all on the side of the vacuum. The exact number of bonds that an atom makes to the substrate is of particular interest since it has been found to affect the bond lengths, as indicated in Fig. 7.2; the fewer the number of nearest neighbors, the smaller the bond lengths to some of them.

This bond-shortening effect is well known and can be explained in several phenomenological ways [7.1]. Surface bonds are shorter than bulk bonds due to greater atomic overlap resulting from the reduced number of nearest neighbors. Equivalently, the electrons belonging to the severed bonds are partly shifted toward the remaining bonds, which reduces the length of those bonds. In another view, the surface electron density attempts to form a smooth surface, as if it possessed a surface tension. The resulting redistribution of charge produces electrostatic forces that draw the surface atoms toward the substrate [7.2].

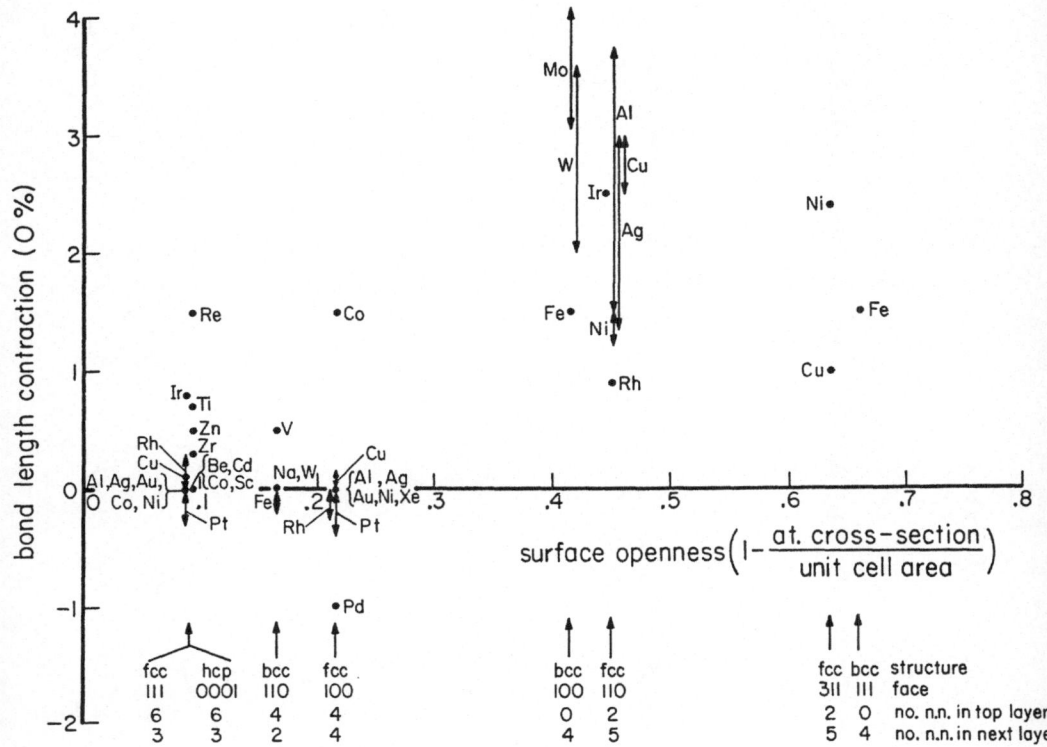

Fig. 7.2. Plot of observed bond-length contractions between surface atoms and underlying atoms versus surface openness (lack of close-packing) for unreconstructed metal surfaces, not including the error bars (uncertainty of 0.5 to 2% of the bond lengths). The atomic cross section used in defining the surface openness is the geometrical cross section of the bulk muffin-tin sphere. For a surface atom, the number of its nearest neighbors (n-n) in the surface plane and in the next layers is indicated for each type of surface. Arrows indicate positions of various crystal faces along the horizontal axis

This bond-length reduction is manifest as a decrease in the topmost interlayer spacing in the case of unreconstructed surfaces. Then the bond lengths between atoms in the topmost layer and the next layer are reduced. The bonds parallel to the surface are usually not reduced since the lattice constant parallel to the surface is fixed by the substrate. If this lattice constant does decrease, a reconstruction occurs, as in the case of Ir(100), Pt(100), Au(100), and Au(111) surfaces, where a quasi-hexagonal top layer forms with reduced interatomic distances. Figure 7.2 shows the bond-length reductions found at clean unreconstructed surfaces, using the most recent results. They are plotted as a function of surface "openness", a measure of the loss of nearest neighbors, to exhibit a correlation between bond-length reductions and surface openness [7.3]. Although the uncertainties in the data are rather large (0.5-2%), it is clear from Fig. 7.2 that not only the number of nearest neighbors matters, but that especially the number of nearest neighbors in the *next* layer is important. This is understandable in terms of interatomic forces, since those nearest neighbors in the topmost layer cannot, to a first approximation, "push" or "pull" top-layer atoms in the direction perpendicular to the surface.

Changes in interlayer spacings are not confined to the topmost interlayer spacing. This has been predicted theoretically [7.4] and confirmed with LEED (Sect. 7.1.2). Such effects should be particularly pronounced at relatively open surfaces, such as fcc(110) metal surfaces. Indeed, several fcc(110) surfaces were investigated initially in terms of only top-layer relaxations, with a somewhat unsatisfactory agreement between theory and experiment. Recent reexaminations of these surfaces, including second- (and even third-) layer relaxations, have improved dramatically the agreement between theory and experiment. Not only have these results confirmed the quality of the LEED theory, but they have also illustrated the attainability of an accuracy of approximately 0.01 Åfor the determination of interlayer spacings.

Apart from contractions in interlayer spacings, no other deviations from the bulk structure have been found at low-Miller-index metal surfaces that are clean and unreconstructed. Such a deviation could be, for example, "registry changes", i.e. lateral displacements of surface atoms. It is not too surprising that surface atoms of fcc(100) surfaces do not "ride up" on top of atoms of the second layer, but other surfaces could conceivably change in more subtle ways. For example, an fcc(111) surface differs from an hcp(0001) surface only in the layer-stacking arrangement: ABCABC... versus ABABAB..., as illustrated in Fig. 7.1. No case has been found to date where an fcc(111) surface involves an hcp-type layer stacking, such as CBCABC..., or where an hcp(0001) surface involves an fcc-like stacking, such as CBA-BAB..., although it is well known that fcc and hcp lattices often have nearly equal total energies.

Furthermore, the top-layer atoms of a bcc(110) surface could easily "roll over" to threefold coordinated hollows, see Fig. 7.1, in order to increase the number of nearest neighbors between a top-layer atom and next-layer atoms from two to three, but this has not been observed.

All unreconstructed surfaces which have the same two-dimensional unit cells as those given by their bulk structures are designated as (1x1) structures. For example, the (111) surface of the fcc metal iridium is designated as the Ir(111)-(1x1) surface. However, it is very important to realize that the appearance of a (1x1) LEED pattern does not necessarily *guarantee* a clean unreconstructed surface. In fact, a (1x1) pattern may occur for impurity-covered surfaces as well as reconstructed surfaces. For example, a stable (1x1) structure is formed when a bcc Fe(100) surface is exposed to eight Langmuirs of oxygen at room temperature [7.5]. The LEED I-V beam profiles from this surface are markedly different from those of the clean surface, tending to confirm the formation of a stable (1x1) oxygen overlayer structure. Dynamical calculations indicate that oxygen atoms are residing in the fourfold hollow sites on the surface. Of course, a disordered overlayer can also preserve the (1x1) pattern, with almost undetectable background due to disorder at low coverages. Another example, that will be discussed later, is the GaAs(110) surface, which shows a (1x1) LEED pattern even though surface reconstruction has occurred. On the other hand, a clean surface may exhibit a LEED pattern other than the (1x1) pattern due to the rearrangement of surface atoms. Numerous examples can be found in this category of reconstruction, and some will be discussed in Sect. 7.2.

7.1.1 The Rh(111) Surface

A rhodium single crystal has a face-centered cubic structure. Therefore, an ideal (111) surface of this crystal has a layer-stacking sequence that can be symbolized as ABCABC... . The observed LEED pattern, as illustrated in Fig. 7.3, indicates that the surface has the two-dimensional periodicity expected on the basis of the bulk structure. The LEED pattern should show a threefold symmetry and a mirror-plane symmetry if the surface is not reconstructed. At normal incidence, the intensities of the (01), (1$\bar{1}$), and ($\bar{1}$0) beams are indeed found to be essentially identical at all energies, as are the intensities of other symmetry-related beams. It is important to note that the indexing of the beams is somewhat arbitrary. At the present time, there is no convention that defines which beam should be indexed as (10) or (01). Therefore, it is quite possible to discover that the (01) beam quoted in one paper may be indexed as the (10) beam in another paper.

Very good agreement between experiment and theory has been achieved by three independent studies [7.6-8]. In the work of Chan et al. [7.6], two different methods, namely the Convolution-Transform method and dynamical calculations, have been applied to analyze the I-V data. Within experimental errors, neither expansion nor contraction of the topmost interlayer spacing was detected. The application of the Convolution-Transform method to analyze LEED I-V data was discussed in detail in Sect. 6.3.2.

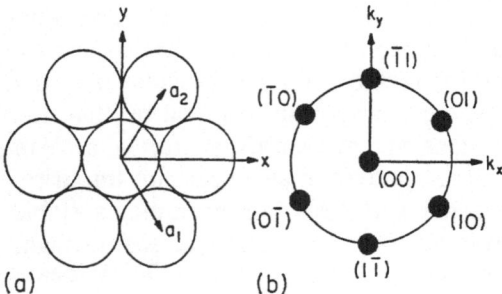

Fig. 7.3. (a) The fcc(111) crystal face; $\vec{a}_1$ and $\vec{a}_2$ are the lattice vectors. (b) The reciprocal space corresponding to (a)

In the dynamical calculations, the perturbation scheme known as Renormalized-Forward Scattering was used for rapid and good convergence. The interaction between the LEED electrons and crystal atoms was represented in the muffin-tin approximation, as described in Sect. 4.4. In this calculation, a band-structure potential, which represents electronic properties of the bulk-crystal atoms, was used, represented by eight phase shifts. In general, the number of phase shifts needed in the calculation increases as the electron energy increases. Since the constant potential between the muffin-tin spheres (i.e. the muffin-tin zero level) does not correspond to any measurable physical parameter, an assumed value of 15 eV was used initially. In the comparison between experimental and theoretical spectra, the inner potential was allowed to vary by a rigid shift of the energy scale. To represent all the inelastic processes, an energy-independent imaginary part of the inner potential of 5 eV was used. In some calculations, this quantity was taken to be energy dependent. The imaginary part of the inner potential is related to the electron-mean-free-path length, which cannot be measured accurately, but it can be estimated from the experimental peak widths. It

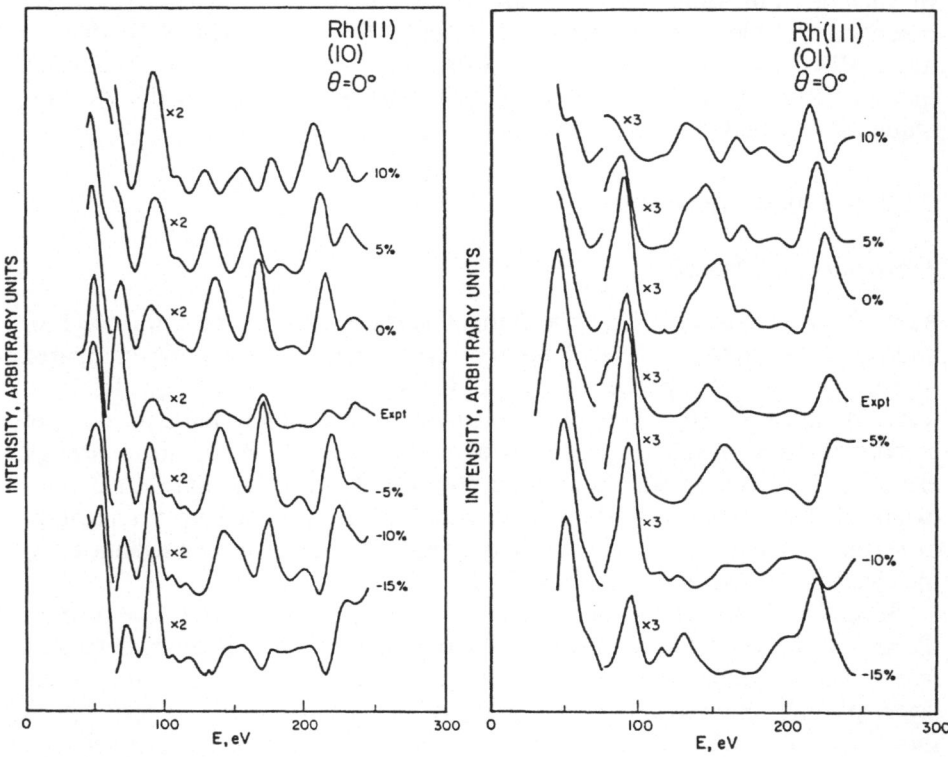

Fig. 7.4. Comparison between the theoretical I-V spectra modified with an inner potential of 10 eV and the experimental I-V spectra for the (01) and (10) beams of Rh(111) at normal incidence. The % parameter refers to expansion (+) or contraction (−) of the topmost rhodium interlayer spacing

can also be varied in the calculations to obtain the best fit between experiment and theory. The thermal vibrations of the crystal atoms reduce the intensities of the scattered beams. Hence, the intensities of the beams decrease as the temperature increases. This temperature effect was included in the calculations through the multiplication of the scattering amplitude of each atom by the Debye-Waller factor. The bulk Debye temperature used was 300 K. To account for the fact that surface atoms have been observed to have larger vibrational amplitudes than corresponding bulk atoms, an enhancement factor of 1.43 was chosen for the surface mean-square vibrational amplitude.

In the calculations, the outer layer spacing of the Rh(111) surface was allowed to relax from −15% (contraction) to +10% (expansion) in steps of 5% of the bulk interlayer spacing. Figure 7.4 shows comparisons between theoretical and experimental I-V spectra for two different beams. The calculated spectra were allowed to shift rigidly in energy. As a result, a modified inner potential of 10 eV was found to give the best fit between experiment and theory. These results agree well with those presented in [7.7,8]. In the work of Van Hove and Koestner, an extra step was taken in the calculation to include four different terminations of the layer stacking, which are described by the stacking sequences: ABCABC...(fcc), ABCBCB...(fcc on hcp), ABACBA...(hcp on fcc) and ABABAB...(hcp). Their results confirmed that the Rh(111) surface is not reconstructed and maintains the stacking sequence of a bulklike fcc(111) surface.

7.1.2 Multilayer Relaxations

Prediction and Detection

As we have seen in Fig. 7.2, "open" metal surfaces such as fcc(110), bcc(100) and bcc(111) exhibit relatively large top-layer spacing contractions, compared to the case of close-packed metal surfaces, such as hcp(0001), fcc(111), fcc(100), and bcc(110). This was explained by Finnis and Heine in terms of a redistribution of valence electrons at the surface which, in turn, through electrostatic forces, affects the atomic positions [7.2]. Landman et al. have extended this type of analysis and predicted a damped oscillatory relaxation of several layer spacings into the bulk, similar to the Friedel oscillations in the electron density near a surface [7.4].

Several LEED analyses have subsequently confirmed the presence of multilayer relaxations with a damped oscillatory character on different metal surfaces. The oscillatory nature is expressed as a contraction of the first interlayer spacing, followed by an expansion of the second interlayer spacing, as shown in Table 7.1. A second-layer expansion has also been observed with High-Energy Ion Scattering for the Cu(110) [7.14] and Ag(110) [7.17] surfaces.

Table 7.1. Results of multilayer-relaxation determinations. Here Δd_{ij} represents the percentage relaxation of the spacing between layers i and j, referred to the bulk value of the layer spacing. Positive (negative) values correspond to expansions (contractions)

Surface	Bulk lattice type	Δd_{12} [%]	Δd_{23} [%]	Δd_{34} [%]	Technique	Reference
Al(110)	fcc	-8.9±1.0	5.9±1.5	2.0	LEED	[7.9]
Al(110)	fcc	-8.4±0.8	4.9±1.0	-1.6±1.1	LEED	[7.10]
V(100)	bcc	-7.0±0.6	1.0±0.6	0 assumed	LEED	[7.11]
Cu(110)	fcc	-10.0	1.9	0 assumed	LEED	[7.12]
Cu(110)	fcc	-8.5±0.6	2.3±0.8	-0.9	LEED	[7.13,14]
Cu(110)	fcc	-5.3	3.3	0 assumed	HEIS	[7.14]
Cu(100)	fcc	-1.1±0.4	1.7±0.6	≥1.0, ≤2.0	LEED	[7.15]
Ag(110)	fcc	-5.7	2.2	0 assumed	LEED	[7.16]
Ag(110)	fcc	-6.0	4.0	0 assumed	HEIS	[7.17]
Re(1010)	hcp	-17.0	≥1.0, ≤2.0	0 assumed	LEED	[7.18]

A relaxation can exist between deeper layers as well, although it is close to the limit of the present resolution of LEED, both in terms of intrinsic distance resolution and in terms of sensitivity to geometry at a depth comparable to the electron-mean-free-path length. Relaxations in the third interlayer spacing have been detected for Al(110) [7.9,10], Cu(110) [7.13,14] and Cu(100) [7.15] by LEED.

Search Through Multi-Dimensional Parameter Space for the Cu(110) Structure

With multilayer relaxation, the search for the optimum geometry involves many parameters that must be fit to the experimental data. This requires a search through a multidimensional parameter space, spanned by the parameters θ_D, V_{oi}, V_o, d_{12}, d_{23}, d_{34}, etc., where d_{ij} is the spacing between layers i and j. How this search is performed will be illustrated here for the case of a careful determination of the Cu(110) structure by Adams et al. [7.13]. The R-factor R2 (6.32), with the prefactor A_2 of (6.34) was used as a measure of the quality of each model structure that was tested.

First, the values of θ_D and V_{oi} were optimized, assuming layer independence of these nonstructural parameters. This was accomplished in several iterative steps, with simultaneous optimization of the first interlayer spacing d_{12}, and with fixed bulklike deeper layer spacings. As a consequence of the very weak correlation between θ_D and V_{oi}, on one hand, and the other parameters, on the other, this initial optimization converges rather well.

Then d_{12}, d_{23}, and V_o were varied, using the optimized values of θ_D and V_{oi} and bulk-like values for d_{34}, d_{45}, etc. For each pair (d_{12},d_{23}), V_o was optimized, yielding the best R-factor value R_2 (d_{12},d_{23}) for that pair. The best overall interpolated combination (d_{12},d_{23}) was found by minimizing R_2, with the reasonable assumption that R_2 is a quadratic function of d_{12} and d_{23} near the minimum.

The resulting optimum parameter values are (the bulk interlayer spacing is 1.278 Å):

$d_{12} = 1.170 \pm 0.008$ Å, i.e. $\Delta d_{12} = -8.5 \pm 0.6\%$,

$d_{23} = 1.307 \pm 0.010$ Å, i.e. $\Delta d_{23} = +2.3 \pm 0.8\%$,

$V_o = 8.9 \pm 0.3$ eV,

$V_{oi} = 2.5 \pm 0.3$ eV,

$\theta_D = 335 \pm 14$ K.

The quoted uncertainties are based on the "error matrix", $\varepsilon = G^{-1}$, where $G_{ij} = \partial^2 R_2 / \partial x_i \partial x_j$, and on standard deviations found in the least-squares fitting [7.13]. Note that the uncertainties in the layer spacings are on the order of 0.01 Å comparable to uncertainties quoted in X-ray crystallographic results for nontrivial structures.

The reliability of these results was tested with respect to variation of the atomic scattering potential. Three different potentials were used, yielding insignificant differences in the optimized structure. Sensitivity to the energy range used (20-360 eV) was also tested and found to be within the uncertainties quoted above. Relaxations of deeper layer spacings were subsequently tested as well, with the previously optimized parameters held fixed. A minor improvement was found with $\Delta d_{34} = -0.9 \pm 0.9\%$ and $\Delta d_{45} = -0.8 \pm 0.9\%$, which the authors consider inconclusive.

7.2 Reconstructed Surfaces

A reconstructed surface is one which has a surface structure that is markedly different from that expected based on the bulk-crystal packing. Reconstructions occur at some clean metal surfaces and more frequently at clean semiconductor surfaces. Surface reconstruction is undoubtedly one of the more interesting and challenging problems in the area of surface structure determination. Advances in the understanding of surface reconstruction will improve our understanding of many fundamental surface phenomena, such as the relationship between the surface energy and the surface structure (the bonding topology of the surface atoms), as well as surface electronic states. To date, the structure of only a few reconstructed surfaces has been determined, with modest success. Some of these reconstructed surfaces are difficult to analyze because the surface unit cells are simply too large, such as the Si(111)-(7x7) surface [7.19]. Furthermore, a number of proposed models are too complicated for any available dynamical LEED program to analyze, i.e. dynamical analyses of these complicated surface structures often exceed the present computer capability. For many reconstructed surfaces, the number of plausible structures is formidable, and this increases the difficulty of searching for the correct structure. However, with the development of better and more efficient methods of computation, such as the Combined-Space Method ([7.20] and Sect. 5.8) and the Quasi-dynamical approximation method [7.21], as well as faster methods of data acquisition (Sect. 2.4), solutions for some of these complicated reconstructed surfaces may be at hand.

Semiconductor Surface Reconstructions

Reconstruction commonly occurs on semiconductor surfaces. The silicon surfaces with low-Miller-index (100) and (111) planes are particularly important, because most MOS (metal-oxide-semiconductor) devices are fabricated from (111)- and (100)-oriented silicon single crystals. Depending on the experimental conditions, either a (2x1) or (7x7) superstructure can be observed on a clean Si(111) surface [7.22,23]. The (2x1) structure, which appears after cleavage, is metastable and can be converted irreversibly to a (7x7) superstructure upon annealing near 473 K. The (7x7) superstructure transforms reversibly into a (1x1) structure at approximately 1163 K [7.24,25]. A buckled-surface structure has been proposed by Haneman for the (2x1) surface [7.26]. In this model, adjacent rows of surface atoms are relaxed and contracted perpendicular to the surface with respect to their ideal positions. The conversion of the (2x1) superstructure by annealing the crystal at elevated temperatures is established by observing the changes in the I-V spectra, in addition to the appearance of seventh-order spots. The substantial changes in the I-V spectra indicate substantial atomic displacements. Despite many years of intensive and repeated efforts with the aid of many different surface-analysis techniques, the structure of Si(111)-(7x7) remains unsolved. This structure has become the most widely known and the most quoted example of surface reconstruction. The number of models proposed for this structure is impressive, including surface-buckling models such as a graphitic top layer [7.19,27] or surface patterns based on a buckled (2x1) reconstruction [7.28], missing-atom models [7.29], "stool" models [7.30], terrace models [7.31] and stacking-fault models [7.32]. There is no sign that the definitive solution is in sight, since every proposed model agrees with some experimental observations but disagrees with others.

The Si(111)-(7x7) structure transforms reversibly into a (1x1) structure at approximately 1163 K. This surface can be stabilized at lower temperatures by a small concentration of impurities, e.g. chlorine [7.22] and tellurium [7.34]. This (1x1) structure was determined by the most recent dynamical calculations [7.35] to involve a contraction of 0.2 Å in the topmost layer spacing (corresponding to a 4% bond-length contraction) and an expansion of 0.075 Å in the next layer spacing (corresponding to a 3.2% bond-length expansion). A laser-annealed (1x1) structure can also be obtained, which is essentially identical to the impurity-stabilized (1x1) structure [7.35].

Some controversy exists concerning the exact structure of the Si(111)-(2x1) surface. However, a LEED result [7.36] and a model calculation [7.37] agree qualitatively with the buckled-top-layer model. In particular, the two studies indicate that the top layer would be buckled by ±22% and ±47%, respectively, of the top layer spacing, which itself would be contracted 8% and 7%, respectively; while the second interlayer spacing would be contracted 10% and expanded 3%, respectively, compared to to bulk values. The Si(100)-(2x1) reconstructed surface, which will be discussed more extensively in Sect. 7.2.2, may be explained by a modified Schlier-Farnsworth model in which every two adjacent rows of first layer atoms are paired to form one asymmetric double row, with additional elastic relaxation of the subsurface down to the fifth layer [7.38,39].

Another extensively studied semiconductor surface is GaAs (110), which will be discussed in detail in Sect. 7.2.3. In this case, the surface is reconstructed without enlarging the surface unit cell, as shown by the observed (1x1) LEED pattern. Dynamical analyses [7.40,41] confirm that the reconstruction is characterized by the outward and inward relaxation of As and Ga atoms, respectively, with respect to their ideal bulk positions.

Clean Metal Surface Reconstructions

Reconstructions are observed on the (100) and (110) surfaces of the metals aluminum [7.42], iridium, platinum, and gold as well as the Au(111) surface. Large superlattices, such as (1x5) or (5x20), are observed on Ir(100), Pt(100), Au(100), and Au(111). These have long been interpreted as due to the top layer adopting a hexagonally close-packed arrangement [7.43]. This has been confirmed for Ir(100) and Pt(100) with LEED intensity calculations [7.44]. An interesting aspect of these structures is that the bonds within the hexagonal layers are shorter than in the bulk. The extent of this contraction is different for the three different metals, which presumably explains the different patterns that are observed. (See also the discussion of this type of pattern in Sect. 3.5). These reconstructions are usually destroyed by a strongly bound adsorbate layer in favor of the unreconstructed geometry, especially if the adsorbate is an electron acceptor. For example, in the case of the Ir(100) surface, the (1x1) structure is stabilized by a small amount of oxygen that is below the detection limit of Auger-Electron Spectroscopy [7.45].

The (110) surfaces of the metals iridium, platinum, and gold exhibit a (1x2) reconstruction. The LEED patterns of these surfaces indicate that the periodicity of the surface in the [001] crystallographic direction is doubled, while the periodicity of the surface in the [1$\bar{1}$0] direction remains the same. Although several models have been proposed for these structures, the missing-row model appears to be the most probable structure. This has been shown in particular for the Ir(110) surface [7.46-48]. The Ir(110)-(1x2) reconstruction can be removed, for example, by the adsorption of a quarter monolayer of oxygen at 850 K on the clean reconstructed (1x2) surface, after which a (1x1) surface structure is formed [7.47,49]. When an additional half monolayer of oxygen is absorbed on the metastable Ir(110)-(1x1) surface, a c(2x2) overlayer structure is formed [7.50]. This two-step process supports the obvious notion that the effect of the adsorbate on the substrate increases with the coverage.

For bcc metals, the (100) surface reconstructs in the case of chromium, molybdenum, and tungsten. A c(2x2) structure forms on Cr(100) [7.51] and W(100) [7.52], and a more complicated but related structure is found on Mo(100). The W(100) structure has been analyzed in some detail, following the observation [7.52] that a p2mg structural symmetry is present. It appears to consist mainly of parallel displacements of top-layer atoms into zigzag rows, although further refinement of this model is required [7.53]. The reconstruction is possibly due to electron-phonon coupling, producing a periodic lattice distortion that is coupled to surface charge-density waves [7.54].

7.2.1 The Ir(110)-(1x2) Reconstructed Surface

It has been shown that a clean Ir(110) surface exhibits a reconstructed (1x2) surface structure [7.46-48] and that an adsorbate-induced (1x1) structure can exist [7.47,49]. The study of the (1x1) surface structure, which is more amenable to a structural analysis by LEED than its reconstructed counterpart, would provide some useful insight concerning the atomic structure of the clean reconstructed (1x2) structure. Consequently, we discuss the (1x1) structure first.

Relationship Between (1x2) and (1x1) Structures

When a clean crystal exhibiting a (1x2) LEED pattern, as shown in Fig. 7.5a, was heated in 5 x 10^{-8} Torr of O_2 at 850 K for approximately two minutes, a (1x1) surface structure was formed, the LEED pattern of which is shown in Fig. 7.5b. The formation of the (1x1) structure is more rapid at higher temperatures. Upon subsequent cooling to room temperature, streaks in the [001] direction (between rows of spots) appeared, as shown in Fig. 7.5c. This is thought to be caused by an excess amount of oxygen on the surface. A stable and sharp (1x1) structure was obtained by subjecting the surface to several Langmuirs of CO and flashing the crystal to 575 K. This (1x1) structure is stable to approximately 1125 K in vacuum, and it is stable to 675 K in 5 x 10^{-8} Torr of CO. Streaks in the [001] direction (linking spots), as shown in Fig. 7.5d, were observed to appear after the crystal was heated above 1125 K in vacuum, or above 675 K in 5 x 10^{-8} Torr of CO for one to two minutes.

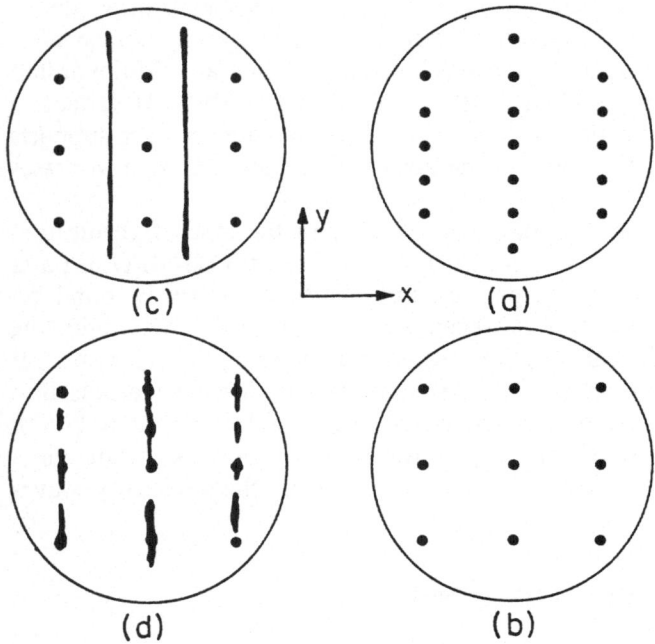

Fig. 7.5. (a) The reciprocal-space lattice of the Ir(110)-(1 × 2) structure. **(b)** The reciprocal-space lattice of the Ir(110)-(1 × 1) structure. **(c)** The reciprocal-space lattice corresponding to (b) with streaks between rows of spots in the [001] direction. **(d)** The reciprocal-space lattice corresponding to (b) with streaks linking spots in the [001] direction

The concentration of surface oxygen which is required to form the (1x1) structure from a clean (1x2) structure was estimated using Auger-Electron Spectroscopy. The peak-to-peak amplitude of the differentiated Auger line [in the dN(E)/dE curve] of the oxygen KLL transition on the (1x1) surface is approximately four times less than that of a (1x2) surface fully saturated with oxygen to monolayer coverage. A monolayer coverage of oxygen is obtained by exposing the clean (1x2) surface to approximately 30 L of oxygen [7.55]. Hence, approximately a quarter monolayer of oxygen is required for the formation of a (1x1) structure from a clean (1x2) structure. One may conclude that the oxygen atoms are distributed randomly and uniformly over the surface. Within the framework of this assumption, the presence of oxygen on the surface would not affect the fine structure of the LEED I-V spectra but could only modulate them. Hence, the presence of oxygen on the surface was neglected in the LEED calculations, which were performed as for a clean, unreconstructed Ir(110)-(1x1) surface structure. Good agreement between experiment and theory suggests that this surface structure is the same as in the bulk, with a topmost interlayer spacing of 1.26 ± 0.05 Å contracted 7.5% with respect to the bulk interlayer spacing.

These results provide useful guidelines for discussing a kinetically activated process for the formation of the (1x1) structure from a clean (1x2) structure. The fact that the formation of the (1x1) structure occurs near 850 K or above indicates that thermal energy is required for the motion of iridium atoms. The presence of the oxygen atoms on the surface must favor the formation of the (1x1) structure due to a reduction of the Gibbs energy of the surface by the rearrangement of the surface atoms.

At temperatures of 850 K, the formation of the (1x1) structure still takes approximately two minutes, which is quite a long time (approximately 10^{15} vibrational periods!) for atomic rearrangements. Upon removing oxygen by CO at 675 K, or increasing the temperature of the crystal to 1125 K, the (1x1) structure does not revert to the (1x2) structure. Rather, the LEED pattern shows the formation of streaks linking spots along the [001] direction as in Fig. 7.5d. Annealing at temperatures above 1600 K is required for the streaky (1x1) structure to return to the (1x2) structure. This fact suggests that a substantial atomic rearrangement has occurred in the transformation of the (1x2) structure to the (1x1) structure.

The (1x2) Surface Structure

Different models have been suggested for the atomic arrangement of an fcc(110) reconstructed surface. The three most common and simplest models proposed for the (1x2) reconstructed surface are the missing-row model, the paired-rows model, and the buckled-surface model. Figure 7.6 shows schematic hard-sphere drawings of these three models. In the missing-row model alternate rows of surface atoms are absent. The paired-rows model suggests that every two adjacent rows of first-layer atoms are paired to form one double row. In the buckled-surface model, adjacent rows of the first layer are relaxed in opposite directions perpendicular to the crystal surface. Since the LEED analysis indicates that the missing-row model represents the most probable structure [7.46-48], it was extended by including slight distortions of the second layer in the form of row pairing and second-interlayer-spacing changes [7.48]. A complete summary of all the models considered with the ranges of geometrical parameters used for each is listed in Table 7.2. Detailed schematic drawings are presented in Fig. 7.7 to show the geometrical parameters used in the different models.

Examples of the comparison between experimentally measured LEED I-V spectra and the best (lowest R-factor) calculated set of I-V spectra for each of the four different models are shown in Figs. 7.8,9 for two integral-order and wo fractional-order beams, respectively. A visual inspection of these figures indicates that the missing-row model is the preferred structure for the (1x2) surface. This qualitative evaluation is confirmed by an R-factor analysis as shown below.

POSSIBLE MODELS FOR THE (1×2) STRUCTURE

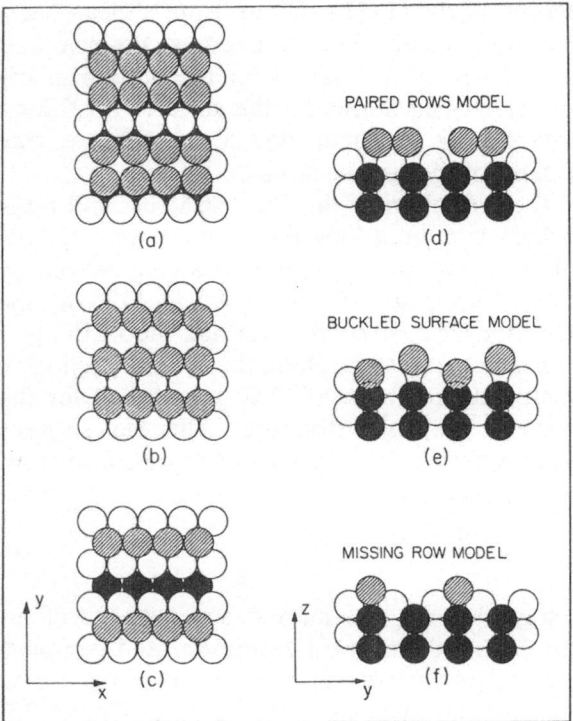

Fig. 7.6a – f. Top views of a hard-sphere representation of (**a**) the paired-row model; (**b**) the buckled-surface model, and (**c**) the missing-row model for fcc-(110)-(1×2) reconstructions. The corresponding side views are shown in (**d – f**), respectively

Table 7.2. Summary of the geometrical parameters used in different models of reconstructed Ir(110)-(1×2). For explanation of the symbols, see Fig. 7.7

Model	Range[a] of Geometrical variables [Å]	Increment in the variables [Å]
Paired-rows model	$d_1 = 1.21\text{-}1.81$	0.15
	$\sigma = 2.95\text{-}3.55$	0.30
Buckled-surface model	$d_\delta = 1.06\text{-}1.51$	0.15
	$2\delta = 0.20\text{-}0.80$	0.20
Missing-row model	$d_1 = 0.91\text{-}1.46$	0.05
Missing-row model with row pairing	$d_1 = 0.76\text{-}1.56$	0.20
	$d_2 = 1.16\text{-}1.76$	0.20
	$\beta = 3.25\text{-}3.55$	0.30

[a] Where more than one parameter is listed, all combinations of the values indicated were evaluated

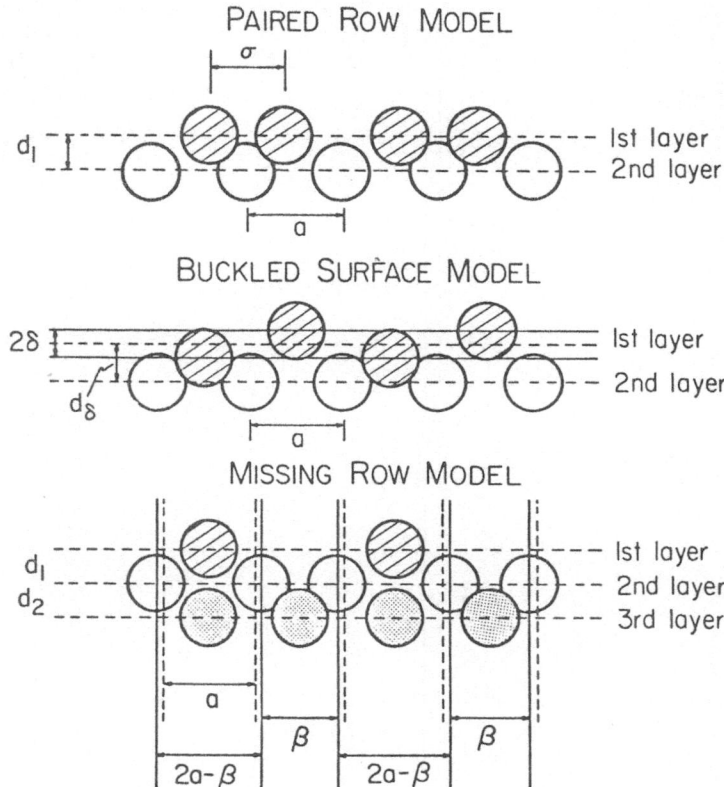

Fig. 7.7. The geometrical parameters used in different Ir(110)-(1 × 2) models where a = 3.58 Å is the normal spacing between adjacent rows of atoms in the [001] crystallographic direction.

The paired-rows model (*top panel*): d_1 is the interlayer spacing between the first and second layers of atoms, σ is the spacing between the adjacent close-packed top rows of atoms which have moved toward each other.

The buckled-surface model (*middle panel*): 2 δ is the spacing between the two topmost layers of atoms which are relaxed in opposite directions perpendicular to the crystal surface, $d_δ$ is the interlayer spacing between the average distance of the buckled top layers of atoms and the second layer of atoms.

The missing-row model with row pairing (*bottom panel*): d_1 is the interlayer spacing between the top and second layers of atoms, d_2 is the interlayer spacing between the second and third layers of atoms, β is the spacing between the adjacent rows of atoms in the second layer in the [001] crystallographic direction. This becomes the missing-row model when β = a with d_2 equal to the bulk interlayer spacing

As is particularly evident in Figs. 7.8,9, the calculated spectra are quite sensitive to even a small movement in the second layer. The sensitivity of the calculated spectra to a variation in geometrical parameters is an important ingredient in the success of any LEED structural analysis. This sensitivity of the calculated spectra enhances the credibility of the results of the analysis.

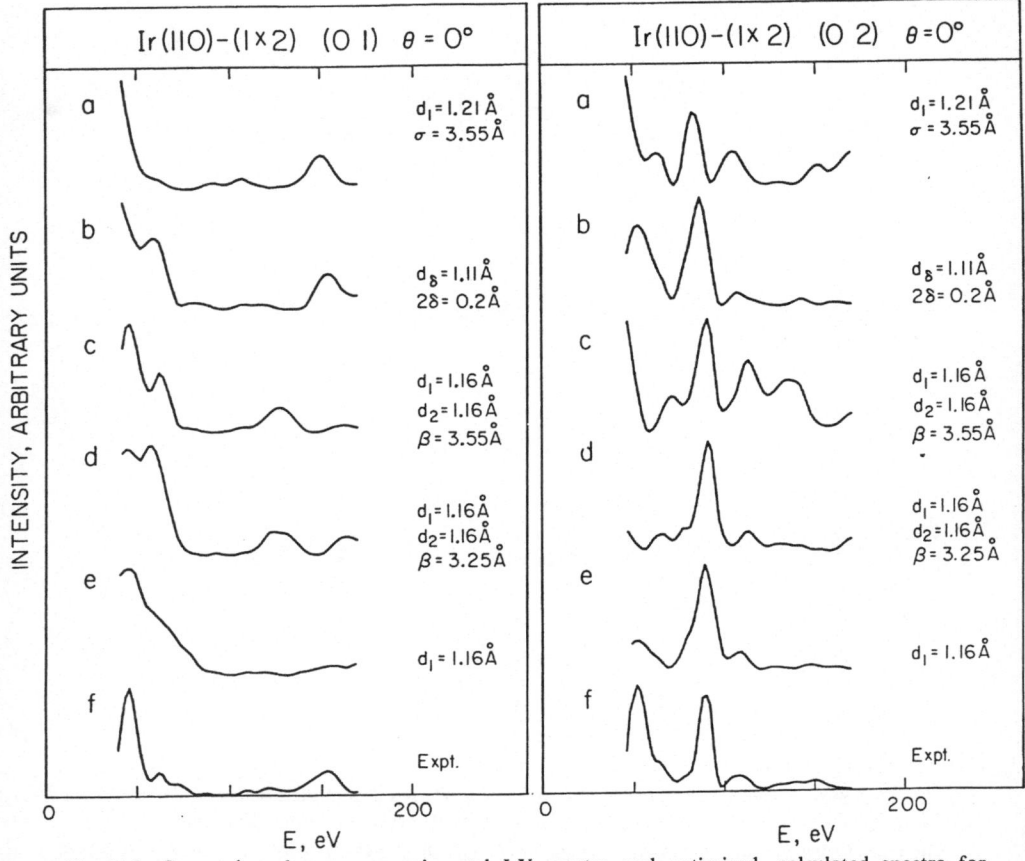

Fig. 7.8. Comparison between experimental I-V spectra and optimized, calculated spectra for different Ir(110)-(1 × 2) models for the (01) and (02) beams, where $\theta = 0°$ defines normal incidence. (a) The paired-rows model with an inner potential $V_0 = 13$ eV. (b) The buckled-surface model with $V_0 = 13$ eV. (c) and (d) The missing-row model with row pairing and with $V_0 = 11$ eV. (e) The missing-row model with $V_0 = 11$ eV. (f) The experimental I-V spectra

The reliability (R-factor) analysis suggested by Zanazzi and Jona [7.56] was used to determine quantitatively the level of agreement between experimental spectra and calculated spectra for all the different models considered. A detailed discussion of the R-factor analysis procedure has been presented in Sect. 6.5. Recall that the interpretation of the value of the Zanazzi-Jona R-factor is that a value below approximately 0.2 indicates high reliability of the proposed model, one near 0.35 indicates a possibly correct structure, while one above approximately 0.5 indicates an incorrect structure [7.56,57]. Thus, a comparison of the R-factors calculated for the different models discriminates among the various models quantitatively.

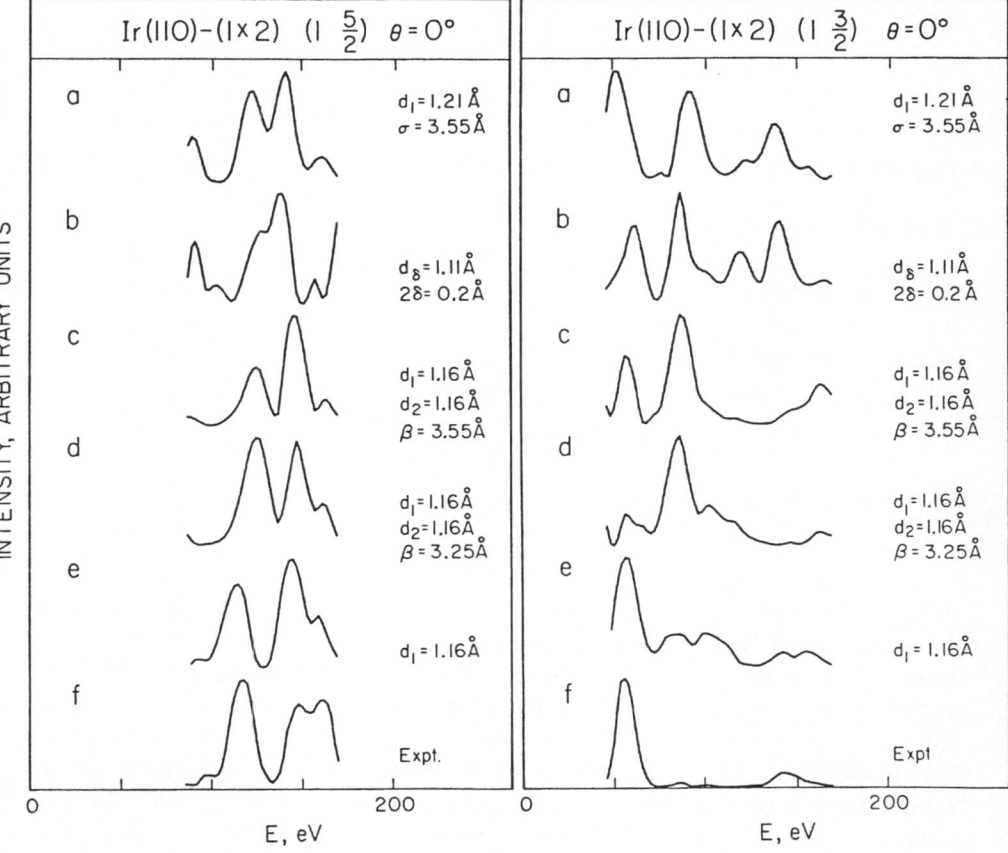

Fig. 7.9. As in Fig. 7.8 for the (1,5/2) and (1,3/2) beams

The R-factors for the optimum structure within each type of surface model for Ir(110)-(1x2) are presented in Table 7.3. The missing-row model with an R-factor of 0.24 is the most probable model among those tested. The top-layer spacing of this optimum structure is contracted by 15% with respect to the bulk interlayer spacing. The two models with row pairing, and the paired-rows model having R-factors of 0.28, 0.27, and 0.29, respectively, could also be regarded as possibly correct structures. The R-factor value of 0.24 for the preferred missing-row structure is only slightly higher than the proposed value of 0.2 for a structure of high reliability.

The Au(110) and Pt(110) Reconstructions

The missing-row model was also considered in two investigations of the clean Au(110) surface which produces a (1x2) LEED pattern at room tem-

Table 7.3. Results of an R-factor analysis for Ir(110)-(1×2)

Model	Inner potential [eV]	Geometrical parameters [Å]	R-factor
Paired-rows model	13	$d_1 = 1.21$ $\sigma = 3.55$	0.29
Buckled-surface model	13	$d_\delta = 1.11$ $2\delta = 0.20$	0.32
Missing-row model	11	$d_1 = 1.16$	0.24
Missing-row model with row pairing	13	$d_1 = 1.16$ $d_2 = 1.16$ $\beta = 3.55$	0.28
	13	$d_1 = 1.16$ $d_2 = 1.16$ $\beta = 3.25$	0.27

perature [7.58,59]. For this surface one of the two LEED studies provided little agreement between experiment and theory [7.58]. Consequently, the missing-row model was rejected as the correct structure for the Au(110)-(1x2) surface based on the fact that the R-factor value for the analysis was greater than 0.6. However, the other independent study indicated that the missing-row model is again the best among the three models tested: the missing-row model, the buckled-surface model and the paired-rows model [7.59]. In this analysis, a modification of the missing-row model consisting of small displacements in the second layer was also included. It was concluded that the missing-row model with a top-layer contraction of 15% of the bulk interlayer spacing, without any atomic displacement in the second layer, is the most probable structure for the Au(110)-(1x2) surface. The overall R-factor for this analysis is 0.28.

Based on the conclusions arrived at from the studies of the (1x2) surfaces of Ir(110) [7.46-48] and Au(110) [7.59], it appears that the missing-row model is certainly not too far removed from the correct structure. This conclusion for Au(110) is supported strongly by results obtained with X-ray Diffraction, Transmission Electron Microscopy and Scanning Tunneling Microscopy [7.59]. Furthermore, this model may also be the correct candidate for the Pt(110)-(1x2) surface, as a LEED study has suggested [7.60].

7.2.2 The Si(100)-(2x1) Reconstructed Surface

The Si(100)-(2x1) surface has been studied extensively by LEED. Many models, such as the surface-dimer model, the vacancy model and the conjugated-chain model, have been proposed over the years for this surface, but only recently has it seemed that the solution is at hand.

The Ideal (1x1) Structure

To facilitate the description of different models for the reconstructed surface, the unreconstructed Si(100)-(1x1) surface is illustrated schematically in Fig. 7.10a, where the open circles represent the atoms in the topmost layer, and the hatched circles represent the atoms in the second layer. A perfect unreconstructed Si(100) surface would exhibit a LEED pattern with twofold rotational symmetry. However, experimentally observed LEED patterns for the reconstructed (2x1), and also for the hydrogen-stabilized (1x1) structure, usually display fourfold rotational symmetry [7.61]. This is because the (100) plane of the diamond lattice can be terminated in two different configurations. The removal of the topmost layer of atoms from the surface exposes the second layer of atoms which is related to the former one by a 90° rotation. The presence of these two kinds of domains in equal concentrations on the surface causes the observation of a LEED pattern of fourfold rotational symmetry.

(1x2) Reconstruction Models

The earliest models proposed for this reconstructed surface are the surface-dimer model and the vacancy model [7.62,63]. The dimer model and the vacancy model are very similar to the paired-row model and the missing-row model, respectively, proposed for the Ir(110)-(1x2) reconstructed surface (Sect. 7.2.7). In the dimer model, shown in Fig. 7.10b, adjacent rows of the first layer of atoms are paired to form double rows in the x direction. In the vacancy model alternate rows of surface atoms are absent, doubling the periodicity in the x or y direction, as shown in Figs. 7.10c and d, respectively.

Comparisons between experimental and theoretical LEED intensities based on the surface-dimer model and the vacancy model were carried out by Jona et al. [7.64]. Their results indicated that neither model yields satisfactory agreement between theory and experiment. Consequently, a conjugate zigzag chain model based on ideas of Seiwatz [7.65] was proposed by Jona et al. [7.64] for the reconstructed Si(100) surface. A schematic diagram of this model is shown in Fig. 7.11. Results of dynamical calculations obtained with this model appear to be superior to those obtained with the surface-dimer model or the vacancy model. Hence, it was concluded that the zigzag chain model was the most promising candidate. However, these

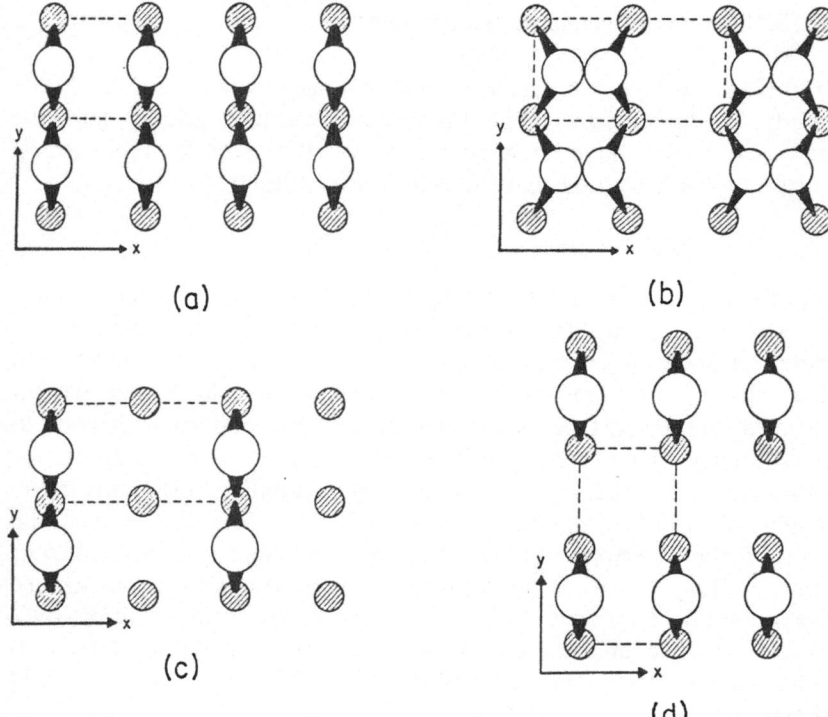

Fig. 7.10a – d. Schematic diagram of the Si(100) surface. The open and hatched circles represent the atoms in the topmost and second layers, respectively. (**a**) Unreconstructed surface. (**b**) Surface-dimer model of Schlier and Farnsworth. (**c**) Vacancy model of Schlier, Farnsworth, and Phillips. (**d**) Alternative vacancy model of Schlier, Farnsworth, and Phillips

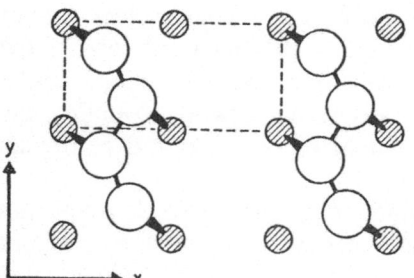

Fig. 7.11. Schematic diagram for the Seiwatz zigzag-chain model of Si(100)-(2 × 1). The open circles represent the atoms in the topmost layer, and the hatched circles represent the atoms in the second layer

conclusions are not in agreement with the results of Ultraviolet Photoelectron Spectroscopy. The surface density of electronic states obtained by self-consistent calculations of the structures of the vacancy model [7.66] and the zigzag-chain model [7.67] is not consistent with the ultraviolet photoemission data [7.68].

Dimer Model with Subsurface Strains

Based on this evidence, a modified dimer model was proposed by Appelbaum and Hamann [7.69] which includes reconstructions extending down to five layers from the surface due to the presence of subsurface strains associated with the reconstruction of the surface layer. A schematic diagram of this model is shown in Fig. 7.12. A calculation of the surface elastic energy, which is dependent on the geometry of the surface atoms, was performed.

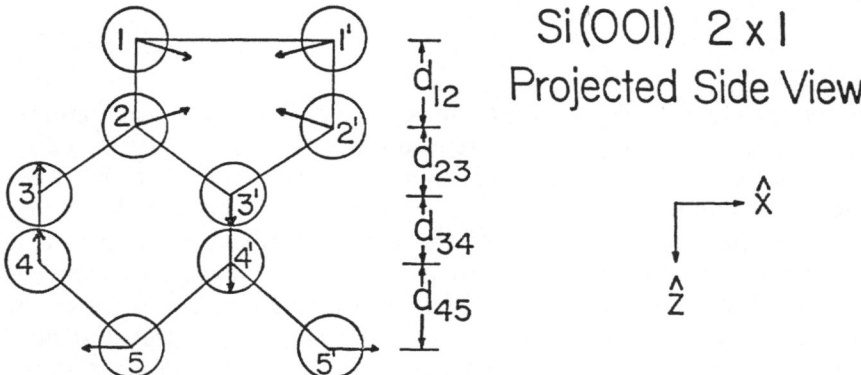

Fig. 7.12. Schematic diagram showing the projected side view of the optimized Appelbaum and Hamann model for the Si(100)-(2×1) surface. The arrow on each atom indicates the direction of displacement of the atom [7.38, 69]

An optimum geometry for the reconstructed surface was achieved by minimizing the surface elastic energy as a function of the geometry of the surface atoms. The surface-dimer model with subsurface reconstruction extending five atomic layers down from the surface appeared to be the most promising candidate for the correct model for the reconstructed Si(100)-(2x1) surface, since kinematic calculations for this structure, carried out by Appelbaum and Hamann [7.69], gave encouraging agreement with the experimental LEED intensities.

The apparent success of the zigzag-chain model proposed by Jona was reviewed by Mitchell and Van Hove [7.70]. It was noted that Jona et al. [7.64] could not execute their computer programs for the precise structural model they proposed. In their calculations, the requirement of distortion in the second layer of the crystal made it necessary to include this distortion down through the deeper layers as well. As a result of this constraint, it is believed that the calculated LEED intensities are useful only for low energies, i.e. below approximately 80 eV. Theoretical LEED intensities based on the zigzag-chain model were calculated using the Quasi-dynamical method

Table 7.4. Surface interlayer spacings and atomic displacements of the optimal Appelbaum-Hamann model for Si(100)-(2×1). The direction of the displacements of atoms are marked in Fig. 7.12

Layer	x [Å]	z [Å]	Interlayer spacing [Å]
1	0.695	0.092	$d_{12} = 1.357$
2	0.119	-0.005	$d_{23} = 1.257$
3	0	-0.130	$d_{34} = 1.207$
4	0	-0.078	$d_{45} = 1.357$
5	-0.032	0	

[7.21]. This method neglects all the multiple scattering within each atomic layer, for which the atoms need not necessarily be coplanar, while including the multiple scattering between the atomic layers by one of the well-established techniques, such as the Renormalized-Forward-Scattering method. The particular advantage of the Quasi-dynamical method is that unit cells with many atoms can be treated efficiently. These quasidynamical studies showed that the results of the calculations based on the zigzag-chain model proposed by Jona et al. appear less promising than expected at first. However, the results of the quasi-dynamical calculations based on the structural model proposed by Appelbaum and Hamann appear to give good results. Later, more complete quasi-dynamical calculations based on the structural model proposed by Appelbaum and Hamann were carried out by Tong and Maldonaldo [7.38]. The structure which gives the best agreement between experiment and theory is shown in Fig. 7.12. The surface interlayer spacings and atomic displacements of this optimal model are summarized in Table 7.4. Examples of the comparisons between the calculated I-V profiles for the optimal model and the experimental data are displayed in Figs. 7.13 and 14. As shown in these figures, better agreement is obtained for the half-order beams than the integral-order beams, because the multiple scattering contributions for the half-order beams are smaller, and, hence, quasi-dynamical results are expected to be more accurate. However, in the later work of Jona et al. [7.71], a full dynamical calculation was performed concluding that, with the Appelbaum and Hamann model [7.69], the agreement between theory and experiment is good only for normal-incidence data. Furthermore, they claimed that this model is not sufficiently good to qualify as the final structure for the Si(100)-(2x1) surface.

Asymmetric Dimer Models

Indeed, more recent theoretical work with energy-minimization model calculations has shown that an asymmetric-dimer model is better [7.37,72]: one of the two paired atoms in a dimer sinks deeper into the surface than the other.

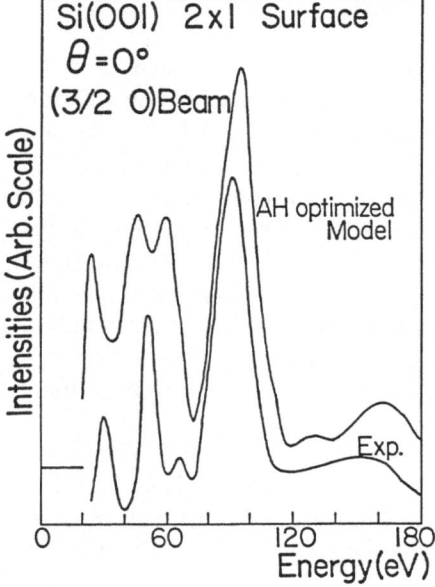

Fig. 7.13. Comparison between the experimental and the quasi-dynamical I-V curves for the optimized Appelbaum and Hamann model for the (3/2,0) beam [7.38]

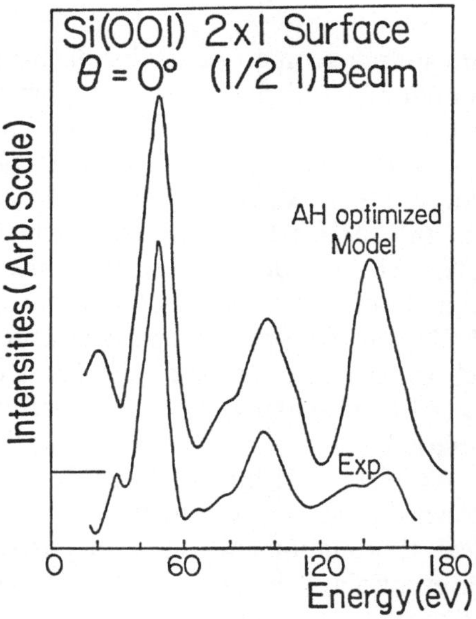

Fig. 7.14. Comparison between the experimental and the quasi-dynamical I-V curves for the optimized Appelbaum and Hamann model for the (1/2,1) beam [7.38]

This model has the virtue of being able to explain the occasional occurrence of a weak c(4x2) pattern through a suitable arrangement of the asymmetrical dimers. The model is also favored by a Medium-Energy Ion Scattering study of this reconstruction [7.73]. The most recent LEED calculations confirm this conclusion, but find that a twisting of the asymmetrical dimer (around an axis perpendicular to the surface) is even better [7.74]. At the same time, subsurface strains are still present in these models.

The history of the Si(100)-(2x1) structure determination teaches at least three lessons. First, LEED is quite sensitive to all atomic positions in the near-surface region and, therefore, all the atomic coordinates must be determined at the same time. Second, the freezing of atomic positions and the assumption of bulklike symmetries can lead one astray. Especially in the case of semiconductor reconstructions, where large atomic displacements from ideal bulklike positions are possible, it is dangerous not to explore such displacements. Third, the interplay with various techniques in addition to LEED can be crucial in delimiting the variety of possible structural models.

7.2.3 The GaAs (110)-(1x1) Reconstructed Surface

The atomic geometry and the electronic structure of the (110) surface of the compound semiconductor GaAs have been studied extensively by different surface spectroscopic techniques, including LEED [7.40,41,75-82] and angle-integrated and angle-resolved photoemission measurements [7.83-87]. The results of both the LEED studies and the photoemission studies indicate that the surface has a structure which is different from the zincblende-like bulk. The reasons for choosing the GaAs (110) surface as an additional example of surface reconstruction are twofold: (1) This example shows that LEED can be applied to surface-structural analysis of compound crystals as well as elemental crystals, and (2) it demonstrates that the observation of a (1x1) LEED pattern is not necessarily the fingerprint of an unreconstructed surface.

As early as 1964, the LEED work of MacRae and Gobeli [7.75] had shown that the arrangement of atoms at the (110) surface of GaAs is not an ideal termination of the bulk structure. The surface unit cell and its corresponding LEED pattern are shown schematically in Fig. 7.15. On one

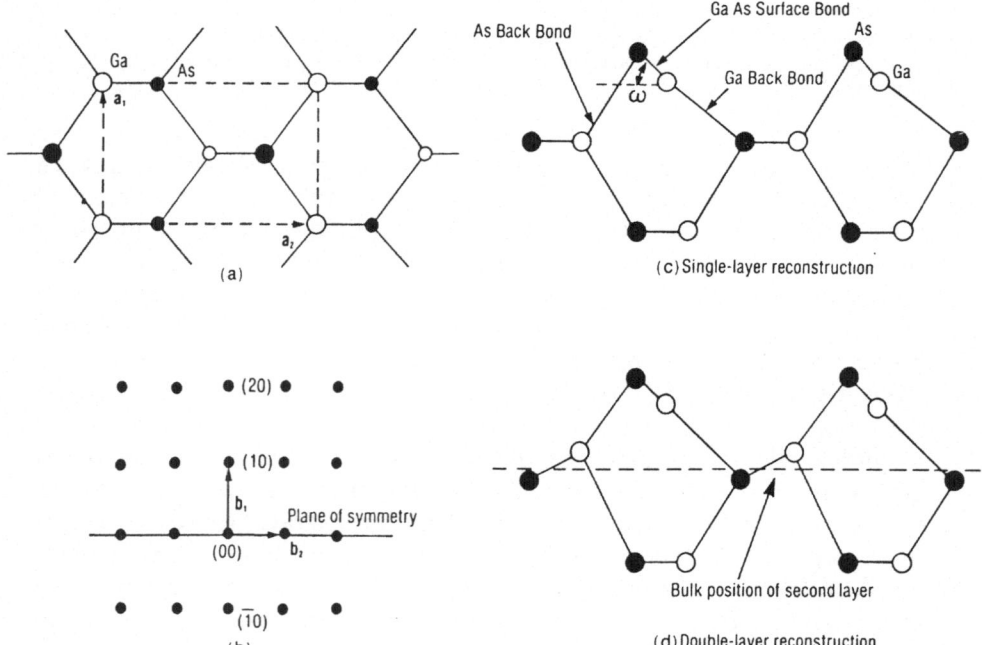

Fig. 7.15. (a) The real-space lattice of the GaAs(110) surface, viewed down onto the surface. The large open and large filled circles represent the top-layer gallium and arsenic atoms, respectively, while the corresponding smaller circles represent the atoms in the second layer. **(b)** The reciprocal lattice corresponding to (a). **(c)** Schematic diagram showing the projected side view of the single-layer reconstruction of GaAs(110)-(1×1). ω is the tilt angle of Ga-As surface bonds. **(d)** Schematic diagram showing the projected side view of the double-layer reconstruction of GaAs(110)-(1×1)

hand, it was found that the dimensions of the two-dimensional unit cell at the surface correspond to those of the substrate, i.e. a (1x1) pattern appeared, in contrast to the situation at most reconstructed surfaces. On the other, however, at normal incidence an extreme asymmetry was found in the intensities of the (hk) and (h$\overline{\text{k}}$) beams, while the intensities of the (10) and ($\overline{1}$0) beams were quite strong. As we shall now discuss, this contradicts the expectations from simple kinematic arguments in the absence of surface reconstruction.

Breaking of Symmetries

If the scattering factors for gallium and arsenic are assumed to be the same, which would yield the diamond lattice, the intensities of the (10) and ($\overline{1}$0) beams at normal incidence would be zero for the ideal surface, due to the presence of a glide plane of symmetry. This interesting case can be illustrated by considering the scattered intensity, which is proportional to the square of the scattering amplitude A(E) (6.11):

$$A(E) = f(E)\sum_n \exp[-M(n)]\alpha(n,E)\sum_j \exp[i\vec{s}\cdot\vec{r}_j(n)]\sum_{l_1,l_2}\exp[i\vec{s}\cdot\vec{\tau}(l_1,l_2)] . \qquad (7.1)$$

Here, E is the energy of the incident electron, exp[-M(n)] is the square root of the Debye-Waller factor of the nth layer, $\alpha(n,E)$ is the attenuation coefficient which is the amplitude at energy E that makes the round trip from the surface to the nth layer and back to the surface, $\vec{r}_j(n)$ is the relative position of the jth basis atom in the nth layer, and $\vec{\tau}(l_1,l_2)$ is the absolute position of the two-dimensional unit cell labeled by the integers l_1 and l_2. The quantity $\vec{s}$ is the momentum transfer to the surface and is given by

$$\vec{s} = \vec{k}_{out} - \vec{k}_o , \qquad (7.2)$$

where $\vec{k}_{out}$ and $\vec{k}_o$ are the wave vectors of the electron after and before the scattering event, respectively. The prefactor f(E) contains all contributions to the scattering other than those included explicitly in the sums.

Equation (7.1) may be applied to obtain the scattering amplitude of the (110) surface of GaAs, assuming that the scattering factors for gallium and arsenic are approximately the same due to their comparable atomic numbers. The unit-cell vectors of the (110) surface of GaAs are given by

$$\vec{a}_1 = a_0(0,1,0) , \quad \text{and} \qquad (7.3)$$

$$\vec{a}_2 = a_0(\sqrt{2},0,0) , \qquad (7.4)$$

where a_0 is the shorter side of the surface unit cell. There are two atoms in the unit cell for each layer, with coordinates given by

$$\vec{r}_j(n) = \vec{\rho}_j(n) + \vec{z}(n) , \qquad (7.5)$$

where $\vec{\rho}_j(n)$ and $\vec{z}(n)$ are the components parallel and perpendicular to the surface, respectively. The components parallel to the surface are

$$\vec{\rho}_1(n) = 0, \quad \vec{\rho}_2(n) = \frac{1}{2}\,\vec{a}_1 + \frac{3}{4}\,\vec{a}_2, \quad n = \text{odd}, \quad \text{and} \tag{7.6}$$

$$\vec{\rho}_1(n) = \frac{1}{2}\,\vec{a}_1 + \frac{1}{2}\,\vec{a}_2, \quad \vec{\rho}_2(n) = \frac{1}{4}\,\vec{a}_2, \quad n = \text{even}, \tag{7.7}$$

where n=1 represents the top layer.

The third sum on the right-hand side of (7.1) can be evaluated to give the two-dimensional Bragg condition for diffraction, i.e.

$$\sum_{l_1,l_2} \exp[i\vec{s}\cdot\vec{r}(l_1,l_2)] = \delta(\vec{s}_\parallel - \vec{g}_{hk}) . \tag{7.8}$$

Here $\vec{s}_\parallel$ is the component of $\vec{s}$ parallel to the surface, and the reciprocal lattice vector $\vec{g}_{hk}$ is

$$\vec{g}_{hk} = h\vec{b}_1 + k\vec{b}_2 , \tag{7.9}$$

where $\vec{b}_1$ and $\vec{b}_2$ are reciprocal unit-cell vectors given by

$$\vec{b}_1 = \frac{2\pi}{a_0}\,(0,1,0) , \quad \text{and} \tag{7.10}$$

$$\vec{b}_2 = \frac{\sqrt{2}\pi}{a_0}\,(1,0,0) . \tag{7.11}$$

From (7.1) the scattering amplitude for the (hk) beam at normal incidence can be obtained as

$$A(E) = f(E)\left| \sum_{n \text{ odd}} \exp[-M(n)]\alpha(n,E)\exp[is_z z(n)]\{1 + \exp[i\pi(h + \frac{3}{2}\,k)]\} \right.$$
$$+ \sum_{n \text{ even}} \exp[-M(n)]\alpha(n,E)\exp[is_z z(n)]$$
$$\left. \{\exp[i\pi(h + k)] + \exp(i\pi k/4)\} \right| , \tag{7.12}$$

where s_z is the component of $\vec{s}$ perpendicular to the surface.

The scattering amplitude A(E) vanishes at all energies only if both h is odd and k=0 mod 8. (There are more extinctions than for normal glide-plane symmetry, which occur only for h odd and k=0, because of the special atomic positions in the unit cell.) In particular, for the (10) or ($\overline{1}$0) beam, the scattering amplitude predicted by (7.12) vanishes. Hence, the observed nonextinction of the (10) or ($\overline{1}$0) beam is attributed to displacements in the position of the surface atoms. However, it is necessary to point out that the occurrence of the odd-numbered (h0) beams in a LEED pattern of a (110) surface of a zincblende structure is not necessarily caused by surface reconstruction. For compounds, such as GaSb or InAs, where the relevant atomic numbers are substantially different, the presence of the odd-numbered (h0) beams in their LEED pattern can be caused by the differences in their atomic scattering factors. Finally, it should be noted that the above symmetry arguments are not affected by multiple scattering.

Historical Development

During the past few years, a number of LEED analyses, including kinematic calculations, some of them employing Constant-Momentum-Transfer Averaging, and dynamical calculations, have been performed to determine the structure of the clean GaAs (110) surface. Table 7.5 summarizes recent LEED results and other studies of the GaAs (110) surface. Kinematic theory was applied by Mark et al. [7.76] to determine the interlayer spacing, assuming an unreconstructed surface. The first dynamical analysis was reported by Lubinsky et al. [7.77] and by Duke et al. [7.78], concluding that the gallium atoms relax inward and the arsenic atoms relax outward relative to the (110) surface plane by a tilt angle, $\omega = 34.8°$, as illustrated in Fig. 7.16. Their results also indicated that the lengths of the Ga-As surface bond,

Table 7.5. Summary of GaAs(110) Surface-Structure Determinations

Technique	Tilt Angle	As Back Bond	Ga Back Bond	Surface Bond	Ref.	Note
Kinematic LEED	0°	<1%	<1%	<1%	[7.76]	
Dynamical LEED	34.8°	0	0	0	[7.77,78]	
Kinematic LEED	17.7°	-9.6%	-22.9%	+1.7%	[7.79]	a
Kinematic LEED	26.4°	+8.0%	+14.8%	+4.0%	[7.79]	b
CMTA LEED	26.4°	+5.2%	-13.7%	+4.0%	[7.80,81]	c
CMTA LEED	26.4°	+5.2%	-13.7%	+4.0%	[7.82]	d
Dynamical LEED	27°	-3.6%	-2.5%	0	[7.40]	
Dynamical LEED	27.4°	-4.0%	+0.25%	0	[7.41]	e
Theory	27.4°	+0.3%	-2.5%	0	[7.88]	f
Theory	27.9°	0	0	0	[7.89]	g

[a] Described as "ionic" structure
[b] Described as "best" kinematic structure
[c] Same structure as line 4, but with additional perpendicular displacements of second-layer atoms by ±0.06 Å
[d] Same structure as line 5, buth with additional perpendicular displacements of third-layer atoms by ±0.03 Å
[e] Same structure as line 7, but with a relative shear of the second layer of 0.12 Å obtained by moving the second-layer arsenic atoms downward 0.06 Å and second-layer gallium atoms upward 0.06 Å from their position in the bulk
[f] Subsurface reconstructions of less than 0.07 Å were also included but did not significantly affect the energy or surface-layer reconstruction
[g] All bond lengths were constrained to maintain the bulk value

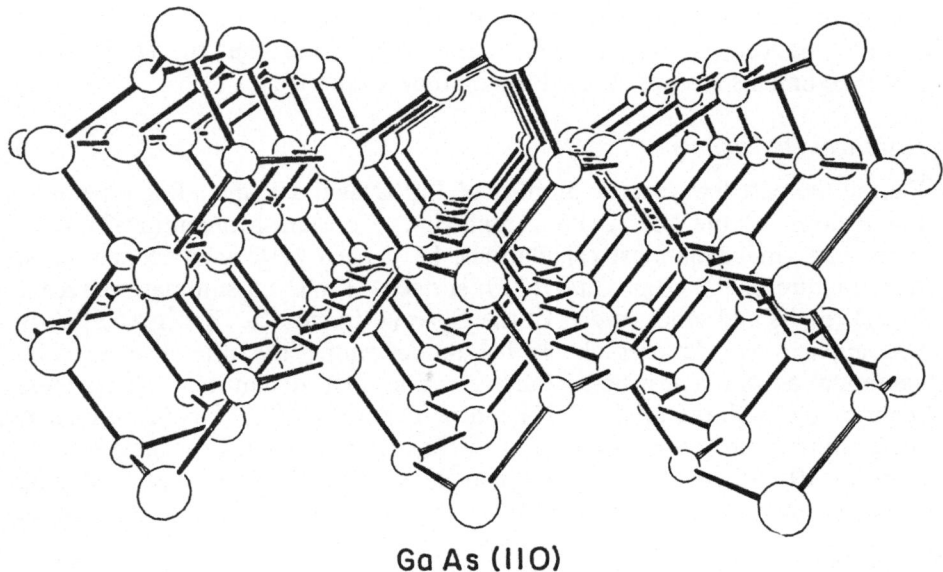

Ga As (110)

Fig. 7.16. Perspective view of single-layer reconstruction of GaAs(110)-(1×1), drawn with the ORTEP plotting program

the gallium back bond and the arsenic back bond are not changed. However, the overall agreement between experiment and theory was far from satisfactory.

Refinement of the earlier kinematic [7.76] and dynamical [7.77,78] results was carried out by Mark et al. [7.79] using kinematic calculations. Applicability of the kinematic calculations was tested by comparing the kinematic and dynamical I-V profiles. It was found that similar prominent features exist in both kinematic and dynamical I-V profiles. The results favored two possible structures: the "best" kinematic structure and the "ionic" structure, which differ in the nature of the assumed atomic scattering potentials, as shown in Table 7.5. The "ionic" and the "best" kinematic structures involve tilt angles ω of 17.7° and 26.4°, respectively. In these two structures, the lengths of the arsenic back bond, the gallium back bond, and the surface bond are not conserved. Even with these two structures there is not a good correspondence between the observed and the calculated spectra, especially for the (11) and $(1\ \bar{1})$ beams. Consequently, further refinement of the structure was undertaken by including perpendicular displacements of the second layer [7.80,81] as shown in Fig. 7.15d, as well as perpendicular displacements of the third layer [7.82]. In these analyses, the Constant-Momentum-Transfer-Averaging method ([7.90], see Sect. 6.2) was used. Comparison between experimental and computed I-V profiles indicated that an improved agreement was achieved with the structure having a tilt angle of 26.4° and second-layer displacements. In their subsequent work [7.82],

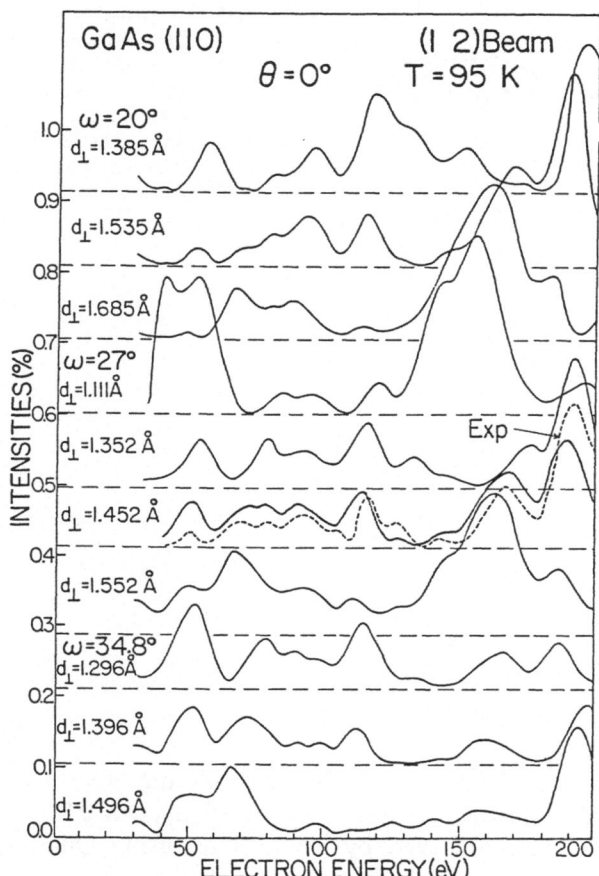

Fig. 7.17. I-V spectra for the (12) beam of GaAs(110)-(1 × 1) corresponding to different tilt angles ω and various heights $d_\perp$ of the surface gallium above the substrate. The experimental spectrum is shown as a dashed line [7.40]

which included third-layer displacements, the improvement claimed is questionable. All kinematic calculations suffer a major disadvantage due to their relative insensitivity to motion of surface atoms parallel to the surface, especially for the specular beam. Results of the Constant-Momentum-Transfer Averaging analysis may also include errors due to incomplete or inaccurate averaging of the data [7.91].

As a result, these proposed models had to be confirmed by a dynamical analysis. Complete and thorough dynamical analyses were carried out by Tong et al. [7.40] and Meyer et al. [7.41]. The results from both studies are identical within the present limits of LEED accuracy. Figures 7.17 and 18 for the (12) and (1$\bar{3}$) beams from the work of Tong et al. [7.40] illustrate the quality of agreement obtained by these results. The best agreement with the experiment was reported for the structure with a tilt angle of 27°, almost no change in the surface bond length, and 3.6% and 2.5% contraction of the arsenic and gallium back bonds, respectively. The two structures which are

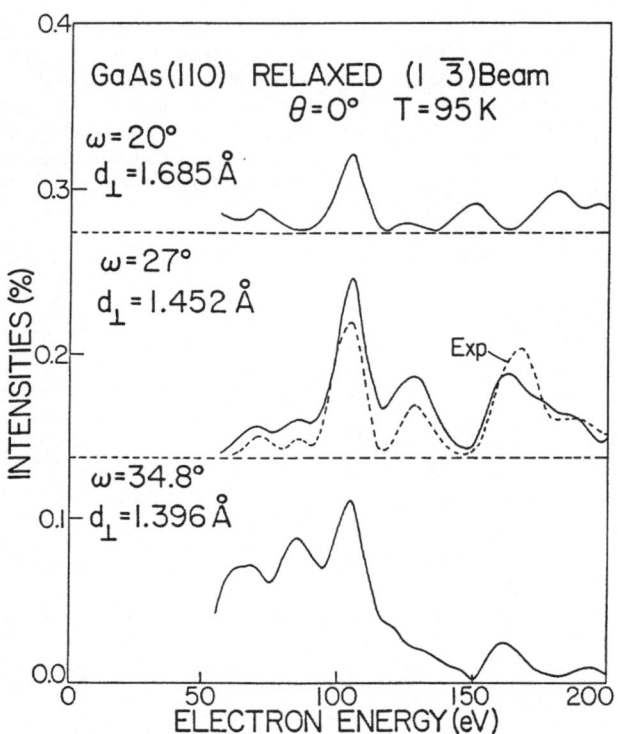

Fig. 7.18. I-V spectra for the $(1\bar{3})$ beam of GaAs(110)-(1×1) corresponding to different tilt angles ω and various heights $d_\perp$ of the surface gallium above the substrate. The experimental spectrum is shown as a dashed line [7.40]

nearly identical (as shown on lines 5 and 6 of Table 7.5), differing only in a perpendicular displacement of atoms in the second and third layer of 0.06 Å and 0.03 Å respectively, were also analyzed by Tong et al. using Quasi-dynamical calculations. Results of these calculations indicate that there was only a small difference between the calculated I-V profiles of these two structures. By contrast, Meyer et al. [7.41] maintained that the second-layer reconstruction is necessary to improve the agreement between experiment and theory. Their reported structure is specified by a tilt angle of 27.4°, no significant change in the surface bond length, and a 4.0% contraction and a 0.25% expansion of the arsenic and gallium back bonds, respectively. In addition, there is a relative shear of the second layer of 0.12 Å obtained by moving the second-layer arsenic atoms downward by 0.06 Å and second-layer gallium atoms upward by 0.06 Å from their positions in the bulk. However, they also agree that their results were insensitive to a small third-layer distortion.

Structures very similar to those suggested by Tong et al. and Meyer et al. were proposed by Chadi [7.88], Goddard et al. [7.89] and Miller et al. [7.92] as a result of theoretical calculations minimizing the total energy. All these studies proposed structures having tilt angles of approximately 27°. In addition, these results indicate that there is little change in bond lengths, as predicted by the various dynamical calculations.

7.3 Adsorbed Atomic Layers

The studies of surface structures of adsorbed atoms and molecules have played an important role in the understanding of a variety of surface phenomena. For example, knowledge concerning the adsite and bond lengths between the adsorbate and the substrate is of relevance in understanding the complex phenomenon of heterogeneous catalysis. Adsorbed atoms or molecules may be ordered periodically to form one or more structures depending on surface coverage and temperature. For example, at a quarter-monolayer coverage, oxygen on Ni(100) produces a p(2x2) pattern. Doubling the oxygen coverage to a half monolayer yields a c(2x2) pattern [7.93]. Often, but by no means always, an overlayer structure possesses the same rotational symmetry as the substrate. More generally, the unit cell of the overlayer structures is the smallest allowed by the atomic or molecular dimension, as well as by the adsorbate-substrate and indirect adsorbate-substrate-adsorbate interactions.

Adsorption Sites

Some of the most commonly occurring geometries of atoms on various metal surfaces are shown in Fig. 7.19. Adatoms have a tendency to occupy sites with the largest coordination number, which are usually sites of high symmetry, as illustrated in Fig. 7.19. An exception is oxygen on several fcc(110) surfaces. In the Ni(110)-p(2x1)-O [7.94] and Ir(110)-c(2x2)-O [7.50] overlayer structures, oxygen has been proposed to reside in the short-bridge site, as shown in Fig. 7.19e, while in the Ag(110)-p(2x1)-O overlayer structure, the long-bridge site is apparently preferred by the oxygen adatoms [7.95]. Both the long-bridge and the short-bridge sites have less than the highest coordination number, although they do have the highest possible symmetry on the fcc(110) surface.

On the fcc(110) surface, the site with the highest coordination number is the centered site, which is chosen by sulfur on the Ni(110) surface [7.95,96]. Oxygen on non-fcc(110) surfaces and other divalent elements (e.g. sulfur, selenium, and tellurium) on various metal surfaces prefer the adsorption site with the largest coordination number. Results of LEED analyses of Ir(111)-(2x2)-O [7.97] and Ni(111)-p(2x2)-O [7.94] overlayer structures agree with these general rules. On these surfaces, it is found that oxygen resides in a threefold hollow site (the site with the largest coordination number), as shown in Fig. 7.19a. It should be noted that there are two different types of threefold sites on the fcc(111) surface, as well as on the hcp(0001) surface. The fcc(111) surface is described by the registry sequence ABCABC... (surface at the left). The two different types of sites on this surface can be described by the registry sequence bABCABC... (the lowercase letter represents the overlayer), which indicates the threefold hollow site directly

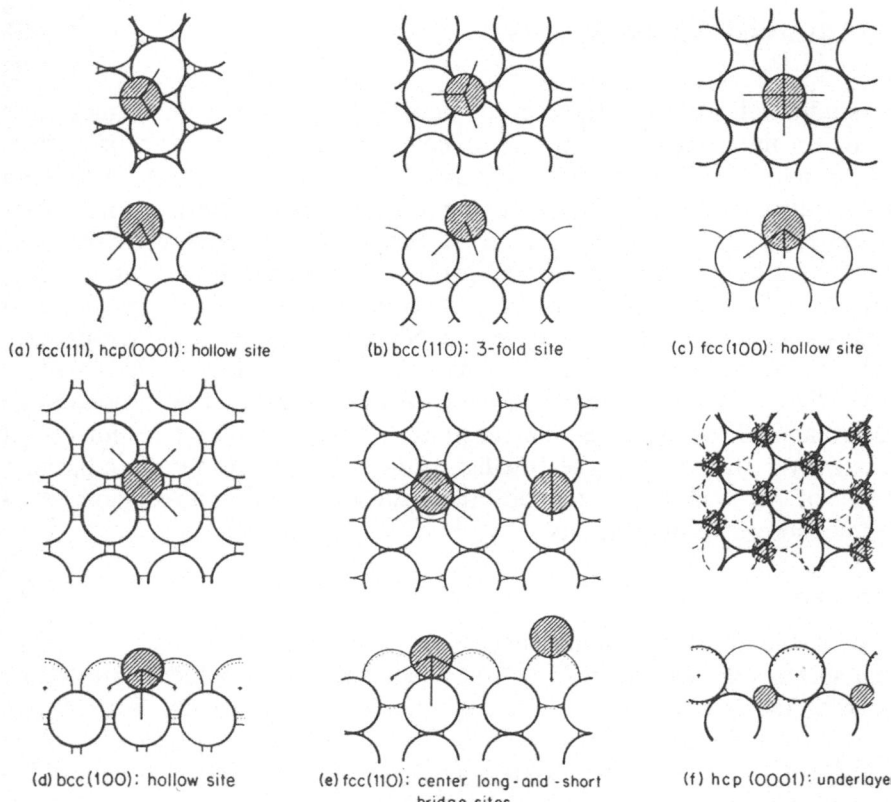

(a) fcc(111), hcp(0001): hollow site

(b) bcc(110): 3-fold site

(c) fcc(100): hollow site

(d) bcc(100): hollow site

(e) fcc(110): center long-and-short bridge sites

(f) hcp(0001): underlayer

Fig. 7.19a – f. Top and side views (in top and bottom sketches of each panel) of atomic adsorption geometries on various metal surfaces. Adsorbates are shown as crosshatched circles. Dotted lines represent clean-surface (relaxed) atomic positions, and arrows show atomic displacements due to adsorption

below which there is another substrate atom in the second layer, and the registry sequence cABCABC..., which represents the threefold hollow site directly below which there is a vacancy in the second layer. On the Ir(111) surface, oxygen was found to reside on the site described by the registry sequence cABCABC..., while on the Ni(111) surface the analysis could not differentiate the two threefold sites for the correct adsorption site. Various adatoms generally prefer the cABCABC... site on fcc(111) surfaces with oxygen and hydrogen being the possible exceptions.

Another interesting overlayer structure is the W(110)-p(2x1)-O on which oxygen chooses the threefold coordinated site, which has only one mirror plane, rather than a site which has two orthogonal mirror planes [7.98,99]. In this case, oxygen prefers a site that has the largest coordination number to one with the highest symmetry, as shown in Fig. 7.19b. On the Fe(100) surface, oxygen is found to reside in the fourfold symmetric site as shown in

Fig. 7.19d [7.5]. The results of a LEED analysis indicate that on this surface oxygen is approximately equidistant from its five neighboring iron atoms, which puts the adatom deep in the hollow site nearly coplanar with the top-layer iron atoms. Once again, the adatom prefers the site with both the highest symmetry and the largest coordination number.

It has been found that nearly all adatoms on metal surfaces reside in only one type of adsorption site at a time, with the exception of Ni(111)-p(2x2)-2H, possibly some oxygen overlayers on fcc(111) surfaces, and incommensurate structures, such as Ag(111)-Xe [7.100]. The Ni(111)-p(2x2)-2H overlayer structure represents an interesting case in which two different types of adsorption sites are occupied simultaneously by the adsorbate. Several models for this overlayer structure have been considered in a LEED analysis [7.101,102], as shown in Fig. 7.20. Two of these structures involve a honeycomb arrangement in which hexagons of atoms are connected at their corners, and each hydrogen atom has three neighboring hydrogen atoms disposed at equal distances, as shown in Figs. 7.20a and b. The structure shown in Fig. 7.20a involves the occupancy of both types of threefold sites (the cABCABC... and bABCABC... sequences), while the structure shown in Fig. 7.20b involves the simultaneous occupancy of both the on-top site and one kind of threefold site. Figure 7.20c shows a surface structure representing "quasi-molecular" chemisorption. In this model, a pair of hydrogen atoms resides in two different types of threefold sites adjacent to one another. The results of an extensive LEED analysis, described in detail in Chap. 8, suggest that the structure in Fig. 7.20a, which involves the two different types of threefold sites, is the most probable structure for the Ni(111)-p(2x2)-2H overlayer structure.

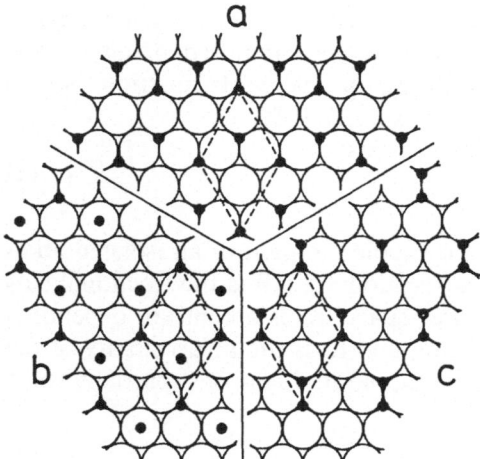

Fig. 7.20. Model geometries for Ni(111)-(2×2)-2H (top view). (a) Honeycomb hollow-hollow arrangement. (b) Honeycomb hollow-top arrangement. (c) Quasi-molecular hollow-hollow arrangement

Subsurface Effect on Adsorption Sites

From the above examples, the effect of the second and deeper substrate layers on the adsorbate seems to be secondary, at least for compact metallic crystals. Nevertheless, it appears that adsorbates favor sites which are the continuation of the bulk structure of the substrate. There are some exceptions, however, e.g. oxygen on the fcc(110) surfaces, hydrogen on the Ni(111) surface, and cadmium on the Ti(0001) surface [7.103,104]. The Ti(0001) surface has an hcp layer-stacking sequence ABABAB..., and the continuation of the bulk structure into the overlayer would imply the sequence bABABAB... . However, the Ti(0001)-p(1x1)-Cd surface adopts a different sequence, which may be represented by cABABABAB... . The choice of this adsorption site may be the result of a repulsion between the cadmium atoms and their second neighbor titanium atoms. Interestingly, both the bulk cadmium structure and the Cd(0001) surface have the hcp stacking sequence ABABAB... .

Bond Lengths and Charge Transfers

Another important parameter that can be obtained from the results of LEED analyses of overlayer structures is the adsorbate-substrate bond length. Figure 7.21 summarizes various results for this adsorbate-substrate bond length determined by LEED. These results agree well with the bond lengths found in appropriate bulk compounds or molecules, as determined by X-ray Diffraction or other methods. A nonstructural quantity which is useful in the understanding of chemisorption is the charge transfer between the adsorbate and the substrate. The charge transfer can be estimated by measuring the change in the work function $\Delta\varphi_w$ upon adsorption. The work function φ_w of a solid is defined as the energy $e\varphi_w$ which is necessary to transfer an electron from the Fermi level to the vacuum level. An adsorbate can change the work function by changing the electronic distribution across the surface, which one may visualize, for convenience, to be composed of an electric dipole at each adsorption site. The change of the work function due to adsorption is given approximately by expression [7.105],

$$\Delta\varphi_w = \frac{-e\mu_o\sigma_a\theta}{\varepsilon_o[1 + K_c\alpha_p(\sigma_a\theta)^{3/2}]} , \qquad (7.13)$$

where e is the electronic charge, μ_o is the dipole moment of an isolated additional dipole, σ_a is the density of the adsorption sites, θ is the fractional surface coverage, ε_o is the permittivity of free space, K_c is a constant depending on the geometry of the overlayer structure, and α_p is the polarizability of the adsorbate. In this simplified picture, the dipole moment of a partially ionic bond is given by

$$\mu_o = q \cdot l_B , \qquad (7.14)$$

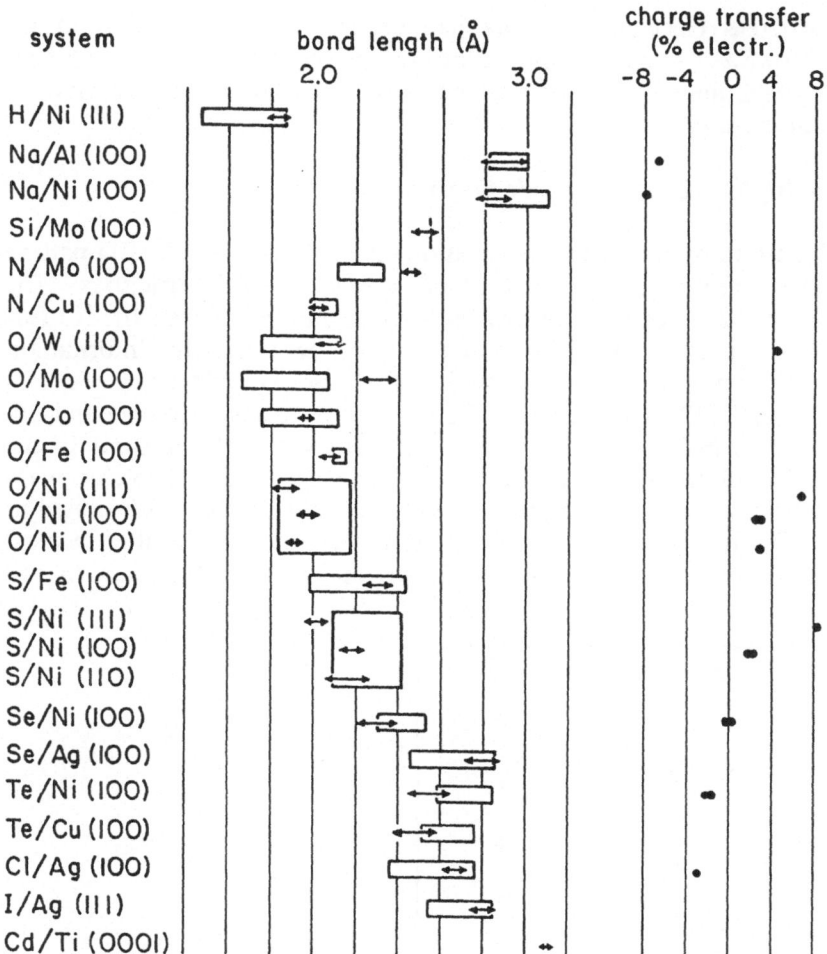

Fig. 7.21. (*Left*) Comparison of adsorbate-substrate bond lengths (arrows showing uncertainty) with equivalent bond lengths in molecules and bulk compounds (blocks extending over range of known values). (*Right*) Induced charge transfers upon adsorption

where q is the charge transfer and l_B is the dipole length, which can be taken as the component of the metal-adatom bond length perpendicular to the surface. Using (7.13,14), but neglecting the small depolarization term $K_c\alpha_p(\sigma_a\theta)^{3/2}$, the charge transfer between the adsorbate and the substrate can be calculated, and typical results are included in Fig. 7.21. Since this model for the change in the work function is quite simple, the absolute magnitude of the charge transfer is not very meaningful, but the trends should be reliable. As one would expect, electron donors like sodium donate charge to the surface, whereas electron acceptors like oxygen extract charge from the surface. However, selenium and tellurium on Ni(100) exhibit a different behavior, which illustrates that the charge redistribution due to adsorption is a rather complicated matter.

7.3.1 The Ir(110)-(2x2)-2S Atomic Overlayer

The symmetry information that is occasionally contained in a LEED pattern is particularly useful for the analysis of complex surface structures. The analysis of the (2x2) pattern formed by the exposure of a clean Ir(110) surface to H_2S is used as an illustration to show how symmetry information may restrict the number of possible surface structures [7.106].

A series of LEED patterns was observed upon exposure of the Ir(110)-(1x2) surface to H_2S at 350 K. A (2x2) pattern with missing spots in the (n+1/2, 0) positions was observed at low exposures and reached its maximum intensity after an exposure of 4 L of H_2S. This pattern is illustrated in Fig. 7.22. The intensity of this LEED pattern was not changed after increasing the temperature of the surface to 1200 K and then cooling back to 350 K.

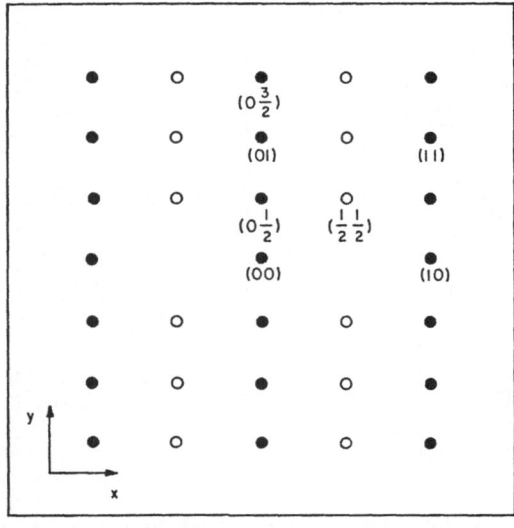

Fig. 7.22. The p1g1-(2 × 2) LEED pattern observed after exposure of Ir(110)-(1 × 2) to 4 L H_2S at 350 K. Solid circles indicate substrate (1 × 2) beams; open circles indicate the "extra" beams due to the (2 × 2) structure

The observation that the spots corresponding to the (1x2) substrate reconstruction remain bright and sharp throughout the growth and the disappearance of the (2x2) pattern at temperatures above 1200 K suggests that the adsorption of sulfur does not cause the Ir(110) surface to lose its (1x2) reconstruction. This, together with the absence of some spots in the (2x2) pattern, allows the structure corresponding to the (2x2) pattern to be delimited.

Restrictions Imposed by the Spot Extinctions

The absence of certain beams at all incident electron energies is indicative of glide-plane symmetry, as has been discussed in Sects. 3.4 and 5.10.3. This symmetry results in destructive interference due to equivalent scatterers within the (2x2) unit cell. Specifically, the extinction of the (n+1/2, 0) beams indicates a surface structure of either p1g1 or p2mg symmetry ([7.107]; see also Fig. 3.9). Both of these symmetries include two glide planes parallel to the x-axis and located within the (2x2) unit cell. A glide plane indicates invariance under the following operation: translation parallel to the plane by one-half the length of the (2x2) unit cell, followed by reflection across the plane. Due to the low symmetry of the (1x2) surface, as illustrated in Fig. 7.23 for the missing-row model for Ir(110)-(1x2) [7.46,48], there is only one possible set of locations for these glide planes. The p1g1 and p2mg symmetries differ in the absence or presence of a twofold rotational axis and in the absence or presence of mirror planes perpendicular to the glide planes, respectively.

For the p1g1 symmetry, there are two atoms within the unit cell, located at (x,y) and (1+x,ȳ) in the units of Fig. 7.23, with x and y such that the glide planes are the only symmetry elements of the structure. A possible p1g1-(2x2) structure is shown in Fig. 7.24. In this particular illustration, the sulfur atoms are prevented from occupying the higher-coordination sites on the "missing rows" by S-S interactions.

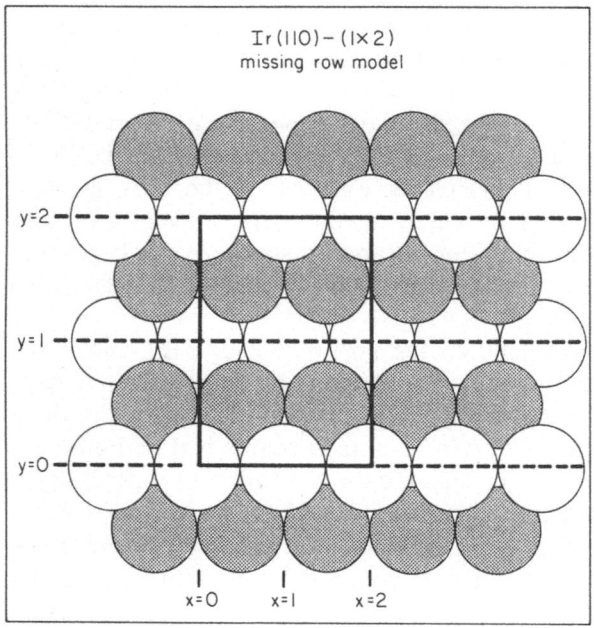

Fig. 7.23. Location of glide planes (*dashed lines*) for a p1g1 or a p2mg-(2×2) unit cell on the Ir(110)-(1×2) surface (missing-row model). Open circles indicate first- and third-layer iridium atoms; shaded circles, second-layer iridium atoms

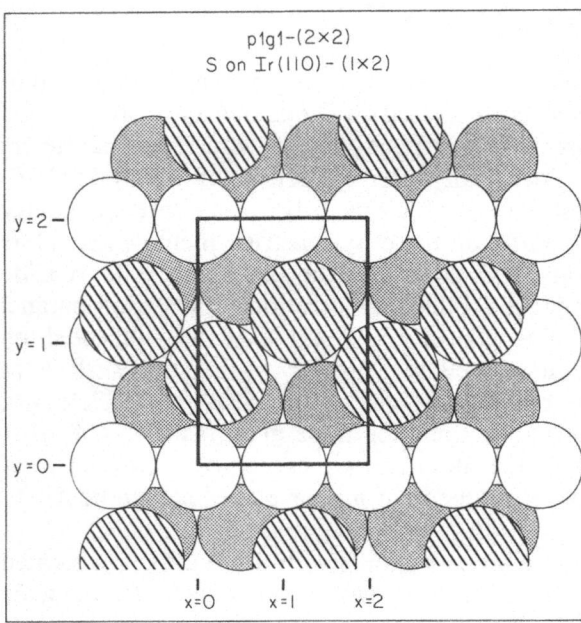

p1g1-(2×2)
S on Ir(110)-(1×2)

Fig. 7.24. A possible p1g1-(2 × 2) structure for sulfur on Ir(110)-(1 × 2). Sulfur atoms (*crosshatched*) are shown with a hard-sphere radius of 1.7 Å. The radius of an iridium atom is 1.36 Å. Sulfur atoms are located at (x, y) and (1 + x, ȳ). Translating the sulfur atoms to (x̄, y), (1 + x̄, ȳ) results in an equivalent (mirror image) domain of the structure shown

Either two or four atoms per unit cell are allowed for the p2mg structure. The four-atom structure (which would correspond to a fractional coverage of unity) is considered unlikely on the basis of the observation that coverages of sulfur higher than that required for the (2x2) structure are possible without significant attenuation of the iridium Auger peaks. In the case of two atoms per unit cell, the sulfur atoms would be located so that a mirror plane parallel to the y-axis runs through each sulfur atom. Two possible p2mg structures are shown in Fig. 7.25. The structures depicted would arise if sulfur atoms with a covalent radius of 1.0 Å [7.108] were adsorbed in the three-fold sites of the terraces of the (111) planes exposed by the missing rows. Other p2mg structures, generated by changing the y-coordinates shown in Figs. 7.25a and b, are also plausible. For instance, a structure similar to that shown in Fig. 7.24 but with a p2mg symmetry would be a reasonable hypothesis.

In an attempt to distinguish between the possibilities of p1g1 and p2mg symmetry, the I-V profiles for two sets of corresponding beams in three quadrants were measured. As shown in Fig. 7.26, the I-V spectra for the three beams of each set are identical. If the (n/2, ±m/2) beams had had different I-V spectra than the (-n/2, ±m/2) beams, then one domain of a p1g1 structure would have predominated on the surface [7.109]. Since the I-V spectra are the same, it is not possible to determine which of the two symmetries is present, since the observed equivalence could be due either to equal representation of the two domains of a p1g1 structure or to a p2mg structure.

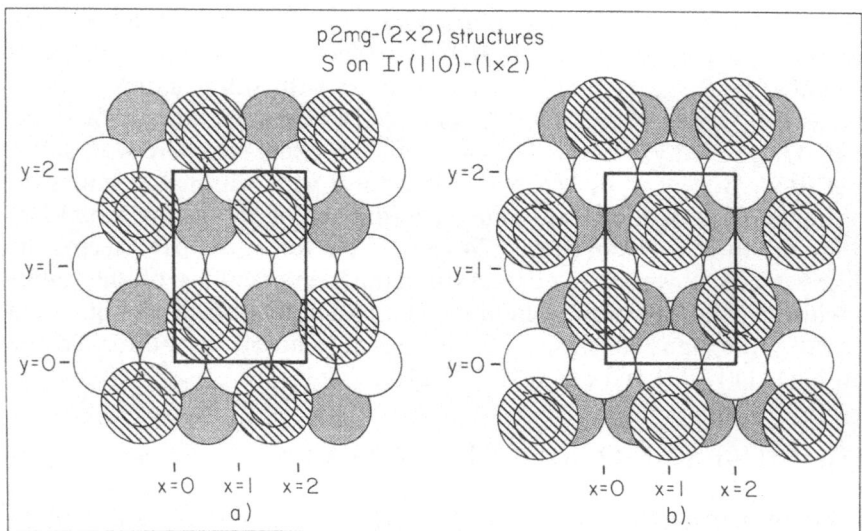

Fig. 7.25a,b. Two possible p2mg-(2 × 2) structures for sulfur on Ir(110)-(1 × 2). The inner circle on the sulfur atoms shows the covalent radius of 1.0 Å. (**a**) Sulfur atoms located in threefold sites formed by two iridium atoms in the top layer and one iridium atom in the second layer. (**b**) Sulfur atoms located in threefold sites formed by one iridium atom in the top layer and two iridium atoms in the second layer

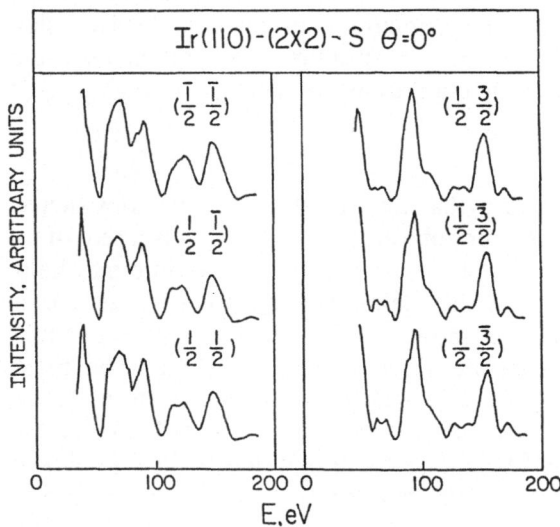

Fig. 7.26. LEED I-V spectra showing the equivalence of corresponding symmetrical beams in three quadrants

In any case, symmetry considerations allow the structure of the (2x2) sulfur overlayer on the reconstructed Ir(110) surface to be limited to one of three possible types, namely (1) a p1g1 structure with sulfur atoms at (x,y), $(1+x,\overline{y})$, x $\neq$ 0, 1/2, or 1, y $\neq$ 0, 1/2, or 1, (2) a p2mg structure with sulfur atoms at $(1/2,y)$, $(3/2,\overline{y})$, y $\neq$ 0, 1/2, or 1, or (3) a p2mg structure with sulfur atoms at (0,y), $(1,\overline{y})$, y $\neq$ 0, 1/2, or 1. Structures with reasonable values for the x- (in the case of the p1g1) and y-coordinates of the sulfur atoms are shown for the three types in Figs. 7.24,25. Of the three, the p1g1 seems the least likely since it would require asymmetrical interactions of the sulfur atoms with neighboring iridium atoms. The structures shown in Fig. 7.25a and b seem quite plausible since sulfur is known to adsorb in a threefold site both on Ni(111) [7.96] and on Ir(111) [7.110].

7.3.2 The Ir(110)-c(2x2)-O and Ir(111)-(2x2)-O Atomic Overlayers

These two overlayer structures of oxygen adatoms on iridium are used to illustrate the effects of coordination number and symmetry on the choice of the adsorption site. In general, an atomic adsorbate appears to choose an adsorption site with the largest coordination number and the highest symmetry as the most favorable bonding geometry. However, under certain circumstances where all these ideal conditions cannot be met, the adsorbate will accept a site which has less favorable conditions. In the example of the Ir(110)-c(2x2)-O overlayer structure, oxygen chooses the short-bridged site because it has a larger coordination number than the on-top and center sites, as well as a more favorable bonding geometry than the long-bridged site. However, in the Ir(111)-(2x2)-O overlayer structure, oxygen chooses the threefold site where all these favorable conditions are met.

The Ir(110)-c(2x2)-O Structure

It was shown in the previous section that an Ir(110)-(1x1) unreconstructed surface is formed when a quarter monolayer of oxygen is adsorbed on a clean Ir(110)-(1x2) reconstructed surface. When an additional half-monolayer of oxygen is adsorbed on the metastable Ir(110)-(1x1) surface, a c(2x2) LEED pattern is observed as shown in Fig. 7.27e [7.50]. Possible real-space models for this structure are shown in Figs. 7.27a-d. The most plausible (high-symmetry) models for an overlayer of adsorbed atoms forming a c(2x2) structure on a substrate with a rectangular unit cell involve the centered, the on-top, the short-bridged, and the long-bridged sites, as shown in Fig. 7.27. In the centered site, the oxygen atoms fit deeply into the hollow, such that they sit on the top of other iridium atoms in the second substrate layer. The distance between the oxygen atom and the iridium atoms in the first layer is too large to form Ir-O bonds. Consequently, the coordination number for oxygen residing in the centered site or in the on-top site is

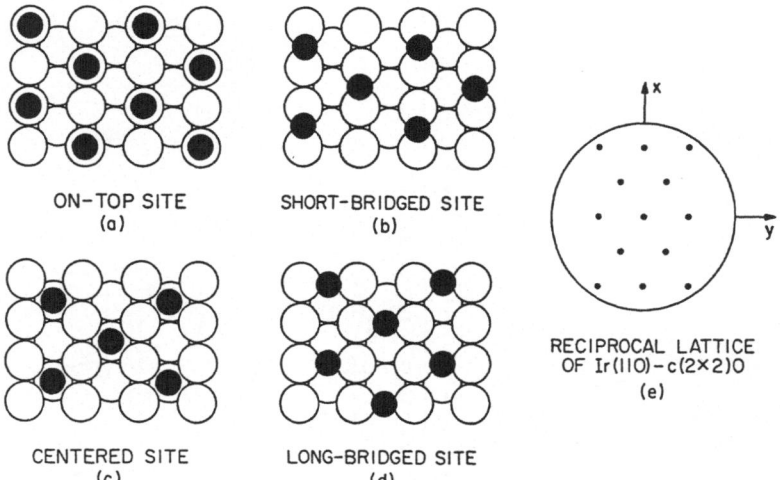

ON-TOP SITE
(a)

SHORT-BRIDGED SITE
(b)

CENTERED SITE
(c)

LONG-BRIDGED SITE
(d)

RECIPROCAL LATTICE
OF Ir(110)-c(2×2)O
(e)

Fig. 7.27a – e. Possible high-symmetry structures of the Ir(110)-c(2 × 2)-O overlayer. Large open circles denote iridium atoms, and small filled circles denote oxygen atoms. (**a**) On-top site. (**b**) Short-bridged site. (**c**) Centered site. (**d**) Long-bridged site. (**e**) Reciprocal-space lattice corresponding to the Ir(110)-c(2 × 2)-O structure

one. From the point of view of the coordination number, these two sites are not favorable. In the dynamical LEED analysis, the centered site was not considered because this site is unlikely to be the correct structure and because the calculation for it is very time-consuming. Hence, calculations have been performed only for the on-top, the short-bridged, and the long-bridged sites, with various physically realistic interlayer spacings between the substrate and the overlayer. An assumed inner potential of 15 eV was used for both the overlayer and the substrate in the dynamical calculations. For the comparison between the theoretical and the experimental I-V spectra, the inner potential was allowed to change by a rigid shift of the energy scale. Examples of comparisons between experimental I-V spectra and the best theoretical spectra for each model, obtained with a modified inner potential of 10 eV, are shown in Figs. 7.28 and 7.29. Comparisons between theory and experiment indicate that the short-bridged site is preferred. For certain beams, the agreement between experiment and theory for the on-top site is acceptable, but the best overall agreement is obtained for the short-bridged site. The long-bridged site is clearly ruled out by the poor quality of agreement between theory and experiment in almost all cases.

It was found that the topmost interlayer spacing of the unreconstructed Ir(110)-(1x1) structure is contracted by approximately 7.5% of the bulk interlayer spacing [7.49]. Therefore, calculations were performed for the short-bridged site model of the oxygen overlayer in which the topmost interlayer spacing of the substrate (d_1) varied from 1.22 Å (-10%) to 1.43 Å ($+5\%$) in

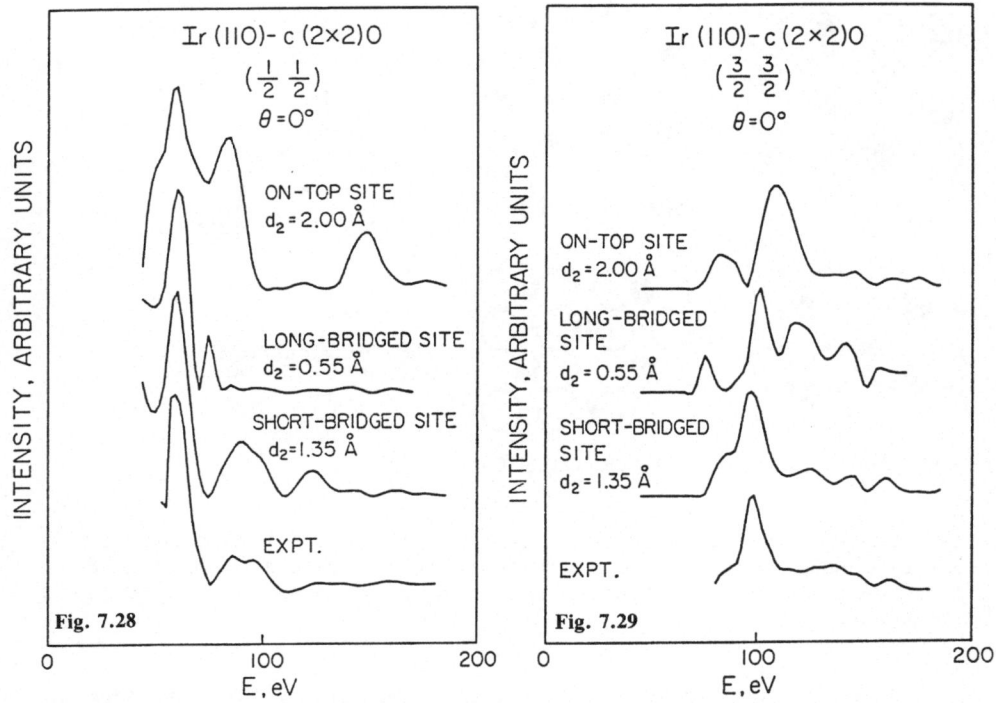

Fig. 7.28. Comparison between the theoretical I-V spectra, modified with an inner potential of 10 eV for each model considered, and the experimental I-V spectrum of the (1/2, 1/2) beam from the Ir(110)-c(2×2)-O surface. $\theta = 0°$ corresponds to normal incidence, and d_2 is the interlayer spacing between the substrate and the oxygen overlayer

Fig. 7.29. As Fig. 7.28, for the (3/2, 3/2) beam

increments of 0.07 Å(5%), while the interlayer spacing between the substrate and the overlayer (d_2) varied from 1.20 Å to 1.50 Å in increments of 0.05 Å Comparisons between experimental and theoretical I-V spectra for d_1 equal to 1.36 Å and 1.29 Å and for d_2 varying between 1.35 Å and 1.50 Å are shown in Figs. 7.30 and 7.31. The results of the calculations indicate that the best agreement is found for values of d_1 and d_2 equal to 1.33 ± 0.07 Å and 1.37 ± 0.05 Å respectively, demonstrating that the contraction of the topmost interlayer spacing of the substrate is essentially removed by the additional adsorbate.

The results of this study, which indicate that the oxygen adatoms in the c(2x2) structure on Ir(110) are residing in the short-bridged sites, are in agreement with the bonding geometry found for the (2x1) oxygen overlayer on Ni(110) [7.94]. The choice of the short-bridged site the long-bridged site, the four-fold site, or the on-top site by the oxygen both on the Ni(110) surface and the Ir(110) surface is in agreement with the results of generalized-

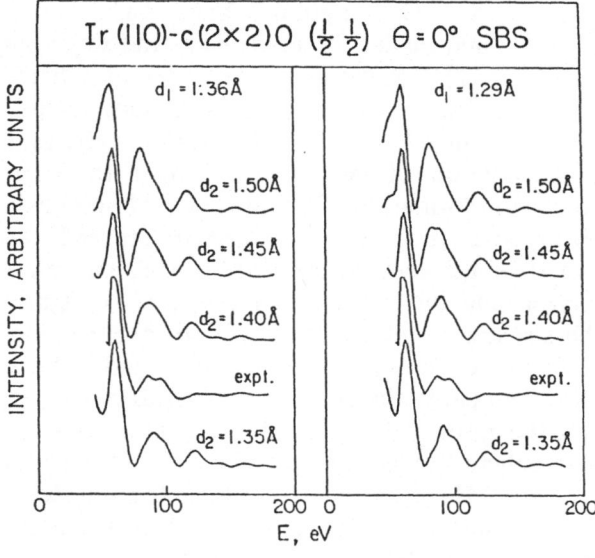

Fig. 7.30. Comparison between theoretical I-V spectra, modified with an inner potential of 10 eV, for the short-bridged site model with several combinations of values of d_1 (the topmost interlayer spacing of the substrate) and d_2 (the interlayer spacing between the substrate and the overlayer), and the experimental I-V spectrum of the (1/2, 1/2) beam from the Ir(110)-c(2×2)-O surface. SBS denotes short-bridged site

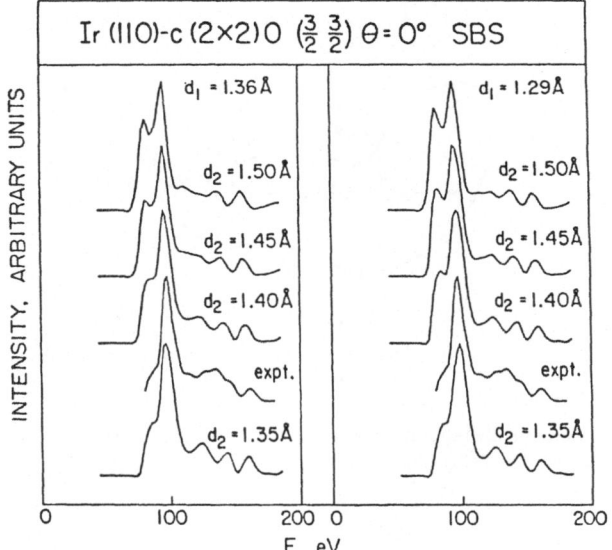

Fig. 7.31. As in Fig. 7.30, for the (3/2, 3/2) beam

valence-bond calculations for oxygen chemisorption on metal clusters [7.111]. If the oxygen atom is located in the on-top site or in the centered site on Ni(110) and/or Ir(110), the chemical bonding will be similar to the bonding between one oxygen atom and one nickel or iridium atom. It was found theoretically that the chemical bond formed between an oxygen atom and two nickel atoms, resembling the short-bridged site, is stronger by 10 kcal/mole than the chemical bond formed between one oxygen atom and one nickel atom, simulating the on-top site.

Moreover, it is known that most X_2O compounds have X-O-X bond angles of approximately 105°, e.g. bond angles for H_2O, F_2O, and Cl_2O are 105°, 103°, and 111°, respectively [7.1]. Within the framework of a hard-sphere model, the bond angles for the short-bridged site and the long-bridged site on the Ir(110) surface are approximately 86° and 148°, respectively, as is shown in Fig. 7.32. Hence, the bonding of oxygen to Ir(110) in the long-bridged site is unfavorable due to the formation of a large Ir-O-Ir bond angle compared to the bond angles found in most X-O-X molecules. Indeed, the LEED calculations indicate that for the short-bridged site model, the inter-layer spacing between the oxygen and the iridium atoms is 1.37 Å slightly less than 1.47 Å which is the spacing calculated from a hard-sphere model assuming a bond length of 2.00 Å i.e. the sum of the covalent radii of the oxygen and iridium atoms. Consequently, the actual bond length of oxygen on Ir(110) is 1.93 Å based on the results of the LEED calculations, rather than 2.00 Å based on the hard-sphere model, and the actual bond angle is approximately 90° rather than 86°. Hence, a contraction of the interlayer spacing between the substrate and the overlayer is reasonable from the point of view of the formation of a more favorable Ir-O-Ir bond angle.

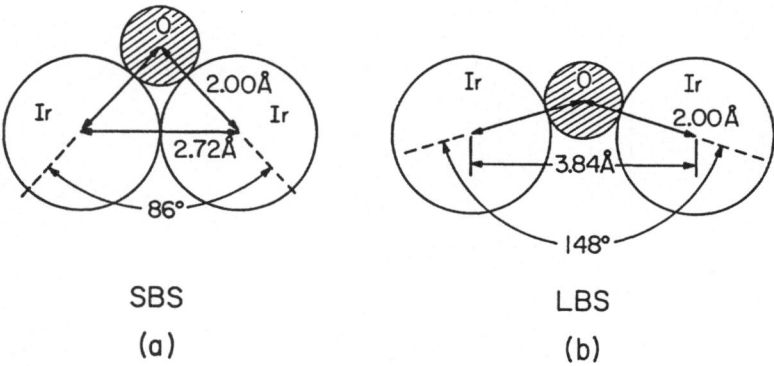

Fig. 7.32a, b. Schematic hard-sphere models showing side views of (a) the short-bridged site (SBS) of Ir(110)-c(2 × 2)-O through the (001) surface and (b) the long-bridged site (LBS) through the (1̄10) surface

The Ir(111)-(2x2)-O Structure: (2x2) or (1x2)?

The (2x2) overlayer on a fcc(111) surface represents an interesting case for a structural analysis. Exposure of a clean Ir(111) surface at room temperature to approximately 20 L of oxygen produces a sharp (2x2) LEED pattern, as shown in Fig. 7.33 [7.97]. This (2x2) pattern can correspond to two different surface structures. One of the possible surface structures for the (2x2) pattern is a p(2x2) array of oxygen adatoms ("p" stands for "primitive"), as shown in Fig. 7.34. This superstructure corresponds to a quarter-monolayer

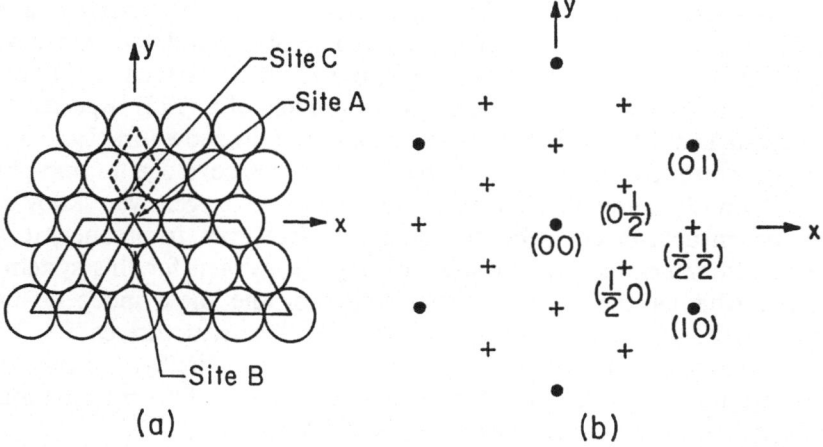

Fig. 7.33. (a) Unit cells of the (1 × 1) substrate for Ir(111)-(2 × 2)-O, and of the p(2 × 2) and (1 × 2) overlayer structures. Site A is the on-top site; site B is the threefold site directly beneath which there is another substrate atom; and site C is the threefold site with a vacancy located beneath it in the second layer. **(b)** Schematic LEED pattern corresponding to a p(2 × 2) superstructure

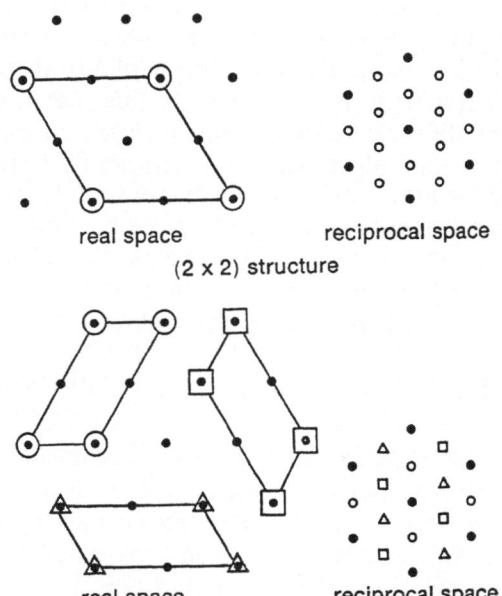

Fig. 7.34. Unit cell of a p(2 × 2) superstructure and its corresponding LEED pattern (*top*); three domains of a (1 × 2) superstructure rotated 120° with respect to one another on an fcc(111) surface and the corresponding LEED pattern (*bottom*)

coverage of oxygen adatoms. The full circles represent the diffraction beams due to the substrate, while the open circles represent the diffraction beams due to the overlayer. Another possible structure for the observed (2x2) pattern consists of three independent, equivalent domains of a (1x2) superstructure, each rotated by 120° with respect to one another, as shown in Fig. 7.34. This superstructure corresponds to a half-monolayer coverage of oxygen adatoms. Obviously if the absolute surface coverage of oxygen were known, these two superstructures could be distinguished. However, in practice, it is quite difficult to determine the absolute coverage of oxygen for this system. Furthermore, whether the p(2x2) superstructure or the three independent domains of a (1x2) superstructure is the correct surface structure cannot be determined by a visual inspection of the LEED pattern. Finally, the precise adsorption site for the oxygen adatoms is also unknown. This information can be determined only by a detailed analysis of I-V spectra.

Adsorption Site

The most symmetric adsites on this surface are the on-top site (site A of Fig. 7.33), and the two different types of threefold sites (sites B and C of Fig. 7.33). Site B represents the threefold hollow site directly below which there is another substrate atom in the second layer, whereas site C represents the threefold site directly below which there is a vacancy in the second layer. Care must be exercised in the calculations of the I-V spectra of the three (1x2) domains rotated by 120° with respect to one another. In this case, the integral-order-beam intensities from different domains are averaged in the calculations, based on the assumption that all domains are sufficiently large to produce mutually incoherent diffraction. Each domain is assumed to be, on the average, larger than the coherence area of the incident beam. Due to the symmetry properties exhibited by the rotated (1x2) domains, the calculated fractional-order beams from each domain do not mix with each other, while the calculated integral-order beams are mixed as follows: $I_{10} = I'_{10} + I'_{0\bar{1}} + I'_{\bar{1}1}$, $I_{01} = I'_{01} + I'_{\bar{1}0} + I'_{1\bar{1}}$, $I_{11} = I'_{11} + I'_{2\bar{1}} + I'_{1\bar{2}} + I'_{\bar{1}\bar{1}} + I'_{\bar{2}1} + I'_{\bar{1}2}$, etc., where I and I' are calculated intensities of the beams after and before averaging, respectively.

Comparisons between two of the ten experimentally measured beams and the optimized, calculated I-V spectra are shown in Figs. 7.35 and 36. The calculated I-V spectra in these figures correspond to the optimum interlayer spacing and inner potential for each of the three adsorption sites tested for both the (2x2) and (1x2) superstructures. A visual inspection of Figs. 7.35,36 shows that site C for either the (2x2) or the (1x2) structure is the preferred adsite. One of the important and interesting features demonstrated in these figures is that the calculated I-V beam profiles for the (2x2) and (1x2) surface structures are very similar. It is impossible to determine the long-range order (as opposed to the short-range order) by a visual inspection of these figures.

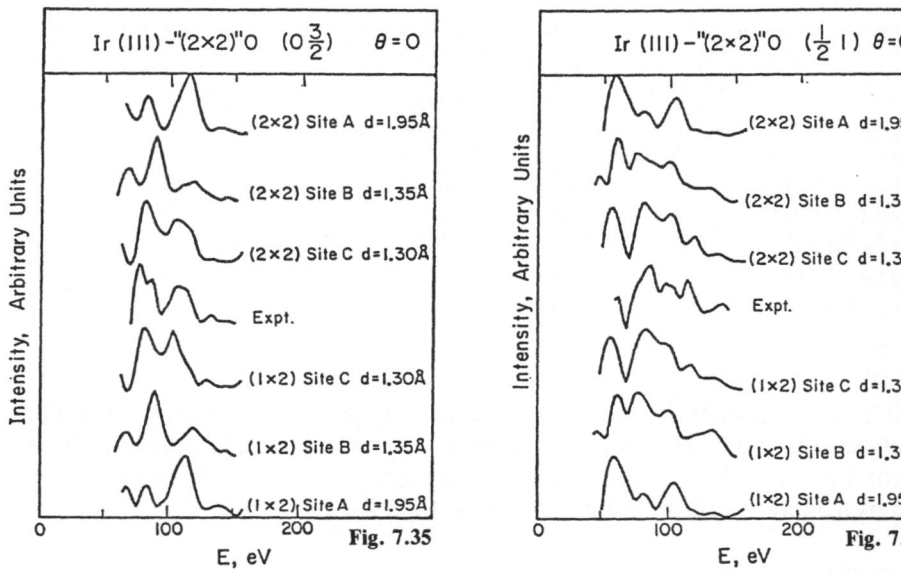

Fig. 7.35. Comparison between theoretical and experimental I-V spectra for the (0,3/2) beam for different proposed structures of Ir(111)-(2×2)-O. The theoretical I-V spectra for site A for both (2×2) and (1×2) structures are modified with an inner potential of 16 eV, and those for site C are modified with an inner potential of 12 eV. θ = 0° defines normal incidence, and d is the interlayer spacing between the oxygen overlayer and the topmost layer of iridium substrate atoms

Fig. 7.36. As in Fig. 7.35, except the (1/2,1) beam

Use of R-Factors

To quantify the comparison between experiment and theory, an R-factor was computed according to the procedure of Zanazzi and Jona [7.56]. The criterion used is that an R-factor near 0.2 indicates high reliability of the tested structure, one near 0.35 indicates a possibly correct structure, while one above approximately 0.5 indicates an incorrect structure [7.56,57]. In Table 7.6, a comparison is shown between the R-factors using ten experimentally measured beams and the optimum, calculated I-V spectra for the three different adsites with p(2x2) and (1x2) superstructures.

Based on the criteria discussed above, the R-factors of 0.22 for site C for both the p(2x2) and the (1x2) surface structures show that the oxygen adatoms occupy the threefold site on Ir(111) with a vacancy in the second substrate layer, forming either a p(2x2) superstructure or three domains of a (1x2) superstructure rotated 120° with respect to one another. The inability of the R-factor analysis to distinguish between the p(2x2) and the (1x2) structures confirms that the calculated I-V spectra for these two structures are very similar. However, it is clear from both a visual inspection and the R-

Table 7.6. Results of an R-factor analysis for Ir(111)-(2×2)-O

	V_o [eV]	d [Å]	R-factor
p(2x2) Site A	10	1.95	0.35
(1x2) Site A	10	1.95	0.34
p(2x2) Site B	16	1.35	0.35
(1x2) Site B	16	1.35	0.34
p(2x2) Site C	12	1.30	0.22
(1x2) Site C	12	1.30	0.22

factor analysis that the correct adsite for oxygen on this surface has been determined. The interlayer spacing between the oxygen adatoms and the iridium surface is 1.30 ± 0.05 Å which corresponds to a bond length between the oxygen and iridium atoms of 2.04 ± 0.08 Å

Comparison of Bond Lengths

The formation of two different types of surface structures on the Ir(111) and Ir(110)-(1x1) surfaces is analogous to two of the many types of complexes formed between iridium and oxygen {e.g. $K_{10}[Ir_3O(SO_4)_9]\cdot 3H_2O$ [7.112] and $K_6H_2[(Ir_2O)_2H_2PO_4)_{10}\cdot(OH)_8]$ [7.113]}. Examples of several inorganic complexes and their bond lengths are presented in Table 7.7. The Ir-O bond lengths in these complexes range from 1.96 to 2.06 Å supporting the LEED results for the Ir-O bond lengths of 1.93 Å and 2.04 Å for the Ir(110)-c(2x2)-O and the Ir(111)-(2x2)-O structures, respectively. Also, as expected, the Ir-O bond lengths for these surface structures compare favorably with the sum of the covalent radii of oxygen and iridium, 2.00 Å [7.118].

Further evidence concerning the covalent character of the bonding between the chemisorbed oxygen and the iridium substrate can be obtained by calculating the charge transfer from the substrate to the overlayer. The

Table 7.7. Complexes between iridium and oxygen and their Ir-O bond lengths

Complex	Ir-O bond length [Å]	Reference
$[Ir(O_2)(PMe_2Ph)_4]B\ Ph_4$ [a]	2.04	[7.114]
$[Ir(O_2)(Ph_2P\ CH_2P\ Ph_2)]ClO_4$	2.06	[7.115]
$[Ir(O_2)(Ph_2P\ CH_2P\ Ph_2)]PF_6$	2.00	[7.115]
$K_3[Ir(C_2O_4)_3]$	1.96	[7.116]
$(NH_4)_4[Ir_3N(SO_4)_6(H_2O)_3]\cdot 3H_2O$	2.005-2.059	[7.117]

[a] Me: methyl; Ph: phenyl

work-function changes $\Delta\varphi_w$ for the Ir(111)-(2x2)-O and Ir(110)-c(2x2)-O structures were measured to be 0.56 eV [7.119] and 0.30 eV [7.55], respectively. The dependence of the change of work function on the coverage is given by (7.13). Neglecting the effect of the depolarizing influence due to surrounding dipoles [the value of $K_c\alpha_p(\sigma_a\theta)^{3/2}$ is estimated to be small] and assuming that the charge is transferred perpendicular to the surface with the center of the top plane of iridium atoms as a reference point, the ionic character of the Ir-O bond is either 0.06 (for $\theta = 1/4$) or 0.03 (for $\theta = 1/2$) for the Ir(111)-(2x2)-O structure, and 0.03 for the Ir(110)-c(2x2)-O structure. These small ionicities of the Ir-O bonds further support the assertion made earlier that the Ir-O bonds of these surface structures are primarily of a covalent character.

7.3.3 The Ti(0001)-(1x1)-N Atomic Underlayer

The study of the Ti(0001)-(1x1)-N structure is unique because it represents the first known underlayer structure [7.120]. This structure is formed by an exposure of 4-5 L of nitrogen on a clean Ti(0001) surface at room temperature. The formation of this (1x1) structure is not evident by observing the LEED pattern visually because this structure has the same unit-cell dimensions as the ideally-terminated bulk structure. However, it can be detected by monitoring changes in the I-V intensities of the diffraction beams and Auger-electron spectra as a function of nitrogen coverage.

The Ti(0001) surface has a hexagonal close-packed stacking sequence ABABAB.... In the search for the correct structure, a number of adsorption sites have been included by Shih et al. [7.120]: the on-top site, the bridged site (the adsorbate lies across the bridge between any two adjacent substrate atoms), and the two types of threefold sites. In addition to these commonly used trial structures, atomic nitrogen was allowed to occupy the interstitial positions underneath the top layer of titanium atoms. There are two types of interstitial positions as shown in Fig. 7.37. In one type of interstitial position, termed the tetrahedral site, a nitrogen atom is placed inside a

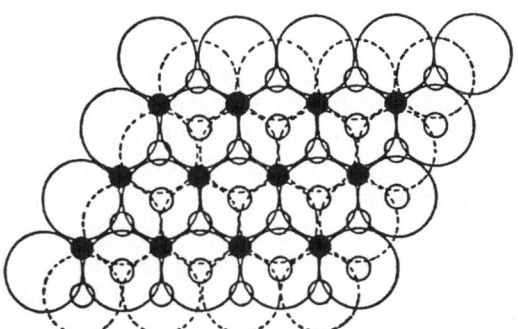

Fig. 7.37. Schematic diagram showing different interstitial positions underneath the top layer of titanium atoms of a Ti(0001) surface. The large open and dashed circles represent the top and the second layers of titanium atoms, respectively. The small open and filled circles represent the tetrahedral and octahedral interstitial sites, respectively. (Two types of tetrahedral sites are shown)

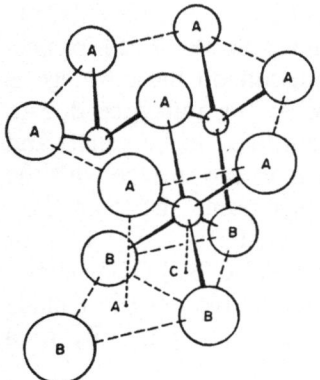

Fig. 7.38. Partial perspective view of the Ti(0001)-(1×1)-N structure, assuming octahedral interstitial sites. The top atomic plane consists of titanium atoms (*large circles*) in A-type positions, while the second titanium layer has atoms in B-type positions. The nitrogen atoms (*small circles*) occupy the octahedral interstitial holes in C-type positions

Table 7.8. Distances in Bulk TiN and Ti(0001)-(1×1)-N [7.120]

	TiN (bulk)	Ti (0001)-(1x1)-N
Ti-Ti distance [Å]	Within (111): 2.998	Within (0001): 2.950
Distance between the first and second Ti layers [Å]	Along [111]: 2.448	Along [0001]: 2.45 ± 0.05
Distance between the first Ti layer and the N plane [Å]	Along [111]: 1.224	1.22 ± 0.05
Ti-N distance [Å]	2.120	2.095

tetrahedron formed by four neighboring titanium atoms. There are two possible tetrahedral sites that must be considered. One requires the nitrogen atom to be located directly below a surface titanium atom, while the other requires the nitrogen atom to be located directly above a titanium atom in the second layer, as shown in Fig. 7.37. In the other interstitial position, the nitrogen is surrounded by six neighboring titanium atoms forming an octahedral structure around it, as shown in Fig. 7.38.

The dynamical LEED analysis indicates that good agreement between experiment and theory is achieved for the model with nitrogen atoms in the octahedral interstitial sites. It was found that nitrogen atoms are equidistant between the first and second layers of titanium atoms, and there is an expansion of 4.6% with respect to the bulk interlayer spacing between the first and second titanium layers. It is interesting that the Ti(0001)-(1x1)-N structure and the bulk structure of the compound TiN show a strong resemblance. The analogy between these two structures is summarized in Table 7.8. This work represents the first structure of this kind determined by LEED. It further enhances the credibility of LEED as a sensitive technique to determine a variety of different classes of surface structures.

7.4 Adsorbed Molecular Layers

Possibly the most important practical application of surface science is to provide an understanding of the elementary steps of heterogeneously catalyzed reactions. Although that goal is far from being achieved, the first steps in that direction are being taken. In particular, the geometrical conformation of molecules at single-crystal surfaces is being actively investigated. This represents an idealized situation for heterogeneous catalysis, but it is nevertheless of fundamental importance. In addition, interesting analogies between surfaces and either inorganic or organometallic cluster complexes add to our understanding of the conformation of molecules adsorbed on metallic surfaces, and such comparisons will be made below.

It is both remarkable and fortunate that many molecules are arranged in well-ordered arrays when adsorbed on various metal surfaces, at least under suitable conditions of temperature and coverage. In fact, the variation of temperature and coverage often generates a succession of different ordered, molecular adsorbate structures that are instructive in terms of the mechanisms of adsorption and chemical reactions. The molecules that have been studied to date in detail with LEED are carbon monoxide and small hydrocarbons.

Carbon Monoxide Adsorption

Carbon monoxide, although it is a small molecule, provides a number of interesting features that qualify it as a prototype for larger molecules. This molecule dissociates on a number of metal surfaces (especially on transition metals toward the left in the Periodic Table), leaving atomic carbon and oxygen both of which can order at suitable coverages, e.g. in a c(2x2) array on Fe(100) [7.121], as indicated in Fig. 7.39. The molecule adsorbs without dis-

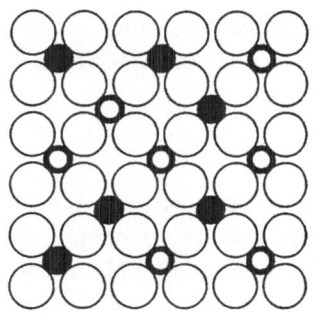

Fe (100) + c(2×2) CO dissociated

Fig. 7.39. Dissociated carbon monoxide on Fe(100) adopts a c(2×2) array, the lattice sites of which are randomly occupied by carbon or oxygen, shown as small circles in hollow sites of the substrate

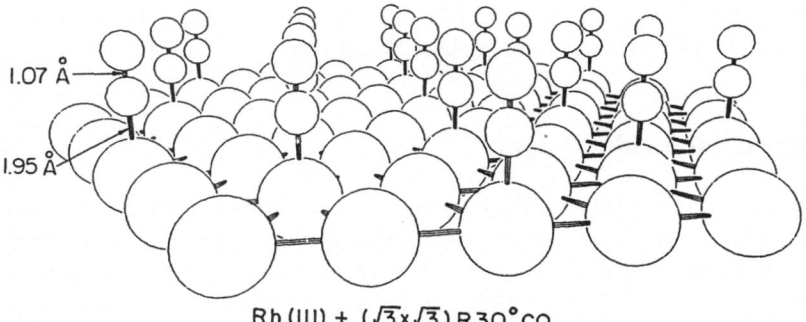

Rh (III) + (√3x√3) R 30° CO

Fig. 7.40. Perspective view (using the ORTEP plotting program) of Rh(111)-($\sqrt{3} \times \sqrt{3}$)R30°-CO. The oxygen ends of the CO molecules point away from the substrate. The sphere radii have no physical significance

sociation on other metal surfaces (further to the right in the Periodic Table), normally binding through its carbon end to the metal substrate. However, unlike the case for most atomic adsorbates, the adsorption site of CO varies from metal to metal, and it can also vary on a given metal as a function of coverage. For example, on Rh(111) at a coverage of approximately one-third of a monolayer, a ($\sqrt{3} \times \sqrt{3}$)R30° structure occurs, in which CO is adsorbed in the on-top site (this is so-called linear bonding since one metal atom, the carbon atom, and the oxygen atom are aligned perpendicular to the surface), as shown in Fig. 7.40 [7.122]. At higher coverages, the ($\sqrt{3} \times \sqrt{3}$)R30° pattern disappears and a (2x2) pattern forms at a fractional coverage of approximately three-quarters. This (2x2) pattern corresponds to a structure in which some CO molecules reside near on-top sites, while others reside at bridge sites, bonded to two rhodium atoms [7.123].

The approximate bonding site of CO molecules may be estimated in many cases from vibration frequencies as measured with Infrared Spectroscopy [7.124] or High-Resolution Electron Energy Loss Spectroscopy [7.125]. However, structural determinations with LEED have been used to confirm the validity of a number of postulated adsites based on the observed CO stretching frequency. In addition, the LEED results have allowed a determination of both bond lengths and bond angles. The results obtained to date by LEED are listed in Table 7.9 where they are compared to similar results known for cluster compounds. The cases of CO on Ni(100) and on Pd(100) will be discussed in greater detail in Sects. 7.4.1 and 2.

Hydrocarbon Adsorption

Small hydrocarbon molecules also provide a variety of features that should be helpful in understanding surface-molecule interactions more generally. For example, it was found that acetylene (C_2H_2), adsorbed on Pt(111) below room temperature, can remain intact while bonding with its axis parallel to

Table 7.9. Molecular CO adsorption structures

Situation	Adsorption Site	Metal-C distance [Å]	C-O distance [Å]	References
Ni(100)-c(2x2)-CO	top	1.72±0.1	1.15±0.1	[7.126]
		1.8±0.1	1.10±0.1	[7.127,128]
		1.7±0.1	1.13±0.1	[7.129]
Ni Clusters	variable	1.75-0.89 (top)	1.04-1.16 (top)	[7.130]
Cu(100)-c(2x2)-CO	top	1.90±0.1	1.15±0.1	[7.127,128]
Cu Clusters	variable	1.79-1.93 (top)^a	1.04-1.16 (top)^a	[7.130]
Pd(100)-($2\sqrt{2} \times \sqrt{2}$)R45°-2CO	bridge	1.93±0.07	1.15±0.1	[7.131]
Pd Clusters	variable	1.85-1.94 (bridge)^b	1.09-1.17 (bridge)^b	[7.130]
Ru(0001)-($\sqrt{3} \times \sqrt{3}$)R30°-CO	top	1.90±0.05	1.16±0.03	[7.132]
Ru Clusters	variable	1.86-1.90 (top)	1.14-1.16 (top)	[7.130]
Ru(111)-($\sqrt{3} \times \sqrt{3}$)R30°-2CO	top	1.95±0.1	1.07±0.1	[7.122]
Rh(111)-(2x2)-3CO	top and bridge	1.94±0.1 2.03±0.07	1.15±0.1 1.15±0.1	[7.123]
Rh Clusters	variable	1.82-1.91 2.00-2.09 (bridge)	1.09-1.17 1.14-1.17 (bridge)	[7.130]

^{a} CuCO distances are extrapolated from RhCO clusters by increasing the Ni radius following bulk values

^{b} CuCO distances are extrapolated from RhCO clusters by increasing the Rh radius following bulk values

the metal surface [7.133]. In addition, at a quarter-monolayer coverage this molecule forms a (2x2) lattice. However, in the presence of sufficient hydrogen and above room temperature, the C_2H_2 molecule transforms to a new species that is now generally agreed to be ethylidyne ($\equiv$C-CH$_3$) in which the C-C axis is perpendicular to the surface, the lower carbon atom bonding strongly to three equidistant metal atoms [7.134], as indicated in Fig. 7.41. Interestingly, ethylene (C_2H_4), when adsorbed on the same clean Pt(111) surface, can again produce a parallel-bonded intact species at lower temperatures, which above room temperature converts to the same ethylidyne ($\equiv$C-CH$_3$) species produced from acetylene, with the loss of one hydrogen atom per molecule.

Parallel behavior, although occurring at lower temperatures, is believed to take place for the same molecules adsorbed on Rh(111) [7.135]. However, by contrast, it is known (from vibrational electron energy loss studies) that

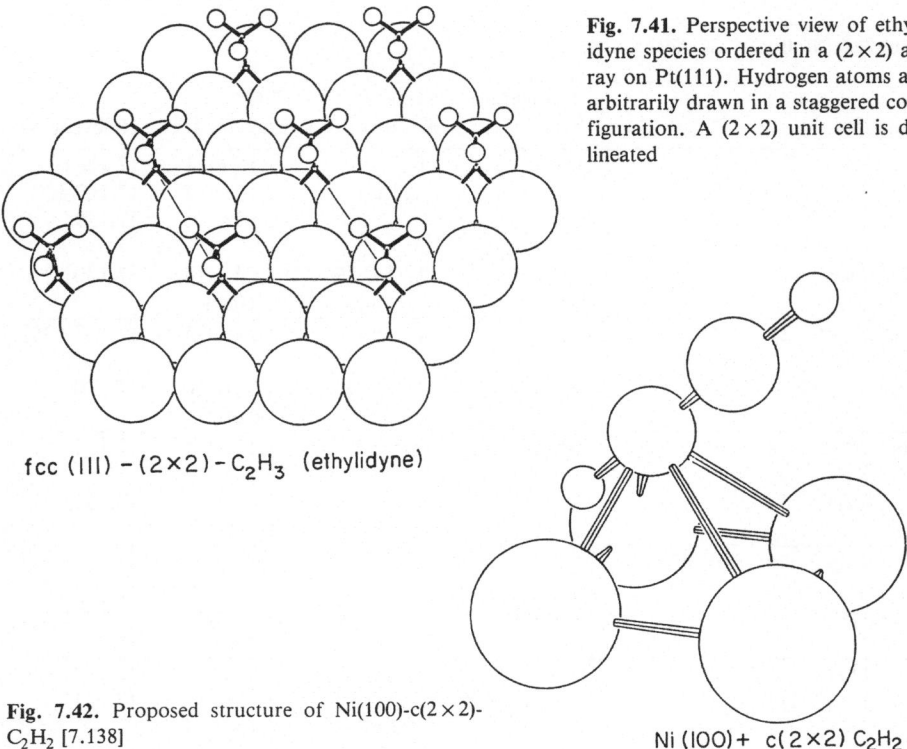

Fig. 7.41. Perspective view of ethylidyne species ordered in a (2×2) array on Pt(111). Hydrogen atoms are arbitrarily drawn in a staggered configuration. A (2×2) unit cell is delineated

fcc (111) – (2×2) – C_2H_3 (ethylidyne)

Fig. 7.42. Proposed structure of Ni(100)-c(2×2)-C_2H_2 [7.138]

Ni (100)+ c(2×2) C_2H_2

these molecules will cleave their C-C bonds on Ni(111) [7.136] and Ru(0001) [7.137] prior to C-H bond cleavage. Such differences are obviously of great importance in understanding catalytic reactions of hydrocarbons.

A full LEED study on a nickel surface was performed for the system Ni(100)-c(2x2)-C_2H_2, which occurs prior to dissociation of the acetylene molecules [7.138]. It was found that the molecules are intact and centered over hollow sites, but with the C-C axis tilted 50° with respect to the surface normal in the [011] direction, as illustrated in Fig. 7.42.

The studies of the adsorption of unsaturated C_2 hydrocarbons on Pt(111) and Rh(111) [hydrocarbons saturated with hydrogen, such as ethane (C_2H_6) are adsorbed considerably less strongly on Pt(111) and Rh(111) and desorb molecularly rather than dissociate as the temperature is increased] have been extended to unsaturated C_3 and C_4 hydrocarbons [7.139,140]. These are found to behave in a very similar way to the C_2 hydrocarbons, with additional effects due to their larger size. For example, the C_3 equivalent of ethylidyne, propylidyne ($\equiv$C-CH$_2$-CH$_3$), has an additional methyl group that on Rh(111) can interact with neighboring methyl groups to form a $(2\sqrt{3}x2\sqrt{3})R30°$ lattice in addition to the (2x2) lattice of the ethylidyne-like base of the molecule [7.140]. This is shown explicitly in Fig. 7.43.

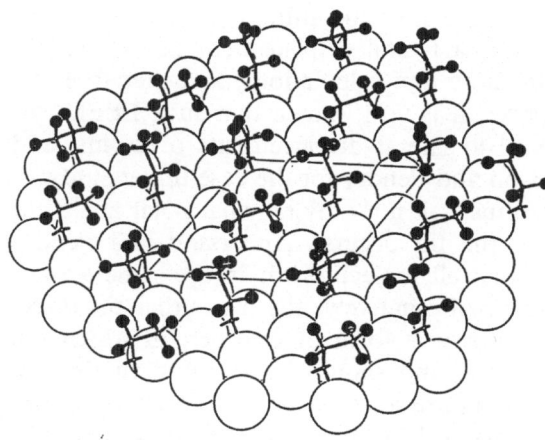

Fig. 7.43. A possible structure of propylidyne ordered on Rh(111). The topmost carbon atoms (and their associated hydrogen atoms) form a $(2\sqrt{3} \times 2\sqrt{3})R30°$ lattice, one unit cell of which is delineated, while the ethylidyne-like bases ($\equiv C\text{-}CH_2\text{-}$) form the (2×2) lattice known for ethylidyne ($\equiv C\text{-}CH_3$)

$fcc(lll) - (2\sqrt{3} \times 2\sqrt{3})R30° - C_3H_5$ (propylidyne)

7.4.1 The Ni(100)-c(2x2)-CO Molecular Overlayer

A half-monolayer of CO on Ni(100) produces a well-defined c(2x2) pattern. This surface was one of the first to be chosen for structural determination of a molecular overlayer by LEED, and it was to teach several practical lessons of importance in subsequent work concerning molecular adsorbates. From vibrational electron energy loss measurements [7.141], it could be expected that CO is adsorbed in the on-top site on this surface, but a frequently asked and unresolved question was whether the molecule was oriented perpendicular to the surface or not.

CO Orientation

The first full LEED study of Ni(100)-c(2x2)-CO was performed by Andersson and Pendry [7.142] using a limited set of experimental data, namely normal-incidence I-V curves for three integral-order beams and for one half-order beam over a range of energies from approximately 10 eV to approximately 115 eV. (Here, as in many other cases, the integral-order beams are less sensitive to the adsorbate structure than the fractional-order beams.) Initially, the C-O layer spacing and the metal-carbon layer spacing were varied, with the conclusion that these spacings are 0.95 $\pm$ 0.1 Å and 1.8 $\pm$ 0.1 Å, respectively.

Given that the C-O bond length could not be very different from the free-molecule value of 1.15 Å (since the C-O bond strength is much larger than the metal-carbon bond strength, and based on a multitude of X-ray Diffraction results for carbonyl cluster compounds), the C-O layer spacing of 0.95 Å would have implied a tilting of the C-O axis by approximately 30° from the surface normal. (Note, however, from Table 7.9 that C-O bond lengths slightly smaller than 1.15 Å are not uncommon in clusters.) One may argue that a tilt would involve a lateral displacement of the oxygen atoms, which was not included in those first calculations, and therefore this conclusion is unwarranted. However, it is well known that such LEED studies have relatively little sensitivity to lateral displacements of atoms, and this was in fact verified by Andersson and Pendry for the case under study.

If a CO molecule is tilted, one must ask not only by what polar angle but also in which azimuthal direction the tilt occurs. Andersson and Pendry assumed in their test calculations that all molecules tilt in the same azimuthal direction (within one domain). Alternatively, the azimuthal tilt direction could be random. However, this is a form of disorder, in which the oxygen atoms have no ordered lattice, and should be manifest in the I-V spectra as a more or less complete disappearance of the oxygen contribution. It can be verified easily, by making a LEED calculation for a Ni(100)-c(2x2)-C surface where the oxygen atoms have been omitted, that this form of disorder is not present to an appreciable extent [7.143]. Thus, all the CO molecules would have to be tilted essentially in the same direction (at least within each domain).

Since the question of the CO tilt was hotly debated, Andersson and Pendry varied many theoretical parameters in their analysis (including later, unpublished work), especially the atomic scattering potentials. However, the tilt angle of 30° remained.

This LEED study was followed by an angle-resolved photoemission investigation of CO adsorbed on the same surface [7.144], based on a three-atom-cluster calculation [7.[145] that predicted the angular distribution of the photoemitted-electron intensity. A peak in photoemitted intensity in the normal direction was interpreted to imply nontilted CO molecules. Another ARUPS study, using a complete two-dimensional surface structure and a more-complete LEED-like theory, favored the on-top site, but left the orientation of the CO molecules uncertain [7.145].

New Experiments and Calculations

Several new LEED studies of the Ni(100)-c(2x2)-CO system were performed soon after these results. In particular, they were spurred by evidence that the original LEED data were unsatisfactory, as determined with a new fast LEED measurement technique based on recording the image on the LEED screen through a vidicon camera [7.147]. It appeared that the CO overlayer deteriorated more easily and quickly under the influence of the incident LEED beam than initially believed, through disordering, through desorption, through molecular dissociation or some combination of all these effects. Apart from these vidicon data, two other new sets of data were obtained [7.126-128] with particular attention to the surface damage problem and with an effort to collect a larger data base. Three independent sets of LEED

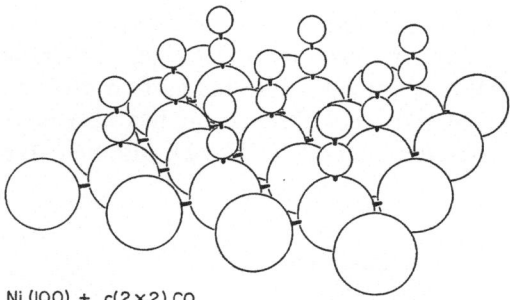

Fig. 7.44. Structure of Ni(100)-c(2×2)-CO [7.126 – 129]

Ni (100) + c(2×2) CO

calculations were performed with independently chosen nonstructural param-eters [7.126-129], which all favored the absence of tilt of the CO molecules, in the sense that the best CO layer spacing was compatible with a CO bond length of approximately 1.15 Å The results of these studies are summarized in Table 7.9 and illustrated in Fig. 7.44.

Therefore, at present there is agreement on the geometry of a half-monolayer of CO adsorbed on Ni(100), with the following lessons for LEED resulting from this determination. First, a sufficiently large data base can be critical in a structural determination. Second, some molecules disorder, desorb, and/or dissociate readily under the influence of the incident electron beam. The damage can be reduced by faster measurement techniques or by a reduction of the incident flux coupled with the insertion of a micro-channelplate, or by reduced exposure to the electron beam, for example by turning the beam off when not needed, or by exposing different parts of the surface in turn. Third, it was noticed that to some extent there is unusual uncertainty in the determination of the distance of the carbon atom from the surface, as if the carbon atom were shadowed by the oxygen atom above it. This shows up clearly as an elongated valley centered on the optimal result in an R-factor plot, where the metal-carbon and carbon-oxygen layer spac-ings are both varied (with constant inner potential), as shown in Fig. 7.45

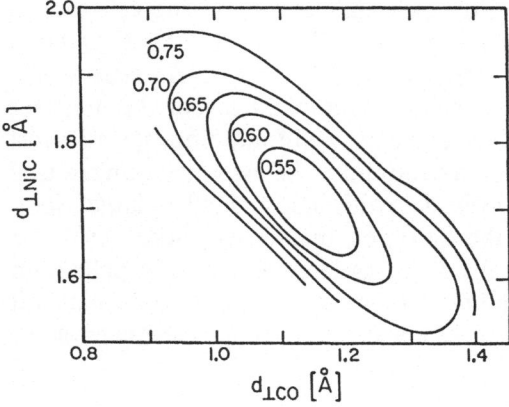

Fig. 7.45. Pendry R-factor contour plot for Ni(100)-c(2×2)-CO [17.128]

[7.128]. The elongation occurs in the direction along which the metal-oxygen distance is constant but the carbon position is variable, indicating that the metal-oxygen distance is well determined, while the carbon position is relatively uncertain. It is significant that this observation has also been made for all other CO adsorption systems studied by LEED, although the exact reason for it remains unclear.

7.4.2 The Pd(100)-(2√2 × √2)R45°-2CO Molecular Overlayer

The ordered structure that carbon monoxide forms at a half-monolayer coverage on Pd(100) is of particular interest from the point of view of structural determination by LEED. We refer to Sect. 3.5.5 for a detailed discussion of the interpretation of the observed (2√2 × √2)R45° pattern and here only summarize the situation and the arguments.

Historical Development

This system was first explored thoroughly by Park and Madden [7.148], who noticed unusual symmetry properties of the LEED pattern, namely the systematic absence of certain spots at all energies when normal incidence was chosen. They recognized that glide-plane symmetry was implied and that this severely restricted the possible surface structures, i.e. two bridge sites per unit cell were necessary to satisfy the glide-plane symmetry, as shown in Sect. 3.5.5. Moreover, unless the CO axes were quite randomly tilted, the CO molecules would most likely be perpendicular to the surface (apart from small thermal vibrations). The bridge sites were in fact later confirmed by vibrational loss measurements [7.148,131]. The major remaining task for a complete LEED analysis was to confirm these deductions and determine the bond lengths. Another need for this LEED analysis arose out of the controversy concerning the tilt angle of CO on Ni(100), which was in progress at the time the CO/Pd(100) work was performed.

The full LEED determination for Pd(100)-(2√2 × √2)R45°-2CO did indeed confirm the bridge sites and the perpendicularity of the CO axis to the surface [7.131], as illustrated in Fig. 7.46. The Pd-C and C-O bond lengths were found to be 1.93 ± 0.07 Å and 1.15 ± 0.1 Å respectively. Compared to corresponding distances in transition-metal carbonyl complexes (cf. Table 7.9), these results are rather satisfactory. In addition, the topmost substrate interlayer spacing was varied and found to be essentially unchanged from the bulk value. It was also checked in this study whether large vibrational amplitudes were present in the adsorbed molecules. Some evidence for oxygen vibrational amplitudes only twice as large as the bulk palladium amplitudes was noted, corresponding to thermally-induced molecular tilt angles of up to approximately 15°. Hence these molecules cannot be regarded as being loosely oriented.

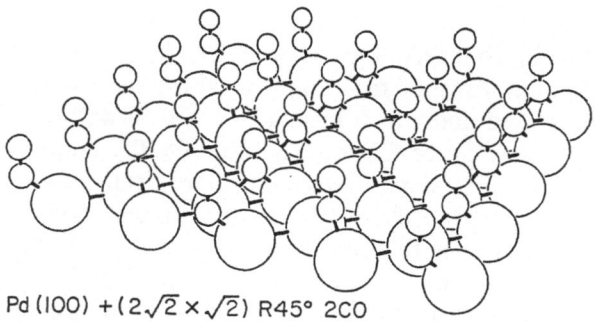

Fig. 7.46. Structure of Pd(100)-$(2\sqrt{2} \times \sqrt{2})$R45°-2 CO [7.131]

Pd (100) + $(2\sqrt{2} \times \sqrt{2})$ R45° 2CO

Comparison Between c(2x2) and $(2\sqrt{2}$ x $\sqrt{2})$R45° Arrangements

It is interesting to compare the c(2x2)-CO structure on Ni(100) with the $(2\sqrt{2} \times \sqrt{2})$R45°-2CO structure on Pd(100), since they have the same coverage but different adsorption sites and different superlattices. We may assume the basic preference of CO for on-top sites on Ni(100) and for bridge sites on Pd(100) at low coverages, with the CO axis perpendicular to the surface. Let us then ask how a half-monolayer of CO molecules can be arranged on these surfaces, assuming that CO-CO intermolecular distances prefer to be at least 3.3 Åwhen possible. Imagine a CO molecule at an on-top site on Ni(100), as shown in the left panel of Fig. 7.47, surrounded by an excluded area the radius of which is approximately 3.3 Å The next available on-top sites are those that generate a c(2x2) array. No other arrangement of on-top sites will maintain a fractional coverage of one-half.

For Pd(100) the first CO molecule is situated in a bridge position (middle panel of Fig. 7.47). The nearest bridge sites outside the excluded area are again those that belong to a c(2x2) lattice. However, there is another choice of bridge sites, as shown in the right panel of Fig. 7.47, which is obtained

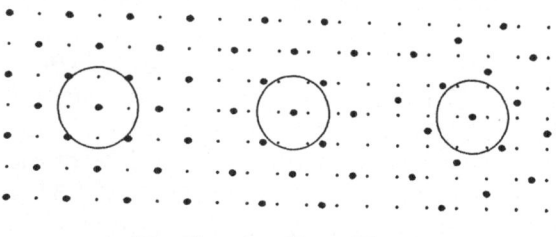

c(2×2) vs $(2\sqrt{2} \times \sqrt{2})$ R45°

Fig. 7.47. Comparison of c(2×2) and $(2\sqrt{2} \times \sqrt{2})$R45° lattices for CO adsorption. Small and large dots represent metal atoms and CO molecules, respectively. (*Left*) Ni(100)-c(2×2)-CO. (*Middle*) imaginary Pd(100)-c(2×2)CO. (*Right*) Pd(100)-$(2\sqrt{2} \times \sqrt{2})$R45°-2CO. Large circles have a radius of 3.3 Å, scaled to the Ni(100) and Pd(100) lattices, representing an excluded area, i.e., the approximate distance of closest possible approach between CO molecules

from the c(2x2) array by sliding alternate rows of molecules parallel to themselves, so that the coverage is not changed. One sees in Fig. 7.47 that this choice results in a quasi-hexagonal array of molecules in which the average intermolecular distance has increased and which has a $(2\sqrt{2} \times \sqrt{2})R45°$ coincidence unit cell with the substrate. Since CO on Pd(100) actually chooses this quasi-hexagonal structure with its larger intermolecular distances, we must conclude that the c(2x2) structure is relatively unfavorable due to repulsive intermolecular interactions over the distances involved, namely the diagonal of the (1x1) square unit cell, which is 3.88 Å for Pd(100) and 3.52 Å for Ni(100) (the larger CO-CO distance in the quasi-hexagonal structure on Pd is 4.34 Å). For comparison, one of the most closely-packed regular CO overlayer structures, that of Rh(111)-(2x2)-3CO, has intermolecular distances of less than 3.1 Å [7.123]. At antiphase-domain boundaries even smaller intermolecular distances may occur [7.149].

7.4.3 Molecular Overlayers of C_2H_2 and C_2H_4 on Pt(111) and Rh(111)

The adsorption of acetylene on Pt(111) was the first molecular structure to be considered for a structural determination by LEED, by Kesmodel et al. [7.133]. It was discovered through the observation of LEED I-V curves that this structure can change drastically from a "metastable" to a "stable" state, both of which show the same (2x2) pattern. The difference was confirmed by UPS measurements [7.150] and later by HREELS data [7.151]. It was also found that ethylene produces both a metastable structure, different from the metastable acetylene structure, and a stable (2x2) structure that is indistinguishable from the stable acetylene structure.

"Metastable" Acetylene Structure

A LEED structural analysis of the (2x2) metastable acetylene species produced a parallel-bonded molecule with both carbon atoms equidistant (2.5 Å) from one platinum atom, and the C-C midpoint sitting near an on-top site, namely 0.25 Å away from an on-top site in the direction of a cABCABC... hollow site, as illustrated in Fig. 7.48 [7.133]. This bonding arrangement is described as π-bonding, since the molecular π-orbital is most involved in the bonding. The C-C bond length is somewhat uncertain and could be of the order of the triple-bond length of 1.20 Å or the double-bond length of 1.33 Å. The hydrogen atoms contribute very little to the I-V curves due to a weak atomic scattering potential, and, consequently, their positions are unknown from LEED (although their presence is established by UPS and HREELS). Subsequent HREELS [7.153] and UPS [7.154,155] interpretations favor partly di-σ, partly π-bonding, as shown in Fig. 7.49, rather than the π-bonding of Fig. 7.48 (the σ bonds are concentrated between individual carbon atoms and the metal substrate). This discrepancy remains to be resolved.

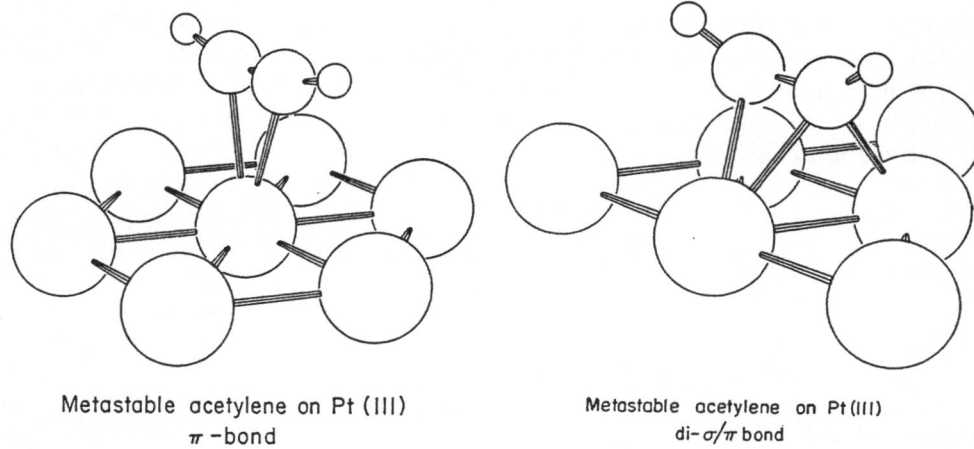

Metastable acetylene on Pt (III)
π -bond

Metastable acetylene on Pt (III)
di-σ/π bond

Fig. 7.48. Proposed structure of Pt(111)-(2 × 2)-C₂H₂, based on LEED [7.133]

Fig. 7.49. Proposed structure of Pt(111)-(2 × 2)-C₂H₂, based on HREELS and UPS [7.153 – 155]

The undetectability of hydrogen is a feature that deserves some comments. First, if hydrogen by itself produces a superlattice as for Ni(111)-(2x2)-2H [7.101,102], shown schematically in Fig. 7.20a, the fractional-order beams, though weak, contain sufficient information to determine the hydrogen positions. However, in the present case, and in all cases involving hydrocarbons, the carbon atoms also contribute to the fractional-order beams and overshadow the contributions from the hydrogen atoms. Second, although one loses valuable information by not being able to determine hydrogen positions, there is a bonus that may outweigh that drawback. There are far fewer structural parameters to fit if the hydrogen atoms can be ignored, and the calculation is appreciably simplified by the neglect of hydrogen, making the "partial" structural determination disproportionately more economical. Often the approximate hydrogen positions in hydrocarbons can be easily guessed from the carbon positions, given C-C bond lengths (which determine the C-C bond orders) and C-C-C or other bond angles determined by LEED. It should be noted that in X-ray Diffraction, hydrogen is also usually undetectable.

Ethylidyne

The structure of the "stable" acetylene (or ethylene) structure on Pt(111) was first analyzed by assuming a parallel-bonded surface species [7.133]. A partly satisfactory geometry was found that involved a different site from that for the "metastable" structure. A subsequent analysis [7.134] extended the search with additional experimental data (in particular, data at larger angles of incidence, θ ~ 30°) to geometries in which the C-C axis is approximately perpendicular to the surface. Most calculations in this work assumed a perpendicular C-C axis, since, as we discussed in the case of Ni(100)-c(2x2)-CO, lateral displacements of atoms have less influence on I-V curves than perpendicular ones. Consequently, the presence of a tilted molecular axis could be detected indirectly in this way, too.

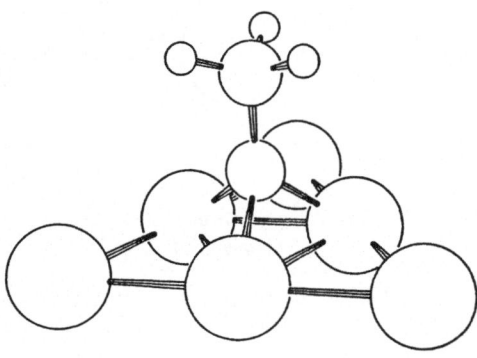

Ethylidyne on Pt (III) and Rh (III)

Fig. 7.50. Detail of ethylidyne species on Pt(111) or Rh(111). The staggered configuration of the hydrogen atoms is arbitrarily chosen for illustration purposes

An improved agreement between experiment and theory was obtained with a metal-carbon layer spacing of 1.20 Å and a C-C spacing of 1.50 Å with the carbon that is bonded to the metal in a cABCABC... threefold coordinated hollow site. The 1.50 Å C-C spacing is compatible with the C-C single-bond length (1.51-1.54 Å in various molecules) for a C-C axis perpendicular to the surface, and not compatible with double- or triple-bond lengths (approximately 1.33 Å and approximately 1.20 Å, respectively) with any tilt angle. By applying the tetrahedral character of sp^3-hybridized carbon to complete the hydrogen complement of the carbon skeleton, one concludes that an ethylidyne species ($\equiv$C-CH$_3$) has formed, as illustrated in Fig. 7.50.

Table 7.10. The structures of ethylidyne on surfaces and in complexes, compared with gas phase C_2H_n

Species	Metal-carbon bond length [Å]	C-C bond length [Å]	Ref.
Pt(111)-(2x2)-CCH$_3$	2.00	1.50	[7.134]
Rh(111)-(2x2)-CCH$_3$	2.03 ± 0.07	1.45 ± 0.10	[7.156]
Co$_3$(CO)$_9$CCH$_3$	1.90 ± 0.02	1.53 ± 0.03	[7.157]
H$_3$Ru$_3$(CO)$_9$CCH$_3$	2.08 ± 0.01	1.51 ± 0.02	[7.158]
H$_3$Os$_3$(CO)$_9$CCH$_3$	2.08 ± 0.01	1.51 ± 0.02	[7.159]
HC≡CH	-	1.20	
H$_2$C=CH$_2$	-	1.33	
H$_3$C-CH$_3$	-	1.54	

In Table 7.10, this structure is compared with trinuclear complexes that contain an ethylidyne ligand, and with an equivalent ethylidyne species found later on the Rh(111) surface [7.156], as well as with C_2 hydrocarbons. Concerning the species on Rh(111), it is curious to note that it was found to reside not in the cABCABC... site, as on Pt(111), but in the other hollow site, bABCABC... .

Initially, the ethylidyne result was somewhat controversial. An ethylidene species (=CH-CH$_3$) was proposed based on HREELS data [7.152], or a vinyl-like species (-CH=CH$_2$) based on UPS data [7.160]. Recently, the issue was settled [7.125] in favor of the ethylidyne species by a reinterpretation of the HREELS data, following a normal-mode analysis [7.161] of the infrared spectrum for the ethylidyne-containing cluster $Co_3(CO)_9(CCH_3)$.

8. Two-Dimensional Order-Disorder Phase Transitions

Thus far, the application of LEED to extract both the symmetry of the two-dimensional unit cell of the surface layer (or overlayer) as well as the basis of this unit cell (i.e. a structural analysis) has been emphasized. It is also possible using elastic LEED to examine order-disorder phase transitions at surfaces, to monitor chemical reactions occurring at surfaces, and to study the formation of ordered domains or "islands" at low coverage. This chapter is concerned with the first topic, while the latter two topics are treated in Chaps. 9 and 10, respectively. In particular, order-disorder phase transitions in chemisorbed overlayers will be considered explicitly here due to their theoretical and practical importance.

However, much fundamental and elegant work has concerned phase transitions in physically adsorbed overlayers as well [8.1], which we shall summarize briefly here. Physical adsorption studies have mainly involved rare gas atoms adsorbed on the graphite basal plane (0001). However, molecules like O_2 and N_2 have been studied on graphite, and the adsorption of rare gases has been investigated on metal surfaces. The principal characteristic of physical adsorption at monolayer coverages is the frequent occurrence of incommensurate lattices (see Sects. 3.5.6 and 6.3.4). Occasionally, commensurate coincidence lattices such as $(\sqrt{3} \times \sqrt{3})R30°$ appear at a suitable coverage. With varying coverage, these systems can exhibit interesting commensurate-incommensurate phase transitions that involve small changes in the lattice parameter and the orientation of the overlayer. In addition, with physically adsorbed molecules (e.g. O_2 and N_2 on graphite), "herringbone" structures may be observed when the molecular axes are oriented at sufficiently low temperatures. Slight heating can destroy the orientational order of these molecular axes without affecting the long-range order of the molecular centers. In all cases of physical adsorption, heating to higher temperatures leads to disordered structures, except when desorption occurs prior to disordering.

8.1 Introduction to Order-Disorder Phase Transitions at Surfaces

8.1.1 Chemisorption and Ordering Principles

Adatom-Adatom Interactions and Ordering

When an atom or molecule adsorbs chemically on a surface, there is generally one location with respect to the surface atoms (i.e. bridge sites, on-top sites, etc.) for which the binding energy is strongest. Thus, a surface of a single crystal represents a periodic array of binding sites for the adsorbed species. It is frequently observed using LEED that atoms or molecules adsorb in ordered overlayers with a periodicity related to, but usually greater than, that of the surface atoms. The fact that not all the available binding sites are occupied indicates that there must be interactions between adspecies that dictate the positions of the adsorbed species with respect to one another. The symmetry of the structure of the overlayer is determined by the qualitative nature of the interactions. A wide variety of types of overlayers have been observed with LEED, and it is often necessary to postulate the existence of interactions that are anisotropic, even within the surface plane, and oscillatory (i.e. changing in sign with distance from a particular adsite) to explain the observations. Such interactions arise as a result of overlap of the oscillatory electron wavefunction in the metal induced by the adsorbed species [8.2,3].

From the preceding discussion, it is apparent that chemically adsorbed overlayers represent a physical realization of the two-dimensional lattice gas that has been widely used as a model in studies of the theory of phase transitions [8.4]. It is surprising that in spite of intense theoretical interest in two-dimensional order-disorder transitions, the first experimental evidence for such a transition, reported by Davisson and Germer in 1927 for the (2x2) structure of oxygen on Ni(111), was largely overlooked for over forty years [8.5]. Order-disorder behavior in chemically adsorbed overlayers was rediscovered by Buchholz and Lagally for the case of oxygen on W(110) in 1975 [8.6]. Since that time there has been a slowly increasing number of experimental reports of order-disorder phenomena in chemically adsorbed systems.

The Ising Model

The theory of two-dimensional order-disorder phenomena has been developed extensively. Until recently, most of the literature has been couched in terms of the Ising model, i.e. spins on a lattice, rather in terms of a lattice gas. As discussed in Sect. 8.3, since the transformation from the spin lattice to lattice gas is well understood, the theoretical results are readily applicable to lattice-gas systems. However, the theory of phase transitions, even for the idealized lattice systems, is quite complicated, therefore, anaytical results have been obtained for only a few very simple systems. These systems include many lattice geometries but are rather limited in the range of interactions between neighboring sites. This is unfortunate since chemically adsorbed overlayers frequently display rather complex structures that must be the result of rather complex sets of interaction energies. In addition, analytical results have been obtained only for spin lattices in zero magnetic field. The analogy between spin lattices and lattice gases is such that a spin lattice in zero magnetic field corresponds to a lattice gas at half-monolayer coverage. Therefore, comparison with theory is limited to only one coverage, whereas experimental phase diagrams of transition temperature as a function of coverage are rich with information concerning the adatom-adatom interactions.

Monte Carlo Simulations

To circumvent the limited availability of analytical results, various approximate methods have been developed. The most successful, in terms of calculating phase diagrams for lattice gases, have been Monte Carlo simulations [8.7], renormalization-group theory [8.8] and finite-size scaling transfer-matrix methods [8.9]. In Monte Carlo simulations, a model of the system of interest is created. Then the system is repeatedly perturbed from its original configuration by changes in the positions of individual atoms in the model, as governed by Boltzmann probabilities, i.e. in a way that simulates real-time changes at a surface. When changes in atomic positions no longer cause significant changes in the thermodynamic quantities of the system, equilibrium has been established, and the properties of interest can be calculated from the known configuration of the atoms. The Monte Carlo technique has the advantage that it is conceptually simple and that it generates physical configurations of the system which can provide insight into the observed types of behavior. The disadvantages of this technique are that it is necessary to use a finite-sized lattice, and that it is extremely demanding in terms of computer time; the latter being more serious than the former.

Renormalization-Group Theory

Renormalization-Group Theory is based on the idea of using a change of scale to reduce the range of correlations in a system. In this way, a difficult many-body problem can be reduced to a tractable few-body problem. Once a solution is found, the scaling transformations are reversed to obtain the solution for the original problem. The scaling hypothesis, if correct anywhere [8.10], is only strictly correct near a critical point. However, Renormalization-Group Theory has been used to calculate phase diagrams even far from a critical point that are in reasonable agreement with diagrams calculated by other methods. Although it is an approximate technique,

Renormalization-Group Theory has the benefit that once the transformations have been formulated, phase diagrams can be calculated rather easily. This allows the effects of changing parameters in the model system to be determined readily, in contrast to Monte Carlo simulations. The related technique of transfer-matrix scaling is described below and discussed in more detail in Sect. 8.4.

Phase Transitions and LEED

It is immediately apparent that LEED, which measures the degree of order in the overlayer, is suitable for monitoring order-disorder transitions. The transitions which have been measured using LEED to date have fallen into two categories. The first category is the second-order phase transitions that occur with increasing temperature at or near the optimum coverage for the overlayer structure [8.11], which is characterized by fractional-order beams. At these coverages, the profile of a fractional-order LEED beam is the sum of a delta function due to the long-range order, and a Lorentzian or similar function due to the short-range order. The long-range order, and thus the intensity due to the long-range order, drops abruptly to zero at the transition temperature, also called the critical temperature T_c. The correlation length, which defines the short-range order, diverges at T_c. The contribution of short-range order to the intensity is therefore a maximum, and the width of the profile is a minimum at T_c. These effects will be convoluted with the instrumental broadening, which must be taken into account before a quantitative analysis of the diffracted-beam profiles can be made.

At lower coverages, first-order transitions from a two-phase coexistence regime to a one-phase regime have been observed using LEED [8.12]. The ordered phase in the two-phase region consists of many small, ordered clusters called islands. As the temperature is increased, the density of the disordered phase increases at the expense of the islands. This leads to a gradually decreasing intensity and increasing width of the fractional-order LEED beams. The two-phase region disappears entirely at the transition temperature, T_c. At this point, there will still be some short-range order in the homogeneous, disordered phase. However, the contribution of this to the LEED intensity will be so small that the transition temperature can be defined adequately as the point at which the LEED beam profile becomes indistinguishable from the background.

8.1.2 Universality, Nonuniversality, Critical Exponents, and Scaling

Statistical fluctuations are more important for two-dimensional compared to three-dimensional systems, and this has a profound ramification within the present context. True long-range two-dimensional order cannot persist at a finite (nonzero) temperature for an isotropic multicomponent order parameter. An example is a lattice gas with a superposed adsorbate density wave of the form $|\zeta(r)|^2$ with

$$\zeta(r) = A \exp[-i(\vec{k}\cdot\vec{r} + \varphi)] \,, \qquad (8.1)$$

where the wavevector $\vec{k}$ is incommensurate with the substrate lattice, and the two components of the order parameter are the amplitude A and the phase φ. At finite temperatures, fluctuations destroy the order, leading to a power law decay of the correlation function of the order parameter. The Fourier transform of the correlation function (the structure factor) then displays power law rather than delta-function singularities at the positions of the Bragg peaks.

In two-dimensional systems, Renormalization-Group Theory predicts that two-component order parameters with cubic anisotropy form a separate universality class of critical phenomena with nonuniversal critical exponents [8.13]. These critical exponents are defined by the temperature variation near the critical temperature, T_c. In particular, the specific heat is proportional to $|T - T_c|^{-\alpha}$, the order parameter is proportional to $|T - T_c|^{\beta}$, the susceptibility is proportional to $|T - T_c|^{-\gamma}$, and the correlation length of the fluctuations of the order parameter is proportional to $|T - T_c|^{-\nu}$ [8.14-16]. The nonuniversality of the critical exponents implies that they depend on the fractional surface coverage θ of the adlayer, for example. This contrasts with the Ising model, the critical exponents of which do not depend on coverage due to its universality class [8.14]. (The critical exponents of the Ising model are given in Table 8.1.) Both of the aforementioned universality classes are possible on a square lattice. For example, the former class applies to a (2x1) superstructure which is known as an XY model with twofold anisotropy, whereas the latter (Ising) class applies to a c(2x2) structure [8.17]. Furthermore, $(\sqrt{3} \times \sqrt{3})R30°$ and (2x2) superstructures on a triangular lattice apparently belong [8.18] to the universality classes of the three- and four-state Potts models [8.19] the critical exponents of which are given also in Table 8.1 [8.20-22].

The importance of correlations in the fluctuations has the unhappy corollary that mean-field approximations are not adequate to describe, even qualitatively, properties of the adlayer such as phase diagrams, isotherms and energies of adsorption [8.23,24]. The inadequacies in the mean-field approx-

Table. 8.1. Critical exponents of various universality classes

Model	α	β	γ	ν
Ising	0	1/8	7/4	1
Three-state Potts	1/3	1/9	13/9	5/6
Four-state Potts	2/3	1/12	7/6	2/3

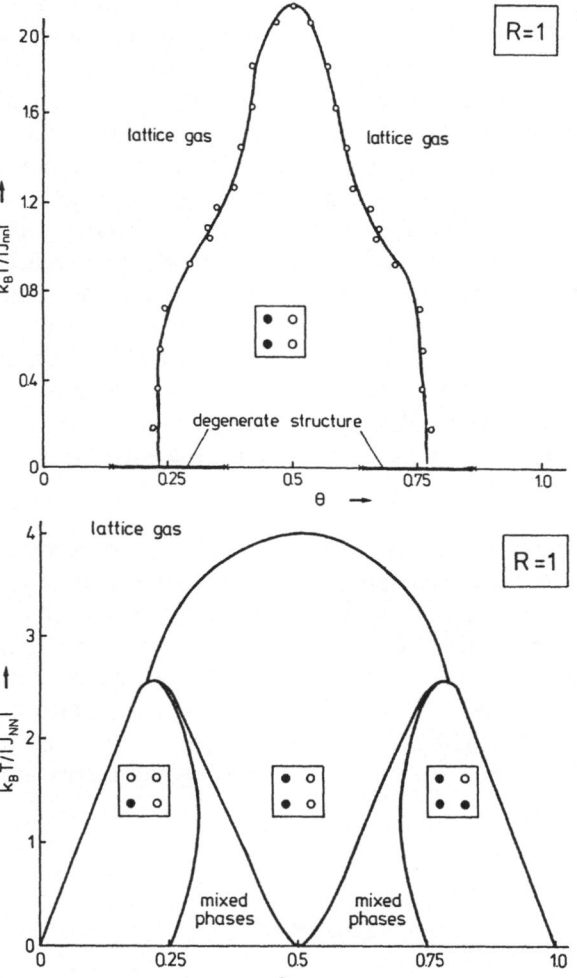

Fig. 8.1. Phase diagram of an adsorbed layer on a square lattice with $R = +1$ according to a Monte Carlo simulation (*top*), and the mean-field approximation (*bottom*) [8.23]

imation are shown clearly in Fig. 8.1 insofar as the calculation of phase diagrams is concerned. The particular case shown in this figure is for a square lattice with repulsive and equal first- and second-neighbor interaction energies (i.e. $R = +1$ as discussed in detail later in this chapter). Not only does the mean-field approximation overestimate the $T - \theta$ domain over which order exists, it is qualitatively incorrect which is obvious by comparing the upper diagram of Fig. 8.1 (due to a reliable Monte Carlo simulation) with the lower one (the mean-field approximation) [8.23,25,26]. For the time being, the ordinate in Fig. 8.1 should be considered simply proportional to temperature, the first-neighbor spin-exchange constant J_{nn} being irrelevant. More physically realistic phase diagrams are discussed in much more detail later in this chapter. In this respect, Monte Carlo simulations have proved to be extremely valuable in describing these two-dimensional systems [8.7,23,27,28].

An active current area of research is the development of more efficient approximate statistical mechanical methods to describe two-dimensional systems, such as Migdal-Kadanoff real-space renormalization-group methods [8.29-31], and, especially, finite-size renormalization-group matrix methods [8.9,25,26,32]. The ordered (2x1) phase indicated in Fig. 8.1 (top) should have critical exponents which are different from the Ising model (Table 8.1) and which are functions both of surface coverage and of adsorbate-adsorbate interaction energies. This has been investigated using a finite-size scaling analysis [8.33] applied to the "data" which were created using a Monte Carlo simulation. The important conclusion of the finite-size scaling analysis is that the order parameter m is a function of a reduced temperature $t \equiv (T_c - T)/T_c$ and the linear dimension of the lattice L in the following way:

$$L^{\beta/\nu} m(t,L) = \tilde{m}(tL^{1/\nu}) .$$ (8.2)

Numerical results are presented in Fig. 8.2. In the absence of second-neighbor (and longer-range) interaction energies, the critical exponents of an Ising model ($\beta = 1/8$ and $\nu = 1$) cause the data points for variable L to lie on a single curve as shown in Fig. 8.2a. As expected, this curve becomes a straight line the slope of which is 1/8 for large values of $tL^{1/\nu}$. If the second-neighbor interaction energy is "turned on" (and, in particular, if the ratio of the second-neighbor to the first-neighbor interaction energy is +1, both interactions being repulsive, as applies to Fig. 8.1), the Ising exponents are

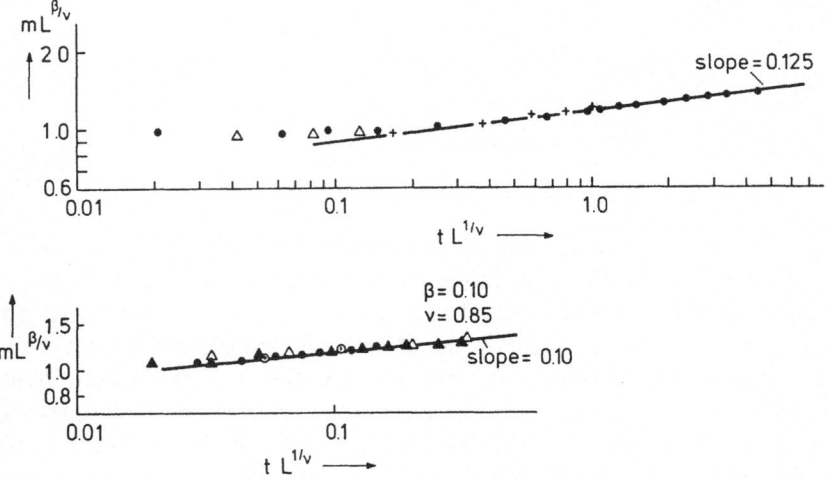

Fig. 8.2a, b. Finite-size scaling description of the order parameter for R = 0 (*top*), along a path of constant $H/k_B T = 1.34$; and (*bottom*) for R = +1 and H = 0. The data are for L = 10 (*full triangles*), 20 (*full circles*), 40 (*open triangles*), 60 (*open circles*), and 80 (*crosses*) [8.23]

no longer appropriate. As may be seen clearly in Fig. 8.2b, better estimates in this case are $\beta = 0.10 \pm 0.2$ and $\nu = 0.85 \pm 0.05$ [8.23].

8.1.3 Applicability to Actual Surfaces

Both the theoretical framework and the experimental ability are now in place for the study of order-disorder phenomena on surfaces, with an evaluation of adsorbate-adsorbate interaction energies, using LEED. In this chapter, three separate investigations related to these topics are considered in detail. These examples are the interactions among hydrogen adatoms on the (111) surface of nickel (Sect. 8.2), the (100) surface of palladium (Sect. 8.3), and the (110) surface of iron (Sect. 8.4). The choice of hydrogen as the adsorbate was dictated by the following reasons. First, hydrogen has been studied extensively on a number of substrate lattices of different symmetries, and theoretical calculations can be used to describe the experimental observations. Also, in the case of hydrogen on Ni(111), Pd(100), and Fe(110), it is clear that the adatoms do not perturb the substrate to any observable extent, i.e. the relaxation of the topmost substrate interlayer spacing is $\lesssim \pm 2\%$, a negligible perturbation. This would not seem surprising since the hydrogen-metal bond energy is in all cases less than 65 kcal/mol, but it should be noted that hydrogen overlayers have been observed to cause reconstruction of the (110) surface of nickel [8.34], the (110) surface of palladium [8.35] and the (100) surface of tungsten [8.36]. Obviously, a lattice-gas description is inappropriate for a substrate that reconstructs in the presence of the overlayer. For this same reason, i.e. possible perturbation of the substrate by the adsorbate, the order-disorder behavior of oxygen on transition metals has been avoided, although a considerable amount of reliable information is available for oxygen adatoms, especially on the (110) surface of tungsten [8.12] and the (111) surface of nickel [8.15,37]. The interaction of hydrogen adatoms on the Ni(111), the Pd(100), and the Fe(110) surfaces serves to illustrate most of the relevant experimental and theoretical issues and, moreover, illustrates these issues on lattices of three fundamentally different types of symmetry: triangular, square, and centered-rectangular.

8.2 The Interaction of Hydrogen with the (111) Surface of Nickel

8.2.1 An Optimum Case

The only ordered overlayer of atomic hydrogen on any surface thus far analyzed successfully with dynamical LEED calculations is the (2x2)-2H superstructure on Ni(111) [8.38,39]. In addition, this system displays an order-disorder phase transition [8.38,39] which has been analyzed theoretically [8.40,41]. On balance, the adsorption of hydrogen on Ni(111) is an optimum case for consideration since good ordering is observed at low substrate temperatures by LEED, rather simple thermal-desorption mass spectra are observed, there is negligible diffusion of the hydrogen into the bulk, the adsorption is entirely dissociative, and there is no reconstruction of the nickel substrate as a consequence of the adsorption of hydrogen.

8.2.2 Experimental Results for Hydrogen Chemisorption on Ni(111)

Hydrogen Coverage and Binding Energy

One of the most important quantities to be determined experimentally is the number of hydrogen atoms chemisorbed on the Ni(111) surface. Since the observed (2x2) LEED superlattice caused by H_2 adsorption cannot be correlated unequivocally with a particular surface coverage θ, there is a need for an independent technique to determine θ: not only could a (2x1) rather than a (2x2) superlattice exist, but in addition the unit cell could contain more than one hydrogen atom. Moreover, the binding energy of an adsorbate, as well as mutual interactions among the adsorbed particles, which come into play especially at higher surface concentrations, supply important information concerning the gas-surface interaction. In principle, this information can be obtained from thermal-desorption mass spectrometric measurements.

A typical sequence of thermal-desorption spectra from Ni(111) surfaces with various coverages of hydrogen adsorbed at 150 K is shown in Fig. 8.3.

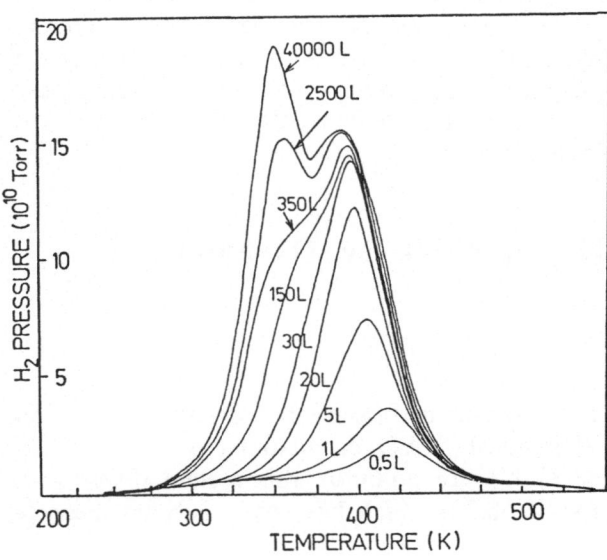

Fig. 8.3. Sequence of thermal-desorption spectra for hydrogen on Ni(111) after adsorption at 150 K, parametric in the hydrogen exposure in Langmuirs [8.39]

Two binding states, β_1 and β_2, can be discerned, which occur (at saturation) at 355 and 383 K, respectively [8.38,42]. By H_2/D_2 exchange experiments, it was shown that both states contain atomic (rather than molecular) hydrogen. The β_2 and β_1 states are filled sequentially and tend to exhibit equal populations at saturation coverage. The maxima of the desorption rates shift to lower temperatures if the hydrogen surface concentration is increased, which is typical for a second-order rate process. An analysis of both the peak maxima as well as the peak shapes as a function of surface coverage according to the method of Chan et al. [8.43] demonstrated that both the activation energy and the preexponential factor of the desorption rate coefficients of the β_1 and β_2 states are almost invariant with surface coverage. For the β_1 state, the activation energy is 19.0 ± 1.5 kcal/mole, and the preexponential factor is 2×10^{-4} cm^2/s; while for the β_2 state, the activation energy is 21.5 ± 1.0 kcal/mole, and the preexponential factor is 10^{-3} cm^2/s. These values are in good agreement with earlier estimates [8.42,44]. From the thermal-desorption spectrum corresponding to surface saturation of hydrogen, a maximum coverage of $\theta \approx 0.8$-0.9 ($\pm 20\%$) is evaluated. Obviously, the β_2 state is filled at $\theta = 0.5$, but not at $\theta = 0.25$. This fact is quite important for the correct interpretation of the LEED pattern.

Adatom-Adatom Interaction Energy and Critical Temperature

Whereas hydrogen adsorption at room temperature causes no noticeable change in the LEED pattern [8.42], striking features are observed when the surface temperature is below 270 K. At these low temperatures, a (2x2) pattern appears [8.45-47]. The new, fractional-order beams are very sharp, but their absolute intensity is at most 1-2% of that of the substrate beams. Furthermore, the intensity of the fractional-order beams depends strongly upon both the coverage and the temperature [8.47], which is the behavior expected for an adsorbed array of interacting adatoms [8.48].

In order to obtain information concerning the interaction energy between adjacent adatoms, it is necessary to monitor the intensity of a given fractional-order beam as a function of θ and T. The relationship between $I_{(1/2,1/2)}$ and the hydrogen coverage (as determined from thermal-desorption mass spectrometry) at a constant temperature (T = 150 K) is shown in Fig. 8.4. No linear increase of $I_{(1/2,1/2)}$ with θ is found, and at $\theta = 0.5$, $I_{(1/2,1/2)}$ goes through a sharp maximum. Upon a further increase of θ, $I_{(1/2,1/2)}$ decreases and approaches zero at $\theta \approx 0.72$. In Fig. 8.5, the temperature dependence of the half-order-beam intensity at various coverages is reproduced. Within the kinematic diffraction approximation, these curves reflect directly the degree of order in the hydrogen overlayer as a function of temperature. In separate low-temperature (120-180 K) measurements, the intensity loss due to vibrational effects (e.g. the Debye-Waller factor) as well as the background intensity due to disorder have been determined; the contributions from both

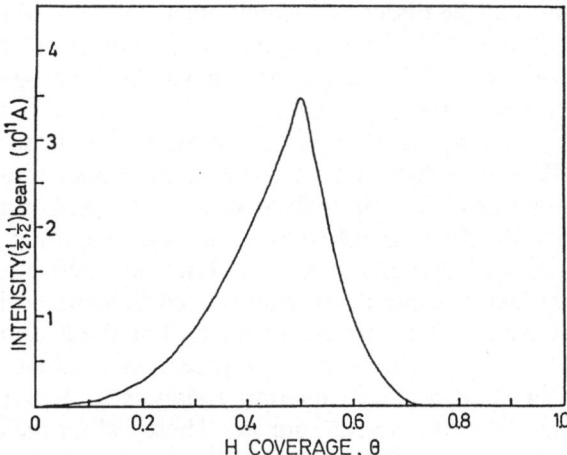

Fig. 8.4. Intensity of the (1/2, 1/2) fractional-order beam as a function of hydrogen coverage on Ni(111) at a temperature of 150 K and an incident electron energy of 63 eV [8.39]

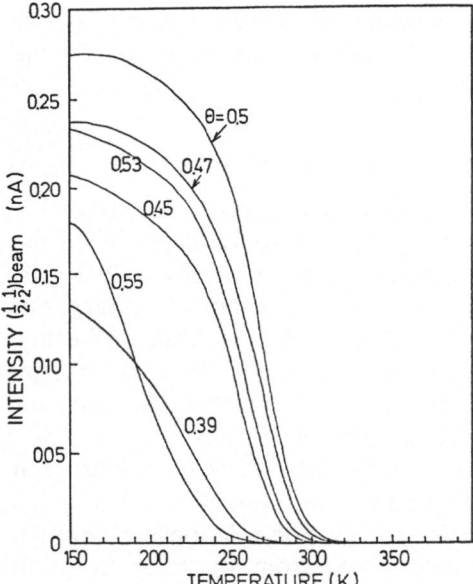

Fig. 8.5. Variation of the intensity of the (1/2, 1/2) fractional order beam with temperature for hydrogen on Ni(111) at various fractional surface coverages. The incident electron energy is 63 eV [8.39]

of these effects have been taken into consideration when presenting the experimental I(T) curves shown in Fig. 8.5. Although dynamic LEED effects will be of only minor importance in this type of intensity measurement and will not alter the qualitative features, the data are not a quantitative measure of the degree of long-range order within an adsorbate layer of infinite size. This is true for the following reasons: (1) The coherence width of the electron beam undergoing diffraction (approximately 100 Å) limits the size of coherently scattering regions, thereby causing a smooth decrease in the I(T) curves above the transition temperature [8.48]; (2) above the long-range order transition, some degree of short-range order may still persist, leading to a broadening of the diffraction beams due to the continuous decrease of the mean, ordered domain size; and (3) the actual measured beam intensity is convoluted with the instrumental response function (caused mainly by the finite aperture width of the Faraday cup). However, the dominant features of Fig. 8.5, in principle, are not affected by these issues. This implies that the transition temperature is highest for $\theta = 0.5$ and that the I(T) curves are not identical for coverages represented by $\theta + \Delta\theta$ and $\theta - \Delta\theta$, i.e. above $\theta = 0.5$ the degree of order decreases more rapidly than below this coverage. This fact can be explained easily within the framework of the structural model developed by the theoretical LEED analysis which is described in more detail in the following sections.

The various transition temperatures have been combined with the corresponding absolute coverage data in a phase diagram, which is shown in Fig. 8.6. The determination of the transition temperatures T_c [which have been taken as the points of inflection of the I(T) curves] is complicated somewhat by the fact that at coverages deviating from 0.5 by $\Delta\theta \approx 0.1$, complete order is no longer achieved in the accessible temperature range, and the order-disorder transitions are rather continuous. From Fig. 8.6, the asymmetry of T_c with respect to $\theta = 0.5$ again becomes quite evident. Linear extrapolation suggests that T_c reaches 0 K at approximately $\theta = 0.65$ and $\theta = 0.2$. The implication of this phase diagram with regard to the forces between adsorbed hydrogen atoms is discussed in more detail in Sects. 8.2.6—8.2.8. It is important to note that model calculations for a lattice gas with half the available sites occupied and with pairwise interactions predict symmetric phase diagrams [8.49,50].

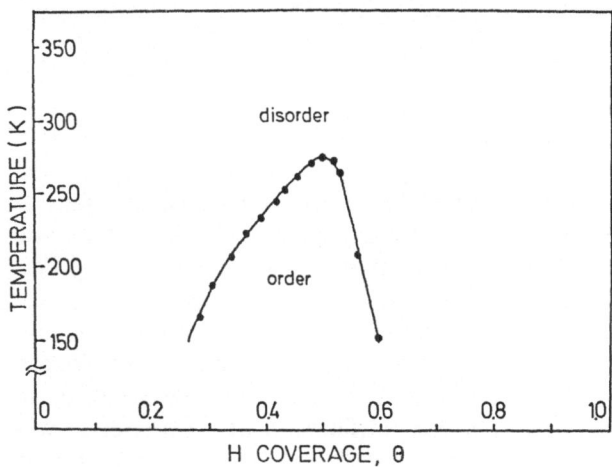

Fig. 8.6. Phase diagram showing the order-disorder transition temperature of hydrogen on Ni(111) as a function of fractional coverage [8.39]

Work-Function Change

In another series of experiments, the work-function change caused by the chemisorption of hydrogen on Ni(111) has been investigated in detail. In a room-temperature study of this system [8.42], a maximum work-function increase of +195 meV was obtained, and the corresponding hydrogen surface coverage may be estimated to be approximately a half monolayer or less, with the hydrogen atoms being in the disordered state. On lowering the temperature, the surface coverage as well as the degree of order within the adsorbate increases. Figure 8.7 shows the work-function change as a function of coverage θ obtained with a Ni(111) surface at 150 K exposed to hydrogen gas. An almost linear increase of $\Delta\varphi$ (indicating a positively polarized adsorbed species) up to approximately 165 meV at $\theta \cong 0.5$ is followed by a slow decrease as the β_1 chemisorption state is filled.

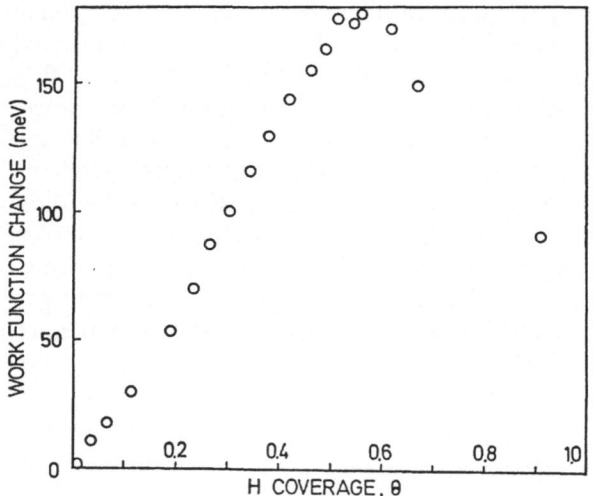

Fig. 8.7. Work-function change for hydrogen adsorption on Ni(111) at 150 K as a function of the absolute fractional coverage [8.39]

8.2.3 Parameters for LEED Analysis

For the analysis of the measured LEED I-V spectra, dynamical formalisms were used, in particular Renormalized-Forward Scattering (Sect. 5.7) and Layer Doubling (Sect. 5.6) [8.51], supplemented by Reverse-Scattering Perturbation (Sect. 5.4) [8.52] when small interlayer spacings occurred (e.g. for underlayers, for intercalation and for overlayers with two atoms per unit cell). The scattering potential for the nickel atoms was one due to Wakoh [8.53], since it has proved to be of good quality in previous analyses of clean [8.54] and adsorbate-covered [8.55] nickel surfaces. Little experience was available to guide the choice of a scattering potential for the chemisorbed hydrogen atoms. From model calculations [8.56], it is expected that chemisorbed hydrogen atoms, at least when not embedded in the surface, are probably only weakly charged so that the free-atom electron distribution is not seriously distorted. Since the work-function change upon hydrogen chemisorption is very small (0.165 eV for a half-monolayer), very little charge transfer is expected on experimental grounds as well, at least in the case of overlayers. Therefore, the simple 1s-hydrogenic wave function, truncated at a radius of 1 a.u., was used to describe the charge distribution. Within the muffin-tin model, scattering potentials were constructed that included Slater exchange [8.57] in addition to the electrostatic term. Five phase shifts were used and shown to be sufficient by test calculations using eight phase shifts.

The muffin-tin constant was set equal to 11 eV below the vacuum level, in accordance with results for clean nickel obtained with the Wakoh potential [8.54] (the muffin-tin constant depends on the choice of muffin-tin potential). This value was used also in the hydrogen layers, since there is no indication in the many LEED studies performed to date that the resulting structure depends on this choice, at least if one uses energies above approximately 40 eV (the lower limit of energy used in these calculations). For the same reason, the reflection by the surface-potential step is neglected. Similarly, the imaginary part of the potential was chosen for both the substrate and the overlayer to have the value used previously with success for nickel [8.54,55], namely it was given a value proportional to $E^{1/3}$, fixed at -3.8 eV for a kinetic energy of 90 eV (relative to the vacuum zero).

All substrate atoms were assigned a Debye temperature of 314 K, which corresponds to the bulk Debye temperature, weakened by a factor of 1.4 to represent an average increase in vibrational amplitudes near the surface. This choice and the thermal state of the hydrogen atoms are discussed in Sect. 8.2.5.

8.2.4 The Geometry of Chemisorbed Hydrogen on Ni(111)

Experimental LEED observations show that the scattering strength of the hydrogen is quite small. The integral-order I-V spectra for the $\theta = 0.5$ hydrogen overlayer are almost indistinguishable from those for the clean surface. In addition to illustrating the weakness of scattering by hydrogen (which was verified by theoretical calculations as well), this observation also shows that the substrate geometry remains essentially unperturbed in the presence of hydrogen. In an analysis of the clean Ni(111) surface, a possible slight contraction of the top interlayer spacing of approximately 1% was found [8.54]. Such a contraction is too small to have any effect on the analysis of the geometry of the hydrogen overlayer. Consequently, the bulk structure is assumed for the substrate. This is especially reasonable since, in general, adlayers tend to cancel any clean surface contraction.

Trial Structures

Little help from previous work was available to give an idea of the chemisorption geometry. Consequently, many possibilities had to be assumed a priori in a trial-and-error search. Although the thermal-desorption mass spectrometric and the work-function-change ($\Delta\varphi$) measurements indicated that the ordered (2x2) hydrogen superstructure corresponds to a half-monolayer coverage, for completeness quarter-monolayer coverages corresponding to p(2x2)-H superstructures were investigated as well. The trial geometries tested included on-top sites, twofold bridging sites, and both types of inequivalent (hcp and fcc) threefold hollow sites. Various perpendicular distances of the hydrogen atoms (including both overlayers and underlayers) were tested with three fundamentally different symmetries of the superstructure: a p(2x2)-H (corresponding to $\theta = 1/4$), three independent domains of (2x1) arrays rotated 120° with respect to one another which gives an apparent (2x2) LEED pattern (corresponding to $\theta = 1/2$), and a (2x2)-2H "honeycomb" structure shown explicitly in Fig. 8.8 (corresponding to $\theta = 1/2$). The latter geometry consists of two interpenetrating hexagonal lattices of hydrogen adatoms. Each hydrogen adatom has three hydrogen neighbors at a distance of 2.88 Å (Figs. 8.8a and 8.8b), which is large compared to the molecular H-H bond length of 0.74 Å

Hydrogen Detection by LEED

Since the integral-order beams for the ordered overlayer of hydrogen on Ni(111) are essentially unaffected by the presence of the chemisorbed hydrogen, only the fractional-order beams can be used in the determination of the adatom positions. It follows, more generally, that to be able to extract

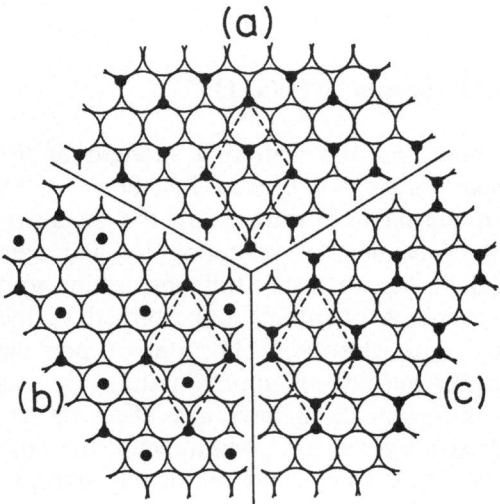

Fig. 8.8a–c. Various Ni(111)-(2×2)-2H geometries (*top view*): (a) honeycomb hollow-hollow arrangement. (b) honeycomb hollow-top arrangement. (c) "Quasi-molecular" hollow-hollow arrangement [8.39]

hydrogen positions from LEED I-V spectra, it is necessary to use those beams that would have vanishing intensities if the hydrogen atoms had vanishing scattering amplitudes, i.e. those beams that are not due to other aspects of the surface geometry. For example, if one adsorbs a p(2x2) overlayer of hydrogen on a surface that already has a p(2x2) superstructure due to non-hydrogen atoms, it will be quite difficult to determine the hydrogen positions. This applies, in particular, to the hydrogen atoms of adsorbed hydrocarbons, so that LEED is not suitable to directly determine hydrogen locations in such adsorbed molecules. However, a determination of the C-C bond length provides the approximate state of hybridization of the individual carbon atoms, and this permits the rather reliable a posteriori placement of the hydrogen atoms in the hydrocarbon skeleton.

Experimental Data Set

Returning to the case of hydrogen on Ni(111), the (½ 0), (0 ½), and (½ ½) beams [and their counterparts rotated about the (00) beam] are available. At normal incidence (chosen to make the computation manageable using symmetry), that leaves three nonequivalent, usable beams a relatively small data set. Other fractional-order beams emerge at higher energies, where the computational effort becomes excessive due to the large number of beams that has to be included in the calculation. Furthermore, at those energies the absolute experimental intensities become rather small compared to the rising background. The lower energy limit was chosen to be 40 eV since the fractional-order beams are difficult to measure below that energy, and the physical parameters of the calculation (especially the hydrogen potential) become uncertain there. The upper energy limit was fixed at approximately 120 eV for two reasons. First, the (½ 0) and (0 ½) beams are in the flank of the (00) beam, which is very intense from approximately 120 eV to over 150 eV, making their measured intensities unreliable. Moreover, as noted above, the computational effort becomes excessive at such energies.

The Optimum Structure

Of all the geometries tested, the hollow-hollow combination of sites shown in Fig. 8.8a, corresponding to a Ni(111)-(2x2)-2H superstructure with $\theta = 1/2$, gave by far the best agreement between experimentally measured and theoretically calculated I-V spectra. The optimized geometry consists of a planar hydrogen overlayer with an H-Ni interlayer spacing d of 1.15 ± 0.1 Å This corresponds to equal Ni-H bond lengths of 1.84 ± 0.06 Å Buckling this overlayer by differentially shifting the inequivalent hydrogen atoms away from and toward the surface by ±0.1 Å caused a degradation in the agreement between experiment and theory. Consequently, no buckling occurs in the honeycomb structure to within ±0.1 Å

The Ni-H bond length of 1.84 Å obtained by LEED for Ni(111) can be compared with other known Ni-H bond lengths. In the NiH diatomic molecule, the value is 1.47 Å [8.58]; in the $H_3Ni_4(C_5H_5)_4$ cluster it is 1.69 Å [8.59]; and the bulk structure of nickel hydride consists of hcp nickel (with lattice constants a = 2.65 Å and c = 2.155 Å) intercalated with hydrogen atoms in octahedral holes, yielding a Ni-H bond length of 1.87 Å [8.60]. As expected for hydrogen chemisorbed in threefold coordination sites on Ni(111), the Ni-H bond length is comparable to that for hydrogen in bulk nickel compounds and longer than for hydrogen located in on-top sites (using the analogous vocabulary of the surface phase).

The following theoretical calculations concerning the chemisorption of hydrogen on nickel are relevant in this discussion. Semiempirical calculations have predicted that both types of threefold coordination sites are energetically favored over other sites [8.61]. However, the calculated barriers to surface diffusion are not sufficiently low to permit the high mobility of hydrogen on the surface that is observed experimentally. Similar results, but with satisfactorily low barriers were obtained with density-functional calculations [8.56]. Using a more sophisticated ab initio self-consistent field-Hartree Fock (SCF-HF) scheme, Upton et al. [8.62,63] have examined the chemisorption of hydrogen on naked clusters of nickel containing either 20 or 28 nickel atoms. With such large clusters, on-top, twofold, threefold, and fourfold binding sites could be examined and compared. For these respective types of sites, the Ni-H bond lengths were found to be 1.49 Å, 1.57 Å, 1.62 Å, and 1.78 Å. The variation in bond length between the two inequivalent types of threefold sites was found to be negligible. Although there is an obvious trend toward increasing bond length with increasing coordination, the calculation suggests a Ni-H bond length in the threefold sites that is approximately 0.2 Å smaller than that found by LEED. This discrepancy may be related to the fact that a single 4s conduction electron was considered in the bonding calculation, the remaining Ar core of electrons and the $3d^9$ band of electrons being modeled by an effective pseudopotential. It may also be related to the lack of adatom-adatom interactions in the cluster calculation.

For the chemisorption of hydrogen on Ni(111), the second substrate layer shows little, if any, effect. In this case, the thermal-desorption mass spectra, the LEED structural determination and theoretical calculations of the chemisorption bond point to the near equality of bond lengths and binding energies at the two inequivalent threefold hollow sites. The work-function behavior, shown in Fig. 8.7, can be explained also on the basis of the honeycomb structural model. Up to θ = 0.5, the hydrogen atoms can occupy chemisorption sites with identical symmetry and nearly identical binding energy. The single dipole moments arising from the formation of individual H-Ni surface complexes simply add linearly, giving the maximum Δφ value at θ = 0.5. Further chemisorption of hydrogen atoms, however, leads to a rapid breakdown of order within the hydrogen overlayer and a significant change of the geometric location of each hydrogen atom, as is indicated by

LEED and thermal-desorption mass spectrometry. It is quite reasonable to associate a different dipole moment (in this case it has to be of a reversed sign) with the new chemisorption sites, and to regard the work-function change beyond $\theta = 0.5$ as being caused by a superposition of both kinds of dipole moments.

8.2.5 Thermal Motion and Disorder in the Hydrogen Overlayer

Discrepancy in LEED Intensities

For the case of hydrogen on Ni(111), a comparison of the calculated absolute intensities for the fractional-order beams with those for the integral-order beams shows that the former are approximately one order of magnitude less intense than the latter at a half-monolayer surface coverage. This comparison quantifies the assertion, made above, concerning the relative weakness of the scattering by hydrogen. Experimentally, however, there is a difference of almost two orders of magnitude between the intensities of the fractional- and integral-order beams. Consequently, there is a discrepancy of one order of magnitude between experiment and theory in the intensities of the fractional-order beams for the Ni(111)-(2x2)-2H superstructure.

Several related explanations for this discrepancy are possible, all of which are concerned with disorder in the adlayer. The first explanation is that there is cancellation of diffraction intensity between adjacent strips of out-of-phase domains. The second explanation is that atomic thermal vibrations may have larger amplitudes than assumed in the calculations. Third, there may be a substantial amount of disorder within ordered regions of the surface. Fourth, and finally, ordered-island formation might take place, in which islands of order occur within a sea of disorder (or "lakes" of disorder occur between regions of order).

Out-of-Phase Domains

It is customary to assume that neighboring adlayer domains diffract incoherently if their average diameter is rather larger than the coherence length of the incident electron beam (on the order of 100 Å), and that coherent diffraction occurs otherwise. However, well-ordered adlayer patches close to each other (relative to the coherence length), but on either side of an out-of-phase domain boundary, will produce canceling contributions to some of the fractional-order beams. Therefore, one expects that complete strips of the adlayer along the domain boundaries, of width comparable to the coherence length, will not contribute to the intensity of the fractional-order beams. Some fractional-order-beam splitting could occur then. However, with the present geometry there are four possibilities of different domain registries, each producing splittings in different directions, so that such an occurrence may well become difficult to observe experimentally. In fact, such beam splittings are not often observed for any adsorbate-substrate combination, presumably because the practical conditions for the occurrence of clear beam splittings are rather stringent: domains should be sufficiently small to produce a significant number of boundaries, while they may not be so small that beam broadening occurs; and they must have a rather uniform size.

Now, if the area of these strips is significant compared to the domain area, which is the case when the domains have dimensions comparable to the coherence length, a considerable fraction of the intensity of the fractional-order beams is lost, while the spot sharpness is not affected significantly. An estimate of this effect is obtained easily by averaging over all possible positions of a square coherence zone relative to an array of square domains, and by assuming that areas of the coherence zone astride a domain boundary cancel. If the domain edge is p times the coherence length, one obtains an intensity weakening by a factor of $[1 - 1/(2p)]^2$. Thus, for $p = 1$ (equal coherence length and domain edge), the intensities of fractional-order beams are multiplied by a factor of 1/4. For a slightly smaller domain edge, the intensities decrease rapidly, e.g. with $p = 3/4$ the reduction factor is 1/9, approximately the factor needed in the present case.

This effect obviously should occur at all surfaces where adatoms are arranged in superlattices. Fractional-order beams should be weakened by the presence of domains, most strongly when the domains are small (on the order of the coherence length). It is believed that this effect plays the most important role in the present context.

Thermal Vibrations

Thermal vibrations is the second possibility mentioned above to explain the discrepancy in the diffraction intensities. The thermal vibrations of the substrate atoms are not noticeably different from those for the clean Ni(111) surface, since the intensities of the integral-order beams are not systematically reduced by the chemisorbed hydrogen. (In fact, increases are observed for certain beams at certain energies.) For this reason, nickel vibrational amplitudes were chosen that are also appropriate for clean Ni(111) calculations. The nickel vibrations therefore cannot be the source of the observed discrepancy in intensities.

An attempt was made to reduce the fractional-order-beam intensities to the experimental values by allowing the hydrogen vibrational amplitudes to increase to several times the vibrational amplitude of the substrate atoms (which is on the order of 0.05 Å at a temperature of 110 K). This was done within the usual Debye-Waller formalism, via "temperature-dependent" phase shifts. Large hydrogen vibrational amplitudes are certainly not unreasonable since the diffusion barriers along the surface are sufficiently low to allow hydrogen to diffuse thermally over the surface at the experimental temperature of 110 K. In addition, the thermal energy at 110 K gives free hydrogen atoms (now considering the chemisorbed atoms as free in directions parallel to the surface) a de Broglie wavelength of approximately 2 Å For comparison, the zero-point motion of hydrogen trapped in the binding potential well (perpendicular to the surface) is on the order of 1.0 eV [8.64], corresponding to an amplitude of vibration perpendicular to the surface of approximately 0.15 Å Vibrations of 0.15 Å produce little effect on the calculated I-V spectra. Even a large increase in vibrational amplitudes parallel to the surface to 0.5 Å would reduce the calculated intensities by only approximately 50% for the fractional-order beams under consideration. This small effect derives from the fact that the Debye-Waller factor is still close to unity, due to the small scattering angle involved for the beams of interest. The hydrogen atoms scatter predominantly in the forward direction since the electrons are turned around mainly by the substrate, as test calculations with a vanishing hydrogen-layer reflectivity have shown.

Lattice Disorder

It appears therefore that very large vibrational amplitudes (at least on the order of the substrate lattice constant) would be required to completely explain the observed weakness of the fractional-order beams. Only then would the Debye-Waller factor affect the intensities sufficiently. Since such large amplitudes are equivalent to hopping of adatoms from site to site, this may be considered to be adsorption-site disorder, i.e. adsorption of some atoms in nonperiodically arranged sites. This is the third possibility advanced to explain the discrepancy in intensities. It is well known that disorder in an adlayer has the effect of decreasing the intensity of fractional-order LEED beams. However, the diffraction spots on the LEED screen are often rendered diffuse by disorder, contrary to the observation in this case. The experimental spots have a sharpness that implies significant order on the scale of at least ten superlattice constants, i.e. on the scale of at least 50 Å This requires domains of at least those dimensions but does not exclude short-range disorder within each domain, such as would occur if a fraction of the adatoms were misplaced in the lattice of each domain. This kind of disorder would be perfectly consistent with the observed high surface mobility of hydrogen. However, to obtain a beam weakening by a factor of ten would require that approximately 2/3 of the atoms are misplaced. In this case, the diffracted intensities would be due to approximately 1/3 of the adatoms, producing intensities weaker by $(1/3)^2$, approximately 0.1, compared to the perfectly ordered case. However, it is difficult to imagine that 2/3 of the adatoms could be misplaced, while long-range order on the scale of at least ten superlattice constants remains among the other 1/3 of the adatoms.

Ordered Islands

The fourth possibility for the weakening of the fractional-order beams by ordered-island formation may play only a marginal role in the case of hydrogen on Ni(111) as well. At a half-monolayer coverage, it would be necessary for 1/3 of the surface to be well ordered in small islands and the rest to be disordered in "lakes" or "seas" to obtain the needed reduction of approximately $(1/3)^2 \approx 0.1$ in the intensities of fractional-order beams. However, it appears that at the experimental temperature of 110 K, the half-monolayer has reached a state of ordering that cannot be improved appreciably by cooling. The evidence is derived from the way in which the intensity of the fractional-order beams increases with decreasing temperature due to better ordering, see Fig. 8.5. At a temperature of 110 K, this intensity has saturated (more precisely, its temperature dependence has become very small). Such behavior argues against the presence of any substantial residual disordered phase on the surface, which therefore cannot explain a tenfold decrease in intensities.

It appears that among the various causes of beam-weakening discussed in this section, cancellations in out-of-phase boundary strips cause the most prominent effect in the present case of hydrogen on Ni(111). More specific statements would require a closer investigation, e.g. an analysis of the diffraction beam profiles [8.63] or optical simulation of the diffraction from test lattices incorporating the various possible forms of disorder.

Effect of Disorder on I-V Curve Structure

A question that must still be answered is whether the presumed adlayer disorder affects the structure of the I-V curves, i.e. whether peaks can be shifted, relative peak heights changed, etc. The most obvious consequence of disorder is a general weakening of the intensity of the LEED beams, this loss of intensity being compensated by an increase of background intensity in all space directions (sometimes mainly in preferential directions due to the retention of some degree of order). One way to describe this intensity weakening is to realize that a particular I-V spectrum gives (in the kinematic limit) the magnitude of a particular Fourier component of the surface structure parallel to the surface (modulated by perpendicular structure). Hence, by observing a particular beam, one focuses on a feature of the surface with a particular periodicity. By definition, however, disorder usually has few or no features with the periodicity defined by the particular beam under consideration and is therefore largely or totally filtered out. Disorder has many Fourier components of different periodicities that deflect intensity into directions away from the beams under consideration. This explains why LEED structural determinations have performed well for many surfaces, most of which are presumably more or less poorly ordered, since only the well ordered part is analyzed. A consequence is, of course, that the structural result of a LEED analysis applies only to that part of the surface which is sufficiently well ordered to generate intensity in the beams that are considered.

It has been found that the main effect of disorder on the structure of I-V spectra (as opposed to simple intensity weakening) arises through multiple scattering [8.66]. In particular, disorder creates nonunique surroundings for surface atoms that otherwise would have identical surroundings. This implies that different surface atoms experience different incident electron wave fields, since multiple scattering makes the wave field incident on a particular atom depend on the geometry of the particular surroundings of that atom. Hence, the I-V spectra become a kind of average over I-V spectra due to different atomic configurations. In the present case, any scattering involving hydrogen atoms is so weak that *multiple* scattering involving hydrogen atoms is especially weak, leaving mainly the kinematic behavior of simple intensity weakening. Moreover, if it is assumed that misplaced adatoms are in the same type of adsorption site as when "properly" positioned, then their immediate surroundings will to a large extent be indistinguishable from the surroundings in the perfectly ordered case. Consequently, the little multiple scattering that may still be present would be largely identical to that of the perfectly ordered case and could not affect the I-V spectra markedly.

Based on these arguments, it seems that the Ni(111)-(2x2)-2H system has a relatively large amount of disorder (as characterized above), that the I-V spectra are largely unaffected in their structure by this disorder, and that the structural determination is reliable for the well ordered part of the surface. Clearly, no statement may be made concerning the position of misplaced hydrogen atoms.

8.2.6 The Order-Disorder Phase Transition and Adatom-Adatom Interaction Energies

A better understanding of the origin of the ordering phenomena which occur with hydrogen atoms on a Ni(111) surface would derive from a knowledge of the adsorbate-induced mutual interactions among the adsorbed atoms. Since experimental evidence for ordering is obtained well below a half-monolayer coverage, see Fig. 8.4, it is evident that indirect long-range interactions between the adsorbed hydrogen atoms must exist. Particular consideration of this kind of interaction has been given to the H/W(100) system. Grimley and Torrini [8.2] as well as Einstein and Schrieffer [8.3] have predicted long-range interactions via the metallic substrate over a distance of up to three lattice constants. The magnitude of these interaction energies, which are assumed to be almost independent of the strength of the hydrogen-metal chemisorption bond, was calculated to range between 1 and 3 kcal/mol (0.05 to 0.15 eV), i.e. less than 5% of that of the chemisorption bond energy.

From Quarter-Monolayer (2x2) to Half-Monolayer (2x2)

The experimentally measured $I(\theta)$ curve, shown in Fig. 8.4, contains information concerning hydrogen-hydrogen interactions on the Ni(111) surface. In this case, the strong decrease in the order-disorder transition temperature as the coverage deviates from $\theta = 1/2$ (see Fig. 8.6) tends to rule out island formation caused by strong attractive interactions. If this were the case at $\theta < 0.5$, the surface would consist of two phases — ordered domains with identical local coverage and bare patches — the relative proportion of which would govern the total coverage θ. In this case, the transition temperature would be independent of coverage for small variations in surface coverage ($\Delta\theta \leqslant 0.15$). At coverages below 0.25 monolayer, little can be inferred from experiment, however, since the fractional-order-beam intensities become so weak that both the $I(\theta)$ curve and the beam sharpness become difficult to evaluate. Thus, one cannot assess whether island formation takes place from these data for $\theta < 0.25$. Above $\theta = 0.25$, Fig. 8.4 shows a quadratic dependence of $I_{(\frac{1}{2},\frac{1}{2})}$ on θ. This indicates that with increasing coverage, hydrogen atoms are added within the existing overlayer rather than at the edge of it. The latter would give a linear $I(\theta)$ plot. It may be that at $\theta = 0.25$ a simple (2x2) arrangement of hydrogen atoms exists and that further adatoms merely fill in the honeycomb overlayer [the (2x2) spot pattern would remain, as observed]. Thus, in Fig. 8.8, first the (2x2) pattern defined by the dashed unit cell (containing only one atom) would form, involving only one type of threefold hollow site (the most favorable one, whichever it is). Then, if one assumes that the hydrogen-hydrogen interaction force is repulsive at the distances involved here, additional adatoms would prefer to choose the other type of hollow site as in Fig. 8.8a. Note that this structure maximizes the hydrogen-hydrogen distance in the overlayer.

Repulsive Hydrogen-Hydrogen Interaction Energy

There are several ways to obtain more quantitative information concerning the repulsive hydrogen-hydrogen interactions on the surface. To a first approximation, the H/Ni(111) system may be treated within the two-dimensional Ising model [8.67]. This model may be transferred to a lattice-gas system assuming that the adsorbed particles occupy surface sites of identical local geometry, irrespective of coverage, degree of order or occupation of neighboring sites. With a square-lattice geometry involving only repulsive interactions between nearest neighbors, the critical temperature T_c drops to zero at $\theta = 0.35$ and $\theta = 0.65$, i.e. for $\Delta\theta = 0.15$. This leads to a completely symmetric phase diagram [8.49]. An analytical expression relates the maximum transition temperature T_c^{max} to the repulsive energy ω_2 via [8.48,68]

$$\omega_2 = 1.76 \ k_B T_c^{max} . \tag{8.3}$$

Other models, e.g. the four-state Potts model, have been employed to describe lattice-gas systems. Domany et al. [8.69] considered phase transitions within a (2x2) structure on a triangular lattice and showed them to be quite continuous (as is true also in the present case). A lattice-gas model can be applied to this system if it is borne in mind that the concentration of threefold sites is twice the concentration of surface nickel atoms, as pointed out by Domany et al. [8.40] and discussed in detail in the next section. The mandatory condition of identical local symmetry can be fulfilled by destruction of the honeycomb overlayer at $\theta > 0.5$. The observed asymmetry of the phase diagram is due in part to the fact that the ordered structure occurs at a coverage of one-quarter of the number of binding sites rather than one-half [8.40]. A weak third-neighbor attractive interaction is probably responsible for increasing the asymmetry. In any case, (8.3) can be used to obtain an approximate estimate of the interaction energy ω_2 of the present system, neglecting the different geometrical configurations. With T_c^{max} equal to 270 K, the value for ω_2 is found to be approximately 0.9 kcal/mol (0.04 eV), which is in quite good agreement with theoretically predicted values for interacting lattice-gas systems. Moreover, it should be mentioned that the thermal-desorption spectra, spanning the entire coverage range, $0 < \theta < 1$, also give clear evidence of an interaction energy of this order of magnitude. In the present system, two desorption states, β_1 and β_2, have been observed which are separated by approximately 2.5 kcal/mol. The β_2 state can be identified with desorption from the identical surface sites fitting into the honeycomb structural model and is saturated near $\theta \cong 0.5$. On the other hand, the β_1 state represents hydrogen atoms desorbing from sites occupied at coverages higher than 0.5, the energy of which is affected particularly by the interaction between hydrogen atoms.

The position of the hydrogen atoms in the β_1 state is not known, but their proximity to hydrogen atoms in the β_2 state results in an interaction energy on the order of 2 kcal/mol as manifest in the thermal-desorption spectra of Fig. 8.3.

Finally, it should be pointed out how the observed asymmetry of the phase diagram of the ordered/disordered hydrogen adsorbate strongly supports the geometrical model derived from the LEED data and presented above. It is very plausible that the removal of a few randomly chosen adatoms from the perfect honeycomb overlayer will not affect the remaining adatoms. On the other hand, the addition of a few adatoms may very well break up the honeycomb arrangement near each additional atom [for example, to produce locally a (1x1) arrangement]. It should be noted that even the eventual formation of a (2x2)-3H structure at $\theta = 0.75$ cannot proceed by simply filling empty sites of the honeycomb structure. In this way, more adatoms are affected above than below a half-monolayer, producing the observed asymmetry about a half-monolayer in the intensities of the fractional-order beams (Fig. 8.4).

8.2.7 A Renormalization-Group Theory Description of the Order-Disorder Transition of Hydrogen on Ni(111)

Domany et al. [8.40] have used Renormalization-group theory to calculate the phase diagram of a lattice gas on a hexagonal lattice and have applied their results to the case of Ni(111)-(2x2)-2H. The calculation assumes that both the first-neighbor (at 1.44 Å) and the second-neighbor (at 2.49 Å) interaction energies ω_1 and ω_2, respectively, are repulsive. This assignment is obviously consistent with the structure of the overlayer at $\theta \leqslant 0.5$ which is shown in Fig. 8.8a. Furthermore, since saturation coverage of hydrogen on Ni(111) corresponds approximately to $\theta = 1$ (Fig. 8.7), the (reasonable) approximation is made that $\omega_2 \ll \omega_1$. In actual fact, this corresponds to assumed infinite repulsive first-neighbor interactions and finite repulsive second-neighbor interactions.

Due to the two inequivalent types of threefold hollow sites in which the hydrogen atoms are bound on the surface, it is appropriate to define the density of the lattice gas n as $\theta/2$ (n is the ratio of the number of adatoms to the total number of threefold hollow sites). Since there are two threefold sites associated with each surface atom, the relationship $n = \theta/2$ follows directly. With these assumptions, the calculated phase diagram, T(n), is asymmetric with respect to $n = 1/4$ (i.e. $\theta = 1/2$), as observed experimentally. As may be seen in Fig. 8.9, however, the experimental phase diagram (Fig. 8.9a) is significantly broader than the theoretical one (Fig. 8.9b). The narrowness of the theoretical phase diagram is typical of calculations which include only repulsive interactions. The inclusion of a three-body repulsive interaction of strength $2\omega_2$ (certainly an overestimate) causes no significant broadening of

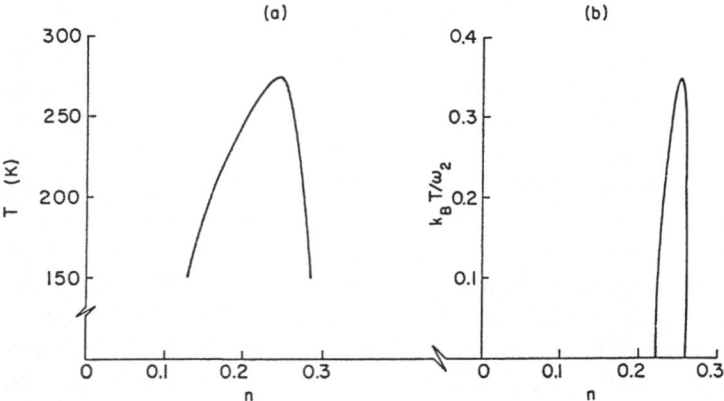

Fig. 8.9. (a) The experimental phase diagram for hydrogen on Ni(111) showing the transition temperature as a function of coverage n (the ratio of hydrogen adatoms to the number of threefold adsites, i.e. $n = \theta/2$). (b) Calculated phase diagram of a hexagonal lattice gas with infinite first- and finite second-neighbor repulsion [8.40]

the calculated phase diagram. This implies that the experimental "width" of the phase diagram is due to attractive interactions and/or third-neighbor interactions, with some contribution to the width of the phase diagram possibly deriving from the way in which it is constructed from the experimental data. This point is discussed fully in Sect. 8.3.

The inclusion of a third-neighbor attractive interaction in the calculation results in qualitative agreement with the experimental phase diagram, but the magnitude of ω_3 has not been estimated. However, ω_2 can be estimated if it is assumed that T_c^{max} (the maximum order-disorder transition temperature) is invariant between the inclusion of ω_1 and ω_2, and the inclusion of ω_1, ω_2, and ω_3. In this case, theoretically (cf. Fig. 8.9b),

$$\omega_2 = 2.84 \, k_B T_c^{max} \,, \tag{8.4}$$

which corresponds to a second-neighbor interaction energy of 1.52 kcal/mol (0.066 eV). This value should be more reliable than the one calculated using (8.3), since the proper triangular lattice has been used here.

8.2.8 A Cluster-Variational Description of the Order-Disorder Transition of Hydrogen on Ni(111)

Kittler and Bennemann [8.41] have used a cluster-variational method to describe the order-disorder phase transition of hydrogen on the (111) surface of nickel. In this calculation, the Helmholtz energy of the adlayer is minimized using an expression due to Kikuchi [8.80] for the configurational entropy. The latter includes three-site cluster correlations with the three sites being the two inequivalent types of threefold hollow sites and the on-top site. For clarity, this is shown explicitly in Fig. 8.10.

There are four input parameters in the calculation involving interaction energies, three of which are adjustable. These three are considered to be in

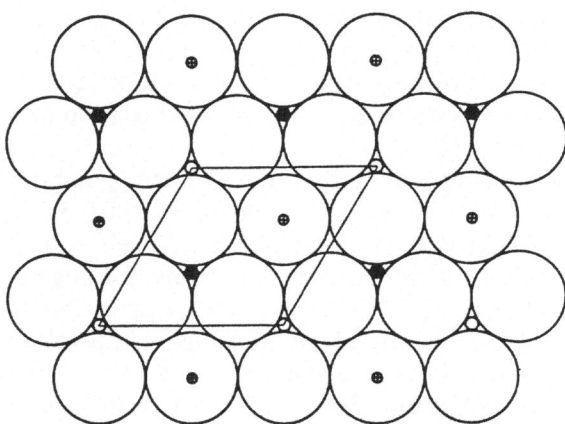

Fig. 8.10. The chemisorption geometry showing the positions of atoms on the Ni(111) surface (*large open circles*), and the three sublattices of possible hydrogen chemisorption sites: on-top sites (*small open circles with crosses*), threefold sites with a second-layer nickel vacancy (*small open circles*), and threefold sites with second-layer nickel atoms (*small filled circles*) [8.41]

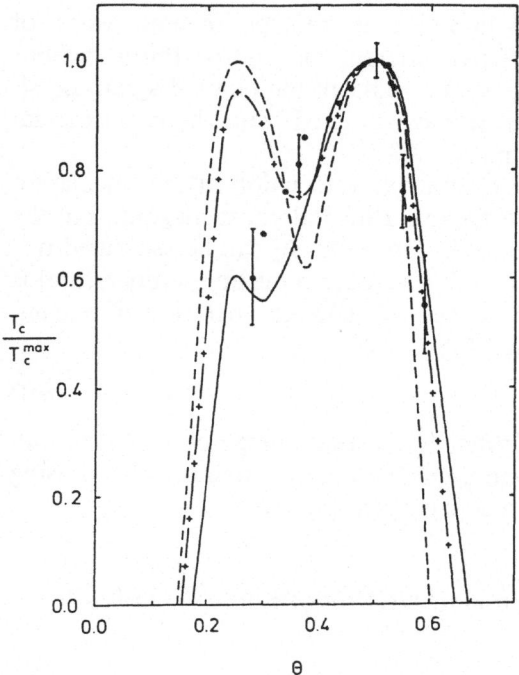

Fig. 8.11. The transition temperature T_c as a function of fractional surface coverage for hydrogen on Ni(111). The dots are experimental values. Calculated results are shown for E = W = 0 and V = 1 (*dashed curve*); for E = 0.034, V = 1 and W = 0 (*cross-dashed curve*); and for E = 0.03, V = 1.115 and W = 0.3 (*solid curve*) [8.41]

units of U, the pair-interaction energy between two hydrogen atoms in the two inequivalent threefold sites of the (nonprimitive) unit cell of Fig. 8.10. The three adjustable parameters are the following: (1) E, the single-atom binding-energy difference between on-top and threefold sites (with the two inequivalent types of threefold sites assumed to have the same intrinsic binding energy); (2) V, the pair-interaction energy between on-top and threefold sites in the superlattice; and (3) W, the three-body interaction energy among the hydrogen atoms on the three inequivalent types of sites in the superlattice. The necessity of including three-body interactions has been emphasized [8.69], but it is difficult to disentangle their effect from, for example, the inclusion of additional (slightly more distant) pair-interaction energies. As may be seen clearly in Fig. 8.11, however, the inclusion of three-body interactions is necessary to suppress the occurrence of a phase transition at θ = 1/4 [involving a p(2x2)-H superlattice].

By comparing theory and experiment, the optimized set of energy parameters is the following: U = 0.037 eV = 0.85 kcal/mole; E = 0.001 eV = 0.023 kcal/mole; V = 0.041 eV = 0.95 kcal/mole; and W = 0.011 eV = 0.25 kcal/mole. With this optimized choice of parameters, reasonably good agreement was achieved between the following calculated and experimentally measured quantities: (1) the variation in the phase transition temperature with surface coverage, i.e. the phase diagram shown in Fig. 8.11; (2) the (relative) intensity variation of the fractional-order LEED beams with surface coverage at various constant temperatures (Fig. 8.4); (3) the (relative) intensity variation of the fractional-order LEED beams with temperature at various constant surface coverages (Fig. 8.5); and (4) the variation in the heat of adsorption with surface coverage at constant temperature. It is perhaps premature to assess the value of this type of calculation. In the meantime, both the values of the energy parameters deduced for hydrogen on Ni(111), as well as their uniqueness, should be regarded with caution.

8.2.9 An Atomic Band Structure Description of Hydrogen on Ni(111)

The following peculiar aspects of hydrogen chemisorbed on Ni(111) should be noted: (1) the thermal energy at 300 K, $kT \sim 0.03$ eV, gives the hydrogen atoms in their motion parallel to the surface (ignoring diffusion barriers) a de Broglie wavelength on the order of 1 Å i.e. on the order of the substrate lattice constant of 2.49 Å; (2) this thermal energy is comparable to observed diffusion barriers; (3) the zero-point energy due to the hydrogen-binding-energy profile perpendicular to the surface is approximately 0.1 eV, which gives the hydrogen atoms a total ground-state energy similar to or larger than the expected diffusion barriers; and (4) a relatively large amount of adsorption disorder is observed at room temperature.

The Hydrogen Atom as a Wave

All this suggests that a more elegant description of the situation might be given by representing the adparticles by quantum-mechanical wave functions that are trapped in a binding-energy well perpendicular to the surface, but that are more or less delocalized along the surface [8.72]. Two obvious analogies come to mind: bound-state resonances which are observed in thermal-atomic-beam scattering from surfaces [8.73], and electronic surface states. For hydrogen on Ni(111), one would have a Schrödinger equation for a particle of mass M equal to the hydrogen mass, with a three-dimensional potential given by the binding energy of the hydrogen to the nickel surface. Energy exchange between the adatom and the substrate, e.g. phonon creation or annihilation, is neglected.

For the sake of exploring the qualitative features of this model, separability of this potential well will be assumed, i.e. motion parallel and perpendicular to the surface are assumed to be independent. First, one must ask what the energy levels are that the Schrödinger equation would predict for chemisorbed hydrogen. The perpendicular profile of the well gives rise to a discrete set of energy levels below the vacuum zero level (the energy levels crowding infinitely densely near the vacuum zero level) and a continuum of levels above that. It should be recalled that the ground state lies typically at $\hbar\omega/2 = 0.1$ eV above the bottom of the well (the total depth of which is on the order of 2.8 eV in this case). In the harmonic approximation, this would imply levels at approximately $E_{\perp n} = (n + 1/2)\hbar\omega = 0.1, 0.3, 0.5, 0.7, \ldots$ eV. To these energies should be added the eigenvalues of the motion parallel to the surface.

Atomic Band Structure

As a result of the periodicity parallel to the surface, one obtains a band structure parallel to the surface that describes the motion of the hydrogen

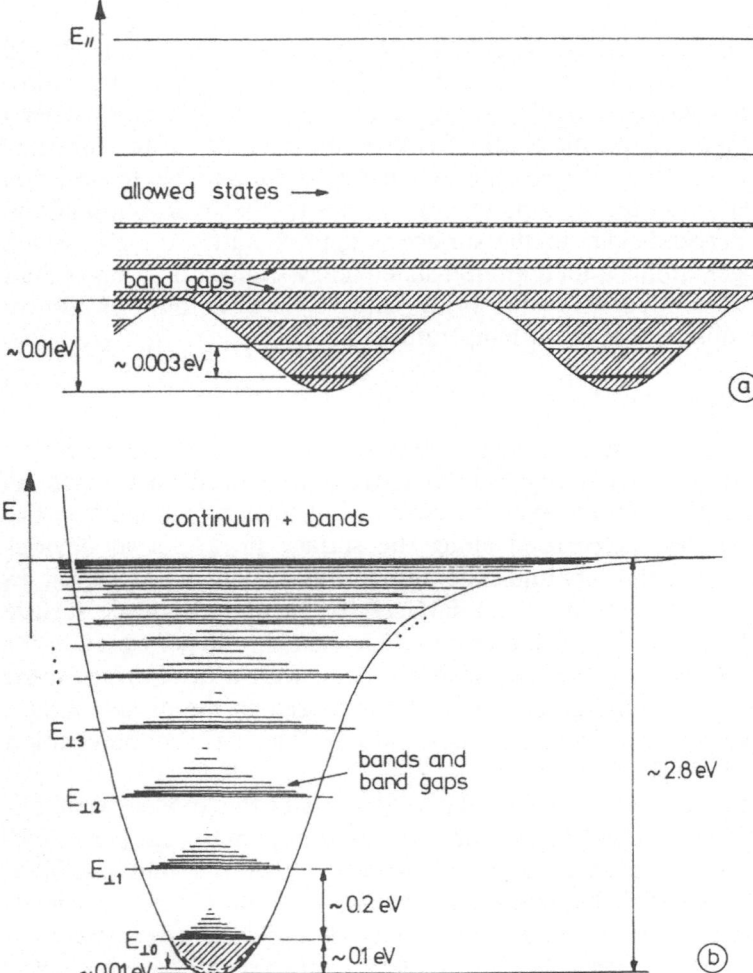

Fig. 8.12. Schematic atomic band structure for hydrogen chemisorbed on a metal surface with low diffusion barriers where the numerical values correspond approximately to hydrogen on Ni(111). (a) Atomic band structure for motion parallel to the surface. (b) Total atomic band structure, including motion perpendicular to the surface [8.39]

adatoms, neglecting their internal structure. As one moves up in energy from the bottom of the deepest well, one finds bands of allowed and bands of forbidden energies, as shown in Fig. 8.12. The energies and widths of these band gaps may be characterized more precisely. The kinetic energy of hydrogen moving parallel to the surface is, in the free-particle limit, $E = \hbar^2 k^2/2M$, where k is the wave number and M is the atomic mass. Band gaps occur when $k = n(\pi/a)$ (n is an integer), where a is the lattice constant, so that electronic band structures may be used with their energies scaled by a factor (m/M) (m is the electronic mass), i.e. for hydrogen, by a factor of 1/1840, to describe at least qualitatively the atomic band structure. Since the scale of band widths is approximately 5 eV in electronic band structures, the corresponding scale in the hydrogenic band structure will be approximately 0.003 eV, i.e. rather less than the expected periodic variations in the potential. Within each allowed band, one can fit a quasi continuum of energy levels, as many levels as there are unit cells on the surface, according to the familiar argument of periodic boundary conditions.

Concerning the band gaps, their widths ΔE_g can be approximated (when they are small) by the corresponding Fourier component V_g of the periodic potential. Compared with electronic band gaps, where V_g can be sizable for large g (due to the sharpness of the interaction potential near the nuclei), it is expected here that V_g will rapidly become quite small for large g, since the potential is rather smooth. This results in a band structure like that shown in Fig. 8.12a, where the periodic potential variations are taken to be approximately 0.01 eV. Below the crests of the periodic potential, slightly broadened, allowed levels (i.e. very large band gaps) occur. These are states that are largely localized at individual sites; above the crests, the band gaps rapidly become smaller. One may estimate from Fig. 8.12a that the hydrogen rms vibrational amplitudes in the ground state are on the order of 1/4 of the nickel lattice constant, i.e. on the order of 0.6 Å in qualitative agreement with earlier estimates.

Combining perpendicular and parallel motions, the energy-level diagram shown in Fig. 8.12b for chemisorbed hydrogen is obtained by simply adding the spectra of each dimension (since separability of motion parallel and perpendicular to the surface was assumed). For each level $E_{\perp n}$ of perpendicular motion, there is a band structure (of the type shown in Fig. 8.12a) extending indefinitely high above it due to parallel motion. Above the second level, these individual band structures overlap.

Distribution of Atoms Among Levels

The occupation of each allowed level, when hydrogen adatoms are present on the surface, must now be considered. If the adatoms are considered to be bosons, Bose-Einstein statistics apply, and a Boltzmann distribution, exp(-E_n/kT), results for the occupation of each level. If the adatoms are fermions, as deuterium might be, Fermi-Dirac statistics would apply, but at the experimental temperature of 110 K, this would make little difference. At zero temperature, all adatoms should be trapped in the lowest-energy sites (but hydrogen-hydrogen interactions may complicate the details of this picture at sufficiently high coverages). For a thermal energy of $kT \approx 0.01$ eV, a spread of the adatoms over many of the lower bands occurs, since these bands are only approximately 0.003 eV apart. Hence, one expects a sizable fraction of the adatoms to be in the localized site-bound states but also a reasonable number of adatoms to be in delocalized states. This conclusion is identical with the earlier picture of a substantial amount of disorder and a large hydrogen mobility.

8.3 The Interaction of Hydrogen with the (100) Surface of Palladium

8.3.1 Significance of the H/Pd(100) System

Thermal-Desorption Mass Spectrometry, work-function measurements, and LEED have provided a rather complete experimental description of the adsorption of hydrogen on the (100) surface of palladium [8.74]. Most important to the discussion here is the observation of a reversible order-disorder phase transition of the c(2x2)-H superstructure which forms at low temperatures on Pd(100). As in the case of hydrogen on Ni(111), this phase transition has been characterized by LEED [8.74]. Indeed, in the case of hydrogen on Pd(100), the situation is slightly more interesting, since more extensive model calculations, corresponding to a lattice gas on a square substrate, have been carried out both with an Ising model and with Monte Carlo simulations [8.27,28]. Consequently, additional microscopic details concerning the interactions among hydrogen adatoms on the Pd(100) surface may be inferred.

Experimentally, the adlayer of hydrogen on Pd(100) is also especially suitable to compare with theoretical calculations. Up to at least a half monolayer of surface coverage, and probably up to a full monolayer (defined to be an equal number of hydrogen adatoms and palladium atoms in the top, unreconstructed layer), only a single type of adsite is occupied on the surface, as judged by Thermal-Desorption Mass Spectrometry and work-function measurements. Furthermore, the hydrogen adatoms are ordered better below the order-disorder transition temperature on Pd(100) than on Ni(111). This assertion is based on the observation that the intensities of the fractional-order LEED beams are approximately 10% of the intensities of the integral-order beams on Pd(100) compared to 1-2% on Ni(111). As discussed in Sect. 8.2, a relative intensity of 10% for the fractional-order LEED beams implies very little disorder within the ordered overlayer of hydrogen. This

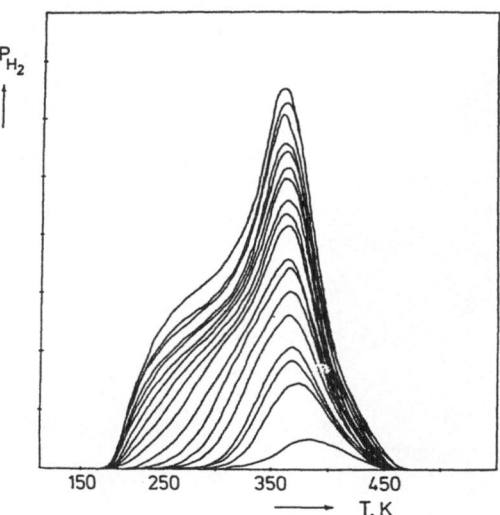

Fig. 8.13. A series of thermal-desorption spectra for hydrogen on Pd(100) after adsorption at 170 K, recorded with a heating rate of 14 K/s. Exposures, corresponding to increasing integrated areas, are the following: 0, 0.1, 0.15, 0.2, 0.3, 0.4, 0.5, 0.6, 0.8, 1.0, 1.2, 1.4, 1.6, 2.0, 2.4, 3.6, 4.4, and 6 L. The "0 L" curve represents the amount of hydrogen that is adsorbed from the residual gas atmosphere while the crystal is cooled to 170 K [8.74]

has the direct implication that the corrugation of the Pd(100) surface (i.e. the barrier to diffusion on the surface) is greater than that of the Ni(111) surface.

8.3.2 An Experimental Characterization of Hydrogen on Pd(100)

Coverage Calibration

As may be seen in Fig. 8.13, thermal-desorption mass spectra reveal the existence of two "states" of hydrogen adsorbed on Pd(100). The thermal-desorption spectra are calibrated to obtain known surface coverages by a separate determination of a half-monolayer coverage (θ = 0.5 adatoms per surface palladium atom). As discussed below, this calibration (or normalization) point is defined by an overlayer that is optimally ordered and corresponds to a c(2x2)-H superstructure. On this basis, the "high-temperature" state of adsorbed hydrogen in the thermal-desorption spectra saturates at θ = 1, and an additional 35% of this total can be adsorbed on the surface at 170 K. This means that the saturation coverage of (atomic) hydrogen on Pd(100) at 170 K in vacuo is θ = 1.35 corresponding to a density of adatoms on the surface of 1.8×10^{15} cm^{-2}. The fractional coverage θ is defined in terms of the number of Pd atoms in the surface layer, and a value of θ greater than unity by no means implies adsorption into a second layer. In fact, each Pd(100) unit cell has two twofold-symmetric bridging sites, one fourfold site and one on-top site associated with it, as may be seen in Fig. 8.14. If all of these high-symmetry sites were occupied, the surface coverage would be four.

The variation in the work function with surface coverage is presented in Fig. 8.15. The physically measured quantity is the change in work function

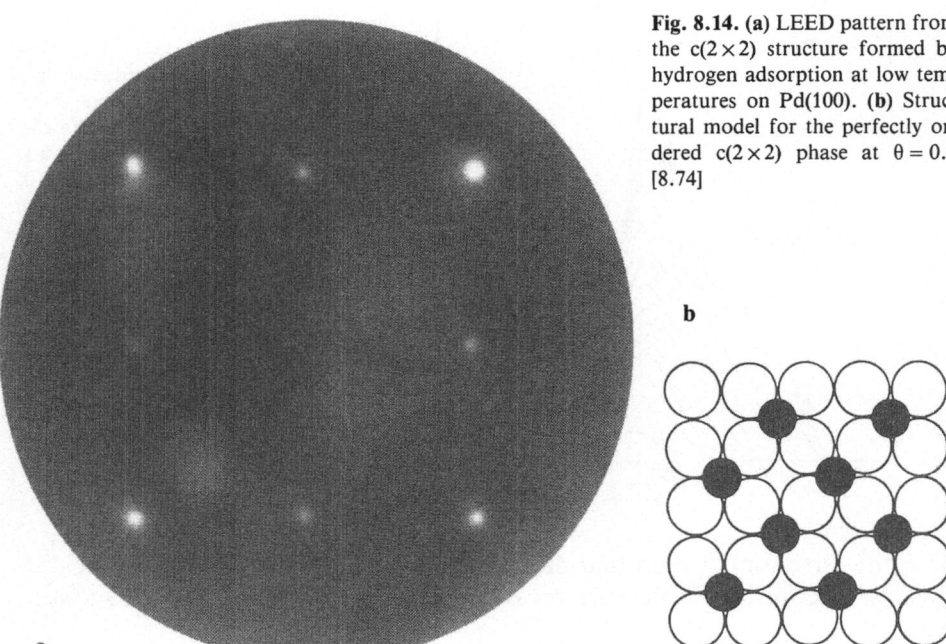

Fig. 8.14. (a) LEED pattern from the c(2×2) structure formed by hydrogen adsorption at low temperatures on Pd(100). **(b)** Structural model for the perfectly ordered c(2×2) phase at θ = 0.5 [8.74]

with exposure of the surface to hydrogen gas. However, the relationship between surface coverage and exposure to H₂ is known quantitatively from the calibrated thermal-desorption spectra of Fig. 8.13. Hence, the change in the work function may be displayed as a function of surface coverage as is done in Fig. 8.15. There are clearly two linear segments of the curve in Fig. 8.15 with the break near θ = 1 (actually at approximately θ = 0.9, but this difference is not significant considering experimental uncertainties). This

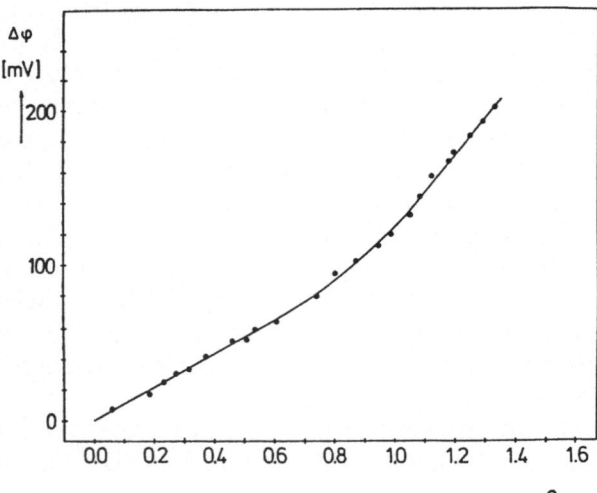

Fig. 8.15. The variation in the work function change with fractional coverage for hydrogen adsorption on Pd(100) [8.74]

suggests that the hydrogen adatoms occupy a single type of adsite at $\theta \leqslant 1$. (More precisely, the average dipole moment of the hydrogen adatoms is constant in this regime of coverage.) This observation is consistent with the thermal-desorption mass spectra of Fig. 8.13 which show only a single desorption "state" at coverages up to $\theta \sim 1$. Both the thermal-desorption spectra and the change in the work function suggest that other sites are occupied at coverages greater than unity. Whether these are in addition to those sites that are occupied at $\theta \leqslant 1$, or whether adsites of a different symmetry are occupied at $\theta > 1$ cannot be determined from these data.

Heat of Adsorption

As may be seen in Fig. 8.15, the change in the work function is a unique (i.e. not a multivalued) function of surface coverage and, consequently, may be used to evaluate the isosteric heat of adsorption q_{st} as a function of surface coverage. A measurement of $\Delta\varphi(T)$, parametric in the total pressure of H_2, yields an equilibrium family of isosteres. From these isosteres, the isosteric heat of adsorption may be determined, via the Clausius-Clapeyron equation, i.e.

$$\left[\frac{\partial \ln p}{\partial (1/T)}\right]_\theta = -q_{st}/k_B , \tag{8.5}$$

since the line $\Delta\varphi = $ constant intersects the family of isosteres at the equilibrium values of pressure and temperature which are appropriate for the coverage corresponding to the (chosen) value of $\Delta\varphi$. Performing this construction for various values of $\Delta\varphi$ (i.e. coverage) allows a determination of $q_{st}(\theta)$, and the results for hydrogen on Pd(100) are shown in Fig. 8.16. The value

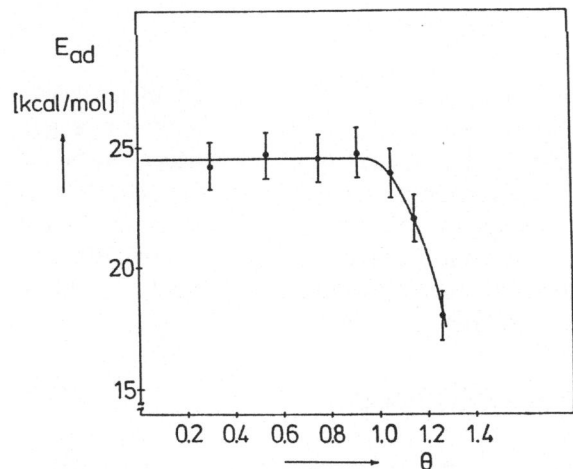

Fig. 8.16. The isosteric heat of adsorption as a function of fractional coverage for hydrogen on Pd(100) [8.74]

of q_{st} in the limit of zero surface coverage, 24.5 kcal/mole (corresponding to a Pd-H bond energy of 64.4 kcal/mole), remains constant up to $\theta \sim 1$, whereas at higher coverages there is a precipitous drop in q_{st} as may be seen clearly in Fig. 8.16. As will be seen later in this section, adatom-adatom interaction energies occur over the coverage range between zero and unity. These effects are not manifest in $q_{st}(\theta)$ in this coverage range, because either the attractive and repulsive interactions cancel, or the net interaction energy is within the experimental uncertainty of the measurement, probably the latter. Likewise, the precipitous decrease in q_{st} at $\theta > 1$ could result from at least three effects. First, there could be a change in the adsite at $\theta = 1$, which intrinsically represents a smaller Pd-H bond energy. Second, repulsive adatom-adatom interaction energies could become of considerable importance as "nearest-neighbor" sites are occupied at $\theta > 1$. Third (and closely related to the first effect), a "surface ligand effect" could be manifest in the decrease in q_{st}. Using the analogy between clusters and surfaces, as more ligands (adatoms) are added to metallic atoms in an inorganic or organometallic cluster (on a surface), the bond strength of each added ligand (adatom) generally becomes progressively weaker. In actual fact, all three of these effects probably combine to yield the observed decrease in q_{st} shown in Fig. 8.16.

Preexponential Factor

This evaluation of the heat of adsorption at equilibrium permits a determination of the preexponential factor of the (second-order) desorption reaction. At equilibrium, the rate of adsorption R_a is equal to the rate of desorption R_d, where

$$R_a = S(\theta)\frac{p_{H_2}}{(2\pi Mk_BT_g)^{1/2}} , \quad \text{and} \tag{8.6}$$

$$R_d = v_d(\theta)(n_s\theta)^2\exp[-E_a(\theta)/k_BT_s] . \tag{8.7}$$

Here, $S(\theta)$ is the coverage-dependent probability of adsorption of hydrogen (which is approximately equal to 0.5 in the limit of low coverage), $p_{H_2}/(2\pi Mk_BT_g)^{1/2}$ is the impingement flux of H_2 onto the surface, n_s is the concentration of sites on the surface (properly normalized to the definition of $\theta = 1$), $E_a(\theta)$ is the activation energy to desorption (which is equal to the isosteric heat of adsorption, the adsorption reaction being unactivated), and $v_d(\theta)$ is the preexponential factor of the desorption-rate coefficient. Equating the expression for R_a and R_d, and solving for v_d gives

$$v_d(\theta) = \frac{S(\theta)}{(n_s\theta)^2} \frac{p_{H_2}}{(2\pi Mk_BT_g)^{1/2}} \exp[E_a(\theta)/k_BT_s] . \tag{8.8}$$

The relationship between surface coverage and gas exposure necessary to obtain $S(\theta)$ (via differentiation) is embodied in the thermal-desorption spectra of Fig. 8.13. The functional $\theta(p_{H_2}, T)$ is known from the isosteres, obtained from a measurement of the change in the work function, and $E_a(\theta)$ is simply $q_{st}(\theta)$. Consequently, it is trivial to derive an expression for $v_d(\theta)$, which, for hydrogen on Pd(100), has a value of 8×10^{-3} cm^2/s in the limit of zero surface coverage, a "normal" value for the preexponential factor of a second-order surface reaction.

Island Formation

At low surface temperature, a well ordered c(2x2)-H LEED pattern is observed on Pd(100), as may be seen in Fig. 8.14a. The variation with surface coverage of the intensity of a fractional-order LEED beam (due entirely to the hydrogen superstructure) is shown in Fig. 8.17. The maximum in the curve occurs at $\theta = 0.5$, by construction, as explained earlier. However, this assumption is eminently reasonable in view of the fact that the intensity of the fractional-order LEED beams is approximately 10% of the intensity of the integral-order beams. Clearly, the c(2x2) superstructure, the optimum coverage of which is $\theta = 1/2$, is very well ordered on the surface. The nearly linear increase in the intensity of the fractional-order LEED beam as a function of coverage for $\theta \leq 1/2$, evident in Fig. 8.17, is indicative of island formation on the surface, even at extremely low coverages of hydrogen. Consequently, hydrogen atoms are added to the edges of ordered domains on the surface. Were the adatoms added within the existing overlayer, the intensity would scale with the square of the coverage rather than linearly with the cov-

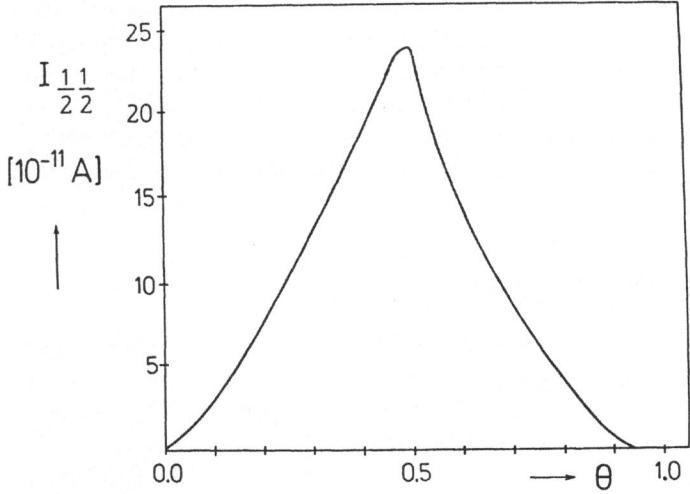

Fig. 8.17. Variation in the intensity of the (1/2, 1/2) beam with fractional coverage for hydrogen on Pd(100). The temperature is 105 K, and the incident electron energy is 52 eV [8.74]

erage as observed. The observation of island formation indicates the occurrence of attractive second-neighbor interactions for hydrogen on this surface, e.g. attractive pairwise interactions between the hydrogen atoms as they are distributed on the lattice shown in Fig. 8.14b. The quadratic decrease in the fractional-order LEED beam intensity with coverage for $\theta >$ 1/2, evident in Fig. 8.17, is a consequence of the rapid destruction of the ordered overlayer by a (relatively) smaller number of "misplaced" hydrogen atoms [as was discussed in Sect. 8.2 for the case of hydrogen on Ni(111)].

Probable Adsorption Site

No structural determination has been made for hydrogen on Pd(100) by LEED (nor by any other technique). The hydrogen adatoms are placed, somewhat arbitrarily, in the fourfold hollow sites in Fig. 8.14b. There are, however, three plausibility arguments one can make to show that the four-fold site is the proper assignment. First, experience suggests that hydrogen is adsorbed in high-coordination sites on metallic surfaces [8.75]. For example, on the (100) surface of nickel, High-Resolution Electron Energy Loss Spectroscopy has shown that hydrogen is adsorbed in fourfold hollow sites (with an unknown interlayer spacing and Ni-H bond length [8.64]). Second, the fourfold hollow sites correspond to octahedral sites which are occupied when hydrogen is dissolved in Pd [8.76]. Indeed, neutron diffraction has verified the existence, at low temperatures, of an ordered bulk phase, the structure of which corresponds to the c(2x2) adphase [8.77,78]. Finally, both thermal-desorption spectra and work-function measurements indicate that a single type of site is occupied up to $\theta = 1$. A value of the fractional coverage of unity would correspond to complete filling of the fourfold hollow sites, whereas the twofold sites would be only half occupied. The same (weak) argument would not exclude the on-top sites, but experience mitigates against this type of site [8.75]. A final decision concerning the binding site and especially the Pd-H bond length must await a full LEED analysis. Fortunately, the precise binding site does not influence any of the arguments in the remainder of this section concerning the observed order-disorder phase transition.

8.3.3 The Order-Disorder Phase Transition

The occurrence of an order-disorder phase transition is shown clearly in the variation of the fractional-order-LEED-beam intensity with temperature at various (constant) surface coverages. These data are presented in Fig. 8.18 and have been corrected for Debye-Waller (thermal vibrational) attenuation by measuring and removing the temperature dependence of the LEED beam intensity at low temperatures for a half-monolayer coverage. If the order-disorder phase transition temperature is assumed to correspond to the

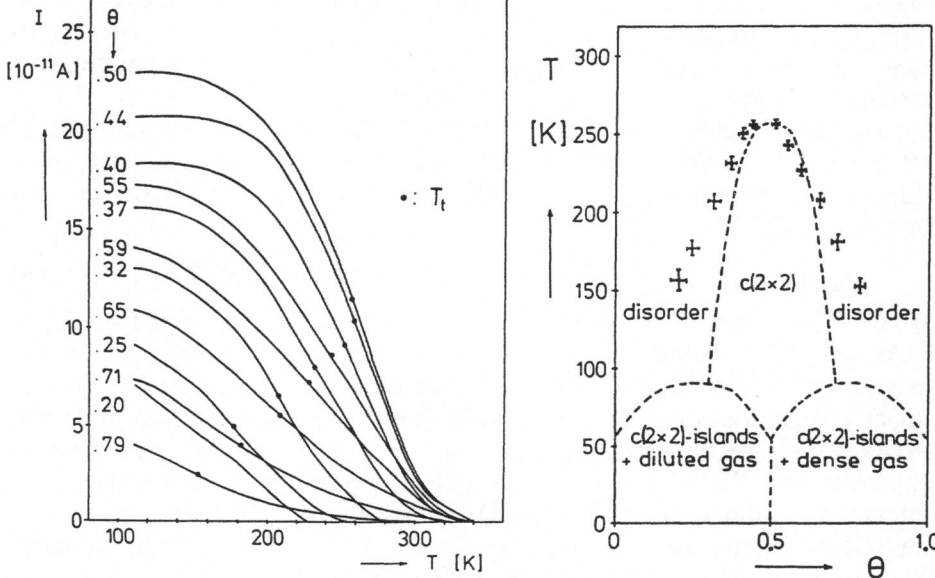

Fig. 8.18. Variation in the intensity of the (1/2, 1/2) beam with temperature (corrected for the effective Debye-Waller factor) at various fractional coverages for hydrogen on Pd(100). The points of inflection, which are indicated by dots, represent the order-disorder transition temperatures. The incident electron energy is 52 eV [8.74]

Fig. 8.19. Experimental phase diagram (*data points*) and theoretical phase diagram for a square lattice-gas model with $\omega_1 = -2\,\omega_2 = 0.51$ kcal/mole [8.49, 74]

inflection point in the I(T) curves of Fig. 8.18 (approximately equal to the temperature at which the intensity is half its maximum value), the phase diagram is given by the data points of Fig. 8.19 (with experimental uncertainties displayed explicitly). The dashed phase diagram corresponds to a lattice-gas calculation (see Sect. 8.3.4) which includes repulsive first-neighbor ($\omega_1 = 0.51$ kcal/mole) and attractive second-neighbor ($\omega_2 = -\omega_1/2 = -0.25$ kcal/mole) interactions, see Sect. 8.3.5 [8.49]. Two features should be noted especially in Fig. 8.19. First, the experimental phase diagram is broader than the calculated one (which assumes only first- and second-neighbor interactions). Also, the experimental phase diagram is not symmetric about $\theta = 1/2$, but rather appears to be approximately symmetric about $\theta \approx 0.47$. The latter is especially important (and will be discussed fully in Sects. 8.3.5-8) since all models involving just first- and second-neighbor interactions (or just first-neighbor interactions) predict a symmetric phase diagram about $\theta = 1/2$. The failure of the two-dimensional Ising model is immediately apparent from the experimental phase diagram shown in Fig. 8.19. The Ising model, which involves only repulsive first-neighbor interactions, predicts that even at T=0, no ordered adphase occurs for $\theta \leq 0.35$ and for $\theta \geq 0.65$. The importance of second-neighbor attractive interactions was already apparent from the island growth mechanism of the overlayer as discussed in Sect. 8.3.2.

The first-neighbor repulsive interaction ω_1 and the second-neighbor attractive interaction ω_2 are approximately related to the maximum two-dimensional critical temperature T_c^{max} [which is 260 K for hydrogen on Pd(100)], by [8.79]

$$\exp(\omega_1/2k_BT_c^{max}) = 2\exp(-\omega_2/2k_BT_c^{max}) + \exp(-\omega_2/k_BT_c^{max}) . \tag{8.9}$$

If $\omega_1 = 8|\omega_2|$, then $\omega_1 \approx 0.77$ kcal/mole; whereas if $\omega_1 = |\omega_2|$, then $\omega_2 \approx -0.39$ kcal/mole. Intuitively (and as verified later in this section), one would expect the ratio of $|\omega_2|$ to ω_1 to be within these rather wide limits. Consequently, it can be concluded, tentatively, that $0.4 \lesssim \omega_1 \lesssim 0.8$ kcal/mole and $-0.4 \lesssim \omega_2 \lesssim -0.1$ kcal/mole, and the ratio $R \equiv |\omega_2|/\omega_1$ lies between approximately 0.1 and unity. This conclusion is tentative because these interaction parameters, while capable of reproducing the coarse features of the phase diagram, cannot reproduce either its broadness or its asymmetry. This issue is discussed in detail in Sects. 8.3.6-8.

8.3.4 The Connection Between the Ising Model and the Lattice-Gas Model

Since detailed calculations have been carried out for a square-lattice gas, comparisons can be made between the theoretical results and the experimental results for hydrogen on Pd(100). In order to appreciate fully the nature of the calculations, however, it is necessary to understand the relationship between the Ising model and the lattice-gas model [8.27,28,49]. The lattice-gas model is an appropriate description of an adlayer that at low temperatures occupies sites of a superstructure that is in registry with the substrate lattice. The results of these model calculations can be compared with experimental order-disorder phase transitions, characterized by LEED, in order to determine (approximately) interatomic or intermolecular interaction energies. This is especially important since the magnitude of the latter is rather small (e.g. compared to adsorbate-substrate binding energies in chemisorbed systems), and they are difficult to calculate theoretically for a "real" as opposed to a "model" system [8.2,3,71,80,81]. The lattice-gas calculations typically involve Monte Carlo simulations [8.82] on a finite lattice [typically a (40x40) lattice] with periodic boundary conditions. The Monte Carlo calculations are of particular value since they simulate the limited homogeneity of the substrate and the finite coherence (or transfer) width of the LEED optics.

The square lattice under consideration is shown in Fig. 8.20 together with the first- (φ_{nn}), second- (φ_{nnn}), and third- (φ_3) neighbor interaction energies, and the three-body interaction energy (φ_t). The sign convention which has

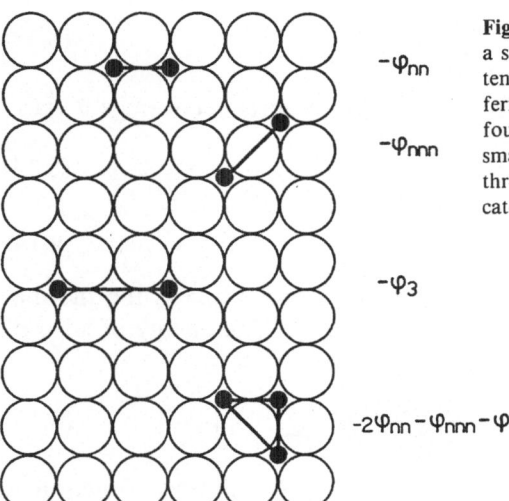

$-\varphi_{nn}$

$-\varphi_{nnn}$

$-\varphi_3$

$-2\varphi_{nn}-\varphi_{nnn}-\varphi_t$

Fig. 8.20. Schematic view of the (100) surface of a substrate (*large open circles*) the periodic potential of which provides a square lattice of preferred adsites which are assumed here to be the fourfold hollow sites. The adatoms are shown as small filled circles, and the various two- and three-body interactions between them are indicated by straight lines [8.27]

been introduced to make a connection with magnetic models should be noted particularly. All the adsorbate-adsorbate interaction energies, as well as the binding energy of the adsorbate to the substrate (ε), are assumed to be independent of both temperature and fractional surface coverage θ.

The magnetic Ising model may be introduced by writing the magnetic field as

$$-H = \frac{1}{2}\,(\varepsilon + \mu) + \frac{1}{4}\,\sum_{j(\neq i)}\varphi_{ij} + \frac{1}{8}\,\sum_{j\neq k(\neq i)}\varphi_{ijk}\,, \qquad (8.10)$$

where μ is the chemical potential of the thermodynamic system. The sums over $j(\neq i)$ extend over all jth neighbors of site i, and the sums over $j\neq k(\neq i)$ extend over all *pairs* of sites j and k of triangles i,j,k *once* with the prefactors related to corrections for overcounting (1/2 and 1/3, respectively).

The effective two- and three-spin exchange constants, J_{ij} and J_t, are related to the adsorbate-adsorbate interaction energies via

$$J_{ij} = \frac{1}{4}\,\varphi_{ij} + \frac{1}{8}\,\sum_{k(\neq i,j)}\varphi_t\,, \quad \text{and} \qquad (8.11)$$

$$J_t = -\frac{1}{8}\,\varphi_t\,. \qquad (8.12)$$

The fact that the two-spin-coupling constant of this Ising model contains a contribution from the three-body interaction (when the latter is "turned on") should be noted especially. This means that the ratios of spin-coupling constants in the Ising model are in general not identical to analogous ratios of interaction energies in the lattice-gas model. By combining (8.10-12), the expression for the magnetic field may be rewritten as

$$-H = \frac{1}{2} (\varepsilon + \mu) + \sum_{j(\neq i)} J_{ij} + \sum_{j\neq k(\neq i)} J_t . \qquad (8.13)$$

In this case the magnetization m is related to the surface coverage θ via

$$m = \frac{1}{N} \sum_{i=1}^{N} <S_i> , \qquad \text{and} \qquad (8.14)$$

$$\theta = (1 - m)/2 , \qquad (8.15)$$

where N is the number of (accessible) lattice sites, and $<S_i>$ is the thermal average of the "spin" of the ith site (where $S_i = \pm 1$). The energy of adsorption U_{ads} is related to the magnetic energy $U_I = <H_I>/N$ by

$$U_{ads} + \varepsilon\theta = U_I + mH + (\frac{1}{2} - m)\left[\sum_{j(\neq i)} J_{ij} + \sum_{j\neq k(\neq i)} J_t \right] . \qquad (8.16)$$

In the definition of the magnetic energy, the Ising Hamiltonian function is given by

$$H_I = -H \sum_{i=1}^{N} S_i - \sum_{i\neq j} J_{ij}S_iS_j - \sum_{i\neq j\neq k} J_tS_iS_jS_k . \qquad (8.17)$$

The symmetry of the Ising Hamiltonian implies that the transformation

$$H,J_t,\{S_i\} \rightarrow -H,-J_t,\{-S_i\} \qquad (8.18)$$

(which leaves the Hamiltonian invariant) leaves (also) the Gibbs energy of the system, the magnetic energy, and other thermodynamic variables unchanged. The replacement of $\{S_i\}$ by $\{-S_i\}$ is equivalent to replacing m by $-m$, (8.14), and θ by $1 - \theta$, (8.15). As a consequence of this symmetry invariance, the phase diagram for $- J_t$ is obtained from that of $+J_t$ by taking its mirror image in $\theta = 1/2$.

It should be noted that if J_t is assumed to be zero, the symmetry properties of these pairwise-interaction models imply that the phase diagram is symmetric, i.e. $T(\theta)$ is symmetric about $\theta = 1/2$, and the adsorption isotherm is antisymmetric, i.e. $\theta(\mu)$ is antisymmetric about the point $\theta = 1/2$ and $\mu = \mu_c$, where μ_c is the chemical potential corresponding to $H = 0$, see (8.13) [8.23]. On the other hand, the total energy of adsorption, $U_{ads} + \varepsilon\theta$ in (8.16), is not symmetric in this case even with J_t equal to zero due to the interaction contribution embodied in the third term on the right-hand side of (8.16). The total energy of adsorption is only symmetric about $\theta = 1/2$ for the special case of

$$\sum_{j(\neq i)} J_{ij} = 0$$

for which $\mu_c = 0$, and all adsorption isotherms intersect at a single point at $\theta = 1/2$.

8.3.5 The Lattice-Gas Model with First- and Second-Neighbor Interactions

The formalism presented in Sect. 8.3.4 was general. Now, the model will be made particular to include only nearest- and next-nearest-neighbor pairwise interactions. In order to make a connection with the experimental results for hydrogen on Pd(100), described in Sects. 8.3.2,3 it will be assumed further that $\varphi_{nn} < 0$ (a repulsive interaction, see Fig. 8.20), $\varphi_{nnn} > 0$ (an attractive interaction), and, of course, $\varphi_3 = \varphi_t = 0$. For $R \equiv J_{nnn}/J_{nn} < 0$, the $T = 0$ (ground) state of this Ising system described by (8.17) is the antiferromagnetic c(2x2) superstructure shown, for example, in Fig. 8.14b.

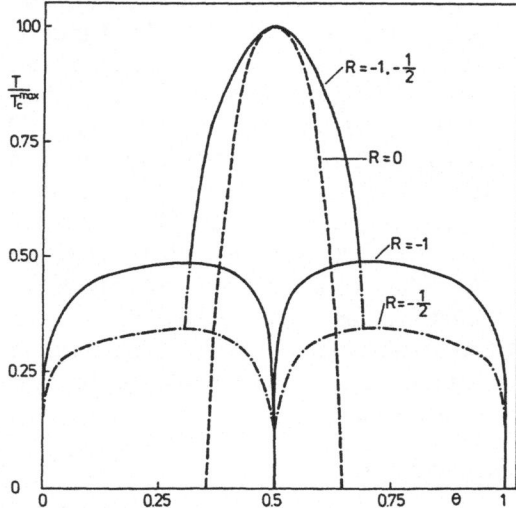

Fig. 8.21. The phase diagram in the temperature-coverage plane for three different values of R [8.27]

Phase diagrams for $R = 0$, $R = -1/2$ and $R = -1$, and with $\varphi_3 = \varphi_t = 0$ are shown in Fig. 8.21. The case of $R = 0$ corresponds to the Ising model on an infinite lattice. For $R = -1/2$ and $R = -1$, Monte Carlo calculations were carried out on finite lattices, generally (40x40) lattices in terms of atomic spacings, with typically a few hundred Monte Carlo steps per spin [8.80]. As will be discussed in Sect. 8.3.8, a value of R between $-1/2$ and -1 applies apparently to the case of hydrogen on the Pd(100) surface. As may be seen in Fig. 8.21, for $R = 0$ there is no tricritical point (or, more precisely, the tricritical temperature T_t is equal to zero). However, for $R \neq 0$ tricritical points appear as is evident from Fig. 8.21. As the magnitude of R increases, T_t approaches T_c^{max}, the maximum order-disorder phase-transition temperature, and the region over which the ordered c(2x2) superstructure exists is broader for $T > T_t$. However, the calculated phase diagram remains symmetric with respect to $\theta = 1/2$, which is not in keeping with the experimental observation (Fig. 8.19).

Finally, it should be noted that values of R < −1 are probably not significant physically, since in that case φ_3 (or J_3) and/or φ_t (or J_t) would almost certainly become important, see Sects. 8.3.6,7. Furthermore, the case of R > 0 is of much more theoretical [8.23,28] than physical interest.

8.3.6 Effects of Three-Body Interactions

As would be expected, the "turning on" of three-body interactions leads to extremely interesting features in the phase diagrams constructed by Monte Carlo simulations. More important is the fact that these new features can be related to the experimental measurements for hydrogen on Pd(100). In addition to the lattice gas, the c(2x2) superstructure and the lattice fluid phases, p(2x2) (at $\theta = 0.25$) and (2x2)-3X (at $\theta = 0.75$) superlattices appear in the phase diagram. The three-body interaction energy φ_3 is related to the three-spin-exchange-coupling constant J_t via (8.12). By analogy with the first- and second-neighbor interaction model, the parameter characterizing the strength of the three-body interaction is defined as $R_t \equiv J_t/J_{nn}$. Recall that the near-neighbor interaction is repulsive which, as defined here, means that $J_{nn} < 0$.

In fact, very small values of R_t can change the phase diagram significantly, and these are precisely the values of R_t that are probably of the most physical significance. Unfortunately, for small values of R_t and nonzero temperature, the Monte Carlo method is less useful due to rather large fluctuations. However, an approximate decoupling scheme can be used in conjunction with the Ising (infinite) lattice, namely,

$$S_iS_jS_k \simeq <S_i>S_jS_k + S_i<S_j>S_k + S_iS_j<S_k>$$
$$- 2<S_i><S_j><S_k> , \tag{8.19}$$

which means that the Ising Hamiltonian of the square lattice becomes

$$H_I \simeq -H \sum_{i=1}^{N} S_i - (J_{nn} + 4J_t<S>) \sum_{(i \neq j)_{nn}} S_iS_j$$
$$- (J_{nnn} + 2J_t<S>) \sum_{(i \neq j)_{nnn}} S_iS_j + 8NJ_t<S>^3 . \tag{8.20}$$

Here, the three summations extend over individual sites, nearest-neighbor sites and next-nearest-neighbor sites, respectively. Underlying (8.20) is the assumption that $<S_i>$ is the same for all lattices, which restricts its applicability to "paramagnetic" phases. To summarize the important manipulations, the approximation of (8.19) has been used to simplify the Ising Hamiltonian of (8.17) to the one given by (8.20). The latter contains only effective pairwise-interaction parameters dependent on the magnetic field via the magnetization which is proportional to $<S>$. As a consequence, the phase diagram is no longer symmetric about $\theta = 1/2$, in agreement with the experimental observations for hydrogen on Pd(100).

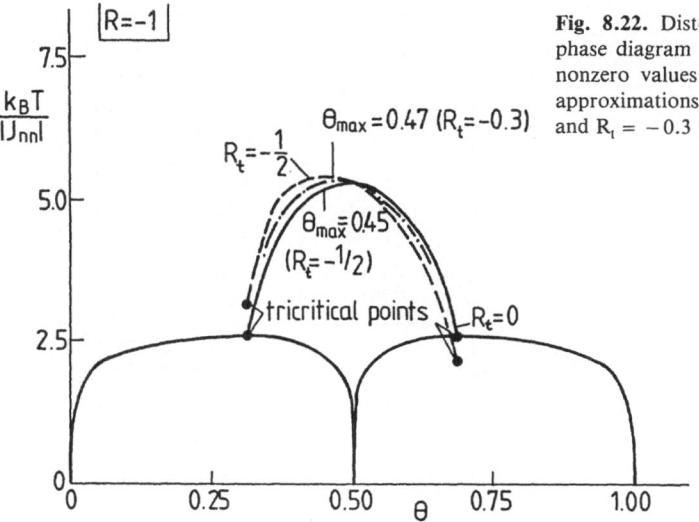

Fig. 8.22. Distortion of the symmetric phase diagram ($R = -1$, $R_t = 0$) due to nonzero values of R_t, calculated from approximations to (8.20), for $R_t = -1/2$ and $R_t = -0.3$ [8.27]

The distortion of the phase diagram due to R_t is shown in Fig. 8.22 for three cases: $R = -1$, $R_t = 0$; $R = -1$, $R_t = -0.3$; and $R = -1$, $R_t = -1/2$. Increasing the magnitude of R_t causes T_c^{max} to shift from $\theta = 1/2$ (i.e. the symmetric phase diagram for $R_t = 0$) to $\theta = 0.47$ (at $R_t = -0.3$) to $\theta = 0.45$ (at $R_t = -0.5$). Consequently, the asymmetry that is observed experimentally can be induced by the turning on of three-body interactions. For values of R_t that are not too small (in magnitude), Monte Carlo methods can be used to calculate the phase diagram. The qualitative results of such calculations are shown in Fig. 8.23 for various values of R_t (in terms of R). For $R/2 < R_t < 0$, the phase diagram is similar, for example, to that shown in Fig. 8.21 but is distorted asymmetrically. For $R_t < R/2$, a (2x2)-3X phase appears, the critical temperature of which occurs at $\theta = 0.75$.

The adsorption energy, given by (8.16) with appropriate modifications to include the three-body interactions [8.27], can be evaluated as a function of surface coverage, with the parameters T, R, and R_t. Examples of such calculations are shown in Fig. 8.24 where the infinite-lattice Ising model (or approximations thereto) applies at $T = 0$ and $T = \infty$, and Monte Carlo simulations have been carried out for intermediate temperatures. In Fig. 8.24a, the case of $R = -1/2$, $R_t = 0$ is shown; whereas the cases of $R = -1$, $R_t = 0$ and $R = -1$, $R_t = -1/2$ are shown in Figs. 8.24b and c, respectively. Experimentally, the isosteric heat of adsorption of hydrogen on Pd(100) was found not to vary significantly for fractional coverages of hydrogen below unity (Fig. 8.16). However, the variation in the adsorption energy that is evident in Fig. 8.24 is not large in view of the fact that $|J_{nn}|$ is certainly a rather small quantity, see (8.11). Indeed, a comparison between calculated and measured adsorption isotherms would appear to constitute a more definitive test of the calculations and/or an evaluation of the interaction parameters

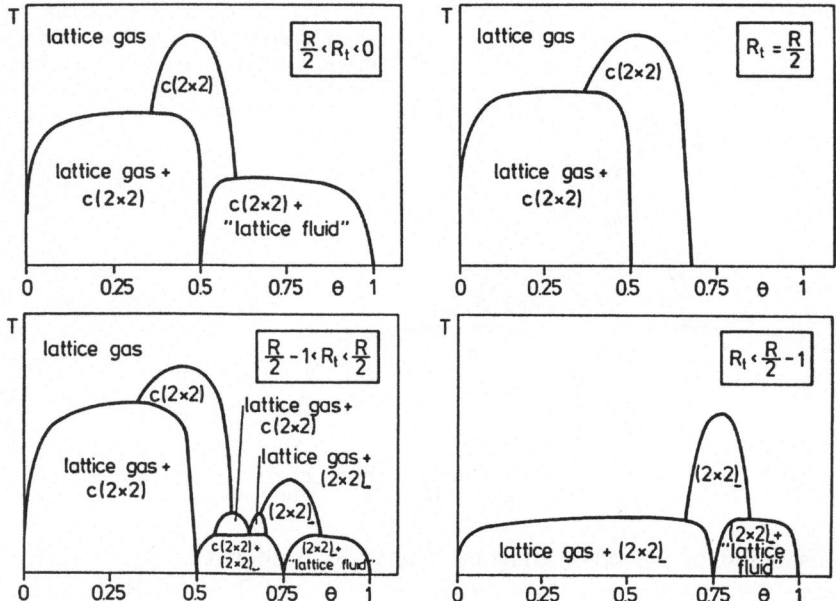

Fig. 8.23. Schematic temperature-coverage phase diagrams for various values of R_t [8.27]

than either the phase diagram or the variation of adsorption energy with sur-face coverage [8.27].

8.3.7 Effects of Third-Neighbor Interactions

Finally, the case of first-, second-, and third-neighbor interactions (with $\varphi_t = 0$) has been considered for the special case of $H = 0$, see (8.10,13). The ground-state ($T = 0$ K) phase diagram of this system, expressed as $R'(R)$ where $R' \equiv J_3/J_{nn}$, is shown in Fig. 8.25. It should be noted especially that for the physically realistic case of $R < 0$ [e.g. repulsive first-neighbor and attractive second-neighbor interactions, forming a c(2x2) superstructure at low temperature], a large and positive value of R' is necessary to induce a phase other than a c(2x2) one. For example, if $R = -1$, then a value of R' > 1 is required. This represents a repulsive and rather large third-neighbor interaction, neither of which would be expected experimentally. However, more extensive calculations are necessary with $H \neq 0$ in order to fully assess the effects of third-neighbor interactions. Tentatively, they appear to be less important than three-body interactions.

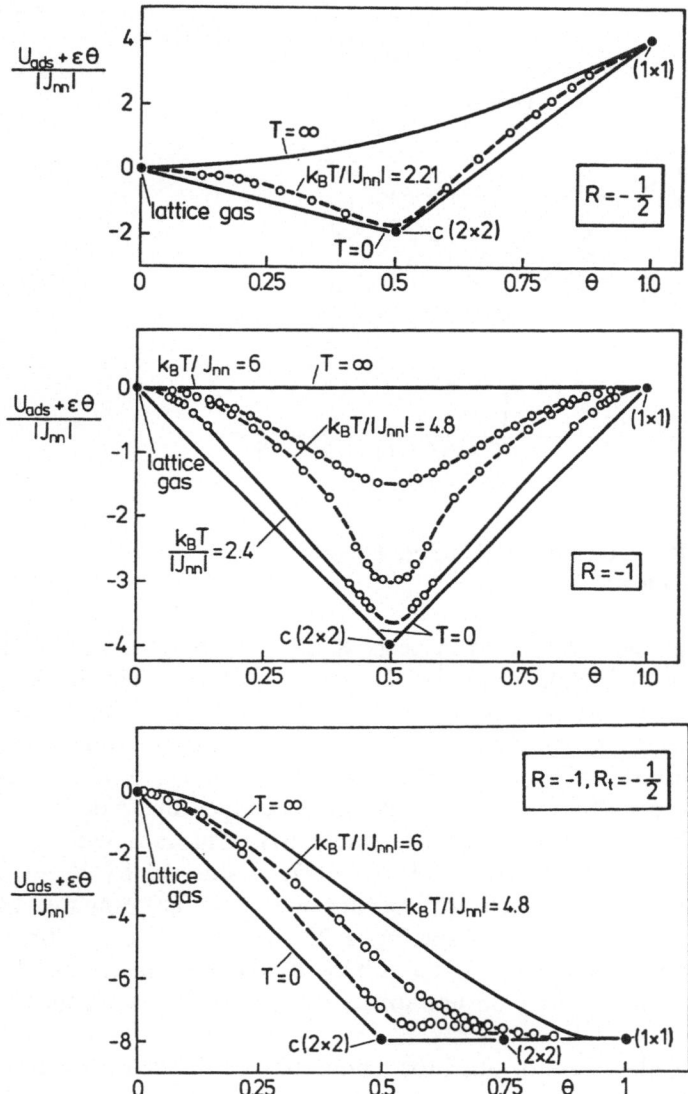

Fig. 8.24. Dimensionless total adsorption energy as a function of coverage for R = − 1/2 and R_t = 0 (*top*); R = −1 and R_t = 0 (*middle*); and R = −1 and R_t = −1/2 (*bottom*). Various temperatures are indicated for each case [8.27]

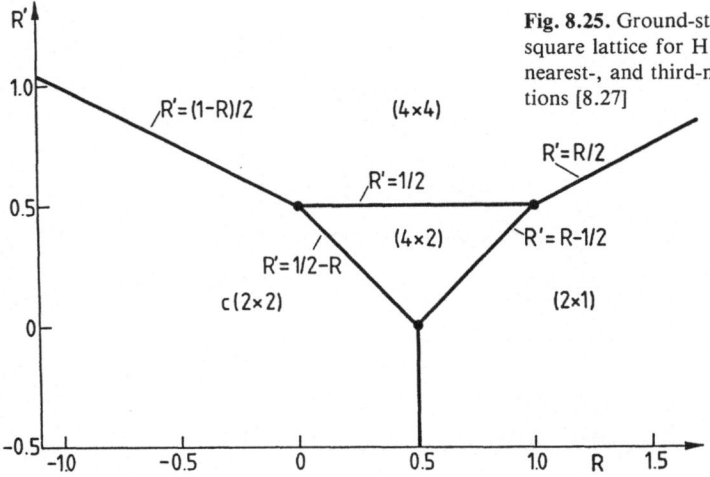

Fig. 8.25. Ground-state phase diagram of the square lattice for H = 0 with nearest-, next-nearest-, and third-nearest-neighbor interactions [8.27]

8.3.8 Comparison Between Experiment and Theory for Hydrogen on Pd(100)

The theoretical calculations discussed in Sects. 8.3.5-7 can be compared with the experimental data of Behm et al. [8.72], which were described in Sect. 8.3.3 and which concern the interaction of hydrogen with the (100) surface of palladium. Although a value of $R_t \approx -0.2$ to -0.3 can reproduce the experimentally observed asymmetry in the phase diagram (i.e. θ at T_c^{max} is not equal exactly to a half-monolayer), it cannot reproduce the wide coverage domain over which the c(2x2) superstructure exists. This is shown explicitly in Fig. 8.26 [8.28] for the case of $R_t = -1/2$. Values of $R_t < -1/2$ would shift the value of θ at which T_c^{max} occurs down too far compared to the experimental value of $\theta = 0.47$, see Fig. 8.22. Hence, although three-body interactions certainly appear to contribute to the asymmetry of the experimental phase diagram, they do not appear to be the sole reason for the broadening.

More insight into the cause of this broadening may be derived from considering the way in which the experimental phase diagram was constructed. As illustrated by the heavy points in Fig. 8.26b, the order-disorder phase-transition temperature at each coverage is assumed to correspond to that temperature at which the intensity of the fractional-order LEED beam has decreased to half its observed value at low temperatures. Within experimental uncertainty, this is equivalent to measuring the inflection points in the I(T) curves. In order to assess this way of constructing the phase diagrams from experimental data, Binder and Landau [8.27] have simulated the LEED data of Fig. 8.26b on a (40x40) lattice. The results are shown in Fig. 8.27a, where $\tilde{m}^2$ is the square of the order parameter of the c(2x2) lattice.

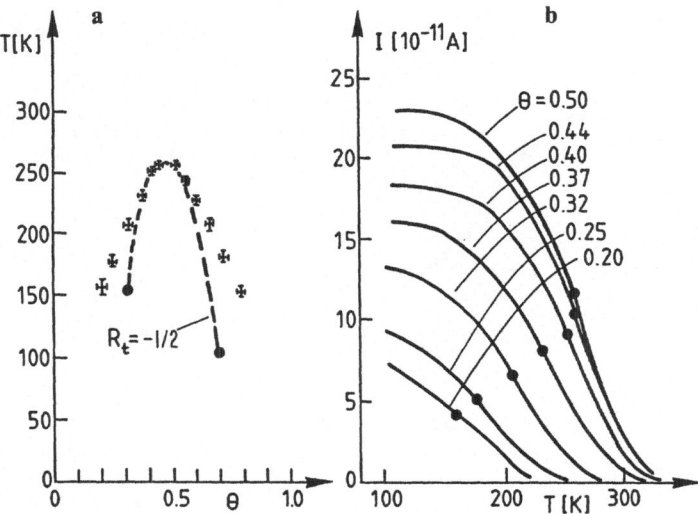

Fig. 8.26. (a) Comparison of the experimental phase diagram for hydrogen on Pd(100) to that of Fig. 8.22. **(b)** Experimental fractional-order LEED beam intensities as a function of temperature at various fractional surface coverages [8.27]

Note that, as required, $\tilde{m}^2 \rightarrow (1 - m)^2$ at low temperature, where $1 - m$ is the total order parameter, which is related to the surface coverage via (8.15). From this relation, it is obvious that at a coverage θ, the fraction of the adlayer that is ordered is $1 - m$, whereas the fraction that is disordered is m. It is then trivial to extract the phase diagram from the computer generated "data" by determining the temperature at which $\tilde{m}^2$ has decreased to half its value at $T = 0$, i.e. $(1 - m)^2$. This corresponds to the order-disorder temperature appropriate for that particular surface coverage. The phase diagram constructed in this way is shown as a solid line in Fig. 8.27b, whereas the phase diagram calculated directly (for $R = -1$, $R_t = 0$) is shown as dashed lines (cf. Fig. 8.22).

Clearly, the way in which the phase diagram is constructed from experimental data leads to additional broadening of the correct phase diagram, whereas three-body interactions lead to an asymmetry in the phase diagram, as observed experimentally. The broadening in the phase diagram of the simulated LEED "data" points up the importance of either the finite homogeneity of the surface or the finite transfer width of the experimental LEED optics, probably the former [8.83]. Experimental measurements on surfaces of high perfection (e.g. as deduced from thermal helium-beam scattering [8.84]) would be most useful to resolve this issue. The three-body interactions result in an asymmetry which shifts the surface coverage corresponding to the maximum critical temperature from 0.5 to 0.47 in agreement with experiment, see Figs. 8.22,26a. As is clear from Fig. 8.22, one would expect

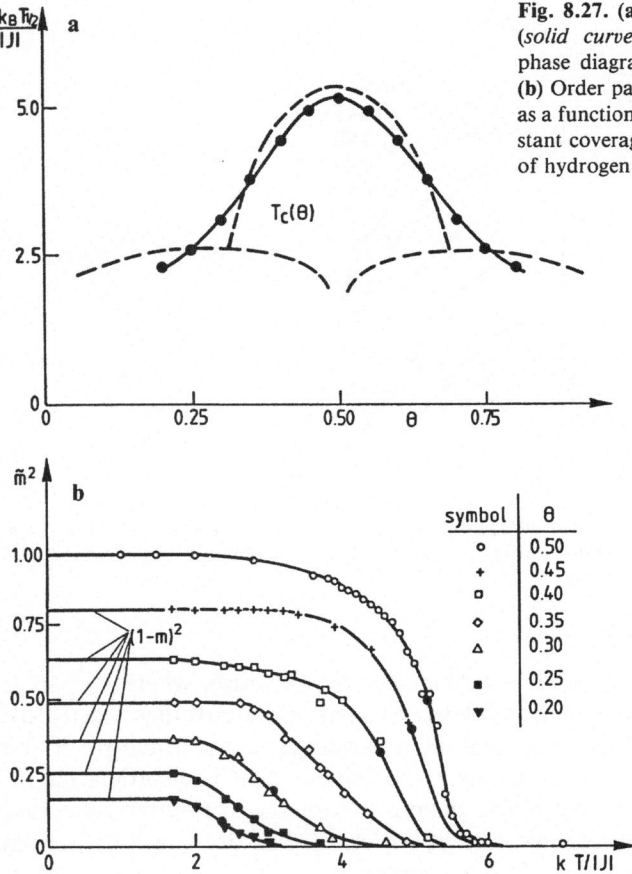

Fig. 8.27. (a) The simulated phase diagram (*solid curve*) compared to the "correct" phase diagram (*dashed curve*) (*top*) [8.27]. **(b)** Order parameter of the c(2 × 2) structure as a function of temperature for various constant coverages, obtained from a simulation of hydrogen on Pd(100)

to observe tricritical points for this set of interaction parameters. The experimental observation of such tricritical points on surfaces of high perfection would be exceedingly valuable in delimiting the values of the various interadsorbate interaction energies.

Additional theoretical calculations and simulations are needed to assess fully the importance of third-neighbor interactions. The observation of island formation at low coverages is an indication (but not a proof) of the importance of third-neighbor attractive interactions, but this depends also on the magnitude of the three-body interactions, as illustrated schematically in Fig. 8.23. On balance, it appears that for hydrogen on Pd(100), in the absence of any third-neighbor interactions, the parameters $-0.8 \lesssim R \lesssim -0.6$ and $-0.2 \lesssim R_t \lesssim -0.3$ provide the best description of the experimental data.

8.4 The Interaction of Hydrogen with the (110) Surface of Iron

8.4.1 Significance of the H/Fe(110) System

As a final example of order-disorder phase transitions on surfaces, the interaction of hydrogen with Fe(110) will be considered. This example is useful for a number of reasons. In Sect. 8.2, order-disorder transitions on a triangular lattice were considered [H on Ni(111)], while in Sect. 8.3, a similar discussion concerned a square lattice [H on Pd(100)]. The Fe(110) surface presents a lattice of a third type of symmetry: the unit cell is a centered rectangle (occasionally termed a "distorted triangular" lattice). Moreover, calculations have been carried out that are relevant to the adsorption of hydrogen on Fe(110) [8.32] which employ both Monte Carlo simulations [8.82] and more recently developed finite scaling transfer-matrix methods [8.85]. The latter calculational scheme appears particularly attractive due to its computational efficiency compared, for example, to more traditional Monte Carlo simulations.

Experimentally, the case of hydrogen on Fe(110) is an attractive system. The adsorption is entirely dissociative with no dissolution into the bulk (i.e. a closed, two-dimensional system) and with no disruption of the Fe(110) substrate (neither an adsorbate-induced relaxation nor reconstruction) [8.86]. Finally, there are two ordered superstructure are formed [8.87], which allow a closer comparison with theoretical calculations [8.32].

8.4.2 An Experimental Characterization of Hydrogen on Fe(110)

The experimental details of the adsorption on Fe(110) will be reviewed only briefly since there is a strong similarity with the results for hydrogen on Ni(111) and Pd(100) discussed in detail in Sects. 8.2,3, respectively. At a temperature of adsorption of 190 K, the saturation coverage θ of hydrogen on Fe(110) is approximately 0.7 adatoms per iron atom in the surface layer, as judged by a comparison of thermal-desorption spectra (normalized for ordered overlayers at fractional coverages $\theta = 0.5$ and $\theta = 0.67$, as described later in this section).

The thermal-desorption spectra show a single peak at coverages below $\theta = 0.5$. The peak temperature, corresponding to the maximum rate of desorption, shifts from approximately 500 K at low coverages to approximately 425 K at a half-monolayer coverage, indicative of second-order desorption kinetics. At coverages above a half monolayer, a second apparent adstate builds in and is manifest as a low-temperature shoulder on the dominant high-temperature peak. This low-temperature "state", as will be discussed subsequently in this section, is apparently due to repulsive adatom-adatom interactions at $\theta > 0.5$, rather than the occupation of a second type of binding site. The heat of adsorption remains constant and equal to 25.5 kcal/mole at coverages up to a half monolayer, implying an iron-hydrogen

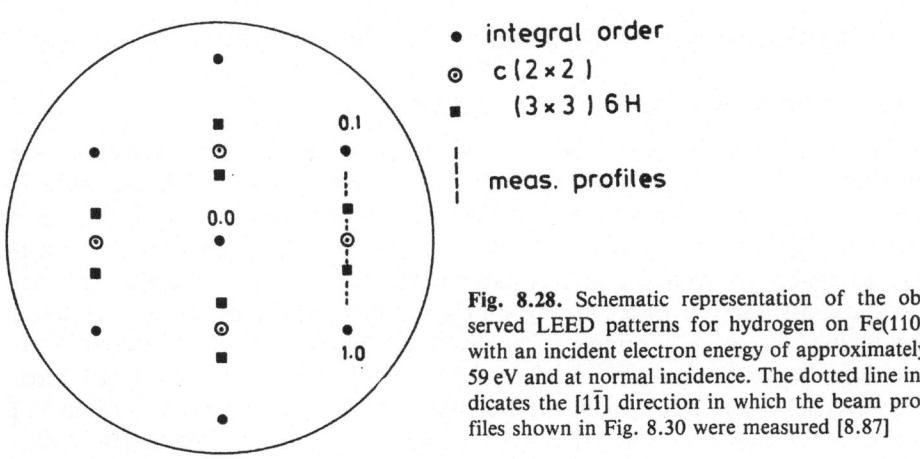

- integral order
- ⊙ c(2×2)
- ■ (3×3)6H

| meas. profiles

Fig. 8.28. Schematic representation of the observed LEED patterns for hydrogen on Fe(110) with an incident electron energy of approximately 59 eV and at normal incidence. The dotted line indicates the [1Ī] direction in which the beam profiles shown in Fig. 8.30 were measured [8.87]

bond energy of 64.8 kcal/mole. This value is virtually identical to that observed for Pd-H, 64.4 kcal/mole, and Ni-H, 62.8 kcal/mole.

At low temperatures and at fractional surface coverages near 1/2 and 2/3, well-ordered LEED patterns corresponding, respectively, to c(2x2)-H and (3x3)-6H superstructures are observed [the primitive lattice in the latter case is (3x1)-2H]. These LEED patterns are drawn schematically in Fig. 8.28. The real-space overlayers to which these LEED patterns correspond are shown in Fig. 8.29. In this figure, the hydrogen adatoms have been placed in the long-bridge sites. A multiple-scattering LEED calculation has suggested that this is the most probable location of the hydrogen adatoms, although the pseudo-threefold sites (which would be occupied by a slight translation of the

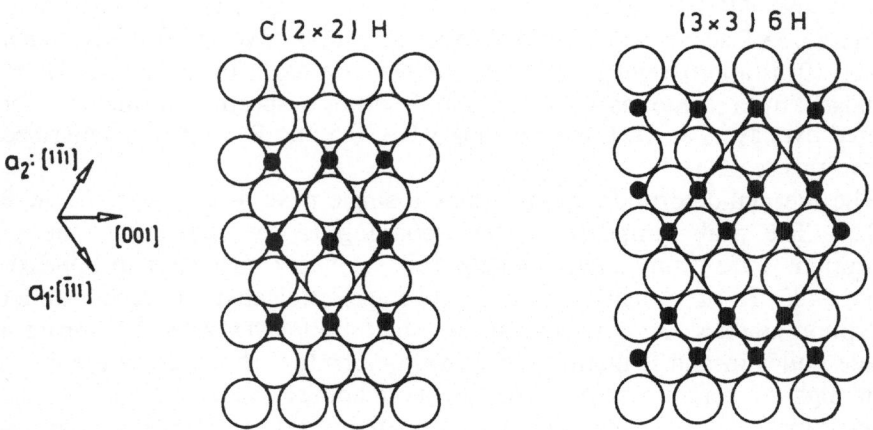

Fig. 8.29. Real-space structural models of the c(2×2)-H and (3×3)-6H overlayers of hydrogen on Fe(110). The full lines indicate the respective unit cells. The long-bridge sites are shown as the hydrogen (*small filled circles*) adsites [8.87]

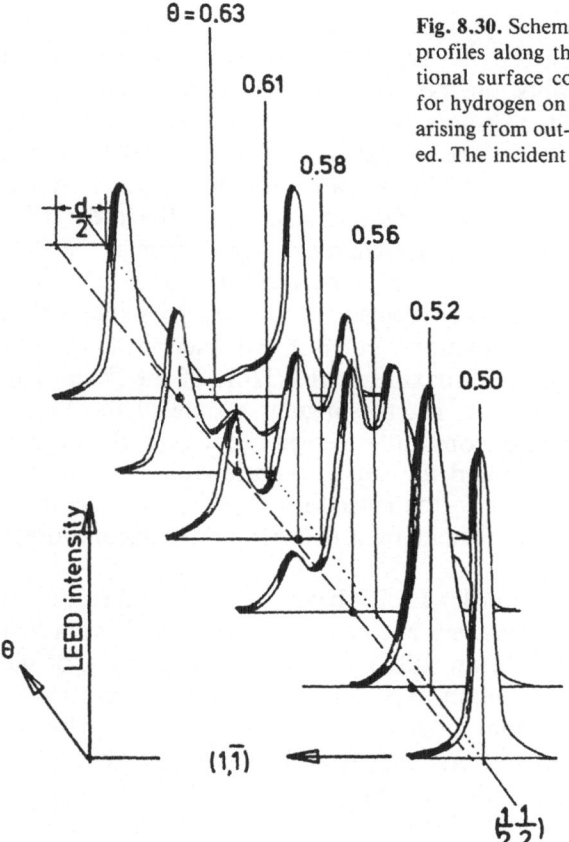

Fig. 8.30. Schematic representation of the angular beam profiles along the [1$\bar{1}$] direction as a function of fractional surface coverage between $\theta = 0.50$ and $\theta = 0.63$ for hydrogen on Fe(110). The position of the split spots arising from out-of-phase domain boundaries is indicated. The incident electron energy is 59 eV [8.87]

hydrogen atoms either in the [1$\bar{1}$] or [$\bar{1}$11] directions in Fig. 8.29, cf. Fig. 7.19b) cannot be excluded completely yet [8.88]. At fractional surface coverages between 0.5 and 0.67, there is a broadening and splitting of the fractional-order LEED beams in the [1$\bar{1}$] crystallographic direction. This direction is indicated by the dashed line in Fig. 8.28, and the effect is shown clearly for the (½ ½) beam in Fig. 8.30. These LEED results are discussed fully in Sect. 8.4.3 in connection with the observed order-disorder phase transitions.

8.4.3 LEED Observations and Order-Disorder Phase Transitions of Hydrogen on Fe(110)

As may be seen in Fig. 8.29, the c(2x2)-H and (3x3)-6H patterns may be indexed also as (2x1)-H and (3x1)-2H, respectively (using nonorthogonal lattice vectors). The (2x1) pattern yields circular fractional-order LEED beams, the angular width of which is limited by the transfer width of the LEED optics, i.e. there are large ordered domains of this structure on the surface (with respect to approximately 100 Å). Moreover, the absolute intensity of the fractional-order LEED beams, on the order of 10% of the integral-order beams, suggests there is very little disorder within these ordered domains. In this sense, hydrogen on Fe(110) is more similar to hydrogen on Pd(100) than hydrogen on Ni(111). The half-order LEED beams, due to the (2x1) super-structure, become visible at fractional surface coverages on the order of a quarter monolayer, and there is an approximately linear increase in the intensity of the half-order beams at coverages between 1/3 and 1/2. This implies either an attractive (or a weakly repulsive) third-neighbor interaction and/or an attractive three-body interaction. For $\theta \geq 0.4$, there is a continu-ous increase in the full-width at half-maximum (FWHM) of the fractional-order LEED beams with temperature, i.e. the angular profile of the LEED beam broadens with temperature at constant coverage. However, the limited homogeneity of both the substrate and the superstructure would not permit the observation of a sudden increase in the FWHM at $T = T_c$. Hence, these LEED data do not permit an assessment of first-order versus second-order phase transitions.

The $I(\theta)$ behavior of the $(\frac{1}{2}, \frac{1}{2})$ order LEED beam in the $[1\bar{1}]$ direction is shown in Fig. 8.30. As the fractional surface coverage increases above a half monolayer and at a temperature below 200 K, there is a broadening of the $(\frac{1}{2}, \frac{1}{2})$ LEED beam in the $[1\bar{1}]$ direction only, followed by a clear splitting of the $(\frac{1}{2}, \frac{1}{2})$ beam at $\theta = 0.53$. As may be seen in Fig. 8.30, third-order beams [due to the (3x1)-2H superstructure] first appear at $\theta = 0.56$, and they coex-ist with the split beams up to $\theta = 0.61$. At $0.62 \leq \theta \leq 0.67$, only the third-order beams are present. At $\theta = 0.56$, the split beams are present at $T \leq$ 220 K, whereas at $\theta = 1/2$ and at $\theta = 2/3$ the half-order and third-order beams, respectively, are stable to higher surface temperatures.

The splitting of the half-order LEED beams, which is observed at $0.53 \leq \theta \leq 0.62$, indicates that out-of-phase domain boundaries exist in the over-layer on the surface [8.89] (see also Sect. 3.57). Indeed, it is easy to see from Fig. 8.29 that the formation of out-of-phase domain boundaries is a natural consequence of the (2x1) → (3x1) transformation as the surface coverage is increased, i.e. as additional rows of hydrogen atoms are added to the over-layer in the [001] direction. An average periodicity of the domain walls gives rise to the beam splitting which is observed in LEED. The fact that LEED selects those domain boundaries that are periodic means that even a rather broad distribution of domain sizes can be observed as resolved, split beams in LEED.

At $0.62 \leq \theta \leq 0.67$, the third-order LEED beams have equivalent angu-lar profiles and intensities to the half-order LEED beams at coverages near θ = 0.5. Consequently, the (3x1)-2H superstructure consists of large domains (linear dimension > 100 Å) of high perfection just as the (2x1) superstructure. At $\theta > 0.67$, the third-order beams broaden in the $[1\bar{1}]$ direction suggesting

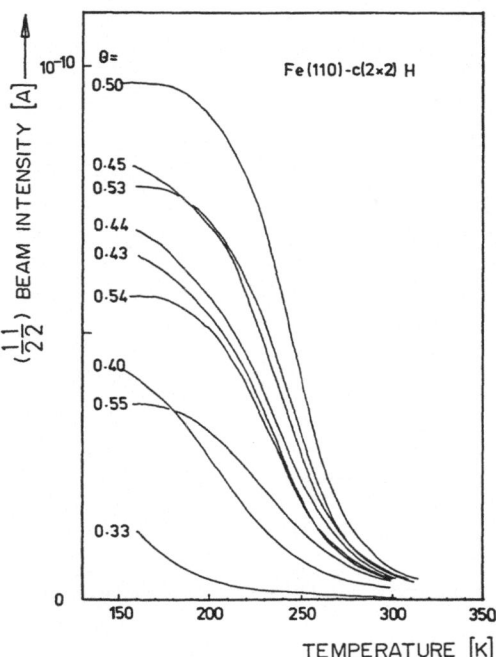

Fig. 8.31. Intensity of the (1/2, 1/2) LEED beam intensity as a function of temperature for various fractional coverages of hydrogen between 0.33 and 0.55 on Fe(110). The incident electron energy is 59 eV [8.85]

an incipient splitting with the formation of additional domain boundaries. Theoretically, a (4x1)-3H LEED pattern would be expected for this lattice near $\theta = 0.75$ [8.32], but such coverages could not be achieved in the experimental work [8.87]. It is thus possible that the broadening of fractional-order LEED beams observed at $\theta > 0.67$ would sharpen into fourth-order beams at coverages only slightly above the maximum thus far examined experimentally, namely, 0.7 monolayer.

Order-disorder phase transitions are observed for both the (2x1)-H and (3x1)-2H phases on Fe(110). For example, the variation in the diffracted half-order-beam intensity with temperature for $0.33 \leqslant \theta \leqslant 0.55$ is shown in Fig. 8.31. Taking the inflection point in these curves as the temperature at each coverage which separates the disordered lattice gas from the ordered (2x1)-H superstructure, results in the phase diagram of Fig. 8.32 for $0.33 \leqslant \theta \leqslant 0.55$. Data similar to those presented in Fig. 8.31, except for the third-order beams, allow the phase diagram for θ near 0.67 to be determined, as is shown by the squares in Fig. 8.32. The region of existence of the out-of-phase domain boundaries is indicated by the hatched regions of Fig. 8.32. The maximum critical temperature of the (2x1)-H phase is approximately 245 K, while the maximum critical temperature of the (3x1)-2H phase is approximately 265 K, as may be seen also in Fig. 8.32. These values are quite close to those observed for the (2x2)-2H phase on Ni(111), 270 K, and the c(2x2)-H phase on Pd(100), 260 K. Consequently, the interaction ener-

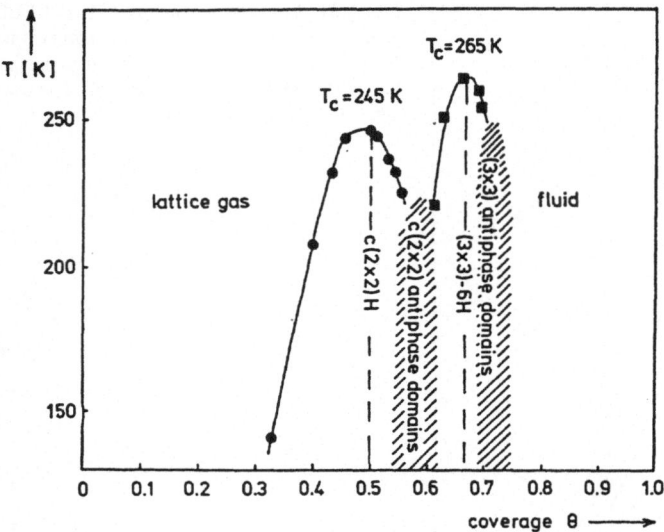

Fig. 8.32. Phase diagram of hydrogen on Fe(110). The filled symbols represent experimental data, and the hatched regions correspond to the regimes where out-of-phase domain boundaries occur [8.87]

gies among hydrogen adatoms on Fe(110) should be similar to those on Ni(111) and Pd(100). This issue is discussed more fully in Sect. 8.4.5.

8.4.4 Theoretical Predictions: A Lattice Gas on a Centered-Rectangular Lattice

A lattice gas on a centered-rectangular lattice with short-range, pairwise-additive interactions, as well as three-body interactions, has been considered in detail by Kinzel et al. [8.32]. Phase diagrams showing (2x1)-X, (3x1)-2X, and (4x1)-2X have been calculated using finite-size scaling transfer-matrix techniques [8.85] and have been verified with Monte Carlo simulations [8.82]. As will be seen subsequently in this section, these calculations provide a deeper understanding of the order-disorder behavior of hydrogen on Fe(110) as described in Sect. 8.4.3. The lattice-gas model used has been discussed in detail in Sect. 8.3.4.

Relevant Interaction Parameters

In this case, the ratios of interaction parameters are defined as

$$R_v \equiv J_v/J_2 \tag{8.21}$$

in terms of spin-exchange constants, or as

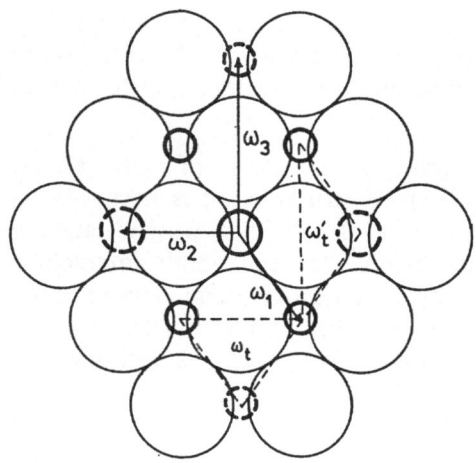

Fig. 8.33. Schematic representation of a (1×1) hydrogen overlayer (*small circles*) on the Fe(110) surface. The arrows indicate the (pairwise) mutual interactions between the hydrogen adatoms: φ_2 = first-neighbor interaction, φ_1 = second-neighbor interaction, and φ_3 = third-neighbor interaction. The dashed lines refer to the three-body interaction φ_t [8.87]

$$R_v^{lg} \equiv \varphi_v/\varphi_2 \tag{8.22}$$

in terms of the interaction energies embodied in the lattice gas analogy to the spin system. In both (8.21,22), v (= 1,3) and t indicate, respectively, second- and third-neighbor pairwise-additive interactions and three-body interactions. As may be seen in Fig. 8.33, each hydrogen atom [e.g. in a (1x1) overlayer] has four first-neighbors at a distance $a\sqrt{3}/2$ = 2.48 Å with an interaction energy φ_1, two second-neighbors at a distance a = 2.86 Å with an interaction energy φ_1, and two third-neighbors at a distance $a\sqrt{2}$ = 4.04 Å with an interaction energy φ_3. These intersite distances refer to the Fe(110) surface, iron having a lattice constant of 2.86 Å The R_v^{lg} may be written in terms of the R_v as follows:

$$R_1^{lg} = \frac{R_1 - 2R_t}{1 - 2R_t} \; , \tag{8.23}$$

$$R_3^{lg} = \frac{R_3}{1 - 2R_t} \; , \quad \text{and} \tag{8.24}$$

$$R_t^{lg} = \frac{2R_t}{1 - 2R_t} \; . \tag{8.25}$$

Under certain conditions, the model presented above either can be solved exactly (for an infinite lattice) or coincides with cases investigated previously, e.g.

1. With only even coupling (i.e. R_t = H = 0, and R_1 or R_3 equal to zero, where H is the magnetic field), this model is an Ising model on a triangular lattice with anisotropic interactions. Such a model has been solved exactly [8.18,25,90,91].

2. With $R_1 = R_3$, this model becomes the square-lattice gas described in Sect. 8.3. This model has been studied extensively both with Monte Carlo simulations [8.23,27], and with finite-size scaling transfer-matrix methods [8.26].

In order to construct the phase diagram representative of hydrogen on Fe(110) (Fig. 8.32) with (2x1)-H and (3x1)-2H structures, it is necessary to have repulsive φ_2 and φ_3 interactions (i.e. $J_2 < 0$ and $R_3 > 0$), and either a weakly repulsive or an attractive φ_1 interaction. The three-body interaction can be either attractive or repulsive, although an attractive one is more sensible physically in the case of hydrogen on Fe(110). Since the calculations involve transfer-matrix scaling, a brief discussion of this topic is necessary.

Transfer-Matrix Scaling

Transfer-matrix scaling calculations are based on infinite strips of finite width N (in units of atomic spacings). The grand partition function of the strip can be written as the trace of the product of a transfer matrix T, i.e. Tr T^∞. The transfer matrix contains Boltzmann weighting factors of adjacent rows only. The largest (positive and real) eigenvalue λ_o of T determines the Gibbs energy of the system via

$$G = - \frac{k_B T}{N} \, ln\lambda_o \, . \tag{8.26}$$

Spin-spin correlations are embodied in λ_1, the second largest eigenvalue (in magnitude), which may be complex. In particular, a correlation length ξ is defined by

$$\xi^{-1} = lln \left(\frac{\lambda_o}{|\lambda_1|} \right) , \tag{8.27}$$

where (8.27) gives $\xi(T,1/N)$. The asymptotic value of ξ (as $N \rightarrow \infty$) is given by [8.92]

$$\xi(t, \frac{1}{N}) = b\xi(b^{1/\nu}t, \frac{b}{N}) , \quad \text{where} \tag{8.28}$$

$$t \equiv \frac{|T - T_c|}{T_c} , \tag{8.29}$$

and b is taken to be N/N' in order to obtain T_c and the critical exponent ν, by comparing $\xi(T,1/N)$ for two strips of width N and N'. Even for quite small values of N (e.g. N = 3 or 4), good estimates of T_c and ν for the infinite system are obtained. In order to describe the case of hydrogen on Fe(110), the strips are chosen to be infinite in the φ_2 direction and finite in the φ_1 direction, with periodic boundary conditions employed in the latter case. The transfer matrix is a $(2^{2N} \times 2^{2N})$ matrix.

Calculated Phase Diagrams

Phase diagrams calculated using these finite-size scaling transfer-matrix techniques for $N/N' = 2/4$ and for various sets of interaction parameters are shown in Fig. 8.34. The phase diagram is symmetric about $\theta = 1/2$ when the three-body interaction is set equal to zero, i.e. $R_t = R_t^{lg} = 0$, as may be seen in Fig. 8.34e. For attractive three-body interactions with $-1/2 \leqslant R_t^{lg} \leqslant -1/3$, the (3x1)-H phase, which is not observed experimentally, is suppressed. This is clearly evident in Figs. 8.34a-d. Since φ_2 and φ_3 are both assumed to be repulsive (R_3, $R_3^{lg} > 0$), neither first-order phase transi-

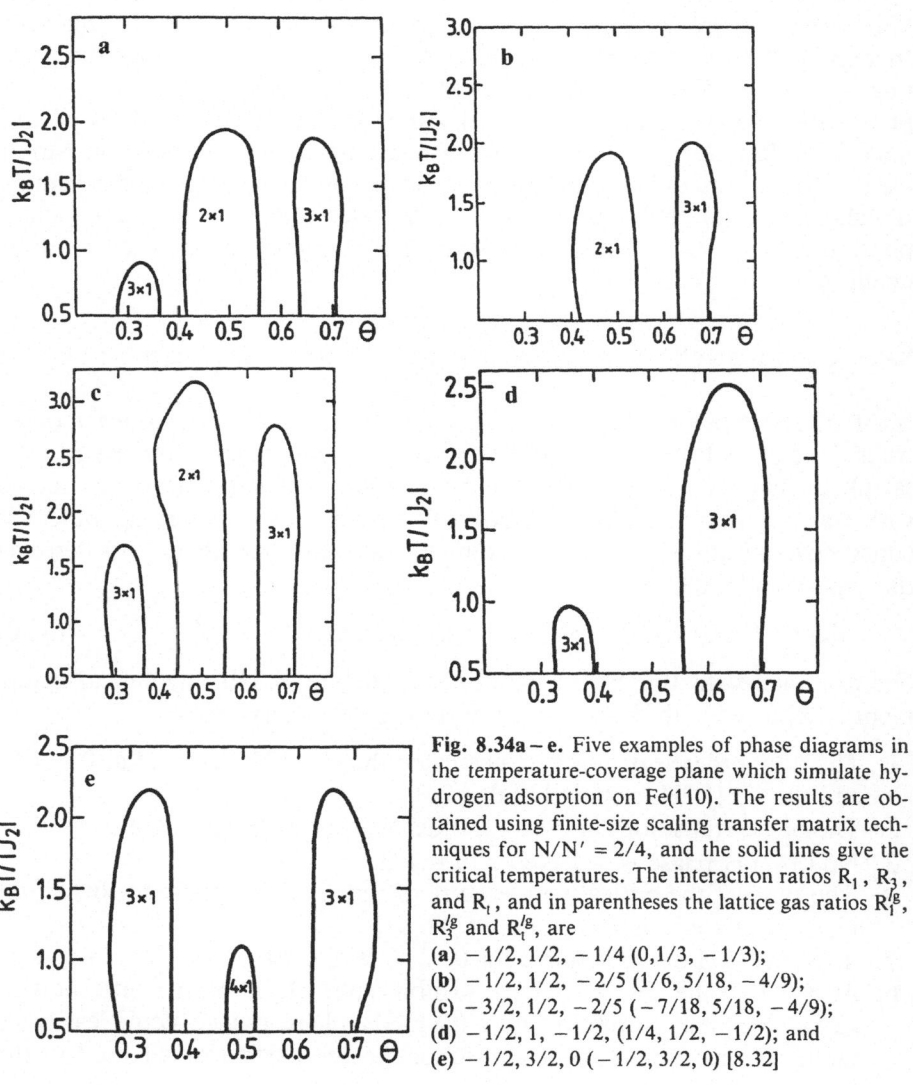

Fig. **8.34a – e.** Five examples of phase diagrams in the temperature-coverage plane which simulate hydrogen adsorption on Fe(110). The results are obtained using finite-size scaling transfer matrix techniques for $N/N' = 2/4$, and the solid lines give the critical temperatures. The interaction ratios R_1, R_3, and R_t, and in parentheses the lattice gas ratios R_1^{lg}, R_3^{lg} and R_t^{lg}, are
(a) $-1/2$, $1/2$, $-1/4$ (0,1/3, $-1/3$);
(b) $-1/2$, $1/2$, $-2/5$ (1/6, 5/18, $-4/9$);
(c) $-3/2$, $1/2$, $-2/5$ ($-7/18$, 5/18, $-4/9$);
(d) $-1/2$, 1, $-1/2$, (1/4, 1/2, $-1/2$); and
(e) $-1/2$, 3/2, 0 ($-1/2$, 3/2, 0) [8.32]

tions nor tricritical points are observed in Fig. 8.34, i.e. there is no tendency for the formation of ordered islands at low fractional surface coverages. Longer-range attractive pair-interaction energies would lead to such features in the phase diagram, but to this point they have not been observed experimentally. The phase diagrams of Figs. 8.34a-c show critical temperatures of the (2x1)-X and (3x1)-2X phases which are of similar magnitude, as observed experimentally for hydrogen on Fe(110). Indeed, the calculated phase diagram of Fig. 8.34b is quite similar to the experimental one shown in Fig. 8.32. The attractive J_1 interaction in Fig. 8.34b results in a repulsive φ_1 interaction due to an attractive three-body interaction ($R_t < 0$) via (8.23). As may be seen in Figs. 8.34b-c, increasing the attractive φ_1 interaction stabilizes (px1) phases generally, and the (2x1) and (3x1) phases, in this case. Increasing the repulsive strength of φ_3 (increasing R_3) suppresses the (2x1) phase at $\theta = 0.5$ (Fig. 8.34d) and results in the formation of a (4x1)-2X phase (Fig. 8.34e). The qualitative features of the phase diagrams of Fig. 8.34, including the second-order phase transitions, were verified via Monte Carlo simulations on a (60x60) lattice (with considerably greater computational effort than that required by the transfer scaling technique). These phase diagrams and the associated critical exponents are discussed in more detail by Kinzel et al. [8.32].

8.4.5 Comparison Between Experiment and Theory for Hydrogen on Fe(110)

Most of the experimental observations concerning the order-disorder phase transitions of hydrogen on Fe(110), which are summarized in the experimental phase diagram of Fig. 8.32, can be reproduced by the lattice-gas model with transfer-matrix scaling calculations. Insofar as the lattice-gas model is concerned, the following set of parameters appears (tentatively) to describe the experimental data best

$$\{R_1, R_3, R_t; R_1^{/g}, R_3^{/g}, R_t^{/g}\} = \{-1/2, 1/2, -2/5; 1/6, 5/18, -4/9\} . \tag{8.30}$$

For example, with this set of parameters, see Fig. 8.34b, theoretical agreement is achieved with the following experimental observations:

1. The only ordered phases that occur are the (2x1)-H, centered at $\theta = 1/2$, and the (3x1)-2H, centered at $\theta = 2/3$.
2. The phase transitions are second-order with no tricritical points.
3. No (3x1)-H phase is observed at $\theta = 1/3$.
4. The ratio of the critical temperature of the (3x1)-2H phase to that of the (2x1)-H phase is approximately 1.08.
5. The (3x1)-2H phase is stable for $0.62 \leqslant \theta \leqslant 0.70$.
6. At $\theta > 0.7$, the (3x1)-2H phase exhibits split LEED beams with the suggested incipient formation of a (4x1)-3H phase (which could easily be made to occur via a slight adjustment to the parameter set used in the model calculations).

The only two obvious disagreements between experiment and theory concern the broadness of the experimental (2x1)-H phase toward low fractional surface coverages (Fig. 8.32), and the absence of beam splitting for $0.53 \leqslant \theta \leqslant 0.62$ in the calculations (due to out-of-phase domain boundaries). As explained in detail in Sect. 8.3, the first disagreement is only apparent and is due to the finite homogeneity of the substrate and hence the superstructure (and, consequently, due to the way in which the LEED data are analyzed experimentally). The second disagreement is an indication of subtle residual imperfections in the (basically correct) theoretical description.

The optimum set of ratios of interaction energies for hydrogen on Fe(110) is given by (8.30). In order to define the individual interaction energies from these ratios, it is necessary to know a priori one of the φ_i. An estimate can be made of the various φ_i by noting that the critical temperature for hydrogen on Ni(111), 270 K, is essentially identical to those for hydrogen on Fe(110). In the former case, the interaction energy at a distance of 2.48 Å on the Ni(111) surface was found to be approximately 1.5 kcal/mole (Sect. 8.2.7). Consequently, it would not be unreasonable to adopt this value for φ_2 for hydrogen on Fe(110), which corresponds to sites which are also separated by 2.48 Å Once this choice is made, the set of interaction energies may be calculated from (8.30), and the result is

$$\{\varphi_1, \varphi_2, \varphi_3, \varphi_t\} = \{0.25, 1.5, 0.4, -0.7\} \tag{8.31}$$

in units of kcal/mole.

To summarize, finite-size scaling transfer-matrix techniques as applied to a lattice-gas model can faithfully and efficiently simulate the observed order-disorder phase transitions which occur in the case of hydrogen chemisorbed on the (110) surface of iron. The calculation yields reliable values for the ratios of the various interadsorbate interaction energies. Assuming the "transferability" of a single parameter φ_2 from hydrogen on Ni(111) at the same interadsorbate separation, allows a determination of φ_1, φ_3, and φ_t for hydrogen on Fe(110) (Fig. 8.33). This set of interaction energies is given by (8.31).

9. Chemical Reactions at Surfaces and LEED

LEED can yield valuable information concerning chemical reactions that take place at surfaces. The crystallographic data obtained by analyzing LEED intensities (Chaps. 7 and 12) provide an important, although indirect, means of access to such information. By identifying the structure of relatively stable atomic and especially molecular species adsorbed on crystalline substrates, one may infer the primary reaction mechanisms and pathways.

For example, acetylene (C_2H_2) adsorbs molecularly on Pt(111) below room temperature, with the C-C axis parallel to the surface, as was indicated in Fig. 7.48 [9.1]. In the presence of additional hydrogen and above room temperature, the acetylene molecules convert to ethylidyne species ($\equiv C-CH_3$), the C-C bond axes of which are perpendicular to the metal surface and which are bonded to three metal atoms in a hollow site [9.2], as was shown schematically in Fig. 7.50. This indicates that, on Pt(111) at least, C-H bond breaking is more likely than C-C bond breaking for acetylene. The same conclusion is found to hold for related molecules such as ethylene (C_2H_4), methylacetylene (C_3H_4), propylene (C_3H_6) and both cis- and trans-2-butene (C_4H_8) [9.3]. It is known, by contrast, that C-C bond breaking is the more likely reaction on Ni(111) [9.4] and on Ru(0001) [9.5]. This behavior can be related to the selectivity and activity of catalytic reactions.

LEED can also monitor surface reactions through the observation of a loss of order that results from the chemical reactions. Since this is a more direct use of LEED than the detailed crystallographic data, we shall devote this chapter to an example of a reaction that has been monitored in this way. It should be stressed that the information obtained with LEED becomes particularly powerful when combined with complementary data due to other techniques (cf. Chap. 11). This is amply demonstrated in the present chapter.

9.1 Monitoring Surface Reactions by LEED

Since LEED is particularly sensitive to overlayers which possess long-range order, it should be possible to monitor surface reactions which destroy this long-range order. This can be achieved, for example, by observing the loss of intensity of the fractional-order LEED beams which are due to one of the reactants on the surface. If, in addition, there is some partial disorder within an otherwise ordered reactant, it should be possible to disentangle the reaction of the ordered part of that reactant (monitored with LEED) from the reaction of all of the reactant (monitored with a probe of the total surface concentration, such as Auger-Electron Spectroscopy or X-ray Photoelectron Spectroscopy). That this is possible in practice has been demonstrated by Yates et al. [9.6] for the reaction between hydrogen and oxygen to form water on the Rh(111) surface. In this case, the adsorbed hydrogen forms no ordered LEED pattern on the surface [9.7], and H_2O desorbs immediately at the temperature of reaction [9.6]. Only oxygen adatoms form an ordered LEED pattern: a p(2x2) superstructure at a fractional surface coverage of 0.25, and three independent (1x2) domains rotated by 120° with respect to one another at $\theta = 0.5$ [9.8]. Both superstructures give rise to an apparent (2x2) LEED pattern but with different intensities of the fractional-order beams. Hence, this system is ideal to monitor the rate of reaction between hydrogen adatoms and ordered oxygen adatoms to produce water, by monitoring the decrease in intensity of the fractional-order oxygen LEED beams upon reaction.

On Rh(111), oxygen adsorbs in a disordered atomic overlayer at approximately 100 K and undergoes an activated ordering reaction beginning near 200 K, with an activation energy of approximately 13.5 kcal/mole [9.8]. Upon further heating, activated disordering begins near 300 K, the activation energy of which is coverage dependent. At essentially saturation coverage of a half-monolayer, this activation energy is approximately 8.2 kcal/mole [9.6,8]. The reaction to form water occurs at conveniently low surface temperatures, above approximately 275 K, and at convenient pressures of hydrogen, for example, between approximately 10^{-8} and 10^{-6} Torr. The (111) surface of rhodium is both unreconstructed and unrelaxed, and hence there are no complications due to a disruption of the rhodium substrate when analyzing the observed rates of the chemical reaction [9.9,10].

9.2 The Adsorption of Oxygen on Rh(111) at 335 K

The coverage-exposure relationship for oxygen on Rh(111) at 335 K is shown in Fig. 9.1. These data were obtained by exposing the crystal to O_2 and then using Auger-Electron Spectroscopy to measure both the rhodium (303-293 eV) and the oxygen (510-497 eV) peak-to-peak amplitudes. The relative coverages θ of chemisorbed oxygen were determined using the O/Rh intensity ratio, correcting the data for their appropriate sensitivity factors and for systematic variations in attenuation of the intensity of the rhodium transition by the overlayer. At saturation coverage of oxygen, a 22% attenuation of the rhodium (293 eV) Auger peak was observed. It was shown that elec-

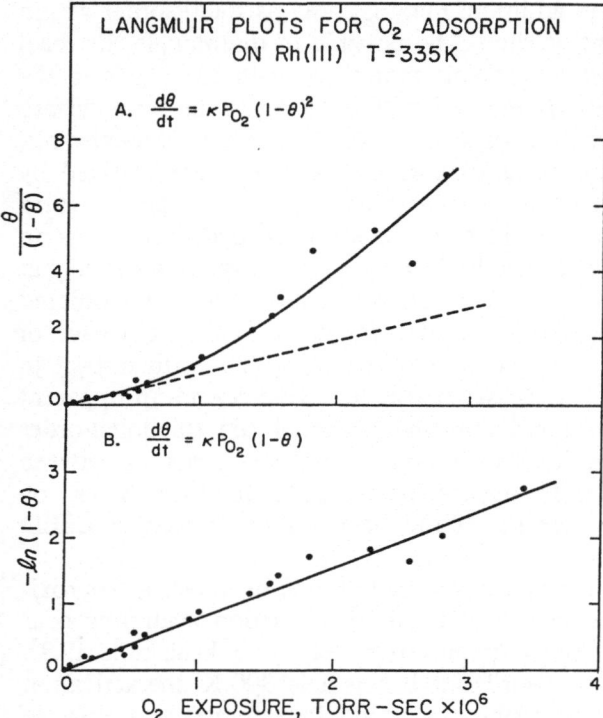

LANGMUIR PLOTS FOR O_2 ADSORPTION
ON Rh(III)   T = 335K

A. $\dfrac{d\theta}{dt} = \kappa P_{O_2} (1-\theta)^2$

B. $\dfrac{d\theta}{dt} = \kappa P_{O_2} (1-\theta)$

O_2 EXPOSURE, TORR–SEC $\times 10^6$

Fig. 9.1. Langmuir plots for O_2 adsorption on Rh(111) at T = 335 K. The oxygen coverage θ was measured by Auger-electron spectroscopy. In the kinetic rate expressions for adsorption, t is the time of exposure at a given pressure of oxygen p_{O_2} and κ is a proportionality constant related to the initial probability of adsorption an the translational velocity of O_2

tron impact on the oxygen overlayer did not cause a measurable loss of the intensity of the oxygen Auger peak in times equivalent to those used in the measurements.

9.2.1 First-Order Langmuir Adsorption

The experimental data have been fitted both to first-order (Fig. 9.1, part B) and second-order (Fig. 9.1, part A) Langmuir models. In this case the rate of adsorption $d\theta/dt$ is

$$\frac{d\theta}{dt} = \kappa p_{O_2}(1 - \theta)^n \ , \tag{9.1}$$

where n = 1 for first-order adsorption kinetics, and n = 2 for second-order adsorption kinetics, i.e. the order is that with respect to the fraction of vacant sites on the surface. In (9.1), κ is equal to $S_{O_2}(0)/(2\pi M_{O_2}k_BT)^{1/2}$, where $S_{O_2}(0)$ is the initial probability of adsorption of oxygen, p_{O_2} is the pressure of oxygen, the coverage has been renormalized so that saturation coverage corresponds to $\theta = 1$, M_{O_2} is the mass of an oxygen molecule, and k_B is the Boltzmann constant. In the case of n = 1, corresponding to the first-order

Langmuir model, the integrated form of (9.1) is

$$-\ln(1 - \theta) = \kappa\varepsilon_{O_2} \ , \tag{9.2}$$

where $\varepsilon_{O_2} = p_{O_2}t$ is the exposure of oxygen. In the case of n = 2, corresponding to the second-order Langmuir model, the integrated form of (9.1) is

$$\frac{\theta}{1 - \theta} = \kappa\varepsilon_{O_2} \ . \tag{9.3}$$

In the constructions of Fig. 9.1, the criteria for acceptance of a proposed model are a linear representation of the data and an intercept of zero, i.e. the origin represents one point on the straight line. A comparison of the first-order and second-order Langmuir models for the rate of adsorption of oxygen clearly favors the first-order model, Fig. 9.1, part B, in view of the nonlinearity of Fig. 9.1, part A.

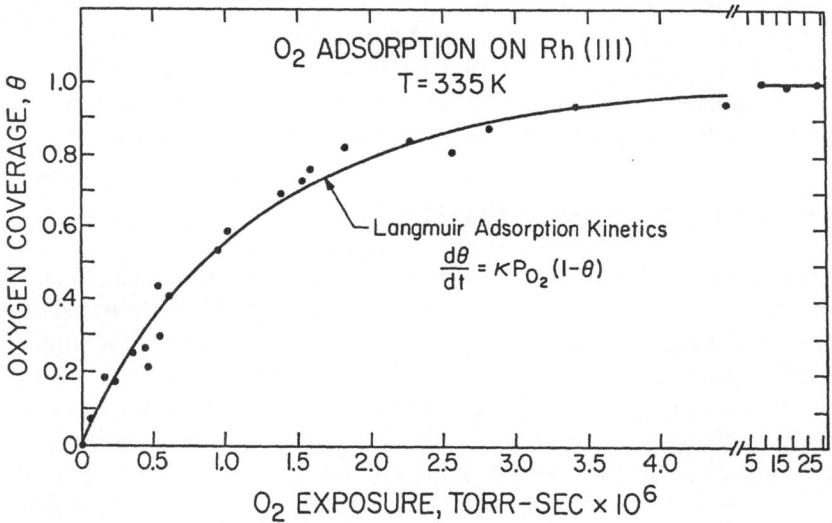

Fig. 9.2. Kinetics of O_2 adsorption on Rh(111) at T = 335 K

A plot of the coverage of oxygen as a function of exposure to oxygen on Rh(111) is shown in Fig. 9.2, where the solid line is a least-squares fit to the data using the first-order Langmuir model. Saturation coverage of oxygen on Rh(111), corresponding to three independent domains of (1x2) super-structures, is 8.0 x 10^{14} atoms/cm^2. This fact allows the initial probability of adsorption of oxygen on Rh(111), or equivalently the probability of adsorption of oxygen on an empty adsite, to be evaluated using the data of Fig. 9.2. This value is found to be approximately 0.9.

The observation at 335 K that the rate of adsorption of oxygen is proportional to $(1 - \theta)$ is unexpected for the dissociative adsorption of a diatomic

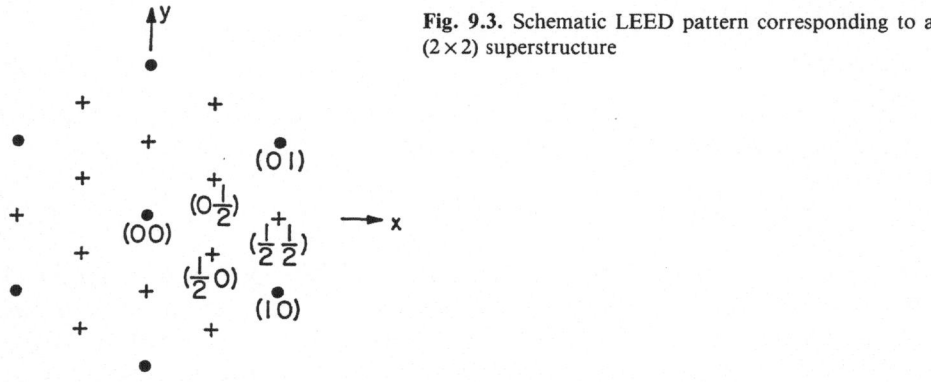

Fig. 9.3. Schematic LEED pattern corresponding to a (2×2) superstructure

molecule into a mobile chemisorbed overlayer, for which a $(1 - \theta)^2$ dependence would be appropriate [9.11]. This implies that the rate-limiting step for the adsorption of oxygen involves the interaction of an oxygen molecule with a single adsite, and the subsequent dissociation is not rate controlling. Although at 335 K the molecular peroxo or superoxo complexes are not sufficiently long-lived to be observed spectroscopically, the lifetime of this intermediate is sufficiently long to control the kinetics of adsorption.

9.2.2 The Structure of Oxygen on Rh(111)

An indexed (2x2) LEED pattern, as is formed by oxygen on Rh(111), is shown in Fig. 9.3. As discussed in Sect. 9.1, this LEED pattern can be caused either by a p(2x2) superstructure or by three independent domains of (1x2) superstructures rotated by 120° with respect to one another. This is illustrated clearly in Fig. 9.4. As oxygen forms a (2x2) overlayer on Rh(111), the intensity of the $(1,\bar{1}/2)$ LEED beam $I_{(1/2bar)}$ was monitored as a function of oxygen exposure, and the result is shown in Fig. 9.5. An initial region of ordering is observed below an exposure of approximately 1.5 L, followed by a region in which the intensity decreases. At exposures above approximately 3 L, a final region of ordering is observed in which a 20-fold increase in the $(1,\bar{1}/2)$ beam intensity is observed. It should be noted, by comparison with the experimental adsorption-kinetics curve, that the final region of ordering occurs over a wide range of exposures very near the saturation coverage of oxygen. This observation implies that a significant cooperative ordering effect takes place in the final stages of chemisorption, which is induced by the addition of a small number of oxygen adatoms to a nearly-saturated oxygen overlayer at this temperature.

The maximum in the LEED beam intensity in Fig. 9.5 at an oxygen exposure of approximately 1.5 L may be associated with an ordered p(2x2) superstructure, whereas the plateau in the intensity at large oxygen exposures may be associated with three domains of (1x2) superstructures. The fact that

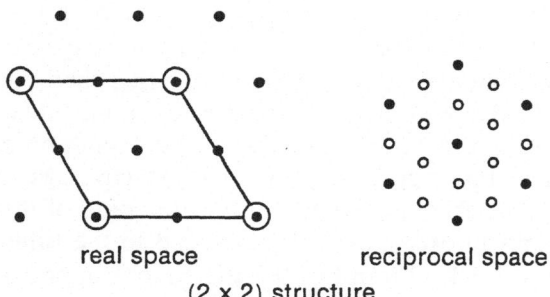

Fig. 9.4. (*Upper panel*) Unit cell of a p(2×2) superstructure and the corresponding LEED pattern. (*Lower panel*) Three domains of a (1×2) superstructure rotated by 120° with respect to one another on an fcc(111) surface and the corresponding LEED pattern

real space reciprocal space

(2 × 2) structure

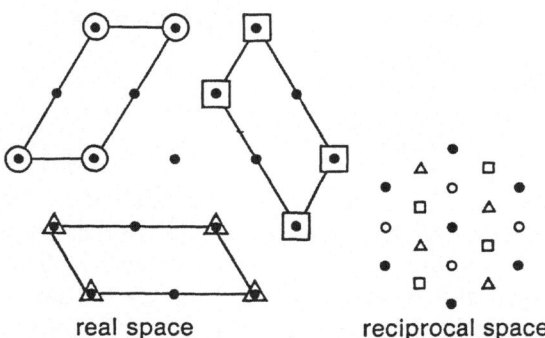

real space reciprocal space

three domains of the (1x2) structure

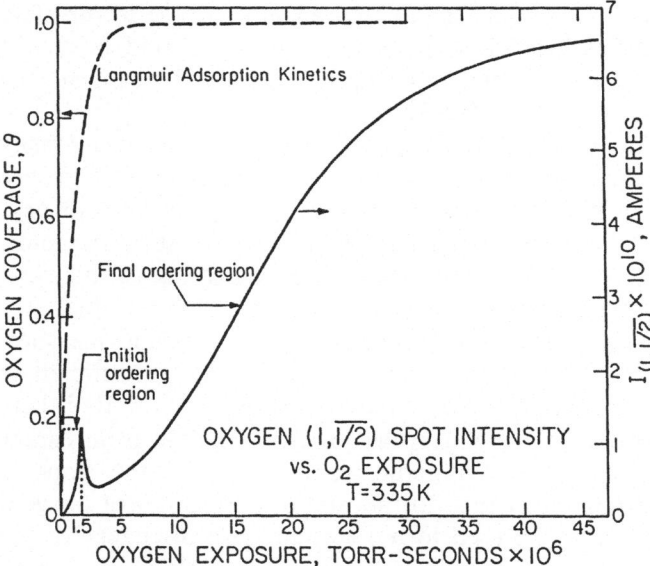

Fig. 9.5. The oxygen (1, $\overline{1/2}$) LEED beam intensity as a function of O_2 exposure at T = 335 K. (The Langmuir adsorption curve is taken from Fig. 9.2.)

the maximum in $I_{(1,\overline{1/2})}$, which is observed at an oxygen exposure of 1.5 L, apparently corresponds to a relative coverage of oxygen somewhat greater than 0.6 (i.e. an absolute coverage somewhat greater than 0.3 rather than the expected 0.25) is because the Faraday cup used to measure the beam intensity does not integrate over the entire beam, and there is a pronounced sharpening of the $(1,1/2)$ beam in this region where the oxygen coverage is varying rapidly with exposure. This effect is manifest also in the ratio of the value of $I_{(1,\overline{1/2})}$ at oxygen exposures in excess of 45 L compared to the value of $I_{(1,\overline{1/2})}$ at an oxygen exposure of 1.5 L. From Fig. 9.5 this ratio is approximately five, whereas a value of four would be expected (based on the fact that the intensity is proportional to the square of the number of ordered scatterers).

9.2.3 LEED Intensity Proportional to Oxygen Coverage

As may be seen in Fig. 9.6, the I-V profile of the $(1,\overline{1/2})$ beam is identical for the p(2x2) superstructure (part A) and the three domains of (1x2) superstructures (part B), apart from the expected differences in absolute intensity. This means that the adsites of the oxygen at $\theta = 0.25$ and at $\theta = 0.5$ are the same. This observation has also been made in the case of oxygen adsorbed on the (111) surface of iridium [9.12]. In this case, by comparison of calculated and measured I-V beam profiles, it was not possible to distinguish a p(2x2) superstructure from three domains of (1x2) superstructures. However, it was possible to determine that on Ir(111) the oxygen adatoms reside in threefold hollow sites with vacancies located beneath them in the second layer (fcc sites), see Sect. 7.3.2. This same conclusion applies also in the case of oxygen on the (111) surface of rhodium [9.13].

The results shown in Fig. 9.7 indicate that the rate of change of intensity of the $(1,1/2)$ beam is directly proportional to the rate of change of surface coverage in the region of final ordering. In Fig. 9.7, the adsorption of oxygen at 6.1×10^{-8} Torr was interrupted in three separate measurements. When p_{O_2} is reduced to zero, the $(1,1/2)$ beam stops increasing in intensity. This demonstrates that the intensification observed in the region of final ordering is not due to a slow, thermally activated reaction. A second measurement carried out with $p_{O_2} = 2.6 \times 10^{-8}$ Torr produces an immediate intensification in $I_{(1,\overline{1/2})}$. The third measurement used $p_{O_2} = 6.1 \times 10^{-8}$ Torr and yielded the rapid rise in intensity expected by extrapolation from earlier times. In these three measurements, indicated in Fig. 9.7, the slope is directly proportional to p_{O_2}, and hence to the rate of adsorption of oxygen. This suggests that $I_{(1,\overline{1/2})}$ may be used to monitor the rate of change of the oxygen coverage in the ordered overlayer. Separate measurements confirmed that there is no effect, whether ordering, disordering or desorption, which can be related to the incident electron LEED beam during the adsorption of oxygen.

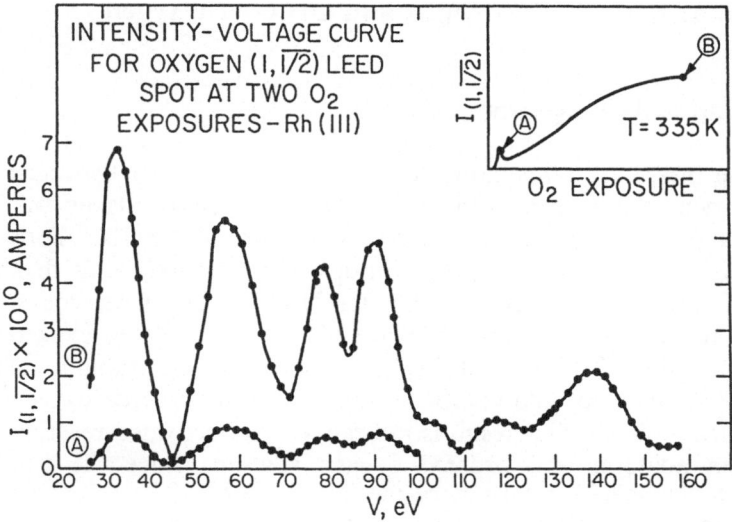

Fig. 9.6. Intensity-voltage curves for the oxygen $(1, \overline{1/2})$ LEED beam at two O_2 exposures on Rh(111). The oxygen exposure at A is approximately 1.5×10^{-6} Torrs, whereas the oxygen exposure at B is approximately 38×10^{-6} Torrs

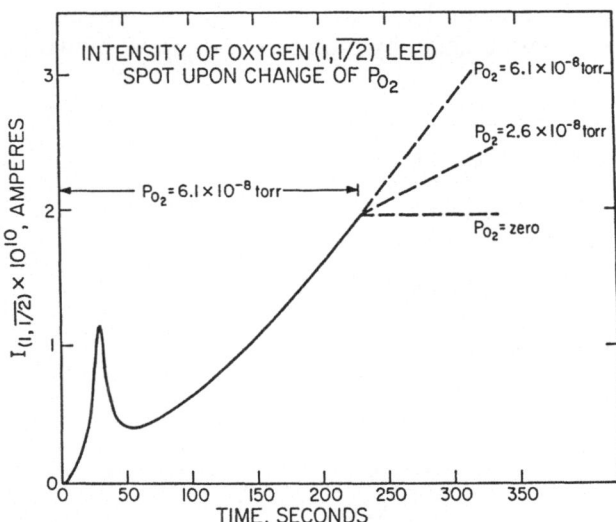

Fig. 9.7. Intensity behavior of the oxygen $(1, \overline{1/2})$ LEED beam upon change in p_{O_2}

9.3 The Reaction Between Hydrogen and Ordered Oxygen on Rh(111)

9.3.1 Reaction Threshold Temperature

To determine the threshold temperature for the reaction between hydrogen and ordered oxygen, the crystal with the ordered oxygen overlayer was cooled to 90 K and hydrogen was admitted at a pressure of 4 x 10^{-7} Torr. The crystal was heated successively to various temperatures and then recooled to 90 K, where the intensity of the $(1,\overline{1/2})$ LEED beam was measured. As shown in Fig. 9.8, the first evidence of a loss of ordered oxygen occurs at approximately 275 K. At higher temperatures, a monotonic decrease in $I_{(1,\overline{1/2})}$ is observed, and complete removal of the ordered oxygen adatoms occurs near 400 K. This result is indicative of an activated reaction between hydrogen adatoms and ordered oxygen adatoms since the two species coexist on the Rh(111) surface below 275 K [9.8]. The threshold temperature, observed near 275 K, suggests an activation energy for the surface reaction of approximately 17 kcal/mole.

9.3.2 First-Order Catalytic Reaction

The kinetics of the reaction between hydrogen and ordered oxygen on Rh(111) may be investigated in more detail by monitoring $I_{(1,\overline{1/2})}$ as shown in

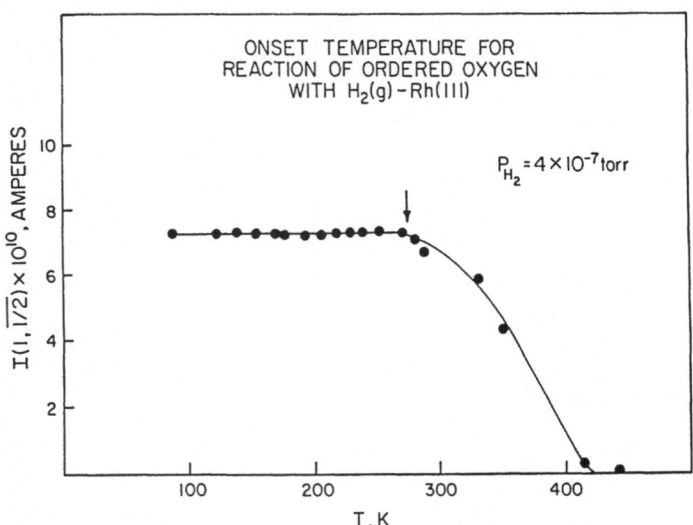

Fig. 9.8. Threshold temperature for the reaction of ordered oxygen with $H_2(g)$. The initial O_2 exposure is 8.2×10^{-6} Torrs at 100 K, followed by heating to 300 K to produce an ordered oxygen overlayer

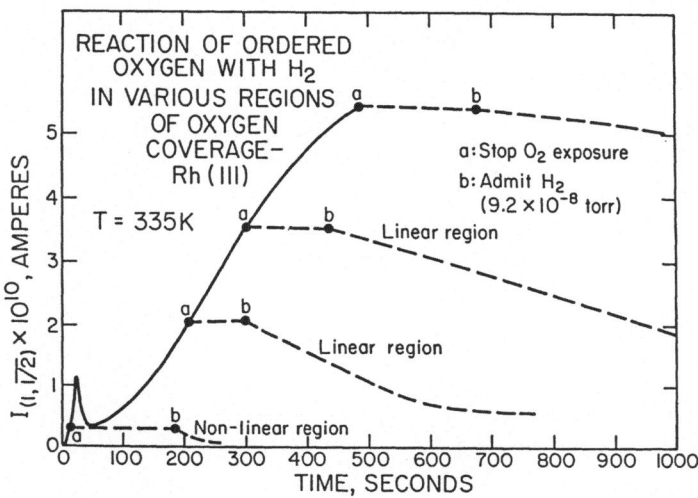

Fig. 9.9. Reaction of ordered oxygen with H_2 in various regions of oxygen coverage at $T = 335$ K on Rh(111). The O_2 exposures were carried out at $p_{O_2} = 6.1 \times 10^{-8}$ Torr to the various points labeled a. H_2 was admitted at the point labeled b

Fig. 9.9. The ordered overlayer of oxygen was produced by exposure to oxygen at $p_{O_2} = 6.1 \times 10^{-8}$ Torr, and the adsorption was interrupted at various points (labeled a) as shown in Fig. 9.9. Following periods of 100-200 s, hydrogen was admitted (at point b), and $I_{(1,\overline{1/2})}$ was measured. The initial slope of the decrease in $I_{(1,\overline{1/2})}$ may be used to determine the rate of depletion of ordered oxygen on Rh(111) as a function of hydrogen pressure, as is illustrated explicitly in Fig. 9.10. In this figure, a sequence of typical measurements is shown for various values of p_{H_2} at 335 K. With identical initial values of $I_{(1,\overline{1/2})}$, each measurement results in a linear region of loss of intensity. In the doubly logarithmic representation of Fig. 9.11, the initial rates of the decrease in intensity are plotted as a function of p_{H_2}. It is clear that the reaction rate is proportional to the first power of p_{H_2} over almost a 100-fold variation in p_{H_2}. (The dashed line in Fig. 9.11 shows the variation in reaction rate with pressure that would be expected if the kinetics were half-order in p_{H_2}.)

9.3.3 Model for the Catalytic Reaction

The mechanism for the formation of water via the reaction of hydrogen with preadsorbed oxygen on Rh(111) is described by the following sequence of elementary reactions:

$$H_2(g) \underset{k_{-1}}{\overset{S_{H_2}F_{H_2}}{\rightleftharpoons}} 2H(a) \ , \tag{9.4}$$

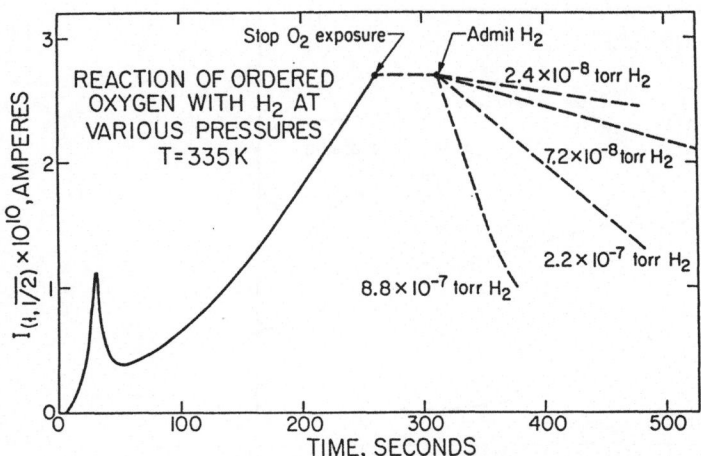

Fig. 9.10. Reaction of ordered oxygen with H_2 at various H_2 pressures and T = 335 K. The O_2 exposure was carried out at $p_{O_2} = 6.1 \times 10^{-8}$ Torr

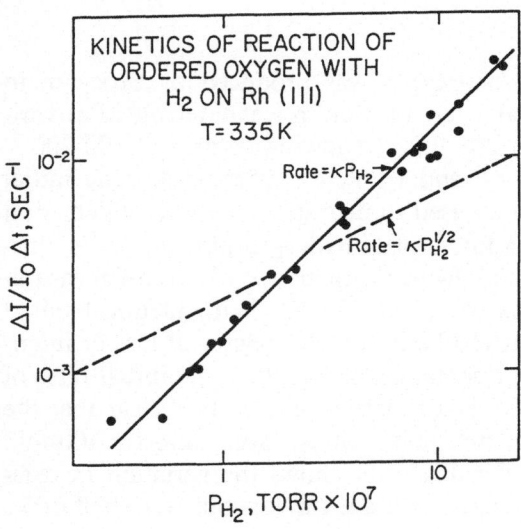

Fig. 9.11. Kinetics of the reaction of ordered oxygen with $H_2(g)$ on Rh(111) at T = 335 K

$$H(a) + O(a) \underset{k_{-2}}{\overset{k_2}{\rightleftarrows}} OH(a) \ , \tag{9.5}$$

$$OH(a) + H(a) \overset{k_3}{\rightarrow} H_2O(a) \ , \quad \text{and} \tag{9.6}$$

$$H_2O(a) \overset{k_4}{\rightarrow} H_2O(g) \ , \tag{9.7}$$

where k_i is the reaction rate coefficient associated with the ith elementary reaction, S_{H_2} is the probability of adsorption of hydrogen, and F_{H_2} is the impingement flux of hydrogen $[p_{H_2}/(2\pi M_{H_2}k_B T)^{1/2}]$. The reaction described by (9.6) is irreversible since water does not dissociate on Rh(111) under the experimental conditions of temperature and pressure considered here, and the reaction described by (9.7) is irreversible since the partial pressure of water in the gas phase is negligible.

Since the flux of hydrogen adsorbed is equal to the sum of the flux of hydrogen desorbed and the flux of water desorbed, we may write

$$S_{H_2}F_{H_2} = k_{-1}\theta_H^2 n_s^2 - n_s \frac{d\theta_O}{dt} , \tag{9.8}$$

where n_s is the concentration of surface sites, and θ_H and θ_O are the fractional surface coverages of hydrogen and oxygen, respectively. Note that the flux of water desorbed is written as $-n_s (d\theta_O)/dt$ since this is the only reaction that leads to the loss of adsorbed oxygen.

Since the rate of desorption of water is rapid under the experimental conditions considered here, we may write the rate of formation of water as

$$R_{H_2O} = - n_s \frac{d\theta_O}{dt} = k_3\theta_{OH}\theta_H n_s^2 . \tag{9.9}$$

Making the reasonable assumption that the concentration of adsorbed hydroxyl groups is at steady-state, we have

$$n_s \frac{d\theta_{OH}}{dt} = k_2\theta_H\theta_O n_s^2 - k_{-2}\theta_{OH}n_s - k_3\theta_{OH}\theta_H n_s^2 \simeq 0 . \tag{9.10}$$

Solving (9.10) for $\theta_{OH}n_s$ gives

$$n_s\theta_{OH} = \frac{k_2\theta_H\theta_O n_s^2}{k_{-2} + k_3\theta_H n_s} , \tag{9.11}$$

and, consequently, the rate of formation of water is given by

$$R_{H_2O} = \frac{k_2 k_3 \theta_H^2 \theta_O n_s^3}{k_{-2} + k_3\theta_H n_s} . \tag{9.12}$$

9.3.4 Activation Energies and Preexponential Factors

In view of (9.12), three separate cases may be considered for the rate of formation of water, namely *(i)* $k_{-2} \ll k_3\theta_H n_s$, *(ii)* $k_{-2} \approx k_3\theta_H n_s$, and *(iii)* $k_{-2} \gg k_3\theta_H n_s$. Since the rate of water formation strictly obeys an Arrhenius construction over wide temperature limits (between 300 and 500 K), it is unlikely that case *(ii)* applies. [This would be possible only if the activation energy for the reverse reaction of (9.5) were equal to the activation energy for the reaction of (9.6).] Considering cases *(i)* and *(iii)*, it is rather more likely that case *(iii)* applies. The activation energy associated with k_{-2} would be expected to be less than the activation energy associated with k_3, resulting in the "preequilibrium" given by (9.5); and, furthermore, $k_2^{(0)}/k_3^{(0)}\theta_H n_s$ would be expected to be greater than unity. For example, using "normal" values for the preexponential factors of the elementary reaction rate coefficients $(k_i^{(0)})$, $k_{-2}^{(0)} \approx 10^{13}s^{-1}$ and $k_3^{(0)} \approx 10^{-3}cm^2/s$, and realizing that the activation energy for the desorption of hydrogen in the presence of coadsorbed oxygen is approximately 12 kcal/mole [9.14], we find that $k_2^{(0)}/k_3^{(0)}\theta_H n_s$ varies between approximately 10^4 and 10^5 for surface temperatures between 300 and 500 K and for a hydrogen pressure of 6 x 10^{-8} Torr. These are the experimental conditions to be considered in detail below. Under these conditions, the equilibrium fractional coverage of hydrogen varies between approximately 2 x 10^{-3} and 10^{-4}. Even at lower temperatures, the value of $k_2^{(0)}/k_3^{(0)}\theta_H n_s$ still exceeds unity by approximately an order of magnitude.

Consequently (9.12) may be written as

$$R_{H_2O} = -n_s \frac{d\theta_0}{dt} = \frac{k_2 k_3}{k_{-2}} \theta_H^2 \theta_0 n_s^3$$

$$= \frac{k_2^{(0)} k_3^{(0)}}{k_{-2}^{(0)}} \theta_H^2 \theta_0 n_s^3 exp[-(E_2 + E_3 - E_{-2})/k_B T] , \qquad (9.13)$$

or

$$-n_s \frac{d\theta_0}{dt} \equiv k_r^{(0)} \theta_H^2 \theta_0 n_s^3 exp(-E_r/k_B T) , \quad where \qquad (9.14)$$

$$k_r = k_r^{(0)} exp(-E_r/k_B T) \qquad (9.15)$$

is the *effective* rate coefficient for the surface reaction producing water. Solving (9.14) for the square of the hydrogen surface coverage gives

$$n_s^2 \theta_H^2 = \frac{-d\theta_0/dt}{k_r^{(0)}\theta_0} exp(E_r/k_B T) . \qquad (9.16)$$

Using (9.16) to eliminate $n_s^2\theta_H^2$ from (9.8) gives

$$-n_s \frac{d\theta_0}{dt} = \frac{S_{H_2}F_{H_2}}{\dfrac{k_{-1}^{(0)}}{k_r^{(0)}\theta_0 n_s} exp[-(E_{-1} - E_r)/k_B T] + 1} , \qquad (9.17)$$

which shows that the rate of water formation (oxygen removal) is proportional to the first power of the hydrogen pressure (F_{H_2}), as observed experimentally.

With the following definitions,

$$\frac{k_{-1}^{(0)}}{k_r^{(0)}} \equiv \hat{k}^{(0)} \quad and \quad E_{-1} - E_r \equiv \hat{E} , \qquad (9.18)$$

(9.17) may be rewritten as

$$\frac{-n_s \dfrac{d\theta_0}{dt}}{S_{H_2}F_{H_2} + n_s \dfrac{d\theta_0}{dt}} = \frac{\theta_O n_s}{\hat{k}^{(0)}} \exp(\hat{E}/k_B T) . \tag{9.19}$$

Defining α as the left-hand side of (9.19), we may write

$$\ln \alpha = \ln \left(\frac{\theta_O n_s}{\hat{k}^{(0)}} \right) + \frac{\hat{E}}{k_B T} . \tag{9.20}$$

Consequently, a plot of $\ln \alpha$ as a function of T^{-1} should yield a straight line, the slope of which is $\hat{E}/k_B$ and the intercept of which is $\ln(\theta_O n_s/\hat{k}^{(0)})$.

9.3.5 Experimental Determination

The experimental approach embodied in Figs. 9.9,10 has also been used to determine the temperature dependence of the reaction between hydrogen and ordered oxygen adatoms on Rh(111). The procedure is illustrated in Fig. 9.12. Oxygen was adsorbed at 317 K until the intensity $I_{(1,\overline{1/2})}$ rose to the same values as in previous measurements, at which point the adsorption was interrupted. Then the temperature of the crystal was adjusted to a constant, higher value, and hydrogen was admitted. At this point it was not possible to monitor the decrease in intensity of the $(1,\overline{1/2})$ beam due to the effects of

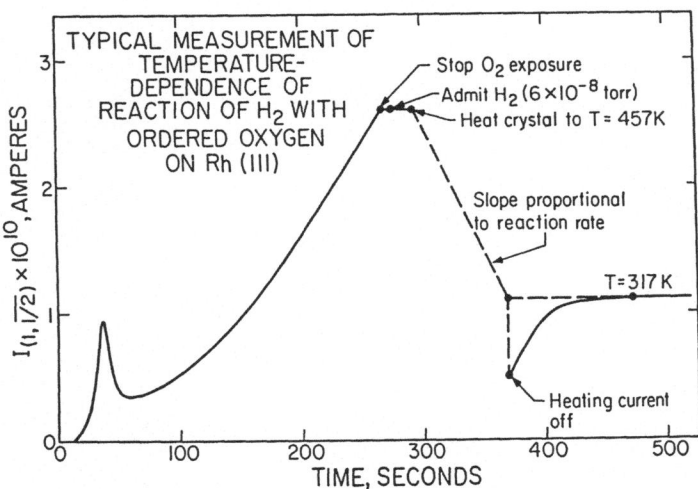

Fig. 9.12. Typical measurement of the temperature dependence of the reaction of hydrogen with ordered oxygen on Rh(111). The dashed lines represent a geometrical construction based upon the experimental data (solid lines)

a magnetic field resulting from the heating current and to a thermal attenuation in the intensity of the LEED beam. Following a convenient reaction time, the heating current was interrupted, and the crystal was cooled rapidly to 317 K, which resulted in the increase in the intensity of the LEED beam beyond 370 s in Fig. 9.12. The geometrical construction shown by the dashed lines was used to deduce the rate of change of the LEED beam during the chemical reaction. Although this method is subject to some error due both to the time required to reach a constant elevated temperature initially and the time required to cool back finally to the initial temperature, these effects are compensating. These sources of error become less serious at temperatures below the temperature of the measurement shown in Fig. 9.12, 457 K, which was the maximum temperature employed.

Based on these data, a plot of $\ln\alpha$ as a function of T^{-1}, as suggested by (9.20), is shown in Fig. 9.13. The slope and intercept of the straight line in Fig. 9.13 yield a value of $\hat{E}$ of -5.4 kcal/mole and a value of $\theta_0 n_s / \hat{k}^{(0)}$ of approximately 120. The value of S_{H_2} used in constructing Fig. 9.13 is 0.4, which is the value appropriate for hydrogen adsorption on a Rh(111) surface

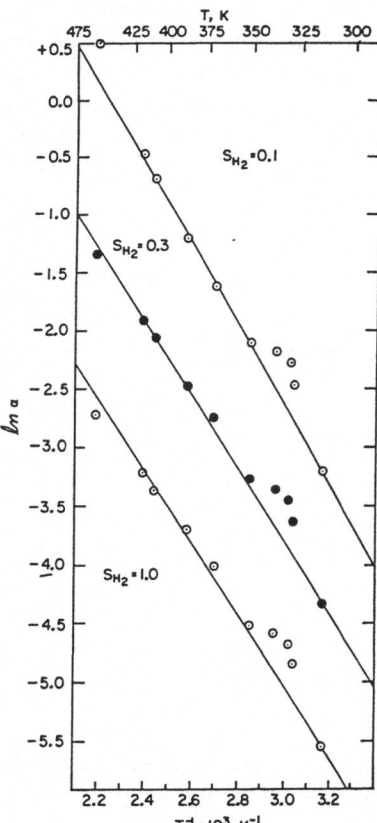

Fig. 9.13. Arrhenius construction for the reaction of hydrogen with an ordered oxygen overlayer on Rh(111)

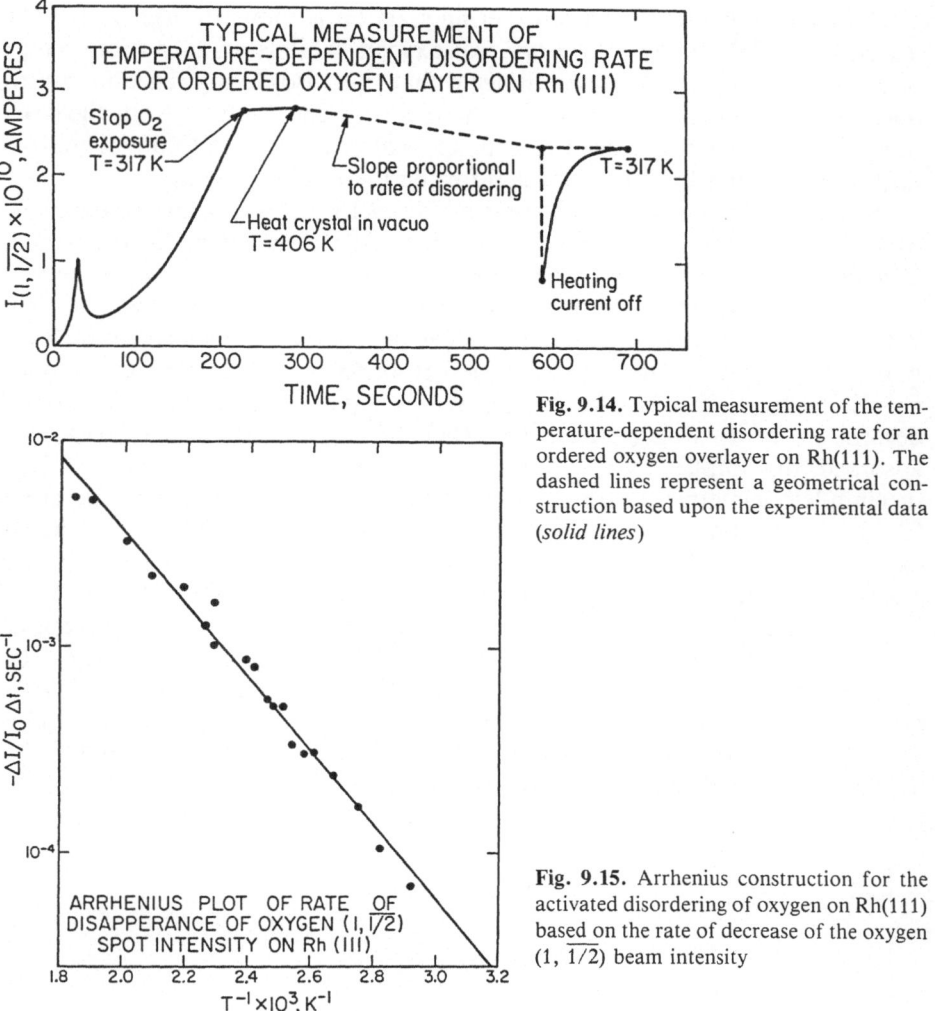

Fig. 9.14. Typical measurement of the temperature-dependent disordering rate for an ordered oxygen overlayer on Rh(111). The dashed lines represent a geometrical construction based upon the experimental data (*solid lines*)

Fig. 9.15. Arrhenius construction for the activated disordering of oxygen on Rh(111) based on the rate of decrease of the oxygen $(1, \overline{1/2})$ beam intensity

with ordered (1x2) overlayers of oxygen adatoms [9.14] [0.65 on clean Rh(111)]. However, if S_{H_2} is varied from 0.1 to unity, $\hat{E}$ varies by less than ± 1 kcal/mole, and $\ln(\theta_0 n_s / \hat{k}^{(0)})$ varies by less than ± 2.3.

Since the ordered oxygen overlayer on Rh(111) undergoes a disordering reaction at temperatures in excess of approximately 300 K [9.8], the loss of intensity in the LEED beam discussed above in connection with the reaction of hydrogen with the ordered overlayer of oxygen contains a contribution from thermal disordering. In order to measure the kinetics of the disordering reaction, a procedure similar to that shown in Fig. 9.12 was employed. This is shown explicitly in Fig. 9.14, where heating in vacuo was carried out to induce the disordering reaction.

An Arrhenius plot of the disordering reaction is shown in Fig. 9.15. The disordering of the ordered overlayer of oxygen exhibits an activation energy of 8.2 kcal/mole. Calculations indicate that this disordering reaction contributes insignificantly compared with the loss of order resulting from the reaction of oxygen with hydrogen to produce water, i.e. $\hat{E}$ is indeed -5.4 ± 0.3 kcal/mole and need not be corrected for the thermal disordering.

Since the Arrhenius construction of Fig. 9.13 yields a single straight line over the entire temperature range examined, the mechanism for the formation of water is invariant under these experimental conditions. However, this does not *necessarily* prove that the mechanism embodied by (9.4-7) is correct. In order to assess this, the values of $\hat{E}$ and $\hat{k}^{(0)}$ deduced from (9.13) may be related to *elementary* reaction-rate coefficients. Since $\hat{E} = E_{-1} - E_r = -5.4$ kcal/mole and E_{-1} is equal to 12 kcal/mole for the desorption of hydrogen from an Rh(111) surface on which oxygen is adsorbed, this implies that E_r is equal to 17.4 kcal/mole. This value of E_r is in complete agreement with the observation that the threshold temperature for the formation of water on Rh(111) is approximately 275 K, as shown explicitly in Fig. 9.8. The fact that E_r is the activation energy of the surface reaction to form water is demonstrated in the energy diagram of Fig. 9.16.

The experimentally derived value of $\theta_0 n_s / \hat{k}^{(0)}$ taken from the intercept of Fig. 9.13 is 120. This implies that $(\theta_0 n_s\, k_2^{(0)} k_3^{(0)} / k_{-1}^{(0)} k_{-2}^{(0)})$ should be equal to 120, see (9.13,14,18). Since $\theta_0 n_s \sim 10^{15} \text{cm}^{-2}$ and normal values of $k_{-1}^{(0)}$, $k_2^{(0)}$, and $k_3^{(0)}$ lie between 10^{-1} and $10^{-2} \text{cm}^2\text{s}^{-1}$, while $k_{-2}^{(0)}$ would be expected to be $10^{12} - 10^{13}\text{s}^{-1}$, one would expect $\theta_0 n_s / \hat{k}^{(0)}$ to lie between 10^{-1} and 10^{+3}. This expectation is entirely consistent with the intercept of Fig. 9.13, namely 120.

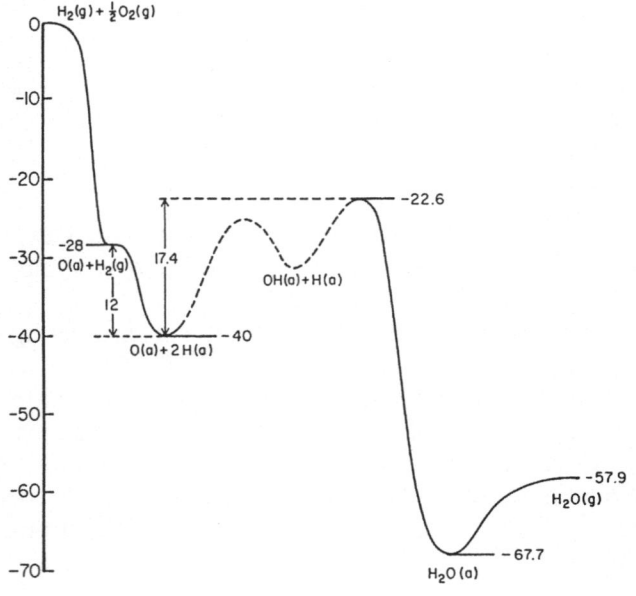

Fig. 9.16. Energy diagram representing the reaction between hydrogen and an ordered oxygen overlayer on Rh(111)

Consequently, the reaction between hydrogen and ordered oxygen on Rh(111) may be described by the elementary reactions given in (9.4-7), the rate expression of (9.19), and the energy diagram of Fig. 9.13. It should be noted that the experimental data imply that the probability of reaction (i.e. the ratio of the reaction rate to the rate of adsorption of hydrogen) is approximately 10^{-2} at 315 K and 0.15 at 460 K (for a hydrogen pressure of 6 x 10^{-8} Torr).

9.4 The Reaction Between Hydrogen and Both Ordered and Disordered Oxygen on Rh(111)

9.4.1 Order-Dependent Kinetics

The use of LEED has been especially informative in determining the details of the reaction of hydrogen with ordered oxygen to produce water on Rh(111). However, at temperatures near 335 K, a mixture of ordered and disordered oxygen adatoms exists on the (111) surface of rhodium [9.8]. As shown in Fig. 9.17, Auger-Electron Spectroscopy has been used to measure the total rate of oxygen depletion via reaction with hydrogen. In this case, as is clear from Fig. 9.18 at the lower values of p_{H_2}, approximate half-order kinetics in p_{H_2} are observed over a tenfold range of p_{H_2}. This is in contrast to the first-order kinetics with respect to p_{H_2} which are observed for the deple-

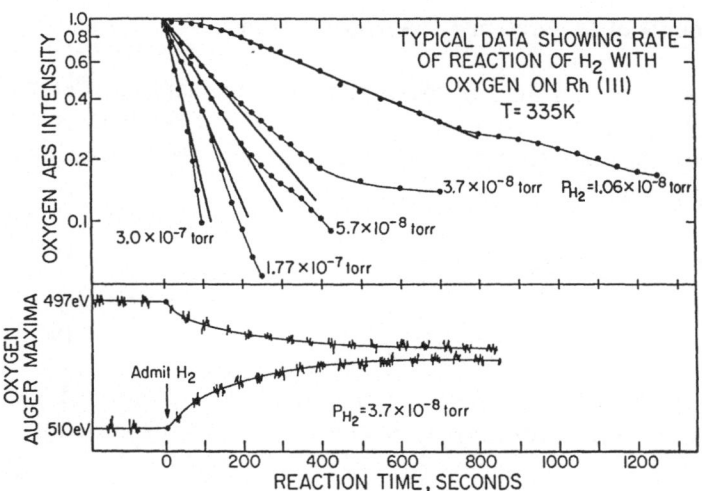

Fig. 9.17. (*Upper panel*) Typical data showing the rate of reaction of H_2(g) with the total oxygen adsorbed on Rh(111) at T = 335 K, as monitored with Auger-Electron Spectroscopy. The straight lines drawn on the semi-logarithmic plots were used to estimate the rate coefficients (first-order in oxygen coverage). (*Lower panel*) Variation in peak-to-peak amplitude in the oxygen Auger intensity as a function of time during exposure to hydrogen at 335 K

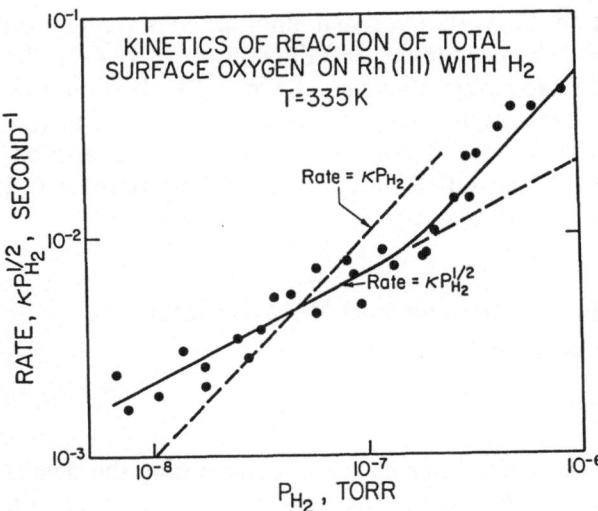

Fig. 9.18. Kinetics of reaction of the total surface oxygen on Rh(111) with $H_2(g)$ at T = 335 K

tion of ordered oxygen adatoms (Fig. 9.11). The reason for the conversion to first-order kinetics above a hydrogen pressure of approximately 2×10^{-7} Torr in Fig. 9.18 evidently lies in the fact that the rate of removal of ordered oxygen is proportional to p_{H_2}, whereas the rate of removal of disordered oxygen is proportional to $p_{H_2}^{1/2}$. Hence, as p_{H_2} increases, a proportionally greater contribution from the reaction involving ordered oxygen occurs.

9.4.2 Relative Amounts of Ordered and Disordered Oxygen

The kinetic measurements indicate the coexistence of appreciable concentrations of both ordered and disordered oxygen at 335 K. If θ_d denotes the fractional surface coverage of disordered oxygen and θ_o denotes the fractional coverage of ordered oxygen, then

$$\frac{d\theta_d}{dt} = -k_d\theta_d p_{H_2}^{1/2} \qquad \text{and} \tag{9.21}$$

$$\frac{d\theta_o}{dt} = -k_o\theta_o p_{H_2} , \tag{9.22}$$

where k_d and k_o are the rate coefficients for the reaction of disordered and ordered oxygen, respectively. From Fig. 9.18, below $p_{H_2} = 2 \times 10^{-7}$ Torr,

$$k_d = 21.5 \text{ s}^{-1} \text{ Torr}^{-1/2} ; \tag{9.23}$$

and from Fig. 9.11,

$$k_o = 1.48 \times 10^4 \text{ s}^{-1} \text{ Torr}^{-1} . \tag{9.24}$$

Combining (9.21-24) gives

$$\frac{d\theta_d/dt}{d\theta_o/dt} = 1.45 \times 10^{-3} p_{H_2}^{-1/2} \frac{\theta_d}{\theta_o} \ . \tag{9.25}$$

Making the reasonable assumption that, at the value of p_{H_2} where the kinetics of the reaction switch from half-order to first-order in p_{H_2}, the rates of reaction of disordered and of ordered oxygen are equal, then θ_d/θ_o may be evaluated. Inserting $p_{H_2} \simeq 2 \times 10^{-7}$ Torr into (9.25), see Fig. 9.18, with $d\theta_d/dt = d\theta_o/dt$ yields

$$\theta_d/\theta_o \simeq 0.3 \ . \tag{9.26}$$

Consequently, at 335 K slightly more than 75% of the oxygen adatoms are ordered, and the remainder are disordered. With LEED it is possible to conveniently separate the reaction of the ordered overlayer from that of the total overlayer.

10. Island Formation of Adspecies and LEED

If the temperature of the surface is sufficiently low, an attractive interaction among adatoms or admolecules which are bound at sites of specific symmetry on the surface will give rise to the formation of ordered islands of the adsorbate. These islands are characterized by a local density greater than the overall density of the adsorbate on the surface. This local density is related to the two-dimensional periodicity within the ordered islands. Evidence for the formation of ordered islands has been observed quite frequently with LEED [10.1-5]. Both the thermodynamic and kinetic (e.g. adsorption, desorption, and surface diffusion) properties of an overlayer that contains islands are determined by the degree of order and the size and relative positioning of the islands [10.6]. This information can be obtained using LEED since the shape (angular profile) of the diffracted electron beams is determined by the size and distribution of ordered regions of scatterers on the surface [10.7-15].

10.1 The Nature of Islands on Surfaces

If the positions of ordered islands in an overlayer are random, then the fractional-order-beam profiles are equal to the weighted sum of the angular profiles of the individual islands on the surface [10.7,15,16]. However, if the positions of the islands are not random but are correlated with one another, this simple result does not hold, and splitting and streaking of the diffracted beams occur due to antiphase (out-of-phase) domains in the overlayer [10.8,9,11,15].

On the W(110) surface, ordered oxygen adatoms cluster into many small islands rather than forming one large island as would be expected from energetic considerations alone, as was deduced by Lu et al. [10.15,17,18]. The formation of small islands evidently arises from limitations on the diffusion of adatoms or admolecules across the surface. In the case of oxygen adatoms on W(110), it appears that steps on the surface may act as barriers to diffusion, isolating the adatoms on particular terraces. Quantitative studies of the size of ordered islands can provide information concerning the limitations of diffusion of adatoms or admolecules across a surface. Furthermore, the finite size of islands can influence strongly the nature of two-dimensional phase transitions [10.19-21]. Thus, detailed information concerning the dimensions of ordered superstructures on surfaces may be crucial to a complete understanding of order-disorder phenomena in chemisorbed overlayers (Chap. 8).

In order to determine the size of islands from LEED measurements, it is necessary to understand the effect of the distribution of the island positions on the LEED beam profiles. In view of this, a model for the distribution of the island positions has been developed, and one- and two-dimensional overlayers based on this model have been generated by Williams and Weinberg [10.16]. The LEED beam profiles for these overlayers have been calculated kinematically to determine the effect of different degrees of correlation between the island positions on the profiles [10.16]. After these important issues are explained in the following section, a detailed example of island formation, namely for CO on Ru(0001), will be presented in Sect. 10.3.

10.2 LEED Beam Profiles for Arrays of Ordered Islands

10.2.1 Distributions of Islands

Computations of LEED beam intensities from model overlayers can be used to determine the amount of information that can be obtained from LEED concerning the structure of the overlayer. We shall illustrate this with an excluded-area model that has been formulated for the formation of overlayers and in which the occurrence of islands is due to attractive interactions among adspecies [10.16]. It is reasonable to assume that at low surface coverages adsorption occurs at lattice sites of random position but identical environment with respect to one another (e.g. bridge sites, on-top sites, etc.). Exceptions to this assumption are molecules with a very small probability of adsorption (the adsorption of which is biased toward defect sites), or those that adsorb via a mobile precursor state. For islands to form, some mobility of the adspecies following adsorption is required. If the adspecies are sufficiently mobile, they will move randomly on the surface until each one either (1) encounters another adparticle and forms a pair (due to an attractive interaction) which has reduced mobility and serves as a nucleus for an island, (2) is trapped at a defect, again forming a nucleus for an island, or (3) encounters a previously nucleated island and joins it. Statistically, those adspecies that adsorb initially near the nucleus of an island are more likely to join the preexisting island than to form a new nucleus. Consequently, once an island begins to form, it acts as a sink for adparticles that adsorb nearby. For this reason, the distribution of the island positions will not be random but will be governed by a weak excluded-area effect, i.e. there will tend to be a minimum distance between the centers of islands. This minimum distance will be related to the mobility of the adspecies (which in turn may be limited

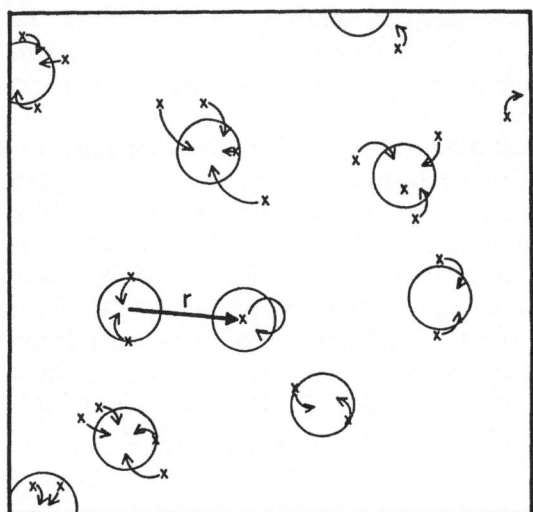

Fig. 10.1. Creation of an overlayer with an excluded area about each island site. The x's represent initial random locations of adatoms. Arrows show the motion of nearby adatoms to a mean site which becomes the island position. Circles show islands centered at the resulting locations. A vector between the nearest-neighbor islands is labeled r

by surface defects). Highly mobile adspecies may move relatively long distances prior to joining an island and thus will form overlayers in which the islands are far apart. If the adparticles are of low mobility, small closely packed islands will result.

A computational algorithm was formulated to create distributions of the island positions based on this excluded-area model. Initially, a large number of islands were placed on a lattice representing the substrate at coordinates chosen using a random-number generator. A section of this lattice, with the initial positions represented by x's, is illustrated in Fig. 10.1. To determine the island positions, this initial set of random adsorption sites was treated in the following way. A subset of the sites was chosen arbitrarily, and the positions of all adparticles within a distance r_{min} of each chosen site were located. The average position of the sites in each group was taken as a new site, and the original positions of the adparticles were eliminated. This procedure was continued until all sites were separated by a distance of at least r_{min}. The "motion" of adparticles to a final nucleation site is illustrated schematically by the arrows in Fig. 10.1. Once the island positions were assigned in this way, islands of arbitrary size, as shown by the circles in Fig. 10.1, could be placed on the lattice to form the overlayer. This algorithm is not meant to simulate the nucleation of islands rigorously, but rather to construct an overlayer conforming to the excluded-area concept discussed above.

Distributions of island positions were created on hexagonal lattices with 4000x4000 sites, for three different values of r_{min}: 70, 35, and 21 substrate-lattice spacings. The hexagonal lattice is chosen to make a connection with the experimental results for CO on the (0001) surface of ruthenium discussed in the next section. Each set of island positions is characterized by the

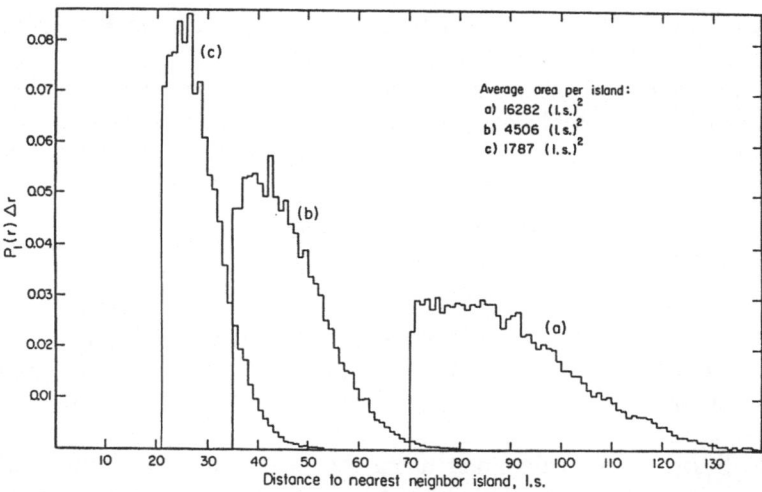

Fig. 10.2. The probability of finding the nearest-neighbor island at a distance r in any direction for overlayers of three densities created on 4000×4000 lattices. (a) $r_{min} = 70$ substrate-lattice spacings (l.s.), number of islands $= 983 \pm 15$; (b) $r_{min} = 35$ l.s., number of islands $= 3551 \pm 25$; (c) $r_{min} = 21$ l.s., number of islands $= 8953 \pm 51$. Each histogram has been normalized so that the area under it integrates to unity

values of distances between nearest-neighbor islands, as shown in Fig. 10.2. The histograms in Fig. 10.2 represent the probability of finding the nearest-neighbor island at a distance r in any direction. Each histogram represents averaged results from 16 to 36 independently created distributions of positions. Curve (a) in Fig. 10.2 is the result for a rather large value of r_{min}, which gives rise to a low density of islands. Curves (b) and (c) represent increasing densities of islands resulting from smaller values of r_{min}. The abrupt cutoff in $P_1(r)\Delta r$ at r_{min} in Fig. 10.2 is not physically realistic. In a real overlayer, a more gradual decrease in $P_1(r)\Delta r$ at small values of r would occur. The use of this sharp cutoff causes a greater degree of long-range order in the positions of the islands than would occur otherwise. Since the order of the island positions, as well as the order within the islands, determines the intensity and angular profile of the LEED beams, the beam profiles for these calculated overlayers will have a greater dependence on the distribution of the island positions than would occur for a physical overlayer.

10.2.2 One-Dimensional Overlayers

To investigate the consequences of an excluded-area distribution of this type for the island positions, LEED beam profiles were calculated for infinite one-dimensional and finite two-dimensional overlayers with the distributions of the island positions shown in Fig. 10.2. In the one-dimensional case, the ordered islands were assigned a structure with one scatterer on every third lattice site. A $(\sqrt{3} \times \sqrt{3})R30°$ superstructure, as applies to CO on Ru(0001), was employed for the islands in the two-dimensional case.

Given the probability distribution for the positions of nearest-neighbor islands, the LEED beam profile can be calculated kinematically for a one-dimensional overlayer [10.9,10]. When all the islands have the same size and structure,

$$I(k) = I_\Gamma(k) + I_\Gamma(k) \left[\frac{F[P_1(r)]}{1 - F[P_1(r)]} + \frac{F^*[P_1(r)]}{1 - F^*[P_1(r)]} \right], \qquad (10.1)$$

where $I_\Gamma(k)$ is the beam profile for a single island of size Γ, $P_1(r)$ is the normalized probability of finding the nearest-neighbor island at a distance r, and F and F^* denote the Fourier transform and its complex conjugate, respectively.

The LEED beam profile calculated using the probability distribution of Fig. 10.2c is shown in Fig. 10.3. The dashed line represents the beam profile for the entire overlayer, whereas the solid line is the profile for a single island. It is clear that placing the islands in the one-dimensional array has very little effect on the beam profile. The beam profiles for arrays based on the distributions of Figs. 10.2a and 10.2b show even smaller (but higher-frequency) deviations from the beam profiles of a single island. These results

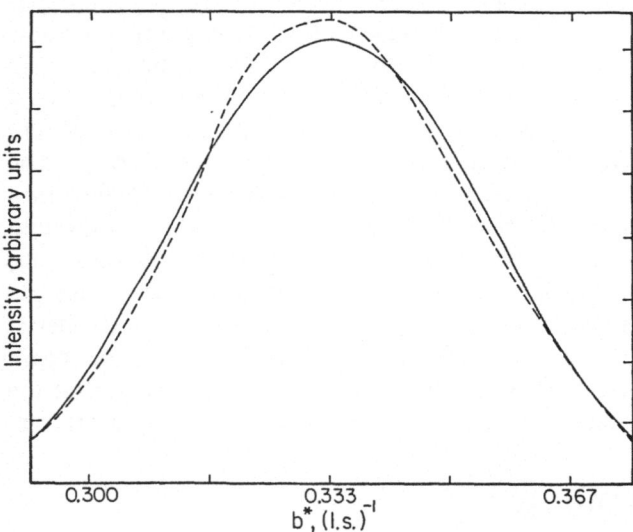

Fig. 10.3. Beam profile for a one-dimensional array of ordered islands based on the distribution of Fig. 10.2, curve (c). All islands are the same size, containing six scatterers, one on every third substrate site. The approximate absolute coverage (the number of adatoms per substrate atom) is $\theta = 0.23$, and the approximate relative coverage (a fully ordered overlayer corresponds to $\theta = 1.0$) is $\theta_r = 0.69$. The dashed line is the profile for the array, and the solid line is the profile for a single island. The reciprocal-space vector b* is in units of reciprocal substrate-lattice spacings, (l.s.)$^{-1}$

are in contrast to results for islands separated by domain boundaries, where beam splitting was found for a similar distribution of nearest-neighbor distances [10.9]. The reason for this difference is that in the case discussed here, the phase (domain) of each island is determined independently of the phase of the adjacent islands. Only if adjacent islands are required to be out-of-phase does beam splitting occur for probability distributions of the width used here.

10.2.3 Two-Dimensional Overlayers

For a two-dimensional overlayer, the effect of the distribution of the island positions on the beam profile should be even weaker than in one dimension due to the additional "degree of freedom", i.e. the probability distribution of finding the nearest-neighbor island in any particular direction on a two-dimensional overlayer will be broader and flatter than the distributions shown in Fig. 10.2. The beam profiles for two-dimensional arrays of islands were calculated directly using the positions of the islands which were created as described in the previous section. The calculations can be carried out quite efficiently since the positions of all the scatterers in the overlayer can be represented as the convolution product

$$L(\vec{r}) = \sum_{\Gamma} j_{\Gamma}(\vec{r})^* l_{\Gamma}(\vec{r}) \; , \tag{10.2}$$

where $L(\vec{r})$ is an array of δ-functions at the position of each scatterer in the overlayer, $j_{\Gamma}(\vec{r})$ is an array of δ-functions at the position of each scatterer within an island of size Γ, $l_{\Gamma}(\vec{r})$ is an array of δ-functions representing the positions of all islands of size Γ, and * represents the convolution product.

The kinematic intensity is the absolute square of the Fourier transform of the positions of the scatterers. Making use of (10.2) and the fact that the Fourier transform of a convolution product is equal to the product of the Fourier transforms, the intensity can be expressed as

$$I(\vec{k}) = |\sum_{\Gamma} F[j_{\Gamma}(\vec{r})] \times F[l_{\Gamma}(\vec{r})]|^2 \; , \tag{10.3}$$

i.e. the total diffracted intensity results from two separable terms. One term arises from the positions of the scatterers within the individual islands, while the second depends only on the absolute positions of the centers of the islands. This separability is convenient in understanding the physical effects that influence diffraction from overlayers with islands.

Constant Island Size

Computations were first performed for overlayers in which all the islands were of the same size. It was found that the beam profiles calculated for individual 4000x4000 lattices were quite noisy since a single lattice is not

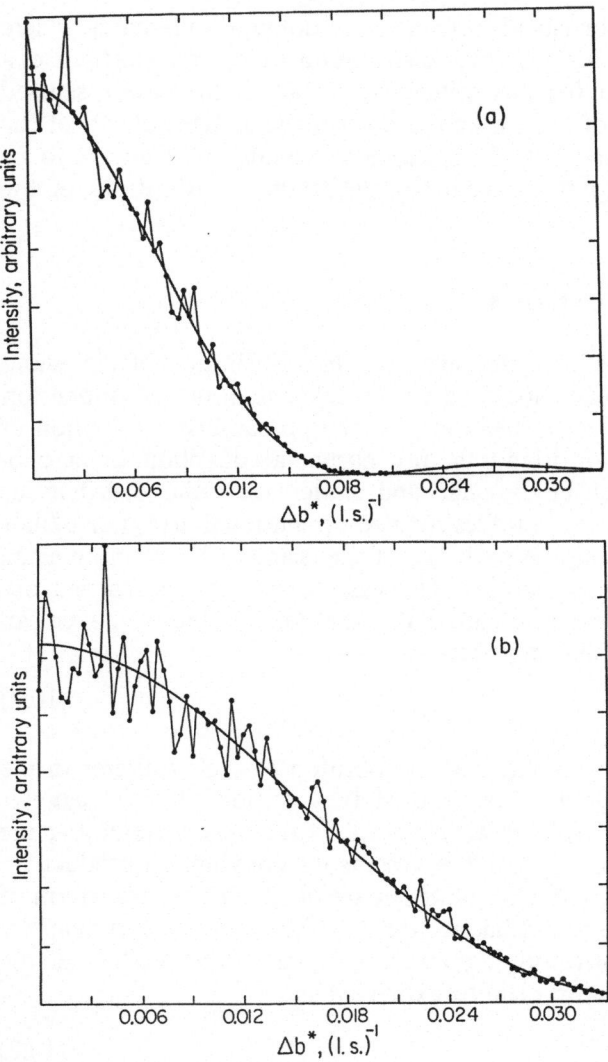

Fig. 10.4a – c. Beam profiles for arrays of uniform island size (*jagged line*) compared with the profile of a single island of the same size (*smooth curve*). (**a**) Average of profiles from 36 overlayers with position distribution as in Fig. 10.2, curve (*a*). Islands are 62 substrate-lattice spacings (l.s.) in diameter; absolute coverage $\theta = 0.071$; relative coverage, cf. Fig. 10.3, $\theta_r = 0.21$. (**b**) Average of 36 in accordance with Fig. 10.2, curve (*b*); islands 31 l.s. in diameter, $\theta = 0.066$ and $\theta_r = 0.20$. (**c**) Average of 16 as for Fig. 10.2, curve (*c*); islands 21 l.s. in diameter, $\theta = 0.071$ and $\theta_r = 0.21$. b* is the $(\sqrt{3} \times \sqrt{3})R30°$ reciprocal-space vector. The units are in terms of reciprocal substrate-lattice spacings, $(l.s.)^{-1}$

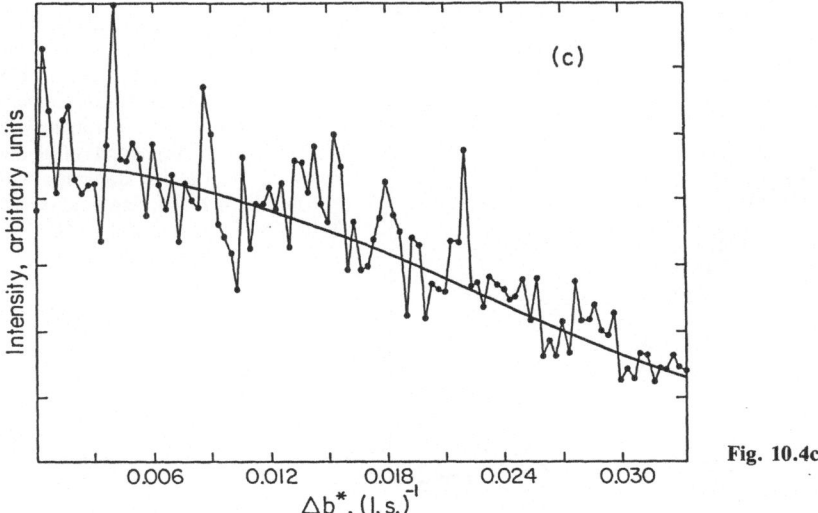

Fig. 10.4c

sufficiently large to represent a true statistical distribution of the positions of islands. A similar effect has been noted in computer simulations of the LEED beam profiles for a finite lattice with random displacements, performed by Henzler [10.12]. To determine the beam profile, it was necessary to average the results for a rather large number (between 16 and 36) of overlayers. The criterion that beam profiles in different directions be the same was used to determine the extent of averaging required.

The results of the computations for overlayers with all islands of the same size are shown in Fig. 10.4. The points in these curves show the calculated beam profiles for three overlayers of which the absolute coverage is approximately 0.07, and the distributions of the positions of the islands correspond to Fig. 10.2. The smooth curves each represent the beam profile for a single island of the size present in that overlayer. The normalization factor used in comparing the profiles of the single island to the profiles of the overlayer is the number of islands on the surface. Within the noise of the computation, the profile of the overlayer is identical to the profile of the single island. This indicates that, in spite of the excluded-area constraint on the positions of the islands, the centers of the islands are acting as a randomly distributed set of scatterers that contribute a homogeneous background of constant intensity [10.12]. The intensity profile due to a single island is then just multiplied by this background, as in (10.3), to give the overall profile of the beam.

Variable Island Size

Beam profiles were calculated also for overlayers containing islands of different sizes. For a given distribution of island sizes, the smallest islands were placed at positions that had the closest-neighbor islands. Typical results

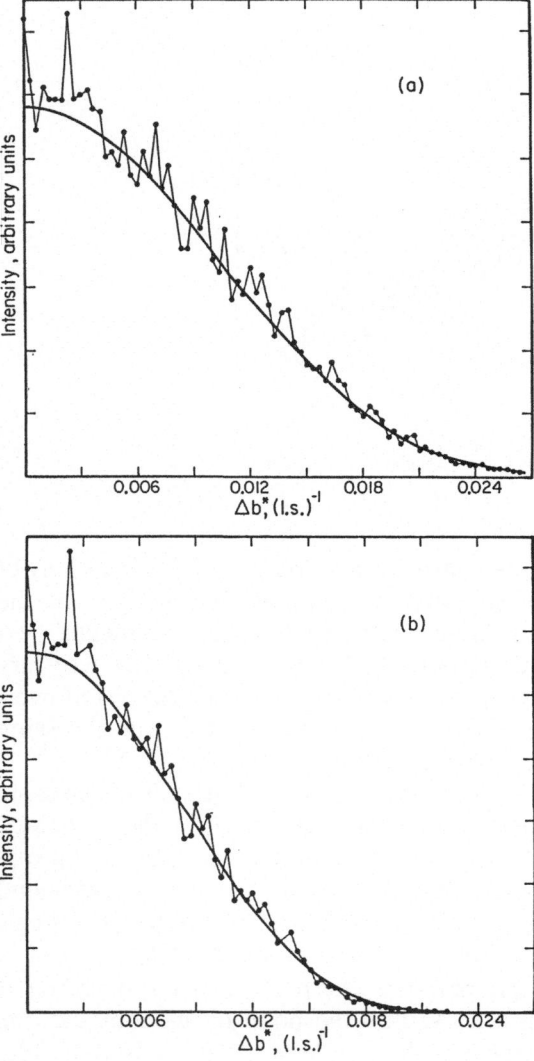

Fig. 10.5a, b. Beam profiles for arrays of mixed island size. The position distributions correspond to those in Fig. 10.2, curve (*a*). (**a**) Islands of diameter 35, 38, 42, 45 and 48 l.s. occurring with equal probability. (**b**) Islands of diameter 48, 55, and 62 l.s. occurring with probabilities 0.2, 0.6, and 0.2, respectively. Conventions as in Fig. 10.4

of this type of calculation are shown in Fig. 10.5, where the points show the result of the computation. The smooth curves in Fig. 10.5 were calculated according to

$$I(\vec{k}) = N\sum_{\Gamma}P(\Gamma)I_{\Gamma}(\vec{k}) \ , \tag{10.4}$$

where N is the number of islands, and $P(\Gamma)$ is the probability of occurrence of an island of size Γ. Assuming that the islands are distributed randomly and that the distributions for islands of different size are independent, (10.4) is exact [10.7,15,16]. With these assumptions, the cross terms in islands of different size that arise from an expansion of (10.3) contribute intensity only in the specular direction [10.16]. As is evident in Fig. 10.5, the distribution of islands does not affect the shape of the beam profile. Due to the abrupt cutoff in $P_1(r)\Delta r$ (cf. Fig. 10.2), the calculated distributions of islands used here have a stronger influence on the beam profile than would a physically more realistic distribution. These results, therefore, can be extrapolated safely to real systems.

The peak intensity of a diffracted beam for an island of size Γ containing N ordered scatterers is N_Γ^2 (provided the island is smaller than the instrumental response width). Consequently, the largest islands will influence the shape of the beam profile most strongly. Empirically, it has been found that the full width at half maximum (FWHM) W of a beam profile for an overlayer of mixed island sizes can be approximated by

$$
W \approx \frac{\sum_\Gamma P(\Gamma)N_\Gamma^2 W_\Gamma}{\sum_\Gamma P(\Gamma)N_\Gamma^2} \; , \tag{10.5}
$$

where N_Γ is the number of scatterers in an island of size Γ, and W_Γ is the FWHM for a beam profile from an island of size Γ. A comparison of the FWHM calculated using the approximation of (10.5) and the FWHM determined from beam profiles calculated according to (10.4) was carried out for 20 arbitrary distributions and ten "ball-in-urn" distributions (Sect. 10.2.4). Rather good agreement was found between the two calculations with the values from (10.5) ranging from 1 to 10% higher than those from (10.4) for values of the FWHM ranging from 0.01 to 0.06, in units of the reciprocal substrate-lattice spacing.

As mentioned above and as is clear from (10.5), an experimental determination of the FWHM will not refer to the mean diameter of the islands on the surface but will be weighted heavily toward the largest islands present. To the extent that the effects of multiple scattering [10.13,22] and especially instrumental limitations [10.23,24] allow, a detailed analysis of the beam profile given by (10.4) can provide information concerning the size distribution of the islands (as a function of surface coverage). In particular, a distribution containing many small islands will give rise to broad wings on the beam profile.

10.2.4 Dependence on Surface Coverage

Given a model for the distribution of island sizes as a function of surface coverage, (10.5) can be used to calculate both the LEED beam profile and the relative kinematic intensity at different surface coverages. For comparison with experiment, the calculated beam profile must be convoluted with the appropriate instrumental response function [10.23,24].

Ball-In-Urn Model

A model for the nucleation and growth of islands must account for the formation of new islands with decreasing frequency as the surface coverage increases, as well as an increasing mean size of the islands with coverage. These characteristics can be described crudely with a ball-in-urn model in which the "urns" are nucleation sites for islands, of which there will be a particular number M, depending both on the nature of the surface and on the adsorbate. During adsorption, the number of adspecies n will be distributed randomly among the M sites. The number of ways in which this can be accomplished is given by

$$\Omega(M,n) = \frac{(M - 1 + n)!}{(M - 1)!n!} \ . \tag{10.6}$$

The number of times that an arbitrary nucleation site is occupied by an island containing l adparticles for all $\Omega(M,n)$ configurations is given by

$$\hat{\Omega}(M,n,l) = M \ \frac{(M - 2 + n - l)!}{(M - 2)!(n - l)!} \ . \tag{10.7}$$

Consequently, the frequency of observing an island of size l is

$$F(l) = \frac{\hat{\Omega}(M,n,l)}{M\Omega(M,n)} \ . \tag{10.8}$$

According to this model, the most frequently occurring size of a single island corresponds to l equal to zero, see Sect. 10.3. The most probable size of an island in which a given adparticle resides, however, is $l \approx n/M$, for n >> M.

Coverage Dependence of Intensity and Angular Profiles

To calculate the coverage dependence of the intensity and angular profiles of the LEED beams according to this ball-in-urn model, a finite number of island sizes (l) was chosen. The probability of occurrence of islands of each size was evaluated using (10.8) for various values of the number of adspecies n. The number of nucleation sites (M) was chosen to be 3600 on a 4000x4000 hexagonal lattice. Once the set of probabilities was determined, (10.4) was used to calculate both the intensity and the angular profile of the LEED beam. The computations were carried out up to an absolute coverage θ of 0.137, at which point adjacent islands on the surface begin to coalesce, and the assumptions embodied in the model for the growth of islands as well as the assumption of circular islands are no longer valid. The optimum fractional surface coverage for the $(\sqrt{3} \times \sqrt{3})R30°$ superstructure is 1/3, and, consequently, at $\theta = 0.137$ somewhat less than half the surface is occupied by islands.

To simulate the effect of instrumental broadening, the calculated beam profiles were convoluted with Gaussian functions of various widths. The

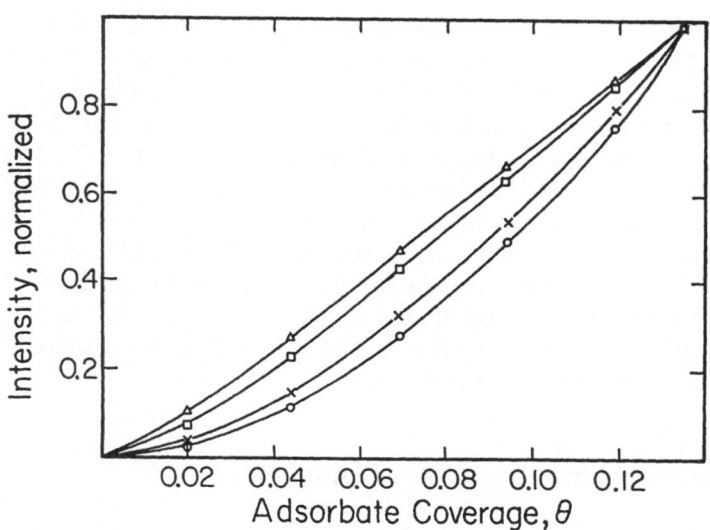

Fig. 10.6. Intensity as a function of coverage for island growth via a ball-in-urn model. (*Circles*) no instrumental broadening; (*crosses*) instrumental width 0.010 (l.s.)$^{-1}$; (*squares*) instrumental width 0.030 (l.s.)$^{-1}$; (*triangles*) instrumental width 0.050 (l.s.)$^{-1}$. Each curve has been normalized independently to unity at $\theta = 0.137$

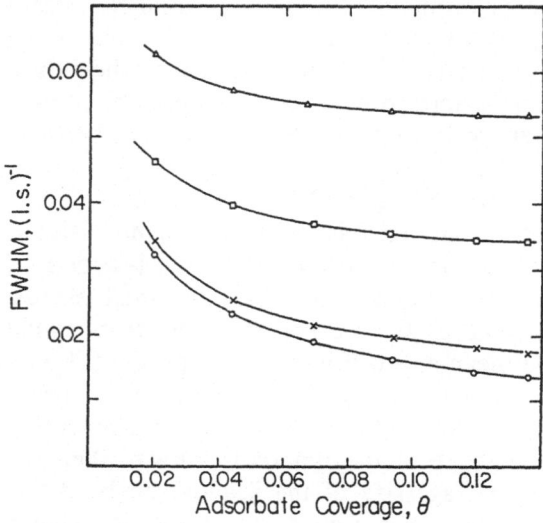

Fig. 10.7. FWHM as a function of coverage for island growth via a ball-in-urn model. Symbols as in Fig. 10.6

property that the beam is isotropic was used to carry out a full two-dimensional convolution [10.25]. The values of the intensity at $\Delta b^* = 0$ [the center of the $(\sqrt{3} \times \sqrt{3})R30°$ fractional-order beam] for different instrumental widths are plotted as a function of surface coverage in Fig. 10.6. Each curve has been normalized independently to unity at $\theta = 0.137$. The corresponding widths of the LEED beams are shown in Fig. 10.7. It is clear from Fig. 10.7 that instrumental broadening is a major barrier to extracting information concerning the overlayer. Even without instrumental broadening, however, (10.4,5) remain only two equations in many unknowns [the set $P(\Gamma)$]. Therefore, a comparison of calculated curves, as in Figs. 10.6 and 7, with experimental observations can serve as a check of a proposed model of island growth but not as a proof of its correctness.

10.2.5 Summary of Theoretical Results for Beam Profiles

For simulated overlayers containing islands, no perturbation of the shape of the LEED beam profiles occurs (streaking, splitting, etc.), even for overlayers for which the mean first-neighbor distance between the centers of islands is as small as 30 lattice spacings. This result, which is independent of any arguments concerning instrumental broadening, is valid so long as there is no correlation between the phases (domains) of different islands. Beam profiles for overlayers containing islands may be described as the sum of the profiles due to the individual islands (10.4). Although the mathematical derivation of this result requires the islands to be positioned randomly, this is a weak constraint. Even overlayers with a rather strong correlation between the positions of nearest-neighbor islands appear to be random insofar as their effect on the beam profile is concerned (cf. Fig. 10.2).

Since the intensity of the beam profile from an individual island scales as the square of the number of scatterers in the island, the largest islands in a distribution will influence the beam profile far more than the small islands. For this reason, *the width of the overall profile will not reflect the mean island diameter.* A determination of the mean size will be model dependent if only the FWHM of the beam profile is known. Analysis of the shape of the beam profile, particularly its wings, can provide additional information concerning the island-size distribution. If the shape of the islands is known, then the size distribution and not just the average size of the islands can be determined directly from the beam profiles as has been demonstrated for islands assumed to be parallelograms [10.26].

10.3 Island Formation in a Real System: CO on Ru(0001)

10.3.1 Conditions of Island Formation

In this section, we discuss in detail the results of a LEED investigation of the formation of islands of CO on Ru(0001), carried out by Williams et al. [10.27]. The fact that CO adsorbs molecularly in the on-top site on Ru(0001) is known from High-Resolution Electron Energy-Loss Spectroscopy (HREELS) [10.28], Infrared Reflection-Absorption Spectroscopy (IRAS) [10.29], and dynamical LEED calculations [10.30]. At fractional surface coverages up to $\theta = 1/3$ (one CO molecule per three ruthenium surface atoms, or 5.26×10^{14} CO molecules/cm^2), the admolecules order into a $(\sqrt{3} \times \sqrt{3})R30°$ superstructure [10.31-33]. The formation of this superstructure, in which nearest-neighbor sites are unoccupied, indicates a repulsive first-neighbor interaction. Results of thermal-desorption [10.1], IRAS [10.29] and especially LEED [10.27] measurements indicate that there is an attractive second-neighbor interaction between CO admolecules, which gives rise to the formation of islands at low temperature and low surface coverage. The $(\sqrt{3} \times \sqrt{3})R30°$ superstructure with first- and second-neighbor interactions J_1 and J_2 is shown in Fig. 10.8.

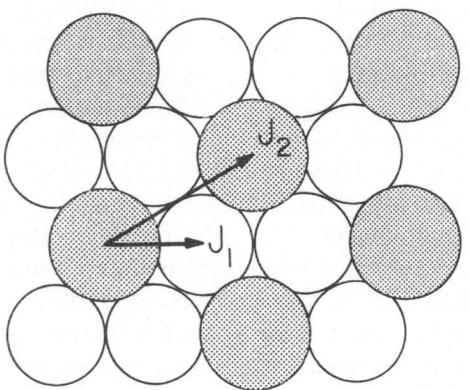

Fig. 10.8. The $(\sqrt{3} \times \sqrt{3})R30°$ superstructure of CO on Ru(0001). Arrows indicate the repulsive first-neighbor interaction J_1 and the attractive second-neighbor interaction J_2

10.3.2 Experimental Results

Angular LEED beam profiles were measured by positioning a Faraday cup, of which the aperture diameter is 0.13 mm, on the center of the beam and varying the energy of the incident electron beam to sweep the profile across the aperture of the cup [10.23]. The Faraday cup contains a channel electron multiplier as a detector and contains also an einzel lens, which is biased negatively in order to accept only those electrons of energy within approximately 0.5 eV of the energy of the incident electron beam. The LEED beam profiles were corrected for the variation in intensity of the incident beam with energy by division by the I-V curve. In making this correction, it was assumed that there was no variation in the width of the beam over the energy range of the beam profile. The energy width (FWHM) of the substrate beams ranged from approximately 1.5 eV at 37 eV to 3.4 eV at 95 eV. The FWHM of the fractional-order overlayer beams ranged from 1.5 eV for the narrowest beam measured to 2.7 eV for the widest. Transformation of the beam profiles functions of energy to functions of wave vector was accomplished using

$$\left(\frac{\partial k_\parallel}{\partial E} \right)_\theta = \frac{1}{2} \cdot \frac{\sin\theta}{\sqrt{150.4}} E^{-1/2} \,, \tag{10.9}$$

where θ is the angle of the diffracted beam with respect to the surface normal, E is the incident electron energy in eV, and $k_\parallel$ is the magnitude of the parallel component of the wave vector in Å^{-1} [10.23]. The bulk value of 2.7058 Å was used for the nearest-neighbor Ru-Ru distance on the surface in calculating the values of $k_\parallel$ at the center of the diffracted beam profiles.

The overlayers of CO were prepared by adsorption at 350 K, followed by cooling to either 100 K or 310 K. This procedure was followed since it has been shown that direct adsorption at low temperatures leads to a large density of defects (domain boundaries) in the overlayer at $\theta = 1/3$ [10.34]. To calibrate the coverage, the fractional-order-LEED-beam intensity was measured as a function of exposure to CO at 330 K. The exposure at which a maximum in intensity occurs represents optimum ordering of the superstructure and, consequently, a coverage of $\theta = 1/3$. The known approximate constancy of the probability of adsorption of CO at room temperature up to $\theta = 1/3$ was then used to relate lower coverages to exposure [10.32,35].

Three distinct sets of LEED measurements were carried out in order to clarify the formation of islands by CO on Ru(0001). First, profiles of the first-order LEED beams of the Ru(0001) substrate were measured as a function of incident electron energy in order to determine both the instrumental response function and the step density of the surface. Second, fractional-order-beam profiles for the $(\sqrt{3} \times \sqrt{3})R30°$ superstructure of CO were measured at a variety of surface coverages at both 100 and 310 K. Finally, the disordering of the $(\sqrt{3} \times \sqrt{3})R30°$ superstructure at temperatures up to 400 K was quantified at three coverages by monitoring the LEED intensity as a function of temperature. Each of these three sets of measurements is described in detail below.

Instrumental Response Function and Surface Step Density

The parameters determining the instrumental response function are the energy spread of the incident electron beam ΔE, the diameter of the aperture of the Faraday cup d, the effective width of the electron beam D, and the angular spread (source extension) of the electron beam γ [10.23,24] (cf. Sect.

2.5). For the instrument with which the behavior of CO on Ru(0001) was monitored, the aperture of the Faraday cup is 0.13 mm, and the true width of the incident electron beam is approximately 1 mm. However, the latter may be modified to a different effective width by a focusing action of the einzel lens of the Faraday cup.

To determine the values of D and γ and to measure the density of steps, the width of the first-order LEED beams of the substrate was measured at energies between 35 and 90 eV. For a surface with a distribution of terraces of different sizes separated by steps, the beam profiles will become broader and narrower with varying energy, as shown in Fig. 10.9 [10.12,13,24,36]. The smallest measured width corresponds to the instrumental width as observed by Wang and Lagally [10.24]. Henzler [10.36] has derived a relationship for the energies at which broadening and narrowing should be observed for the (0001) surface of an hcp lattice with steps of height equal to the lattice constant along the hexagonal axis (4.28 Å for ruthenium). These energies are indicated by arrows in Fig. 10.9, together with the experimentally-determined values of the width of the beam. Clearly, the measured values are consistent with a model of the surface containing a distribution of steps of height 4.28 Å. The degree of broadening of the beam profile is determined by the average distance between steps. The relative reduced width (the deconvoluted FWHM divided by the value of $k_{\parallel}$ for the beam) of the broadened beams shown in Fig. 10.9 is 1.0 ± 0.3%. Depend-

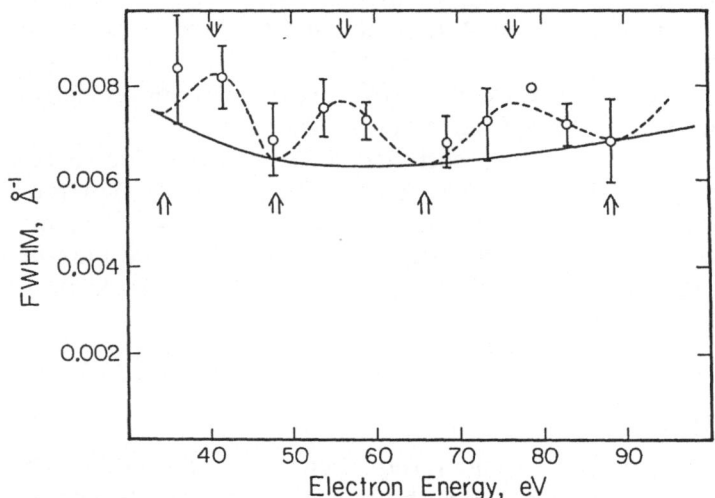

Fig. 10.9. FWHM of the first-order diffraction beams of the clean Ru(0001) surface as a function of energy. Error bars represent the standard deviation determined from repeated measurements. Arrows represent the energies at which maxima (*down-arrows*) and minima (*up-arrows*) are expected in the width [10.36]. The solid curve is the instrumental width. The dashed curve is drawn empirically to show the broadening

ing on the model used for the distribution of terrace sizes, this indicates terraces of width of 100 lattice spacings (270 Å) [10.13] or 50 lattice spacings (135 Å) [10.14].

The minima in the measured widths in Fig. 10.9 represent the width of the instrumental response function. Using the known values of ΔE and d, these minima were used to determine the values of D and γ. The optimized values for these parameters were found to be D = 0.05 mm and γ = 0.006 radian. The small value of the effective width indicates that the einzel lens in the Faraday cup discriminates against electrons not moving orthogonally to the detector. The instrumental resolution is limited by the energy spread ΔE at low electron energies and by the source extension γ, at higher energies.

Since the fractional-order beams associated with the $(\sqrt{3} \times \sqrt{3})R30°$ superstructure occur at a smaller angle than the substrate beams for the same energy of the incident electrons, the width of the instrumental response function is narrower for the fractional-order beams. Using the experimentally determined values for ΔE, d, D, and γ, the width of the instrumental response function of the first fractional-order beam is 0.0060 Å^{-1} at 28 eV and 0.0057 Å^{-1} at 49 eV with an uncertainty of ± 0.0006 Å^{-1}.

Beam Profiles of the CO Overlayer

Beam profiles of the fractional-order beams due to the CO superstructure were measured at 100 K and at 310 K for a range of surface coverages as described earlier in this section. Profiles measured at 100 K at fractional surface coverages of 1/3 and 1/10 are shown in Fig. 10.10. At θ = 1/3, the FWHM of the beam profile is 0.0064 Å^{-1}, only slightly broader than the instrumental response width; whereas at θ = 1/10, the FWHM is 0.0132 Å^{-1}. This increase in the width indicates that the CO admolecules are present in islands on the surface, i.e. in ordered regions of limited extent.

The measured beam profiles were corrected for broadening due to the instrumental response function by a Fourier-transform-deconvolution procedure. Park et al. have shown that the measured profile $I_m(k)$ is the convolution product of the instrumental response function $T(k)$ and the true beam profile $I_t(k)$ [10.23], therefore, the true profile can be recovered from the measured profile via

$$I_t(k) = F^{-1} \left[\frac{F[I_m(k)]}{F[T(k)]} \right], \tag{10.10}$$

where F and F^{-1} are the forward and reverse Fourier transforms, respectively. The average of two fractional-order-beam profiles that has the correct FWHM (0.006 Å^{-1}) was used for the instrumental response function $T(k)$, and the measured beam profiles were symmetrized by averaging about their center prior to deconvolution. Since the application of (10.10) is quite sensitive to truncation of the beam profiles and to noise, the profiles were smoothed and their wings extended prior to taking the Fourier transforms.

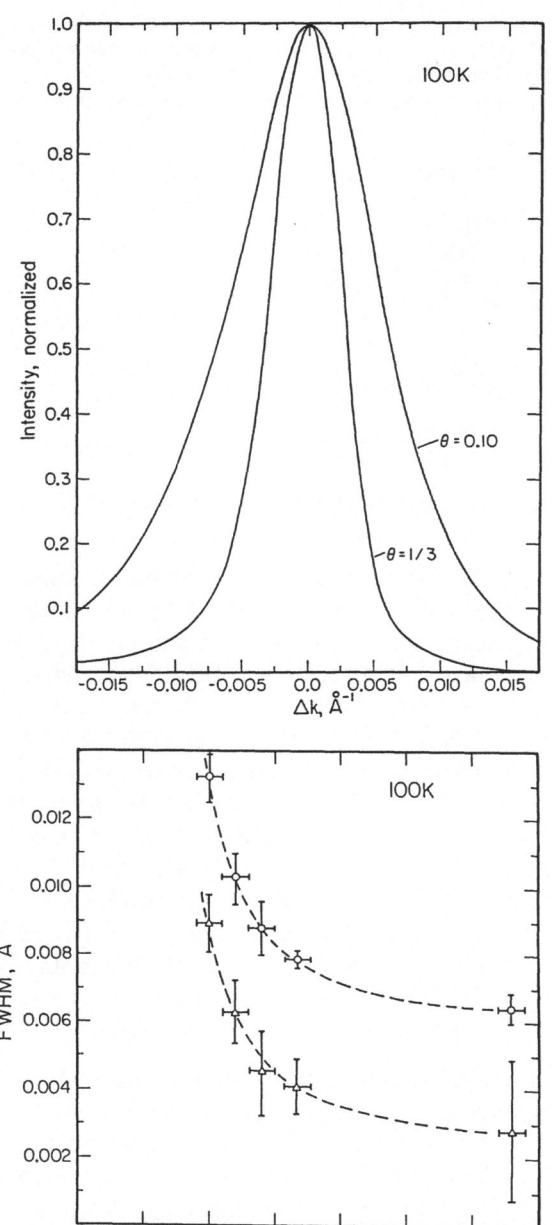

Fig. 10.10. Averaged beam profiles for the $(\sqrt{3} \times \sqrt{3})R30°$ superstructure of CO on Ru(0001) at absolute coverages $\theta = 1/3$ and $\theta = 0.10$, measured at 100 K. The profile for $\theta = 1/3$ is the average of nine measured profiles, that at $\theta = 0.10$ is the average of five. Measurements were made with an incident electron energy of 28 eV. The center of the fractional-order-beam profile, $\Delta k = 0$, is at $k_{||} = 0.2464 \, Å^{-1}$

Fig. 10.11. FWHM of a fractional-order-beam profile for CO on Ru(0001) as a function of coverage, at 100 K. Circles indicate measured widths at an incident electron energy of 28 eV. Triangles indicate widths corrected for the instrumental response function by deconvolution. Error bars on the FWHM are the standard deviations determined from repeated measurements. Error bars on the coverage are estimated

The widths of beam profiles measured at 100 K at fractional surface coverages between 1/10 and 1/3 are shown in Fig. 10.11 before (top curve) and after (bottom curve) deconvolution. The width increases as the coverage decreases, indicating that smaller islands form at lower coverages. Values of the widths shown in Fig. 10.11, as well as widths measured at 310 K, are

Table 10.1. Widths of beam profiles in $\mathring{A}^{-1}$ for the $(\sqrt{3} \times \sqrt{3})R30°$ superstructure of CO on Ru(0001) at different coverages and temperatures. $FWHM_m$ and σ_m are the measured width and standard deviation. $FWHM_t$ and σ_t are the width and standard deviation following deconvolution to correct for the instrumental response function. Values for $\theta = 1/3$ at 100 K and 310 K are combined since they are identical

T [K]	θ	$FWHM_m$	σ_m	$FWHM_t$	σ_t
100	1/3	0.0064	0.0005	0.0028	0.0021
	1/6	0.0079	0.0002	0.0041	0.0008
	0.14	0.0088	0.0008	0.0046	0.0012
	0.14[a]	0.0090	0.0002	0.0049	0.0007
	0.12	0.0103	0.0007	0.0063	0.0010
	0.10	0.0132	0.0007	0.0090	0.0009
310	0.20	0.0068	0.0006	0.0030	0.0017
	1/6	0.0089	0.0005	0.0058	0.0009
	0.14	0.0155	0.0015	0.0115	0.0016

[a] Value measured at 49 eV incident energy. All other values measured at 28 eV incident energy

listed in Table 10.1. The standard deviation in the widths of the deconvoluted profiles was calculated using the error-propagation equation appropriate for the deconvolution of two Gaussian functions. At a fractional surface coverage of 1/3, the optimum coverage for the $(\sqrt{3} \times \sqrt{3})R30°$ superstructure, there is no difference between the FWHM at 100 K and that at 310 K. At lower coverages, e.g. $\theta = 1/6$ and $\theta = 0.14$, the FWHM increases with temperature indicating a decrease in the size of the islands due to the loss of CO admolecules from the islands. At $\theta = 0.12$, the beam profile is sufficiently weak and broad at 310 K not to be measurable. At lower coverages, no intensity due to the ordered superstructures can be observed. The good agreement between the deconvoluted widths of the beam profiles measured at 28 eV and at 49 eV indicates that the use of the kinematic (single-scattering) approximation is entirely adequate for the analysis of these data.

Temperature Dependence of the CO Overlayer

Changes in the CO overlayer with temperature were monitored first by measuring the temperature dependence of the intensity of the beam profile at its center. The results for three different surface coverages are shown in Fig. 10.12. At $\theta = 1/3$, there is only an approximately 20% decrease in intensity between 100 K and 400 K. At $\theta = 1/6$ and $\theta = 0.14$, however, there is a much more dramatic decrease in intensity with increasing temperature.

Some decrease in intensity with temperature is expected due to the displacement of CO admolecules from their optimum positions on the surface as a result of vibrational motion (positional disorder). The expected variation in intensity is given by (4.63), which requires knowledge of the mean-square displacements $<(\Delta \bar{r})^2>$. The relevant vibrations in this case have fre-

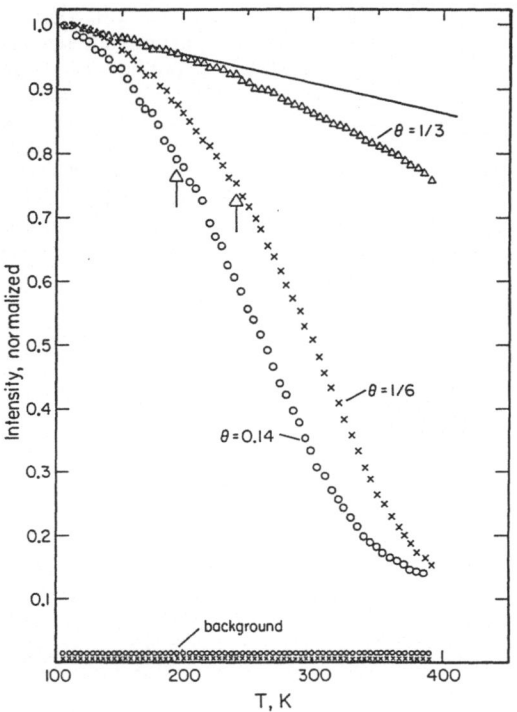

Fig. 10.12. The variation with temperature of the intensity at the center of the fractional-order-beam profile for $\theta = 1/3$, $1/6$, and 0.14 for CO on Ru(0001). The variation of the background intensity at $\theta = 1/6$ (*crosses*) and $\theta = 0.14$ (*circles*) is also shown. Arrows indicate the temperatures at which the beams begin to broaden. The continuous curve is the calculated decrease in intensity due to the frustrated translational motion of the CO molecules parallel to the surface, with a frequency of $45\ \text{cm}^{-1}$, using a Debye-Waller factor

quencies of $2021\ \text{cm}^{-1}$ for the carbon-oxygen stretching mode, $445\ \text{cm}^{-1}$ for the ruthenium-carbon stretching mode (the frustrated translational mode perpendicular to the surface), between 400 and $600\ \text{cm}^{-1}$ for the frustrated rotational motion, and between 34 and $126\ \text{cm}^{-1}$ for the frustrated translational motion parallel to the surface [10.28,29,37-39]. Of these, only the latter is sufficiently low in frequency to cause a measurable change in the mean displacement of CO between 100 and $400\ \text{K}$. The mean-square displacement parallel to the surface of a two-dimensional harmonic oscillator is given by

$$<(\Delta \vec{r}_{\parallel})^2> \ = \ \frac{h}{8\pi^2 m_a \nu} \left\{ \frac{1}{2} + [\exp(h\nu/k_B T) - 1]^{-1} \right\} , \qquad (10.11)$$

where ν is the vibrational frequency, and m_a is the mass of the CO admolecule (which is treated as a single particle). The variation in intensity with temperature was calculated from (4.63) and (10.11) for a range of values of ν. The calculated variation in intensity is shown by the continuous curve in Fig. 10.12 for a value of ν equal to $45\ \text{cm}^{-1}$. Up to approximately $200\ \text{K}$, the observed decrease in intensity at $\theta = 1/3$ can be attributed to vibrational motion. Above $200\ \text{K}$, an additional type of disorder occurs, i.e. site disorder in which CO admolecules occupy incorrect sites in the lattice with respect to the $(\sqrt{3} \times \sqrt{3})R30°$ superstructure. It is likely that site disorder at

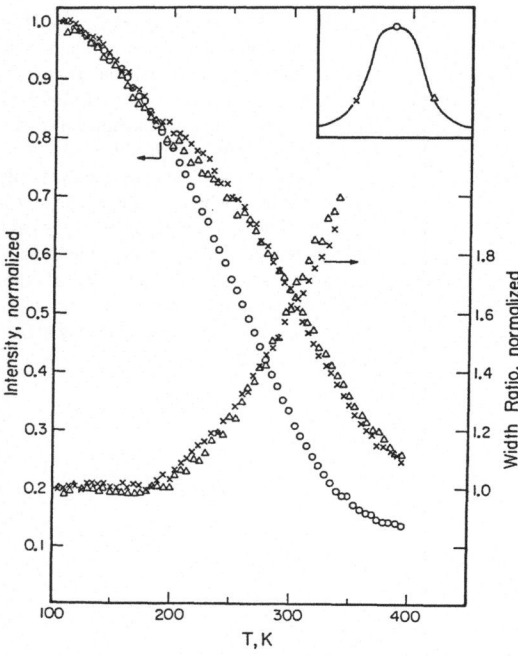

Fig. 10.13. The variation with temperature of the intensity at three different positions on the beam profile for $\theta = 0.14$ for CO on Ru(0001). The position for each curve is illustrated schematically in the inset. Each curve has been normalized independently to unity at 100 K. The ratio of the curves measured in the wings of the profile to the curve measured at the center is also shown for the same points

$\theta = 1/3$ begins at boundaries of domains, only becoming prevalent throughout the overlayer at high temperatures.

It is known that the beam profiles at $\theta = 1/6$ and $\theta = 0.14$ broaden with increasing temperature (Table 10.1). To monitor this change in shape, the intensity at different points on the profile was measured as a function of temperature. As a profile broadens, the intensity in the wings of the profile will decrease less rapidly than the intensity at the center, as illustrated for $\theta = 0.14$ in Fig. 10.13. In this figure, three intensity-temperature curves are plotted together; one was measured at the center of the profile and the other two at quarter-maximum intensity with each curve normalized independently. The slower decrease in intensity at the quarter-maximum points, which is indicative of beam broadening, is apparent in Fig. 10.13. The ratio of the intensity at quarter-maximum to that at the center, i.e. the width ratio, shows the broadening with temperature even more clearly in Fig. 10.13. The onset of an increase in the width ratio is abrupt, occurring at a temperature of 195 ± 5 K for $\theta = 0.14$. This same behavior is observed at $\theta = 1/6$, with the onset of broadening at 240 ± 10 K. The temperatures at which the width ratio begins to change are indicated with arrows in Fig. 10.12. At $\theta = 1/3$, the width ratio is constant up to 400 K confirming the previous observation that the FWHM of the beam at $\theta = 1/3$ is the same at 100 K and at 310 K.

10.3.3 Analysis and Discussion of Results

An analysis and discussion of the results presented in the preceding section will be concerned with island sizes, the distribution of island sizes, the mechanism of the formation of islands, and the relationships between these issues and order-disorder phenomena. In principle, the mechanism of the formation of islands determines the distribution of island sizes, as discussed by Lagally et al. [10.40]. There are two possible reasons for the formation of small islands. The first is the presence of steps which act as a barrier to surface diffusion, trapping adspecies on the terraces on which they are adsorbed initially [10.15]. In this case, the size distribution of the islands will be determined by the size distribution of the terraces. Except at quite low surface coverages, this step-limited model requires that there be a constant number of islands which vary in size directly with coverage. A second possible reason for the formation of small islands is a limited diffusion distance of the adspecies. In this diffusion-limited model, adspecies which are adsorbed initially near one another merge to form small islands. Once formed, the configuration with many small islands may represent a local minimum in the Helmholtz energy, with an activation barrier to the formation of a single large island.

10.3.3a The Step-Limited Model of Island Formation

In practice, size distributions have been predicted only for the step-limited model of island formation [10.13,15]. The experimental beam widths for CO on Ru(0001) were analyzed using these two distributions as well as three semiempirical size distributions. To calculate beam profiles for comparison with experiment, given a distribution of sizes $P(l)$, we used an expression equivalent to (10.4), namely

$$I(k) = N\sum_l P(l)I_l(k) \ , \tag{10.12}$$

where N is the number of islands, $P(l)$ is the probability of occurrence of an island containing l admolecules, and $I_l(k)$ is the beam profile due to a single island with l admolecules. A set of 34 of the $I_l(k)$ profiles were calculated for values of l ranging from 59 to 4955, with the admolecules arranged in round islands. The 34 island sizes were chosen to represent constant increments in the value of the diameter of the islands. Consequently, the increments in l are smaller for smaller values of l, where the width of the beam profile varies more rapidly with the size of the islands. The summation in (10.12) was carried out over this set of values of l, with $P(l)$ replaced by $P(l)\Delta l$, for each of five different size distributions, as described below.

Geometrical Distribution of Terrace Widths

First, Lu et al. [10.14,15] have developed a "geometrical distribution" for terrace widths Γ which is given by

$$P(\Gamma) \propto (1 - \Gamma)^{\Gamma-1}\gamma \ , \tag{10.13}$$

where γ is the probability of encountering a step between two surface atoms in a given direction, and Γ is the width of the terrace in units of the number of surface atoms. An analysis of the step distribution of the Ru(0001) surface on which CO was adsorbed indicates that the mean terrace width is 50 ruthenium atoms, see Sect. 10.3.2, corresponding to a value of γ equal to 0.02. If the distribution of island sizes is step-limited, the size l of an island will be determined by the overall coverage θ and the size Γ of the terrace on which it resides through

$$l = \theta\Gamma^2 \ , \tag{10.14}$$

where l is the number of admolecules in the island. It is assumed that, on average, terraces will have uniform widths in two dimensions. Using this model, the calculated FWHM of the CO beam profile varies only slightly (from 0.0040 Å^{-1} to 0.0042 Å^{-1}) between $\theta = 1/6$ and $\theta = 1/10$. It is clear that using a constant average terrace width, this model cannot account for the rapid change in the FWHM with coverage at intermediate coverages that

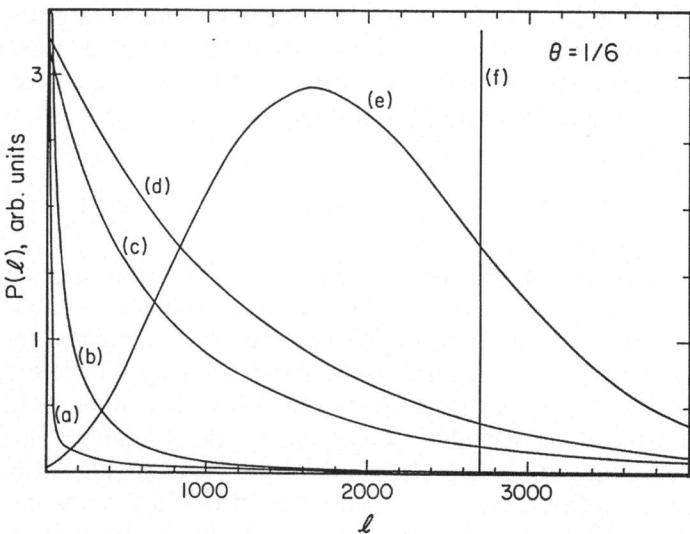

Fig. 10.14. Probability of observing an island containing l admolecules as a function of l for (*a*) geometrical distribution, (*b*) Henzler's distribution, (*c*) empirical distribution (described in text), (*d*) ball-in-urn distribution, (*e*) distance distribution, and (*f*) delta-function distribution. Each distribution gives rise to a beam profile of width 0.0041 Å^{-1} as observed for $\theta = 1/6$ at 100 K for CO on Ru(0001)

is observed experimentally (Table 10.1 and Fig. 10.11). The experimental values can be reproduced only if the mean terrace width is allowed to vary substantially with surface coverage. Curve (a) in Fig. 10.14 shows the distribution of island sizes (with $\gamma = 0.022$) that gives the correct FWHM for $\theta = 1/6$ at 100 K.

Henzler's Distribution of Terrace Widths

A second distribution, due to Henzler [10.13], has been proposed for the distribution of terrace widths and is given by

$$\Gamma P(\Gamma) \propto \Gamma \exp\left[-\left(\frac{\Gamma}{w}\right)^E - \left(\frac{w}{\Gamma}\right)^E \right], \tag{10.15}$$

where E is approximately equal to 0.8, and w is an adjustable parameter determining the mean terrace width. Based on this distribution, the mean terrace width on the Ru(0001) surface was found to be 100 ruthenium atoms, as discussed in Sect. 10.3.2. A value of w equal to 40 in (10.15) gives this mean terrace width. The distributions of island sizes were evaluated from (10.15) with (10.14) used to relate l to Γ. As in the case of the geometrical distribution, the requirement of islands limited by the size of terraces results in a very slow variation in the FWHM with coverage between $\theta = 1/6$ and $\theta = 1/10$, the range of the values of the FWHM lying between 0.0036 Å^{-1} at $\theta = 1/6$ and 0.0037 Å^{-1} at $\theta = 1/10$. Consequently, a fit to the experimental data requires a different mean terrace width at each surface coverage. The distribution that corresponds to the correct width of the beam profile at $\theta = 1/6$ is shown in curve (b) of Fig. 10.14 and corresponds to a value for w of 17 and a mean terrace width of 41 ruthenium atoms.

Ball-In-Urn Distribution of Terrace Widths

A third distribution for the size of islands is the "ball-in-urn distribution" described in detail in Sect. 10.2.4. In this case, P(l) is proportional to F(l) given by (10.8). As expected, a different number of nucleation sites (M) had to be used for each coverage to fit the experimental data with this distribution. This is consistent with the previous results using the first two distribution functions, i.e. the experimental values are incompatible with a model that requires a fixed number of islands at all surface coverages. Curve (d) in Fig. 10.14 shows the ball-in-urn distributions (for M equal to 2100 and n equal to 2,666,667 on a 4000x4000 ruthenium lattice) which gives the value of the FWHM measured for $\theta = 1/6$ at 100 K.

Distance Distribution of Terrace Widths

A computer simulation based on a simple model of diffusion-limited island nucleation was used to generate probability distributions for the nearest-neighbor distance between centers of islands r that are Gaussian in r, see Sect. 10.2. This "distance distribution" is the fourth distribution to be considered. The width of the distributions increases approximately linearly with the mean distance r_0. Distributions of island sizes were generated using this model assuming that the number of molecules in an island is related to the distance to its nearest-neighbor by $l = \theta r^2$. A decreasing value of r_0 with decreasing surface coverage was required to fit the measured values of the FWHM, indicating that the model used to generate this distribution is too simple to describe the overlayer correctly. Curve (e) in Fig. 10.14 represents the distance distribution with $r_0 = 106$ ruthenium atoms which reproduces the measured FWHM for $\theta = 1/6$ at 100 K.

Delta-Function Distribution of Terrace Widths

For comparison with the other distributions, a fifth distribution was considered in which there is only one island size. This "delta-function distribution" is shown by curve (f) of Fig. 10.14 for $\theta = 1/6$ at 100 K. It is demonstrated clearly in Fig. 10.14 that knowledge of the width of a beam profile alone is insufficient to determine the island sizes. The use of the delta-function distribution to analyze the size does not give the mean size in any sense but, at best, an upper limit to the mean size (Sect. 10.2).

Although knowledge of the coverage dependence of the beam width can be used to test specific models of the size distribution, models that require a fixed number of islands fail to fit the experimental data since they cannot generate the rapid change in FWHM with coverage that is observed. This indicates that neither a step-limited model nor a defect-nucleation model is correct for CO on Ru(0001). Although the distance distribution also seems to be incorrect, since it requires larger diffusion distances at larger coverages, this distribution results from a rather crude simulation, and hence a diffusion-limited model for the formation of islands cannot be excluded.

10.3.3b Dissolution of Islands

It is clear from Figs. 10.12,13 that no change occurs in the width of the beam profiles at low surface coverage with increasing temperature, until the intensity has decreased by approximately 20%. Since the variation in intensity with temperature at $\theta = 1/3$ indicates that there is little site disorder within the $(\sqrt{3} \times \sqrt{3})R30°$ superstructure below approximately 200 K, the loss of intensity must be due to the loss of CO admolecules from the islands. Intuitively, it might be expected that variations in intensity and FWHM would occur simultaneously, i.e. as CO admolecules leave the islands, the islands become smaller and the FWHM larger. Although this effect has been observed for oxygen adatoms on W(110) [10.17], it is possible that the size distribution of the islands could modify this expectation. Since the intensity of a beam profile is proportional to the square of the number of admolecules in the island, large islands dominate overwhelmingly in determining the observed beam profile, cf. (10.12). On the other hand, since the FWHM is inversely proportional to the diameter of the island, the FWHM changes more rapidly with size for small islands than for large islands. Consequently, if small islands dissolve totally, the FWHM will decrease. However, if large islands lose a fraction of their admolecules, the intensity will decrease, and the FWHM will increase slightly. It is possible that for the correct distribution of island sizes, these two influences on the FWHM will cancel until the intensity has decreased appreciably. This possibility was tested for CO on Ru(0001) using a rather crude model of disordering for both a step-limited and a non-step-limited model of island formation, as described next.

Ideal Two-Dimensional Gas

The dissolved phase of CO on Ru(0001) was treated as an ideal two-dimensional gas and the $(\sqrt{3} \times \sqrt{3})R30°$ phase as an ideal two-dimensional solid. The chemical potentials of the two phases were calculated and equated in order to determine the number density of disordered CO admolecules θ_g as a function of temperature. The result is given by [10.27]

$$\theta_g = \left(\frac{2\pi m_a k_B T}{h^2} \right) \frac{q_{i,g}}{q_{i,s}} \exp\left(\frac{6J_2}{2k_B T} \right) , \tag{10.16}$$

where $q_{i,g}$ and $q_{i,s}$ are the internal partition functions of a single CO admolecule in the "gas" and the "solid" phases, respectively, and J_2 is the interaction energy between CO admolecules in second-neighbor adsites (Fig. 10.8). The values of $q_{i,g}$ and $q_{i,s}$ were assumed to be the same except for the minor difference due to the different carbon-oxygen stretching vibrational frequencies [10.28,29].

However, it is necessary to modify (10.16) slightly in view of the fact that for an island of finite size, the total interaction energy will be less than $6J_2 N_s/2$ since the admolecules at the edge of an island have a coordination less than six. Here, N_s is the number of admolecules in the two-dimensional solid. The number of admolecules at the edge of an island will be proportional to the square root of the number in the island, so that the total energy of interaction may be written as

$$E = \frac{6J_2 N_s}{2} - cJ_2\sqrt{N_s} , \tag{10.17}$$

where c is a constant that accounts for the coordination of the admolecules at the edge of an island as well as the proportionality of $\sqrt{N_s}$ to the number at the edge. Using this modified value of E in the partition function of the solid gives a modified value of the density of disordered admolecules of CO (i.e. those in the "gas" phase on the surface). The result may be written as

$$\hat{\theta}_g = \theta_g \exp(-cJ_2/k_B T\sqrt{N_s}) . \tag{10.18}$$

Step-Limited Model

In a step-limited model of island formation, each island is located on a terrace that is isolated from all the other islands. Consequently, (10.18) can be used directly to calculate the two-dimensional gas-phase density as a function of temperature on each terrace. Once $\hat{\theta}_g$ is known, the change in the size of the island follows immediately, and the beam profile can be calculated as a function of temperature. In order to be compared with experiment, the calculated beam profiles were convoluted numerically with a Gaussian instrumental response function of width 0.006 $\overset{\circ}{A}{}^{-1}$ for the distribution functions of Fig. 10.14. Although the calculation was quite successful in reproducing the sudden onset in the change of the width ratio (Fig. 10.13), the intensity did not decrease by more than 10% prior to a change in the width ratio for any of the size distributions. Moreover, only a 10 K difference between the intensity curves at $\theta = 1/6$ and $\theta = 0.14$ was calculated, compared to the observed difference of 30 K or more (Fig. 10.12).

Non-Step-Limited Model

If the growth and disappearance of the islands are not step-limited, all islands in the overlayer should be considered when calculating the partition function of the two-dimensional solid. In order to model this mathematically, the overall two-dimensional "gas" density was evaluated from (10.16). There it was assumed that all islands lose CO admolecules from their edges at the same rate, with a correction term for the higher energy of smaller islands, see (10.18), until the correct overall two-dimensional gas density was attained. Since the inclusion of this energy correction term causes small islands to lose admolecules from their edges at a greater rate than large islands, the omission of this term causes a distribution to resemble a distribution with a slightly larger mean island size. When beam profiles were calculated as for the step-limited model, it was found that for different distributions of island sizes, the onset of the change of the width ratio was shifted to higher temperatures as the mean island size became smaller. This is in agreement with the qualitative arguments presented earlier concerning the relative effects on the FWHM for a loss of CO admolecules from small and large islands. For the ball-in-urn distribution [Fig. 10.14, curve (d)], the intensity decreased by 10% prior to the observation of beam broadening. For the geometrical distribution [Fig. 10.14, curve (a)], the beam profile actually became narrower with increasing temperature as the large number of small islands in that distribution "disordered" preferentially. Consequently, a distribution intermediate in shape between these two was sought.

Empirical Distribution

An empirical distribution of the form

$$P(l) \propto (1 - b)^{l^{3/4}} , \tag{10.19}$$

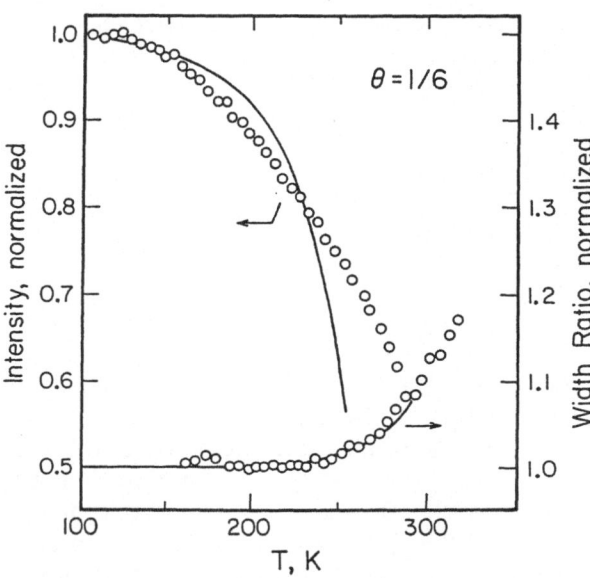

Fig. 10.15. Comparison of calculated (*solid curves*) and experimental (*circles*) intensity and width ratio as functions of temperature. An empirical size distribution (see text) and a value of $J_2 = -1.28$ kcal/mole were used in the calculation. The calculated intensity has been multiplied by the Debye-Waller factor

where b is an empirical constant, was found to give the correct relative behavior of intensity and width with temperature. The distribution used at θ = 1/6 is shown in curve (c) of Fig. 10.14. The calculated intensity and width ratio for $J_2 = -1.28$ kcal/mole is compared with the experimental data in Fig. 10.15. The calculation reproduces the delayed onset of broadening quite successfully, and the more rapid increase in width with temperature at θ = 0.14 than at θ = 1/6 is predicted as well. However, as in the case of the step-limited model, the observed difference in temperature between the two intensity curves is not reproduced exactly. The best fit to both sets of data (θ = 1/6 and θ = 0.14) occurs with $J_2 = -1.20$ kcal/mole, which places the calculated curves for θ = 1/6 approximately 10 K too low and those for θ = 0.14 approximately 10 K too high in temperature.

It was not possible to fit the experimental intensity and width-ratio curves using a step-limited model of island growth in which the density of the disordered phase about an island is due to the loss of CO admolecules from that island alone. If the number of CO admolecules lost from each island is proportional to the number of molecules at the edge of the island, a very reasonable approximation, it is possible to fit the experimental curves with only one model for the island-size distribution. Consequently, curve (c) of Fig. 10.14 represents the physical distribution of island sizes although (10.19), used to describe it, is an empirical expression.

Mean Island Sizes

The mean island sizes were determined by varying the parameter b in (10.19) to reproduce the beam profiles corresponding to the widths measured

Table 10.2. Mean island size and mean island diameter for the correct distribution described in the text and for the delta-function distribution. The value b is the parameter used in (10.19)

	Correct distribution			Delta-function Distribution	
θ	b	$\bar{l}$	$\bar{d}$[Å]	l	d [Å]
1/6	0.008	1020	138	2700	255
0.14	0.014	515	96	1720	218
0.12	0.019	325	77	1100	163
0.10	0.032	160	54	560	115

at 100 K. The results are compared in Table 10.2 with values calculated using the delta-function distribution [Fig. 10.14, curve (f)]. In carrying out these calculations, it was assumed that the beam width measured at 100 K is the narrowest width that occurs at each coverage, an assumption that was shown to be correct experimentally for $\theta = 0.14$ and $\theta = 1/6$. Extrapolating the observed behavior at $\theta = 1/6$ and $\theta = 0.14$, it appears that the onset of broadening is near 100 K at $\theta = 0.12$ and is likely to be below 100 K at $\theta = 0.10$. Consequently, the calculated island sizes are correct for $\theta = 1/6$, 0.14, and 0.12, but the one for $\theta = 0.10$ is likely to be somewhat small.

10.3.4 Summary of Island Formation Properties for CO/Ru(0001)

Measurements of fractional-order-LEED-beam profiles as a function of CO coverage were made at low temperatures on Ru(0001). These beam profiles increased rapidly in width with decreasing coverage, indicating that smaller islands are present at lower coverages. The coverage dependence of the beam width is inconsistent with a step-limited model of island formation, suggesting that island formation is diffusion-limited in this system. The disordering of the islands was monitored via changes in the LEED beam intensity and width, and the disordering was found to be influenced strongly by the finite size of the islands. By modeling the island dissolution, it was found that only one distribution of island sizes is consistent with the observed behavior, and, as a consequence, a unique determination of the island size as a function of coverage was possible. Presumably, these observations are rather general and are not limited to the particular case of CO on Ru(0001).

11. The Future of LEED

Thus far in this book, we have sketched the developments in the use of LEED up to the present. Despite the evolution that has occurred over the years, the technique of LEED retains many untapped potentials. Some applications have had to await a sufficient experimental or theoretical development of the technique, e.g. the case of the study of less stable and more complex surface structures. Other applications of LEED have not yet received the attention they deserve, even though the technique can already be applied to them. For example, oxides, magnetic materials, alloys, disordered structures, and phase transitions present many unanswered questions that can be studied with the present state of the art.

In this chapter, we shall discuss the prospects for the future of LEED from four different points of view. First, present experimental limitations and possible technological improvements as well as new applications will be discussed. Second, present theoretical limitations will be addressed, as well as theoretical developments that are already taking place or can be forecast. Third, we shall consider the specific topic of structural determination and the prospects for its development. Fourth, and finally, it is important to take a broader view and to relate LEED to the many other surface-sensitive techniques that can be used to study surfaces in general, and surface structure in particular.

11.1 Experimental Outlook

11.1.1 Improvements in Experimental Techniques

Low Energies

At low kinetic energies, for example below 20 eV, LEED measurements usually become inaccurate. One of the primary causes is residual magnetic fields that distort the electron trajectories between gun, sample, and detector. The routine use of Helmholtz coils and/or μ-metal shielding has reduced the magnetic-field effects sufficiently for kinetic energies above approximately 20 eV. Although it is certainly possible to carry out LEED measurements at lower kinetic energies, in general this is not necessary. At these low kinetic energies, few data are measurable since few beams emerge from the surface, and there is too great a sensitivity to the scattering potential. Exceptions to this generalization concern studies of the effect of the surface-potential step, as well as other aspects of the scattering potential at a surface.

Angle of Incidence and Surface Orientation

To set the angle of incidence accurately, a good sample manipulator is required. The commercial manipulators that are commonly used are not quite sufficiently accurate to orient the sample reproducibly within approximately 0.5°, as is required for detailed structural studies. Normal incidence can be determined relatively easily with such an accuracy by observing the symmetries in the spot intensities (at least when the surface structure has such symmetries), but other incidence directions are not so easily calibrated. More accurate goniometers have been tested and have shown a much higher degree of reliability in setting the incidence angle [11.1]. This appears to be a desirable experimental improvement, but it raises another issue which must be addressed at the same time, namely that of the proper orientation of the surface plane with respect to the bulk crystallographic planes of the sample (i.e. that of the cutting angle of the sample). The present approach to surface orientation could be improved easily to yield better results. An accuracy of approximately 0.1° should be attained routinely with sufficient care [11.2], rather than the generally accepted accuracy which gives a surface orientation within 0.5° or 1° of the desired crystallographic plane.

Electron-Beam-Induced Damage

Electron-beam-induced damage is a frequent problem, especially with molecular surfaces, whether they be submonolayers, monolayers, or thicker films of adsorbed molecules, or even surfaces of molecular crystals. The current cure is usually to reduce the exposure to the incident electron beam. In the photographic and vidicon methods of recording LEED data, the intensities are measured rapidly to allow a reduced beam exposure. Alternatively, field-emitter tips may be used to reduce the incident electron current drastically, by up to six orders of magnitude, thus eliminating the problem [11.3]. In that case, the weaker diffracted beams can be amplified through microchannel plates and measured by means of the traditional LEED screen or new position-sensitive detectors. Moreover, small channel electron multipliers may be placed in the rotatable Faraday-cup collectors, and this permits a reduction from four to six orders of magnitude in the incident electron-beam flux [11.4].

Surface Charging

The use of low incident currents also solves another problem, namely the charging of poorly conducting samples. This is most valuable when studying molecular films, molecular crystals, and surfaces of oxides or other insulators, including semiconductors.

Surface Preparation

The improvements described above will contribute markedly to the reproducibility of LEED measurements. Another aspect of reproducibility concerns the state of the surface in terms of cleanliness, flatness, adlayer ordering, and the stability of the overlayer, for example. Effective cleaning techniques must be developed on a case-by-case basis to remove impurities from the sample surface. In some instances Auger-Electron Spectroscopy (AES) is not effective in assuring low impurity levels, for example for combinations of elements for which Auger lines overlap. In these cases other techniques should be applied, for example X-ray Photoelectron Spectroscopy (XPS). In some special cases, for example when hydrogen is the adsorbate, neither AES nor XPS would be useful. Laser annealing may be an effective solution to the impurity problem, and it seems to improve the substrate flatness greatly as well. The substrate perfection and overlayer ordering can be assessed by monitoring the spot widths obtained with electron guns that produce a larger coherence length (transfer width). Annealing until LEED beams have a sharpness corresponding to a transfer width on the order of 1000 Å should be achievable without difficulty [11.5], so long as the diffusion of bulk impurities to the surface is avoided.

Unstable Overlayers

Chemical conversion of surface species, i.e. the stability of the overlayer, is a more delicate problem. Each case must be addressed individually to establish the conversion rates which dictate the stability of a given surface structure. Examples of currently problematic surface structures are oxygen on Ni(100) at a fractional coverage of one half, and oxygen on Al(111) at a fractional coverage of unity. Different measurements often yield different structures for these surfaces, and the cause is likely to be a conversion from an overlayer state to an underlayer state, i.e. the penetration of oxygen into the metal lattice (see the references in Chap. 12). Molecular overlayers can be even more delicate in this respect due to the many possibilities of structural change, for example dissociation or other surface-mediated chemical reactions, in addition to the possibility of either different ordering arrangements or disordering. Impurity-stabilized states must also be considered.

11.1.2 New Experimental Directions

New Materials

We have already mentioned the extension of LEED measurements to less stable overlayers and nonconducting surfaces. Many other kinds of surfaces have not been investigated in very much detail yet, although the basic techniques have been available for some time. These include coadsorbed overlayers, where one adsorbate affects the structure of the other; surfaces of magnetic materials and magnetic monolayers, which can be studied in particular with spin-polarized LEED, and the surface properties of which may be quite different from the bulk properties (for example, the issue of magnetically "dead" surface layers) [11.6]; alloy surfaces, some of which order into simple lattices, while others are disordered, and the surface composition of which may be quite different from the bulk composition; liquid surfaces [11.7]; liquid monolayers on solid substrates; multilayer oxide growth; epitaxial growth of one material on another [11.8]; and multilayer films deposited on inert substrates such as graphite or MoS_2 to study low-dimensional properties of such films.

Domain and Island Sizes

A technical development that should be within grasp is the possibility of varying the coherence length of the incident electron beam intentionally (i.e. the instrumental transfer width). By varying this quantity, it should be possible to measure both the size and the distribution of domains and islands of adsorbates at surfaces. This would be most valuable in studying adsorbate-adsorbate interactions and in investigating phase transitions at surfaces.

I-V Curves as Fingerprints

We should mention new possibilities offered by the advent of fast intensity-measurement techniques, such as the vidicon and the position-sensitive detectors. The ability to acquire complete LEED intensity curves in a matter of minutes or less allows the convenient use of intensity curves independently of a calculational interpretation. For example, an I-V curve can serve as the unique fingerprint of a particular surface under specified conditions, just as an Auger spectrum characterizes the surface composition. The I-V curves are quite sensitive to the surface structure, especially if it is ordered, and therefore they can be used to ensure that a desired surface condition has been reproduced in an experiment.

Moreover, the fast measurement techniques make it possible to control a LEED experiment better since one can immediately reject unsatisfactory intensity data due, for example, to improper crystal orientation and preparation.

Mutual Comparison of Experimental Intensities

An important application of measured intensity data is the comparison of such data, not with theory, but with equivalent data measured for another surface structure. One may thereby, for example, prove that different preparation methods yield the same structure, or that different molecules eventually produce the same surface species. In this way, if one structure has been determined by a comparison with LEED calculations, a sequence of others may be determined by simple comparisons without additional calculations.

An issue that should be explored in detail is to what extent identical intensity curves necessarily imply identical structures. For example, an extremely disordered surface layer has a minimal effect on the I-V curves of the substrate. Therefore, if one part of an adsorbed molecule has no well-defined orientation because it is free to rotate about the fixed atom to which it is bonded, then that molecular segment will be disordered and will not affect the I-V curves. In this example, only the ordered segment of the adsorbed molecule is monitored by the beam intensities, and any statement concerning the identity of structures applies only to that ordered segment. Such considerations have proved to be very useful in recent structural determinations of adsorbed molecules [11.9].

Benefits of Large Data Bases

Finally, it should be mentioned that the fast acquisition of data can result in the availability of large collections of intensity data for any given surface structure. These are very useful in refining conventional structure determinations since more data can then be fit with calculated curves, resulting in a higher accuracy of the structural determination. Large data collections could also lead to better data averaging, as in the Constant-Momentum-Transfer-Averaging scheme. The result could be more "kinematic" data which are susceptible to an easier interpretation.

11.2 Theoretical Outlook

11.2.1 Survival of the Kinematic Theory

Complex Surface Structures

For several reasons, the convenient kinematic theory of LEED would be expected to remain popular. The LEED pattern, especially when it is complex, can be interpreted advantageously within the kinematic framework. This applies, for example, to ordered surface structures with large unit cells containing several molecules, or to modulated structures due to a lattice mismatch between the surface layer and the bulk. It also applies, in particular, to disordered structures. Much information concerning short-range ordering and long-range disordering phenomena is available in the diffuse "background" of the LEED pattern. Important developments have taken place over the years in this respect [11.10-12], but much more remains to be done.

Laser Simulation

An interesting possibility for further development is the expanded use of laser simulations of LEED patterns [11.13,14]. At present, this approach is limited largely by the inability to compare spot intensities from LEED and laser diffraction directly, due to different "atomic" form factors in the two cases and the influence of multiple scattering in LEED. However, corrections for the difference in form factors could be included, and the effect of multiple scattering could be reduced by suitable averaging (over the energy, for example). Then the use of laser simulation would become more quantitative, leading to a more detailed interpretation of complex structures. Of course, the above remarks apply not only to laser simulation, but also to the straightforward simulation of the LEED pattern by kinematic calculations. The latter is already practiced with disordered systems [11.11], where the atomic form factor and multiple scattering are of less concern.

Preliminary Kinematic Structural Search

Another reason for the continued use of the kinematic theory is the constant need to reduce the computational effort in the interpretation of LEED intensities. There are a number of obvious possibilities. One is simply to apply the kinematic rather than the dynamical formalism in a conventional structural search. The reduced computational expense allows a search over many trial structures and thereby the elimination of a large number of otherwise plausible structures [11.15] (see also Sect. 6.1). The neglect of multiple scattering, of course, reduces the reliability of the results. However, the use of a large data base (obtained easily with the new high-speed data-acquisition systems) should be quite beneficial, since it should average out to some extent the effects of multiple scattering. This approach is especially attractive with materials in which multiple scattering is relatively weak, namely low-atomic-number and low-density materials (for example, silicon and organic surface layers). In any case, after a kinematic search for the surface structure, it will remain desirable to evaluate and refine the fewer remaining acceptable structures with more accurate dynamical calculations.

Kinematic Conditions

Another possibility is to search specifically for experimental conditions where multiple scattering is weak. High energies and high temperatures are useful in this regard through a reduction in the atomic backscattering strength, especially due to the Debye-Waller factor. However, it must not be forgotten that forward multiple scattering remains strong, and many systems of interest could be sensitive to temperature.

11.2.2 Partial Multiple Scattering

A compromise can often be found between a kinematic calculation and a fully self-consistent dynamical calculation. For example, an adsorbed over-layer may produce little internal multiple scattering if its density is low or if its atomic number is low. Consequently, a kinematic calculation *within* that layer is justified, which can be combined with a dynamical calculation throughout the substrate. The multiple scattering between the overlayer and the substrate can then be included to convergence [11.16-18]. Alternatively, the multiple scattering within each layer (surface or bulk) may be neglected, yielding kinematic diffraction matrices for each layer. In this case, the individual layers can still be stacked together, with the inclusion of full multiple scattering between them. The justification for this approach, which has been called Quasi-dynamical [11.16-18], lies in the preponderance of forward scattering from one layer to the next, as opposed to "sideways" scattering to atoms in the same plane parallel to the surface. Predictably, this approach works best for weakly scattering materials, but it has been shown to be effective in a crude structural search of strongly scattering surfaces [11.17].

A different approach to neglecting multiple-scattering contributions adopts the point of view that multiple scattering is strongest between nearest-neighbor atoms. Consequently, it should be a reasonable approximation to neglect multiple scattering between atoms separated by more than one or two nearest-neighbor distances [11.18]. This situation arises especially with non-close-packed materials (e.g. molecular surface layers), where the number of atoms in each shell of equidistant neighbors is relatively small. For example, in an overlayer of carbon monoxide molecules which are adsorbed perpendicularly to a metallic surface, each carbon atom has only one nearest neighbor (an oxygen atom at a small distance of approximately 1.15 Å) and only one or a few next-nearest neighbors (metal atoms at a distance of approximately 2.0 Å); the oxygen atoms are separated even further from their next-nearest neighbors. Similar considerations apply within a hydrocarbon overlayer, with neglect of the hydrogen atoms.

This partial-multiple-scattering approach has been applied successfully in a few test cases [11.18] (it is related to the cluster approach designed by Pendry [11.19]), but its execution can be improved considerably. Possible extensions include the calculation and storage (for repeated use) of the self-consistent scattering properties of individual constituent clusters of atoms. Arbitrary molecules could then be assembled efficiently from these small clusters using a kinematic formalism, but including exact near-neighbor multiple scattering through the stored scattering properties of the clusters.

11.2.3 Developments in the Dynamical Theory

A major factor that will govern the direction of developments in the dynamical LEED theory is the continued concern with reducing the computational effort. At present, an experimentalist can solve a surface structure by dynamical LEED calculations without much preparation or outside help only for the simpler structures, namely for low-Miller-index surfaces of metals, with or without simple overlayers, with one atom per unit cell.

Computational costs and computer core size limits are the obstacles. To reduce the cost and core size for more complicated structures generally requires some understanding and modification of the computer programs. Certainly, continuing improvements in the computer hardware will help greatly in making more complex surface structures more easily accessible to computation; however, the researcher's desires inevitably increase rather more rapidly than does the capacity of computers, with the result that the search for computational efficiency remains essential.

As pointed out in Sect. 5.15, a number of specific issues need improvement in LEED theory. Some of these concern the accuracy of the physical parameters describing the scattering, together with the underlying theoretical assumptions (e.g. the muffin-tin potential, the imaginary part of the potential, the Debye-Waller factor, and the Debye temperature). This accuracy is becoming increasingly important in connection with the higher accuracy being claimed for simple surface structures, for example clean metal surfaces.

Perhaps the most troublesome quantity is the scattering potential, for which no individual construction scheme appears to be optimal for all chemical elements. In some cases, nonsphericity of the ion cores may have to be included, particularly in non-dense overlayers and at lower kinetic energies. Special care may be needed with the 5d-transition-metal potentials, because they give relatively poor agreement between theory and experiment. The imaginary part of the potential may have to be made layer dependent and direction dependent, again especially at the lower energies. Finally, the Debye-Waller factor should be refined to include the effects of correlated vibrations, especially in molecules.

A specific need for improvement concerns the ability to perform LEED calculations for stepped surfaces, since no current formalism can treat an arbitrarily stepped (or kinked) surface. The problem lies in the small layer spacings, which adversely affect calculations carried out either with a plane-wave expansion or with a spherical-wave expansion.

The potential for considerable improvement also exists in the area of large unit cells. We have mentioned previously the use of partial multiple scattering. Other developments that have been proposed are described in Sects. 11.2.3a-c.

11.2.3a Coherent Kinematic Summation of Amplitudes over Different Local Configurations

One part of a large unit cell is imagined to extend along the surface with a smaller period and to encompass the entire surface. The smaller unit cell yields an affordable calculation. The other subparts of the large unit cell are treated similarly in independent calculations, and the amplitudes diffracted by these different parts are added coherently. An example is the

$$Pt(100)\text{-}\begin{bmatrix} 14 & 1 \\ 1 & 5 \end{bmatrix}$$

surface reconstruction [11.16], which is thought to be a slight distortion of the (1x5) quasi-hexagonal reconstruction that exists on Ir(100). The large

$$\begin{bmatrix} \underline{14} & 1 \\ 1 & 5 \end{bmatrix}$$

unit cell is imagined to consist of 14 strips of (1x5) structure. Essentially, 14 calculations for the simpler (1x5) structure are performed rather than one calculation for the large unit cell, and the 14 results are added coherently for each beam. A requirement with this approach is that the small unit cell yield beams that correspond directly to some beams due to the large unit cell. The next method also starts from this idea.

11.2.3b Reduced Unit Cell

Of the many diffraction spots generated by a large unit cell, often some can be generated equally well by a small unit cell with the same contents tele-scoped together [11.21]. For example, a (2x2) pattern due to a (2x2) adsor-bate structure contains spots that could be due to a (2x1) superstructure of the same adsorbate with a doubled coverage. For the intensities of the common spots, a calculation with the (2x1) periodicity can be a good approximation to the (2x2) calculation, with a corresponding reduction in the computing effort. Thereby, a structural determination can be achieved using a suitable subset of the experimental data base.

11.2.3c Asymptotic Regime

As the unit-cell size and the number of inequivalent atoms within the unit cell become very large, the magnitude of the computation of present calcula-tional schemes scales either as the square or the cube of the unit-cell area and the number of atoms (Sect. 5.8). Essentially, this scaling law is dictated by the number of atoms within a few mean-free-path lengths of any given atom (more precisely by the number of multiple-scattering paths that need to be included). We have seen already, in Sect. 11.2.3a, an example of an approximate scheme that scales only as the first power of the unit-cell area. Furthermore, the approximate cluster approach mentioned in Sect. 11.2.2 satisfies this scaling law. Another example is the calculational method dev-ised by Moritz et al. for disordered surfaces [11.12], see Sect. 5.14.3. These methods become exact, and retain the advantages of the first-power scaling, in the limit where the neighborhood within which multiple scattering is cal-culated correctly becomes larger than the mean-free-path length. Pendry has discussed this "asymptopia" in some detail [11.22]. One conclusion he reaches is that it should be beneficial to shift the focus from I-V to I-Θ and I-ϕ curves. The greatest computational effort, namely that needed to describe the multiple scattering, need not be repeated if one maintains a constant energy while changing the electron-beam orientation. In this case, the first-power scaling can be realized with a cluster approach. Periodicity is no longer required, and therefore disordered as well as ordered surfaces can be treated within the same scaling law (although ordered surfaces, of course, will contain symmetries that are computationally advantageous).

11.2.4 New Directions

Real Part of the Scattering Potential

We have mentioned the continuing interest in molecular overlayers. When this type of overlayer becomes thicker, its influence on LEED intensities increases due to the reduced contribution from the substrate. Therefore, the physical parameters describing the molecular overlayer (especially the scattering potential, the imaginary part of the potential and the thermal parameters) become more critical. Current LEED theory has been developed with compact, high-density metal surfaces in mind, and a number of aspects may have to be revised when molecular surfaces are considered. For the real part of the scattering potential, the greater departure from spherically-symmetrical ion cores has already been mentioned. The simplest solution to this problem is to employ kinetic energies that are large compared to the aspherical deviations, so that the latter can be ignored. However, it is not obvious what the best radius for the muffin-tin spheres is. Indeed, overlapping spheres may have advantages over non-overlapping spheres.

Another unresolved issue is that of the muffin-tin zero level, which can be defined as the average value of the potential outside the muffin-tin spheres. Since the potential that is to be averaged is strongly position dependent in a molecular assembly, the constant muffin-tin zero can be a poor approximation. In particular, it is not clear where this potential should be interrupted, i.e. where the potential step to the vacuum zero level should be positioned, and whether this step can still be considered to occur in a plane parallel to the surface rather than along a corrugated surface that more closely follows the contour of the molecules. Again, high kinetic energies will minimize the effects of poor approximations. Nevertheless, a well-characterized muffin-tin zero is important to align peaks properly in the theory and the experiment (fitting the muffin-tin zero to experiment will probably remain the most convenient approach).

Imaginary Part of the Scattering Potential

Concerning the imaginary part of the potential, representing the loss of elastic current due to inelastic effects, a fundamental question is whether its value for molecular surfaces is comparable to that of conventional (i.e. metallic) surfaces. Experimental determinations of the electronic mean-free-path length in molecular solids are scarce and inconclusive. Values substantially larger as well as substantially smaller than those customary for metallic surfaces have been reported for similar molecular materials [11.23]. A limited set of LEED I-V curves has been measured for a multilayer of ice condensed on a metallic surface [11.24], and it shows essentially the same peak widths as I-V curves of conventional surfaces; this also applies to graphite, which in some ways is representative of molecular solids [11.25]. From these observations, we tentatively conclude that the universality of the mean free path versus kinetic energy curve (Fig. 4.10) is maintained for molecular surfaces.

A separate issue is whether the imaginary part of the potential can still be treated as a single position- and orientation-independent number. Presumably this will depend strongly on the homogeneity and isotropy of the molecular layer on the scale of 5 to 10 Å Graphite could serve as a simple structure to investigate these questions.

Thermal Motions

Thermal motions at a molecular surface can be quite different from those at a compactly bonded surface. Structural groups (e.g. terminal groups), bonded only to one or a few anchoring points, can oscillate as a whole, or a molecule can rotate as a whole around a fixed center, so that neighboring atoms may have high correlations in their large thermal motions. This is an important issue in the presence of multiple scattering. Multiple scattering internal to a rigidly vibrating or rotating structural group should be calculated as if there were no collective motion, since it is the relative atomic positions that are important in that case. On the other hand, the remaining scattering that relates such a group to its environment should include a conventional Debye-Waller factor or other similar device which does include that collective motion. However, at the same time, there will normally be some relative thermal motion internal to a rigid group (rapid bond stretching and bond bending) which requires, for its internal multiple scattering, a Debye-Waller factor that includes this internal relative motion. The cluster approach discussed above is well suited to such a treatment of thermal effects [11.19].

Magnetic Materials

A different kind of surface that deserves much more attention than it has received thus far is that of magnetic materials [11.6]. It appears that the dynamical theory of spin-polarized LEED is sufficiently advanced at present to permit its application to the understanding of surface magnetism. Several studies have already confirmed this point of view and have shown that much remains to be learned concerning surface magnetism itself. This subject could also help illuminate the complicated nature of bulk magnetism and contribute to the understanding of catalysis, for example in the case of the adsorption of hydrogen atoms, since they have spins that can interact with the spins of the substrate.

Disordered Surfaces

Another direction with great potential is that of disordered surfaces. Although the kinematic formalism is adequate in a number of cases, e.g. for the understanding of the diffuse LEED pattern due to a disordered overlayer, there remain situations that require a multiple-scattering formalism within disordered structures. This applies especially to dense materials where multiple scattering between near neighbors is strong, e.g. at disordered alloy surfaces, at liquid surfaces and at noncrystalline oxide surfaces. Theoretical techniques are available at present [11.10-12], but they require excessively large computational efforts. Suitable statistical averaging methods are now the most-needed development.

11.3 Progress in Structural Determination

11.3.1 Degree of Completeness of Structural Determinations

The major application of LEED will undoubtedly remain the structural determination of surfaces and overlayers, although various other uses will continue to be developed. Despite the serious complications due to multiple scattering and the growing competition from other techniques, LEED has remained the most productive tool for surface-structure analysis. This competition will be discussed in more detail in Sect. 11.4.

A structural determination can be carried out at many different levels of completeness, for example ranging from nothing more than the two-dimensional periodicity (as deduced directly from a LEED pattern) to the complete description of all atomic coordinates and possibly any form of disorder (such as thermal vibrations, defects, and other non-periodicities). With the growing variety of types and complexities of surfaces that will be studied and with the growing variety of calculational schemes that will be applied, one should expect an increase in the number of incomplete structural determinations. This will be most pronounced with complex structures, where for instance one part of an adsorbed molecule may be structurally characterized, while others (e.g. hydrogen positions or disordered structural groups and coadsorbates) are not. In this connection, it should be recognized that a complete structural determination may occasionally not only be overambitious but also unnecessary. One may well be satisfied with just one part of the structural information (e.g. the symmetry alone, the bond lengths alone, or the coordination numbers alone).

Another factor favoring incomplete structural determinations is the making of assumptions that reduce the number of adjustable structural parameters which are to be fit to the experimental data. An example of such an assumption, which has already been applied frequently, is that of structural symmetries corresponding to the symmetries observed in the LEED pattern and spot intensities (e.g. high-symmetry adsorption sites). In principle, one can safeguard against erroneous assumptions of symmetry by experimentally isolating individual domains on the surface (Sect. 3.4). Clearly, it would be worth pursuing efforts to achieve this routinely in practice. Other kinds of structural assumptions can occasionally be verified with results from other surface techniques. For example, any buckling of a close-packed top layer, as in the reconstructions of Ir, Pt, and Au(100) surfaces [11.20], or in epitaxial growth on a substrate [11.14] can be examined by helium diffraction. Alternatively, the tilting of an adsorbed molecule away from the surface normal may be investigated by vibrational spectroscopic measurements, either with High-Resolution Electron Energy-Loss Spectroscopy or Infrared Reflection-Absorption Spectroscopy.

One must accept the fact that no structural determination can ever be considered final, whether with LEED or any other technique. The question becomes one of the desired level of confidence and the desired level of accuracy, both of which depend to a large extent on the use one wishes to make of the result. For example, at one extreme, the determination of deviations from the bulk layer spacing at a metal surface requires a high accuracy (since the deviations are small) and a high confidence level (since an inadequate atomic scattering potential could render the results incorrect). At the other extreme, in the case of molecularly adsorbed benzene, one might be satisfied with the knowledge as to whether it is lying flat on the surface and at what height it lies, since these are the primary structural quantities that affect the chemistry of benzene on a surface. The adsorption site may be of secondary importance, and thus one may desire a lower accuracy and a lower confidence level in determining the adsorption geometry of benzene.

11.3.2 R-Factors and Structural Search Techniques

The use of R-factors will become increasingly widespread in the structural determination of all kinds of surfaces by LEED. With the simple surfaces, a high structural accuracy can be expected, which can be obtained only with an unchanging sensitivity of R-factors. At the other extreme of complexity, even when only an approximate structural result is desired, R-factors are needed to make the large required number of comparisons between theoretical and experimental intensity curves, due to the many trial structures that must be considered, and due to the large number of beams frequently involved.

Unfortunately, too many R-factor definitions are available at present (Sect. 6.5). This reflects differences in opinion concerning how various features of intensity curves should be weighted relative to one another. (Nevertheless, as we pointed out in Sect. 6.5.5, the diversity of R-factors can be put to advantage by observing whether their individual structural predictions agree or disagree.) No R-factor has thus far become generally accepted as the superior one. In this regard, improvement is needed, if only for the sake of uniformity. This would at least allow different structural results to be compared in terms of a common figure of merit (although we also noted conceptual difficulties in making such comparisons in Sect. 6.5.3).

Nevertheless, additional R-factors may be beneficial. For example (Sect. 6.5.5), an R-factor that measures the disagreement in structural predictions between different beam R-factors or between different R-factor definitions would be valuable to discriminate against local minima corresponding to incorrect structures. R-factors have been developed explicitly to compare I-V curves. However, LEED comparisons can also be made between I-Θ and between I-φ curves. Consequently, there is a need to explore the adequacy of existing R-factors when applied to such curves. Possibly, adaptations of existing R-factors will be needed, or entirely different R-factors may prove preferable, since I-Θ and I-φ curves often have a different peak-and-valley structure than I-V curves.

However, it is in the area of structural-search strategies that the greatest developments are needed. Let us assume that an R-factor is used in such strategies as the basic measure of the level of correctness of a trial structure. Then the problem is to *predict* the next trial structure from the R-factors of previous trial structures in such a way as to provide an iteration scheme that approaches the correct structure most rapidly. The emphasis here is on minimizing the number of trial structures that need to be considered in the process, since the LEED calculation for each trial structure is by far the dominant cost in this operation. The matter is complicated further by the fact that some structural parameters (e.g. layer spacings) can be changed much more inexpensively than others. One may also wish to include constraints representing independent information concerning known bond lengths and/or bond angles, for example. An iterative search should take such factors into account. An effective structural-search strategy is needed badly at present, especially in the increasingly frequent case where many ($\gtrsim 3$) structural parameters have to be determined.

Next we discuss a number of specific ideas for structural-search strategies that need further development to become effective in determining surface structures [11.26].

11.3.2a Projection Improvement

If a reflected beam $\bar{g}$ involves a momentum transfer $\Delta \vec{k}_{\bar{g}}$, and the atomic positions are denoted by $\vec{r}_j$, then in the kinematic limit the reflection amplitude in beam $\bar{g}$ will be proportional to the structure factor

$$\sum_j \exp(i\Delta\vec{k}_{\bar{g}}\cdot\vec{r}_j) \ .$$

Therefore, it is only the *projection* of the position vectors onto the direction of the momentum transfer of any particular beam that is of concern (in the kinematic limit). Thus, each beam samples a different projection of the surface geometry, and the R-factor of each beam measures the quality of the corresponding projection.

The knowledge of how good or how bad the various projections of a given structure are can be used to predict an improvement of the structure. For example, if the three coordinates of one atom (per unit cell) are to be determined, one could use several beams to provide several projections of the position of this atom. If the R-factor for one of these beams is small, the implication is that the projection onto the corresponding momentum transfer is relatively good and should not be changed significantly. Thus the position of the atom should be varied in (or nearly in) the plane perpendicular to this momentum transfer. (However, a large change in $\Delta\vec{r}$ such that $\Delta\vec{k}_{\bar{g}}\cdot\Delta\vec{r}$ is a multiple of 2π is also acceptable. Such a change may be needed if one has initiated the search with a very wrong structure, which would be manifest as a lack of coincidence of beam R-factor minima.) A bad R-factor for another beam implies that the position of the atom should be varied with a substantial component parallel to the momentum transfer of this beam, but avoiding $\Delta\vec{k}_{\bar{g}}\cdot\Delta\vec{r} \cong n2\pi$ for any integer n. Consequently, an improved atomic position can be predicted, given a sufficient number of beams.

Let us consider the particular case of the determination of the positions of two atoms per unit cell. Then it is easy to show in the kinematic limit that the calculated intensity in a given beam will remain constant only if neither of the two atoms changes its projection $\Delta\vec{k}_{\bar{g}}\cdot\Delta\vec{r}$ (modulo 2π) onto the momentum transfer of this beam. In other words, one cannot compensate the change of the projection of one atom by a change in the projection of the other atom. A bad projection can be improved by moving one or both atoms in a way that can be determined if a sufficient number of different projections (beams) are used.

With three or more atoms, it is possible (in the kinematic limit) to leave the diffracted intensity in one beam unchanged by suitable combinations of atomic displacements, so that more freedom of movement is present. However, a sufficient number of independent beams will provide strong restrictions on possible displacements. Furthermore, additional information is often available and can be incorporated to restrict any excess freedom of movement. For example, one may have calculated R-factors for more than one trial geometry, so that some trends and slopes of R-factors with certain structural changes are available. It may be useful also to include constraints at this point, such as known bond lengths and bond angles. It remains to formulate and test a detailed algorithm to perform the improvement of atomic positions by this method.

11.3.2b Functional Fitting of R-Factors

If R-factor values are obtained for a certain set of trial geometries of a surface, one may fit those values with an appropriate simple function of the
unknown atomic positions. The minimum (or minima) of this function can
then be determined by established methods to provide an estimate of the
correct geometry. Iteration can refine the result.

In the neighborhood of the optimum geometry, the R-factor has the
shape of a paraboloid, the minimum of which can be found easily. For
example, if two unknown parameters are to be optimized, the intersection of
two straight lines determines the position of the minimum R-factor, as discussed in Sect. 6.5.5. A graphical solution is then possible to find the intersection. In general, one obtains a positive definite problem that leads to a set
of linear equations in the variables, which can be solved by standard
methods.

A paraboloid is suitable when a number of R-factor values sufficiently
close to a minimum have been obtained. Farther away from a minimum, an
R-factor can have a very different behavior, including an oscillatory shape,
and higher-order polynomial functions are needed to fit it. The minimum
(or minima) of the polynomial function can still be found numerically, using
for example the methods described below, where the fitted function rather
than the original R-factor is evaluated in order to reduce the computational
effort. Alternatively, or in an early stage of the structural search, kinematic
approximations may be used in the LEED calculations.

11.3.2c Steepest Descent

A well-known and efficient numerical method for searching for a minimum
of a function starts with a trial position and follows the direction of the
steepest descent. This direction is given by the negative of the gradient of
the function. Thus for an R-factor as a function of the atomic coordinates,
the partial derivatives of the R-factor with respect to these coordinates are
needed at every geometry that is considered in the course of the search.
Since these partial derivatives can involve substantial computations ($N + 1$
LEED calculations at each geometry, where N is the number of unknown
atomic coordinates), it may be preferable to use a fitted functional form of
the R-factor, as described in Sect. 11.3.2b. Furthermore, approximations in
the LEED theory may be used to reduce the computational effort involved in
the method of steepest descent.

11.3.2d Least Squares

Finally, we mention the method of least squares. Let $R_{\bar{g}}$ be an R-factor for
beam $\bar{g}$, evaluated for a certain geometry, and let $\partial R_{\bar{g}}/\partial r_i$ be its partial
derivative with respect to the unknown atomic coordinate r_i, also evaluated
for that geometry. Corrections Δr_i to improve that geometry can be obtained

by minimizing the "residuals" $\delta_{\bar{g}}$ in the linearized equation

$$R_{\bar{g}} + \sum_i \left(\frac{\partial R_{\bar{g}}}{\partial r_i} \right) \Delta r_i = \delta_{\bar{g}} \ .$$

The number of these equations (equal to the number of beams used) should be at least equal to the number of unknown coordinates. By the method of least squares, the values of Δr_i can be calculated. If desired, the method can be iterated to convergence.

As with the method of steepest descent, the computational effort can be reduced greatly by fitting $R_{\bar{g}}$ with an appropriately simple function or by suitable approximations in the LEED theory.

It is obvious from the foregoing discussion that many methods are available for systematizing structural determinations by LEED. Only experience will dictate which methods (or combinations thereof) are the most effective. In addition, the choice may very well depend on the type of surface under consideration.

11.4 Leed vs. Other Surface-Sensitive Techniques

The future of LEED depends greatly on the further development and implementation of other surface-sensitive techniques, which relate directly or indirectly to surface-structure determinations. There is an important competitive element involved, and there may well be a better technique to determine a particular surface structure than LEED. However, we must also consider the complementarity of various techniques. A combination of two or more techniques is clearly more effective than any one technique alone. This is especially true in view of the practical limitations to which most techniques are subject. The issue of the optimum approach is far from settled, and it is therefore possible to give only a preliminary view of its possible evolution. The major reasons for the present uncertainty are the large number of techniques that have been able to provide some surface-structural information, together with the fact that most of these techniques are only in their initial development stage. In addition, totally new techniques are added to the list each year, while the older ones are being extended to new capabilities.

First we shall address, individually, the primary techniques that are sensitive to surface structure on the atomic scale. Surface structure will be taken in a broad sense to include sensitivity to surface topography, composition, and electronic structure. It should be borne in mind that many other techniques are possible, but to date they have not been developed sufficiently. For classification purposes, the techniques are ordered according to the nature of the probe(s) used: ion, atom, neutron, electron, and photon. Theoretical simulation of surfaces is also included, since it has contributed significantly to surface-structure determination. The various techniques will be compared directly in Sects. 11.4.2 and 11.4.3. For convenience, a listing of a number of acronyms encountered in the surface-science literature as well as their meaning is provided in Appendix A. This listing should be useful both when reading the following section, and more generally.

11.4.1 Individual Techniques

Ion-Scattering Spectroscopy (ISS)

In ISS, ions such as H^+, He^+ and Ar^+ are scattered by a surface, and their energy distribution is measured [11.27]. During the scattering process, the ions lose energy to the surface atoms. The collision is usually so rapid (at kinetic energies on the order of 1 keV to 1 MeV) that a binary-collision model is a good description of the interaction. It is then easy to relate the energy loss ΔE to the mass M of the surface atoms involved via

$$\frac{E - \Delta E}{E} = \frac{1}{(1 + M/m)^2} \left\{ \cos\Theta \pm \left[\left(\frac{M}{m} \right)^2 - \sin^2\Theta \right]^{1/2} \right\}^2 ,$$

where E is the incident energy, m the ion mass, and Θ the laboratory scattering angle. Thus, different peaks in the energy-loss spectrum occur for collisions from surface atoms with different masses M, allowing a qualitative determination of the chemical composition.

Depending on energy and incidence direction (including channeling and blocking effects, as discussed below under MEIS and HEIS), the depth resolution of ISS can vary from less than a monolayer to approximately 300 Å Quantitative evaluations of the chemical composition of surfaces may be achieved in special cases, but these are generally hampered by uncertainties in the ion-atom scattering potential (especially at the lower energies), the possibility of multiple scattering, and the ever-present question of the depth distribution of the individual species.

Low-Energy Ion Scattering (LEIS)

Ions with low kinetic energies (≤ 2 keV) are highly surface sensitive since they do not penetrate beyond the first layer of surface atoms [11.27]. Mutual shadowing of surface atoms is exploited to investigate their mutual positions. Although the physics of the scattering process at the lower kinetic energies is not well understood, uncertainties in position determinations on the order of 0.1 Å can be achieved. However, even when the uncertainties are much larger, it is possible to make important observations, such as whether adsorbed atoms are embedded among substrate atoms or are situated in more exposed positions.

Medium- and High-Energy Ion Scattering (MEIS and HEIS) or Rutherford Backscattering Spectrometry (RBS)

Medium-energy (approximately 50-500 keV) and high-energy ($\geq$500 keV) ions are obtained with ion accelerators [11.27,28]. At these energies, the cross sections for ion scattering by surface atoms become very small, allowing deep penetration into the bulk. Surface sensitivity is maintained, however, by utilizing channeling (penetration along open channels between nuclei in the bulk crystalline structure) and observing the blocking of this channeling by surface atoms, the positions of which deviate from the bulk positions. (In the case of adsorbates, the energy loss can also serve to select surface information, as described under ISS above.) The channeling leads to a kind of "triangulation", in which the directions of the lines connecting pairs of surface atoms are identified by observing shadowing of one atom of each pair by the other atom. These directions provide sufficient information to determine the relative positions of the surface atoms. At high energies, the well-understood Rutherford scattering is the dominant mechanism, simplifying the interpretation.

A general problem with ion scattering is that the thermal vibrations of the surface atoms complicate the interpretation. Occasionally, computer simulations of the scattering are made to disentangle the structural from the thermal effects. In principle, accuracies on the order of 0.02 Å are possible in the determination of atomic positions.

Secondary-Ion Mass Spectrometry (SIMS)

The technique of SIMS analyzes the mass and angular distributions of charged clusters of atoms ejected during ion bombardment of a surface, using incident ions with kinetic energies on the order of 1 keV [11.29]. The knowledge of the cluster masses allows the chemical composition of the surface to be investigated. The depth resolution of SIMS is on the order of a few atomic layers.

A quantitative analysis of the surface composition requires a comparison of the experimental yield of the individual clusters with corresponding yields obtained theoretically. This may be done by numerical simulation of the complex collision process, but the accuracy of the result cannot yet be ascertained. The accuracy of the composition analysis depends to some extent on such poorly known factors as the interatomic potential, ionization cross sections, and quantum-mechanical corrections to a treatment based on classical trajectories. In addition, preferential sputtering of one or more atomic species on the surface, causing changes in the SIMS spectra as a function of time, compounds the difficulty of obtaining quantitative surface compositions.

Once the theory of the SIMS process is understood properly, this technique should be capable of analyzing some aspects of the surface geometry, such as the registries of adsorbates on a substrate and whether molecules adsorb associatively or dissociatively.

Field-Ion Microscopy (FIM)

In FIM, a hemispherical sample tip is imaged by allowing a gas (usually helium, but also hydrogen and neon among others) to ionize at the surface of the tip under the influence of a strong applied electric field [11.30]. The field also projects the ions onto a screen that can be photographed. The ionization probability depends strongly on the local field variations induced by the atomic structure of the surface. Protruding atoms generate appreciably stronger ionization than atoms embedded in close-packed atomic planes and, consequently, produce individual bright spots on the screen. The imaging by the ions from the sample tip to the screen occurs with very little motion tangential to the tip surface, especially at low temperatures (a temperature of approximately 4 K is often used for this reason), allowing a resolution of 2-3 Å The use of small-radius tips (500-2000 Å) is needed to produce the high field required for ionization, and this is also responsible for the immense magnification of this microscope. The tip surface is imaged directly with a magnification of approximately 10^7.

Only a limited class of materials withstands the strong electric field at the sample tip without field desorption of the surface atoms. As a consequence, the more refractory transition metals have been studied primarily. Covalently bonded adsorbates are especially affected by field desorption, greatly restricting the range of usable substrate-adsorbate systems that can be studied by FIM. Furthermore, the properties of surfaces under high electric field conditions may differ from those of the field-free state. On the other hand, real-time observation of the motion of individual atoms is possible. In this regard, FIM has been very helpful in understanding the properties of metal surfaces and metal-on-metal surfaces, including alloy surfaces, defect structures, reconstruction, thermal disordering, adatom-adatom interactions, two-dimensional cluster formation and evolution, and atomic surface diffusion.

Concerning the analysis of the detailed geometrical surface structure, FIM does not provide the depth information required to investigate accurately the coordination of surface atoms to underlying atoms (i.e. layer registries cannot be determined) or to measure bond lengths.

Electron-Stimulated Desorption Ion Angular Distributions (ESDIAD)

Structural information is provided by ESDIAD through the preferential direction of emission of (atomic or molecular) ions desorbed by electron impact [11.31]. It appears that these ions are generally ejected from the surface in the direction in which the (broken) adsorption bond was oriented just prior to its rupture. An angular spread of approximately $\pm 10°$ is typical for the emission direction, corresponding presumably in part to thermal vibrations of the adsorbed species and in part to the detailed mechanism of electron-stimulated desorption. For example, the desorbing ion can interact with its image charge in the metallic substrate, an effect which becomes more significant at grazing angles of desorption. This technique has found its major application in the determination of molecular orientation at a surface, i.e. short-range order. One can obtain both the polar and the azimuthal directions of bonds with respect to the surface normal and with respect to the substrate orientation in the surface plane, respectively.

Ion-Neutralization Spectroscopy (INS)

In INS, slow, positively charged noble-gas ions (typically He^+) are allowed to neutralize at a surface by capturing surface electrons [11.32]. The energy liberated is transferred to other electrons in the near-surface region, which can leave the surface and be detected. The probability of this two-electron process involves the self-convolution of the surface density of occupied electronic states, the density of final states, and matrix elements both for electron tunneling to the ion and for ejection of the detected electrons. The procedure is to extract the surface density of occupied states for energies between the Fermi level and approximately 10 eV below it, by deconvolution from the measured emission probabilities. Then one may predict the relative atomic positions from this information concerning the electronic structure of the surface. For example, adsorbate-induced peaks will occur that depend on the adsorbate and its position, as in UPS. This technique is primarily sensitive to the outermost atoms of the surface, in particular adsorbates, since the emitted electrons originate from this region only. The difficulties in deconvoluting and interpreting the density-of-states information have limited the use of INS, however.

Surface Penning Ionization (SPI), Penning Ionization Electron Spectroscopy (PIES) or Metastable Deexcitation Spectroscopy (MDS)

A metastable helium atom incident (with thermal kinetic energy) onto a surface allows a surface electron to tunnel into the unoccupied helium-1s level, enabling the excited helium electron to be emitted and detected [11.33]. This occurs with energy and angular distributions characteristic of the electronic structure and geometry of the surface, in a way similar to photoemission. Whereas in photoemission the excitation occurs via a dipolar photon-electron interaction, in SPI it is an electron-electron interaction proportional to $|\vec{r}_1 - \vec{r}_2|^{-1}$ that takes place. Consequently, the selection rules are different. A further difference lies in the surface sensitivity. The photoemission excitation is sensitive to the entire region sampled by the initial state of the electron to be excited, while the SPI excitation occurs at the outer edge of the outermost surface atoms. Similarities exist between SPI and INS but SPI but is intrinsically easier to interpret, largely due to the absence of a deconvolution of the surface density of occupied electronic states. Furthermore, SPI appears to be even more surface-sensitive than INS.

Atomic Scattering and Diffraction

Helium atoms with thermal energy on the order of 0.02 eV have an associated de Broglie wavelength of approximately 1 Å and can therefore diffract from surfaces [11.34]. Three different regimes of the scattering process can be distinguished. There is usually strong specular reflection of the helium atoms from the surface, because many surfaces present a rather flat electron-density profile parallel to the surface. There is also rainbow scattering that results in the appearance of multiple peaks. This type of scattering may be viewed as the classical limit of diffraction since it is due to scattering of atoms by the waviness of the electron-density profile. The third type of scattering is diffraction due to the surface periodicity, giving rise to well-defined diffraction beams, the intensity of which in the lowest-order approximation is proportional to the amplitude of the surface waviness (corrugation). Beam intensities give information concerning the corrugation of the "classical turning point" surface, which corresponds to a surface of low, constant charge density several Ångstroms from the nuclei. The technique has been used to obtain the gross surface structure, which can be used to discriminate between various types of structural models for the surface. To obtain information concerning parameters such as bond angles and bond lengths requires further refinement of the theory.

Thermal-Desorption Mass Spectrometry (TDMS), Flash Desorption or Temperature-Programmed Desorption (TPD)

The technique of TDMS measures the rate of desorption of adsorbed atoms or molecules from surfaces as the temperature of the substrate is increased, often linearly. This is usually repeated at a succession of different coverages of the adsorbate [11.35,36]. One or more peaks appear in the desorption rate versus temperature curves. Each peak may correspond to a particular bonding state, since different temperatures are required to induce desorption from different states. Therefore it is possible to monitor with TDMS how many different bonding states are occupied at any given adsorbate coverage, and also to estimate their relative populations. One often finds that with increasing coverage, the first state (which may correspond to a specific adsorption site) is populated, and then a second state (perhaps another site) begins filling in.

The positions and relative intensities of the desorption peaks as a function of initial coverage can be analyzed with appropriate models. In this way, one obtains good estimates for the binding energies, i.e. the activation energies of desorption, as well as the preexponential factors of the desorption-rate coefficients. A more detailed analysis involves measuring the widths and positions of the desorption peaks [11.36].

Frequently, the surface coverage can be obtained from the thermal-desorption spectra, since a suitable integration over the spectra gives the concentration of the adsorbate that is desorbed. This is an especially important method for measuring hydrogen coverage since rather few other techniques are sensitive to hydrogen.

The reliability of TDMS is somewhat limited by the fact that the effect on the desorption reaction due to mutual-interaction energies among adsorbates could be substantial and is poorly understood. Therefore this technique is used primarily to examine the surface composition, and its evolution with coverage and temperature, as well as the number of distinct species or adsorption sites, and the kinetics of adsorption and desorption. The ease of the experiment explains its universal use, together with Auger-Electron Spectroscopy, for surface characterization.

Neutron Diffraction

Neutrons have a very small cross section for interaction with other matter and can therefore be applied in surface analysis only in special circumstances, namely in materials with a high surface-to-volume ratio [11.37]. Therefore a commonly used substrate is graphite (often exfoliated or otherwise of imperfect crystallinity), with adsorption on as many of the atomic planes as possible. Thermal neutrons have convenient de Broglie wavelengths for diffraction by crystals and allow structural studies to be carried out for ordered adsorbates, as in the case of bulk materials. One major advantage of neutrons is their sensitivity to hydrogen atoms.

Low-Energy Electron Diffraction (LEED) and Spin-Polarized LEED (SPLEED)

We shall summarize the principles of LEED (and SPLEED) here in order to allow a direct comparison to be made with the other techniques discussed in this section.

Both LEED and SPLEED rely on the use of electrons with kinetic energies that correspond to de Broglie wavelengths on the order of 1 Å The surface sensitivity is dictated by the inelastic mean-free-path length of electrons in solids, which is typically on the order of 5-10 Å This approach has been most successful with ordered surface structures. Observation of the LEED pattern on a luminescent screen has become a primary surface-characterization tool, both because surface ordering is quite sensitive to improper preparation, and because the degree of ordering is recognized easily in the diffraction pattern.

The intensities of the diffraction beams (spots) contain detailed information concerning the relative atomic positions at the surface (bond lengths and bond angles). Unlike the diffraction of X-rays, neutrons, and high-energy electrons, however, multiple-scattering of the diffracted electrons in LEED must generally be taken into account when extracting this structural information, leading to enhanced computational costs. It follows also that a structural determination must proceed by trial and error. On the other hand, LEED data are usually measured in large quantities compared to other techniques, so that relatively more information is available for the structural determination.

Structural determinations by LEED have been applied to many kinds of surfaces of moderate complexity, and the boundary of complexity is being pushed back constantly. The strength of LEED is its high sensitivity to the detailed surface structure. Therein also lies a weakness: in the trial-and-error approach, it is often difficult to propose the next trial structure on the basis of previous, unsuccessful trial structures.

The accuracy in determining atomic positions varies from approximately 0.02 Å for layer spacings of the simplest surfaces to approximately 0.2 Å for coordinates parallel to the surface in the case of more complex surfaces. Hydrogen cannot normally be detected by LEED, unless it forms its own separate superlattice on a surface.

In SPLEED, an additional variable is measured, namely the spin polarization of the incident and scattered electrons. This quantity not only provides information concerning relativistic effects and magnetic surface structures, but it also shows greater sensitivity to structural surface details than "intensity-only" LEED.

Medium-Energy Electron Diffraction (MEED) and Reflection High-Energy Electron Diffraction (RHEED)

The differences between LEED, MEED, and RHEED lie in the range of energies used [11.38]. While in LEED low kinetic energies of approximately 20-500 eV are employed, in RHEED higher kinetic energies on the order of 1-10 keV are used, with MEED bridging the intermediate energy range. Higher energies imply longer mean-free-path lengths (approximately 20-100 Å at RHEED energies). The surface sensitivity is maintained by using grazing angles of incidence and emergence at the higher energies. However, grazing angles of incidence require a very good large-scale flatness of the surface. The multiple scattering present in LEED is also present at the higher energies, and corresponding dynamical theories have been developed. However, a severe difficulty is the lack of accurate experimental data to interpret. The required calculations are of a complexity and size comparable to those for LEED. It appears that MEED and RHEED should ultimately provide a capability equivalent to that of LEED for surface-structure determinations.

Electron Microscopy (EM)

In EM, very high energy electrons ($\geq$50 keV) are diffracted by a sample and then focused electromagnetically into an image [11.39.40]. Various modes of operation have been developed. In Transmission Electron Microscopy (TEM, at an energy of approximately 100 keV), the image can show clearly resolved detail in the 2-3 Å range. This image constitutes a two-dimensional projection of the sample structure. If the sample is sufficiently thin (on the order of 1000 Å), monolayer steps can be observed and single, heavy-element atoms on specially prepared substrates have been resolved. Single-atom resolution is also possible with Scanning Transmission Electron Microscopy (STEM), when the incident beam is focused to a size of 2-3 Å and secondary electrons are detected.

Bond lengths and bond angles are not likely to be determined very accurately in the foreseeable future with EM. The strength of EM is rather the capability to exhibit morphology, i.e. the general atomic conformation and coordination numbers. Indeed, TEM has recently confirmed the correctness of the "missing-row" model for the (1x2) and (1x3) reconstructions of the Au(110) surface [11.40]. In addition, the diffusion of surface atoms and adatom-adatom interactions can be analyzed.

Present developments include the combination of EM with ultrahigh vacuum. As an example, clean-surface phase transitions have been observed with imaging based on fractional-order diffraction beams, which can provide surface sensitivity. Problems to be overcome are relatively large electron-beam induced damage and the presence of strong magnetic fields that could affect the surface structure.

Scanning Tunneling Microscopy (STM)

In STM, a metal tip with an extremely small radius of curvature ($\leq$10 Å) is positioned sufficiently close to a surface that the resistance for vacuum tunneling of electrons between the tip and the surface is measurably small [11.41]. The surface is then scanned in its plane by the tip, while the distance between the tip and the surface (approximately 5 Å) is adjusted to maintain a constant tunneling resistance. The result is a contour map of the electronic density of the surface (at the Fermi level), which indicates the nuclear positions. The lateral resolution of STM is $\leq$5 Å and it provides a real-space representation of the surface structure. The technique has confirmed the "missing-row" picture of the reconstruction of the (1x2) and (1x3) Au(110) surfaces, and it has provided insight into the nature of the (7x7) reconstruction of the Si(111) surface. Furthermore, a rather simple one-electron, asymptotic tunneling picture appears adequate to describe the limited experimental data available.

High-Resolution Electron Energy-Loss Spectroscopy (HREELS)

Electrons scattered by surfaces can lose energy to the surface in various ways. One of these involves excitation of the vibrational modes of atoms and molecules on the surface, and this effect is measured in HREELS [11.42,43]. A beam of electrons with kinetic energy on the order of 5 eV is used. The energy spread should be at most approximately 10 meV (which corresponds to 80 cm^{-1}), compared with ≥ 100 meV in most LEED experiments. This electron beam excites the vibrational modes of the different chemical bonds present at the surface (including phonons), which typically have frequencies up to 500 meV (i.e. up to 4000 cm^{-1}). These are manifest as separate loss peaks in an energy scan performed by a suitable energy analyzer. Consequently, HREELS can be used to effectively identify the adsorbed species and their chemical state, by comparison with known vibration frequencies in gauge molecules.

The energy scan can be carried out at specular reflection or at off-specular reflection. In the specular mode, "dipole scattering" dominates, which obeys an approximate selection rule. To be detected clearly, vibrational modes must have a component of their dipole derivative perpendicular to the surface. This feature is valuable in determining the approximate orientation of molecules or individual bonds relative to the surface plane.

Hydrogen may be detectable at coverages much less than a monolayer with HREELS. In addition, the isotopic shifts due to the different masses of isotopes may also be observed, and this helps greatly in assigning specific vibrational modes to the observed frequencies.

One can determine some aspects of the surface structure by fitting atomic positions and force constants to measured vibrational frequencies. This requires a normal-mode analysis in the more complex cases. This approach has heretofore been applied only to simple cases such as hydrogen and CO adsorption.

A more complete structural analysis can be performed by interpreting HREELS I-V curves, i.e. the HREELS intensity (at a given angle) as a function of the primary energy for constant loss energy [11.43]. This proceeds as in ARPES by computed simulation as a function of trial structures, using a LEED-like multiple-scattering formalism.

Inelastic Electron Tunneling Spectroscopy (IETS)

IETS, like HREELS, measures vibrational frequencies by detecting electron energy losses [11.44]. However, the experimental geometry is very different compared to HREELS. The electrons, the energy losses of which are to be measured, tunnel from a conducting substrate (such as aluminum) to a second metal electrode (typically lead) across a thin insulating layer (aluminum oxide in the case of an aluminum substrate). The species of interest are adsorbed or reacted on the oxide surface and then "matrix isolated" with lead to complete the sandwich structure. The specimen, after preparation in a suitable vacuum, involves no vacuum and is cooled in liquid helium to obtain the desired resolution (which is proportional to temperature) of approximately 2 meV. The current across the insulator/adspecies layer is then measured as a function of a bias voltage applied across this junction. The current increases abruptly, though only slightly, whenever a new vibrational mode opens a new channel for tunneling. A second-derivative measurement renders these discontinuities considerably more apparent, i.e. they appear as peaks in the "tunneling spectrum".

The information that can be extracted is essentially the same as that in HREELS and IRAS. The ease of measurement is somewhat offset by the restrictions imposed by the junction, in particular in the choice of substrate materials and crystallinity. However, the technique is ideally suited to studying oxide surfaces, including supported metallic or organometallic clusters on oxides.

Work-Function Change or Contact-Potential Difference (CPD) Measurements

The work function is the energy that an electron at the Fermi level of a solid must be given to reach the level of zero kinetic energy in the vacuum [11.45]. It is due to the interaction of an electron with the charges of the surface, through electrostatic, exchange, and correlation effects. Any change of the dipole component perpendicular to the surface of the charge distribution at the surface will change the work function (by up to 2-3 eV in some cases). The work function is a sensitive function of the presence and position of an adsorbed layer on a surface. Therefore, work-function measurements have become a sensitive technique for monitoring the state of the surface, usually as a function of coverage.

The work function can be measured by, among other methods, photoemission, since the work function appears as a clearly distinguishable threshold energy. Changes in the work function are often measured by a Kelvin probe, which uses a vibrating capacitor, or by retarding-potential techniques.

However, the work function and its change are complicated functions of the surface composition and structure. They depend on the detailed electronic distribution, as governed by the charge transfer between an adsorbate and a substrate, by competition for the charge transfer between adsorbates, and by dipole-dipole interactions between adsorbates. Although exceptions exist, the sign of the work-function change for atomic adsorption is generally that implied by the relative electronegativity of the adsorbate with respect to the substrate, as one would expect. Furthermore, in atomic adsorption it appears that large atoms (such as alkali atoms) produce large work-function changes, mainly because of the size effect. Many organic molecules adsorbed on metal surfaces produce a decrease in the work function, indicating the net transfer of electrons from the molecules to the substrate.

Auger-Electron Spectroscopy (AES)

In AES surface atoms are ionized (usually by incident electrons) in their deep-lying electronic levels, and these levels are then filled by electrons from higher-lying levels [11.46]. The excess energy is either emitted as X-rays or transferred to electrons in other electronic levels, which are then emitted and detected at energies characteristic of the particular levels involved. The second process is called the Auger effect, and the electrons that are detected are called Auger electrons. The detection usually integrates over a large solid angle.

Since the level energies depend sensitively on the chemical identity of the atoms, the emitted electrons have energies characteristic of the chemical identity of the emitting species. Therefore, each species has a particular fingerprint in AES, which makes this technique ideal for the determination of the surface chemical composition. As a consequence, AES is used routinely to monitor surface cleanliness in most surface-analysis work. Impurity concentrations on the order of 1% of a monolayer are detectable (except for hydrogen and helium, to which AES is not sensitive for lack of suitable energy levels). Similarly, the surface composition of alloys and other compounds can be investigated with AES.

Such applications, however, do not distinguish clearly between atoms adsorbed on and atoms incorporated in the substrate, since the signal is derived from a surface layer of thickness approximately equal to the electronic mean free path, which is on the order of several atomic layers. The observed signal may be due either to a small concentration of adsorbed atoms or to a larger concentration of atoms incorporated in the first few layers of the substrate. This ambiguity, coupled with uncertainties in the absolute yield of the AES excitation process for different species, makes a quantitative analysis of atomic concentrations at surfaces difficult and frequently unreliable.

Angle-Resolved Auger-Electron Spectroscopy (ARAES)

In ARAES, the emitted Auger-electron current is measured at selected emission directions [11.47]. Under these conditions, diffraction effects become important, as in ARPES. These diffraction effects are manifest as variations in the ARAES intensity with kinetic energy and with angle of detection, and they are sensitive to the surface structure. Several attempts to reproduce ARAES intensity variations by theoretical calculations have shown the feasibility of this approach, but a more detailed description of the atomic Auger process than available at present is a prerequisite. The capacity of ARAES for structural determination could be comparable to that of ARPES, taking into account the advantage of not needing the technical complexity of synchrotron radiation.

Extended Appearance-Potential Fine Structure (EAPFS) and Surface Extended-Energy-Loss Fine Structure (SEELFS)

In EAPFS and SEELFS, electrons with kinetic energy on the order of 1 keV are directed at the surface [11.48]. These electrons create core holes, and one measures the intensity of electron or X-ray emission due to deexcitation of the core hole. This intensity oscillates in the EAPFS mode as the incident energy is varied above any core-hole excitation energy ("edge"), and this extended fine structure yields structural information in the same manner as EXAFS and SEXAFS. Actually, it is the first or second derivative of the emitted intensity with respect to the energy that shows the useful fine structure. These derivatives are obtained by detecting the first or second harmonic in the intensity due to a periodic modulation of the incident intensity. This approach is called Appearance-Potential Spectroscopy, or APS. Extended fine structure can be obtained also at constant incident energy by varying the monitored energy loss. This approach is taken in SEELF.

In EAPFS, the collection of soft X-rays bears the name Soft X-ray Appearance-Potential Spectroscopy (SXAPS) [11.49], while the collection of all secondary electrons is called Auger-electron APS (AEAPS) [11.50]. If one measures the decrease in elastic electron yield above a core excitation threshold, one obtains Disappearance-Potential Spectroscopy (DAPS) [11.51].

Compared to EXAFS and SEXAFS, the matrix element giving the excitation probability is more complicated in EAPFS. In general, the latter requires the use of more than one angular-momentum eigenstate, i.e. more than one phase shift. (These phase shifts must be calculated for EAPFS, whereas the EXAFS and SEXAFS phase shifts can be "transfered" from other systems.) Another complication is any diffraction of the incident electron beam by an ordered lattice, since such diffraction can generate undesirable fine structure of its own. This difficulty is reduced in the SXAPS mode and eliminated in the SEELFS mode. Damage due to the incident electron beam may also be a problem at some surfaces. On the other hand, compared to EXAFS and SEXAFS, it is quite easy to vary the incident beam energy, while the beam intensity is also strong.

In terms of accuracy of surface-structure determination, it appears that EAPFS should be comparable to SEXAFS for determining near-neighbor bond lengths (i.e. approximately 0.02-0.07 Å), especially at disordered surfaces.

X-Ray Photoelectron Spectroscopy (XPS)

In XPS, photoelectrons ejected by incident X-rays are detected, integrated over emission angle, and analyzed with respect to kinetic energy [11.52]. X-rays can be generated with a number of different anode materials. The most common laboratory X-ray anodes are aluminum and magnesium. The

Al Kα (1486.6 eV) and Mg Kα (1253.6 eV) lines, which are used frequently, have an inherent energy line width on the order of 1 eV. Thus, the energy resolution of X-ray photoelectron spectra is limited to approximately this value in the absence of a reliable deconvolution scheme to account for this instrumental broadening. However, in high-resolution XPS, the inherent energy line width can be reduced to less than 0.5 eV with the use of mono-chromators, at the expense of lowering the incident photon intensity (and consequently, the resulting photoelectron current).

For surface studies, final kinetic energies of the photoemitted electrons are on the order of 100 to 1000 eV. The photoemitted electrons have ener-gies characteristic of the deep-core-level energies from which they were excited and consequently can be used to identify the chemical elements present at the surface. Therefore XPS has also been known historically as Electron Spectroscopy for Chemical Analysis (ESCA). However, these core-level energies also depend on the electronic state of the emitting atoms and therefore on their environment. Chemical shifts occur that are of the order of one to several eV, and these chemical shifts can be used to distinguish different states of the same atomic species. In this way it is possible, for example, to distinguish between adatoms that are adsorbed in an overlayer, adatoms that have penetrated into the first substrate layer, and adatoms that have penetrated deeper into the substrate.

Ultraviolet Photoelectron Spectroscopy (UPS)

As in XPS, photoemitted electrons are detected in UPS [11.53]. However, in this case ultraviolet radiation such as HeI (21.2 eV), HeII (40.8 eV), NeI (16.7 eV), and NeII (26.9 eV) is used rather than X-rays. These lower pho-ton energies give rise to emission only from valence levels, conduction levels, and other weakly bound levels. These levels are strongly sensitive to chemical-bonding effects. Thus, the interaction between an adsorbate and a substrate can be characterized in terms of which orbitals are involved. Furthermore, the dissociation of adsorbed molecules and the nature of the resulting adsorbed fragments can be identified. Even the orientation of adsorbed molecules can be estimated. For example, it was found by UPS that the CO molecule bonds to many metallic surfaces through the carbon end rather than the oxygen end, since the 5σ orbital (which is centered mainly on the carbon atom) is markedly more perturbed by the chemisorp-tion than the 4σ orbital (which is centered to a greater extent on the oxygen atom).

Angle-Resolved Photoelectron Spectroscopy (ARPES: ARXPS and ARUPS)

The angular integration performed in XPS and UPS to a large extent aver-ages out diffraction effects undergone by the photoemitted electrons, thereby removing sensitivity to direct geometrical structure. These diffraction effects are enhanced by angle-resolved measurements, so that geometrical information concerning the detailed surface structure becomes available in this case [11.54]. The diffraction of the photoemitted electrons through the surface lattice is very similar to LEED.

The use of a surface atom as the "electron gun" has both its advantages and its disadvantages. In ARXPS, one can select which chemical element acts as the electron source. Only short-range order is required (but no orientational disorder should be present), and the diffraction is somewhat more kinematic than in LEED. On the other hand, the photoemission process in the individual atoms must be calculated, introducing new parameters. This is especially complicated in ARUPS, where the initial electron state is normally a delocalized orbital extending over a number of atoms. Another disadvantage of ARPES is the need for synchrotron radiation. Otherwise, the analysis of ARPES data is very similar to that of LEED data of the same energy. In particular, the structural accuracy that can be achieved is comparable. At higher than the usual LEED energies (i.e. above approximately 200 eV), the ARPES data become relatively more kinematic, and it is possible that a Fourier transformation, similar to that used in SEXAFS and EAPFS, can be applied. This would lead to accuracies of approximately 0.02 Å in the technique known as Angle-Resolved Photoelectron Fine Structure (ARPEFS).

Surface Extended X-Ray-Absorption Fine Structure (SEXAFS)

In SEXAFS, incident X-rays of variable energy can eject electrons with sufficiently low energies that surface sensitivity is guaranteed [11.55]. Alternatively, ejected ions, which desorb as a result of atomic deexcitation, can be counted. It is possible to focus attention on particular chemical elements, for example, those of an adsorbate, by considering only Auger electrons with an energy characteristic of those elements. One measures the electron yield above an excitation threshold over an energy range of hundreds of eV, but excluding energies within approximately 50 eV of the threshold, where multiple scattering of the emitted electrons can occur.

The yield of the angle integrated electrons or ions is modulated as a function of incident energy due to interference between the electron wave leaving the emitting atom and the wave that is backscattered from the neighboring atoms. By Fourier analysis of this modulation, the interatomic distances can be extracted with an accuracy that is often quoted to be approximately 0.02 Å It is also possible to determine the approximate number of neighbors at the individual distances, thereby specifying the adsorption site through the coordination number. In addition, by varying the polarization plane of the incident X-rays with respect to the surface plane, it is occasionally possible to obtain bond orientations, perhaps within 10°.

An advantage of SEXAFS over some other techniques is that only short-range order is required, and even orientational disorder is permissible. However, there must be some limit to the number of different environments that a given element may have at any one time on a surface, so that only a small number of interatomic distances exists in a given experimental spectrum. Another advantage of SEXAFS is that the phase shift which is needed to interpret the data is "transferable" from other known samples, for example from bulk crystals. By contrast, in EAPFS for example, phase shifts are not available experimentally and must be computed.

X-Ray-Absorption Near-Edge Structure (XANES) or Near-Edge X-Ray-Absorption Fine Structure (NEXAFS)

The technique of XANES is concerned with the near-edge fine structure in X-ray absorption that SEXAFS avoids; hence its alternative name NEXAFS [11.56]. The multiple-scattering effects present in this regime imply not only that dynamical-scattering calculations are needed, but more importantly that there is direct sensitivity to bond angles and the general near-neighbor conformation. Only short-range order is necessary with XANES, and although orientational ordering presumably is beneficial, it is not required. The multiple scattering can be treated in a LEED-like fashion, using a cluster approach. One may expect an accuracy in structural determinations on the order of 0.1 Å

X-Ray Diffraction (XRD)

X-rays have a mean-free-path length on the order of 10000 Å or more in most metallic materials. To obtain surface sensitivity, one can concentrate on materials that contain suitable internal surfaces. For example one could use graphite, other layered compounds, or crystallized arrays of clusters, where each cluster represents an identical area of a surface with some adsorbate or with some reconstruction. (Clearly, Neutron Diffraction and EXAFS may be used in the same manner.)

However, there has been progress in the use of X-ray Diffraction to determine the structure of conventional single-crystal surfaces. One idea is to use fluorescent scattering from X-ray standing waves near a Bragg total-reflection condition [11.57]. The fluorescence gives sensitivity to surface adatoms if they have a different atomic number than the bulk atoms. The total-reflection condition constructs a strong standing wave that modulates the fluorescence when the nodal and antinodal planes of the standing wave pass through the adatom centers (during tilting of the crystal). This allows the distance of the adatom to the crystallographic planes of the substrate to be determined. By switching to Bragg total-reflection from other crystallographic planes, the position of adatoms can be determined with respect to each such crystallographic plane, which is sufficient to establish the complete set of coordinates of the adatoms.

The accuracy of position determination with this fluorescent XRD approach is at present approximately 0.02 Å It must be pointed out, however, that this technique cannot determine the coordinates of the topmost substrate atoms, so that adatom-substrate bond lengths depend on some assumption concerning the position of these substrate atoms.

In another approach, surface sensitivity is obtained by the measurement of fractional-order diffraction beams of a surface superlattice [11.58,59]. However, since the perfect substrate does not diffract kinematically into those beams, only the internal structure of the deviating top layers can be obtained. Using this approach, it has been shown that the Au(110)-(1x2) reconstructed surface follows the "missing-row" model with lateral pairing displacements in the second layer of 0.122 ± 0.017 Å[11.59].

In XRD, sufficient diffracted intensity is achieved only by using high-atomic-number surfaces with rotating anodes in the laboratory, or alternatively, by using synchrotron radiation.

Extended X-Ray-Absorption Fine Structure (EXAFS)

We have discussed SEXAFS above, which is the surface-sensitive variant of EXAFS. In EXAFS, the X-ray absorption (or sometimes the fluorescence) is measured above an absorption edge, and its oscillation with wavelength (i.e. the fine structure) is used to determine near-neighbor bond lengths [11.60]. This method has been very successful in bulk materials, whether crystalline or amorphous. To obtain surface information, as in the case of neutron and X-ray Diffraction, one can use layered materials such as graphite, or crystals of clusters representing surfaces. In particular, the former has been applied to physical adsorption on Grafoil. Bond length accuracies on the order of 0.05-0.1 Åhave been reported.

Infrared Reflection-Absorption Spectroscopy (IRAS)

Absorption of infrared radiation by characteristic vibrations at a surface can be used to obtain information concerning that surface, by comparison with known absorption frequencies in molecules of known structure, as in the case of HREELS and IETS [11.61,62]. Surface sensitivity can be obtained with small particles and thin films or with a multiple-reflection arrangement. A difficulty is the weak signal due to the very small cross section for the photon-vibrational coupling (approximately 10^6 times smaller than for vibrational excitation by electrons). The accessible energy losses are somewhat restricted on the low-frequency side, so that only a limited set of vibrations can be detected. On the other hand, the resolution is over an order of magnitude superior to that attainable with HREELS. This technique has been particularly successful in studies of adsorbed carbon monoxide due to the large dynamical dipole of the latter. In this case single-reflection IR absorption was used to monitor the adsorption on single-crystal surfaces from coverages below 10^{-2} monolayer up to saturation [11.62] . One great advantage of IRAS is that it may be carried out while the surface is subjected to high gas pressures or in the presence of (nonabsorbing) liquids. A dipole-selection rule is in effect, which permits only vibrational modes with significant components of their dynamical dipoles perpendicular to the surface to be detected. This allows the orientation of vibrational modes (i.e. chemical bonds) to be estimated.

Other Approaches

None of the experimental techniques mentioned above, together with their supporting theories, has been able singlehandedly to solve surface structures completely. Help has often come from first-principles calculations based on theoretical models that predict surface structures. Usually the calculations are made to fit some experimental quantities such as electronic-level energies (including densities of states) measured with UPS and XPS, atomic and molecular binding energies measured with TDMS, vibrational frequencies from HREELS and IRAS, or work-function changes. Some calculations, however, minimize the total energy self-consistently, yielding the optimum structure without experimental input. There is considerable hope that these types of calculations will prove to be very useful in determining (or confirming) surface structures.

At the other extreme of an approach to structural determination is the comparison of results obtained at surfaces with results obtained in other environments, such as in bulk materials, gas-phase molecules, and either inorganic or organometallic clusters. As an elementary example, one can assume that many bond lengths are transferable from one environment to another, especially when the immediate neighborhoods of the bonds in question are very similar. Organometallic clusters provide a specific example. In these clusters, a rather small number of metal atoms form a core to which organic molecules can be attached and examined structurally by X-ray crystallography or other methods. Analogous bonding of the same organic molecules to a single-crystal surface of the same metal might be expected. Especially with the use of vibrational frequencies, it is possible to identify surface species by analogy with the same species in a cluster. Empirical structural determinations of this type will proceed in parallel with the ab initio self-consistent calculations.

11.4.2 Comparisons Between Surface-Sensitive Techniques

Particular features of the various techniques mentioned in the previous section are summarized in Table 11.1. This table provides a compact overview (at the price of oversimplification) of the present capabilities and limitations of the different techniques. Some explanations of the entries in the table are in order.

In Table 11.1, a "variable" depth sensitivity refers to a strong change of the penetration of the probe as energy, angle of incidence, or angle of emergence are varied. For SIMS and ESDIAD, the entries represent typical depths at which bonds are broken in the experimental process. "Region with superlattice" refers to the case where surface sensitivity is obtained by measuring the intensities of superlattice beams only. However, recall that other possibilities for obtaining surface sensitivity are possible, as with high-surface-area layered substrates. By "overall conformation" we mean the approximate surface structure in terms of the bonding arrangement. This includes the state of adsorbed molecular species (i.e. whether they are physically adsorbed, chemisorbed, dissociated, or not dissociated), the identity of adsorbed molecular species (i.e. the chemical composition and the particular isomer), which atom(s) of a molecular species bond(s) to the substrate, approximately what orientation the species has adopted, whether an adatom has penetrated the substrate, and whether a surface is atomically smooth, rumpled, or grooved. For several techniques, the limit on the atomic number Z is quoted as ≥ 1, corresponding to the fact that very light elements (especially hydrogen) are detectable only under special circumstances.

We shall now attempt to classify these techniques according to their capabilities for detailed structural determination. The divisions between classes are not meant to be strict, and the classes are clearly not mutually exclusive.

Techniques Directly Sensitive to Atomic Coordinates

This category can be subdivided as follows:

i. Techniques sensitive to three coordinates: LEED, MEED, RHEED, MEIS, HEIS, ARPES, ARAES, XANES, HREELS (if done with a normal-mode analysis or I-V curve interpretation), SEXAFS (when contrasting results are obtained with polarized radiation), Neutron Diffraction, and X-ray Diffraction. Of these, several techniques are mainly sensitive to layer spacings (because they involve angle-resolved measurements), namely LEED, MEED, RHEED, ARPES, ARAES, and HREELS (with I-V curve interpretation), while SEXAFS with polarization is most sensitive to bond lengths and orientations.

ii. Techniques sensitive to bond lengths only: SEXAFS (without polarization dependence), EXAFS, and EAPFS.

iii. Techniques sensitive to bond orientation only: ESDIAD, HREELS (when applying only dipole selection rules) and IRAS (with selection rules).

iv. Techniques sensitive to overall conformation only: FIM, ESDIAD, SPI, Atomic Diffraction, EM, HREELS (with selection rules only), IETS, STM, UPS, and IRAS.

v. Techniques sensitive only to the low-electron-density contour of the surface: LEIS, Atomic Diffraction, SPI, and STM.

Table 11.1. Comparison of capabilities and limitations of surface-sensitive techniques. Crosses indicate

Technique	Probe		Depth sensitivity [Å]	Bond length accessibility	Length accuracy [Å]	Bond angle accessibility	Long-range order required
	in	out					
ISS	Ion	Ion	variable	–	–	–	–
LEIS	Ion	Ion	<1	some	0.5	some	. –
MEIS, HEIS	Ion	Ion	>1	×	0.02	×	–
SIMS	Ion	Ion	5	–	–	–	–
FIM	E field	Ion	<1	–	1	–	–
ESDIAD	electron	Ion	2	–	–	×	–
INS	Ion	atom, molecule	5	–	–	–	
SPI	atom	electron	<1	–	–	–	–
Atomic diffraction	atom	atom	<1	some	0.5	some	×
TDS	neat	atom, molecule	2	–	–	–	–
Neutron diffraction	neutron	neutron	region with superlattice	×	0.1	×	×
LEED, SPLEED, MEED, RHEED	electron	electron	≥5	×	0.02 – 0.1	×	×
EM	electron	electron	region with superlattice	–	1	–	–
HREELS (loss spectrum only)	electron	electron	5	·	–	–	–
HREELS (with vibrational mode analysis)	electron	electron	5	×	0.05	×	–
HREELS (I-V curves)	electron	electron	5	×	0.1	×	–
IETS	electron	electron	5	–	–	–	–
STM	electron	electron	1	–	–	–	–
Work function change	photon	electron	2	–	–	–	–
AES	electron	electron	5	–	–	–	–
ARAES	electron	electron	5	×	0.05	×	–
EAPFS	electron	electron	5	×	0.02	–	–
XPS	photon	electron	≥5	–	–	–	–
UPS	photon	electron	5	–	–	–	–
ARPES	photon	electron	5	×	0.02 – 0.1	×	–
SEXAFS	photon	electron	5	×	0.02	some	–
XANES	photon	photon	5	×	0.1	×	–

strong capabilities

Accessibility of overall conformation	Multiple scattering involved	Limit on atomic number Z	Accessibility of chemical composition	Accessibility of electronic structure	UHV required
some	–	–	×	–	×
some	some	–	×	–	×
some	–	–	×	–	×
some	–	–	×	–	×
×	–	≥13	–	–	×
×	–	–	×	–	×
–	–	–	×	×	×
×	–	–	×	×	×
×	some	–	–	–	×
some	–	–	×	–	×
some	–	–	–	–	–
some	×	≥1	–	–	×
×	some	≥1	some	–	×
×	–	–	×	–	×
×	–	–	×	–	×
some	×	–	some	–	×
×	–	–	×	–	–
×	–	–	some	some	×
some	–	–	–	some	×
–	–	≥3	×	some	×
some	×	≥3	some	–	×
some	–	≥3	×	some	×
some	–	≥1	×	×	×
×	–	≥1	×	×	×
some	variable	≥1	×	×	×
some	–	≥1	some	–	×
some	×	≥1	–	–	×

continued on next page

Table 11.1 (continued)

Technique	Probe		Depth sensitivity [Å]	Bond length accessibility	Length accuracy [Å]	Bond angle accessibility	Long-range order required
	in	out					
XRD	photon	photon	region with super-lattice	×	0.02	×	×
EXAFS	photon	photon	≫10	×	0.02	some	–
IRAS	photon	photon	>2	–	–	–	–
Theoretical simulation	–	–	any	×	0.02 – 0.20	×	variable

vi. Techniques sensitive to foreign atoms only (i.e. not sensitive to the geometry within the substrate surface): SEXAFS and EAPFS.

Techniques Directly Sensitive to the Electronic Structure of Surfaces

i. Techniques sensitive to core levels: XPS and ARXPS.

ii. Techniques sensitive to valence levels: UPS, ARUPS, SPI, and INS.

Techniques Directly Sensitive to the Spatial Electronic Distribution at Surfaces

i. Techniques sensitive to the electronic distribution perpendicular to the surface: STM and work-function-change measurements.

ii. Techniques sensitive to the electronic distribution parallel to the surface: LEIS, Atomic Diffraction, SPI, and STM.

Techniques Directly Sensitive to the Chemical Composition of Surfaces

i. Nondestructive techniques: AES, ISS, and XPS.

ii. Destructive techniques: TDMS and SIMS.

Techniques Directly Sensitive to Surface Vibrations

i. Techniques measuring vibrational frequencies: HREELS, IRAS and IETS.

11.4.3 Complementary and Competitive Techniques

The above comparisons among the various surface-sensitive techniques make it quite clear which of them complement each other and which of them compete with one another. It cannot be overstressed, however, that no single

Accessibility of overall conformation	Multiple scattering involved	Limit on atomic number Z	Accessibility of chemical composition	Accessibility of electronic structure	UHV required
some	–	≥ 3	–	–	–
some	–	≥ 1	–	–	–
×	–	≥ 1	×	–	–
×	not applicable	–	given	×	–

technique can determine a surface structure by itself. Complementary information is always required. For example, LEED is not very useful without information concerning chemical composition as determined by AES, XPS, or other methods. In the case of molecular adsorption, some identification of the molecular species present on the surface is also most useful. This can be obtained, for example, with HREELS, IRAS, or UPS. Insofar as the future is concerned, there is no sign that any technique will be able to avoid this need for complementary information. That is the reason why we have included so many non-LEED techniques in our discussion in Sects. 11.4.1 and 2.

The main competitors of LEED at present are MEIS, HEIS, ARPES, and SEXAFS. The need for an ion accelerator or synchrotron radiation to use these techniques will always constitute a severe drawback. However, they benefit in general from a greater ease of interpretation. This situation may result ultimately in these techniques being used to examine a limited set of relatively important structures, while more routine surface crystallography would be based on LEED.

However, we must make a distinction between different kinds of surfaces. ARPES and SEXAFS are not so well suited as LEED and MEIS or HEIS to studying clean surfaces, with their possible reconstructions and layer-spacing relaxations. For clean surfaces, ARPES involves an initial electronic state delocalized over many atoms in many layers, which removes the benefits due to photoexcitation from an orbital localized on a single adatom. Similarly, SEXAFS involves many subtly different bond lengths that are difficult to extricate. With reconstructions and layer-spacing changes of some complexity, MEIS and HEIS also have some difficulty. Although MEIS and HEIS can determine that some atoms are out of alignment with others, it may be difficult to decide which are the atoms that are out of alignment, and, quantitatively, by how much.

The most direct competition among LEED, MEIS, HEIS, ARPES, and SEXAFS occurs, at present, in the case of simple atomic adsorbate structures. All of these techniques are well suited to this kind of structure. Nevertheless, it has been amply demonstrated that such issues as whether adatom penetration into the substrate occurs, are not *easily* settled with any technique.

For more complex adsorbates, especially for adsorbed molecules, the capabilities of these techniques again diverge somewhat. For example, whenever an adsorbed molecule contains two or more inequivalent atoms of the same atomic number (such as several carbon atoms in a hydrocarbon), MEIS, HEIS, ARPES, and SEXAFS cannot distinguish between these atoms, while LEED can. On the other hand, in principle, ARPES can determine the structure layer-by-layer, from the substrate outward, by considering photoemission from one layer at a time, and this represents a great convenience. In some cases, the LEED result may be considered as an overdetermination-namely when only some structural parameters are desired, such as bond lengths obtained from SEXAFS at the expense of bond angles. Thus, one may forecast a divergence in the application of the various techniques for structure determinations, with specialization to certain types of surfaces. Clearly, the level of experimental or theoretical sophistication brought to bear on a problem is a sensitive function of the level of information that is required.

Other factors of greater sociopolitical complexity also play a role in this competition among the various techniques, such as manpower needs (LEED relies relatively more on theoreticians than other techniques) and funding patterns (techniques that depend on sophisticated equipment are relatively more subject to vagaries in funding). On the other hand, the very diversity of techniques enhances each of them individually through the inevitable competitive process. Cross-checks are provided which cause reexamination of previous results, an escalation of accuracy is set in motion, and the capabilities of each technique are extended, when possible, to match those of others.

12. Reference List and Table for Surface Structures

The following references generally contain surface structure analyses based on LEED intensities and many detailed structure determinations by other techniques, including model calculations (non-LEED results are labeled clearly as such). This list has been updated to mid-1984, but we do not claim it to be complete, even for LEED work.

The classification of references is by the value of the atomic number Z of the substrate (or by pairs of atomic numbers for compound solids) listed in order of increasing atomic number, with a subclassification by atomic number for adsorbates. The particular crystallographic orientations involved are given by their Miller indices, and the superlattice is indicated as well for ordered surfaces, (1x1) being the default structure. If the superlattice notation is preceded by a +, an overlayer is present. Otherwise, a reconstruction is implied. A + which is not followed by a superlattice notation indicates an undetermined adsorbate unit cell. The stoichiometry is added where useful, as in MoS_2 (the presence of a + indicates an adsorbate, its absence a compound). The acronyms denoting non-LEED techniques referred to in this bibliography are defined in Appendix A.

A comprehensive table of results of surface crystallography for clean surfaces and atomic and molecular adsorption follows the surface structure reference list.

Surface Structure References

The results are all from LEED unless otherwise specified. Listing is by chemical element (or combinations thereof) in order of increasing atomic number.

----3-9--LiF----

3.9.1   G. E. Laramore, A. C. Switendick: Phys. Rev. B **7**, 3615 (1973) - (100)

----4--Be----

4.1   J. A. Strozier, R. O. Jones: Phys. Rev. B **3**, 3228 (1971) - (0001)

4.2   J. A. Strozier, R. O. Jones: Phys. Rev. Lett. **25**, 516 (1970) - (0001)

----6--C----

6.1   N. J. Wu, A. Ignatiev: Bull. Am. Phys. Soc. **27**, 310 (1982) - (0001)

6.2   W. S. Yang, J. Sokolov, P. Jona, P. M. Marcus: Bull. Am. Phys. Soc. **27**, 311 (1982) - (100)(1x1), (100)(2x1) and (111)(1x1)

6.3   N. J. Wu, A. Ignatiev: J. Vac. Sci. Techno. **20**, 896 (1982) - (0001)
6.4   N. J. Wu, A. Ignatiev: Phys. Rev. B **25**, 29 3 (1982) - (0001)
6.5   W. S. Yang, J. Sokolov, F. Jona, P. M. Marcus: Solid State Commun.
        41, 191 (1982) - (111)(1x1)
6.6   D. Vanderbilt, S. G. Louie, J. Vac. Sci. Te nol. B **1**, 723 (1983)
        - (111)(1x1), (2x1) (Model Calculation)
6.7   D. Vanderbilt, S. G. Louie: Bull. Am. Phy Soc. **29**, 546 (1984)
        - (111)(2x1) (Model Calculation)

----6-1--C-H----
6.1.1   G. Vidali, M. W. Cole, W. H. Weinberg, W. A. Steele: Phys. Rev.
          Lett. **27**, 118 (1983) - (111)+(1x1) (Atom Diffraction)

----6-8--C-O----
6.8.1   M. F. Toney, R. D. Diehl, S. C. Fain, Phys. Rev. B **30**, 1115 (1984)
          - (0001) + incommensurate O_2
6.8.2   H. Yo, S. C. Fain: Bull. Am. Phys. Soc. **29**, 222 (1984) - (0001)+CO

----6-18--C-Ar----
6.18.1   C. G. Shaw, S. C. Fain, M. D. Chinn, M. F. Toney: Surface Sci. **97**,
           128 (1980) - (0001) + incommensurate

----6-19--C-K----
See 6.1 - (0001) + disordered
6.19.1   N. J. Wu, A. Ignatiev: Solid State Commun. **46**, 59 (1983)
           - (0001)+intercalate
6.19.2   N. J. Wu, A. Ignatiev: Phys. Rev. B **28**, 7288 (1983)
           - (0001)+intercalate

----6-36--C-Kr----
6.36.1   C. Bouldin, E. A. Stern: Phys. Rev. B **25**, 3462 (1982)
           - (0001)+ (EXAFS)

----11--Na----
11.1   P. M. Echenique: J. Phys. C **9**, 3193 (1976) - (110)
See 11.8.1 - (110)
11.2   S. A. Lindgren, J. Paul, L. Wallden, P. Westrin: J. Phys. C **15**, 6285
         (1982) - (0001)

----11-8--Na-O----
11.8.1   S. Andersson, J. B. Pendry, P. M. Echenique: Surf. Sci. **65**,
           539 (1977) - $Na_2O(111)$

----12-8--MgO----
12.8.1   K. O. Legg, M. Prutton, C. Kinniburgh: J. Phys. C **7**, 4236 (1974)
           - (100)

12.8.2   C. G. Kinniburgh: J. Phys. C **8**, 2382 (1975) - (100)
12.8.3   C. G. Kinniburgh: J. Phys. C **9**, 2695 (1976) - (100)
12.8.4   M. R. Welton-Cook, W. Berndt: J. Phys. C **15**, 5691 (1982) - (100)
12.8.5   T. Urano, T. Kanaji, M. Kaburagi: Surf. Sci. **134**, 109 (1983) - (100)

----13--Al----
13.1   S. Y. Tong, T. N. Rhodin, Phys. Rev. Lett. **26**, 711 (1971) - (100)
13.2   D. W. Jepsen, P. M. Marcus, F. Jona: Phys. Rev. Lett. **26**,
        1365 (1971) - (100)
13.3   C. B. Duke, G. E. Laramore, B. W. Holland, A. M. Gibbons:
        Surf. Sci. **27**, 523 (1971) - (110)
13.4   D. W. Jepsen, P. M. Marcus, F. Jona: Phys. Rev. B **5**, 3933 (1972)
        - (100)
13.5   Groupe d'Etude des Surfaces: Surf. Sci. **32**, 297 (1972) - (100)
13.6   G. E. Laramore, C. B. Duke: Phys. Rev. B **5**, 267 (1972)
        - (100),(111),(110)
13.7   M. R. Martin, G. A. Somorjai: Phys. Rev. B **7**, 3607 (1973)
        - (111),(110)
13.8   S. Y. Tong, T. N. Rhodin, R. H. Tait: Phys. Rev. B **8**, 430 (1973)
        - (100),(110)
13.9   M. W. Finnis, V. Heine: J. Phys. F **4** , L37 (1974) - (111),(100),(110)
        (Model Calculation)
13.10   R. H. Tait, S. Y. Tong, T. N. Rhodin: Phys. Rev. Lett. **28**, 553
        (1972) - (100),(110)
13.11   V. Hoffstein, D. S. Boudreaux: Phys. Rev. Letters **25**, 512 (1970)
        - (100)
13.12   D. S. Boudreaux, V. Hoffstein: Phys. Rev. B **3**, 2447 (1971)
        - (111),(110)
13.13   D. Aberdam, R. Baudoing, C. Gaubert, E. G. McRae: Surf. Sci. **57**,
        715 (1976) - (100),(110)
13.14   D. W. Jepsen, P. M. Marcus, F. Jona: Phys. Rev. B **6**, 3684 (1972)
        - (100),(110),(111)
13.15   N. Masud, C. G. Kinniburgh, J. B. Pendry: J. Phys. C **10**, 1 (1977)
        - (100),(110) (MEED)
See 13.11.2 - (100)
13.16   Groupe d'Etude des Surfaces: Surf. Sci. **62**, 567 (1977) - (100),(110)
See 28.11 - (110)
13.17   A. Bianconi, R. Z. Bachrach: Phys. Rev. Lett. **42**, 104 (1978)
        - (111),(100)(EXAFS)
13.18   C. W. B. Martinson, S. A. Flodstrom, J. Rundgren, P. Westrin:
        Surf. Sci. **89**, 102 (1979) - (111)
13.19   F. Jona, D. Sondericker, P. M. Marcus: J. Phys. C **13**, L155 (1980)
        - (111)
See 13.8.1 - (111)

13.20   C.-M. Chan, S. L. Cunningham, M. A. Van Hove, W. H. Weinberg:
         Surf. Sci. **67**, 1 (1977) - (110)
See 13.8.7 - (111)
13.21   H. B. Nielsen, D. L. Adams: J. Phys. **C** 15, 615 (1982) - (111)
See 13.8.12 - (111)
13.22   H. B. Nielsen, J. N. Andersen, L. Petersen, D. L. Adams: J. Phys. C
         15, L1113 (1982) - (110)
13.23   J. R. Noonan, H. L. Davis, F. Jona: Bull. Am. Phys. Soc. **28**, 572
         (1983) - (110)
13.24   R. N. Barnett, U. Landman, C. L. Cleveland: Phys. Rev. B **27**, 6534
         (1983) - (110) (Model Calculation)
13.25   J. N. Andersen, H. B. Nielsen, L. Peterson, D. L. Adams: J. Phys. C,
         to be published - (110)
13.26   N. Masud, R. Baudoing, D. Aberdam, C. Gaubert: Surf. Sci. **133**,
         580 (1983) - (100)(MEED)

----13-8--Al-O----
13.8.1   H. L. Yu, M. C. Munoz, F. Soria: Surf. Sci. **94**, L184 (1980)
          - (111)+(1x1)
13.8.2   L. I. Johansson, J. Stohr: Phys. Rev. Lett. **43**, 1882 (1979)
          - (111)+(1x1) (SEXAFS)
See 13.18 - (111)+(1x1).
13.8.3   M. L. Den Boer, T. L. Einstein, W. T. Elam, R. L. Park, L. D.
          Roelofs, G. E. Laramore: Phys. Rev. Lett. **44**, 496 (1980)
          - (100)+ (EAPFS)
13.8.4   F. Jona, P. M. Marcus: J. Phys. C **13**, L477 (1980) - (111)+(1x1)
13.8.5   L. Kleinman, K. Mednick: Phys. Rev. B **23**, 4960 (1981)
          - (111)+(1x1) (Model Calculation)
13.8.6   D. Norman, S. Brennan, R. Jaeger, J. Stohr: Surf. Sci. **105**, L297
          (1981) - (111)+ (SEXAFS)
13.8.7   R. Payley, J. A. Ramsey: J. Phys. C **13**, 505 (1980) - (111)+(1x1)
13.8.8   R. Z. Bachrach, G. V. Hansson, R. S. Bauer: Surf. Sci. **109**, L560
          (1981) - (111)+(PES,SEXAFS)
13.8.9   D. S. Wang, A. J. Freeman, H. Krakauer: Phys. Rev. B **24**, 3104
          (1981) - (111)+(Model Calculation)
13.8.10   F. Soria, V. Martinez, M. C. Munoz, J. L. Sacedon: Phys. Rev. B **24**,
          6926 (1981) - (111)+(1x1)
13.8.11   M. E. Schwartz: Appl. Surf. Sci. **11/12**, 157 (1982)
          - (111)+ (Model Calculation)
13.8.12   J. Neve, J. Rundgren, P. Westrin: J. Phys. C **15**, 4391 (1982)
          - (111)+(1x1)
13.8.13   R. L. Strong, B. Firey, F. W. De Wette, J. L. Erskine: Phys. Rev. B
          26, 3483 (1982) - (111)+(1x1) (HREELS)
13.8.14   I. P. Batra, O. Bisi: Surf. Sci. **123**, 283 (1982) - (111)+(Model
          Calculation)

13.8.15   V. Martinez, F. Soria, M. C. Muñoz, J. L. Sacedon:
          Surf. Sci. **128**, 424 (1983) - (111)+(1x1)
13.8.16   D. M. Bylanderm, L. Kleinman: Phys. Rev. B **28**, 523 (1983)
          - (111)+(Model Calculation)

----13-11--Al-Na----
13.11.1   B. A. Hutchins, T. N. Rhodin, J. E. Demuth: Surf. Sci. **54**, 419
          (1976) - (100)+c(2x2)
13.11.2   M. Van Hove, S. Y. Tong, N. Stoner: Surf. Sci. **54**, 259 (1976)
          - (100)+c(2x2)

----13-15--Al-P----
13.15.1   A. Kahn, C. R. Bonapace, C. B. Duke, A. Paton: J. Vac. Sci.
          Technol. B **1**, 613 (1983) - (110)
13.15.2   C. B. Duke, A. Paton, A. Kahn, C. R. Bonapace: Phys. Rev. B **28**,
          852 (1983) - (110)
13.15.3   C. B. Duke, A. Paton, A. Kahn: Bull. Am. Phys. Soc. **29**, 221
          (1984) - (110)

----14--Si----
14.1   S. J. White, D. P. Woodruff: J. Phys. C **9**, L451 (1976) - (100)(2x1)
14.2   H. D. Shih, F. Jona, D. W. Jepsen, P. M. Marcus: Bull. Am. Phys.
          Soc. **22**, 357 (1977) - (111)
14.3   H. D. Shih, F. Jona, D. W. Jepsen, P. M. Marcus: Phys. Rev. Lett. **37**,
          1622 (1976) - (111)
14.4   F. Jona, H. D. Shih, D. W. Jepsen, P. M. Marcus: Bull. Am. Phys.
          Soc. **22**, 357 (1977) - (100)(2x1)
14.5   F. Jona, H. D. Shih, A. Ignatiev, D. W. Jepsen, P. M. Marcus: J.
          Phys. C **10**, L67 (1977) - (100)(2x1)
14.6   A. Ignatiev, F. Jona, M. Debe, D. E. Johnson, S. J. White, D. P.
          Woodruff: J. Phys. C **10**, 1109 (1977) - (100)(2x1)
14.7   S. J. White, D. P. Woodruff: Surf. Sci. **64**, 131 (1977) - (100)
14.8   P. Mark: in Proc 7th Int. Vac. Cong. & 3rd Int. Conf. Sol. Surf.,
          Vienna (1977), ed. and publ. by R. Dobrozemsky,
          F. Rüdenauer, F. P. Viehböck, A. Breth (Vienna, 1977) p. 533
          - (111)(7x7)
14.9   P. Mark, J. D. Levine, S. H. McFarlane: Phys. Rev. Lett. **38**, 1408
          (1977) - (111)(7x7)
14.10   J. A. Appelbaum, G. A. Baraff, D. R. Hamann, H. D. Hagstrum,
          T. Sakurai: Surf. Sci. **70**, 654 (1978) - (100)(2x1)
14.11   J. D. Levine, S. H. McFarlane, P. Mark: Phys. Rev. B **16**, 5415
          (1977) - (111)(7x7)
14.12   K. A. R. Mitchell: Surf. Sci. **64**, 797 (1977) - (100)(2x1)
14.13   J. D. Levine, P. Mark, S. H. McFarlane: J. Vac. Sci. Technol. **14**,
          878 (1977) - (111)(7x7)

14.14   S. J. White, D. P. Woodruff: Surf. Sci. **63**, 254 (1977) - (100)

14.15   J. A. Appelbaum, D. R. Hamann: Surf. Sci. **74**, 21 (1978)
         - (100)(2x1)

14.16   K. A. R. Mitchell, M. A. Van Hove: Surf. Sci. **75**, 147L (1978)
         - (100)(2x1)

14.17   S. Y. Tong, A. L. Maldonado: Surf. Sci. **78**, 459 (1978) - (100)(2x1)

14.18   P. P. Auer, W. Mönch: Surf. Sci. **80**, 45 (1979) - (111)(1x1),(2x1)

14.19   S. J. White, D. P. Woodruff, B. W. Holland, R. S. Zimmer: Surf. Sci.
         74, 34 (1978) - (100)(1x1)

14.20   T. D. Poppendieck, T. C. Ngoc, M. B. Webb: Surf. Sci. **75**, 287
         (1978) - (100)c(4x2)

14.21   L. C. Snyder, Z. Wasserman: Surf. Sci. **77**, 52 (1978)
         - (111)(Model Calculation)

14.22   D. J. Chadi: Phys. Rev. Lett. **41**, 1062 (1978) - (111)(2x1)
         (Model Calculation)

14.23   D. J. Chadi: J. Vac. Sci. Technol. **16**, 1290 (1979) - (100)(2x1),c(4x2)
         (Model Calculation)

See 31.33.12 - (111)(2x1)

14.24   D. J. Miller, D. Haneman: J. Vac. Sci. Technol. **16**, 1270 (1979)
         - (111)(7x7)

14.25   W. Moench, P. P. Auer, R. Feder: J. Vac. Sci. Technol. **16**, 1286
         (1979) - (111)(2x1)

See 74.1.4 - (100)(2x1)

14.26   W. S. Verwoerd: Surf. Sci. **99**, 581 (1980) - (100)(2x1) (Model
         Calculation)

14.27   I. Stensgaard, L. C. Feldman, P. J. Silverman: Surf. Sci. **102**, 1
         (1981) - (100)(2x1) (HEIS)

14.28   R. J. Culbertson, L. C. Feldman, P. J. Silverman: Phys. Rev. Lett.
         45, 2043 (1980) - (111)(7x7) (HEIS)

14.29   H. L. Davis, J. R. Noonan, C. W. White, D. M. Zehner: Bull. Am.
         Phys. Soc. 26, 350 (1981) - (111)(1x1)

14.30   W. S. Yang, J. C. Fernandez, H. D. Shih, F. Jona, P. M. Marcus:
         Bull. Am. Phys. Soc. **26**, 351 (1981) - (100)(2x1)

14.31   J. Ihm, M. L. Cohen, D. J. Chadi: Phys. Rev. B **21**, 4592 (1980)
         - (100)(2x1) (Model Calculation)

14.32   D. W. Jepsen, H. D. Shih, F. Jona, P. M. Marcus: Phys. Rev. B **22**,
         814 (1980) - (111)(1x1)

14.33   M. J. Cardillo: Phys. Rev. B **23**, 4279 (1981) - (111)(7x7)
         (Atom Diffraction)

14.34   R. M. Tromp, R. G. Smeenk, F. W. Saris: Phys. Rev. Lett. **46**, 939
         (1981) - (100)(2x1) (MEIS)

14.35   D. M. Zehner, J. R. Noonan, H. L. Davis, C. W. White: J. Vac. Sci.
         Technol. **18**, 852 (1981) - (111)(1x1)

14.36   D. J. Chadi: J. Vac. Sci. Technol. **18**, 856 (1981)
         - (111)(1x1),(2x1),(7x7) (Model Calculation)

14.37   R. M. Tromp, R. G. Smeenk, F. W. Saris: Solid State Commun. **39**, 755 (1981) - (100)(2x1) (MEIS)

14.38   A. Redondo, W. A. Goddard, T. C. McGill, G. T. Surratt: Solid State Commun. **20**, 733 (1976) - (111)(1x1) (Model Calculation)

14.39   M. T. Yin, M. L. Cohen: Phys. Rev. B **24**, 2303 (1981) - (100)(2x1) (Model Calculation)

14.40   D. J. Chadi: Phys. Rev. Lett. **43**, 43 (1979) - (100)(2x1) (Model Calculation)

14.41   D. J. Chadi: J. Vac. Sci. Technol. **16**, 1290 (1979) - (100)(2x1) (Model Calculation)

14.42   J. A. Appelbaum, G. A. Baraff, D. R. Hamann: Phys. Rev. Lett. **35**, 729 (1975) - (100)(2x1) (Model Calculation)

14.43   J. A. Appelbaum, G. A. Baraff, D. R. Hamann: Phys. Rev. B **14**, 588 (1976) - (100)(2x1) (Model Calculation)

14.44   E. G. McRae: Bull. Am. Phys. Soc. **27**, 270 (1982) - (111)(7x7)

14.45   P. M. Marcus, J. W. Connolly: Bull. Am. Phys. Soc. **27**, 271 (1982) - (100)(2x1)

14.46   W. S. Yang, F. Jona, P. M. Marcus: Solid State Commun. **43**, 847 (1982) - (100)(2x1)

14.47   E. G. McRae, C. W. Caldwell: Phys. Rev. Lett. **46**, 1632 (1981) - (111)(7x7)

14.48   J. E. Northrup, J. Ihm, M. L. Cohen, Phys. Rev. Lett. **47**, 1910 (1981) - (111)(2x1) (Model Calculation)

14.49   K. C. Pandey: Phys. Rev. Lett. **47**, 1913 (1981) - (111)(2x1) (Model Calculation)

14.50   D. J. Chadi, R. Del Sole: J. Vac. Sci. Technol. **21**, 319 (1982) - (111)(1x1),(2x1) (Model Calculation)

14.51   J. E. Northrup, M. L. Cohen: J. Vac. Sci. Technol. **21**, 333 (1982) - (111)(2x1) (Model Calculation)

14.52   A. Redondo, W. A. Goddard: J. Vac. Sci. Technol. **21**, 344 (1982) -(100)(2x1) (Model Calculation)

14.53   J. E. Northrup, M. L. Cohen: Phys. Rev. Lett. **49**, 1349 (1982) - (111)(2x1) (Model Calculation)

14.54   S. J. White, D. C. Frost, K. A. R. Mitchell: Solid State Commun. **42**, 763 (1982) - (100)(2x1)

14.55   R. M. Tromp, E. J. van Loenen, M. Iwami, F. W. Saris: Solid State Commun. **44**, 971 (1982) - (111)(1x1) (MEIS)

14.56   N. Garcia, J. M. Soler: Surf. Sci. **122**, 535 (1982) - (111)(7x7) (Atom Diffraction)

14.57   G. Binnig, H. Rohrer, C. H. Gerber, E. Weisel: Phys. Rev. Lett. **50**, 120 (1983) - (111)(7x7) (STM)

14.58   R. Feder: Solid State Commun. **45**, 51 (83) - (111)(2x1)

14.59   K. C. Pandey: Bull. Am. Phys. Soc. **28**, 479 (1983) - (111)(2x1),(7x7) (Model Calculation)

14.60   R. M. Tromp: Bull. Am. Phys. Soc. **28**, 479 (1983) - (111),(1x1),(7x7)
         (HEIS)
14.61   E.G. McRae: Bull. Am. Phys. Soc. **28**, 480 (1983) - (111)(7x7)
14.62   W. S. Yang, F. Jona, P. M. Marcus: Surf. Sci. (1983) - (100)(2x1)
14.63   F. J. Himpsel: Phys. Rev. B. **27**, 7782 (1983) - (111)(7x7)
         (Model Calculation)
14.64   P. M. Petroff, R. J. Wilson: Phys. Rev. Lett. **51**, 199 (1983)
         - (111)(7x7)(TEM)
14.65   J. Ihm, D. H. Lee, J. D. Joannopoulos, A. N. Berker: J. Vac. Sci.
         Technol. B **1**, 705 (1983) - (100)(Model Calculation)
14.66   O. H. Nielsen, R. M. Martin, D. J. Chadi, K. Kunz: J. Vac. Sci.
         Technol. B **1**, 714 (1983) - (111)(2x1)(Model Calculation)
14.67   W. S. Yang, F. Jona, P. M. Marcus: J. Vac. Sci. Technol. B **1**, 718
         (1983) - (100)(2x1)
14.68   M. Aono, R. Souda, C. Oshima, Y. Ishizawa: Phys. Rev. Lett. **51**, 801
         (1983) - (111)(7x7)(LEIS)
14.69   W. S. Yang, F. Jona, P. M. Marcus: Phys. Rev. B **28**, 2049 (1983)
         - (100)(2x1)
14.70   E. G. McRae: Phys. Rev. B **28**, 2305 (1983) - (111)(7x7)
14.71   P. A. Bennett, L. C. Feldman, Y. Kuk, E. G. McRae, J. E. Rowe:
         Phys. Rev. B **28**, 3656 (1983) - (111)(7x7)
14.72   R. M. Tromp, L. Smit, J. F. van der Veen: Phys. Rev. Lett. **51**, 1672
         (1983) - (111)(2x1)(MEIS)
14.73   W. S. Yang, F. Jona: Phys. Rev. B **28**, 1178 (1983) - (111)(1x1)
14.74   G. J. R. Jones, B. W. Holland: Solid State Commun. **46**, 651 (1983)
         - (100)(2x1)
14.75   R. M. Tromp, R. G. Smeenk, F. W. Saris, D. J. Chadi: Surf. Sci. **133**,
         137 (1983) - (100)(2x1)(MEIS)
14.76   W. S. Yang, F. Jona: Solid State Commun. **48**, 377 (1983)
         - (111)(7x7)
14.77   H. Liu, M. R. Cook, F. Jona, P. M. Marcus: Phys. Rev. B **28**, 6137
         (1983) - (111)(2x1)
14.78   L. Pauling, Z. S. Herman: Phys. Rev. B **28**, 6154 (1983) - (100)(2x1)
         (Model Calculation)
14.79   I. P. Batra, F. J. Himpsel, P. M. Marcus, R. Tromp, M. R. Cook,
         F. Jona, H. Liu: Bull. Am. Phys. Soc. **29**, 233 (1984) - (111)(2x1)
14.80   J. E. Northrup, M. L. Cohen: Bull. Am. Phys. Soc. **29**, 264 (1984)
         - (111)(7x7)(Model Calculation)
14.81   P. Chiaradia, A. Cricenti, S. Selci, G. Chiarotti: Phys. Rev. Lett. **52**,
         1145 (1984) - (111)(2x1)(Optical Reflection)
14.82   M. A. Olmstead, N. M. Amer: Phys. Rev. Lett. **52**, 1148 (1984)
         - (111)(2x1)(Photothermal Displacement Spectroscopy)
14.83   J. E. Northrup, M. L. Cohen: Phys. Rev. B **29**, 1966 (1984)
         - (111)(2x1) (Model Calculation)

14.84   K. Higashiyama, S. Kono, H. Sakurai, T. Sagawa: Solid State
         Commun. **49**, 253 (1984) - (111)(7x7)(ARXPS)

----14-1--Si-H----

14.1.1   S. J. White, D. P. Woodruff, B. W. Holland, R. S. Zimmer: Surf. Sci.
         68, 457 (1977) - (100)+(1x1)
See 14.10
See 14.1
See 14.19
See 74.1.4
14.1.2   R. M. Tromp, R. G. Smeenk, F. W. Saris: in Proc. 4th Int. Conf. Sol.
         Surf. & 3rd Eur. Conf. Surf. Sci., Cannes (1980), ed. by D. A.
         Degras, M. Costa, Suppl. "Le Vide, les Couches Minces", No. 204
         (Societe Francaise du Vide, Paris 1980) p. 943 - (100)+(1x1)
         (MEIS)
14.1.3   R. M. Tromp, R. G. Smeenk, F. W. Saris: Surf. Sci. **104**, 13 (1981)
         - (100)+(1x1) (HEIS)
See 14.34 - (100)+(2x1) (MEIS)
14.1.4   W. S. Verwoerd: Surf. Sci. **108**, 153 (1981) - (100)+(2x1)H,(1x1)2H
         (Model Calculation)
14.1.5   W. M. Gibson, T. Narusawa, K. Kinoshita, L. C. Feldman:
         Bull. Am. Phys. Soc. **27**, 271 (1982) - (100)+(1x1) (HEIS)
14.1.6   D. J. Chadi: Bull. Am. Phys. Soc. **28**, 537 (1983) - (111)+(1x1)
         (Model Calculation)

----14-8--Si-O----

14.8.1   I. P. Batra, P. S. Bagus, K. Herman: Phys. Rev. Lett. **52**, 384 (1984)
         - (100)+(Model Calculation)

----14-13--Si-Al----

14.13.1 W. S. Yang, H. D. Shih, F. Jona, P. M. Marcus: Bull. Am. Phys. Soc.
         27, 270 (1982) - (111)+($\sqrt{3} \times \sqrt{3}$)R30°

----14-17--Si-Cl----

See 14.53.1 - (111)(7x7)+(SEXAFS)
14.17.1 G. B. Bachelet, M. Schlüter: Bull. Am. Phys.
         Soc. **28**, 537 (1983) - (111)+(Model Calculation)
14.17.2 B. N. Dev, K. C. Mishra, W. M. Gibson, T. P. Das: Bull. Am. Phys.
         Soc. **28**, 537 (1983) - (111)+(Model Calculation)
14.17.3 G. B. Bachelet, M. Schlüter: J. Vac. Sci. Technol. B **1**, 726 (1983)
         - (111)+(Model Calculation)
14.17.4 P. H. Citrin, J. E. Rowe, P. Eisenberger: Phys. Rev. B **28**, 2299
         (1983) - (111)+(SEXAFS)
14.17.4 G. B. Bachelet, M. Schlüter: Phys. Rev. B **28**, 2302 (1983)
         - (111)+(Model Calculation)

----14-32--Si-Ge----
See 14.17.4 - (111)+(SEXAFS)
See 14.17.5 - (111)+(Model Calculation)
14.32.1   Y.-N. Xu, K.-M. Zhang, X.-D. Xie: Solid State Commun. **47**, 93
          (1983) - (111)+(Model Calculation)

----14-35--Si-Br----
14.35.1   J. A. Golovchenko, J. R. Patel, D. R. Kaplan, P. L. Cowan, M. J.
          Bedzyk, Phys. Rev. Lett. **49**, 560 (1982)
          - (111)+(X-ray Diffraction)
14.35.2   G. Materlik, A. Frahm, M. J. Bedzyk: Phys. Rev. Lett. **52**, 441
          (1984) - (111)+(XRD)

----14-47--Si-Ag----
14.47.1   M. Saitoh, F. Shoji, K. Oura, T. Hanawa: Jap. J. Appl. Phys. **19**,
          L421 (1980) (111)+($\sqrt{3} \times \sqrt{3}$)R30° (LEIS)
14.47.2   Y. Terada, T. Yoshizuka, K. Oura, T. Hanawa: Surf. Sci. **114**, 65
          (1982) - (111)+(3x1), ($\sqrt{3} \times \sqrt{3}$)R30°
14.47.3   J. Stöhr, R. Jaeger, G. Rossi, T. Kendelewicz, I. Lindau: Surf. Sci.
          134, 813 (1983) - (111)(7x7)+(SEXAFS)

----14-52--Si-Te----
See 14.53.1 - (111)(7x7)+, (111)+($\sqrt{3} \times \sqrt{3}$)R30° (SEXAFS)
See 14.53.2 - (111)(7x7)+ (SEXAFS)

----14-53--Si-I----
14.53.1   P. H. Citrin: Bull. Am. Phys. Soc. **27**, 333 (1982)
          - (111)(7x7)+(SEXAFS)
14.53.2   P. H. Citrin, P. Eisenberger, J. E. Rowe: Phys. Rev. Lett. **48**,
          802 (1982) - (111)(7x7)+(SEXAFS)

----20-8--Ca-O----
20.8.1   M. Prutton, J. A. Ramsey, J. A. Walker, M. R. Welton-Cook:
          J. Phys. C **12**, 5271 (1979) - (100)

----21--Sc----
21.1   S. Tougaard, A. Ignatiev, D. L. Adams: in Proc. 4th Int. Conf. Sol.
          Surf. & 3rd Eur. Conf. Surf. Sci., Cannes (1980), ed. by D. A.
          Degras, M. Costa, Suppl. "Le Vide, les Couches Minces", No. 204
          (Societe Francaise du Vide, Paris 1980) p. 677 - (100)+(1x1)(MEIS)

----22--Ti----
22.1   H. D. Shih, F. Jona, D. W. Jepsen, P. M. Marcus: J. Phys. C **9**, 1405
          (1976) - (0001)

----22-1--Ti-H----
22.1.1   P. Cremaschi, J. C. Whitten: Phys. Rev. Lett. **46**, 1242 (1981)
           - (0001) + (Model Calculation)

----22-6-8--Ti-C-O----
22.6.8.1   H. D. Shih, F. Jona, D. W. Jepsen, P. M. Marcus: J. Vac. Sci.
           Technol. **15**, 596 (1978) - (0001)+
22.6.8.2   C. R. Fischer, L. A. Burke, J. L. Whitten: Phys. Rev. Lett. **49**, 344
           (1982) - (0001) + (Model Calculation)

----22-7--Ti-N----
22.7.1   H. D. Shih, F. Jona, D. W. Jepsen, P. M. Marcus: Phys. Rev. Lett.
           36, 798 (1976) - (0001)+(1x1)
22.7.2   H. D. Shih, F. Jona, D. W. Jepsen, P. M. Marcus: Surf. Sci. **60**,
           445 (1976) - (0001)+(1x1)

----22-16--Ti-S----
22.16.1   B. Lau, B. J. Mrstik, S. Y. Tong, M. A. Van Hove: unpublished
           - $TiS_2(0001)$

----22-34--Ti-Se----
See 22.16.1 - $TiSe_2(0001)$

----22-48--Ti-Cd----
22.48.1   H. D. Shih, F. Jona, D. W. Jepsen, P. M. Marcus: Commun. Phys.
           1, 25 (1976) - (0001)+(1x1)
22.48.2   H. D. Shih, F. Jona, D. W. Jepsen, P. M. Marcus: Phys. Rev. B **15**,
           5550 (1977) - (0001)+(1x1)
22.48.3   H. D. Shih, F. Jona, D. W. Jepsen, P. M. Marcus: Phys. Rev. B **15**,
           5561 (1977) - (0001)+(1x1)

----23--V----
23.1   P. W. Davies, R. M. Lambert: Surf. Sci. **95**, 571 (1980) - (100)(5x1)
23.2   D. L. Adams, H. B. Nielsen: Surf. Sci. **107**, 305 (1981);
         ibid. **116**, 598 (1982) - (110)
23.3   P. W. Davies, R. M. Lambert: Surf. Sci. **107**, 391 (1981) - (100)(5x1)
23.4   V. Jensen, J. N. Andersen, H. B. Nielsen, D. L. Adams: Surf. Sci.
         116, 66 (1982) - (100)(1x1)

----26--Fe----
26.1   R. Feder, G. Gafner: Surf. Sci. **57**, 45 (1976) - (110)
26.2   R. Feder: Phys. Status Solidi **58**, K137 (1973) - (100)
26.3   H. D. Shih, F. Jona, D. W. Jepsen, P. M. Marcus: Bull. Am. Phys.
         Soc. **22**, 357 (1977) - (111)

26.4   K. O. Legg, F. Jona, D. W. Jepsen, P. M. Marcus: J. Phys. C **10**, 937
        (1977) - (100)
26.5   R. Storbeck: Krist. Tech. **11**, 969 (1976) - (100),(110),(111)
26.6   H. D. Shih, U. Bardi, F. Jona: Bull. Am. Phys. Soc. **23**, 391 (1978)
        - (110)
26.7   J. Sokolov, H. D. Shih, U. Bardi, F. Jona, D. W. Jepsen,
        P. M. Marcus: Bull. Am. Phys. Soc. **25**, 327 (1980) - (211)·
26.8   H. D. Shih, F. Jona, U. Bardi, P. M. Marcus: J. Phys. C **13**, 3801
        (1980) - (110)
26.9   H. D. Shih, F. Jona, D. W. Jepsen, P. M. Marcus: Surf. Sci. **104**, 39
        (1981) - (111)
26.10   J. Sokolov, H. D. Shih, U. Bardi, F. Jona, P. M. Marcus:
        Solid State Commun. **48**, 739 (1983) - (211)
26.11   J. Sokolov, F. Jona, P. M. Marcus: Bull. Am. Phys. Soc. **29**, 220
        (1984) - (211),(310),(210)

----26-6--Fe-C----
26.6.1 S. Polizzi, F. Antonangeli, G. Chiarello, M. de Crescenzi: Surf. Sci.
        136, 555 (1984) - Polycrystalline Fe_3C (SEELFS)

----26-6-8--Fe-C-O----
26.6.8.1   F. Jona, K. O. Legg, H. D. Shih, D. W. Jepsen, P. M. Marcus:
           Phys. Rev. Lett. **40**, 1466 (1978) - (100)+c(2x2)

----26-7--Fe-N----
26.7.1 R. Imbihl, R. J. Behm, G. Ertl, W. Moritz: Surf. Sci. **123**, 129
        (1982) - (100)+c(2x2)

----26-8--Fe-0----
26.8.1 K. O. Legg, F. Jona, D. W. Jepsen, P. M. Marcus: J. Phys. C **8**, L492
        (1975) - (100)+(1x1)
26.8.2 A. B. Anderson: Phys. Rev. B **16**, 900 (1977) - (100)+(Model
        Calculation)
26.8.3 K. O. Legg, F. Jona, D. W. Jepsen, P. M. Marcus: Phys. Rev. B **16**,
        5271 (1977) - (100)+(1x1)
26.8.4 J. Sokolov, F. Jona, P. M. Marcus: Bull. Am. Phys. Soc. **26**, 225
        (1981) - (211)+(2x1)

----26-16--Fe-S----
26.16.1   K. O. Legg, F. Jona, D. W. Jepsen, P. M. Marcus: Surf. Sci. **66**, 25
          (1977) - (100)+c(2x2)
26.16.2   R. Feder, H. Viefhaus: to be published - (100)+c(2x2)
26.16.3   P. M. Marcus, D. W. Jepsen, H. D. Shih, F. Jona: Bull. Am. Phys.
          Soc. **26**, 225 (1981) - (110)+(2x2)

26.16.4   H. D. Shih, F. Jona, D. W. Jepsen, P. M. Marcus: Phys. Rev. Lett.
           46, 731 (1981) - (110)+(2x2)

----27--Co----
27.1   B. W. Lee, R. Alsenz, A. Ignatiev, M. A. Van Hove: Bull. Am. Phys.
        Soc. **22**, 357 (1977) (0001),(111)
27.2   M. Maglietta, E. Zanazzi, F. Jona, D. W. Jepsen, P. M. Marcus:
        Bull. Am. Phys. Soc. **22**, 355 (1977) - (100)
27.3   R. Alsenz, B. W. Lee, A. Ignatiev, M. A. Van Hove: Solid State
        Commun. **25**, 641 (1978) - (0001),(111)
27.4   M. Maglietta, E. Zanazzi, F. Jona, D. W. Jepsen, P. M. Marcus:
        Appl. Phys. **15**, 409 (1978) - (100)
27.5   B. W. Lee, R. Alsenz, A. Ignatiev, M. A. Van Hove: Phys. Rev. B
        17, 1510 (1978) - (0001),(1.11)
27.6   G. L. P. Berning, G. P. Alldredge, P. E. Viljoen: Surf. Sci. **104**, L225
        (1981) - (0001)
27.7   G. L. P. Berning, P. E. Viljoen, G. P. Alldredge, M. A. Van Hove:
        So. Afr. J. Phys. **4**, 77 (1981) - (0001)
27.8   M. Welz, W. Moritz, D. Wolf: Surf. Sci. **125**, 473 (1983) - (11$\bar{2}$0)

----27-8--Co-O----
27.8.1   M. Maglietta, E. Zanazzi, U. Bardi, F. Jona, D. W. Jepsen,
          P. M. Marcus: Surf. Sci. **77**, 101 (1978) - (100)+c(2x2)
27.8.2   A. Ignatiev, B. W. Lee, M. A. Van Hove: in Proc 7th Int. Vac. Cong.
          & 3rd Int. Conf. Sol. Surf., Vienna (1977), ed. and publ. by
          R. Dobrozemsky, F. Rüdenauer, F. P. Viehböck, A. Breth
          (Vienna, 1977) p. 2435 - (111)(7x7) - CoO(111)
27.8.3   R. C. Felton, M. Prutton, S. P. Tear, M. R. Welton-Cook: Surf. Sci.
          88, 474 (1979)
          - CoO(100)

----27-16--Co-S----
27.16.1   M. Maglietta: Solid State Commun. **43**, 395 (1982) - (100)+c(2x2)

----28--Ni----
28.1   J. B. Pendry: J. Phys. C **2**, 2283 (1969) - (111)
28.2   J. E. Demuth, S. Y. Tong, T. N. Rhodin: J. Vac. Sci. Technol. **9**,
        639 (1972) - (100)
28.3   G. E. Laramore: Phys. Rev. B **8**, 515 (1973) - (100),(111)
28.4   J. E. Demuth, P. M. Marcus, D. W. Jepsen: Phys. Rev. B **11**, 1460
        (1975) - (100),(111),(110)
See 13.10 - (100),(110)
See 29.4 - (100)
28.5   S. Y. Tong, L. L. Kesmodel: Phys. Rev. B **8**, 3753 (1973)
        - (100),(110)

28.6   S. Y. Tong: Solid State Commun. **16**, 91 (1975) - (100)
28.7   T. C. Ngoc, M. G. Lagally, M. B. Webb: Surf. Sci. **35**, 117 (1973)
        - (111)
28.8   J. E. Demuth, T. N. Rhodin: Surf. Sci. **42**, 261 (1974)
        - (100),(111),(110)
28.9   C. Gaubert, Y. Gauthier: in Proc 7th Int. Vac. Cong. & 3rd Int.
        Conf. Sol. Surf., Vienna (1977), ed. and publ. by R. Dobrozemsky,
        F. Rüdenauer, F. P. Viehböck, A. Breth
        (Vienna, 1977) p. 2427 - (111)(7x7)
28.10   R. Feder: Phys. Rev. B **15**, 1751 (1977) - (111)
28.11   C.-M. Chan, S. L. Cunningham, M. A. Van Hove, W. H. Weinberg:
        Surf. Sci. **67**, 1 (1977) - (110)
See 28.8.12 and 28.8.13 - (110)
See 28.8.15 - (100)
See 13.20   - (110)
See 28.8.24 - (111) (HEIS)
28.12   W. T. Moore, S. J. White, D. C. Frost, K. A. R. Mitchell: Surf. Sci.
        116, 253,261 (1982) - (311)
28.13   Y. Gauthier, R. Baudoing, C. Gaubert, L. Clarke: J. Phys. C **15**,
        3223,3231,3249 (1982) - (110)
28.14   R. Feder, S. F. Alvarado, E. Tamura, E. Kisker: Surf. Sci. **127**, 83
        (1983) - (100)(SPLEED)
28.15   R. Feidenhansl, J. E. Sorensen, I. Stensgaard: Surf. Sci. **134**, 329
        (1983) - (110)(HEIS)
28.16   J. W. M. Frenken, R. G. Smeenk, J. F. van der Veen: Surf. Sci. **135**,
        147 (1983) - (100)(HEIS)
28.17   E. Törnqvist, E. D. Adams, M. Copel, T. Gustafsson, W. R. Graham,
        J. Vacuum Sci. Technol. A **2**, 939 (1984) - (110)(MEIS)

----28-1--Ni-H----
28.1.1 J. E. Demuth: J. Colloid Interface Sci. **58**, 184 (1977) - (110)+(2x1)
28.1.2 M. A. Van Hove, G. Ertl, W. H. Weinberg, K. Christmann, R. J.
        Behm: Bull. Am. Phys. Soc. **22**, 356 (1977) - (111)+(2x2)
28.1.3 M. A. Van Hove, G. Ertl, W. H. Weinberg, K. Christmann, R. J.
        Behm: in Proc 7th Int. Vac. Cong. & 3rd Int. Conf. Sol. Surf.,
        Vienna (1977), ed. and publ. by R. Dobrozemsky, F. Rüdenauer,
        F. P. Viehböck, A. Breth (Vienna, 1977) p. 2427 - (111)(7x7)
28.1.4 M. A. Van Hove, G. Ertl, K. Christmann, R. J. Behm,
        W. H. Weinberg: Solid State Commun. **28**, 373 (1978)
        - (111)+(2x2)
28.1.5 G. Casalone, M. G. Cattania, M. Simonetta, M. Tescari:
        Chem. Phys. Letters **61**, 36 (1979) - (111)+(2x2)
28.1.6 K. Christmann, R. J. Behm, G. Ertl, M. A. Van Hove,
        W. H. Weinberg: J. Chem. Phys. **70**, 4168 (1979) - (111)+(2x2)

28.1.7   R. J. Behm, K. Christmann, G. Ertl, M. A. Van Hove,
W. H. Weinberg: Surf. Sci. **89**, 403 (1979) - (111)+(2x2)

28.1.8   T. Engel, K. H. Rieder: Surf. Sci. **109**, 140 (1981)
- (110)+(Atom Diffraction)

28.1.9   T. F. Koetzle, R. K. McMullan, R. Bau, D. W. Hart, R. G. Teller,
D. L. Tipton, R. D. Wilson: Adv. Chem. Ser. **167**, 61 (1978)
- $H_3Ni_4(Cp)_4$ cluster (Neutron Diffraction)

28.1.10   T. H. Upton, W. A. Goddard, C. F. Melius: J. Vac. Sci. Technol.
16, 531 (1979) - (Model Calculation)

28.1.11   K. H. Rieder: Phys. Rev. B **27**, 7799 (1983) - (110)+(Atom
Diffraction)

See 28.6.4 - (100)+(1x1)(Atom Diffraction)

28.1.12   P. Nordlander, S. Holloway, J. K. Norskov: Surf. Sci. **136**, 59 (1984)
- (111)+,(110)+(Model Calculation)

28.1.13   L. Olle, A. M. Baro: Surf. Sci. **137**, 607 (1984) - (110)+(HREELS)

----28-6--Ni-C----

28.6.1   M. A. Van Hove, S. Y. Tong: Surf. Sci. **52**, 673 (1975) - (100)+(2x2)

28.6.2   J. F. Onuferko, D. P. Woodruff, B. W. Holland: Surf. Sci. **87**, 357
(1979) - (100)+(2x2)

28.6.3   R. Rosei, M. de Crescenzi, F. Sette, C. Quaresima, A. Savoia, P.
Perfetti: Phys. Rev. B **28**, 1161 (1983) - (111)+(1x1)(SEELFS)

28.6.4   K. H. Rieder, H. Wilsch: Surf. Sci. **131**, 245 (1983)
- (100)+(2x2)(Atom Diffraction)

28.6.5   L. Papagno, L. S. Caputi: Phys. Rev. B **29**, 1483 (1984)
- (110)+Graphite (SEELFS)

----28-6-1--Ni-C-H----

28.6.1.1   G. Casalone, M. G. Cattania, M. Simonetta: Surf. Sci. **103**, L121
(1981) - (100)+c(2x2)C_2H_2

28.6.1.2   G. Casalone, M. G. Cattania, F. Merati, M. Simonetta: Surf. Sci.
120, 171 (1982) - (111)+(2x2)C_2H_2

28.6.1.3   R. G. Carr, T. K. Sham: Bull. Am. Phys. Soc. **29**, 552 (1984)
- (111)+C_2H_6,C_2H_4,C_2H_2,C_6H_6,C_6H_{12} (NEXAFS)

----28-6-8--Ni-C-O----

28.6.8.1   S. Andersson, J. B. Pendry: Surf. Sci. **71**, 75 (1978)
- (100)+c(2x2)

See 28.16.5 - (100)+c(2x2) (ARUPS)

28.6.8.2   K. Horn, A. M. Bradshaw, K. Jacobi: Surf. Sci. **72**, 719 (1978)
- (100)+c(2x2) (ARUPS)

28.6.8.3   M. A. Passler, A. Ignatiev: Bull. Am. Phys. Soc. **24**, 469 (1979)
-(100)+c(2x2)

28.6.8.4   M. A. Passler, A. Ignatiev, F. Jona, D. W. Jepsen, P. M. Marcus:
Phys. Rev. Letters **43**, 360 (1979) - (100)+c(2x2)

28.6.8.5   S. Andersson, J. B. Pendry: Phys. Rev. Lett. **43**, 363 (1979)
          - (100)+c(2x2)
28.6.8.6   L.-G. Petersson, S. Kono, N. F. T. Hall, C. S. Fadley, J. B. Pendry:
          Phys. Rev. Lett. **42**, **42**, 1545 (1979) - (100)+c(2x2) (ARXPS)
28.6.8.7   S. Andersson, J. B. Pendry: J. Phys. C **13**, 3547 (1980)
          - (100)+c(2x2)
28.6.8.8   S. Y. Tong, A. Maldonado, C. H. Li, M. A. Van Hove: Surf. Sci. **94**,
          73 (1980) - (100)+c(2x2)
28.6.8.9   P. J. Orders, S. Kono, J. T. Lloyd, K. A. Thompson, C. S. Fadley:
          Bull. Am. Phys. Soc. **26**, 258 (1981) - (100)+c(2x2) (ARXPS)
28.6.8.10 S. D. Kevan, R. F. Davis, D. H. Rosenblatt, J. G. Tobin,
          M. G. Mason, D. A. Shirley: Bull. Am. Phys. Soc. **26**, 258 (1981)
          - (100)+c(2x2), (111)+($\sqrt{3} \times \sqrt{3}$)R30° (ARNPD)
28.6.8.11 K. Heinz, E. Lang, K. Mueller: Surf. Sci. **87**, 595 (1979)
          - (100)+c(2x2)
28.6.8.12 S. D. Kevan, R. F. Davis, D. H. Rosenblatt, J. G. Tobin, M. G.
          Mason, D. A. Shirley, C. H. Li, S. Y. Tong: Phys. Rev. Lett. **46**,
          1629 (1981) - (100)+c(2x2), (111)+($\sqrt{3} \times \sqrt{3}$)R30° (ARNPD)
28.6.8.13 J. Stöhr, R. Jaeger: Phys. Rev. B **26**, 4111 (1982) - (100)+
          (NEXAFS)

----28-7--Ni-N----
See 28.6.8.13 - (100)+N_2 (NEXAFS)

----28-7-8--Ni-N-O----
28.7.8.1   M. A. Passler, T. H. Lin, A Ignatiev: J. Vac. Sci. Technol. **18**,
          481 (1981) - (100) + c(2x2)
See 28.6.8.13 - (100) + (NEXAFS)
28.7.8.2   C. W. Bauschlicher, P. S. Bagus: J. Chem. Phys. **80**, 944 (1984)
          - (Model Calculation)

----28-8--Ni-O----
28.8.1   H. D. Hagstrum, G. E. Becker: J. Chem. Phys. **54**, 1015 (1971)
          - (100)+c(2x2)(INS)
28.8.2   S. Andersson, B. Kasemo, J. B. Pendry, M. A. Van Hove:
          Phys. Rev. Lett. **31**, 595 (1973) - (100)+c(2x2)
28.8.3   J. E. Demuth, D. W. Jepsen, P. M. Marcus, Phys. Rev. Lett. **31**,
          540 (1973) - (100)+c(2x2)
28.8.4   H. H. Brongersma, J. B. Theeten: Surf. Sci. **54**, 519 (1976)
          - (100)+c(2x2)(LEIS)
28.8.5   M. Van Hove, S. Y. Tong: J. Vac. Sci. Technol. **12**, 230 (1975)
          - (100)+p(2x2)
See 28.1.1 - (110)+(2x1)
28.8.6   W. Heiland, E. Taglauer: J. Vac. Sci. Technol. **9**, 620 (1972)
          - (100)+c(2x2) (LEIS)

28.8.7   P. M. Marcus, J. E. Demuth, D. W. Jepsen: Surf. Sci. **53**, 501 (1975)
          - (100),(111),(110)
28.8.8   J. E. Demuth, T. N. Rhodin: Surf. Sci. **45**, 249 (1974)
          - (LEED experiment)
28.8.9   S. Walch, W. A. Goddard: Surf. Sci. **75**, 609 (1978)
          - (100)+(Model Calculation)
28.8.10   C. H. Li, J. W. D. Connolly: Surf. Sci. **65**, 700 (1977)
          - (100)+(Model Calculation)
28.8.11   W. Heiland, H. G. Schaeffler, E. Taglauer: Surf. Sci. **35**, 381
          (1973) - (110)+ (LEIS)
28.8.12   J. F. van der Veen, R. G. Smeenk, R. M. Tromp, F. W. Saris:
          Ned. Tijdschr. Vacuumtech. **2/3/4**, 284 (1978) - (110)+ (MEIS)
28.8.13   W. C. Turkenburg, R. G. Smeenk, F. W. Saris: Surf. Sci. **74**, 181
          (1978) - (110)+ (MEIS)
28.8.14   J. F. van der Veen, R. G. Smeenk, R. M. Tromp, F. W. Saris: Surf.
          Sci. **79**, 212 (1979) - (110)+ (MEIS)
28.8.15   G. Hanke, E. Lang, K. Heinz, K. Müller: Surf. Sci. **91**, 551 (1980)
          - (100)+c(2x2), p(2x2)
28.8.16   L. G. Petersson, S. Kono, N. F. T. Hall, S. Goldberg, J. T. Lloyd,
          C. S. Fadley, J. B. Pendry: Mat. Sci. Eng. **42**, 111 (1980)
          - (100)+ (ARXPS)
28.8.17   C. G. Kinniburgh, J. A. Walker: Surf. Sci. **63**, 274 (1977) - NiO(100)
28.8.18   J. A. Walker, C. G. Kinniburgh, J. A. D. Matthew: Surf. Sci. **68**, 221
          (1977) - NiO(100)
28.8.19   M. Prutton, J. A. Walker, M. R. Welton-Cook, R. C. Felton, J. A.
          Ramsey: Surf. Sci. **89**, 95 (1979) - NiO(100)
28.8.20   M. R. Welton-Cook, M. Prutton: J. Phys. C **13**, 3993 (1980)
          - NiO(100)
28.8.21   D. H. Rosenblatt, J. G. Tobin, M. G. Mason, R. F. Davis,
          S. D. Kevan, D. A. Shirley, C. H. Li, S. Y. Tong:
          Phys. Rev. B **23**, 3828 (1981) - (100)+c(2x2) (ARNPD)
See 28.16.10 - (100)+c/p(2x2) (ARXPS)
28.8.22   T. S. Rahman, J. E. Black, D. L. Mills: Phys. Rev. Lett. **46**, 1469
          (1981) - (100)+c/p(2x2) (Model Calculation)
28.8.23   R G Smeenk, R M Tromp, F W Saris: Surf. Sci. **107**, 429 (1981)
          - (110)+(2x1) (HEIS)
28.8.24   T. Narusawa, W. M. Gibson, E. Tornqvist: Phys. Rev. Lett. **47**, 417
          (1981) - (111)+(2x2), ($\sqrt{3} \times \sqrt{3}$)R30° (HEIS)
28.8.25   A. B. Anderson: J. Chem. Phys. **66**, 2173 (1977)
          - (100)+(Model Calculation)
28.8.26   J. A. van der Berg, L. K. Verhey, D. G. Armour: Surf. Sci. **91**, 218
          (1980) - (110)+(2x1) (LEIS)
28.8.27   T. Narusawa, W. M. Gibson, E. Tornqvist: Surf. Sci. **114**, 331
          (1982) - (111)+(2x2), ($\sqrt{3} \times \sqrt{3}$)R30° (HEIS)

28.8.28   J. E. Black, T. S. Rahman, D. L. Mills: J. Vac. Sci. Technol. **20**,
567 (1982) - (100)+c/p(2x2) (Model Calculation)

28.8.29   T. H. Upton, W. A. Goddard: Phys. Rev. Lett. **46**, 1635 (1981)
- (100)+c/p(2x2) (Model Calculation)

28.8.30   J. Stöhr, R. Jaeger, T. Kendelewicz: Phys. Rev. Lett. **49**,
142 (1982) - (100)+c/p(2x2) (SEXAFS)

28.8.31   S. Y. Tong, K. H. Lau: Phys. Rev. B **25**, 7382 (1982)
- (100)+c(2x2)

28.8.32   D. Norman, J. Stöhr, R. Jaeger, P. Durham, J. B. Pendry:
Phys. Rev. Lett. **51**, 2052 (1983) - (100)+ (XANES)

28.8.33   I. P. Batra, J. A. Barker: Bull. Am. Phys. Soc. **28**, 463 (1983)
- (100)+c/p(2x2) (Atom Diffraction)

28.8.34   W. N. Mei, S. Y. Tong: Bull. Am. Phys. Soc. **28**, 470 (1983)
- (100)+c(2x2) (ARPES)

28.8.35   J. E. Demuth, N. J. DiNardo, G. S. Cargill III: Phys. Rev.
Lett. **50**, 1373 (1983) - (100)+c(2x2)

28.8.36   M. de Crescenzi, F. Antonangeli, C. Bellini, R. Rosei: Phys. Rev.
Lett. **50**, 1949 (1983) - (100)+c(2x2)(SEELFS)

28.8.37   S. Y. Tong, W. M. Kang, D. H. Rosenblatt, J. G. Tobin, D. A.
Shirley: Phys. Rev. B **27**, 4632 (1983) - (100)+c(2x2)(ARPES)

28.8.38   K. H. Rieder: Phys. Rev. B **27**, 6978 (1983)
- (100)+c/p(2x2) (Atom Diffraction)

28.8.39   J. A. Barker, I. Batra: Phys Rev. B **27**, 3138 (1983)
- (100)+c/p(2x2) (Atom Diffraction)

28.8.40   J. M. Szeftel, S. Lehwald, H. Ibach, T. S. Rahman, J. E. Black,
D. L. Mills: Phys. Rev. Lett. **27**, 2688 (1983)
- (100)+c(2x2) (Model Calculation)

28.8.41   K. H. Rieder: Surf. Sci. **128**, 325 (1983) - (100)+c/p(2x2)
(Atom Diffraction)

See 29.8.8 - (100)+(2x2)

28.8.42   J. W. M. Frenken, J. F. van der Veen, G. Allan: Phys. Rev. Lett. **51**,
1876 (1983) - (100)+c(2x2) (HEIS)

28.8.43   D. Norman, J. Stöhr, R. Jaeger, P. J. Durham, J. B. Pendry:
Phys. Rev. Lett. **51**, 2052 (1983) - (100)+c/p(2x2)(NEXAFS)

28.8.44   A. M. Baro, G. Binnig, H. Rohrer, Ch. Gerber, E. Stoll, A. Baratoff:
F. Salvan: Phys. Rev. Lett. **52**, 1304 (1984) - (110)+(2x1)(STM)

See 28.16 - (100)+c/p(2x2)(HEIS)

28.8.45   C. W. Bauschlicher and P. S. Bagus, Phys. Rev. Lett. **52**, 200
(1984) - (100)+c/p(2x2) (Model Calculation)

----28-11--Ni-Na----

28.11.1   S. Andersson and J. B. Pendry, J. Phys. C **6**, 601 (1973)
- (100)+c(2x2)

28.11.2   S. Andersson and J. B. Pendry, J. Phys. C **5**, L41 (1972)
- (100)+c(2x2)

28.11.3   S. Andersson and J. B. Pendry, Solid State Commun. **16**, 563 (1975)
          - (100)+c(2x2)
28.11.4   J. E. Demuth, D. W. Jepsen and P. M. Marcus, J. Phys. C **8**, L25
          (1975) - (100)+c(2x2)
28.11.5   N. V. Smith, H. H. Farrell, M. M. Traum, D. P. Woodruff,
          D. Norman, M. S. Woolfson, B. W. Holland: Phys. Rev. B **21**,
          3119 (1980). - (100)+c(2x2) (ARUPS)

----28-11-16--Ni-Na-S----
28.11.16.1 S. Andersson, J. B. Pendry: J. Phys. C **9**, 2721 (1976)
          - (100)+c/p(2x2)

----28-13--Ni-Al----
See 29.79.1 - Ni$_3$Al(111)

----28-16--Ni-S----
See 28.8.1 - (100)+c(2x2)
See 28.8.3 - (100)+c(2x2)
28.16.1   C. B. Duke, N. O. Lipari, G. E. Laramore, J. B. Theeten: Solid State
          Commun. **13**, 579 (1973) - (100)+c(2x2)
28.16.2   J. E. Demuth, D. W. Jepsen, P. M. Marcus: Phys. Rev. Lett. **32**,
          1182 (1974) - (100),(110),(111)
28.16.3   Groupe d'Etude des Surfaces: Surf. Sci. **48**, 577 (1975)
          - (100)+c(2x2)
28.16.4   H. D. Hagstrum, G. E. Becker: J. Vac. Sci. Technol. **14**, 369
          (1977) - (100)+c/p(2x2) (INS)
See 28.8.5 - (100)+p(2x2)
See 28.8.7 - (110),(111),(100)
See 28.8.8 - Experiment
See 28.9 - (100)+c(2x2)
28.16.5   C. H. Li, S. Y. Tong: Phys. Rev. Lett. **40**, 46 (1978)
          - (100)+c(2x2)(ARUPS)
28.16.6   J. B. Theeten, H. H. Brongersma: Rev. Physique Appl. **11**, 57 (1976)
          - (100)+c(2x2) (LEIS)
28.16.7   J. F. van der Veen, R. M. Tromp, R. G. Smeenk, F. W. Saris:
          Surf. Sci. **82**, 468 (1979)
          - (110)+c(2x2) (MEIS)
28.16.8   S. P. Walch, W. A. Goddard: Surf. Sci. **72**, 645 (1978)
          - (100)+ (Model Calculation)
28.16.9   Y. Gauthier, D. Aberdam, R. Baudoing: Surf. Sci. **78**, 339 (1978)
          - (100)+c(2x2)
28.16.10 R. E. Connelly, N. F. T. Hall, P. J. Orders, C. S. Fadley, S. Kono,
          L.-G. Peterson: Bull. Am. Phys. Soc. **26**, 259 (1981) - (100)+c(2x2)
          (ARXPS)

See 28.8.21 - (100)+c(2x2) (ARNPD)
28.16.11   D. H. Rosenblatt, S. D. Kevan, J. G. Tobin, R. F. Davis, M. G.
           Mason, D. R. Denley, D. A. Shirley, Y. Huang, S. Y. Tong:
           Phys. Rev. B **26**, 1812 (1982) - (110)+c(2x2)(ARNPD)
28.16.12   S. Brennan, J. Stöhr, R. Jaeger: Phys. Rev. B **24**, 4871 (1981)
           - (100)+c(2x2)(SEXAFS)
28.16.13   J. Stöhr, R. Jaeger, S. Brennan: Surf. Sci. **117**, 503 (1982)
           - (100)+c(2x2)(SEXAFS)
28.16.14   P. J. Orders, C. S. Fadley: Phys. Rev. B **27**, 781 (1983)
           - (100)+c(2x2)(NPD)
28.16.15   J. J. Barton, C. C. Bahr, Z. Hussain, S. W. Robey, L. E. Klebanoff,
           D. A. Shirley: J. Vac. Sci. Technol. A **2**, 847 (1984)
           - (100)+c(2x2)(ARPEFS)
28.16.16   E. L. Bullock, C. S. Fadley, P. J. Orders: Phys. Rev. B **28**, 4867
           (1983) - (100)+c/p(2x2)(ARXPS)
28.16.17   R. Baudoing, E. Blanc, C. Gaubert, Y. Gauthier: Surf. Sci. **128**, 22
           (1983) - (110)+c(2x2)(ARAES)

----28-29--Ni-Cu----
28.19.1   M. Abujodeh, P. P. Vaishnava, P. A. Montano: Bull. Am. Phys. Soc.
           29, 221 (1984) - (100)+(1x1)

----28-34--Ni-Se----
See 28.8.3 - (100)+c(2x2)
See 28.8.5 - (100)+p(2x2)
See 28.8.7 - (100)+c/p(2x2)
See 28.8.8 - (100)+c/p(2x2)
See 28.16.4 - (100)+c/p(2x2)
See 28.16.10 - (100)+c(2x2) (ARXPS)
See 28.16.11 - (110)+c(2x2) (ARNPD)
See 28.16.11 - (111)+(2x2) (ARNPD)
28.34.1   D. H. Rosenblatt, S. D. Kevan, J. G. Tobin, R. F. Davis,
           M. G. Mason, D. A. Shirley, J. C. Tang, S. Y. Tong:
           Phys. Rev. B **26**, 3181 (1982) - (100)+c(2x2) (ARNPD)

----28-52--Ni-Te----
28.52.1   J. E. Demuth, D. W. Jepsen, P. M. Marcus: J. Phys. C **6**, L307
           (1973) - (100)+c(2x2)
See 28.16.4 - (100)+c(2x2)/c(2x4)
See 28.8.3 - (100)+c(2x2)
See 28.8.5 - (100)+p(2x2)
See 28.8.7 - (100)+c/p(2x2)
See 28.8.8 - (100)+c/p(2x2)
See 28.11.5 - (100)+c(2x2) (ARUPS)

28.52.2   J. K. Lang, K. D. Jamison, F. B. Dunning, G. K. Walters,
            M. A. Passler, A. Ignatiev, E. Tamaru, R. Feder:
            Surf. Sci. **123**, 247 (1982) - (100)+c(2x2) (PLEED)

----29--Cu----

29.1   J. B. Pendry: J. Phys. C **4**, 2514 (1971) - (100)
29.2   G. Capart: Surf. Sci. **26**, 429 (1971) - (100)
29.3   P. M. Marcus, D. W. Jepsen, F. Jona: Surf. Sci. **31**, 180 (1972) - (100)
29.4   D. L. Adams, U. Landman: Phys. Rev. B **15**, 3775 (1977) - (100)
29.5   G. E. Laramore20: Phys. Rev. B **9**, 1204 (1974) - (100),(111)
29.6   G. G. Kleiman, J. M. Burkstrand: Surf. Sci. **50**, 493 (1975) - (100)
29.7   J. B. Pendry: *Low Energy Electron Diffraction* (Academic, London,
            1974) - (100)
29.8   D. C. Frost, K. A. R. Mitchell, W. T. Moore, R. W. Streater, P. R.
            Watson: in Proc. 7th Int. Vac. Cong. & 3rd Int. Conf. Sol. Surf.,
            Vienna (1977), ed. and publ. by R. Dobrozemsky, F. Rüdenauer,
            F. P. Viehböck, A. Breth (Vienna, 1977) p. 2403 - (311)
29.9   H. L. Davis: to be published - (110)
29.10   P. R. Watson, F. R. Shepherd, D. C. Frost, K. A. R. Mitchell:
            Surf. Sci. **72**, 562 (1978) - (111)
29.11   R. W. Streater, W. T. Moore, P. R. Watson, D. C. Frost,
            K. A. R. Mitchell: Surf. Sci. **72**, 744 (1978) - (311)
29.12   J. R. Noonan, H. L. Davis, L. H. Jenkins: J. Vac. Sci. Technol. **15**,
            619 (1978) - (110)
29.13   H. L. Davis, J. R. Noonan, L. H. Jenkins: Surf. Sci. **83**, 559 (1979)
            - (110)
29.14   H. L. Davis, J. R. Noonan: Bull. Am. Phys. Soc. **25**, 328 (1980)
            - (100)
29.15   J. R. Noonan, H. L. Davis: in Proc. 4th Int. Conf. Sol. Surf. & 3rd
            Eur. Conf. Surf. Sci., Cannes (1980), ed. by D. A. Degras, M. Costa,
            Suppl. "Le Vide, les Couches Minces", No. 204 (Societe Francaise
            du Vide, Paris 1980) p. 1100 - (100)
29.16   A. J. Algra, S. B. Luitjens, E. P. Th. M. Suurmeijer, A. L. Boers:
            Physics Letters **75A**, 496 (1980) - (410) (LEIS)
29.17   J. R. Noonan, H. L. Davis: Surf. Sci. **99**, L424 (1980) - (110)
29.18   A. J. Algra, S. B. Luitjens, E. P. Th. M. Suurmeijer, A. L. Boers:
            Surf. Sci. **100**, 329 (1980) - (410) (LEIS)
29.19   M. N. Read, D. N. Lowy: Surf. Sci. **107**, L313 (1981) - (111) (LEPD)
29.20   S. P. Tear, K. Röll, M. Prutton: J. Phys. C **14**, 3297 (1981) - (111)
29.21   J. R. Noonan, H. L. Davis: Bull. Am. Phys. Soc. **27**, 237 (1982)
            - (100)
29.22   H. L. Davis, J. R. Noonan: J. Vac. Sci. Technol. **20**, 842 (1982)
            - (100)
29.23   F. Jona, D. W. Jepsen, P. M. Marcus, I. J. Rosenberg, A. H. Weiss,
            K. F. Canter: Solid State Commun. **36**, 957 (1980) - (111) (LEPD)

29.24   D. L. Adams, H. B. Nielsen, J. N. Andersen, I. Stensgaard, R.
         Feidenhansl, J. E. Sorensen: Phys. Rev. Lett. **49**, 669 (1982)
         - (110) (LEED+HEIS)
29.25   H. L. Davis, J. R. Noonan: Surf. Sci. **126**, 245 (1983) - (100),(110)
29.26   B. Salanon, G. Armand, J. Perreau, J. Lapujoulade: Surf. Sci. **127**,
         135 (1983) - (110)(Atom Diffraction)
29.27   A. H. Weiss, I. J. Rosenberg, K. F. Canter, C. B. Duke, A. Paton:
         Phys. Rev. B **27**, 867 (1983) - (111),(100)(LEPD)
29.28   I. Stensgaard, R. Feidenhansl, J. E. Sorensen: Surf. Sci. **128**, 281
         (1983) - (110)(HEIS)
29.29   D. L. Adams, H. B. Nielsen, J. N. Andersen: Surf. Sci. **128**, 294
         (1983) - (110)
29.30   J. Neve, P. Westrin, J. Rundgren: J. Phys. C **16**, 1291 (1983) - (111)
29.31   S. A. Lindgren, L. Wallden, J. Rundgren, P. Westrin: Phys. Rev. B
         29, 576 (1984) - (111)

----29-6-8--Cu-C-O----
See 28.6.8.5 - (100)+c(2x2)
See 28.6.8.7 - (100)+c(2x2)

----29-7--Cu-N----
29.7.1   J. M. Burkstrand, G. G. Kleiman, G. G. Tibbetts, J. C. Tracy:
         J. Vac. Sci. Technol. **13**, 291 (1976) - (100)+c(2x2)(CMTA-LEED)
29.7.2   J. M. Burkstrand, S. Y. Tong, M. A. Van Hove: unpublished
         - (100)+c(2x2)

----29-8--Cu-O----
29.8.1   L. McDonnell, D. P. Woodruff, K. A. R. Mitchell: Surf. Sci. **45**, 1
         (1974) - (100)+c(2x2)
29.8.2   J. F. Onuferko, D. P. Woodruff: Surf. Sci. **95**, 555 (1980)
         - (100)+c(2x2),(100)+c($\sqrt{2} \times 2\sqrt{2}$)R45°
29.8.3   J. Lapujoulade, Y. Le Cruer, M. Lefort, Y. Lejay, E. Maurel:
         Phys. Rev. B **22**, 5740 (1980) - (110)+(2x1) (Atom Diffraction)
29.8.4   S. Kono, S. M. Goldberg, N. F. T. Hall, C. S. Fadley: Phys. Rev. B
         22, 6085 (1980) - (100)+c(2x2) (ARXPS)
29.8.5   A. G. J. de Wit, R. P. N. Bronckers, Th. M. Hupkens, J. M. Fluit:
         Surf. Sci. **90**, 676 (1979) - (110)+(2x1) (LEIS)
29.8.6   J. G. Tobin, L. E. Klebanoff, D. H. Rosenblatt, R. F. Davis,
         E. Umbach, A. G. Baca, C. C. Bahr, D. A. Shirley, Y. Huang,
         W. M. Kang, S. Y. Tong: Bull. Am. Phys. Soc. **27**, 240 (1982)
         - (100)+c(2x2) (ARNPD)
29.8.7   J. G. Tobin, L. E. Klebanoff, D. H. Rosenblatt, R. F. Davis,
         E. Umbach, A. G. Baca, C. C. Bahr, D. A. Shirley, Y. Huang,
         W. M. Kang, S. Y. Tong: Phys. Rev. B **26**, 7076 (1982)
         - (100)+c(2x2),+c($\sqrt{2} \times 2\sqrt{2}$)R45° (ARNPD)

29.8.8   H. Richter, U. Gerhardt: Phys. Rev. Lett. **51**, 1570 (1983)
          - (100)+c(2x2)
29.8.9   R. Feidenhansl, I. Stensgaard: Surf. Sci. **133**, 453 (1983)
          - (110)+(2x1)(HEIS)
29.8.10   U. Döbler, K. Baberschke, J. Haase, A. Puschmann: Phys. Rev. Lett.
          52, 1437 (1984) - (110)+(2x1)(SEXAFS)

----29-13--Cu-Al----
29.13.1   D. F. Ogletree, M. A. Van Hove, G. A. Somorjai, R. J. Baird:
          Bull. Am. Phys. Soc. **29**, 222 (1984) - $Cu_3Al(111)$

----29-16--Cu-S----
See 28.16.14 - (100)+p(2x2) (ARPEFS)
See 28.16.18 - (100)+p(2x2) (ARPEFS)

----29-17--Cu-Cl----
29.17.1   D. R. Hamann, P. H. Citrin, L. F. Mattheiss, J. E. Rowe: Bull. Am.
          Phys. Soc. **27**, 239 (1982) - (100)+c(2x2)(SEXAFS)
29.17.2   D. Westphal, A. Goldmann, F. Jona, P. M. Marcus: Solid State
          Commun. **44**, 685 (1982) - (100)+c(2x2)
29.17.3   P. H. Citrin, D. R. Hamann, L. F. Mattheis, J. E. Rowe: Phys. Rev.
          Letters **49**, 1712 (1982) - (100)+c(2x2) (SEXAFS)
29.17.4   F. Jona, D. Westphal, A. Goldmann, P. M. Marcus: Bull. Am. Phys.
          Soc. **28**, 537 (1983) - (100)+c(2x2)
29.17.5   F. Jona, D. Westphal, A. Goldmann, P. M. Marcus: J. Phys. C **16**,
          3001 (1983) - (100)+c(2x2)
29.17.6   F. Jona, P. M. Marcus: Phys. Rev. Lett. **50**, 1823 (1983)
          - (100)+c(2x2)
29.17.7   P. H. Citrin, D. R. Hamann, L. F. Mattheiss, J. E. Rowe:
          Phys. Rev. Lett. **50**, 1824 (1983) - (100)+c(2x2)

----29-28--Cu-Ni----
29.28.1   S. P. Tear, K. Röll: J. Phys. C **15**, 5521 (1982) - (111)+(1x1)

----29-52--Cu-Te----
29.52.1   A. Salwen, J. Rundgren: Surf. Sci. **53**, 523 (1975) - (100)+p(2x2)
29.52.2   F. Comin, P. H. Citrin, P. Eisenberger, J. E. Rowe: Phys. Rev. B **26**,
          7060 (1982) - (100)+(2x2),(111)+(2√3 × √3)R30° (SEXAFS)

----29-53--Cu-I----
See 47.53.3 - (111)+(√3 × √3)R30° (SEXAFS)
29.53.1   P. H. Citrin: Bull. Am. Phys. Soc. **25**, 383 (1980) - (100)+(2x2),(110)
          +c(2x2),(110) + compressed c(2x2) (SEXAFS)
29.53.2   P. H. Citrin, P. Eisenberger, R. C. Hewitt: Phys. Rev. Lett. **45**, 1948
          (1980) - (100),(111)+(SEXAFS)

----29-55--Cu-Cs----
29.55.1   S. A. Lindgren, L. Wallden, J. Rundgren, P. Westrin, J. Neve:
           Phys. Rev. B **28**, 6707 (1983) - (111)+(2x2)

----29-79--Cu-Au----
29.79.1   D. Sondericker, F. Jona, P. M. Marcus: Bull. Am. Phys. Soc. **29**,
           222 (1984) - $Cu_3Au(111)$

----29-82--Cu-Pb----
29.82.1   W. Hoesler, W. Moritz: Surf. Sci. **117**, 196 (1982)
           - $(100)+c(5\sqrt{2} \times \sqrt{2})R45°$

----30--Zn----
30.1   W. N. Unertl, H. V. Thapliyal: J. Vac. Sci. Technol. **12**, 263 (1975)
           - (0001)

----30-8--Zn-O----
30.8.1   C. B. Duke, A. R. Lubinsky: Surf. Sci. **50**, 605 (1975)
           - (0001)(Zn termination),(000$\bar{1}$)(O termination)
30.8.2   A. R. Lubinsky, C. B. Duke, S. C. Chang, B. W. Lee, P. Mark:
           J. Vac. Sci. Technol. **13**, 189 (1976) - (10$\bar{1}$0),(11$\bar{2}$0)
See 31.33.2 - (10$\bar{1}$0),(11$\bar{2}$0)
30.8.3   C. B. Duke, A. R. Lubinsky, S. C. Chang, B. W. Lee, P. Mark:
           Phys. Rev. B **15**, 4865 (1977) - (10$\bar{1}$0)

See 31.33.16 - (10$\bar{1}$0)

----30-16--Zn-S----
30.16.1   C. B. Duke, R. J. Meyer, A. Paton, A. Kahn, J. Carelli, J. L. Yeh:
           J. Vacuum Sci. Technol. **18**, 866 (1981) - (110)
See 31.15.4 - (110)

----30-34--Zn-Se----
30.34.1   C. B. Duke, A. R. Lubinsky, M. Bonn, G. Cisneros, P. Mark:
           J. Vac. Sci. Technol. **14**, 294 (1977) - (110)
See 31.33.3 - (110)
See 31.33.5 - (110)
See 31.33.10 - (110)
See 31.33.12 - (110)
See 13.15.3 - (110)

----30-52--Zn-Te----
See 31.33.16 - (110)
30.52.1   R. J. Meyer, C. B. Duke, A. Paton, E. So, J. L. Yeh, A. Kahn,
           P. Mark: Phys. Rev. B **22**, 2875 (1980) - (110)

30.52.2   C. B. Duke, A. Paton, A. Kahn: J. Vac. Sci. Technol. A **1**, 672
        (1983) - (110)

----31-15--Ga-P----
31.15.1   C. B. Duke, A. Paton, J. Carelli, A. Kahn: Bull. Am. Phys. Soc. **26**,
        352 (1981) - (110)
31.15.2   C. B. Duke, A. Paton, W. K. Ford, A. Kahn, J. Carelli: Phys. Rev. B
        24, 562 (1981) - (110)
31.15.3   B. W. Lee, R. K. Ni, N. Masud, X. R. Wang, D. C. Wang, M. Rowe:
        J. Vac. Sci. Technol. **19**, 294 (1981) - (110)
31.15.4   C. B. Duke, A. Paton, A. Kahn: J. Vac. Sci. Technol. A **2**, 515
        (1984) - (110)

----31-33--Ga-As----
31.33.1   A. R. Lubinsky, C. B. Duke, B. W. Lee, P. Mark: Phys. Rev. Lett.
        36, 1058 (1976) - (110)
31.33.2   C. B. Duke, A. R. Lubinsky, B. W. Lee, P. Mark: J. Vac. Sci.
        Technol. **13**, 761 (1976) - (110)
31.33.3   P. Mark, G. Cisneros, M. Bonn, A. Kahn, C. B. Duke, A. Paton,
        A. R. Lubinsky: J. Vac. Sci. Technol. **14**, 910 (1977) - (110)
31.33.4   B. J. Mrstik, M. A. Van Hove, S. Y. Tong: Bull. Am. Phys. Soc. **22**,
        356 (1977) - (110)
31.33.5   P. Mark: in Proc 7th Int. Vac. Cong. & 3rd Int. Conf. Sol. Surf.,
        Vienna (1977), ed. and publ. by R. Dobrozemsky,
        F. Rüdenauer, F. P. Viehböck, A. Breth
        (Vienna, 1977) p. 533 - (110)
31.33.6   S. Y. Tong, A. R. Lubinsky, B. J. Mrstik, M. A. Van Hove:
        Phys. Rev. B **17**, 3303 (1978) - (110)
31.33.7   A. Kahn, G. Cisneros, M. Bonn, P. Mark, C. B. Duke, Surf. Sci. **71**,
        387 (1978) - (110)
31.33.8   S. Y. Tong, M. A. Van Hove, B. J. Mrstik: in Proc 7th Int. Vac.
        Cong. & 3rd Int. Conf. Sol. Surf., Vienna (1977), ed. and publ.
        by R. Dobrozemsky, F. Rüdenauer, F. P. Viehböck, A. Breth
        (Vienna, 1977) p. 2407 - (110) + As, (100)
31.33.9   B. J. Mrstik, M. A. Van Hove, S. Y. Tong: Bull. Am. Phys. Soc. **23**,
        390 (1978) - (100)
31.33.10   D. J. Chadi: Phys. Rev. Lett. **41**, 1062 (1978) - (110)
        (Model Calculation)
31.33.11   A. Kahn, E. So, P. Mark, C. B. Duke: J. Vac. Sci. Technol. **15**, 580
        (1978) - (110)
31.33.13   R. J. Meyer, C. B. Duke, A. Paton, A. Kahn, E. So, J. L. Yeh,
        P. Mark: Phys. Rev. B **19**, 5194 (1979) - (110)
31.33.14   C. B. Duke, R. J. Meyer, A. Paton, P. Mark, A. Paton, E. So,
        J. L. Yeh: J. Vac. Sci. Technol. **16**, 1253 (1979) - (110)
31.33.15   R. J. Meyer, C. B. Duke, A. Paton: Surf. Sci. **97**, 512 (1980) - (110)

31.33.16   C. B. Duke, R. J. Meyer, P. Mark: J. Vac. Sci. Technol. **17**, 971
           (1980) - (110)
31.33.17   B. J. Mrstik, S. Y. Tong, M. A. Van Hove: J. Vac. Sci. Technol. **16**,
           1258 (1979) - (110)
31.33.18   C. A. Swarts, T. C. McGill, W. A. Goddard: Surf. Sci. **110**, 400
           (1981) - (110) (Model Calculation)
31.33.19   N. Masud: J. Phys. C **15**, 3209 (1982) - (110)
31.33.20   C. B. Duke, S. C. Richardson, A. Paton, A. Kahn: Surf. Sci. **127**,
           L135 (1983) - (110)
31.33.21   G. Xu, Y. Huang, S. R. Wang, W. N. Mei, S. Y. Tong, B. W. Lee:
           Bull. Am. Phys. Soc. **28**, 458 (1983) - (111)(2x2)
See 14.59 - (110)(1x1)(Model Calculation)
31.33.22   H. J. Gossmann, W. M. Gibson, T. Itoh, L. C. Feldman: J. Vac. Sci.
           Technol. A **1**, 1059 (1983) - (110)(HEIS)
31.33.23   S. Y. Tong, G. Xu, W. N. Mei: J. Vac. Sci. Technol. A **2**, 863
           (1984) - (111)(2x2)
See 13.15.3 - (110)

----31-33-13--Ga-As-Al----
31.33.13.1 C. B. Duke, A. Paton, R. J. Meyer, L. J. Brillson, A. Kahn,
           D. Kanani, J. Carelli, J. L. Yeh, G. Margaritondo, A. D. Katnani:
           Phys. Rev. Lett. **46**, 440 (1981) - (110)+(1x1)
31.33.13.2 J. R. Chelikowsky, D. J. Chadi, M. L. Cohen: Phys. Rev. B **23**,
           4013 (1981) (110)+(Model Calculation)
31.33.13.3 A. Kahn, D. Kanani, J. Carelli, J. L. Yeh, C. B. Duke, R. J. Meyer,
           A. Paton: L. Brillson: J. Vac. Sci. Technol. **18**, 792 (1981)
           - (110)+(1x1)
31.33.13.4 A. Kahn, J. Carelli, D. Kanani, C. B. Duke, A. Paton, L. Brillson:
           J. Vac. Sci. Technol. **19**, 331 (1981) - (110)+(1x1),0-8.5 Monolayers
31.33.13.5 J. Ihm, J. D. Joannopoulos: J. Vac. Sci. Technol. **21**, 340 (1982)
           - (110)+(1x1)(Model Calculation)
31.33.13.6 A. Kahn, J. Carelli, D. L. Miller, S. P. Kowalczyk: J. Vac. Sci.
           Technol. **21**, 380 (1982) - (110)
31.33.13.7 J. Ihm, J. D. Joannopoulos: Phys. Rev. B **26**, 4429 (1982)
           - (110)+ (Model Calculation)
31.33.13.8 S. L. Richardson, C. B. Duke, A. Paton, A. Kahn: Bull. Am. Phys.
           Soc. **28**, 458 (1983) - (110)+

----31-33-51--Ga-As-Sb----
31.33.51.1 A. Kahn, J. Carelli, C. B. Duke, A. Paton, W. K. Ford: J. Vac. Sci.
           Technol. **20**, 775 (1982) - (110)+(1x1)
31.33.51.2 C. B. Duke, A. Paton, W. K. Ford, A. Kahn, J. Carelli:
           Phys. Rev. B **26**, 803 (1982) - (110)+(1x1)

----31-51--Ga-Sb----
31.51.1   R. S. Williams, T. M. Buck, G. H. Wheatley: Bull. Am. Phys. Soc.
            26, 352 (1981) - (100)(LEIS)
See 30.52.2 - (110)
31.51.2   C. B. Duke, A. Paton, A. Kahn: Phys. Rev. B **27**, 3436 (1983)
            - (110)

----32--Ge----
32.1   P. Eisenberger, W. L. Marra: Phys. Rev. Lett. **46**, 1081 (1981)
            - (100)(2x1)(X-ray Diffraction)
32.2   J. C. Fernandez, W. S. Yang, H. D. Shih, F. Jona, D. W. Jepsen,
            P. M. Marcus: J. Phys. C **14**, L55 (1981) - (100)(2x1)
32.3   Y. Kuk, L. C. Feldman, P. J. Silverman: Bull. Am. Phys. Soc. **27**,
            270 (1982) - (100) c(4x2) (HEIS)

----32-14--Ge-Si----
See 14.32.1 - (111)+ (Model Calculation)

----32-17--Ge-Cl----
See 14.53.1 - (111)+(2x8) (SEXAFS)
See 14.17.1 - (111)+ (Model Calculation)
See 14.17.3 - (111)+ (Model Calculation)

----32-52--Ge-Te----
See 14.53.1 - (111)+(2x2) (SEXAFS)
See 14.53.2 - (111)+(2x2) (SEXAFS)

----32-53--Ge-I----
See 14.53.1 - (111)+ (SEXAFS)
See 14.53.2 - (111)+ (SEXAFS)

----40--Zr----
40.1   W. T. Moore, P. R. Watson, D. C. Frost, K. A. R. Mitchell:
            J. Phys. C **12**, L887 (1979) - (0001)

----41-34--Nb-Se----
See 42.16.2 - $NbSe_2$(0001)
See 42.16.3 - $NbSe_2$(0001)
See 42.16.4 - $NbSe_2$(0001)
See 42.16.5 - $NbSe_2$(0001)

----42--Mo----
42.1   A. Ignatiev, F. Jona, H. D. Shih, D. W. Jepsen, P. M. Marcus:
            Phys. Rev. B **11**, 4787 (1975) - (100)

42.2   L. J. Clarke: in Proc 7th Int. Vac. Cong. & 3rd Int. Conf. Sol. Surf.,
        Vienna (1977), ed. and publ. by R. Dobrozemsky, F. Rüdenauer,
        F. P. Viehböck, A. Breth (Vienna, 1977) A2725 - (100)
42.3   T. E. Felter, R. A. Barker, P. J. Estrup: Phys. Rev. Lett. **38**, 1138
        (1977) - (100)
42.4   L. J. Clarke: Surf. Sci. **91**, 131 (1980) - (100)
42.5   C. Noguera, D. Spanjaard, D. Jepsen, Y. Ballu, C. Guillot, J. Lecante,
        J. Paigne, Y. Petroff, R. Pinchaux, P. Thiry, R. Cinti:
        Phys. Rev. Lett. **38**. 1171 (1977) - (100)
43.6   G. Gafner: Surf. Sci. **19**, 9 (1970) - (110)
42.7   L. Morales de la Garza, L. J. Clarke: J. Phys. C **14**, 5391 (1981)
        - (110)

----42-7--Mo-N----
42.7.1 A. Ignatiev, F. Jona, D. W. Jepsen, P. M. Marcus: Surf. Sci. **49**, 189
        (1975) - (100)+c(2x2)

----42-8--Mo-O----
42.8.1 L. J. Clarke: in Proc 7th Int. Vac. Cong. & 3rd Int. Conf. Sol. Surf.,
        Vienna (1977), ed. and publ. by R. Dobrozemsky, F. Rüdenauer,
        F. P. Viehböck, A. Breth (Vienna, 1977) A2725 - (100)+p(2x1)

----42-14--Mo-Si----
42.14.1 A. Ignatiev, F. Jona, D. W. Jepsen, P. M. Marcus: Phys. Rev. B **11**,
        4780 (1975) - (100)+(1x1)

----42-16--Mo-S----
42.16.1 B. J. Mrstik, S. Y. Tong, R. Kaplan, A. K. Ganguly: Solid State
        Commun. **17**, 755 (1975) - $MoS_2(0001)$
42.16.2 S. Y. Tong, M. Van Hove, B. J. Mrstik, R. Kaplan, T. Reinecke:
        J. Vac. Sci. Technol. **13**, 188 (1976) - $MoS_2(0001)$
42.16.3 M. A. Van Hove, S. Y. Tong, M. H. Elconin: Surf. Sci. **64**, 85
        (1977) - $MoS_2(0001)$
42.16.4 B. J. Mrstik, R. Kaplan, T. L. Reinecke, M. A. Van Hove,
        S. Y. Tong: Phys. Rev. B **15**, 897 (1977) - $MoS_2(0001)$
42.16.5 B. J. Mrstik, R. Kaplan, T. L. Reinecke, M. Van Hove, S. Y. Tong:
        Nuovo Cimento **38B**, 387 (1977) - $MoS_2(0001)$
42.16.6 L. J. Clarke: Surf. Sci. **102**, 331 (1981) - (100)+c(2x2)

----44--Ru----
44.1   G. Michalk, W. Moritz, H. Pfnür, D. Menzel: Surf. Sci. **129**, 92
        (1983) - (0001)

----44-6-8--Ru-C-O----
See 44.1 - $(0001)+(\sqrt{3} \times \sqrt{3})R30°$

----45--Rh----

45.1   D. C. Frost, K. A. R. Mitchell, F. R. Shepherd: in Proc 7th Int. Vac.
       Cong. & 3rd Int. Conf. Sol. Surf., Vienna (1977), ed. and publ.
       by R. Dobrozemsky, F. Rüdenauer, F. P. Viehböck,
       A. Breth (Vienna, 1977) A2725 - (100),(111)
45.2   K. A. R. Mitchell, F. R. Shepherd, P. R. Watson, D. C. Frost:
       Surf. Sci. **64**, 737 (1977) - (100)
45.3   P. R. Watson, F. R. Shepherd, D. C. Frost, K. A. R. Mitchell:
       Surf. Sci. **72**, 562 (1978) - (100)
45.4   F. R. Shepherd, P. R. Watson, D. C. Frost, K. A. R. Mitchell:
       J. Phys. C **11**, 4591 (1978) - (111)
45.5   D. C. Frost, S. Hengrasmee, K. A. R. Mitchell, F. R. Shepherd,
       P. R. Watson: Surf. Sci. **76**, L585 (1978) - (110)
45.6   C.-M. Chan, P. A. Thiel, J. T. Yates, W. H. Weinberg: Surf. Sci. **76**,
       296 (1978) - (111)
45.7   S. Hengrasmee, K. A. R. Mitchell, P. R. Watson, S. J. White:
       Can. J. Phys. **58**, 200 (1980) - (100),(111),(110)
45.8   M. A. Van Hove, R. J. Koestner: in *Determination of Surface
       Structure by LEED*, ed. by P. M. Marcus, F. Jona (Plenum, New
       York, London, 1984) 357 - (111)

----45-6-1--Rh-C-H----

45.6.1.1 M. A. Van Hove, R. J. Koestner, G. A. Somorjai: Bull. Am. Phys.
          Soc. **26**, 225 (1981) - (111)+(2x2)C_2H_3
45.6.1.2 M. A. Van Hove, R. J. Koestner, G. A. Somorjai: J. Vac. Sci.
          Technol. **20**, 886 (1982) - (111)+(2x2)C_2H_3,
          (111)+($2\sqrt{3} \times 2\sqrt{3}$)R30° C_3H_5
45.6.1.3 M. A. Van Hove, R. J. Koestner, G. A. Somorjai: Surf. Sci. **121**, 321
          (1982) - (111)+(2x2)C_2H_3
45.6.1.4 A. Gavezzotti, M. Simonetta, M. A. Van Hove, G. A. Somorjai:
          Surf. Sci. **122**, 292 (1982) - (111)+(2x2), ($2\sqrt{3}\times2\sqrt{3}$)R30° C_3H_5,
          C_4H_7 (Model Calculation)
45.6.1.5 M. A. Van Hove, R. F. Lin, R. J. Koestner, G. A. Somorjai:
          Bull. Am. Phys. Soc. **28**, 538 (1983) - (111)+$\left(\begin{smallmatrix}3&1\\1&3\end{smallmatrix}\right)C_6H_6$
45.6.1.6 M. A. Van Hove, R. F. Lin, R. J. Koestner, B. E. Koel,
          J. E. Crowell, M. Mate, G. A. Somorjai: Vac. **33**, 860 (1983)
          - (111)+$\left(\begin{smallmatrix}3&1\\1&3\end{smallmatrix}\right)C_6H_6$
45.6.1.7 M. A. Van Hove, R. F. Lin, G. A. Somorjai: Phys. Rev. Lett. **51**,
          778 (1983) - (111) + $\left(\begin{smallmatrix}3&1\\1&3\end{smallmatrix}\right)C_6H_6$

----45-6-8--Rh-C-O----

45.6.8.1 R. J. Koestner, M. A. Van Hove, G. A. Somorjai: Surf. Sci. **107**,
          439 (1981) - (111) + ($\sqrt{3} \times \sqrt{3}$)R30°CO

45.6.8.2   M. A. Van Hove, R. J. Koestner, G. A. Somorjai: Bull. Am. Phys. Soc. **27**, 238 (1982) - (111)+(2x2)3CO

45.6.8.3   M. A. Van Hove, R. J. Koestner, G. A. Somorjai: Phys. Rev. Lett. **50**, 903 (1983) - (111)+(2x2)3CO

45.6.8.4   M. A. Van Hove, R. J. Koestner, G. A. Somorjai: Surf. Sci. **129**, 482 (1983) - (111) + (2x2)3CO

----45-16--Rh-S----

45.16.1   S. Hengrasmee, P. R. Watson, D. C. Frost, K. A. R. Mitchell: Surf. Sci. **87**, L249 (1979) - (100)+(2x2)

45.16.2   S. Hengrasmee, P. R. Watson, D. C. Frost, K. A. R. Mitchell: Surf. Sci. **92**, 71 (1980) - (110)+c(2x2)

----46--Pd----

46.1   R. J. Behm, K. Christmann, G. Ertl, M. A. Van Hove, P. A. Thiel, W. H. Weinberg: Surf. Sci. **88**, L59 (1979) - (100)

46.2   R. J. Behm, K. Christmann, G. Ertl, M. A. Van Hove: J. Chem. Phys. **73**, 2984 (1980) - (100)

46.3   Y. Kuk, L. C. Feldman, P. J. Silverman: Phys. Rev. Lett. **50**, 511 (1983) - (111)(HEIS)

----46-1--Pd-H----

46.1.1   K. H. Rieder, M. Baumberger and W. Stocker, Phys. Rev. Letters **51**, 1799 (1983) - (110)+(2x1)/(1x2) (Atom Diffraction)

----46-6-1--Pd-C-H----

46.6.1   L. L. Kesmodel, J. A. Gates: Bull. Am. Phys. Soc. **26**, 425 (1981) - (111) + (2x2)C_2H_3 (HREELS)

46.6.2   J. A. Gates, L. L. Kesmodel: Surf. Sci. **120**, L461 (1982) - (111)+C_2H_4

----46-6-8--Pd-C-O----

See 46.1 - (100)+($2\sqrt{2} \times \sqrt{2}$)R45°CO
See 46.2 - (100)+($2\sqrt{2} \times \sqrt{2}$)R45°CO

----46-16--Pd-S----

46.16.1   W. Berndt, R. Hora, M. Scheffler: Surf. Sci. **117**, 188 (1982) - (100)+c(2x2)

----46-79--Pd-Au----

See 46.3 - (111)+ (HEIS)

----47--Ag----
See 13.4 - (100)
47.1   D. W. Jepsen, P. M. Marcus, F. Jona: Phys. Rev. B 8, 5523 (1973)
        - (100)
47.2   E. Zanazzi, F. Jona, D. W. Jepsen, P. M. Marcus: J. Phys. C 10, 375
        (1977) - (110)
47.3   W. Moritz: PhD Thesis, University of Munich (1976) - (100),(110)
47.4   F. Forstmann: Jpn. J. Appl. Phys., Suppl. 2 Part 2, 657 (1974) - (111)
See 28.7 - (111)
47.5   N. Stoner: PhD Thesis, University of Wisconsin, Milwaukee (1976)
        - (111)
47.6   F. Soria, J. L. Sacedon, P. M. Echenique, D. Titterington: Surf. Sci.
        68, 448 (1977) - (111)
47.7   M. Maglietta, E. Zanazzi, F. Jona, D. W. Jepsen, P. M. Marcus: 3287
        (1977) - (110)
47.8   M. Alff: PhD Thesis, University of Munich (1976) - (110)
See 28.11 - (110)
47.9   M. Alff, W. Moritz: to be published - (110)
47.10   J. R. Noonan, H. L. Davis: Bull. Am. Phys. Soc. 25, 327 (1980)
        - (110)
See 29.15 - (110)
See 13.20 - (110)
47.11   J. R. Noonan, H. L. Davis: Bull. Am. Phys. Soc. 26, 224 (1981)
        - (110)
47.12   R. J. Culbertson, L. C. Feldman, P. J. Silverman, H. Boehm: Bull.
        Am. Phys. Soc. 26, 226 (1981); Phys. Rev. Lett. 47,
        657 (1981)- (111) (HEIS)
47.13   S. P. Withrow, R. J. Culbertson, J. H. Barrett: Bull. Am. Phys. Soc.
        28, 572 (1983) - (110) (HEIS)
47.14   Y. Kuk, L. C. Feldman: Bull. Am. Phys. Soc. 29, 390 (1984)
        - (110) (HEIS)

----47-8--Ag-O----
47.8.1 E. Zanazzi, M. Maglietta, U. Bardi, F. Jona, D. W. Jepsen,
        P. M. Marcus: in Proc 7th Int. Vac. Cong. & 3rd Int. Conf. Sol.
        Surf., Vienna (1977), ed. and publ. by R. Dobrozemsky,
        F. Rüdenauer, F. P. Viehböck, A. Breth (Vienna, 1977) 2447
        - (110)+(1x2)
47.8.2 W. Heiland, F. Iberl, E. Taglauer, D. Menzel: Surf. Sci. 53, 383
        (1975) - (110)+(LEIS)
47.8.3 E. Zanazzi, M. Maglietta, U. Bardi, F. Jona, P. M. Marcus: J. Vac.
        Sci. Technol. A 1, 7 (1983) - (110)+(1x2)

----47-17--Ag-Cl----
47.17.1   E. Zanazzi, F. Jona, D. W. Jepsen, P. M. Marcus: Phys. Rev. B 14,
        432 (1976) - (100) + c(2x2)

47.17.2   H. S. Greenside, D. R. Hamann: Phys. Rev. B **23**, 4879 (1981)
    - (100)+c(2x2) (Model Calculation)
47.17.3   M. J. Cardillo, G. E. Becker, D. R. Hamann, L. Mattheiss,
    J. A. Serri, L. Whitman: Bull. Am. Phys. Soc. **27**, 238 (1982)
    - (100)+c(2x2) (Atom Diffraction)
See 29.17.6 - (100)+c(2x2)
See 29.17.7 - (100)+c(2x2)
47.17.4   M. J. Cardillo, G. E. Becker, D. R. Hamann, J. A. Serri,
    L. Whitman, L. F. Mattheiss: Phys. Rev. B **28**, 494 (1983)
    - (100)+c(2x2) (Atom Diffraction)

----47-34--Ag-Se----
47.34.1   A. Ignatiev, F. Jona, D. W. Jepsen, P. M. Marcus: Surf. Sci. **40**, 439
    (1973) - (100)+c(2x2)

----47-53--Ag-I----
47.53.1   F. Forstmann, W. Berndt, P. Buttner: Phys. Rev. Lett. **30**, 17 (1973)
    - (111) + ($\sqrt{3} \times \sqrt{3}$)R30°
47.53.2   J. D. Head, K. A. R. Mitchell, L. Noodleman: Surf. Sci. **61**, 661
    (1977) - (111)+(Model Calculation)
See 47.4   - (111)+($\sqrt{3} \times \sqrt{3}$)R30°
47.53.3   P. H. Citrin, P. Eisenberger, R. C. Hewitt: Surf. Sci. **89**, 28 (1979)
    - (111) + ($\sqrt{3} \times \sqrt{3}$)R30° (SEXAFS)
47.53.4   M. Maglietta, E. Zanazzi, U. Bardi, D. Sondericker, F. Jona:
    Surf. Sci. **123**, 141 (1982) - (111) + ($\sqrt{3} \times \sqrt{3}$)R30°

----47-54--Ag-Xe----
47.54.1   P. I. Cohen, J. Unguris, M. B. Webb: Surf. Sci. **58**, 429 (1976)
    - (111) + incommensurate
47.54.2   N. Stoner, M. A. Van Hove, S. Y. Tong, M. B. Webb: Phys. Rev.
    Lett. **40**, 243 (1978)
    - (111) + incommensurate
47.54.3   M. B. Webb, P. I. Cohen: Crit. Rev. Solid State Sci. **6**, 253 (1976)
See 47.5   - (111)+incommensurate

----47-79--Ag-Au----
See 47.6   - (111)+(1x1)
See 47.12   - (111)+(1x1) (HEIS)
47.79.1   L. C. Feldman, R. J. Culbertson, P. J. Silverman: J. Vac. Sci.
    Technol. **20**, 368 1982) - (111)+(1x1) (HEIS)
47.79.2   Y. Kuk, L. C. Feldman, P. J. Silverman: Bull. Am. Phys. Soc.
    28, 260 (1983) - (110)+(1x1)

----48--Cd----
See 22.48.1 - (0001)

----48-52--Cd-Te----
48.52.1   C. B. Duke, A. Paton, W. K. Ford, A. Kahn, G. Scott: Phys. Rev. B
           24, 3310 (1981) - (110)
48.52.2   C. B. Duke, A. Paton, W. K. Ford, A. Kahn and G. Scott: J. Vac.
           Sci. Technol. **20**, 778 (1982) - (110)
See 13.15.3 - (110)

----49-15--In-P----
49.15.1   C. B. Duke, R. J. Meyer, A. Paton, J. C. Tsang, J. L. Yeh, A. Kahn:
           Bull. Am. Phys. Soc. **25**, 328 (1980) - (110)
See 31.33.16 - (110)
49.15.2   R. J. Meyer, C. B. Duke, A. Paton, J. C. Tsang, J. L. Yeh, A. Kahn,
           P. Mark: Phys. Rev. B **22**, 6171 (1980) - (110)
49.15.3   S. P. Tear, M. R. Welton-Cook, M. Prutton, J. A. Walker: Surf. Sci.
           99, 598 (1980) - (110)

----49-33--In-As----
49.33.1   C. B. Duke, A. Paton, A. Kahn, C. R. Bonapace: Phys. Rev. B **27**,
           6189 (1983) - (110)

----49-51--In-Sb----
See 31.33.16 - (110)
49.51.1   R. J. Meyer, C. B. Duke, A. Paton, J. L. Yeh, J. C. Tsang, A. Kahn,
           P. Mark: Phys. Rev. B **21**, 4740 (1980) - (110)
See 13.15.3 - (110)

----51-52-34--Sb-Te-Se----
51.52.34.1 R. L. Benbow, M. R. Thuler, Z. Hurych, K. H. Lau, S. Y. Tong:
           Phys. Rev. B **28**, 4160 (1983) - Sb_2Te_2Se basal plane (ARUPS)

----52--Te----
52.1   R. J. Meyer, W. R. Salanek, C. B. Duke, A. Paton, C. H. Griffiths,
           L. Kovnat, L. E. Meyer: Phys. Rev. B **21**, 4542 (1980) - (10$\bar{1}$0)

----54--Xe----
54.1   A. Ignatiev, J. B. Pendry, T. N. Rhodin: Phys. Rev. Lett. **26**, 189
           (1971) - (100)
54.2   A. Ignatiev, T. N. Rhodin, S. Y. Tong, B. I. Lundqvist, J. B. Pendry:
           Solid State Commun. **9**, 1851 (1971) - (111)

----73--Ta----
73.1   A. Titov, W. Moritz: Surf. Sci. **123**, L709 (1982) - (100)

----74--W----
74.1   R. Feder: Phys. Rev. Lett. **36**, 598 (1976) - (100)

74.2   M. A. Van Hove, S. Y. Tong: Surf. Sci. **54**, 91 (1976) - (100),(110)

74.3   R. Feder: Phys. Stat. Sol. (b) **62**, 135 (1974) -(100),(110)

74.4   B. W. Lee, A. Ignatiev, S. Y. Tong, M. Van Hove: J. Vac. Sci.
        Technol. **14**, 291 (1977) - (100)

See 74.8.2 - (110)

74.5   R. W. Strayer, W. Mackie, L. W. Swanson: Surf. Sci. **34**, 225 (1973)
        - (110),(100),(111)(Work Function)

74.6   R. Feder: Surf. Sci. **63**, 283 (1977) - (100)

74.7   M. K. Debe, D. A. King: Phys. Rev. Lett. **39**, 708 (1977) - (100)

74.8   L. C. Feldman, R. L. Kauffman, P. J. Silverman, R. A. Zuhr,
        J. H. Barrett: Phys. Rev. Lett. **39**, 38 (1977) - (100) (HEIS)

74.9   J. Kirschner, R. Feder: Surf. Sci. **79**, 176 (1979) - (100)

74.10   M. K. Debe, D. A. King, F. S. Marsh: Surf. Sci. **68**, 437 (1977)
        - (100)

74.11   M. K. Debe, D. A. King: J. Phys. C **10**, L303 (1977) - (100)

74.12   T. E. Felter, R. A. Barker, P. J. Estrup: Phys. Rev. Lett. **38**, 1138
        (1977) - (100)c(2x2)

74.13   M. Kalisvaart, T. W. Riddle, F. B. Dunning, G. K. Walter:
        paper presented at 37th Phys. Elect. Conf. Stanford (1977) - (100)

74.14   R. A. Barker, P. J. Estrup, F. Jona, P. M. Marcus: Solid State
        Commun. **25**, 375 (1978) - (100)c(2x2)

74.15   J. Kirschner, R. Feder: Verh. Deutsch. Phys. Ges. **2**, 557 (1978)
        - (100)

74.16   I. Stensgaard, L. C. Feldman, P. J. Silverman: J. Vac. Sci. Technol.
        16, 492 (1979) - (100)c(2x2)

74.17   M. N. Read, G. J. Russell: Surf. Sci. **88**, 95 (1979) - (100)

74.18   P. Heilmann, K. Heinz, K. Müller: Surf. Sci. **89**, 84 (1979) - (100)

74.19   G.-C. Wang, R. J. Celotta, D. T. Pierce: Bull. Am. Phys. Soc. **25**,
        328 (1980) - (100)

74.20   L. J. Clarke, L. Morales de la Garza: Surf. Sci. **99**, 419 (1980)
        - (100)(1x1)

74.21   F. S. Marsh, M. K. Debe, D. A. King: J. Phys. C **13**, 2799 (1980)
        - (100)(1x1)

74.22   R. Feder, J. Kirschner: Surf. Sci. **103**, 75 (1981) - (100)(1x1)

74.23   M. A. Stevens, G. J. Russell: Surf. Sci. **104**, 354 (1981) - (100)(1x1)

74.24   J. A. Walker, M. K. Debe, D. A. King: Surf. Sci. **104**, 405 (1981)
        - (100)c(2x2)

74.25   G.-C. Wang, R. J. Celotta, D. T. Pierce: Surf. Sci. **119**, 479 (1982)
        - (100)(1x1)(CMTA PLEED)

74.26   A. J. Melmed, W. R. Graham: Appl. Surf. Sci. **11/12**, 470 (1982)
        - (100) (Review)

74.27   M. Posternak, H. Krakauer, A. J. Freeman: Phys. Rev. B **25**, 755
        (1982) - (100)(Model Calculation)

74.28   M. K. Debe, D. A. King: J. Phys. C **15**, 2257 (1982) - (100)(1x1)

74.29   D. P. Woodruff: Surf. Sci. **122**, L653 (1982) - (100) c(2x2)
74.30   P. J. Estrup, L. D. Roelofs, S. C. Ying: Surf. Sci. **123**, L703 (1982)
         - (100) (Discussion)
74.31   A. J. Melmed, W. R. Graham: Surf. Sci. **123**, L706 (1982)
         - (100) (Discussion)
74.32   J. P. Bourdin, G. Treglia, M. C. Desjonqueres, J. P. Ganachaud,
         D. Spanjaard: Solid State Commun. **47**, 279 (1983)
         - (110)(Model Calculation)
74.33   H. L. Davis, G.-C. Wang: Bull. Am. Phys. Soc. **29**, 221 (1984) - (211)

----74-1--W-H----
74.1.1   D. W. Bullett, M. L. Cohen: J. Phys. C **10**, 2083 (1977) - (100)+(1x1)
         (Model Calculation)
74.1.2   A. Ignatiev, B. W. Lee, M. A. Van Hove: Bull. Am. Phys. Soc. **23**,
         391 (1978) -(100)+(1x1)
74.1.3   R. F. Willis: Surf. Sci. **89**, 457 (1979) - (100)+c(2x2)H,(1x1)2H
         (HREELS)
74.1.4   L. C. Feldman: Bull. Am. Phys. Soc. **25**, 403 (1980)
         - (100)+(1x1) (HEIS)
74.1.5   G. B. Blanchet, J. DiNardo, F. Greuter, E. W. Plummer: Bull. Am.
         Phys. Soc. **26**, 424 (1981) - (110)+(2x1) (ELS+UPS)
74.1.6   W. Ho, R. F. Willis, E. W. Plummer: Phys. Rev. B **21**, 4202 (1980)
         - (100)+(1x1) (HREELS)
74.1.7   M. A. Passler, A. Ignatiev, B. W. Lee, D. Adams, M. A. Van Hove:
         in *Determination of Surface Structure by LEED*, ed. by
         P. M. Marcus, F. Jona (Plenum, New York, London, 1984) 233
         - (100)+(1x1)
74.1.8   M. A. Passler, A. Ignatiev: Bull. Am. Phys. Soc. **27**, 237 (1982)
         - (100)+(1x1)2H
74.1.9   G.-C. Wang, J. Unguris, D. T. Pierce, R. J. Celotta: Surf. Sci. **114**,
         L35 (1982) - (100)+c(2x2)H, +(1x1)2H (PLEED)
See 28.1.12 - (110)+(100)+ (Model Calculation)

----74-6-8--W-C-O----
See 74.7.2 - (100)+c(2x2)CO

----74-7--W-N----
74.7.1   K. Griffiths, C. Kendon, D. A. King, J. B. Pendry: Phys. Rev. Lett.
         46, 1584 (1981) - (100)+c(2x2) Islands
74.7.2   K. Griffiths, D. A. King, G. C. Aers, J. B. Pendry: J. Phys. C **15**, 4921
         (1982) - (100)+c(2x2)

----74-8--W-O----
74.8.1   M. A. Van Hove, S. Y. Tong: Phys. Rev. Lett. **35**, 1092 (1975)
         - (110)+(2x1)

74.8.2   M. G. Lagally, J. C. Buchholz, G.-C. Wang: J. Vac. Sci. Technol. **12**, 213 (1975) - (110)+(2x1)
See 42.16.3 - (110)+(2x1)
74.8.3   H. Niehus, E. Bauer: Surf. Sci. **47**, 222 (1975) - (110)+(LEIS)
74.8.4   H. Niehus, S. Prigge, E. Bauer: Verh. Deutsch. Phys. Ges. **9**, 1008 (1976) - (100)+ (LEIS)
74.8.5   S. Prigge, H. Niehus, E. Bauer: Surf. Sci. **65**, 141 (1977) - (100)+ (LEIS)
74.8.6   E. Bauer, H. Poppa, Y. Viswanath: Surf. Sci. **58**, 517 (1976) - (100)
74.8.7   R. J. Smith, S. Fine, N. Holland, M. W. Kim: Bull. Am. Phys. Soc. **28**, 470 (1983) - (110)+(2x1) (HEIS)

----75--Re----
75.1   H. L. Davis, D. M. Zehner: Bull. Am. Phys. Soc. **24**, 468 (1979) - (0001)
75.2   H. L. Davis, D. M. Zehner: J. Vac. Sci. Technol. **17**, 190 (1980) - (0001)

----77--Ir----
77.1   C.-M. Chan, S. L. Cunningham, M. A. Van Hove, W. H. Weinberg: Bull. Am. Phys. Soc. **22**, 357 (1977) - (111)
77.2   C.-M. Chan, S. L. Cunningham, M. A. Van Hove, W. H. Weinberg, S. P. Withrow: Surf. Sci. **66**, 394 (1977) - (111)
77.3   C.-M. Chan, S. L. Cunningham, K. L. Luke, W. H. Weinberg, S. P. Withrow: Surf. Sci. **78**, 15 (1978) - (110)(1x1)
77.4   C.-M. Chan, M. A. Van Hove, W. H. Weinberg, E. D. Williams: Surf. Sci. **91**, 440 (1980) - (110)(1x2)
77.5   M. A. Van Hove, R. J. Koestner, G. A. Somorjai: Bull. Am. Phys. Soc. **24**, 468 (1979) - (100)(1x5)
77.6   M. A. Van Hove, R. J. Koestner, P. C. Stair, J. P. Biberian, L. L. Kesmodel, I. Bartos, G. A. Somorjai: Surf. Sci. **103**, 189, 218 (1981) - (100)(1x5)
77.7   C.-M. Chan, M. A. Van Hove, W. H. Weinberg, E. D. Williams: J. Vac. Sci. Technol. 16, 642 (1979) - (110)(1x2)
77.8   C.-M. Chan, M. A. Van Hove, W. H. Weinberg, E. D. Williams: Solid State Commun. **30**, 47 (1979) - (110)(1x2)
77.9   E. Lang, W. Grimm, K. Heinz: Surf. Sci. **117**, 169 (1982) - (100)(1x1)
77.10   K. Heinz, G. Besold: Surf. Sci. **125**, 515 (1983) - (100)(1x1)
77.11   E. Lang, K. Müller, M. A. Van Hove, R. J. Koestner, G. A. Somorjai: Surf. Sci. **127**, 347 (1983) - (100)(1x5)

----77-8--Ir-O----
77.8.1   C.-M. Chan, K. L. Luke, M. A. Van Hove, W. H. Weinberg, S. P. Withrow: Surf. Sci. **78**, 386 (1978) - (110)+c(2x2)

77.8.2   C.-M. Chan, W. H. Weinberg: J. Chem. Phys. **71**, 2788 (1979)
- (111)+(2x2) or (2x1)

----77-16--Ir-S----

77.16.1   C.-M. Chan, W. H. Weinberg: J. Chem. Phys. **71**, 3988 (1979)
- (111)+($\sqrt{3} \times \sqrt{3}$)R30°

----78--Pt----

78.1   L. L. Kesmodel, G. A. Somorjai: Phys. Rev. B **11**, 630 (1975) - (111)

78.2   L. L. Kesmodel, P. C. Stair, G. A. Somorjai: Surf. Sci. **64**, 342 (1977)
- (111)

78.3   J. A. Davies, D. P. Jackson, J. B. Mitchell, P. R. Norton,
R. L. Tapping: Phys. Lett. **54A**, 239 (1975) - (111) (HEIS)

78.4   J. F. van der Veen, R. G. Smeenk, F. W. Saris: in Proc 7th Int. Vac.
Cong. & 3rd Int. Conf. Sol. Surf., Vienna (1977), ed. and publ.
by R. Dobrozemsky, F. Rudotdotdenauer, F. P. Viehböck,
A. Breth (Vienna, 1977) 2515 - (111)(MEIS)

78.5   E. Bogh, I. Stensgaard: in Proc 7th Int. Vac. Cong. & 3rd Int. Conf.
Sol. Surf., Vienna (1977), ed. and publ. by R. Dobrozemsky,
F. Rüdenauer, F. P. Viehböck, A. Breth
(Vienna, 1977) A2757 - (111)(HEIS)

78.6   R. Feder: Surf. Sci. **68**, 229 (1977) - (100)(1x1), (111)

78.7   J. A. Davies, D. P. Jackson, N. Matsunami, P. R. Norton,
J. U. Andersen: Surf. Sci. **78**, 274 (1978) - (111)(HEIS)

78.8   J. F. van der Veen, R. G. Smeenk, R. M. Tromp, F. W. Saris:
Surf. Sci. **79**, 219 (1979) - (111) (MEIS)

See 77.5   - (100)$\begin{bmatrix} 14 & 1 \\ -1 & 5 \end{bmatrix}$

78.9   P. Heilmann, K. Heinz, K. Muller: Surf. Sci. **83**, 487 (1979)
- (100)$\begin{bmatrix} 14 & 1 \\ -1 & 5 \end{bmatrix}$

78.10   P. R. Norton, J. A. Davies, D. P. Jackson, N. Matsunami: Surf. Sci.
85, 269 (1979) - (100)(1x1)

78.11   D. L. Adams, H. B. Nielsen, M. A. Van Hove: Bull. Am. Phys. Soc.
24, 468 (1979) - (111)

78.12   D. L. Adams, H. B. Nielsen, M. A. Van Hove: Phys. Rev. B **20**,
4789 (1979) - (111)

78.13   P. Bauer, R. Feder, N. Muller: Surf. Sci. **99**, L395 (1980) - (111)

78.14   D. L. Adams, H. B. Nielsen, M. A. Van Hove, A. Ignatiev: Surf. Sci.
104, 47 (1981) - (110)(2x1)

78.15   R. Feder, H. Pleyer, P. Bauer, N. Muller: Surf. Sci. **109**, 419 (1981)
- (111)

78.16   J. A. Davies, T. E. Jackman, D. P. Jackson, P. R. Norton: Surf. Sci.
109, 20 (1981) - (100)(1x1) (HEIS)

See 77.9   - (100)(1x1)

78.17   A. M. Lahee, W. Allison, R. F. Willis, K. H. Rieder: Surf. Sci. **126**,
654 (1983) - (110)(1x2) (Atom Diffraction)

78.18   D. Tomanek, H.-J. Brocksch, K. H. Bennemann: Surf. Sci. **138**, L129 (1984) - (110)(1x2) (Model Calculation)

----78-1--Pt-H----

78.1.1   J. A. Barker, I. P. Batra, N. Garcia, K. H. Rieder: Bull. Am. Phys. Soc. **28**, 278 (1983) - (111)+(1x1) (Atom Diffraction)

78.1.2   D. J. Auerbach, I. P. Batra, J. A. Barker: Bull. Am. Phys. Soc. **28**, 539 (1983) - (111) + (1x1) (Atom Diffraction)

78.1.3   I. P. Batra: Surf. Sci. **137**, L97 (1984) - (111)+(1x1)(Atom Diffraction)

78.1.4   I. P. Batra, J. A. Barker, D. J. Auerbach: J. Vac. Sci. Technol. A **2**, 943 (1984) - (111)+(1x1) (Atom Diffraction)

----78-6-1--Pt-C-H----

78.6.1.1   L. L. Kesmodel, P. C. Stair, R. C. Baetzold, G. A. Somorjai: Phys. Rev. Letters **36**, 1316 (1976) - (111)+(2x2)C_2H_2,C_2H_4

78.6.1.2   L. L. Kesmodel, R. C. Baetzold, G. A. Somorjai: Surf. Sci. **66**, 299 (1977) - (111)+(2x2)C_2H_2,C_2H_4

78.6.1.3   A. B. Anderson, A. T. Hubbard: Surf. Sci. **99**, 384 (1980) - (111)+C_2H_2 (Model Calculation)

78.6.1.4   A. Gavezotti, M. Simonetta: Surf. Sci. **99**, 453 (1980) - (111)+C_2H_2 (Model Calculation)

78.6.1.5   L. L. Kesmodel, L. H. Dubois, G. A. Somorjai: J. Chem. Phys. **70**, 2180 (1979) - (111)+(2x2)C_2H_3

78.6.1.6   M. R. Albert, L. G. Sneddon, W. Eberhardt, F. Greuter, T. Gustafsson, E. W. Plummer: Surf. Sci. **120**, 19 (1982) - (111)+C_2H_3(ARUPS)

78.6.1.7   R. J. Koestner, J. C. Frost, P. C. Stair, M. A. Van Hove, G. A. Somorjai: Surf. Sci. **116**, 85 (1982) - (111) + (2x2), (2√3×2√3)R30° C_3H_5, C_4H_7

See 45.6.1.4 - (111)+(2x2),(2√3×2√3)R30° C_3H_5, C_4H_7(Model Calculation)

----78-6-8--Pt-C-O----

78.6.8.1   R. Brooks, N. V. Richardson, D. A. King: Surf. Sci. **117**, 434 (1982) - (100)+c(4x2) (ARUPS)

78.6.8.2   P. Hoffman, S. R. Bare, N. V. Richardson, D. A. King: Solid State Commun. **42**, 645 (1982) - (111),(110)+(ARUPS)

----78-79--Pt-Au----

78.79.1   J. W. A. Sachtler, M. A. Van Hove, J. P. Biberian, G. A. Somorjai: Surf. Sci. **110**, 19 (1981) - (100)+

----79--Au----

See 47.6   - (111)

79.2   J. F. Wendelken, D. M. Zehner: Surf. Sci. **71**, 178 (1979) - (111)
79.3   D. Wolf, H. Jagodzinski, W. Moritz: Surf. Sci. **77**, 265, 283 (1978)
        - (110)(2x1)
See 78.6   - (100)(1x1)
79.4   R. Feder, N. Muller, D. Wolf: Zeitschr. f. Phys. B **28**, 265 (1977)
        - (110)(1x1)
79.5   W. Moritz, D. Wolf: Surf. Sci. **88**, L29 (1979) - (110)(2x1)
79.6   R. Feder, W. Moritz: Surf. Sci. **77**, 505 (1978) - (111)
79.7   J. R. Noonan, H. L. Davis: J. Vac. Sci. Technol. **16**, 587 (1979)
        - (110)
79.8   S. H. Overbury, W. Heiland, D. M. Zehner, S. Datz, R. S. Thoe:
        Surf. Sci. **109**, 239 (1981) - (110)(2x1)
See 77.9   - (100)(1x1)
79.9   M. Manninen, J. K. Norskov, C. Umrigar: Surf. Sci. **119**, L393
        (1982) - (110) (Atom Diffraction)
79.10   I. K. Robinson: Phys. Rev. Lett. **50**, 1145 (1983) - (110)(2x1)
        (X-ray Diffraction)
79.11   I. K. Robinson, W. Toy: Bull. Am. Phys. Soc. **28**, 572 (1983)
        - (110)(2x1) (X-ray Diffraction)
79.12   K. H. Rieder, T. Engel, R. H. Swendsen, M. Manninen: Surf. Sci.
        127, 223 (1983) - (100)(1x5) (Atom Diffraction)
79.13   L. D. Marks: Phys. Rev. Lett. **51**, 1000 (1983) - (110)(2x1)(EM)
79.14   G. Binnig, H. Rohrer: in Proc. 9th Int. Vac. Cong. & 5th Int. Conf.
        Sol. Surf., Madrid (1983), ed. by J. L. de Segovia (Assoc. Esp.
        del Vacio, Madrid, 1983) p. 77 - (110),(100)(STM)
79.15   G. Binnig, H. Rohrer, Ch. Gerber, E. Weibel: Surf. Sci. **131**, L379
        (1983) - (110)(2x1) (STM)
79.16   L. D. Marks, V. Heine, D. J. Smith: Phys. Rev. Lett. **52**, 656 (1984)
        - (111)(EM)
79.17   Y. Kuk, L. C. Feldman, I. K. Robinson: Surf. Sci. **138**, L168 (1984)
        - (110)(1x2) (HEIS)

----79-47--Au-Ag----
See 47.79.1 - (111)+(1x1) (HEIS)

----79-78-Au-Pt----
See 78.79.1 - (100)+

Table 12.1. A comprehensive list of results of surface crystallography for clean surfaces and atomic and molecular adsorption, classified by structural type of substrate surface (alphabetical order is used within classes, considering only the letters of the chemical species). First the cases of atomic adsorption on metal substrates are listed, in order of decreasing closepackedness of the metal surfaces. These are then followed by molecular adsorption systems. Finally, other materials are listed. For metal substrates, the second column indicates relative bond length changes between first and second layers of the clean surface or the substrate and the corresponding layer spacing changes [between square brackets], referred to the bulk values: expansions and contractions have positive and negative signs, respectively (values close to 0% are usually quoted as 0%, generally when no variation away from 0% was tried in LEED calculations). In the adsorption site description, the stacking sequence for fcc(111) and hcp (0001) is indicated by the familiar ABCABC... or ABABAB... notation, lower-case letters being used for adsorbates. Adsorption bond lengths are compared with bond lengths known from other sources for molecules and bulk compounds (to be found in the standard structure and crystallographic tables). Analytical methods are abbreviated as follows: KLEED, DDLEED, QDLEED, DLEED and SPLEED for Kinematical, Double-Diffraction (an approximation), Quasi-Dynamical, Dynamical and Spin-Polarized LEED; GVB, EH and Xα for Generalized-Valence-Bond, Extended Huckel and Xα cluster calculations. The reference numbers correspond with the comprehensive list which precedes this table. (Abbreviations, ad.: adsorbate; b.l.: bond length; sub.: substrate; l.s.: layer spacing.)

a. Clean Metals and Atomic Adsorbates

Surface type and surface	Topmost substrate bond length [and layer spacing] relaxation	Adsorption site	Adsorbate-to-substrate layer spacing [Å]	Adsorbate-to-substrate bond length [Å]	Equivalent bond lengths for non-surfaces [Å]	Method	Ref.	Comments
fcc(111)								
Al(111)	0% [0%]					DLEED	13.6	
	1.5% [15%]					DLEED	13.14	
	0% [0%]					DLEED	13.7	
	−1% [−3%]					KLEED	29.4	
	−5% [−8%]					SEXAFS	13.17	Deconvolution
	0% [0%]					DLEED	13.18	
	0.7±0.4% [2.2±1.5%]					DLEED	13.19	
	0±1% [0±3%]					DLEED	13.8.1	
	0±1.5% [0±5%]					DLEED	13.8.7	
Al(111)-(1×1) O	0% [0%]	b or cABC...	$0.70^{+0.10}_{-0.15}$	1.79±0.05	1.86−1.97	SEXAFS	13.8.2	
	0% [0%]	cABC...	1.33±0.08	2.12±0.05		DLEED	13.18	
	0% [0%]	cABC...	1.46±0.05	2.21±0.03		DLEED	13.8.1	1.46 Å applies to all Al-O layer spacings
		cAcBC...	1.46±0.05	2.21±0.03		DLEED	13.8.1	

Surface								
Ag(111)	0% [0%]	cABC...	1.6±0.07	2.3±0.04		DLEED	13.8.7	c or bABC site assumed for spacing determination
			0.60±0.10	1.76±0.03		SEXAFS	13.8.6	Low pressure; possibly undissociated O_2, top O not determined[f]
		b or cABC...	1.33±0.16	2.22±0.1		SEXAFS	13.8.8	
		b or cABC...	0.98±0.1	1.92±0.05		SEXAFS	13.8.8	High pressure
		b or cABC...	0.7	1.79		Model calc.	13.8.9	
	0% [0%]					KLEED	28.7	Averaging
	0% [0%]					DLEED	47.4	
	0% [0%]					DLEED	47.5	
	0% [0%]					DLEED	47.6	
	0% [0%]					HEIS	47.12	
Ag(111)-(1×1) Au	0% [0%]	cABC...	2.36	2.88	2.88	DLEED	47.6	
						HEIS	47.12	Qualitative confirmation of Ref. 47.6
Ag(111)-($\sqrt{3}\times\sqrt{3}$)R30°I	0% [0%]	cABC...	2.25	2.80	2.54 – 2.85	DLEED	47.53.1	
		b or cABC...	2.34	2.87±0.02		SEXAFS	47.53.3	
Ag(111)-incommensurate Xe	0% [0%]	variable	3.5±0.1	variable	3.53	KLEED	47.54.1	Averaging
	0% [0%]	variable	3.55±0.1	variable		DLEED	47.54.2	
Au(111)	0% [0%]					DLEED	47.6	Clean surface reconstructs
Co(111)	−1.5% [−5%]					DLEED	27.1	High-temperature phase
	0% [0%]					DLEED	27.3, 5	

Table 12.1 (continued)

Surface	Sub. b.l. [l.s.] relaxation	Adsorption site	Ad.-sub. l.s. [Å]	Ad.-sub. b.l. [Å]	Equiv. non-surf. b.l. [Å]	Method	Ref.	Comments
Cu(111)	0% [0%]					DLEED	29.5	
	−1.3% [−4%]					DLEED	29.10	Positron diffraction
	0% [0%]					DLEPD	25.19	
	−0.1±0.3% [−0.3±1.0%]					DLEED	29.20	
Ir(111)	−0.8% [−2.5%]					DLEED	77.1, 2	
Ir(111)-(√3×√3) R30°S	0% [0%]	cABC...	1.65	2.28	2.26 − 2.44	DLEED	77.16.1	
Ni(111)	0% [0%]					DLEED	28.1	
	0% [0%]					DLEED	28.3	
	0% [0%]					KLEED	28.7	Averaging
	0% [0%]					DLEED	28.4	
	0% [0%]					DLEED	28.10	
	0% [0%]					HEIS	28.8.24	
Ni(111)-(2×2) 2H	0% [0%]	b & cABC...	1.15±0.10	1.84±0.06	1.47 − 1.87	DLEED	28.1.3, 28.1.6	Honeycomb overlayer lattice
		b or cABC	1.1	1.81		DLEED	28.1.5	(2×1) lattice assumed
					1.63	Cluster calc.	28.1.10	
					1.69 ± 0.1	ND	28.1.9	Neutron diffraction $H_3Ni_4(Cp)_4$ cluster
Ni(111)-(2×2) O	0% [0%]	b or cABC...	1.20±0.10	1.88±0.06	1.84 − 2.18	DLEED	28.8.7	No O determination
	2% [7%]					HEIS	28.8.24	
Ni(111)-(√3×√3) R30° O	2% [7%]					HEIS	28.8.24	No O determination

	0% [0%]	cABC...	1.40 ± 0.1	2.02 ± 0.06	2.10 – 2.23			
Ni(111)-(2×2) S	0% [0%]					DLEED	28.8.7, 28.16.2	
Pt(111)	0% [0%]					DLEED	78.1	
	0% [0%]					DLEED	78.2	
	0.5% [1.5%]					HEIS	78.4	
	0% [0%]					HEIS	78.5	
	0% [0%]					SPLEED	78.6	
	0.5 ± 0.15% [1.3 ± 0.4%]					HEIS	78.7	
	0.5 ± 0.3% [1.5 ± 1%]					MEIS	78.8	
	0.3 ± 1.5% [1 ± 4%]					DLEED	78.11, 12	
	0% [0%]					SPLEED	78.13	
	0.15 ± 0.3% [0.5 ± 1%]					SPLEED	78.15	
Rh(111)	0 ± 0.3% [0 ± 1%]					DLEED	45.1	
	−0.3 ± 0.6% [−1.2 ± 2%]					DLEED	45.4	
	0.5 ± 1.5% [1.5 ± 5%]					DLEED & KLEED	45.6	
Xe(111)	0% [0%]					DLEED	54.1, 2	
hcp(0001) Be(0001)	0% [0%]					DLEED	4.1, 2	
Cd(0001)	0% [0%]					DLEED	22.48.1	
Co(0001)	−1.5% [−5%]					DLEED	27.1	Low-temperature phase
	0% [0%]					DLEED	27.3, 5	
	0% [0%]					DLEED	27.6	
Re(0001)	−1.5% [−5%]					DLEED	75.1	
Sc(0001)	0% [0%]					DLEED	21.1	

Table 12.1 (continued)

Surface	Sub. b.l. [l.s.] relaxation	Adsorption site	Ad.-sub. l.s. [Å]	Ad.-sub. b.l. [Å]	Equiv. non-surf. b.l. [Å]	Method	Ref.	Comments
Ti(0001)	−0.7% [−2%]					DLEED	22.1	
Ti(0001)-(1×1)Cd	0% [0%]	cABA...	2.57±0.05	3.08±0.03	3.01	DLEED	22.48.1 − 3	
Ti(0001)-(1×1)N	1% [5%]	AcABA...	1.22±0.05	2.095±0.03	2.12	DLEED	22.7.1, 2	N underlayer
Zn(0001)	−0.5% [−2%]					KLEED	30.1	Averaging
Zr(0001)	−0.3±0.6% [−1±2%]					DLEED	40.1	
bcc(110) Fe(110)	0% [0%] 0.2±0.7% [0.5±2%]					SPLEED DLEED	26.1 26.6, 8	
Fe(110)-(2×2)S	0% [0%]	center	1.43	2.17 & 2.36	1.99 − 2.44	DLEED	26.16.3, 4	Top Fe atoms move parallel to surface by 0.2 Å and 0.12 Å; slight buckling possible
Na(110)	0% [0%]					DLEED	11.1, 11.8.1	
V(110)	−0.5±0.3% [−1.5±1%]					DLEED	23.2	
W(110)	0% [0%] 0% [0%] 0% [0%]					DLEED KLEED DLEED	74.3 74.8.2 74.2	Averaging
W(110)+(2×1)O	0% [0%] 0% [0%]	3-fold	1.25±0.1	2.08±0.07	1.75 − 2.12	KLEED DLEED	74.8.2 74.8.1	Averaging, no O determination

Structure	Coverage [%]	Site	d	d	Range	Method	Ref.	Comments
fcc(100) **Al(100)**	0% [0%]					DLEED	13.1	
	0% [0%]					DLEED	13.2	
	0% [0%]					DLEED	13.4	
	0% [0%]					DLEED	13.5	
	0% [0%]					DLEED	13.6	
	0% [0%]					DLEED	13.14	
	0% [0%]					Model calc.	13.8	
	0% [0%]					DLEED	13.9	
	0% [0%]					DLEED	13.11.2	Deconvolution
	0% [0%]					KLEED	29.4	
	0% [0%]					MEED	13.15	
	0% [0%]					SEXAFS	13.16	
	0% [0%]						13.17	
Al(100)-c(2×2) Na	0% [0%]	hollow	2.08 ± 0.12	2.90 ± 0.09	2.82 – 3.00	DLEED	13.11.2	
	0% [0%]	hollow	2.05 ± 0.1	2.86 ± 0.07		DLEED	13.11.1	
Ag(100)	0% [0%]					DLEED	13.4	
	0% [0%]					DLEED	47.1	
	0% [0%]					DLEED	47.3	
Ag(100)-c(2×2) Cl	0% [0%]	hollow	1.72 ± 0.1	2.67 ± 0.06	2.36 – 2.77	DLEED	47.17.1	Cl substitutes for Ag atoms in top layer
	−2% [5%]		−0.24	2.89 & 2.63		Model calc.	47.17.2	
Ag(100)-c(2×2) Se	0% [0%]	hollow	1.91 ± 0.1	2.80 ± 0.07	2.46 – 2.86	DLEED	47.34.1	
Au(100) (1×1)	0% [0%]					SPLEED	78.6	Metastable surface
Co(100)	−1.5% [−5%]					DLEED	27.2	
	−1.5% [−4%]					DLEED	27.4	
Co(100)-c(2×2) O	0% [0%]	hollow	0.80	1.94	2.12	DLEED	27.8.1	

Table 12.1 (continued)

Surface	Sub. b.l. [l.s.] relaxation	Adsorption site	Ad.-sub. l.s. [Å]	Ad.-sub. b.l. [Å]	Equiv. non-surf. b.l. [Å]	Method	Ref.	Comments
Cu(100)	0% [0%]					DLEED	29.1	
	0% [0%]					DLEED	29.2	
	0% [0%]					DLEED	29.3	
	0% [0%]					DLEED	13.4	
	0% [0%]					DLEED	29.5	
	0% [0%]					DLEED	29.7	
	0% [0%]					KLEED	29.6	
	−0.15 ± 0.3% [−0.5 ± 1%]					KLEED	29.4	Averaging
	−0.1 ± 0.3% [−0.1 ± 0.3%]					DLEED	29.14	Deconvolution
	[−0.3 ± 0.8%]					DLEED	29.15	
Cu(100)- c(2×2) N	0% [0%]	hollow	1.45 ± 0.04	2.32 ± 0.03	1.993 − 2.11	KLEED	29.7.1	Averaging; layer spacing of 0.90 Å is second choice
	0% [0%]	hollow	0.90 ± 0.10	2.02 ± 0.05		DLEED	29.7.2	Poor agreement between theory and experiment
Cu(100)- c(2×2) O	0% [0%]	bridge	1.4	1.97	1.81 − 2.1	KLEED	29.8.1	Averaging; reconstruction occurs
	0% [0%]	hollow	0.0	1.81		DLEED	29.8.2	Poor agreement between theory and experiment
						ARXPS	29.8.4	
Cu(100)- (2×2) Te	0% [0%]	hollow	1.70 ± 0.15	2.48 ± 0.1	2.51 − 2.76	DLEED	29.52.1	
Ir(100)- (1×5)						DLEED	77.5, 6	Quasi-hexagonal top layer with two-bridge registry, buckled; contracted bond lengths parallel to surface
Ni(100)	0% [0%]					DLEED	13.10	
	0% [0%]					DLEED	28.3	
	0% [0%]					DLEED	28.5	
	0% [0%]					DLEED	28.4	
	0% [0%]					DLEED	28.6	

Structure	Coverage	Site			Range	Method	Ref.	Comments
	0% [0%]					KLEED	29.4	Deconvolution
	0% [0%]					DLEED	28.9	
	0% [0%]					DLEED	28.8.15	
Ni(100)-(2×2) 2C	4% [8.5%]	hollow	0.1 ± 0.1	1.803 ± 0.015		DLEED	28.6.1	Position of C unknown
	5.5% [11.4%]					DLEED	28.6.2	Ni: parallel movement by 0.35 ± 0.05; glide plane symmetry (p4g)
Ni(100)-c(2×2) Na	0% [0%]	hollow	2.86	3.35	2.80 – 3.10	DLEED	28.11.1, 2	
	0% [0%]	hollow	2.23 ± 0.1	2.84 ± 0.08		DLEED	28.11.4	Cancels Refs. 28.11.1, 2
	0% [0%]	hollow	2.23	2.84		DLEED	28.11.3	
	0% [0%]	hollow	2.23 ± 0.1	2.84 ± 0.08		ARUPS	28.11.5	
Ni(100)-c(2×2) O	0% [0%]	hollow	1.5 ± 0.1	2.31 ± 0.06	1.84 – 2.18	DLEED	28.8.2	Result retracted
	0% [0%]	hollow	0.9 ± 0.1	1.98 ± 0.05		DLEED	28.8.3, 7	
		hollow	0.9 ± 0.2	1.98 ± 0.10		LEIS	28.8.4	
	0% [0%]	hollow	0.9	1.98		DLEED	28.8.15	
	0% [0%]	hollow	0.9 ± 0.04	1.98 ± 0.02		ARNPD	28.8.21	
	0% [0%]	hollow	0.8	1.93		ARXPS	28.8.16	
Ni(100)-(2×2) O	0% [0%]	hollow	0.9 ± 0.1	1.98 ± 0.05	1.84 – 2.18	DLEED	28.8.5	
	0% [0%]	hollow	0.9 ± 0.1	1.98 ± 0.05		DLEED	28.8.7	
		hollow	0.75	1.90		Xα cluster	28.8.10	
			0.96	2.01		GVB cluster	28.8.9	
	0% [0%]	hollow	0.8	1.93		ARXPS	28.8.16	
		bridge	1.20 ± 0.10	2.13 ± 0.05		EH	28.8.25	
Ni(100) c(2×2) S	0% [0%]	hollow	1.30 ± 0.1	2.19 ± 0.06	2.10 – 2.23	DLEED	28.8.3, 28.16.2	
		hollow	1.70 ± 0.1	2.45 ± 0.07		DLEED	28.16.1	
		hollow	1.30	2.19		LEIS	28.16.6	
	0% [0%]	bridge	1.30 ± 0.1	2.19 ± 0.1		DLEED	28.9, 28.16.9	
	0% [0%]	hollow	1.30	2.19		ARUPS	28.16.5	
	0% [0%]	hollow	1.30 ± 0.04	2.19 ± 0.02		ARNPD	28.8.21	

Table 12.1 (continued)

Surface	Sub. b.l. [l.s.] relaxation	Adsorption site	Ad.-sub. l.s. [Å]	Ad.-sub. b.l. [Å]	Equiv. non-surf. b.l. [Å]	Method	Ref.	Comments
Ni(100)-(2×2) S	0% [0%]	hollow	1.30±0.1	2.19±0.06	2.10−2.23	DLEED	28.8.5	
	0% [0%]	hollow	1.30±0.1	2.19±0.06		DLEED	28.8.7	
		hollow	1.33	2.21		GVB cluster	28.16.8	
		hollow	1.31	2.20		EH	28.8.25	
Ni(100)-c(2×2) Se	0% [0%]	hollow	1.45±0.1	2.28±0.06	2.31−2.53	DLEED	28.8.3	
Ni(100)-(2×2) Se	0% [0%]	hollow	1.55±0.1	2.34±0.07	2.31−2.53	DLEED	28.8.5	
	0% [0%]	hollow	1.55±0.1	2.34±0.07		DLEED	28.8.7	
		hollow	1.47	2.29		EH	28.8.25	
Ni(100)-c(2×2) Te	0% [0%]	hollow	1.90±0.1	2.58±0.07	2.54−2.85	DLEED	28.8.3	
	0% [0%]	hollow	1.90±0.1	2.58±0.07		DLEED	28.52.1	
	0% [0%]	hollow	1.9	2.58		ARUPS	28.11.5	
Ni(100)-(2×2) Te	0% [0%]	hollow	1.80±0.1	2.52±0.07	2.54−2.85	DLEED	28.8.5	
	0% [0%]	hollow	1.90±0.1	2.58±0.07		DLEED	28.8.7	
Ni(100)-c(2×2) S-c(2×2) Na	0% [0%]	S: hollow Na: hallow	NiS: 1.30 SNa: 1.13	NiS: 2.19 NiNa: 2.99 SNa: 2.76	S-Na: 2.73−3.38	DLEED	28.11.16.1	
Ni(100)-c(2×2) S-(2×2) Na						DLEED	28.11.16.1	Same sites and distances as for Ni(100)−c(2×2) S−c(2×2) Na
Ni(100)-(2×2) S-c(2×2) Na						DLEED	28.11.16.1	Same sites and distances as for Ni(100)−c(2×2) S−c(2×2) Na, Na bridged between S's
Ni(100)-c(2×2) NO	0% [0%]	hollow	0.8	2.28		DLEED	26.7.8.1	Dissociated N and O fill one c(2×2) sublattice at random (1/4 monolayer of NO)

Surface	Method	Ref.				Site	% [%]	Comments
Pd(100)	DLEED	46.1, 2					1 ± 1% [2.5 ± 2.5%]	Metastable phase
Pt(100)-(1×1)	DLEED HEIS	78.6 78.10					0% [0%] 0.2 ± 0.2% [0.5 ± 0.5%]	
Pt(100)-(14,1/1̄,5)	DLEED	77.5						Quasi-hexagonal, slightly rotated, buckled top layer; contracted bond lengths parallel to surface
Rh(100)	DLEED DLEED DLEED	45.1, 2 45.3 45.7					1 ± 1.5% [3 ± 5%] 0 ± 1% [0 ± 3%] 0.25 ± 0.3% [0.75 ± 1%]	Later retracted
Rh(100)-(2×2) S	DLEED	45.16.1	2.23 − 2.38	2.30	1.29	hollow	0% [0%]	
Xe(100)	KLEED	54.1					0% [0%]	
bcc(100) Fe(100)	SPLEED DLEED DLEED	26.2 26.8.1 26.4	2.46 − 2.70				0% [0%] − 1.5% [− 4%] − 0.7% [− 1.4%]	Surface not quite clean
Fe(100)-(1×1) O	DLEED DLEED EH	26.8.1 26.8.3 26.8.2	2.09 − 2.15	2.07 & 2.09 2.02 & 2.08 2.07 & 2.09	0.53 ± 0.06 0.48 0.53 ± 0.06	hollow hollow hollow	3% [7.5%] 3% [7.5%]	O closest to 2nd layer Fe atoms
Fe(100)-c(2×2) S	DLEED DLEED	26.16.1 26.16.2	1.99 − 2.44	2.3 2.33	1.09 ± 0.05 1.15 ± 0.05	hollow hollow	0% [0%] 0% [0%]	
Fe(100)-c(2×2) CO	DLEED	26.6.8.1	2.09 − 2.15	2.02 & 2.08	0.48	hollow	0% [0%]	Dissociated CO; C and O randomly occupy sites in a c(2×2) array, with equal spacings to substrate
Mo(100)	DLEED DLEED DLEED	42.1 42.2 42.5 42.3, 4					− 4% [− 11.5%] − 3.3% [− 10%] − 4.1% [− 13%] − 3.1% ± 0.6% [− 9.5 ± 2%]	Surface resonance analysis

Table 12.1 (continued)

Surface	Sub. b. l. [l. s.] relaxation	Adsorption site	Ad.-sub. l. s. [Å]	Ad.-sub. b. l. [Å]	Equiv. non-surf. b. l. [Å]	Method	Ref.	Comments
Mo(100)-c(2×2) N	0% [0%]	hollow	1.02±0.1	2.45±0.05	2.11 – 2.33	DLEED	42.7.1	
Mo(100)-(2×1) O	0% [0%]	hollow	0.70	2.275 & 2.33	1.66 – 2.07	DLEED	42.8.1	
Mo(100)-(1×1) Si	0% [0%]	hollow	1.16±0.1	2.51±0.05	2.53	DLEED	42.14.1	
Mo(100)-c(2×2) S	0% [0%]	hollow	1.04	2.45		DLEED	42.16.6	
V(100)	−2% [−7%]					DLEED	23.4	
W(100) (1×1)	−3.3% [−10%]					SPLEED	74.1	
	−2% [−6%]					DLEED	74.2	
	−3.6±0.6% [−11±2%]					DLEED	74.4	
	−2% [−5%]					DLEED	74.7	
	−2% [−6%]					HEIS	74.8	6% is upper limit for contraction
	−2±0.5% [−5.5±1.5%]					SPLEED	74.15	
	−2% [−5%]					SPLEED	74.9	
	−3.3±0.6% [−10±2%]					DLEED	74.18	
	−2% [−5%]					KLEED	74.19	Averaging
	−2.2±0.6% [−6.7±2%]					DLEED	74.20	
	−3±0.5% [−8±1.5%]					DLEED	74.21	
W(100)-c(2×2)	[−3±3%]					DLEED	74.14	Parallel W displacements of 0.15 – 0.30 Å giving zigzag rows of atoms with pmg symmetry. 1/2 of surface atoms displaced parallel to surface as Ref. 74.14 with parallel displacements ~ 0.16 Å
	[−6%]					HEIS	74.16	
						DLEED	74.24	

System	Displacement	Site			Method	Ref.	Notes
W(100)-c(2×2) H		bridge	1.90	2.15 ± 0.05	HREELS	74.1.3	Top W layer buckled with parallel displacements
W(100)-(1×1)2H	−1% [−3%]	bridge	1.32	2.05 ± 0.05	HREELS	74.1.3	Top W layer unreconstructed
		bridge	1.35	1.92	DLEED	74.1.7	
fcc(110)							
Al(110)	−3% [−10%]				DLEED	13.6	
	−3% [−10%]				DLEED	13.14	
	−3.75 ± 0.75%				DLEED	13.7	
	[−12.5 ± 2.5%]						
	−3% [−9%]				DLEED	13.16	Convolution-transform
	−1.5% [−4%]				KLEED	13.20	
	−1.5% [−5%]				MEED	13.15	
Ag(110)	−1.85% [−6.5%]				DLEED	47.3	
	−2.15% [−7%]				DLEED	47.2	
	−3% [−10%]				DLEED	47.7	
	−2.5% [−8%]				DLEED	47.9	
	−2.15% [−7%]				KLEED	28.11	Convolution-transform
	−1.55% [−6%]				DLEED	47.10	
	−1.9% [−6.6%]				DLEED	29.15	
	−1.35% [−5.7%]				DLEED	47.11	[+2.2%] in 2nd layer spacing
Ag(110)-(2×1) O					LEIS	47.8.2	Probably long-bridge
					DLEED	47.8.1	Probably long-bridge
Au(110)-(2×1)	−4.5% [−15%]				DLEED	79.5	Missing row model
					DLEED	79.7	No satisfactory model found
Cu(110)	−3 ± 1%				DLEED	29.12, 13	
	[−10 ± 2.5%]						
	−2.5 ± 1%				DLEED	29.17	
	[−8 ± 3%]						
Ir(110)-(2×1)	−5% [−15%]				DLEED	77.4, 7, 8	Missing row model

Table 12.1 (continued)

Surface	Sub. b.l. [l.s.] relaxation	Adsorption site	Ad.-sub. l.s. [Å]	Ad.-sub. b.l. [Å]	Equiv. non-surf. b.l. [Å]	Method	Ref.	Comments
Ir(110)-(1×1)	-2.5% $[-7.5\%]$					DLEED	77.3	Impurity stabilized
Ir(110) c(2×2) O	0% [0%]	short bridge	1.37±0.05	1.93±0.07	1.96−2.06	DLEED	77.8.1	1/4 monolayer of randomly positioned O present
Ni(110)	-1.5% $[-5\%]$ -1.5% $[-5\%]$ $-1.2\pm0.3\%$ $[-4\pm1\%]$ $0.3\pm0.3\%$ $[1\pm1\%]$					DLEED KLEED MEIS	28.4 28.11 28.8.14	Convolution transform
						MEIS	28.8.14	1/3 monolayer of randomly positioned O present
Ni(110)-(2×1) H	-2.5% $[-8\%]$					DLEED	28.1.1	Probably (2×1) reconstruction: row pairing
Ni(110)-(2×1) O	0% [0%]	short bridge	1.46±0.05	1.92±0.04	1.84−2.18	DLEED LEIS	28.1.1, 28.8.7 28.8.26	Missing row model (across ridges of clean surface); O position uncertain
						MEIS	28.8.23	Missing row model; O position undetermined
Ni(110)-c(2×2) S	0% [0%] 1.5±0.75% [6±3%]	center center long bridge	0.93±0.1 0.87±0.03 1.04	2.17±0.1 & 2.35±0.04 2.11±0.03 & 2.33±0.01 2.04	2.10−2.23	DLEED MEIS GVB cluster	28.8.7, 28.16.2 28.16.7 28.16.8	2.17 bond length to 2nd layer Ni
Pt(110)-(2×1)						DLEED	78.14	Missing row model slightly favored

Surface type and surface	Topmost substrate bond length [and layer spacing] relaxation	Adsorption site	Adsorbate-to-substrate layer spacing [Å]	Adsorbate-to-substrate bond length [Å]	Equivalent bond lengths for non-surfaces [Å]	Method	Ref.	Comments
Rh(110)	−0.9±0.7% [−2.7±2%]			2.115 & 2.45	2.23 − 2.38	DLEED	45.5	
Rh(110)-c(2×2) S	0% [0%]	hollow	0.77			DLEED	45.16.2	
bcc(111) Fe(111)	−1.5% [−15%] −1.5±0.4% [−15.4±4%]					DLEED DLEED	26.3 26.9	
bcc(211) Fe(211)	[−10%]					DLEED	26.7	2nd layer spacing [+10%]
fcc(311) Cu(311)	−1% [−5%]					DLEED	29.8, 11	
Ni(311)	−2.4% [−14%]					DLEED	28.12	

b. Molecular Adsorbates

Surface type and surface	Topmost substrate bond length [and layer spacing] relaxation	Adsorption site	Adsorbate-to-substrate layer spacing [Å]	Adsorbate-to-substrate bond length [Å]	Equivalent bond lengths for non-surfaces [Å]	Method	Ref.	Comments
Cu(100)-c(2×2) CO	0% [0%]	top	CuC: 1.9±0.1 CO: 1.15±0.1	1.9±0.1 1.15±0.1		DLEED	28.6.8.5,7	CO axis perpendicular to surface
Ni(100)-c(2×2) C_2H_2	0% [0%]	midpoint of C-C over hollow	Ni-midpoint of C-C: 2.02	Ni-closest C: 2.2		DLEED	28.6.1.1	C-C axis filted 50° from normal in [110] azimuth; C-C bond length 1.2 Å; no H included in calculations
Ni(100)-c(2×2) CO	0% [0%]	top	NiC: 1.8±0.1 CO: 0.95±0.1	1.8±0.1 1.15	1.75 − 1.869 1.04 − 1.16	DLEED	28.6.8.1	CO axis tilted ~30° from normal
	0% [0%]	top	NiC: 1.72 CO: 1.15	1.72 1.15		DLEED	28.6.8.3,4	CO axis perpendicular to surface
	0% [0%]	top	NiC: 1.8±0.1 CO: 1.1±0.1	1.8±0.1 1.1±0.1		DLEED	28.6.8.5,7	CO axis perpendicular to surface
	0% [0%]	top	NiC: 1.7±0.1 CO: 1.13±0.1	1.7±0.1 1.13±0.1		DLEED	28.6.8.8	CO axis perpendicular to surface

Table 12.1 (continued)

Surface	Sub. b.l. [l.s.] relaxation	Adsorption site	Ad.-sub. l.s. [Å]	Ad.-sub. b.l. [Å]	Equiv. non-surf. b.l. [Å]	Method	Ref.	Comments
Pd(100)-(2$\sqrt{2}\times\sqrt{2}$)R45° 2 CO	0% [0%]	bridge	PdC: 1.36±0.1, CO: 1.15±0.1	1.93±0.07, 1.15±0.1		DLEED	46.1, 2	CO axis perpendicular to surface
Pt(111)-(2×2) C_2H_2	0% [0%]	near top	PtC: 2.43	2.50		DLEED	76.6.1.1	Site uncertain; C–C axis parallel to surface; H position uncertain; C–C distance roughly 1.2 Å
Pt(111)-(2×2) C_2H_3	0% [0%]	ccABC...	PtC: 1.20, CC: 1.50	2.00, 1.50	1.95 – 2.13, 1.50 – 1.53	DLEED	78.6.1.5	Ethylidyne species; C–C axis perpendicular to surface; H ignored in calculation
Rh(111)-(2×2) C_2H_3	0% [0%]	bbABC...	RhC: 1.31±0.1, CC: 1.45±0.1	2.03±0.07, 1.45±0.1	1.90 – 2.08, 1.50 – 1.53	DLEED	45.6.1.1	Ethylidyne species; C–C axis perpendicular to surface; H ignored in calculation
Rh(111)-($\sqrt{3}\times\sqrt{3}$)R30° CO	0% [0%]	top	RhC: 1.95±0.1, CO: 1.07±0.1	1.95±0.1, 1.07±0.1	1.82 – 1.91, 1.09 – 1.17	DLEED	45.6.8.1	CO axis perpendicular to surface

c. Semiconductors and Compounds

Surface	Surface structure, including bond length [and layer spacing] relaxations, relative to bulk	Method	Refs.	Comments
CaO(100)	Bulk NaCl structure with [−1%] top spacing contraction; buckling less than [±1%]	DLEED	20.8.1	
CdTe(110)	As GaAs(110) with 30.5° rotation, Cd and Te backbonds expanded 0.87%, contracted −4.54%, resp.	DLEED	48.52.1	
CoO(111)	O-terminated polar face of NaCl structure; top layer contraction −5% [−15%]	DLEED	27.8.2	
CoO(100)	Unrelaxed bulk NaCl structure within [±5%]	DLEED	27.8.3	
GaAs(100)	Bulk structure with As termination; no relaxation	DLEED	31.33.8	As termination forced by molecular beam epitaxy
GaAs(110)	Zincblende structure with top Ga and As atoms rotated into, resp., out of surface (keeping about constant mutual bond lengths, rotated by projected angle of 27°); Ga and As backbonds contracted by −2.5% and −3.6%, resp.	DLEED	31.33.1−7, 11, 13−17	Only selected results are quoted here
	As DLEED results	Model calc.	31.33.10, 18	
GaAs(110)-(1×1) Al	Al first substitutes for Ga in 2nd layer; top layer 0° rotated, cf. GaAs(110), contracted [−5%]; at higher coverage Al also substitutes for Ga in 3rd layer, then in 1st layer, ending with pure AlAs(110) similar to clean GaAs(110)	DLEED	31.33.13.1, 3, 4	
	Al overlayer single-bonded to Ga (bond lengths GaAl = 2.95 Å); angle AlGaAs = angle AsGaAs; top GaAs plane rotations by 21° (As out, Ga in)	Model calc.	31.33.13.2	
GaAs(110)-(1×1) As	Substrate has unrelaxed bulk structure; As bonded to surface Ga as in bulk (possible small bond angle change)	QDLEED	31.33.8	
GaP(110)	As GaAs(110) with 25±3° rotation; Ga and P backbonds contracted −2.5±1° and −3±1°, resp.	DLEED	31.15.2	
	As GaAs(110) with 27° rotation; bulk bond lengths	DLEED	31.15.3	
Ge(100)	As Si(100) (2×1) with asymmetric dimer; bond length changes within 5% of bulk value	DLEED	32.2	

Table 12.1 (continued)

Surface	Surface structure, including bond length [and layer spacing] relaxations, relative to bulk	Method	Refs.	Comments
InP(110)	As GaAs(110) with 30.4° rotation. In and P backbonds contracted −3% and −3.5%, resp.	DLEED	49.15.2	
	Qualitatively same result as Ref. 49.15.2	DLEED	49.15.3	
InSb(110)	As GaAs(110) with 28.8° rotation; In and Sb backbonds expanded 1% and contracted −4.5%, resp.	DLEED	49.51.1	
MgO(100)	Unrelaxed bulk NaCl structure within [±5%]	DLEED	12.8.1 − 3	
MoS₂(0001)	Layer compound cleaved between two 3-plane layers; top contraction by −1.6% [−4.7%]; first Van der Waals spacing contracted [−3%]	DLEED	42.16.1 − 5	
Na₂O(111)	Fluorite structure terminated between two Na layers; no relaxation	DLEED	11.8.1	
NbSe₂(0001)	As MoS₂(0001) but top contraction −0.2% [−0.6%]; first Van der Waals spacing contracted [−1.4%]	DLEED	42.16.2 − 5	
NiO(100)	Unrelaxed bulk NaCl structure within [±5%]	DLEED	28.8.17 − 20	
Si(111)-"(1 × 1)"	Bulk structure with −2% [−15%] top contraction	DLEED	14.3	Impurity stabilized
	Bulk structure with 0.8% [5%] and 1% [1%] expansions in two topmost layer spacings, resp.	DLEED	14.18	Annealed at conversion temperature
	Bulk structure with −1% [−6%] top contraction	Model calc.	14.21	
	Bulk structure with −0.33% [−2%] top contraction	DLEED	14.32	
	Bulk structure with −4±0.4% [−25.5±2.5%] top contraction, 3.2±1.5% [3.2±1.5%] 2nd layer expansion	DLEED	14.35	
	Bulk structure with −1.5% [−10%] top contraction	Model calc.	14.38	

Only selected papers are included here

Structure	Description	Method	Ref.
Si(111)-(2×1)	Top layer contracted −1% [−8%], buckled ±3% [±22%]; 2nd layer spacing contracted −10% [−10%]	DDLEED	14.25
	Top layer contracted −1% [−7%], buckled ±10% [±47%]; 2nd layer spacing expanded 3% [3%]	Model calc.	14.36
Si(100)-(2×1)	Symmetric dimer (atom-pairing) model with relaxations down several layers	Model calc. & KLEED	14.15
	Confirmation of above model	QDLEED	14.16, 17
	Asymmetric (buckled) dimer with subsurface relaxations	Model calc.	14.40, 41
	Confirmation of above model (Ref. 14.40, 41)	DLEED	14.30
	Confirmation of above model (Ref. 14.40, 41)	MEIS	14.37
	Further refinement of geometry	Model calc.	14.39
Si(100)-(2×1) H	Clean surface reconstruction maintained; H occupies danglin bonds	Model calc.	14.42, 43
Si(100)-(1×1) 2H	Ideal bulk Si(100) structure found acceptable; H positions unknown	DLEED	14.19
	Ideal bulk Si(100) structure with [−6±3%] top spacing contraction; H positions unknown	MEIS	14.34, 14.1.2, 3
Te(10$\overline{1}$0)	Polymeric trigonal bulk structure retains 3-strand chains parallel to surface, but surface chains rotate 15.5°, moving top-layer atoms inward by [−0.21 Å] and 2nd layer atoms outward by [0.46 Å]	DLEED	52.1
TiS$_2$(0001)	As MoS$_2$, but top contraction by −1.7% [−5%], first Van der Waals spacing contracted [−5%]	DLEED	22.16.1
TiSe$_2$(0001)	As MoS$_2$, but top expansion by 1.7% [5%], first Van der Waals spacing contracted [−5%]	DLEED	22.16.1
ZnO(000$\overline{1}$)	Unreconstructed Zn-terminated wurtzite structure with top contraction by −3% [−25%]	DLEED	30.8.1

Table 12.1 (continued)

Surface	Surface structure, including bond length [and layer spacing] relaxations, relative to bulk	Method	Refs.	Comments
ZnO(10$\bar{1}$0)	Unreconstructed wurtzite structure, top Zn and O pulled somewhat into surface	DLEED	30.8.2, 3 31.33.2, 16	
ZnS(110)	Unreconstructed zincblende structure within [±3%], ±2° rotation (cf. GaAs(110))	DLEED	30.16.1	
ZnSe(110)	Zincblende structure reconstructed as GaAs(110)	DLEED	30.34.1 30.33.3, 5, 10, 12	
ZnTe(110)	As GaAs(110) with 33.2° rotation; Zn and Te backbonds; contracted −0.34% and expanded 2.72%, resp.	DLEED	30.52.1, 31.33.16	

Appendix A: Acronyms of Techniques Related to Surface Science

Acronym	Meaning
AEAPS	Auger-Electron Appearance-Potential Spectroscopy
AES	Auger-Electron Spectroscopy
AMEFS	Auger-Monitored Extended Fine Structure
APS	Appearance-Potential Spectroscopy
APFIM	Atom-Probe Field-Ion Microscopy
ARAES	Angle-Resolved Auger-Electron Spectroscopy
ARNPD	Angle-Resolved Normal Photoelectron Diffraction
ARPEFS	Angle-Resolved Photoelectron Fine Structure
ARPES	Angle-Resolved Photoelectron Spectroscopy
ARSES	Angle-Resolved Secondary-Electron Spectroscopy
ARUPS	Angle-Resolved Ultraviolet-Photoelectron Spectroscopy
ARXPD	Angle-Resolved X-Ray Photoelectron Diffraction
ARXPS	Angle-Resolved X-Ray Photoelectron Spectroscopy
CPD	Contact-Potential Difference (work-function change)
DAPS	Disappearance-Potential Spectroscopy
EAPFS	Extended Appearance-Potential Fine Structure
EDAX	Energy Dispersive X-Ray Analysis
EELS	Electron Energy-Loss Spectroscopy
EID	Electron-Impact Desorption
ELEED	Elastic Low-Energy Electron Diffraction
ELNES	Electron-Energy Loss Near-Edge Structure
ELS	Energy-Loss Spectroscopy

EM	Electron Microscopy
ESCA	Electron Spectroscopy for Chemical Analysis
ESD	Electron-Stimulated Desorption
ESDIAD	Electron-Stimulated Desorption Ion Angular Distributions
ESR	Electron-Spin Resonance
EXAFS	Extended X-Ray-Absorption Fine Structure
EXELFS	Extended Electron-Energy-Loss Fine Structure
FEED	Field-Emission Energy Distribution
FEM	Field-Emission Microscopy
FIM	Field-Ion Microscopy
HEED	High-Energy Electron Diffraction
HEIS	High-Energy Ion Scattering
HREELS	High-Resolution Electron Energy-Loss Spectroscopy
ICISS	Impact-Collision Ion-Scattering Spectroscopy
IETS	Inelastic Electron Tunneling Spectroscopy
IID	Ion-Impact Desorption
ILEED	Inelastic Low-Energy Electron Diffraction
INS	Ion-Neutralization Spectroscopy
IPE	Inverse Photoemission
IRAS	Infrared Reflection-Absorption Spectroscopy
ISS	Ion-Scattering Spectroscopy
LEED	Low-Energy Electron Diffraction
LEIS	Low-Energy Ion Scattering
LEPD	Low-Energy Positron Diffraction
LID	Laser-Induced Desorption
LIF	Laser-Induced Fluorescence
MDS	Metastable Deexcitation Spectroscopy
MEED	Medium-Energy Electron Diffraction
MEIS	Medium-Energy Ion Scattering
MPI	Multi-Photon Ionization

NEXAFS	Near-Edge X-Ray-Absorption Fine Structure
NIS	Neutron Inelastic Scattering
NMA	Nuclear Microanalysis
NMR	Nuclear Magnetic Resonance
NPD	Normal Photoelectron Diffraction
OPD	Off-Normal Photoelectron Diffraction
PED	Photoelectron Diffraction
PES	Photoelectron Spectroscopy
PhD	Photoelectron Diffraction
PIES	Penning Ionization Electron Spectroscopy
PLEED	Polarized Low-Energy Electron Diffraction
PSD	Photon-Stimulated Desorption
PSDIAD	Photon-Stimulated Desorption Ion Angular Distributions
RBS	Rutherford Backscattering Spectroscopy
RHEED	Reflection High-Energy Electron Diffraction
SAES	Scanning Auger-Electron Spectroscopy
SEE	Secondary-Electron Emission
SEELFS	Surface Extended-Energy-Loss Fine Structure
SEM	Scanning Electron Microscopy
SERS	Surface-Enhanced Raman Scattering
SEXAFS	Surface Extended X-Ray-Absorption Fine Structure
SIMS	Secondary-Ion Mass Spectroscopy
SPI	Surface Penning Ionization
SPIES	Surface Penning Ionization Electron Spectroscopy
SPLEED	Spin-Polarized Low-Energy Electron Diffraction
SSIMS	Static Secondary-Ion Mass Spectroscopy
STEM	Scanning Transmission Electron Microscopy
STM	Scanning Tunneling Microscopy
SXAPS	Soft X-Ray Appearance-Potential Spectroscopy
SXPS	Soft X-Ray Photoelectron Spectroscopy

TDMS	Thermal-Desorption Mass Spectroscopy
TDS	Thermal-Desorption Spectroscopy
TED	Transmission Electron Diffraction
TEM	Transmission Electron Microscopy
THEED	Transmission High-Energy Electron Diffraction
TPD	Temperature-Programmed Desorption
TPR	Temperature-Programmed Reaction
UPS	Ultraviolet-Photoelectron Spectroscopy
VLEED	Very-Low-Energy Electron Diffraction
XAES	X-Ray-Stimulated Auger-Electron Spectroscopy
XANES	X-Ray-Absorption Near-Edge Structure
XPD	X-Ray Photoelectron Diffraction
XPS	X-Ray Photoelectron Spectroscopy
XRD	X-Ray Diffraction

Appendix B: A Computer Program to Determine the Angle of Incidence in LEED

Two of the geometrical parameters that are necessary to specify a LEED I-V curve are the incidence angle θ (the angle of the incident beam with respect to the surface normal) and the azimuthal angle φ (the angle between an arbitrarily chosen axis in the surface plane and the projection of the incident beam direction on the surface). These angles can be obtained from photographs of the LEED pattern, as described in Sect. 2.6. This involves a computation for which a program is presented in this Appendix.

The program reproduced below can be run without any additional auxiliary subroutines. It is written in standard Fortran language. An example of the input data can be found following the program listing. It should be noted that a pair of initially guessed values for θ and φ is needed for the input data, and the solutions are weakly dependent on these initially guessed values. The output data can be found following the listing of the input data. The first part of the output data is a more detailed listing of the input data. The second part is the results (for brevity, a middle section of the second part has been omitted). Spot combinations for which $\delta\theta$ or $\delta\varphi$ exceeds the allowable maximum errors are omitted in the calculation for the averaged values of θ and φ, see Sect. 2.6.3. The regular and weighted averages (Sect. 2.6.3) are computed and listed at the end.

Variables whose values the user may wish to change are (see the first group of executable statements)

ITLIM = allowed number of iterations,

TLIM = limit of accuracy,

TERR = the maximum allowable error in determining the value of θ due to the uncertainty in measuring the position of the spot,

PERR = the maximum allowable error in determining the value of φ due to the uncertainty in measuring the position of the spot.

```
C        PROGRAM ANGLE(INPUT,OUTPUT,TAPE5=INPUT,TAPE6=OUTPUT)
C    PROGRAM ANGLE
C    PURPOSE=   TO USE LEED PHOTOGRAPH ANGLE MEASUREMENTS TO DETERMINE
C               THE ANGLE OF INCIDENCE OF THE ELECTRON BEAM
C
        DIMENSION HEAD(20),A1(2),A2(2),B1(2),B2(2),GX(2),GY(2),ANG(2)
        DIMENSION BH(15),BK(15),EPS(15),THETA(105),PHI(105),ERAD(15)
        DIMENSION X(2),D(2),F(2),P(2,2)
        DIMENSION ER1(5),ER2(5)

        DIMENSION DTH(105),DPH(105),ETH(4),EPH(4)
        DIMENSION NOMIT(50,2)
        COMMON AK,GX,GY,ANG
        DATA ER1/0.0,1.0,-1.0,-1.0,1.0/
        DATA ER2/0.0,-1.0,1.0,-1.0,1.0/
C
C    FORMAT STATEMENTS
C
  101 FORMAT(20A4)
  102 FORMAT(2F8.3)

  104 FORMAT(I3)
  105 FORMAT(2F5.1,F10.1)
  111 FORMAT(80X,*IT=*,I2,5X,*F =*,2E10.2,5X,*D =*,2F6.2)
  112 FORMAT(20X,I3,2I10,2F15.2)
  113 FORMAT(1H1,10X,20A4)
  114 FORMAT(/,10X,*A1 =*,2F8.4,/,10X,*A2 =*,2F8.4)
  115 FORMAT(/,10X,*BEAM ENERGY =*,F6.1,* EV*,/,10X,
     1          *INITIAL ANGLE GUESS= THETA =*,F6.1,/,31X,*PHI   =*,F6.1)
  116 FORMAT(//,10X,*INPUT DATA,      I*,8X,*BH*,8X,*BK*,9X,*EPS*,/)
  117 FORMAT(24X,I2,2F10.1,F12.1)
  118 FORMAT(//,10X,*RESULTS*,/,20X,*IND*,5X,*SPOT1*,5X,
     1          *SPOT2*,10X,*THETA*,11X,*PHI*,/)
  119 FORMAT(26X,2F10.0,F12.2,F15.2)
  120 FORMAT(48X,2F15.3)
  121 FORMAT(/,10X,*REGULAR AVERAGE*,5X,*THETA =*,F6.2,* +/-*,F6.3,
     1          10X,*PHI =*,F7.2,* +/-*,F6.3)
  122 FORMAT(/,10X,*WEIGHTED AVERAGE*,4X,*THETA =*,F6.2,* +/-*,F6.3,
     1          10X,*PHI =*,F7.2,* +/-*,F6.3)
  123 FORMAT(//,10X,*SUMMARY*,/,10X,20A4,/,10X,*NUMBER OF SPOTS  =*,
     1          I3,/,10X,*NUMBER OF COMBS. =*,I3)
  124 FORMAT(2X,14H**** OMIT ****)
  125 FORMAT(10X,*NUMBER OF OMITS  =*,I3,/,12X,*OMITTED COMBS. =*,
     1          2I3,49(/,28X,2I3))
C
C    CONSTANTS
C
        RAD=1.745329E-2
        PI=3.1415926
        CONS=0.5123298
        ITLIM=10
        TLIM=1.0E-6
        TLIM=1.0E-5
        TERR=2.5
        PERR=4.0
        IF(ER1(2).GT.1.1) TERR=4.0
        IF(ER2(3).GT.1.1) PERR=6.0
C
C    READ EVERYTHING
C    HEAD =      HEADING CARD USED TO LABEL THE DATA AND OUTPUT (20A4)
C    A1,A2 =     TWO-DIMENSIONAL SURFACE UNIT CELL VECTORS IN ANGSTROMS
C                FORMAT OF EACH IS (2F8.3)
```

```
C   ENERG =      ENERGY OF THE INCIDENT ELECTRON RELATIVE TO VACUUM
C                IN ELECTRON VOLTS  (F8.3)
C
C   TH1,PH1 =    INITIAL GUESS OF INCIDENT ANGLE THETA AND PHI IN DEGREES
C                GUESS SHOULD BE FAIRLY CLOSE (2F8.3)
C   NDAT =       NUMBER OF DATA POINTS READ (I3)
C   BH(I),BK(I) = (H,K) VALUES FOR THE I-TH DATA POINT
C   EPS(I) =     ANGLE EPSILON IN DEGREES FOR I-TH DATA POINT
C                FORMAT FOR DATA POINT IS (2F5.1,F10.1)
C
      READ(5,101) HEAD
      READ(5,102) A1
      READ(5,102) A2
      READ(5,102) ENERG
      READ(5,102) TH1,PH1
      READ(5,104) NDAT
      WRITE(6,113) HEAD
      WRITE(6,114) A1,A2
      WRITE(6,115) ENERG,TH1,PH1
      WRITE(6,116)
      DO 10 I=1,NDAT
      READ(5,105) BH(I),BK(I),EPS(I)
      WRITE(6,117) I,BH(I),BK(I),EPS(I)

   10 ERAD(I)=EPS(I)*RAD
      WRITE(6,118)
C
C   CALCULATE RECIPROCAL LATTICE VECTORS AND K VECTOR MAGNITUDE
C
      TVA=ABS(A1(1)*A2(2)-A1(2)*A2(1))
      ATV=2.0*PI/TVA
      B1(1)= A2(2)*ATV
      B1(2)=-A2(1)*ATV
      B2(2)= A1(1)*ATV
      B2(1)=-A1(2)*ATV
      AK=CONS*SQRT(ENERG)
C
C   LOOPOVER COMBINATION OF TWO SPOTS
C
      IND=0
      NER=0
      IL=NDAT-1
      DO 55 I=1,IL
      JS=I+1
      DO 55 J=JS,NDAT
      IND=IND+1
C
C   LOOP ON ERRORS IN ANG
C
      DO 50 IERR=1,5
C
C   ASSIGN ANGLE FROM PHOTO AND DETERMINE RECIPROCAL VECTORS
C
      X(1)=TH1*RAD
      X(2)=PH1*RAD
      ANG(1)=ERAD(I) + ER1(IERR)*RAD
      GX(1)=BH(I)*B1(1)+BK(I)*B2(1)
      GY(1)=BH(I)*B1(2)+BK(I)*B2(2)
      ANG(2)=ERAD(J) + ER2(IERR)*RAD
      GX(2)=BH(J)*B1(1)+BK(J)*B2(1)
      GY(2)=BH(J)*B1(2)+BK(J)*B2(2)
```

```
C
C   NEWTONS METHOD LOOP
C
        IT=0
        GO TO 20
     15 IT=IT+1
        IF(ABS(D(1)).GT.0.17) D(1)=SIGN(0.17,D(1))
        IF(ABS(D(2)).GT.0.17) D(2)=SIGN(0.17,D(2))
        X(1)=X(1)+D(1)
        X(2)=X(2)+D(2)
        IF(IT.GT.ITLIM) GO TO 25
     20 CALL FN(X,F,P)
        PJB=P(1,1)*P(2,2)-P(1,2)*P(2,1)
        IF(PJB.EQ.0.0) GO TO 30

        D(1)=(P(1,2)*F(2)-P(2,2)*F(1))/PJB
        D(2)=(P(2,1)*F(1)-P(1,1)*F(2))/PJB
        TST=D(1)**2+D(2)**2
        IF(TST.LT.TLIM) GO TO 35
        GO TO 15
     25 D1=D(1)/RAD
        D2=D(2)/RAD
        WRITE(6,111) IT,F,D1,D2

        IND=IND-1
        NER=NER+1
        NOMIT(NER,1)=I
        NOMIT(NER,2)=J
        WRITE(6,124)
        GO TO 55
     30 D(1)=1.0E-2
        D(2)=-1.0E-2
        GO TO 15
     35 CONTINUE
C
C   STORE RESULTS
C
        TH=X(1)/RAD
        PH=X(2)/RAD
        IF(IERR.NE.1) GO TO 45
        THETA(IND)=TH

        PHI(IND)=PH
        WRITE(6,112) IND,I,J,TH,PH
        GO TO 50
     45 WRITE(6,119) ER1(IERR),ER2(IERR),TH,PH
        IEX=IERR-1
        ETH(IEX)=ABS(TH-THETA(IND))
        EPH(IEX)=ABS(PH-PHI(IND))
     50 CONTINUE
        DTH(IND)=AMAX1(ETH(1),ETH(2),ETH(3),ETH(4),0.1)
        DPH(IND)=AMAX1(EPH(1),EPH(2),EPH(3),EPH(4),0.1)
        WRITE(6,120) DTH(IND),DPH(IND)
        IF(DTH(IND).GT.TERR) GO TO 52
        IF(DPH(IND).GT.PERR) GO TO 52
        GO TO 55
     52 CONTINUE
        IND=IND-1
        NER=NER+1
        NOMIT(NER,1)=I
        NOMIT(NER,2)=J
        WRITE(6,124)
     55 CONTINUE
```

```
C
C   CALCULATE AVERAGES AND STANDARD DEVIATIONS
C
        TAVE=0.0
        PAVE=0.0
        TWAV=0.0
        PWAV=0.0
        TSIG=0.0
        PSIG=0.0
        TWT=0.0
        PWT=0.0
C
        DO 60 I=1,IND

        TAVE=TAVE+THETA(I)
        W1=DTH(I)**2
        TWT=TWT+1.0/W1
        TWAV=TWAV+THETA(I)/W1
        PAVE=PAVE+PHI(I)
        W2=DPH(I)**2
        PWT=PWT+1.0/W2
        PWAV=PWAV+PHI(I)/W2
     60 CONTINUE
C
        TAVE=TAVE/FLOAT(IND)
        PAVE=PAVE/FLOAT(IND)
        TWAV=TWAV/TWT

        PWAV=PWAV/PWT

C
        DO 65 I=1,IND

        TSIG=TSIG+(THETA(I)-TAVE)**2
        PSIG=PSIG+(PHI(I)-PAVE)**2
     65 CONTINUE
C
        TSIG=SQRT(TSIG/FLOAT(IND))
        PSIG=SQRT(PSIG/FLOAT(IND))
        TWIG=SQRT(1.0/TWT)
        PWIG=SQRT(1.0/PWT)
C
        WRITE(6,123) HEAD,NDAT,IND
        WRITE(6,125) NER,((NOMIT(I,1),NOMIT(I,2)),I=1,NER)
        WRITE(6,121) TAVE,TSIG,PAVE,PSIG
        WRITE(6,122) TWAV,TWIG,PWAV,PWIG
C
        STOP
        END

        SUBROUTINE FN(X,F,P)
C
C   THIS SUBROUTINE EVALUATES THE FUNCTION F(THETA,PHI,BH,BK) FOR
C   TWO CHOICES OF THE INDEX PAIR (BH,BK).  ALSO OBTAINED AND STORED IN
C   P ARE THE PARTIAL DERIVATIVES OF F WITH RESPECT OT THETA AND PHI.
C
        REAL KCIX,KCIY,KCFX,KCFY,KDUM,KCFZ,KLFX,KLFY
        DIMENSION X(2),F(2),P(2,2),GX(2),GY(2),EPS(2)
        COMMON AK,GX,GY,EPS
C
        ST=SIN(X(1))
        SP=SIN(X(2))
        CT=COS(X(1))
        CP=COS(X(2))
```

```
C
C   INITIAL K IN CRYSTAL FRAME

C
      KCIX=ST*CP*AK

      KCIY=ST*SP*AK

C
C   DERIVATIVES OF FINAL K IN CRYSTAL FRAME
C
      DKCFXT =-CT*CP*AK
      DKCFXP=-ST*SP*AK
      DKCFYT= CT*SP*AK
      DKCFYP= ST*CP*AK
C
C   LOOP FOR THE TWO EVALUATIONS OF THE SAME EQUATION
C
      DO 50 I=1,2
C
C   FINAL K IN CRYSTAL FRAME
C
      KCFX=KCIX + GX(I)
      KCFY=KCIY + GY(I)
      KDUM=AK**2 - KCFX**2 - KCFY**2
      KCFZ=-SQRT(ABS(KDUM))
C
      DKCFZT = -(KCFX*DKCFXT + KCFY*DKCFYT)/KCFZ
      DKCFZP = -(KCFX*DKCFXP + KCFY*DKCFYP)/KCFZ
C
C   FINAL K AND DERIVATIVES IN LAB FRAME

C
      KLFX=-SP*KCFX+CP*KCFY
      KLFY=-CT*CP*KCFX-CT*SP*KCFY+ST*KCFZ
      DKLFXT=-SP*DKCFXT+CP*DKCFYT
      DKLFXP=-SP*DKCFXP+CP*DKCFYP-CP*KCFX-SP*KCFY

      DKLFYT=ST*CP*KCFX+ST*SP*KCFY+CT*KCFZ
     1      -CT*CP*DKCFXT-CT*SP*DKCFYT+ST*DKCFZT
      DKLFYP=CT*SP*KCFX-CT*CP*KCFY
     1      -CT*CP*DKCFXP-CT*SP*DKCFYP+ST*DKCFZP
C
C   ASSIGN FUNCTION AND DERIVATIVES
C
      TEP=TAN(EPS(I))
      F(I)=KLFX+TEP*KLFY
      P(I,1)=DKLFXT+TEP*DKLFYT
      P(I,2)=DKLFXP+TEP*DKLFYP
C
   50 CONTINUE
      RETURN
      END
```

Input data:
```
IRIDIUM(111) SPOT POSITIONS
   2.710                    A1
   1.355    2.347           A2
 245.0                      ENERG
    10.0   -90.0            TH1,PH1
13                          NDAT
  1.0   0.0    24.0
  1.0   1.0    46.0
  0.0   1.0     0.0
 -1.0   0.0   -46.0
 -1.0  -1.0   -26.0
  0.0  -1.0    -1.0
  2.0   1.0    55.0
  2.0   2.0    83.0
  2.0   3.0   115.0
  1.0   2.0   100.0
  1.0   3.0   145.0
 -2.0   1.0  -116.0
 -2.0  -1.0   -56.0
```

Output data:
```
        IRIDIUM(111) SPOT POSITIONS

    A1 =   2.7100  -0.
    A2 =   1.3550   2.3470

    BEAM ENERGY =  245.0 EV
    INITIAL ANGLE GUESS= THETA =   10.0
                         PHI   =  -90.0
```

INPUT DATA,	I	BH	BK	EPS
	1	1.0	0.	24.0
	2	1.0	1.0	46.0
	3	0.	1.0	0.
	4	-1.0	0.	-46.0
	5	-1.0	-1.0	-26.0
	6	0.	-1.0	-1.0
	7	2.0	1.0	55.0
	8	2.0	2.0	83.0
	9	2.0	3.0	115.0
	10	1.0	2.0	100.0
	11	1.0	3.0	145.0
	12	-2.0	1.0	-116.0
	13	-2.0	-1.0	-56.0

RESULTS	IND	SPOT1	SPOT2	THETA		PHI	
	1	1	2	12.80		-83.45	
		1.	-1.	13.38		-87.19	
		-1.	1.	12.09		-79.55	
		-1.	-1.	13.00		-81.43	
		1.	1.	12.62		-85.43	
					.705		3.898
	2	1	3	15.61		-90.00	
		1.	-1.	14.72		-90.53	
		-1.	1.	16.53		-89.29	
		-1.	-1.	17.19		-90.78	
		1.	1.	14.32		-89.51	
					1.580		.776
	3	1	4	13.47		-84.96	
		1.	-1.	13.12		-86.55	
		-1.	1.	13.81		-83.27	
		-1.	-1.	13.28		-82.14	
		1.	1.	13.65		-87.85	
					.346		2.892

```
              4         1         5         14.50        -87.38
                       1.       -1.         13.40        -87.56
                      -1.        1.         15.60        -87.22
                      -1.       -1.         14.48        -84.76
                       1.        1.         14.51        -90.00
                                                    1.108         2.624
              5         1         6         14.64        -87.69
                       1.       -1.         12.69        -85.65
                      -1.        1.         16.85        -90.00
                      -1.       -1.         14.76        -85.36
                       1.        1.         14.51        -90.00
                                                    2.213         2.332
              6         1         7         -4.24        -43.79
                       1.       -1.          9.40        -77.54
                      -1.        1.         -7.30        -35.83
                      -1.       -1.         -3.44        -45.07
                       1.        1.         -5.38        -41.39
                                                   13.643        33.742
**** OMIT ****
              6         1         8          3.80        -62.72
                       1.       -1.          9.61        -78.03
                      -1.        1.         -3.16        -45.73
                      -1.       -1.          2.93        -59.58
                       1.        1.          4.91        -66.85
                                                    6.953        16.995
**** OMIT ****

**** OMIT ****
              6         1        10         11.49        -80.26
                       1.       -1.         12.63        -85.34
                      -1.        1.          9.99        -74.87
                      -1.       -1.         10.85        -76.80
                       1.        1.         12.01        -83.78
                                                    1.503         5.387
**** OMIT ****
              6         1        11         18.60        -97.36
                       1.       -1.         15.35        -92.13
                      -1.        1.         22.48       -103.47
                      -1.       -1.         22.07       -102.41
                       1.        1.         16.24        -94.42
                                                    3.875         6.111
**** OMIT ****
              6         1        12         14.27        -86.82
                       1.       -1.         13.55        -87.62
                      -1.        1.         15.00        -85.90
                      -1.       -1.         14.65        -85.12
                       1.        1.         13.90        -88.48
                                                     .738         1.698
              7         1        13         14.01        -86.21
                       1.       -1.         13.20        -87.04
                      -1.        1.         14.78        -85.41
                      -1.       -1.         14.15        -84.02
                       1.        1.         13.73        -88.48
                                                     .802         2.271
              8         2         3         13.19        -90.00
                       1.       -1.         12.92        -90.35
                      -1.        1.         13.46        -89.60
                      -1.       -1.         13.51        -90.41
                       1.        1.         12.88        -89.66
                                                     .321          .408
              9         2         4         13.19        -90.00
                       1.       -1.         12.90        -90.00
                      -1.        1.         13.49        -90.00
                      -1.       -1.         13.20        -84.66
                       1.        1.         13.20        -95.34
                                                     .298         5.338
```

**** OMIT ****

9	2	5	13.22		-90.47	
	1.	-1.	12.85		-89.13	
	-1.	1.	13.65		-92.15	
	-1.	-1.	13.36		-87.68	
	1.	1.	13.12		-93.45	
				.432		2.976
10	2	6	13.06		-87.79	
	1.	-1.	12.63		-85.66	
	-1.	1.	13.49		-90.00	
	-1.	-1.	13.23		-85.56	
	1.	1.	12.90		-90.00	
				.430		2.237
11	2	7	13.18		-89.81	
	1.	-1.	12.71		-86.54	
	-1.	1.	13.65		-92.81	
	-1.	-1.	13.48		-88.78	
	1.	1.	12.94		-90.63	
				.469		3.266
12	2	8	13.29		-91.65	
	1.	-1.	12.83		-88.82	
	-1.	1.	13.75		-94.65	
	-1.	-1.	13.63		-91.74	
	1.	1.	13.02		-91.81	
				.461		2.999
13	2	9	13.24		-90.01	
	1.	-1.	12.78		-88.11	
	-1.	1.	13.63		-91.73	
	-1.	-1.	13.54		-89.84	
	1.	1.	12.91		-90.05	
				.456		1.899
14	2	10	13.19		-89.21	
	1.	-1.	12.61		-85.25	
	-1.	1.	13.68		-92.76	
	-1.	-1.	13.57		-90.49	
	1.	1.	12.82		-88.05	
				.582		3.959
15	2	11	13.16		-89.56	
	1.	-1.	12.81		-88.46	
	-1.	1.	13.59		-90.81	
	-1.	-1.	13.52		-89.51	
	1.	1.	12.89		-89.82	
				.424		1.246
16	2	12	13.19		-89.24	
	1.	-1.	12.85		-89.24	
	-1.	1.	13.51		-89.37	
	-1.	-1.	13.43		-87.90	
	1.	1.	12.95		-90.69	
				.341		1.448
17	2	13	13.12		-88.82	
	1.	-1.	12.80		-88.30	
	-1.	1.	13.45		-89.35	
	-1.	-1.	13.30		-86.74	
	1.	1.	12.96		-90.87	
				.331		2.082
18	3	4	13.19		-90.00	
	1.	-1.	12.92		-89.65	
	-1.	1.	13.46		-90.40	
	-1.	-1.	12.88		-90.34	
	1.	1.	13.51		-89.59	
				.321		.408
19	3	5	13.39		-90.00	
	1.	-1.	12.66		-89.68	
	-1.	1.	14.32		-90.49	
	-1.	-1.	12.44		-90.30	
	1.	1.	14.72		-89.47	
				1.334		.528

20	3	6	-9.48	-90.00	
	1.	-1.	-24.16	-92.90	
	-1.	1.	9.61	-90.00	
	-1.	-1.	-3.29	-88.67	
	1.	1.	9.61	-90.00	
			19.089		2.899

**** OMIT ****

20	3	7	13.24	-90.00	
	1.	-1.	13.62	-89.58	
	-1.	1.	12.85	-90.34	
	-1.	-1.	13.90	-90.45	
	1.	1.	12.63	-89.68	
			.663		.449
21	3	8	12.83	-90.00	
	1.	-1.	13.05	-89.64	
	-1.	1.	12.59	-90.31	
	-1.	-1.	13.26	-90.38	
	1.	1.	12.42	-89.71	
			.428		.381
22	3	9	13.27	-90.00	
	1.	-1.	13.47	-89.60	
	-1.	1.	13.05	-90.36	
	-1.	-1.	13.86	-90.44	
	1.	1.	12.73	-89.67	
			.587		.444
23	3	10	13.33	-90.00	
	1.	-1.	13.41	-89.60	
	-1.	1.	13.25	-90.38	
	-1.	-1.	13.56	-90.41	
	1.	1.	13.11	-89.63	
			.225		.413
24	3	11	13.50	-90.00	
	1.	-1.	13.63	-89.58	
	-1.	1.	13.33	-90.39	
	-1.	-1.	14.30	-90.49	
	1.	1.	12.77	-89.67	
			.805		.490
25	3	12	12.88	-90.00	
	1.	-1.	12.64	-89.68	
	-1.	1.	13.10	-90.36	
	-1.	-1.	12.35	-90.29	
	1.	1.	13.46	-89.60	
			.577		.402
26	3	13	12.74	-90.00	
	1.	-1.	12.34	-89.72	
	-1.	1.	13.12	-90.37	
	-1.	-1.	12.16	-90.27	
	1.	1.	13.37	-89.61	
			.632		.393
27	4	5	13.19	-90.86	
	1.	-1.	13.68	-87.03	
	-1.	1.	12.62	-94.57	
	-1.	-1.	13.00	-88.93	
	1.	1.	13.38	-92.81	
			.573		3.832
28	4	6	13.32	-87.77	
	1.	-1.	13.75	-85.47	
	-1.	1.	12.90	-90.00	
	-1.	-1.	13.18	-85.57	
	1.	1.	13.49	-90.00	
			.422		2.302

```
29      4        7        13.20       -89.87
        1.      -1.       13.59       -89.12
       -1.       1.       12.88       -90.43
       -1.      -1.       13.07       -87.59
        1.       1.       13.38       -91.97
                                 .394           2.276
30      4        8        13.13       -91.07
        1.      -1.       13.43       -91.01
       -1.       1.       12.83       -91.15
       -1.      -1.       12.95       -89.26
        1.       1.       13.36       -93.13
                                 .304           2.067
31      4        9        13.20       -89.85
        1.      -1.       13.55       -89.87
       -1.       1.       12.90       -90.04
       -1.      -1.       12.99       -88.57
        1.       1.       13.47       -91.37
                                 .355           1.520
32      4       10        13.26       -89.59
        1.      -1.       13.53       -90.26
       -1.       1.       12.97       -88.88
       -1.      -1.       13.05       -87.59
        1.       1.       13.46       -91.53
                                 .289           1.994
33      4       11        13.21       -89.63
        1.      -1.       13.52       -89.43
       -1.       1.       12.91       -89.85
       -1.      -1.       12.98       -88.70
        1.       1.       13.46       -90.55
                                 .308            .925
34      4       12        13.29       -89.03
        1.      -1.       13.67       -87.35
       -1.       1.       12.85       -90.91
       -1.      -1.       13.01       -88.82
        1.       1.       13.59       -89.20
                                 .446           1.877
35      4       13        13.29       -88.29
        1.      -1.       13.75       -85.35
       -1.       1.       12.82       -91.29
       -1.      -1.       13.06       -87.48
        1.       1.       13.54       -89.08
                                 .472           3.008
36      5        6        14.21       -87.70
        1.      -1.       16.41       -85.12
       -1.       1.       12.54       -90.00
       -1.      -1.       14.15       -85.40
        1.       1.       14.51       -90.00
                             2.203           2.573
37      5        7        13.31       -90.21
        1.      -1.       14.17       -90.86
       -1.       1.       12.64       -89.72
       -1.      -1.       13.17       -88.20
        1.       1.       13.47       -92.25
                                 .858           2.035
38      5        8        13.07       -90.88
        1.      -1.       13.70       -92.01
       -1.       1.       12.52       -90.06
       -1.      -1.       12.88       -89.03
        1.       1.       13.32       -92.97
                                 .631           2.094
39      5        9        13.39       -90.34
        1.      -1.       14.08       -91.08
       -1.       1.       12.69       -89.59
       -1.      -1.       13.02       -88.63
        1.       1.       13.73       -91.95
                                 .700           1.712
```

```
            40        5         10          13.39        -90.33
                      1.        -1.         13.79        -91.79
                     -1.         1.         12.96        -88.81
                     -1.        -1.         13.17        -88.22
                      1.         1.         13.59        -92.29
                                                 .435          2.105
            41        5         11          13.49        -90.08
                      1.        -1.         14.28        -90.59
                     -1.         1.         12.69        -89.57
                     -1.        -1.         12.99        -88.71
                      1.         1.         13.98        -91.33
                                                 .795          1.361
            42        5         12          17.76        -78.14
                      1.        -1.         24.67        -55.62
                     -1.         1.         10.29        -96.26
                     -1.        -1.         13.17        -88.47
                      1.         1.         24.51        -56.70
                                                7.478         22.518
**** OMIT ****
            42        5         13           -.02       -124.06
                      1.        -1.         -7.88       -145.14
                     -1.         1.         17.12        -76.38
                     -1.        -1.          6.18       -106.94
                      1.         1.          -.01       -125.04
                                               17.144         47.681
**** OMIT ****
            42        6          7          12.54        -87.84
                      1.        -1.         13.76        -90.00
                     -1.         1.         11.40        -85.87
                     -1.        -1.         12.36        -85.70
                      1.         1.         12.74        -90.00
                                                1.219          2.162
            43        6          8          12.23        -87.86
                      1.        -1.         13.15        -90.00
                     -1.         1.         11.35        -85.88
                     -1.        -1.         11.99        -85.77
                      1.         1.         12.51        -90.00
                                                 .918          2.136
            44        6          9          12.27        -87.86
                      1.        -1.         13.65        -90.00
                     -1.         1.         10.95        -85.95
                     -1.        -1.         11.73        -85.81
                      1.         1.         12.88        -90.00
                                                1.383          2.139
            45        6         10          12.93        -87.80
                      1.        -1.         13.49        -90.00
                     -1.         1.         12.37        -85.70
                     -1.        -1.         12.70        -85.65
                      1.         1.         13.18        -90.00
                                                 .558          2.195
            46        6         11          11.90        -87.89
                      1.        -1.         13.94        -90.00
                     -1.         1.          9.99        -86.11
                     -1.        -1.         10.95        -85.95
                      1.         1.         13.03        -90.00
                                                2.043          2.107
            47        6         12          13.97        -87.72
                      1.        -1.         12.48        -90.00
                     -1.         1.         15.26        -85.28
                     -1.        -1.         14.53        -85.39
                      1.         1.         13.27        -90.00
                                                1.482          2.433
            48        6         13          13.46        -87.76
                      1.        -1.         12.24        -90.00
                     -1.         1.         14.77        -85.30
                     -1.        -1.         13.71        -85.48
                      1.         1.         13.24        -90.00
                                                1.304          2.464
```

49	7	8	10.31	−81.37
	1.	−1.	14.74	−96.18
	−1.	1.	.02	−54.03
	−1.	−1.	9.71	−78.35
	1.	1.	11.00	−84.64
			10.285	27.343
**** OMIT ****				
49	7	9	13.61	−90.85
	1.	−1.	10.55	−83.25
	−1.	1.	16.49	−98.51
	−1.	−1.	14.74	−92.63
	1.	1.	12.67	−89.60
			3.068	7.660
**** OMIT ****				
49	7	10	13.34	−90.01
	1.	−1.	14.24	−94.54
	−1.	1.	12.28	−85.31
	−1.	−1.	13.03	−87.47
	1.	1.	13.66	−92.68
			1.063	4.706
**** OMIT ****				
49	7	11	13.25	−89.75
	1.	−1.	11.79	−87.08
	−1.	1.	14.58	−92.15
	−1.	−1.	13.91	−90.06
	1.	1.	12.66	−89.58
			1.463	2.672
50	7	12	13.09	−89.55
	1.	−1.	12.64	−89.69
	−1.	1.	13.61	−89.16
	−1.	−1.	13.29	−88.21
	1.	1.	12.95	−90.67
			.512	1.339
51	7	13	12.99	−89.22
	1.	−1.	12.49	−89.24
	−1.	1.	13.50	−89.20
	−1.	−1.	13.00	−87.67
	1.	1.	12.99	−90.78
			.511	1.556
52	8	9	12.18	−87.68
	1.	−1.	10.70	−83.59
	−1.	1.	13.63	−91.73
	−1.	−1.	12.48	−87.47
	1.	1.	11.96	−88.08
			1.481	4.090
**** OMIT ****				
52	8	10	14.32	−95.36
	1.	−1.	15.04	−100.13
	−1.	1.	13.20	−90.10
	−1.	−1.	14.11	−93.48
	1.	1.	14.46	−97.61
			1.117	5.253
**** OMIT ****				
52	8	11	12.44	−88.61
	1.	−1.	11.62	−86.85
	−1.	1.	13.22	−90.24
	−1.	−1.	12.68	−88.29
	1.	1.	12.21	−88.96
			.823	1.764
53	8	12	12.85	−90.07
	1.	−1.	12.50	−89.97
	−1.	1.	13.20	−90.16
	−1.	−1.	12.91	−89.12
	1.	1.	12.79	−91.02
			.349	.949
54	8	13	12.79	−89.85
	1.	−1.	12.38	−89.57
	−1.	1.	13.19	−90.15
	−1.	−1.	12.74	−88.49
	1.	1.	12.85	−91.21
			.407	1.364

55	9	10	13.38	-90.23
	1.	-1.	13.87	-92.28
	-1.	1.	12.86	-88.27
	-1.	-1.	13.38	-89.41
	1.	1.	13.38	-91.06
			.519	2.053
56	9	11	12.90	-89.22
	1.	-1.	10.94	-85.93
	-1.	1.	14.54	-91.93
	-1.	-1.	13.17	-88.95
	1.	1.	12.65	-89.51
			1.959	3.288
57	9	12	13.08	-89.59
	1.	-1.	12.68	-89.59
	-1.	1.	13.46	-89.59
	-1.	-1.	13.08	-88.76
	1.	1.	13.08	-90.41
			.392	.824
58	9	13	12.95	-89.33
	1.	-1.	12.50	-89.20
	-1.	1.	13.41	-89.48
	-1.	-1.	12.83	-88.21
	1.	1.	13.09	-90.45
			.456	1.123
59	10	11	13.28	-89.72
	1.	-1.	12.92	-88.62
	-1.	1.	13.62	-90.76
	-1.	-1.	13.34	-89.18
	1.	1.	13.23	-90.26
			.360	1.102
60	10	12	13.21	-89.31
	1.	-1.	12.99	-88.96
	-1.	1.	13.43	-89.67
	-1.	-1.	13.21	-88.49
	1.	1.	13.21	-90.13
			.221	.820
61	10	13	13.12	-88.82
	1.	-1.	12.84	-88.17
	-1.	1.	13.43	-89.70
	-1.	-1.	13.04	-87.54
	1.	1.	13.24	-90.29
			.316	1.468
62	11	12	13.12	-89.50
	1.	-1.	12.69	-89.57
	-1.	1.	13.53	-89.44
	-1.	-1.	13.06	-88.81
	1.	1.	13.18	-90.19
			.425	.694
63	11	13	12.96	-89.30
	1.	-1.	12.47	-89.29
	-1.	1.	13.45	-89.34
	-1.	-1.	12.76	-88.40
	1.	1.	13.18	-90.19
			.491	.898
64	12	13	12.67	-90.40
	1.	-1.	9.92	-97.15
	-1.	1.	15.39	-83.41
	-1.	-1.	11.87	-91.26
	1.	1.	13.61	-89.15
			2.747	6.992

**** OMIT ****

```
SUMMARY
IRIDIUM(111) SPOT POSITIONS
NUMBER OF SPOTS   = 13
NUMBER OF COMBS.  = 63
NUMBER OF OMITS   = 15
   OMITTED COMBS. =   1   7
                      1   8
                      1   9
                      1  10
                      1  11
                      2   4
                      3   6
                      5  12
                      5  13
                      7   8
                      7   9
                      7  10
                      8   9
                      8  10
                     12  13
```

REGULAR AVERAGE   THETA = 13.24 +/- .553   PHI = -89.13 +/-

WEIGHTED AVERAGE   THETA = 13.18 +/- .058   PHI = -89.81 +/-

List of Major Symbols

$a(L_1,L_2,L_3)$	Clebsch-Gordan or Gaunt coefficients
$\vec{a}_1,\vec{a}_2$	Substrate basis vectors
$\vec{a}_1^*,\vec{a}_2^*$	Reciprocal substrate basis vectors
$\vec{b}_1,\vec{b}_2$	Superlattice basis vectors
$\vec{b}_1^*,\vec{b}_2^*$	Reciprocal superlattice basis vectors
$C^l(l'm',l''m'')$	Clebsch-Gordan or Gaunt coefficients
E	Kinetic energy in vacuum
f	Atomic scattering amplitude
F	Fourier transform
$G^{ij},\bar{G}^{ij}$	Green's function
$\vec{g}$	Two-dimensional reciprocal lattice vector
$h_l^{(1)},h_l^{(2)}$	Hankel functions of the 1st and 2nd kinds
$h = 2\pi\hbar$	Planck's constant
h	Miller index, as in (hk) beam or (hkl) plane
I	Intensity
I	Unit matrix
i	Imaginary unit: $\sqrt{-1}$
i	Miller-Bravais index, as in (hkil)
j_l	Bessel function
J	Adatom-adatom interaction
k_B	Boltzmann constant
k	Miller index, see h
k	Wave number $k = |\vec{k}|$
$\vec{k}$	Electron wavevector

$\vec{k}^i = \vec{k}_0 = \vec{k}_0^+$	Wave vector incident from vacuum
$\vec{k}^s = \vec{k}_{out} = \vec{k}_{\bar{g}}^-$	Wave vector emergent into vacuum
$\vec{k}_{\bar{g}}^\pm$	Wave vector of beam $\vec{g}^\pm$
l	Miller index, see h
l	Angular momentum
l_{max}	Cutoff angular momentum in partial-wave expansion
m	Mass of electron
$m, \tilde{m}$	Order parameter
m_a	Mass of atom or molecule
M	Exponent in Debye-Waller factor
$M = \begin{pmatrix} m_{11} & m_{12} \\ m_{21} & m_{22} \end{pmatrix}$	Matrix notation for superlattices
$M^* = \begin{pmatrix} m_{11}^* & m_{12}^* \\ m_{21}^* & m_{22}^* \end{pmatrix}$	Matrix notation for reciprocal superlattices
$M_{\bar{g}'\bar{g}}^{\pm\pm}$	Layer diffraction matrix element
$\vec{p}$	Linear momentum
P	Patterson function
P_l	Legendre polynomials
$P_{\bar{g}}^\pm, p_{\bar{g}}^\pm$	Plane-wave propagators
$\vec{r}$	General position vector
$r_{\bar{g}'\bar{g}}^{\pm\mp}$	Layer reflection matrix element
$\vec{s}$	Momentum transfer
S	Structure factor
$t_{\bar{g}'\bar{g}}^{\pm\pm}$	Layer transmission matrix element
t	Reduced temperature
t	t-matrix for single atom
t_l	t-matrix element for single atom
T	Temperature
T_c	Critical temperature

T^i	t-matrix for a layer consisting of several Bravais-lattice planes of atoms
V_o	Inner potential or muffin-tin zero level (>0)
x,y	Coordinates parallel (∥) to the surface
Y_{lm}	Spherical harmonic function
z	Coordinate perpendicular ($\perp$) to the surface (pointing into the surface)
α	Layer-to-layer electron attenuation coefficient
β	Critical exponent
Γ	Island size
δ_l	Phase shift
θ	Polar angle of incidence or emergence
θ	Fractional surface coverage
θ_D	Debye temperature
θ_D^{eff}	Effective Debye temperature
ν	Frequency
ν	Critical exponent
τ	t-matrix for a Bravais-lattice plane of atoms
φ	Azimuthal angle of incidence or emergence
φ_w	Work function
ω_i	i-th neighbor adatom-adatom interaction energy ($i = 1,2,...$)

References

Chapter 1

1.1   C. J. Davisson, C. H. Kunsman: Science **54**, 522 (1921)

1.2   H. E. Farnsworth: Phys. Rev. **20**, 358 (1922); Proc. Nat. Acad. Sci. U.S.A. **8**, 251 (1922)

1.3   H. E. Farnsworth: Phys. Rev. **25**, 41 (1925)

1.4   H. E. Farnsworth: Phys. Rev. **27**, 413 (1926); ibid. **31**, 405,414,419 (1928)

1.5   L. de Broglie: C. R. Acad. Sci. **177**, 517,548,630 (1923); These de doctorat, Masson (Paris) (1924); Phil. os. Mag. **47**, 446 (1924); Ann. Phys. (Paris) **3**, 22 (1925)

1.6   M. von Laue: Kön. Bay. Ak. **1912**, p. 203

1.7   W. L. Bragg: Proc. Cambridge Philos. Soc. **17**, 43 (1913)

1.8   R. K. Gehrenbeck: Phys. Today **31**, 34 (1978); in *Fifty Years of Electron Diffraction*, ed. by P. Goodman (D. Reidel, Dordrecht 1981) p. 12

1.9   C. J. Davisson, L. H. Germer: Nature (London) **119**, 558 (1927)

1.10   C. J. Davisson, L. H. Germer: Phys. Rev. **29**, 908 (1927)

1.11   C. J. Davisson, L. H. Germer: Phys. Rev. **30**, 705 (1927)

1.12   G. P. Thompson, A. Reid: Nature (London) **119**, 890 (1927)

1.13   O. Stern: Naturwissenschaften **17**, 391 (1929)

        I. Estermann, O. Stern: Z. Phys. **61**, 95 (1930)

        I. Estermann, R. Frisch, O. Stern: Z. Phys. **73**, 348 (1931)

1.14   T. H. Johnson: Phys. Rev. **35**, 1299 (1930); ibid. **37**, 847 (1931)

1.15   H. Halban, P. Preiswerk: C. R. Acad. Sci. **203**, 73 (1936)

        D. P. Mitchell, P. N. Powers: Phys. Rev. **50**, 486 (1936)

1.16   C. J. Davisson, L. H. Germer: Proc. Nat. Acad. Sci. U.S.A. **14**, 317 (1928); ibid. **14**, 619 (1928)

1.17   C. G. Darwin: Philos. Mag. **27**, 315, 675 (1914)

1.18   H. Bethe: Naturwissenschaften **16**, 333 (1928)

1.19   H. Bethe: Naturwissenschaften **15**, 786 (1927); in *Fifty Years of Electron Diffraction*, ed. by P. Goodman (D. Reidel, Dordrecht 1981) p. 73

1.20   H. Bethe: Ann. Phys. (Leipzig) **87**, 55 (1928)

1.21   P. P. Ewald: Ann. Phys. (Leipzig) **54**, 519 (1917)

1.22   P. M. Morse: Phys. Rev. **35**, 1310 (1930)

1.23   H. Mark. R. Wierl: Naturwissenschaften **18**, 778 (1930)

1.24   This development is described in various articles in *Fifty Years of Electron Diffraction*, ed. by P. Goodman (D. Reidel, Dordrecht 1981)

1.25   W. Ehrenberg: Philos. Mag **18**, 878 (1934)

1.26   E. J. Scheibner, L. H. Germer, C. D. Hartman: Rev. Sci. Instrum. **31**, 112 (1960)

1.27   J. J. Lander, F. Unterwald, J. Morrison: Rev. Sci. Instrum. **33**, 784 (1962)

1.28   R. L. Park, H. E. Farnsworth, Rev. Sci. Instrum. **35**, 1592 (1964)

1.29   H. E. Farnsworth: Phys. Rev. **33**, 1069 (1929); ibid. **34**, 679 (1929); ibid. **36**, 1799 (1930)

1.30   H. E. Farnsworth: Adv. Catal. **9**, 493 (1957); ibid. **15**, 31 (1964)

1.31   H. E. Farnsworth: Phys. Rev. **49**, 605 (1936)

1.32   H. E. Farnsworth, R. E. Schlier, T. H. George, R. M. Burger: J. Appl. Phys. **26**, 252 (1955); ibid. **29**, 1150 (1958)

1.33   R. E. Schlier, H. E. Farnsworth: in *Semiconductor Surface Physics*, ed. by H. Kingston (University of Pennsylvania Press, Philadelphia 1956) p. 3

1.34   R. E. Schlier, H. E. Farnsworth: J. Chem. Phys. **30**, 917 (1959)

1.35   J. J. Lander: Prog. Solid State Chem. **2**, 26 (1965)

1.36   P. J. Estrup: in *The Structure and Chemistry of Solid Surfaces*, ed. by G. A. Somorjai, (Wiley, New York 1969) Chap. 19

1.37   R. L. Park, J. E. Houston, D. G. Schreiner: Rev. Sci. Instrum. **42**, 60 (1971)

1.38   E. G. McRae: C. W. Caldwell: Surf. Sci. **2**, 409 (1964)

1.39   E. G. McRae: J. Chem. Phys. **45**, 3258 (1966)

1.40   E. G. McRae: Surf. Sci. **11**, 479, 492 (1968)

1.41   D. S. Boudreaux, V. Heine: Surf. Sci. **8**, 426 (1967)

1.42   P. M. Marcus, D. W. Jepsen: Phys. Rev. Lett. **20**, 925 (1968)

P. M. Marcus, D. W. Jepsen, F. Jona: Surf. Sci. **17**, 442 (1969)

1.43   J. L. Beeby: J. Phys. C **1**, 82 (1968)

1.44   B. W. Holland: Surf. Sci. **28**, 258 (1971)

1.45   C. B. Duke, C. W. Tucker, Jr.: Surf. Sci. **15**, 231 (1969); Phys. Rev. Lett. **23**, 1163 (1969)

1.46   J. C. Slater: Phys. Rev. **51**, 840 (1937)

1.47   K. Moliere: Ann. Phys. (Leipig) **34**, 461 (1939)

1.48   H. Yoshioka: J. Phys. Soc. Jpn. **12**, 618 (1957)

1.49   K. Hirabayashi: J. Phys. Soc. Jpn. **24**, 846 (1968)

1.50   E. R. Jones, J. R. McKinney, M. B. Webb: Phys. Rev. **151**, 476 (1966)

1.51   C. B. Duke, J. R. Anderson, C. W. Tucker, Jr.: Surf. Sci. **19**, 117 (1970)

1.52   K. Kambe: Z. Naturforsch. **22a**, 322,422 (1967); ibid. **23a**, 1280 (1968)

1.53   C. B. Duke, G. E. Laramore: Phys. Rev. B **2**, 4765,4783 (1970)

1.54   C. B. Duke, G. E. Laramore, V. Metze: Solid State Commun. **8**, 1189 (1970)

1.55   J. B. Pendry: J. Phys. C **2**, 1215,2273,2283 (1969); J. Phys. C **4**, 2501,2514 (1971)

1.56   J. B. Pendry: J. Phys. C **4**, 3095 (1971)

1.57   J. B. Pendry: *Low-Energy Electron Diffraction* (Academic, London, 1974)

1.58   M. A. Van Hove, J. B. Pendry: J. Phys. C **8**, 1362 (1975)

1.59   S. Y. Tong, T. N. Rhodin: Phys. Rev. Lett. **26**, 711 (1971)

1.60   S. Y. Tong: Prog. Surf. Sci. **7**, 1 (1975)

1.61   R. H. Tait, S. Y. Tong, T. N. Rhodin: Phys. Rev. Lett. **28**, 553 (1972)

S. Y. Tong, T. N. Rhodin, R. H. Tait: Phys. Rev. B **8**, 421,430 (1973); Surf. Sci. **34**, 457 (1973)

1.62   D. W. Jepsen, P. M. Marcus, F. Jona: Phys. Rev. Letters **26**, 1365 (1971); Phys. Rev. B **5**, 3933 (1972)

1.63   V. Hoffstein, D. S. Boudreaux: Phys. Rev. Lett. **25**, 512 (1970)

1.64   F. Hoffmann, H. P. Smith, Jr.: Phys. Rev. Lett. **19**, 1472 (1967); Phys. Rev. B **1**, 2811 (1970)

1.65   R. O. Jones, J. A. Strozier, Jr.: Phys. Rev. Lett. **22**, 1186 (1969)

       J. A. Strozier, R. O. Jones: Phys. Rev. Lett. **25**, 516 (1970); Phys. Rev. B **3**, 3228 (1971)

1.66   K. Hirabayashi, Y. Takeishi: Surf. Sci. **4**, 150 (1966)

       K. Hirabayashi: Surf. Sci. **28**, 621 (1971)

1.67   S. Andersson: Surf. Sci. **18**, 325 (1969); ibid. **19**, 21 (1970)

1.68   H. B. Lyon, G. A. Somorjai: J. Chem. Phys. **44**, 3707 (1966)

       R. M. Goodman, H. H. Farrell, G. A. Somorjai: J. Chem. Phys. **48**, 1046 (1968)

1.69   M. Lagally, T. C. Ngoc, M. B. Webb: Phys. Rev. Lett. **26**, 1557 (1971)

1.70   R. S. Zimmer, B. W. Holland: J. Phys. C **8**, 2395 (1975)

1.71   S. Y. Tong, M. A. Van Hove: Phys. Rev. B **16**, 1459 (1977)

1.72   M. A. Van Hove, S. Y. Tong: *Surface Crystallography by LEED*, Springer Ser. Chem. Phys., Vol. 2 (Springer, Berlin, Heidelberg 1979).

1.73   N. Masud, J. B. Pendry: J. Phys. C **9**, 1833 (1976)

       N. Masud, C. G. Kinniburgh, J. B. Pendry: J. Phys. C **10**, 1 (1977)

1.74   G. A. Somorjai, M. A. Van Hove: *Adsorbed Monolayers on Solid Surfaces*, Structure and Bonding, Vol. 38 (Springer, Berlin, Heidelberg 1979)

       R. J. Koestner, M. A. Van Hove, G. A. Somorjai: Chemtec **13**, 376 (1983); J. Phys. Chem. **87**, 203 (1983)

Chapter 2

2.1   M. A. Van Hove, S. Y. Tong, *Surface Crystallography by LEED*, Springer Ser. Chem. Phys., Vol. 2 (Springer, Berlin, Heidelberg 1979).

2.2   C.-M. Chan, W. H. Weinberg: J. Chem. Phys. **71**, 2788 (1979). [E.g., the Ir(111)-(2x2)O overlayer structure requires off-normal incident beam data to distinguish between a p(2x2) structure or three independent domains of (1x2) structures rotated 120° with respect to one another.]

2.3   J. Larscheid, J. Kirschner: Rev. Sci. Instrum. **49**, 1486 (1978)

2.4   M. K. Debe: Rev. Sci. Instrum. **47**, 39 (1976)

2.5   R. Chapman, D. L. Blair: Rev. Sci. Instrum. **48**, 939 (1977)

2.6   R. C. Unwin, K. Horn, P. Geng: private communication.

2.7   R. R. Wilson: Rev. Sci. Instrum. **12**, 91 (1941)

2.8   P. Feulner, D. Menzel: private communication.

2.9   C. J. Davisson, L. H. Germer: Phys. Rev. **30**, 705 (1927)

2.10   H. E. Farnsworth: Phys. Rev. **34**, 679 (1929)

2.11   F. Jona: Discuss. Faraday Soc. **60**, 210 (1975)

2.12   S. P. Weeks, C. D. Ehrlich, E. W. Plummer: Rev. Sci. Instrum. **48**, 190 (1977)

2.13   C.-M. Chan, W. H. Weinberg: J. Chem. Phys. **71**, 3988 (1979)

2.14   P. C. Stair, T. J. Kaminska, L. L. Kesmodel, G. A. Somorjai: Phys. Rev. B **11**, 623 (1975)

2.15   P. Heilmann, E. Lang, K. Heinz, K. Müller: Appl. Phys. **9**, 247 (1976)

2.16   D. C. Frost, K. A. R. Mitchell, F. R. Shepherd, P. R. Watson: J. Vac. Sci. Technol. **13**, 1196 (1976)

2.17   T. N. Tommet, G. B. Olszewski, P. A. Chadwick, S. L. Bernasek: Rev. Sci. Instrum. **50**, 147 (1979)

2.18   E. Lang, P. Heilmann, K. Heinz, K. Müller: Appl. Phys. **19**, 287 (1979)

2.19   P. C. Stair: Rev. Sci. Instrum. **51**, 132 (1980)

2.20   J. E. Houston, R. L. Park: Surface Sci. **21**, 209 (1970)

2.21   R. L. Park, J. E. Houston, D. G. Schreiner: Rev. Sci. Instrum. **42**, 60 (1971)

2.22   G. C. Wang, M. G. Lagally: Surf. Sci. **81**, 69 (1979)

2.23   D. G. Welkie, M. G. Lagally: Appl. Surf. Sci. **3**, 272 (1979)

2.24   M. Henzler: in *Electron Spectroscopy for Surface Analysis*, ed. by H. Ibach, Topics Curr. Phys., Vol. 4 (Springer, Berlin, Heidelberg 1977)

2.25   M. Henzler: Appl. Phys. **9**, 11 (1976)

      C. S. McKee, M. W. Roberts, M. L. Williams: Adv. Colloid Interface Sci. **8**, 29 (1977)

2.26   R. L. Park: in *The Structure and Chemistry of Solid Surface*, ed. by G. A. Somorjai (Wiley, New York 1969)

2.27   J. M. Burkstrand: Rev. Sci. Instrum. **44**, 774 (1973), and private communication

2.28   G. E. Thomas, W. H. Weinberg: unpublished.

2.29   J. R. Noonan: private communication.

2.30   S. L. Cunningham, W. H. Weinberg: Rev. Sci. Instrum. **49**, 7 (1978)

2.31   G. P. Price: Rev. Sci. Instrum. **51**, 605 (1980)

2.32   A. C. Sobrero, W. H. Weinberg: Rev. Sci. Instrum. **53**, 1566 (1982)

2.33   H. C. Clark, R. Herman: Phys. Rev. **139**, A860 (1965)

2.34   R. E. Allen, F. W. de Wette: Phys. Rev. **188**, 1320 (1969)

2.35   G. E. Laramore, C. B. Duke: Phys. Rev. B **2**, 4765,4783 (1970)

2.36   C.-M. Chan, P. A. Thiel, J. T. Yates, Jr., W. H. Weinberg: Surf. Sci. **76**, 296 (1978)

2.37   C.-M. Chan, E. D. Williams, W. H. Weinberg: Surf. Sci. **82**, L577 (1979)

2.38   D. T. Quinto, B. W. Holland, W. D. Robertson: Surf. Sci. **32**, 139 (1972)

2.39   D. J. Cheng, R. F. Wallis, C. Megerle, G. A. Somorjai: Phys. Rev. B **12**, 5599 (1975)

2.40   D. Tabor, J. M. Wilson, T. J. Bastow: Surf. Sci. **26**, 471 (1971)

2.41   A. U. MacRae: Surf. Sci. **2**, 522 (1964)

2.42   D. Tabor, J. M. Wilson: Surf. Sci. **20**, 203 (1970)

2.43   R. M. Goodman, H. H. Farrell, G. A. Somorjai: J. Chem. Phys. **48**, 1046 (1968)

2.44   R. J. Reid: Surf. Sci. **29**, 623 (1972)

2.45   P. J. Estrup: in *The Structure and Chemistry of Solid Surfaces*, ed. by G. A. Somorjai (Wiley, New York 1969)

2.46   G. A. Somorjai, H. H. Farrell: Adv. Chem. Phys. **20**, 293 (1971)

2.47   H. B. Lyon, G. A. Somorjai: J. Chem. Phys. **44**, 3707 (1966)

2.48   R. Bastasz, C. A. Colmenares, R. L. Smith: Surf. Sci. **67**, 45 (1977)

2.49   V. S. Sundaram, W. D. Robertson: Surf. Sci. **55**, 324 (1976)

2.50   W. Göpel, G. Neuenfeldt: Surf. Sci. **55**, 362 (1976)

Chapter 3

3.1   S. Trajmar: Acc. Chem. Res. **13**, 14 (1980)

3.2   J. S. Schilling, M. B. Webb: Phys. Rev. B **2**, 1665 (1970)

3.3   M. Henzler: Appl. Surf. Sci. **11/12**, 450 (1982)

3.4   J. B. Pendry: *Low-Energy Electron Diffraction* (Academic, London 1974)

   M. A. Van Hove, S. Y. Tong: *Surface Crystallography by LEED*, Springer Ser. Chem. Phys., Vol. 2 (Springer, Berlin, Heidelberg 1979)

3.5   *International Tables for X-Ray Crystallography* (Kynoch, Birmingham, England 1952)

3.6   C. Kittel: *Introduction to Solid State Physics*, 5th ed. (Wiley, New York 1976)

3.7   E. A. Wood: Bell Syst. Tech. J. XLIII, 541 (1964); J. Appl. Phys. **35**, 1306 (1964)

3.8   S. M. Davis, G. A. Somorjai: in *Encyclopaedia of Materials Science and Engineering*, ed. by M. D. Bever (Pergamon, New York 1982)

   G. A. Somorjai: *Chemistry in Two Dimensions: Surfaces* (Cornell University Press, Ithaca, NY 1981)

3.9   J. F. Nicholas: *An Atlas of Models of Crystal Surfaces* (Gordon and Breach, New York 1965)

3.10   B. Lang, R. W. Joyner, G. A. Somorjai: Surf. Sci. **30**, 454 (1972)

3.11   M. A. Van Hove, G. A. Somorjai: Surf. Sci. **92**, 489 (1980)

3.12   A comprehensive list of observed LEED patterns is given in G. A. Somorjai, M. A. Van Hove: *Adsorbed Monolayers on Solid Surfaces*, Structure and Bonding, Vol. 38 (Springer, Berlin, Heidelberg 1979); and in D. G. Castner, G. A. Somorjai: Chem. Rev. **79**, 233 (1979)

3.13   M. K. Debe, D. A. King: Phys. Rev. Lett. **39**, 708 (1977); J. Phys. C **10**, L303 (1977)

3.14   D. Dahlgren, J. C. Hemminger: Surf. Sci. **109**, L513 (1981)

3.15   M. A. Van Hove, R. J. Koestner, P. C. Stair, J. P. Biberian, L. L. Kesmodel, I. Bartos, G. A. Somorjai: Surf. Sci. **103**, 189,218 (1981)

3.16   R. L. Park, H. H. Madden: Surf. Sci. **11**, 188 (1968)

3.17   A. M. Bradshaw, F. M. Hoffmann: Surf. Sci. **72**, 513 (1978)

3.18   R. J. Behm, K. Christmann, G. Ertl, M. A. Van Hove, P. A. Thiel, W. H. Weinberg: Surf. Sci. **88**, L59 (1979)

R. J. Behm, K. Christmann, G. Ertl, M. A. Van Hove: J. Chem. Phys. **73**, 2984 (1980)

3.19   D. M. Zehner: private communication

3.20   D. G. Fedak, N. A. Gjostein: Surf. Sci. **8**, 77 (1967)

3.21   M. D. Chinn, S. C. Fain, Jr.: J. Vac. Sci. Technol. **14**, 314 (1977); Phys. Rev. Lett. **39**, 146 (1977)

C. G. Shaw, S. C. Fain, Jr., M. D. Chinn: Phys. Rev. Lett. **41**, 955 (1978)

3.22   P. H. Holloway, J. B. Hudson: Surf. Sci. **43**, 123 (1974)

G. Dalmai-Imelik, J. C. Bertolini, J. Rousseau: Surf. Sci. **63**, 67 (1977)

D. F. Mitchell, P. B. Sewell, M. Cohen: Surf. Sci. **61**, 355 (1976)

3.23   J. E. Houston, R. L. Park: Surf. Sci. **21**, 209 (1970)

3.24   J. C. Tracy: J. Chem. Phys. **56**, 2748 (1972)

J. C. Tracy, P. W. Palmberg: J. Chem. Phys. **51**, 4852 (1969)

3.25   K. Horn, J. Pritchard: Surf. Sci. **55**, 701 (1976)

S. Andersson, Solid State Commun. **21**, 75 (1977)

3.26   J. P. Biberian, M. A. Van Hove: Surf. Sci. **118**, 443 (1982)

H. Ibach, D. L. Mills: *Electron Energy Loss Spectroscopy and Surface Vibrations* (Academic, New York 1982)

3.27   W. P. Ellis: in *Optical Transforms*, ed. by H. S. Lipton (Academic, New York 1972)

D. G. Fedak, T. E. Fischer, W. D. Robertson: J. Appl. Phys. **39**, 5658 (1968)

Chapter 4

4.1   R. W. James: *The Optical Principles of the Diffraction of X-Rays* (Cornell University Press, Ithaca, NY 1965)

4.2   H. Lipson, W. Cochran: *The Determination of Crystal Structures* (Bell, London 1953)

4.3   M. J. Buerger: *X-Ray Crystallography* (Wiley, New York 1942)

4.4   G. E. Bacon: *Neutron Diffraction* (Clarendon, Oxford 1955)

4.5   B. K. Vainshtein: *Structure Analysis by Electron Diffraction* (Macmillan, New York 1964).

4.6   P. B. Hirsch, A. Howie, R. B. Nicholson, D. W. Pashley, M. J. Whelan, *Electron Microscopy of Thin Crystals* (Butterworths, London 1965)

4.7   C. G. Darwin: Philos. Mag. **27**, 315,675 (1914)

4.8   J. B. Pendry: *Low Energy Electron Diffraction* (Academic, London 1974)

4.9   S. Y. Tong: Prog. Surf. Sci. **7**, 1 (1975)

4.10   D. W. Jepsen, P. M. Marcus, F. Jona: Phys. Rev. B **5**, 3933 (1972)

4.11   A. Ignatiev, J. B. Pendry, T. N. Rhodin: Phys. Rev. Lett. **26**, 189 (1971)

4.12   D. L. Adams, H. B. Nielsen, M. A. Van Hove: Phys. Rev. B **20**, 4789 (1979)

4.13   A. Bagchi, C. B. Duke: Phys. Rev. B **6**, 2956 (1972)

        J. M. Burkstrand, F. M. Propst: J. Vac. Sci. Technol. **9**, 731 (1972)

4.14   H. Yoshioka: J. Phys. Soc. Jpn. **12**, 618 (1957)

4.15   M. N. Read, D. N. Lowry: Surf. Sci. **107**, L313 (1981)

4.16   J. C. Slater: *Insulators, Semiconductors and Metals* (McGraw-Hill, New York 1967)

        J. C. Slater: Phys. Rev. **81**, 385 (1951)

        W. Kohn, L. J. Sham: Phys. Rev. **140**, A1133 (1965)

4.17   P. M. Echenique, D. J. Titterington: J. Phys. C **10**, 625 (1977)

        R. J. Meyer, C. B. Duke, A. Paton: Surf. Sci. **97**, 512 (1980)

4.18   T. Loucks: *Augmented Plane Wave Method* (Benjamin, New York 1967)

4.19   V. L. Moruzzi, J. F. Janak, A. R. Williams: *Calculated Electronic Properties of Metals* (Pergamon, New York 1978).

4.20   P. M. Morse, H. Feshbach: *Methods of Theoretical Physics*, Vol. II (McGraw-Hill, New York 1953) p. 1574

4.21   A. Messiah: *Quantum Mechanics*, Vol. 1 (North-Holland, Amsterdam 1970).

4.22   D. Tabor, J. M. Wilson, T. J. Bastow: Surf. Sci. **26**, 477 (1971)

4.23   R. J. Reid: Surf. Sci. **29**, 623 (1972)

4.24   D. W. Jepsen, P. M. Marcus, F. Jona: Surf. Sci. **41**, 223 (1974)

4.25   K. Heinz, K. Müller: "LEED Intensities - Experimental Progress and New Possibilities of Surface Structure Determination", in *Springer Tracts Mod. Phys.*, Vol. 91 (Springer, Berlin, Heidelberg 1982) p. 1

Chapter 5

5.1   P. P. Ewald: *Fifty Years of X-Ray Diffraction* (Oosthoek, Utrecht 1962)

    L. V. Azaroff: *Elements of X-Ray Crystallography* (McGraw-Hill, New York 1968)

5.2   G. E. Bacon: *Neutron Diffraction* (Clarendon Press, Oxford 1955)

5.3   B. K. Vainshtein: *Structure Analysis by Electron Diffraction* (Pergamon, New York 1964)

5.4   P. B. Hirsch, A. Howie, R. B. Nicholson, D. W. Pashley, M. J. Whelan: *Electron Microscopy of Thin Crystals* (Butterworth, London 1965)

5.5   J. Korringa: Physica **13**, 392 (1947)

    W. Kohn, N. Rostoker: Phys. Rev. **94**, 1111 (1954)

5.6   K. Kambe: Z. Naturforsch. **22a**, 322,422 (1967); ibid. **23a**, 1280 (1968)

5.7   S. Y. Tong: Prog. Surf. Sci. **7**, 1 (1975)

5.8   J. L. Beeby: J. Phys. C **1**, 82 (1968)

5.9   J. B. Pendry: *Low Energy Electron Diffraction* (Academic, London 1974)

5.10   N. Stoner, M. A. Van Hove, S. Y. Tong: in *Characterization of Metal and Polymer Surfaces*, Vol. 1, ed. by L. H. Lee (Academic, New York 1977) p. 299

5..11   C. B. Duke, C. W. Tucker, Jr.: Surf. Sci. **15**, 231 (1969)

5.12   E. G. McRae: J. Chem. Phys. **45**, 3258 (1966)

5.13   D. W. Jepsen, P. M. Marcus, F. Jona: Phys. Rev. B **5**, 3933 (1972)

5.14   S. Andersson, J. B. Pendry: J. Phys. C **13**, 3547 (1980)

5.15   J. C. Slater: *Quantum Theory of Atomic Structure* (McGraw-Hill, New York 1960)

5.16   R. S. Zimmer and B. W. Holland, J. Phys. C **8**, 2395 (1975).

5.17   D. L. Adams: J. Phys. C **14**, 789 (1981); in Proc. Conf. on Determination of Surface Structure by LEED (Plenum, New York 1985)

5.18   S. Y. Tong, M. A. Van Hove: Phys. Rev. B **16**, 1459 (1977)

5.19   D. S. Boudreaux, V. Heine: Surf. Sci. **8**, 426 (1967)

5.20   D. W. Jepsen: Phys. Rev. B **22**, 5701 (1980)

5.21   Many observed examples are collected in G. A. Somorjai, M. A. Van Hove: *Adsorbed Monolayers on Solid Surfaces*, Structure and Bonding, Vol. 38 (Springer, Berlin, Heidelberg 1979)

5.22   B. W. Holland: Surf. Sci. **28**, 258 (1971)

5.23   C. B. Duke, G. E. Laramore: Phys. Rev. B **2**, 4765,4783 (1970)

5.24   D. Tabor, J. M. Wilson, T. J. Bastow: Surf. Sci. **26**, 471 (1971)

5.25   N. D. Lang, W. Kohn: Phys. Rev. B **3**, 1215 (1971)

5.26   J. C. Inkson: Surf. Sci. **28**, 69 (1971)

5.27   J. A. Appelbaum, D. R. Hamann: Phys. Rev. B **6**, 1122 (1972)

5.28   P. J. Jennings, G. L. Price: Surf. Sci. **93**, L124 (1980)

5.29   R. E. Dietz, E. G. McRae, R. L. Campbell: Phys. Rev. Lett. **45**, 1280 (1980)

5.30   F. Forstmann: in *Photoemission and Electronic Properties of Surfaces*, ed. by B. Feuerbacher, B. Fitton, R. F. Willis (Wiley, London 1978)

5.31   J. M. Baribeau, J. D. Carette: Phys. Rev. B **23**, 6201 (1981)

5.32   E. G. McRae, M. L. Kane: Surf. Sci. **108**, 435 (1981)

5.33   P. M. Echenique, J. B. Pendry: J. Phys. C **11**, 2065 (1978)

5.34   J. Rundgren, G. Malmström, J. Phys. C **10**, 4671 (1977)

5.35   D. T. Pierce, R. J. Coletta: in *Advances in Electronic and Electron, Physics*, Vol. 56, ed. by C. Marton (Academic, New York 1981)

5.36   R. Feder, J. Kirschner: Surf. Sci. **103**, 75 (1981)

5.37   E. L. Garwin, D. T. Pierce, H. C. Siegmann: Helv. Phys. Acta **47**, 393 (1974)

       E. L. Garwin, R. E. Kirby: in Proc. 7th Int. Vac. Congr. and 3rd Int. Conf. Solid Surfaces, Vienna 1977, ed. and publ. b. R. Dobrozemsky, F. Rüdenauer, F. P. Viehböck, A. Breth (Vienna, 1977) p. 533 - (111)(7x7)

5.38   P. J. Jennings: Surf. Sci. **20**, 18 (1970)

5.39   J. Kessler: *Polarized Electrons*, Texts and Monographs in Physics (Springer, Berlin, Heidelberg 1976)

5.40   S. W. Wang: Solid State Commun. **36**, 847 (1980)

5.41   N. Masud, C. G. Kinniburgh, J. B. Pendry: J. Phys. C **10**, 1 (1977)

N. Masud: J. Phys. C **13**, 6359 (1980)

5.42   W. Moritz, H. Jagodzinski, D. Wolf: Surf. Sci. **77**, 233,249 (1978)

5.43   W. Moritz: Proc. Conf. on Determination of Surface Structure by LEED (Plenum, New York 1985)

5.44   J. M. Ziman: Solid State Phys. **26**, 1 (1971)

5.45   M. L. Cohen: Phys. Today **32**, 40 (1979)

5.46   A. P. Jauho, J. W. Wilkins, M. Cohen, R. P. Merrill: Proc. Conf. on Determination of Surface Structure by LEED (Plenum, New York 1985)

5.47   P. M. Echenique, D. J. Titterington: J. Phys. C **10**, 625 (1977)

5.48   C. H. Li, S. Y. Tong: Phys. Rev. B **19**, 1769 (1979)

D. H. Rosenblatt, J. G. Tobin, M. G. Mason, R. F. Davis, S. D. Kevan, D. A. Shirley, C. H. Li, S. Y. Tong: Phys. Rev. B **23**, 3828 (1981)

5.49   J. B. Pendry: J. Phys. C **8**, 2413 (1975)

B. W. Holland: J. Phys. C **8**, 2679 (1975)

5.50   H. L. Davis, T. Kaplan: Solid State Commun. **19**, 595 (1976)

5.51   S. Y. Tong, C. H. Li, D. L. Mills: Phys. Rev. Lett. **44**, 407 (1980)

5.52   G. Aers, T. B. Grimley, J. B. Pendry, K. L. Sebastian: J. Phys. C **14**, 3995 (1981)

5.53   P. H. Citrin, P. Eisenberger, R. C. Hewitt: Surf. Sci. **89**, 28 (1979)

5.54   G. E. Laramore, T. L. Einstein, L. D. Roelofs, R. L. Park: Phys. Rev. B **21**, 2108 (1980)

5.55   H. L. Davis, J. R. Noonan, L. H. Jenkins: Surf. Sci. **83**, 559 (1979)

5.56   R. J. Meyer, C. B. Duke, A. Paton, A. Kahn, E. So, J. L. Yen, P. Mark: Phys. Rev. B **19**, 5194 (1979)

5.57   R. Feder: Solid State Commun. **21**, 1091 (1977)

5.58   N. V. Smith, H. H. Farrell, M. M. Traum, D. P. Woodruff, D. Norman, M. S. Woolfson, B. W. Holland, Phys. Rev. B **21**, 3119 (1980)

5.59   E. Zanazzi, F. Jona: Surf. Sci. **62**, 61 (1977)

5.60   D. J. Spanjaard, D. W. Jepsen, P. M. Marcus: Phys. Rev. B **15**, 1728 (1977)

5.61   J. B. Pendry: Surf. Sci. **57**, 679 (1976)

5.62   J. Rundgren, A. Salwen: Comput. Phys. Commun. **9**, 312 (1975)

5.63   M. A. Van Hove, S. Y. Tong: *Surface Crystallography by LEED*, Springer Ser. Chem. Phys., Vol. 2 (Springer, Berlin, Heidelberg 1979).

5.64   C. H. Li, S. Y. Tong, D. L. Mills: Phys. Rev. B **21**, 3057 (1980)

Chapter 6

6.1   C. J. Davisson, L. H. Germer: Phys. Rev. **30**, 705 (1927)

6.2   L. H. Germer, A. U. MacRae, C. D. Hartman: J. Appl. Phys. **32**, 2432,2923 (1962)

6.3   J. J. Lander, J. Morrison: J. Chem. Phys. **37**, 729 (1962); J. Appl. Phys. **34**, 1403 (1963)

6.4   R. Seiwatz: Surf. Sci. **2**, 473 (1964)

6.5   G. Gafner: Surf. Sci. **2**, 534 (1964)

6.6   A. U. MacRae, G. W. Gobeli: J. Appl. Phys. **35**, 1629 (1964)

6.7   A. Ignatiev, J. B. Pendry, T. N. Rhodin: Phys. Rev. Lett. **26**, 189 (1971)

6.8   K. Christmann, G. Ertl, O. Schober: Surf. Sci. **40**, 61 (1973)

6.9   K. Christmann, G. Ertl, T. Pignet: Surf. Sci. **54**, 365 (1976)

6.10   K. Christmann, G.Ertl: private communication

6.11   M. A. Van Hove: unpublished

6.12   J. A. Appelbaum, D. R. Hamann: Surf. Sci. **74**, 21 (1978)

6.13   S. Andersson, B. Kasemo: Surf. Sci. **25**, 273 (1971)

6.14   R. M. Stern, S. Sinharoy: Surf. Sci. **33**, 131 (1972)

6.15   P. Mark, S. C. Chang, W. F. Creighton, B. W. Lee: Crit. Rev. Solid State Sci. **5**, 189 (1975)

6.16   M. G. Lagally, T. C. Ngoc, M. B. Webb: Phys. Rev. Lett. **26**, 1557 (1971)

6.17   J. B. Pendry: J. Phys. C **5**, 2567 (1972)

6.18   T. C. Ngoc, M. G. Lagally, M. B. Webb: Surf. Sci. **35**, 117 (1973)

562   References

6.19   J. E. Demuth, P. M. Marcus, D. W. Jepsen: Phys. Rev. B 11, 1460 (1975)

6.20   R. Feder: Phys. Rev. B 15, 1751 (1977)

6.21   W. N. Unertl, M. B. Webb, Surf. Sci. 59, 373 (1976)

6.22   L. McDonnell, D. P. Woodruff, K. A. R. Mitchell: Surf. Sci. 45, 1 (1975)

6.23   J. H. Onuferko, D. P. Woodruff: Surf. Sci. 91, 400 (1980)

6.24   D. P. Woodruff: Discuss. Faraday Soc. 60, 218 (1976)

6.25   D. Aberdam, R. Baudoing, C. Gaubert, E. G. McRae, Surf. Sci. 57, 715 (1976)

6.26   C. G. Darwin: Philos. Mag. 27, 675 (1914)

6.27   D. Aberdam, R. Baudoing, C. Gaubert: Surf. Sci. 52, 125 (1973)

6.28   P. P. Ewald: *Fifty Years of X-ray Diffraction* (Oosthoek, Utrecht 1962)

6.29   L. V. Azaroff: *Elements of X-ray Crystallography* (McGraw-Hill, New York 1968)

6.30   C. W. Tucker, Jr.: Surf. Sci. 2, 516 (1964)

6.31   C. W. Tucker, Jr.: J. Appl. Phys. 37, 3013 (1966)

6.32   C. W. Tucker, Jr., C. B. Duke: Surf. Sci. 29, 237 (1972)

6.33   T. A. Clark, R. Mason, M. Tescari: Surf. Sci. 30, 553 (1972)

6.34   T. A. Clark, R. Mason, M. Tescari, Proc. Roy. Soc. Lond. A 331, 321 (1972).

6.35   T. A. Clark, R. Mason, M. Tescari, Surf. Sci. 40, 1 (1973)

6.36   S. Andersson, J. B. Pendry: Solid State Commun. 16, 563 (1975)

6.37   J. E. Demuth, D. W. Jepsen, P. M. Marcus: J. Phys. C 13, L25 (1975)

6.38   J. C. Buchholz, M. G. Lagally, M. B. Webb: Surf. Sci. 41, 248 (1974)

6.39   P. I. Cohen, J. Unguris, M. B. Webb: Surf. Sci. 58, 429 (1976)

6.40   S. L. Cunningham, C.-M. Chan, W. H. Weinberg: Phys. Rev. B 18, 1537 (1978)

6.41   C.-M. Chan, S. L. Cunningham, M. A. Van Hove, W. H. Weinberg: Surf. Sci. 67, 1 (1977)

6.42   C.-M. Chan, S. L. Cunningham, M. A. Van Hove, W. H. Weinberg, S. P. Withrow, Surf. Sci. 66, 394 (1977)

6.43   U. Landman, D. L. Adams: J. Vac. Sci. Technol. **11**, 195 (1974)

6.44   D. L. Adams, U. Landman: Phys. Rev. Lett. **33**, 585 (1974)

6.45   D. L. Adams, U. Landman, J. C. Hamilton: J. Vac. Sci. Technol. **12**, 206 (1975)

6.46   D. L. Adams, U. Landman: Phys. Rev. B **15**, 3775 (1977)

6.47   C.-M. Chan, P. A. Thiel, J. T. Yates, Jr., W. H. Weinberg, Surf. Sci. **76**, 296 (1978)

6.48   J. E. Demuth, P. M. Marcus, D. W. Jepsen: Phys. Rev. B **11**, 1460 (1978)

6.49   C. B. Duke, G. E. Laramore, B. W. Holland, A. M. Gibbons: Surf. Sci. **27**, 523 (1971)

        G. E. Laramore, C. B. Duke: Phys. Rev. B **5**, 267 (1972)

6.50   W. Moritz: PhD Thesis, University of Munich (1976)

        M. Alff: PhD Thesis, University of Munich (1976)

6.51   M. Maglietta, E. Zanazzi, F. Jona, D. W. Jepsen, P. M. Marcus: J. Phys. C **10**, 3287 (1977)

6.52   C.-M. Chan, S. L. Cunningham, M. A. Van Hove, W. H. Weinberg: unpublished.

6.53   R. V. Southwell: *Relaxation Methods in Engineering Science* (Oxford University Press, Oxford 1964)

6.54   J. Unguris, L. W. Bruch, E. R. Moog, M. B. Webb: Surf. Sci. **87**, 415 (1979)

6.55   N. Stoner, M. A. Van Hove, S. Y. Tong, M. B. Webb: Phys. Rev. Lett. **40**, 243 (1978)

6.56   T. C. Ngoc, M. G. Lagally, M. B. Webb: Surf. Sci. **35**, 117 (1973)

6.57   C. G. Shaw, S. C. Fain, Jr., M. D. Chinn, M. F. Toney: Surf. Sci. **97**, 128 (1980)

6.58   S. J. White, D. C. Frost, K. A. R. Mitchell: Surf. Sci. **108**, L435 (1981)

6.59   S. Andersson, B. Kasemo, J. B. Pendry, M. A. Van Hove: Phys. Rev. Lett. **31**, 595 (1973)

6.60   J. E. Demuth, D. W. Jepsen, P. M. Marcus: Phys. Rev. Lett. **31**, 540 (1973)

6.61   S. Y. Tong, K. H. Lau: Phys. Rev. B **25**, 7382 (1982)

564   References

6.62   T. H. Upton, W. A. Goddard: Phys. Rev. Lett. **46**, 1635 (1981)

6.63   J. E. Demuth, D. W. Jepsen, P. M. Marcus: Solid State Commun. **13**, 1311 (1973)

P. M. Marcus, J. E. Demuth, D. W. Jepsen: Surf. Sci. **53**, 501 (1973)

6.64   M. A. Van Hove, S. Y. Tong: Phys. Rev. Lett. **35**, 1092 (1975)

6.65   M. A. Van Hove, S. Y. Tong, M. H. Elconin: Surf. Sci. **64**, 85 (1977)

6.66   A. Ignatiev, F. Jona, D. W. Jepsen, P. M. Marcus: LEED 7 Seminar Notes, Am. Phys. Soc. Meeting, San Diego, California, March 19-21, 1973 (unpublished)

6.67   U. Landman, D. L. Adams: Surf. Sci. **51**, 149 (1975)

6.68   E. Zanazzi, F. Jona: Surf. Sci. **62**, 61 (1977)

6.69   P. R. Watson, F. R. Shepherd, D. C. Frost, K. A. R. Mitchell: Surf. Sci. **72**, 562 (1978)

6.70   G. G. Kleiman, J. M. Burkstrand: Solid State Commun. **21**, 5 (1977)

6.71   D. L. Adams, H. B. Nielsen, M. A. Van Hove: Phys. Rev. B **20**, 4789 (1979)

6.72   J. B. Pendry: J. Phys. C **13**, 937 (1980)

6.73   J. Philip, J. Rundgren: in Proc. Conf. Determination of Surface Structure by LEED (Plenum, New York 1985)

6.74   M. A. Van Hove, R.J. Koestner: in Proc. Conf. Determination of Surface Structure by LEED (Plenum, New York 1985)

6.75   R. J. Koestner, M. A. Van Hove, G. A. Somorjai: Surf. Sci. **107**, 439 (1981)

6.76   M. A. Van Hove, R. J. Koestner, G. A. Somorjai: Surf. Sci. **121**, 321 (1982)

6.77   F. Jona: in Proc. Conf. Determination of Surface Structure by LEED (Plenum, New York 1985)

6.78   R. Z. Bachrach, G. V. Hansson, R. S. Bauer: Surf. Sci. **109**, L560 (1981)

6.79   F. Jona, D. Sondericker, P. M. Marcus: J. Phys. C **13**, L155 (1980)

6.80   R. E. Walpole, R. H. Myers: *Probability and Statistics for Engineers and Scientists* (MacMillan, New York 1972)

Chapter 7

7.1   L. Pauling: *The Nature of the Chemical Bond*, 3rd ed. (Cornell University Press, Ithaca NY 1960).

7.2   M. W. Finnis, V. Heine: J. Phys. F **4**, L37 (1974)

7.3   M. A. Van Hove: Surf. Sci. **81**, 1 (1979.

7.4   U. Landman, R. N. Hill, M. Mostoller: Phys. Rev. B **21**, 448 (1980)

7.5   K. O. Legg, F. Jona, D. W. Jepsen, P. M. Marcus: J. Phys. C **8**, L492 (1975)

7.6   C.-M. Chan, P. A. Thiel, J. T. Yates, Jr., W. H. Weinberg: Surf. Sci. **76**, 296 (1978)

7.7   K. A. R. Mitchell, F. R. Shepherd, P. R. Watson, D. C. Frost: Surf. Sci. **64**, 737 (1977)

7.8   M. A. Van Hove, R. J. Koestner: in Proc. Conf. on Structure Determination by LEED (Plenum, New York 1985)

7.9   J. R. Noonan, H. L. Davis, F. Jona: Bull. Am. Phys. Soc. **28**, 572 (1983)

7.10   H. B. Nielsen, J. N. Andersen, L. Petersen, D. L. Adams: J. Phys. C **15**, L1113 (1982)

7.11   V. Jensen, J. N. Andersen, H. B. Nielsen, D. L. Adams: Surf. Sci. **116**, 66 (1982)

7.12   H. L. Davis, J. R. Noonan: Surf. Sci. **126**, 245 (1983)

7.13   D. L. Adams, H. B. Nielsen, J. N. Andersen: Surf. Sci. (in press)

7.14   D. L. Adams, H. B. Nielsen, J. N. Andersen, I. Stensgaard, R. Feidenhans'l, J. E. Sorensen: Phys. Rev. Lett. **49**, 669 (1982)

7.15   H. L. Davis, J. R. Noonan: J. Vac. Sci. Technol. **20**, 842 (1982)

7.16   J. R. Noonan and H. L. Davis, Bull. Am. Phys. Soc. **26**, 224 (1981)

7.17   S. P. Withrow, R. J. Culbertson, J. H. Barrett: Bull. Am. Phys. Soc. **28**, 572 (1983)

7.18   H. L. Davis, D. M. Zehner: J. Vac. Sci. Technol. **17**, 190 (1980)

7.19   P. Mark, J. D. Levine, S. H. McFarlane: Phys. Rev. Lett. **38**, 1408 (1977)

7.20   S. Y. Tong, M. A. Van Hove: Phys. Rev. B **16**, 1459 (1977)

7.21   S. Y. Tong, M. A. Van Hove, B. J. Mrstik: in Proc. 7th Int. Vacuum Congress and 3rd Int. Conference on Solid Surfaces, Vienna, 2407 (1977)

7.22 J. J. Lander, G. W. Gobeli, J. Morrison: J. Appl. Phys. **34**, 2298 (1963)

7.23 F. Bauerle, W. Mönch, M. Henzler: J. Appl. Phys. **43**, 3917 (1972)

7.24 J. J. Lander: Surf. Sci. **1**, 125 (1964)

7.25 J. V. Florio, W. D. Robertson: Surf. Sci. **22**, 459 (1970)

7.26 D. Haneman: Phys. Rev. **121**, 1093 (1961)

7.27 J. D. Levine, P. Mark, S. H. McFarlane: J. Vac. Sci. Technol. **14**, 878 (1977)

7.28 J. D. Levine, S. H. McFarlane, P. Mark: Phys. Rev. B **16**, 5415 (1977)

7.29 J. J. Lander: in *Progress in Solid State Chemistry*, Vol. 2, ed. by H. Reiss (Pergamon, Oxford(1965) p. 26

7.30 D. J. Miller, D. Haneman: J. Vac. Sci. Technol. **16**, 1270 (1979)

7.31 M. J. Cardillo: Phys. Rev. B **23**, 4279 (1981)

7.32 E. G. McRae: Bull. Am. Phys. Soc. **28**, 480 (1983)

7.33 J. V. Florio, W. D. Robertson: Surf. Sci. **24**, 173 (1971)

7.34 H. D. Shih, F. Jona, D. W. Jepsen, P. M. Marcus: Phys. Rev. Lett. **37**, 1622 (1976)

7.35 D. M. Zehner, J. R. Noonan, H. L. Davis, C. W. White: J. Vac. Sci. Technol. **18**, 852 (1981)

7.36 W. Mönch, P. P. Auer, R. Feder: J. Vac. Sci. Technol. **16**, 1286 (1979)

7.37 D. J. Chadi: J. Vac. Sci. Technol. **18**, 856 (1981)

7.38 S. Y. Tong, A. L. Maldonaldo: Surf. Sci. **78**, 459 (1978)

7.39 W. S. Yang, F. Jona, P. M. Marcus: Solid State Commun. **43**, 847 (1982)

7.40 S. Y. Tong, A. R. Lubinsky, B. J. Mrstik, M. A. Van Hove: Phys. Rev. B **17**, 3303 (1978)

7.41 R. J. Meyer, C. B. Duke, A. Paton, A. Kahn, E. So, J. L. Yeh, P. Mark: Phys. Rev. B **19**, 5194 (1979)

7.42 R. Ludeke, G. Landgren: Phys. Rev. Lett. **47**, 875 (1981)

7.43 D. G. Fedak, N. A. Gjostein: Surf. Sci. **8**, 77 (1967)

7.44 M. A. Van Hove, R. J. Koestner, P. C. Stair, J. P. Biberian, L. L. Kesmodel, I. Bartos, G. A. Somorjai: Surf. Sci. **103**, 189,218 (1981)

7.45   T. N. Rhodin, G. Broden: Surf. Sci. **60**, 466 (1976)

7.46   C.-M. Chan, M. A. Van Hove, W. H. Weinberg, E. D. Williams: Solid State Commun. **29**, 47 (1979)

7.47   C.-M. Chan, K. L. Luke, M. A. Van Hove, W. H. Weinberg, E. D. Williams: J. Vac. Sci. Technol. **16**, 642 (1979)

7.48   C.-M. Chan, M. A. Van Hove, W. H. Weinberg, E. D. Williams: Surf. Sci. **91**, 440 (1980)

7.49   C.-M. Chan, S. L. Cunningham, K. L. Luke, W. H. Weinberg, S. P. Withrow: Surf. Sci. **78**, 15 (1978)

7.50   C.-M. Chan, K. L. Luke, M. A. Van Hove, W. H. Weinberg, S. P. Withrow: Surf. Sci. **78**, 386 (1978)

7.51   G. Gewinner, J. C. Peruchetti, A. Jaegle, R. Riedinger: Phys. Rev. Lett. **43**, 935 (1979)

7.52   M. K. Debe, D. A. King: Phys. Rev. Lett. **39**, 708 (1977)

7.53   M. K. Debe, D. A. King: Surf. Sci. **81**, 193 (1979)

7.54   E. Tosatti: Solid State Commun. **25**, 637 (1978)

7.55   J. L. Taylor, D. E. Ibbotson, W. H. Weinberg: Surf. Sci. **79**, 349 (1979)

7.56   E. Zanazzi, F. Jona: Surf. Sci. **62**, 61 (1977)

7.57   F. Jona: J. Phys. C **11**, 4271 (1978)

7.58   J. R. Noonan, H. L. Davis: J. Vac. Sci. Technol. **16**, 587 (1979)

7.59   W. Moritz, D. Wolf: Surf. Sci. **88**, L29 (1979)

7.60   D. L. Adams, H. B. Nielsen, M. A. Van Hove, A. Ignatiev: Surf. Sci. **104**, 47 (1981)

7.61   S. J. White, D. P. Woodruff: Surf. Sci. **63**, 254 (1977)

7.62   R. E. Schlier, H. E. Farnsworth: *Semiconductor Surface Physics* (University of Pennsylvania Press, Philadelphia 1957).

7.63   J. C. Phillips: Surf. Sci. **40**, 459 (1973)

7.64   F. Jona, H. D. Shih, A. Ignatiev, D. W. Jepsen, P. M. Marcus: J. Phys. C **10**, L67 (1977)

7.65   R. Seiwatz: Surf. Sci. **2**, 473 (1964)

7.66   J. A. Appelbaum, G. A. Baraff, D. R. Hamann: Phys. Rev. Lett. **35**, 729 (1975); Phys. Rev. B **14**, 588 (1976)

7.67   G. P. Kerker, S. G. Louie, M. L. Cohen: Phys. Rev. B **17**, 706 (1978)

568   References

7.68   J. E. Rowe: Phys. Lett. **46A**, 400 (1974)

7.69   J. A. Appelbaum, D. R. Hamann: Surf. Sci. **74**, 21 (1978)

7.70   K. A. R. Mitchell, M. A. Van Hove: Surf. Sci. **75**, L147 (1978)

7.71   F. Jona, H. D. Shih, D. W. Jepsen, P. M. Marcus: J. Phys. C **12**, 455 (1978)

7.72   M. T. Yin, M. L. Cohen: Phys. Rev. B **24**, 2303 (1981)

7.73   R. M. Tromp, R. G. Smeenk, F. W. Saris: Solid State Commun. **39**, 755 (1981)

7.74   W. S. Yang, F. Jona, P. M. Marcus: Surf. Sci. (in press)

7.75   A. U. MacRae, G. W. Gobeli: J. Appl. Phys. **35**, 1629 (1964)

7.76   P. Mark, S. C. Chang, W. R. Creighton, B. Lee: CRC Crit. Rev. Solid State Sci. **5**, 189 (1975)

7.77   A. R. Lubinsky, C. B. Duke, B. W. Lee, P. Mark: Phys. Rev. Lett. **36**, 1058 (1976)

7.78   C. B. Duke, A. R. Lubinsky, B. W. Lee, P. Mark: J. Vac. Sci. Technol. **13**, 761 (1976)

7.79   P. Mark, G. Cisneros, M. Bonn, A. Kahn, C. B. Duke, A. Paton, A. R. Lubinsky, J. Vacuum Sci. Technol. **14**, 910 (1977)

7.80   A. Kahn, E. So, P. Mark, C. B. Duke, R. J. Meyer: J. Vac. Sci. Technol. **15**, 580 (1978)

7.81   A. Kahn, G. Cisneros, M. Bonn, P. Mark: Surf. Sci. **71**, 387 (1978)

7.82   A. Kahn, E. So, P. Mark, C. B. Duke, R. J. Meyer: J. Vac. Sci. Technol. **15**, 1223 (1978)

7.83   K. C. Pandey, J. L. Freeouf, D. E. Eastman: J. Vac. Sci. Technol. **14**, 904 (1977)

7.84   K. C. Pandey: J. Vac. Sci. Technol. **15**, 440 (1978)

7.85   J. A. Knapp, D. E. Eastman, K. C. Pandey, F. Patella: J. Vac. Sci. Technol. **15**, 1252 (1978)

7.86   D. J. Chadi: J. Vac. Sci. Technol. **15**, 631,1244 (1978); Phys. Rev. B **18**, 1800 (1978)

7.87   A. Huijser, J. Van Laar, T. V. Van Rooy: Phys. Lett. **65A**, 337 (1978)

7.88   D. J. Chadi: Phys. Rev. Lett. **41**, 1062 (1978)

7.89   W. A. Goddard, J. J. Barton, A. Redondo, T. C. McGill: J. Vac. Sci. Technol. **15**, 1274 (1978)

7.90   M. G. Lagally, T. S. Ngoc, M. B. Webb: Phys. Rev. Lett. **26**, 1557 (1971)

7.91   C. B. Duke, D. L. Smith: Phys. Rev. B **5**, 4730 (1972)

7.92   D. J. Miller, D. Haneman: Surf. Sci. **82**, 102 (1979)

7.93   J. E. Demuth, T. N. Rhodin: Surf. Sci. **45**, 249 (1974)

7.94   P. M. Marcus, J. E. Demuth, D. W. Jepsen: Surf. Sci. **53**, 501 (1975)

7.95   E. Zanazzi, M. Maglietta, U. Bardi, F. Jona, D. W. Jepsen, P. M. Marcus: in Proc. 7th Int. Vacuum Congress and 3rd Int. Conference on Solid Surfaces, Vienna, 2447 (1977)

7.96   J. E. Demuth, D. W. Jepsen, P. M. Marcus: Phys. Rev. Lett. **32**, 1182 (1974)

7.97   C.-M. Chan, W. H. Weinberg: J. Chem. Phys. **71**, 2788 (1979)

7.98   M. A. Van Hove, S. Y. Tong: Phys. Rev. Lett. **35**, 1092 (1975)

7.99   M. A. Van Hove, S. Y. Tong, M. H. Elconin: Surf. Sci. **64**, 85 (1977)

7.100   N. Stoner, M. A. Van Hove, S. Y. Tong, M. B. Webb: Phys. Rev. Lett. **40**, 243 (1978)

7.101   M. A. Van Hove, G. Ertl, K. Christmann, R. J. Behm, W. H. Weinberg: Solid State Commun. **28**, 373 (1978)

7.102   K. Christmann, R. J. Behm, G. Ertl, M. A. Van Hove, W. H. Weinberg: J. Chem. Phys. **70**, 4168 (1979)

7.103   H. D. Shih, F. Jona, D. W. Jepsen, P. M. Marcus: Commun. Phys. **1**, 25 (1976)

7.104   H. D. Shih, F. Jona, D. W. Jepsen, P. M. Marcus: Phys. Rev. B **15**, 5550, 5561 (1977)

7.105   J. Topping: Proc. R. Soc. London, Ser. A **114**, 67 (1927)

7.106   E. D. Williams, C.-M. Chan, W. H. Weinberg: Surf. Sci. **81**, L309 (1979)

7.107   *International Tables for X-Ray Crystallography*, Vol. I, Kynoch Press, Birmingham, England (1952)

7.108   F. Tuinstra: *Structural Aspects of the Allotropy of Sulfur and the Other Divalent Elements* (Uitgeverij Waltman, Delft 1967)

7.109   M. K. Debe, D. A. King: Phys. Rev. Lett. **39**, 708 (1977)

7.110   C.-M. Chan, W. H. Weinberg: J. Chem. Phys. **71**, 3988 (1979)

7.111   W. A. Goddard, S. P. Walch, A. K. Rappe, T. H. Upton: J. Vac. Sci. Technol. **14**, 416 (1977)

7.112   M. Delepine: Ann. Chim. (Paris) **4**, 1131 (1959)

7.113   N. K. Pshemtsyn, S. I. Ginzburg, L. G. Salskaya: Zh. Nerorg. Khim. **5**, 832 (1960); Russ. J. Inorg. Chem. **5**, 399 (1960)

7.114   M. J. Nolte, E. Singleton: Acta Crystallogr. Sect. B **32**, 1838 (1976)

7.115   M. Laing, M. J. Nolte, E. Singleton: Chem. Commun., 660 (1975)

7.116   P. Herpin: Bull. Soc. Fr. Mineral. Cristallogr. **81**, 201 (1958)

7.117   M. Ciechanowicz, W. P. Griffith, D. Pawson, A. C. Skapski, M. J. Cleare: Chem. Commun., 876 (1971)

7.118   J. A. Dean (ed.): *Lange's Handbook of Chemistry*, 12th ed. (McGraw-Hill, New York 1979

7.119   V. P. Ivanov, G. K. Boreskov, V. I. Savchenko, W. F. Egelhoff, Jr., W. H. Weinberg: Surf. Sci. **61**, 207 (1976)

7.120   H. D. Shih, F. Jona, D. W. Jepson, P. M. Marcus: Surf. Sci. **60**, 445 (1976)

7.121   F. Jona, K. O. Legg, H. D. Shih, D. W. Jepsen, P. M. Marcus: Phys. Rev. Lett. **40**, 1466 (1978)

7.122   P. A. Thiel, E. D. Williams, J. T. Yates, Jr., W. H. Weinberg: Surf. Sci. **84**, 54 (1979)

   R. J. Koestner, M. A. Van Hove, G. A. Somorjai, Surf. Sci. **107**, 439 (1981)

7.123   M. A. Van Hove, R. J. Koestner, G. A. Somorjai: Phys. Rev. Lett. **50**, 903 (1983)

7.124   M. L. Hair: *Infrared Spectroscopy in Surface Chemistry* (Dekker, New York 1967)

   A. M. Bradshaw, J. Pritchard: Surf. Sci. **17**, 372 (1969)

   H. G. Tompkins, R. G. Greenler: Surf. Sci. **28**, 194 (1971)

7.125   H. Ibach, D. L. Mills: *Electron Energy Loss Spectroscopy and Surface Vibrations* (Academic, New York 1982)

7.126   M. A. Passler, A. Ignatiev, F. Jona, D. W. Jepsen, P. M. Marcus: Phys. Rev. Lett. **43**, 360 (1979)

7.127   S. Andersson, J. B. Pendry: Phys. Rev. Lett. **43**, 363 (1979)

7.128   S. Andersson, J. B. Pendry: J. Phys. C **13**, 3547 (1980)

7.129   S. Y. Tong, A. Maldonado, C. H. Li, M. A. Van Hove: Surf. Sci. **94**, 73 (1980)

7.130 P. Chini, G. Longoni, V. G. Albano: Adv. Organomet. Chem. **14**, 285 (1976)

7.131 R. J. Behm, K. Christmann, G. Ertl, M. A. Van Hove, P. A. Thiel, W. H. Weinberg: Surf. Sci. **88**, L59 (1979)

R. J. Behm, K. Christmann, G. Ertl, M. A. Van Hove: J. Chem. Phys. **73**, 2984 (1980)

7.132 G. Michalk, W. Moritz, H. Pfnür, D. Menzel: Surf. Sci. **129**, 92 (1983)

7.133 L. L. Kesmodel, R. C. Baetzold, G. A. Somorjai: Surf. Sci. **66**, 299 (1977)

7.134 L. L. Kesmodel, L. H. Dubois, G. A. Somorjai: J. Chem. Phys. **70**, 2180 (1979)

7.135 L. H. Dubois, D. G. Castner, G. A. Somorjai: J. Chem. Phys. **72**, 5234 (1980)

7.136 J. E. Demuth, H. Ibach: Surf. Sci. **78**, L238 (1978)

S. Lehwald, H. Ibach: Surf. Sci. **89**, 425 (1979)

7.137 P. M. George, N. R. Avery, W. H. Weinberg, F. N. Tebbe: J. Am. Chem. Soc. **105**, 1393 (1983)

7.138 G. Casalone, M. G. Cattania, M. Simonetta: Surf. Sci. **103**, L121 (1981)

7.139 R. J. Koestner, J. C. Frost, P. C. Stair, M. A. Van Hove, G. A. Somorjai: Surf. Sci. **116**, 85 (1982)

7.140 M. A. Van Hove, R. J. Koestner, G. A. Somorjai: J. Phys. Chem. **87**, 203 (1983)

7.141 S. Andersson: Solid State Commun. **21**, 75 (1977)

7.142 S. Andersson, J. B. Pendry: Surf. Sci. **71**, 75 (1978)

7.143 M. A. Van Hove: unpublished.

7.144 C. L. Allyn, T. Gustafsson, E. W. Plummer: Chem. Phys. Lett. **47**, 127 (1977); Solid State Commun. **28**, 85 (1978)

7.145 J. W. Davenport: Phys. Rev. Lett. **36**, 945 (1976)

7.146 C. H. Li, S. Y. Tong: Phys. Rev. Letters **40**, 46 (1978)

7.147 K. Heinz, E. Lang, K. Müller: Surf. Sci. **87**, 595 (1979)

7.148 R. L. Park, H. H. Madden: Surf. Sci. **11**, 188 (1968)

7.149 A. M. Bradshaw, F. M. Hoffmann: Surf. Sci. **72**, 513 (1978)

7.150   J. P. Biberian, M. A. Van Hove: Surf. Sci. **118**, 443 (1982); ibid. **122**, 600 (1982)

7.151   W. J. Lo, Y. W. Chung, L. L. Kesmodel, P. C. Stair, G. A. Somorjai: Solid State Commun. **22**, 335 (1977)

7.152   H. Ibach, S. Lehwald: J. Vac. Sci. Technol. **15**, 407 (1978)

7.153   H. Ibach, H. Hopster, B. Sexton: Appl. Phys. **14**, 21 (1977)

7.154   J. E. Demuth: Surf. Sci. **84**, 315 (1979)

7.155   T. E. Felter, W. H. Weinberg: Surf. Sci. **103**, 265 (1981)

7.156   R. J. Koestner, M. A. Van Hove, G. A. Somorjai: Surf. Sci. **121**, 321 (1982)

7.157   P. W. Sutton, L. F. Dahl: J. Am. Chem. Soc. **89**, 261 (1967)

7.158   G. M. Scheldrick, J. P. Yesinowski: J. Chem. Soc. Dalton Trans. 873 (1975)

7.159   J. P. Yesinowski, D. Bailey: J. Organomet. Chem. **65**, 627 (1974)

7.160   J. E. Demuth: Surf. Sci. **80**, 367 (1979)

7.161   M. H. Howard, S. F. A. Kettle, I. A. Oxton, D. B. Powell, N. Sheppard, P. Skinner: J. Chem. Soc., Faraday Trans. **77**, 397 (1981)

Chapter 8

8.1   S. C. Fain: "Low-Energy Electron Diffraction Studies of Physically Adsorbed Films" in *Chemistry and Physics of Solid Surfaces IV*, ed. by R. Vanselow, R. Howe, Springer Ser. Chem. Phys., Vol. 20 (Springer, Berlin, Heidelberg 1982) p. 203

8.2   T. B. Grimley, M. Torrini: J. Phys. C **6**, 868 (1973)

8.3   T. L. Einstein, J. R. Schrieffer: Phys. Rev. B **7**, 3629 (1973)

8.4   T. D. Lee, C. N. Yang: Phys. Rev. **87**, 410 (1952)

8.5   C. Davisson, L. H. Germer: Phys. Rev. **30**, 705 (1927)

8.6   J. C. Buchholz, M. G. Lagally: Phys. Rev. Lett. **35**, 442 (1975)

8.7   D. P. Landau: "Applications in Surface Physics", in *Monte Carlo Methods in Statistical Physics*, ed. by K. Binder, Topic Curr. Phys., Vol. 7 (Springer, Berlin, Heidelberg 1979) p. 337

8.8   C. Domb, M. S. Green (eds.): *Phase Transitions and Critical Phenomena*, Vol. 6 (Academic, New York 1977)

8.9   M. P. Nightingale: Physica **83A**, 561 (1976)

8.10   H. E. Stanley: *Introduction to Phase Transitions and Critical Phenomena* (Oxford University Press, New York 1971)

8.11   L. D. Roelofs, A. R. Kortan, T. L. Einstein, R. L. Park: Phys. Rev. Lett. **46**, 1465 (1981)

8.12   M. G. Lagally, T.-M. Lu, G.-C. Wang: in *Ordering in Two Dimensions*, ed. by S. Sinha (Elsevier, Amsterdam 1980).

8.13   J. V. Jose, L. P. Kadanoff, S. Kirkpatrick, D. R. Nelson: Phys. Rev. B **16**, 1217 (1977)

8.14   M. E. Fisher: Rev. Mod. Phys. **46**, 597 (1974)

8.15   L. D. Roelofs, A. R. Kortan, T. L. Einstein, R. L. Park: J. Vac. Sci. Technol. **18**, 492 (1981)

8.16   L. D. Roelofs: Appl. Surf. Sci. **11/12**, 425 (1982)

8.17   S. Krinsky, D. Mukamel: Phys. Rev. B **16**, 2313 (1977)

8.18   E. Domany, M. Schick, J. S. Walker, R. B. Griffiths: Phys. Rev. B **18**, 2209 (1978)

8.19   R. B. Potts: Proc. Cambridge Philos. Soc. **48**, 106 (1952)

8.20   M. P. M. den Nijs: J. Phys. A **12**, 1857 (1979)

8.21   B. Nienhuis, E. K. Riedel, M. Schick: J. Phys. A **13**, L189 (1980)

8.22   R. B. Pearson: Phys. Rev. B **22**, 2579 (1980)

8.23   K. Binder, D. P. Landau: Phys. Rev. B **21**, 1941 (1980)

8.24   Y. Saito: Prog. Theor. Phys., Suppl. 69 (1981)

8.25   W. Kinzel, M. Schick: Phys. Rev. B. **23**, 3435 (1981)

8.26   W. Kinzel, M. Schick: Phys. Rev. B. **24**, 324 (1981)

8.27   K. Binder, D. P. Landau: Surf. Sci. **108**, 503 (1981)

8.28   K. Binder, W. Kinzel, D. P. Landau: Surf. Sci. **117**, 232 (1982)

8.29   L. P. Kadanoff: Ann. Phys. (N.Y.) **100**, 359 (1976)

8.30   A. N. Berker: in *Ordering in Two Dimensions*, ed. by S. Sinha (North-Holland, Amsterdam 1980)

8.31   M. Schick: in *Phase Transitions in Surface Films*, ed. by J. G. Dash, S. Ruvalds (Plenum, New York 1980)

8.32   W. Kinzel, W. Selke, K. Binder: Surf. Sci. **121**, 13 (1982)

8.33   M. E. Fisher: in *Critical Phenomena*, ed. by M. S. Green (Academic Press, New York 1971)

574   References

8.34   J. E. Demuth: J. Colloid Interface Sci. **58**, 184 (1977)

8.35   R. J. Behm, V. Penka, M. G. Cattania, K. Christmann, G. Ertl: J. Chem. Phys. **78**, 7486 (1983)

8.36   P. J. Estrup: J. Vac. Sci. Technol. **16**, 635 (1979)

8.37   R. L. Park, T. L. Einstein, A. R. Kortan, L. D. Roelofs: in *Ordering in Two Dimensions*, ed. by S. K. Sinha (North-Holland, Amsterdam 1980) p. 17

8.38   M. A. Van Hove, G. Ertl, K. Christmann, R. J. Behm, W. H. Weinberg: Solid State Commun. **28**, 373 (1978)

8.39   K. Christmann, R. J. Behm, G. Ertl, M. A. Van Hove, W. H. Weinberg, J. Chem. Phys. **70**, 4168 (1979)

8.40   E. Domany, M. Schick, J. S. Walker: Solid State Commun. **30**, 331 (1979)

8.41   R. C. Kittler, K. H. Bennemann: Solid State Commun. **32**, 403 (1979)

8.42   K. Christmann, O. Schober, G. Ertl, M. Neumann: J. Chem. Phys, **60**, 4528 (1974)

8.43   C.-M. Chan, R. Aris, W. H. Weinberg: Appl. Surf. Sci. **1**, 360 (1978)

8.44   H. Rinne: Thesis, University of Hannover (1974)

8.45   J. C. Bertolini, G. Dalmai-Imelik: Colloq. Int. CNRS 135 (1969)

8.46   G. Casalone, M. G. Cattania, M. Simonetta, M. Tescari: Surf. Sci. **72**, 739 (1978)

8.47   R. J. Behm, K. Christmann, G. Ertl: Solid State Commun. **25**, 763 (1978)

8.48   G. Doyen, G. Ertl, M. Plancher: J. Chem. Phys. **62**, 2957 (1975)

8.49   K. Binder, D. P. Landau: Surf. Sci. **61**, 577 (1976)

8.50   B. Mihura, D. P. Landau: Phys. Rev. Lett. **38**, 977 (1977)

8.51   J. B. Pendry: *Low-Energy Electron Diffraction* (Academic, New York 1974)

8.52   M. A. Van Hove, S. Y. Tong: *Surface Crystallography by LEED*, Springer Ser. Chem. Phys., Vol. 2 (Springer, Berlin, Heidelberg 1979)

8.53   S. Wakoh: J. Phys. Soc. Jpn. **20**, 1894 (1965)

8.54   J. E. Demuth, P. M. Marcus, D. W. Jepsen: Phys. Rev. B **11**, 1460 (1975)

8.55 See, for example, M. A. Van Hove, S. Y. Tong: J. Vac. Sci. Technol. **12**, 230 (1975)

8.56 S. W. Wang, W. H. Weinberg: Surf. Sci. **77**, 14 (1978)

8.57 J. C. Slater: *Insulators, Semiconductors and Metals* (McGraw-Hill, New York 1967)

8.58 A. Heimber: Z. Phys. **105**, 56 (1937)

8.59 T. F. Koetzle, R. K. McMullen, R. Bau, D. W. Hart, R. G. Teller, D. L. Tipton, R. D. Wilson: Advan. Chem. **167**, 61 (1978)

8.60 W. Büssem, F. Gross: Z. Phys. **87**, 778 (1934)

8.61 G. Doyen, G. Ertl: J. Chem. Phys. **68**, 5417 (1978)

8.62 T. H. Upton, W. A. Goddard: Phys. Rev. Lett. **42**, 472 (1979)

8.63 T. H. Upton, W. A. Goddard, C. F. Melius: J. Vac. Sci. Technol. **16**, 531 (1979)

8.64 S. Andersson: Chem. Phys. Lett. **55**, 185 (1978)

8.65 D. Wolf, H. Jagodzinski, W. Moritz: Surf. Sci. **77**, 283 (1978)

8.66 H. Jagodzinski, W. Moritz, D. Wolf: Surf. Sci. **77**, 223 (1978)

8.67 E. Ising: J. Phys. **31**, 253 (1925)

8.68 T. L. Hill: *Statistical Mechanics* (McGraw-Hill, New York 1956)

8.69 E. Domany, M.Schick, J. S. Walker: Phys. Rev. Lett. **38**, 1148 (1977)

8.70 R. Kikuchi: Phys. Rev. **81**, 988 (1951)

8.71 T. L. Einstein: CRC Crit. Rev. Solid State Mater. Sci. **7**, 261 (1977)

8.72 J. G. Dash: *Films on Solid Surfaces* (Academic, New York 1975) Chap. 2

8.73 H. Hoinkes: Rev. Mod. Phys. **52**, 933 (1980)

8.74 R. J. Behm, K. Christmann, G. Ertl: Surf. Sci. **99**, 320 (1980)

8.75 M. A. Van Hove: in *The Nature of the Surface Chemical Bond*, ed. by T. N. Rhodin, G. Ertl (North-Holland, Amsterdam 1979) Chap. 4

8.76 J. E. Worsham, M. K. Wilkinson, C. G. Shull: J. Phys. Chem. Solids **3**, 303 (1957)

8.77 I. S. Anderson, C. J. Carlisle, D. K. Ross: J. Phys. C **11**, 15 (1978)

8.78 I. S. Anderson, D. K. Ross, C. J. Carlisle: Phys. Lett. **68A**, 249 (1978)

8.79 C. Fan, F. Y. Wu: Phys. Rev. **179**, 560 (1979)

8.80   T. B. Grimley: Proc. Phys. Soc., London **90**, 571 (1967); ibid. **92**, 776 (1967)

8.81   V. Hartung: Z. Phys. B **32**, 307 (1979)

8.82   K. Binder (ed.): *Monte Carlo Methods in Statistical Physics*, Topics Curr. Phys., Vol. 7 (Springer, Berlin, Heidelberg 1979)

8.83   B. Poelsema, R. L. Palmer, G. Mechtersheimer, G. Comsa: Surf. Sci. **117**, 60 (1982)

8.84   T. Engel, K.-H. Rieder: "Structural Studies of Surfaces with Atomic and Molecular Beam Diffraction", in Springer Tracts Mod. Phys., Vol. 91 (Springer, Berlin, Heidelberg 1982) p. 55

8.85   B. Derrida, L. de Seze, J. Vannimenus: in *Lecture Notes in Physics*, Vol. 149 (Springer, Berlin, Heidelberg 1979) p. 46

8.86   F. Bozso, G. Ertl, M. Grunze, M. Weiss: Appl. Surf. Sci. **1**, 103 (1977)

8.87   R. Imbihl, R. J. Behm, K. Christmann, G. Ertl, T. Matsushima: Surf. Sci. **117**, 257 (1982)

8.88   R. J. Behm, R. Imbihl, K. Christmann, G. Ertl, W. Moritz: to be published

8.89   J. E. Houston, R. L. Park: Surf. Sci. **21**, 209 (1970)

8.90   R. M. F. Houtappel: Physica **16**, 425 (1950)

8.91   M. Schick: in *Phase Transitions in Surface Films*, ed. by J. G. Dash, J. Ruvalds (Plenum, New York 1979) p. 65

8.92   M. E. Fisher, M. N. Barber: Phys. Rev. Lett. **28**, 1516 (1972)

Chapter 9

9.1   L. L. Kesmodel, R. C. Baetzold, G. A. Somorjai: Surf. Sci. **66**, 299 (1977)

9.2   L. L. Kesmodel, L. H. Dubois, G. A. Somorjai: J. Chem. Phys. **70**, 2180 (1979)

9.3   R. J. Koestner, J. C. Frost, P. C. Stair, M. A. Van Hove, G. A. Somorjai: Surf. Sci. **117**, 491 (1982)

9.4   G. Casalone, M. G. Cattania, F. Merati, M. Simonetta: Surf. Sci. **120**, 171 (1982)

9.5   P. M. George, N. R. Avery, W. H. Weinberg, F. N. Tebbe: J. Am. Chem. Soc. **105**, 1393 (1983)

9.6   J. T. Yates, Jr., P. A. Thiel, W. H. Weinberg: Surf. Sci. **82**, 45 (1979)

9.7   J. T. Yates, Jr., P. A. Thiel, W. H. Weinberg: Surf. Sci. **84**, 427 (1979)

9.8   P. A. Thiel, J. T. Yates, Jr., W. H. Weinberg: Surf. Sci. **82**, 22 (1979)

9.9   C.-M. Chan, P. A. Thiel, J. T. Yates, Jr., W. H. Weinberg: Surf. Sci. **76**, 296 (1978)

9.10   F. R. Shepherd, P. R. Watson, D. C. Frost, K. A. R. Mitchell: J. Phys. C **11**, 4591 (1978); in Proc. 3rd Internat. Conf. Solid Surfaces, Vienna (1977), p. A2725

9.11   D. O. Hayward, B. M. V. Trapnell: *Chemisorption* (Butterworth, London 1964)

9.12   C.-M. Chan, W. H. Weinberg: J. Chem. Phys. **71**, 2788 (1979)

9.13   A. C. Sobrero, W. H. Weinberg: unpublished

9.14   P. A. Thiel, J. T. Yates, Jr., W. H. Weinberg: Surf. Sci. **90**, 121 (1979)

Chapter 10

10.1   H. Pfnür, P. Feulner, H. A. Engelhardt, D. Menzel: Chem. Phys. Lett. **59**, 481 (1978)

10.2   F. M. Hoffmann, A. Ortega, H. Pfnür, D. Menzel, A. M. Bradshaw: J. Vac. Sci. Technol. **17**, 239 (1980)

10.3   M. G. Lagally, T.-M. Lu, G.-C. Wang: in *Ordering in Two Dimensions*, ed. by S. Sinha (Elsevier, Amsterdam 1980); and references therein

10.4   L. D. Roelofs, T. L. Einstein, P. E. Hunter, A. R. Kortan, R. L. Park, R. M. Roberts, J. Vac. Sci. Technol. **17**, 231 (1980)

10.5   A. Crossley, D. A. King: Surf. Sci. **95**, 131 (1980)

10.6   M. G. Lagally, T.-M. Lu, D. G. Welkie: J. Vac. Sci. Technol. **17**, 223 (1980)

10.7   J. C. Tracy, J. M. Blakely: in *The Structure and Chemistry of Solid Surfaces*, ed. by G. A. Somorjai (Wiley, New York 1969)

10.8   C. S. McKee, D. L. Perry, M. W. Roberts: Surf. Sci. **39**, 176 (1973)

10.9   J. E. Houston, R. L. Park: Surf. Sci. **21**, 209 (1970)

10.10 J. E. Houston, R. L. Park: Surf. Sci. **26**, 269 (1971)

10.11 G. Ertl, J. Küppers: *Low Energy Electrons and Surface Chemistry* (Verlag Chemie, Weinheim 1974)

10.12   M. Henzler: "Electron Diffraction and Surface Defect Structure", in *Electron Spectroscopy for Surface Analysis*, ed. by H. Ibach (Springer, Berlin, Heidelberg 1977) p. 117

10.13   M. Henzler: Surf. Sci. **73**, 240 (1978)

10.14   T.-M. Lu, S. R. Anderson, M. G. Lagally, G.-C. Wang: J. Vac. Sci. Technol. **17**, 207 (1980)

10.15   T.-M. Lu, G.-C. Wang, M. G. Lagally: Surf. Sci. **107**, 494 (1981)

10.16   E. D. Williams, W. H. Weinberg: Surf. Sci. **109**, 574 (1981)

10.17   G.-C. Wang, T.-M. Lu, M. G. Lagally: J. Chem. Phys. **69**, 479 (1978)

10.18   T.-M. Lu, G.-C. Wang and M. G. Lagally, Surface Sci. **92**, 133 (1980).

10.19   M. E. Fisher: in *Proceedings of the Enrico Fermi International School of Physics* (Academic, New York 1971)

10.20   D. P. Landau: Phys. Rev. B **13**, 2997 (1976)

10.21   S. Ostlund, A. N. Berker: Phys. Rev. Lett. **42**, 843 (1979)

10.22   W. Moritz: in *Electron Diffraction 1927-77*, ed. by P. J. Dobson, J. B. Pendry, C. J. Humphreys (The Institute of Physics, London 1977) p. 261

10.23   R. L. Park, J. E. Houston, D. G. Schreiner: Rev. Sci. Instrum. **42**, 60 (1971)

10.24   G.-C. Wang, M. G. Lagally: Surf. Sci. **81**, 69 (1979)

10.25   E. D. Williams, W. H. Weinberg: in Proc. Fourth Int. Conf. Solid Surfaces, Cannes (1980) p. 311

10.26   L.-H. Zhao, T.-M. Lu, M. G. Lagally: Appl. Surf. Sci. **11/12**, 634 (1982)

10.27   E. D. Williams, W. H. Weinberg, A. C. Sobrero: J. Chem. Phys. **76**, 1150 (1982)

10.28   G. E. Thomas, W. H. Weinberg: J. Chem. Phys. **70**, 1437 (1979)

10.29   H. Pfnür, F. M. Hoffmann, A. Ortega, D. Menzel, A. M. Bradshaw: Surf. Sci. **93**, 431 (1980)

10.30   G. Michalk, W. Moritz, H. Pfnür, D. Menzel: Surf. Sci. **129**, 92 (1983)

10.31   J. T. Grant, T. W. Haas: Surf. Sci. **21**, 76 (1970)

10.32   T. E. Madey, D. Menzel: Jpn. J. Appl. Phys., Suppl. 2, Pt. 2, 229 (1974)

10.33   E. D. Williams, W. H. Weinberg: Surf. Sci. **82**, 93 (1979)

10.34   P.Feulner, H. A. Engelhardt, D. Menzel: Appl. Phys. **15**, 355 (1978)

10.35   J. C. Fuggle, T. E. Madey, M. Steinkilberg, D. Menzel: Surf. Sci. **52**, 521 (1975)

10.36   M. Henzler: Surf. Sci. **22**, 12 (1970)

10.37   T. E. Madey: Surf. Sci. **79**, 575 (1979)

10.38   N. V. Richardson, A. M. Bradshaw: Surf. Sci. **88**, 255 (1979)

10.39   C. O. Quicksall, T. G. Spiro: Inorg. Chem. 7, 2365 (1968)

10.40   M. G. Lagally, G.-C. Wang, T.-M. Lu: CRC Crit. Rev. Solid State Mater. Sci. **1**, 233 (1978)

Chapter 11

11.1   L. de Bersuder: Rev. Sci. Instrum. **45**, 1569 (1974)

11.2   B. Poelsema, R. L. Palmer, G. Mechtersheimer, G. Comsa: Surface Sci. **117**, 60 (1982)

11.3   M. D. Chin, S. C. Fain, Jr.: J. Vac. Sci. Technol. **14**, 314 (1977)

11.4   S. P. Weeks, C. D. Ehrlich, E. W. Plummer: Rev. Sci. Instrum. **48**, 190 (1977)

E. D. Williams: PhD Thesis, California Institute of Technology (1981)

11.5   M. Henzler: Appl. Surf. Sci. **11/12**, 450 (1982)

11.6   D. T. Pierce, R. J. Celotta: in *Advances in Electronic and Electron Physics*, Vol. 56, ed. by C. Marton (Academic, New York 1981

S. W. Wang: Solid State Commun. **36**, 847 (1980)

R. Feder: J. Phys. C **14**, 2049 (1982)

11.7   J. S. Schilling, M. B. Webb: Phys. Rev. B **2**, 1665 (1970)

11.8   J. P. Biberian, G. A. Somorjai: J. Vac. Sci. Technol. **16**, 2073 (1979)

J. W. A. Sachtler, M. A. Van Hove, J. P. Biberian, G. A. Somorjai: Surf. Sci. **110**, 19 (1981)

11.9   M. A. Van Hove, R. J. Koestner, G. A. Somorjai: J. Vac. Sci. Technol. **20**, 886 (1982)

11.10   C. B. Duke, G. E. Laramore: Phys. Rev. B **2**, 4765,4783 (1970)

E. D. Williams, W. H. Weinberg, Surf. Sci. **109**, 574 (1981)

11.11   E. D. Williams, W. H. Weinberg, A. C. Sobrero: J. Chem. Phys. **76**, 1150 (1982)

11.12   W. Moritz, H. Jagodzinski, D. Wolf: Surf. Sci. **77**, 233,249 (1978)

11.13   W. P. Ellis: in *Optical Transforms*, ed. by H. S. Lipson (Academic, London 1972)

   M. A. Van Hove, R. J. Koestner, P. C. Stair, J. P. Biberian, L. L. Kesmodel, I. Bartos, G. A. Somorjai: Surf. Sci. **103**, 189 (1981)

11.14   W. P. Ellis, T. N. Taylor: Surf. Sci. **123**, 77 (1982)

11.15   P. Mark, G. Cisneros, M. Bonn, A. Kahn, C. B. Duke, A. Paton, A. R. Lubinsky: J. Vac. Sci. Technol. **14**, 910 (1977)

   M. A. Van Hove, S. Y. Tong: unpublished

11.16   S. Y. Tong, M. A. Van Hove, B. J. Mrstik: Proc. 7th IVC & 3rd ICSS, Vienna (1977), p. 2407

   K. A. R. Mitchell, M. A. Van Hove: Surf. Sci. **75**, 147L (1978)

11.17   S. J. White, D. C. Frost, K. A. R. Mitchell: Solid State Commun. **42**, 763 (1982)

11.18   M. A. Van Hove, G. A. Somorjai: Surf. Sci. **114**, 171 (1982)

11.19   S. Andersson, J. B. Pendry: J. Phys. C **13**, 3547 (1980)

11.20   M. A. Van Hove, R. J. Koestner, P. C. Stair, J. P. Biberian, L. L. Kesmodel, I. Bartos, G. A. Somorjai: Surf. Sci. **103**, 218 (1981)

11.21   W. S. Yang, F. Jona, P. M. Marcus: Bull. Am. Phys. Soc. **27**, 239 (1982)

   M. A. Van Hove, R. F. Lin, G. A. Somorjai: to be published

11.22   J. B. Pendry: Proc. Conf. on Determination of Surface Structure by LEED, Plenum, New York (1985)

11.23   C. R. Brundle, H. Hopster, J. D. Swalen: J. Chem. Phys. **70**, 5190 (1979)

11.24   L. E. Firment: PhD Thesis, University of California, Berkeley (1977)

11.25   N. J. Wu, A. Ignatiev: Phys. Rev. B **25**, 2983 (1982)

11.26   M. A. Van Hove, R. J. Koestner: Proc. Conf. on Determination of Surface Structure by LEED, Plenum, New York (1985)

11.27   W. Heiland, E. Taglauer: Surf. Sci. **68**, 96 (1977)

   F. W. Saris, J. F. van der Veen: Proc. 7th Int. Vac. Congr. & 3rd Int. Conf. Solid Surfaces, Vienna (1977), p. 2503

11.28   L. C. Feldman, R. L. Kauffman, P. J. Silverman, R. A. Zuhr, J. H. Barrett: Phys. Rev. Lett. **39**, 38 (1977)

R. M. Tromp, E. J. van Loenen, M. Iwami, F. W. Saris: Solid State Commun. **44**, 971 (1982)

11.29   A. Benninghoven: Surf. Sci. **35**, 427 (1973)

B. J. Garrison, N. Winograd, D. E. Harrison, Jr.: Phys. Rev. B **18**, 6000 (1978)

11.30   E. W. Müller, Z. Phys. **136**, 131 (1951); in *Advances in Electronics and Electron Physics*, Vol. XIII, ed. by L. Marton (Academic, New York 1960)

E. W. Müller, T. T. Tsong: *Field Ion Microscopy* (American Elsevier, New York 1969)

G. Ehrlich: Surf. Sci. **63**, 422 (1977)

T. T. Tsong, R. Casanova: Phys. Rev. Lett. **47**, 113 (1981)

11.31   T. E. Madey, J. T. Yates, Jr.: Chem. Phys. Lett. **51**, 77 (1978); Surf. Sci. **76**, 397 (1978)

T. E. Madey: Surf. Sci. **79**, 575 (1979)

T. E. Madey, J. T. Yates, Jr., A. M. Bradshaw, F. M. Hoffmann: Surf. Sci. **89**, 370 (1979)

11.32   H. D. Hagstrum: Phys. Rev. **150**, 495 (1966)

G. E. Becker, H. D. Hagstrum: J. Vac. Sci. Technol. **10**, 31 (1973)

H. D. Hagstrum, G. E. Becker: J. Vac. Sci. Technol. **14**, 369 (1977)

11.33   P. D. Johnson, T. A. Delchar: Surf. Sci. **77**, 400 (1978)

H. Conrad, G. Ertl, J. Küppers, S. W. Wang, K. Gerard, H. Haberland: Phys. Rev. Lett. **42**, 1082 (1979)

S. W. Wang: J. Vac. Sci. Technol. **20**, 600 (1982)

F. Bozso, C. P. Hanrahan, J. Arias, H. Metiu, J. T. Yates, Jr., R. M. Martin: Surf. Sci. **128**, 197 (1983)

11.34   W. H. Weinberg: Adv. Colloid Interface Sci. **4**, 301 (1975)

M. W. Cole, D. R. Frankl: Surf. Sci. **70**, 585 (1978)

H. Hoinkes: Rev. Mod. Phys. **52**, 933 (1980)

M. J. Cardillo: Annu. Rev. Phys. Chem. **32**, 331 (1981)

T. Engel, K. H. Rieder: "Structural Studies of Surfaces with Atomic and Molecular Beam Diffraction", in *Structural Studies of Surfaces*, Springer Tracts Mod. Phys. Vol. 91 (Springer, Berlin, Heidelberg 1982) p. 55

11.35   P. A. Redhead: Vacuum **12**, 203 (1962)

L. A. Petermann: in Progress in Surface Science, Vol. 3, ed. by S. G. Davison (Pergamon, Oxford 1972)

D. A. King: Surf. Sci. **47**, 384 (1975)

11.36   C.-M. Chan, R. Aris, W. H. Weinberg: Appl. Surf. Sci. **1**, 360 (1978)

11.37   J. K. Kjems, L. Passell, H. Taub, J. G. Dash, A. D. Novaco: Phys. Rev. B **13**, 1446 (1976)

J. P. McTague, M. Nielsen, L. Passell: Crit. Rev. Solid State Sci. **8**, 125 (1979)

11.38   J. F. Menadue: Acta Crystallogr. Sect. A **28**, 1 (1972)

N. Masud, J. B. Pendry: J. Phys. C **9**, 1833 (1976)

N. Masud, C. G. Kinniburgh, J. B. Pendry: J. Phys. C **10**, 1 (1977)

11.39   A. V. Crewe, J. Wall, J. Langmore: Science **168**, 133 (1970)

J. M. Cowley: *Diffraction Physics* (North-Holland, Amsterdam 1975)

M. S. Isaacson, J. Langmore, N. W. Parker, D. Kopf, M. Utlaut: Ultramicroscopy **1**, 359 (1976)

N. Osakabe, Y. Tanishiro, K. Yagi, G. Honjo: Surf. Sci. **97**, 393 (1980); ibid. **109**, 353 (1981)

11.40   L. D. Marks: Phys. Rev. Lett. **51**, 1000 (1983)

11.41   G. Binnig, H. Rohrer, Ch. Gerber, E. Weibel: Appl. Phys. Lett. **40**, 178 (1982); Phys. Rev. Lett. **49**, 57 (1982); ibid. **50**, 123 (1983); Surf. Sci. **131**, L379 (1983)

J. Tersoff, D. R. Hamann: Phys. Rev. Lett. **50**, 1998 (1983)

N. Garcia, C. Ocal, F. Flores: Phys. Rev. Lett. **50**, 2002 (1983)

11.42   F. M. Propst, T. C. Piper: J. Vac. Sci. Technol. **4**, 53 (1967)

H. Ibach, D. L. Mills: *Electron Energy Loss Spectroscopy and Surface Vibrations* (Academic, New York 1982)

W. H. Weinberg: in *Methods of Experimental Surface Physics*, ed. by R. L. Park, M. G. Lagally (Academic, New York 1983)

11.43   C. H. Li, S. Y. Tong, D. L. Mills: Phys. Rev. B **21**, 3057 (1980)

11.44   J. Lambe, R. C. Jaklevic: Phys. Rev. **165**, 821 (1968)

P. K. Hansma: Phys. Rep. **30C**, 145 (1977)

W. H. Weinberg: Ann. Rev. Phys. Chem. **29**, 115 (1978)

T. Wolfram (ed.), *Inelastic Electron Tunneling Spectroscopy* (Springer-Verlag, New York 1978); *Tunneling Spectroscopy*, ed. by P. K. Hansma (Plenum, New York 1982)

11.45   C. Herring, M. H. Nichols: Rev. Mod. Phys. **21**, 185 (1949)

N. D. Lang, W. Kohn: Phys. Rev. B **3**, 1215 (1971)

J. Topping: Proc. R. Soc. London, Ser. A **114**, 67 (1927)

J. Hölzl, F. K. Schulte: "Work Function of Metals", in *Solid State Physics*, Springer Tracts Mod. Phys., Vol. 85 (Springer, Berlin, Heidelberg 1979) p. 1

11.46   P. W. Palmberg: in *Electron Spectroscopy*, ed. by D. A. Shirley (North-Holland, Amsterdam 1972)

T. A. Carlson: *Photoelectron and Auger Spectroscopy* (Plenum, New York 1975)

11.47   H. L. Davis, T. Kaplan: Solid State Commun. **19**, 595 (1976)

J. W. Gadzuk: Surf. Sci. **60**, 76 (1976)

11.48   P. I. Cohen, T. L. Einstein, W. T. Elam, Y. Fukuda, R. L. Park: Appl. Surf. Sci. **1**, 538 (1978)

T. L. Einstein: Appl. Surf. Sci. **11/12**, 42 (1982)

G. E. Laramore: Surf. Sci. **81**, 43 (1979)

11.49   R. L. Park: Surf. Sci. **48**, 80 (1975)

11.50   R. L. Gerlach: Surf. Sci. **28**, 648 (1971)

11.51   J. Kirschner, P. Staib, Phys. Lett. **42A**, 335 (1973)

11.52   C. R. Brundle, A. D. Baker (eds.): *Electron Spectroscopy: Theory, Techniques and Applications* (Academic, London 1978)

11.53   D. E. Eastman, J. K. Cashion: Phys. Rev. Lett. **27**, 1520 (1971)

A. M. Bradshaw, L. S. Cederbaum, W. Domcke: Structure and Bonding, Vol. 24 (1975)

E. W. Plummer: in *Photoemission and the Electronic Properties of Surfaces*, ed. by B. Feuerbacher, B. Fitton, R. F. Willis (Wiley, London, in press)

D. A. Shirley (ed.): *Electron Spectroscopy* (North-Holland, Amsterdam 1972)

11.54   J. W. Gadzuk: Phys. Rev. B **10**, 5030 (1974)

J. B. Pendry: Surf. Sci. **57**, 679 (1976)

A. Liebsch, Phys. Rev. Lett. **38**, 248 (1977)

K. Jacobi, M. Scheffler, K. Kambe, F. Forstmann: Solid State Commun. **22**, 17 (1977)

C. H. Li, S. Y. Tong, Phys. Rev. Lett. **40**, 46 (1978)

L. G. Petersson, S. Kono, N. F. T. Hall, S. Goldberg, J. T. Lloyd, C. S. Fadley, J. B. Pendry: Mater. Sci. Eng. **42**, 111 (1980)

D. H. Rosenblatt, J. G. Tobin, M. G. Mason, R. F. Davis, S. D. Keran, D. A. Shirley, C. H. Li, S. Y. Tong: Phys. Rev. B **23**, 3828 (1981)

C. S. Fadley: in *Progress in Surface Science*, ed.by S. G. Davison, (Pergamon, New York, 1983)

11.55   P. A. Lee: Phys. Rev. B **13**, 5261 (1976)

U. Landman, D. L. Adams: Proc. Natl. Acad. Sci. U.S.A. **73**, 2550 (1976)

P. A. Lee, P. H. Citrin, P. Eisenberger, P. M. Kincaid: Rev. Mod. Phys. **53**, 769 (1981)

L. I. Johansson, J. Stöhr: Phys. Rev. Lett. **43**, 1882 (1979)

J. Stöhr, R. Jaeger, S. Brennan: Surf. Sci. **117**, 503 (1982)

11.56   P. J. Durham, J. B. Pendry, C. H. Hodges: Comput. Phys. Commun. **25**, 193 (1982)

G. N. Greaves, P. J. Durham, G. Diakun, P. Quinn: Nature (London) **294**, 139 (1981)

J. B. Pendry: Comments Solid State Phys., to be published.

11.57   J. A. Golovchenko, J. R. Patel, D. R. Kaplan, P. L. Cowan, M. J. Bedzyk: Phys. Rev. Lett. **49**, 560 (1982)

11.58   P. Eisenberger, W. C. Marra: Phys. Rev. Lett. **46**, 1081 (1981)

11.59   I. K. Robinson: Phys. Rev. Lett. **50**, 1145 (1983)

11.60   P. A. Lee, G. Beni: Phys. Rev. B **15**, 2862 (1977)

J. Stöhr, L. Johansson, I. Lindau, P. Pianetta: Phys. Rev. B **20**, 664 (1979)

P. H. Citrin, P. Eisenberger, R. C. Hewitt: Phys. Rev. Lett. **45**, 1948 (1980)

C. Bouldin, E. A. Stern: Phys. Rev. B **25**, 3462 (1982)

11.61   M. L. Hair: *Infrared Spectroscopy in Surface Chemistry* (Marcel Dekker, New York 1967)

A. M. Bradshaw, J. Pritchard: Surf. Sci. **17**, 372 (1969)

H. G. Tompkins, R. G. Greenler: Surf. Sci. **28**, 194 (1971)

11.62   F. M. Hoffmann: Surf. Sci. Rep., to be published.

Subject Index